Merrill
Geometry
Applications and Connections

GLENCOE
McGraw-Hill

New York, New York   Columbus, Ohio   Mission Hills, California   Peoria, Illinois

Send all inquiries to:

Glencoe/McGraw-Hill
936 Eastwind Drive
Westerville, OH 43081

ISBN: 0-02-824000-6 (Student's Edition)
ISBN: 0-02-824001-4 (Teacher's Wraparound Edition)

Printed in the United States of America.
11 12 13 14   071/043   03 02 01 00 99

AUTHORS

Gail F. Burrill is the chairperson of the mathematics department at Whitnall High School, Greenfield, Wisconsin. Ms. Burrill obtained her N.B.S. in Mathematics from Marquette University and her M.S. degree in Mathematics from Loyola University. She is active in professional organizations and is a frequent speaker at national conferences. Ms. Burrill received a Wisconsin State Presidential Award for Excellence in Teaching Mathematics and Science in 1985. She is a past president of the Wisconsin Mathematics Council and has received the Wisconsin Distinguished Mathematics Educator Award. She is a member of the Board of Directors of the National Council of Teachers of Mathematics and the Chair of the Adolescence and Young Adulthood Mathematics Standards Committee for the National Board for Professional Teaching Standards. Ms. Burrill is a coauthor of *Merrill Algebra Essentials*.

Timothy D. Kanold is the Mathematics-Science Chairman and mathematics teacher at Adlai Stevenson High School, Prairie View, Illinois. Mr. Kanold obtained his B.S. degree in Mathematics Education and his M.S. degree in Mathematics from Illinois State University. He also holds a C.A.S. degree from the University of Illinois. Mr. Kanold is active in numerous professional mathematical organizations for which he frequently speaks, and is currently a member of the Regional Services Committee of the National Council of Teachers of Mathematics. He also served as one of seventeen members for NCTM's Professional Standards for Teaching Mathematics commission. He is the 1986 National Presidential Awardee for Excellence in Mathematics Teaching for Illinois and is a past-president of the Council for Presidential Awardees of Mathematics. He is the author of a chapter on effective classroom teaching practices in NCTM's 1990 yearbook and coauthor of *Merrill Informal Geometry*.

Jerry J. Cummins is a teacher and Mathematics Department Chairman for Lyons Township High School, LaGrange, Illinois. Mr. Cummins obtained his B.S. degree in Mathematics Education and M.S. degree in Educational Administration and Supervision from Southern Illinois University. He also holds an M.S. degree in Mathematics Education from the University of Oregon. Mr. Cummins has spoken at many local, state, and national mathematics conferences, and is a past member of the Regional Services Committee of the National Council of Teachers of Mathematics. He received an Illinois State Presidential Award for Excellence in Teaching of Mathematics in 1984. Mr. Cummins is a coauthor of *Merrill Informal Geometry* and Merrill's *Programming in BASIC*.

Lee E. Yunker is a teacher and chairman of the Mathematics Department at West Chicago Community High School, West Chicago, Illinois. Mr. Yunker obtained his B.S. degree from Elmhurst College and his M.Ed. in Mathematics from the University of Illinois, Urbana, Illinois. Mr. Yunker frequently speaks or conducts workshops on a variety of topics. Recently, he participated in the first U.S./Korea Seminar on a Comparative Analysis of Mathematics Education in the United States and Korea at the Seoul National University. Mr. Yunker is a past member of the Board of Directors of both the National Council of Teachers of Mathematics and the National Council of Supervisors of Mathematics. Mr. Yunker is a State Presidential Award Winner for Excellence in Mathematics Teaching. He is a coauthor of *Merrill Advanced Mathematical Concepts*, and *Fractals for the Classroom: Strategic Activities, Volumes One and Two*, co-published by NCTM and Springer-Verlag.

CONSULTANT

Alan G. Foster
Former Chairperson of Mathematics Department
Addison Trail High School
Addison, Illinois

REVIEWERS

Richard Albright
Mathematics Department Chairperson
Hempfield Area Senior High School
Greensburg, Pennsylvania

Rosemary Aragon
Mathematics Teacher
El Rancho High School
Pico Rivera, California

John Bisbikis
Director: Mathematics & Science
Reavis High School
Burbank, Illinois

Jennifer Carmen
Mathematics Teacher
Spring Valley High School
Columbia, South Carolina

Tina Cindea
Mathematics Teacher
Hoover High School
North Canton, Ohio

Patricia Dandridge
Mathematics Teacher
Armstrong High School
Richmond, Virginia

Charles DiGruttolo
Mathematics Department Chairperson
North Allegheny School District
Wexford, Pennsylvania

Mary J. Dubsky
Mathematics Instructional Specialist
Baltimore City Public Schools
Baltimore, Maryland

Vivian P. Fernandez
Mathematics Teacher
H. B. Plant High School
Tampa, Florida

Dianne Foster
Mathematics Teacher
Fairley High School
Memphis, Tennessee

Carol Green
Mathematics Teacher
Hoover High School
North Canton, Ohio

Dennis W. Hodges
Mathematics Teacher
Poly High School
Riverside, California

Dr. Larry L. Houser
Supervisor of Mathematics
Carroll County Public Schools
Westminster, Maryland

David Howell
Mathematics Supervisor
New Haven Public Schools
New Haven, Connecticut

Table of Contents

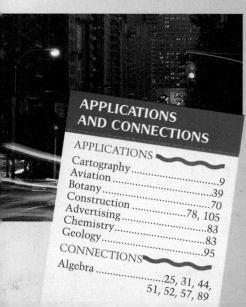
APPLICATIONS AND CONNECTIONS

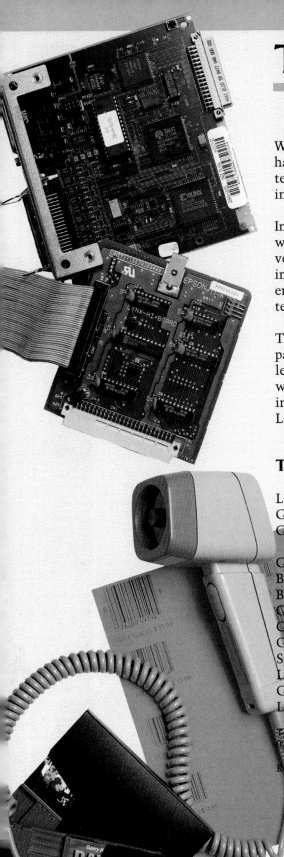

Technology

What would your life be like without technology? Imagine not having computer games, scanners at the store, microwaves and televisions with digital inputs, or hand-held calculators. It's hard to imagine! Technology plays a very large role in our lives.

In fact, calculators and computers have become so essential that it would be rare for your life not to be affected by these amazingly versatile machines. They are vital to businesses, education, industry, retail, and research just to name a few. Even law enforcement agencies and the sports industry rely heavily on technology to perform many important tasks.

The Technology pages in this text let you use technology to explore patterns, make conjectures, and discover mathematics. You will learn to use programs written in the BASIC computer language as well as computer software and spreadsheets. You will also investigate mathematical concepts using graphing calculators and LOGO.

Technology

The Geometric Supposer was developed by Education Development Center, Newton, MA 02160, and published by Sunburst Communications, Inc.

Keystrokes are provided for both Casio and Texas Instruments graphing calculators.

3 Parallels

116

4 Congruent Triangles

162

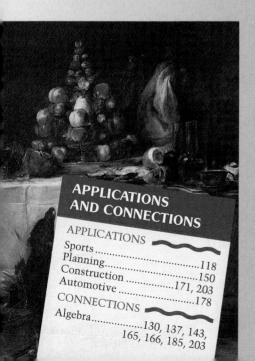

APPLICATIONS AND CONNECTIONS

5 Applying Congruent Triangles

214

6 Quadrilaterals

264

APPLICATIONS
AND CONNECTIONS

Special Features

Why did the study of mathematics cost Hypatia, one of the first known women mathematicians, her life? Read the **History Connection** on page 29 to find out. These features contain information about real people from the past and present and from many different cultures who have had a great influence on what you study in geometry today.

When am I ever going to use geometry? It may be sooner than you think. You'll find geometry in many of the subjects you study in school. In the **Fine Arts Connection** on page 141, you'll see how geometry is used to make unusual drawings.

How can working together with my classmates help me solve problems? The fun, but challenging, problems presented in each **Cooperative Learning Project** gives you an opportunity to cooperate, not compete, with other students. The **Developing Reasoning Skills** features also give you a chance to apply your abilities in logic.

History Connections

Connections

7 Similarity 306

8 Right Triangles and Trigonometry 358

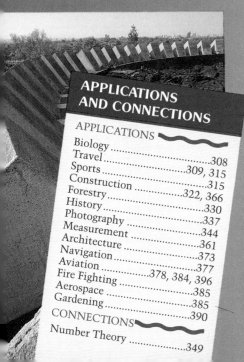

APPLICATIONS AND CONNECTIONS

Algebra Review

"Why is there algebra review in a geometry book?" "I thought after I took algebra last year, I was finished with it, at least for awhile." These are things you may ask or say when you see the pages in this book entitled Algebra Review.

Geometry and algebra are very closely connected. You can't have one without the other! For example, the solutions of a linear equation in algebra corresponds to the graph of the equation on a coordinate plane in geometry. You can use algebra to explore what happens to the volume of a rectangular prism if one or two of its dimensions is doubled.

The Algebra Review pages will not only help you connect algebra and geometry; they will also help you maintain your algebraic and problem-solving skills. Algebra Reviews are provided on pages 66–67, 160–161, 262–263, 356–357, 462–463, 570–571, and 678–679.

9 Circles

10 Polygons and Area

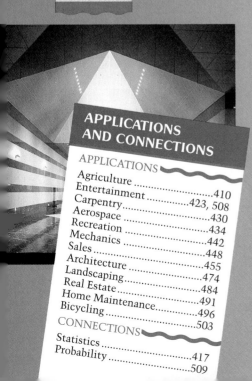

APPLICATIONS AND CONNECTIONS

APPLICATIONS

CONNECTIONS

College Entrance Exam Preview

Have you thought about what you want to do after you graduate from high school? There are many things to consider—college, a full-time job, combination of part-time job and college, military service, and so on. If you want to go to college, you need to begin thinking about what you would like to major in and what college or university you will attend.

To get into most colleges and universities, you need to take the SAT (Scholastic Aptitude Test) or the ACT (American College Test). Approximately 40% of the questions on both the SAT and the ACT are related to geometry. To help you practice for geometry questions as well as the other questions on these tests and other similar tests, you can use the College Entrance Exam Previews. These are provided after every other chapter in this text. They are on pages 114–115, 212–213, 304–305, 406–407, 522–523, and 618–619.

11 Surface Area and Volume 524

12 More Coordinate Geometry 572

13 Loci and Transformations

620

APPLICATIONS AND CONNECTIONS

Symbols

h	altitude*	$\rightarrow$	is mapped onto
$\angle$	angle	$m\angle A$	measure of $\angle A$
a	apothem*	$m\overset{\frown}{AB}$	measure of arc AB
$\approx$	approximately equal to	$\sqrt{}$	nonnegative square root
$\overset{\frown}{AB}$	minor arc with endpoints A and B	(x, y)	ordered pair
$\overset{\frown}{ACB}$	major arc with endpoints A and B	(x, y, z)	ordered triple
		$\parallel$	is parallel to
A	area of a polygon or circle surface area of a sphere*	$\nparallel$	is not parallel to
B	area of base of a prism, cylinder, pryamid, or cone*	$\square$	parallelogram
		P	perimeter*
b	base of a triangle, parallelogram, or trapezoid*	$\perp$	is perpendicular to
		π	pi
$\odot P$	circle with center P	n-gon	polygon with n sides
C	circumference*	r	radius of a circle*
$\cong$	is congruent to	$\overrightarrow{PQ}$	ray with endpoint P passing through Q
$\leftrightarrow$	corresponds to	$\overline{RS}$	segment with endpoints R and S
cos	cosine		
$\circ$	degree	s	side of a regular polygon*
d	diameter of a circle* distance*	$\sim$	is similar to
		sin	sine
AB	distance between points A and B*	ℓ	line ℓ length of a rectangle* slant height*
$=$	equals, is equal to		
$\neq$	is not equal to	m	slope
$>$	is greater than	tan	tangent
A'	the image of preimage A	T	total surface area*
$<$	is less than	Δ	triangle
L	lateral area*	$\overrightarrow{AB}$	vector from A to B
$\overleftrightarrow{DE}$	line containing points D and E	V	volume*

* indicates that this is the symbol for the measure of the item listed.

INSIDE YOUR BOOK

Understanding the Lesson

Each lesson is organized into lessons to make learning manageable. Each lesson begins with an application followed by a well-developed mathematical concept that is explained using several examples. At the end of each lesson, you will complete a variety of exercises.

Objectives tell you exactly what you should be able to do after studying the lesson and completing the exercises.

Interesting math-related trivia and historical facts, presented in **FYI**—"for your information"—enhance the relevance of the mathematics content.

To help you understand each new concept, **example** problems that illustrate the concept are completely worked out. Many are based on real-world applications.

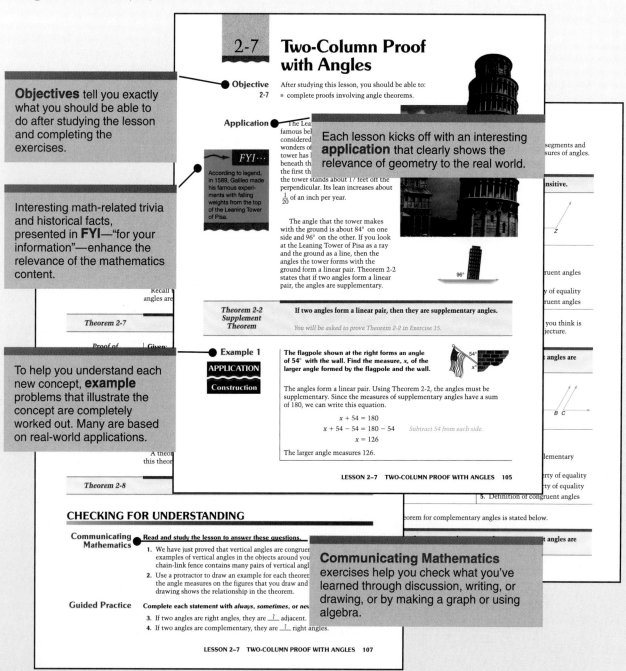

2-7 Two-Column Proof with Angles

Objective
2-7

After studying this lesson, you should be able to:
- complete proofs involving angle theorems.

Application

The Lea... famous bel... considered ... wonders of ... tower has ... beneath th... the first th... the tower stands about 17 feet off the perpendicular. Its lean increases about $\frac{1}{20}$ of an inch per year.

Each lesson kicks off with an interesting **application** *that clearly shows the relevance of geometry to the real world.*

FYI...
According to legend, in 1589, Galileo made his famous experiments with falling weights from the top of the Leaning Tower of Pisa.

The angle that the tower makes with the ground is about 84° on one side and 96° on the other. If you look at the Leaning Tower of Pisa as a ray and the ground as a line, then the angles the tower forms with the ground form a linear pair. Theorem 2-2 states that if two angles form a linear pair, the angles are supplementary.

Theorem 2-2 **Supplement** **Theorem**	If two angles form a linear pair, then they are supplementary angles.

You will be asked to prove Theorem 2-2 in Exercise 15.

Example 1

APPLICATION

Construction

The flagpole shown at the right forms an angle of 54° with the wall. Find the measure, x, of the larger angle formed by the flagpole and the wall.

The angles form a linear pair. Using Theorem 2-2, the angles must be supplementary. Since the measures of supplementary angles have a sum of 180, we can write this equation.

$$x + 54 = 180$$
$$x + 54 - 54 = 180 - 54 \quad \text{Subtract 54 from each side.}$$
$$x = 126$$

The larger angle measures 126.

LESSON 2–7   TWO-COLUMN PROOF WITH ANGLES   105

CHECKING FOR UNDERSTANDING

Communicating Mathematics

Read and study the lesson to answer these questions.

1. We have just proved that vertical angles are congruen... examples of vertical angles in the objects around you... chain-link fence contains many pairs of vertical angl...

2. Use a protractor to draw an example for each theorem... the angle measures on the figures that you draw and... drawing shows the relationship in the theorem.

Communicating Mathematics exercises help you check what you've learned through discussion, writing, or drawing, or by making a graph or using algebra.

Guided Practice

Complete each statement with *always, sometimes,* or *nev...*

3. If two angles are right angles, they are __?__ adjacent.
4. If two angles are complementary, they are __?__ right angles.

LESSON 2–7   TWO-COLUMN PROOF WITH ANGLES   107

2

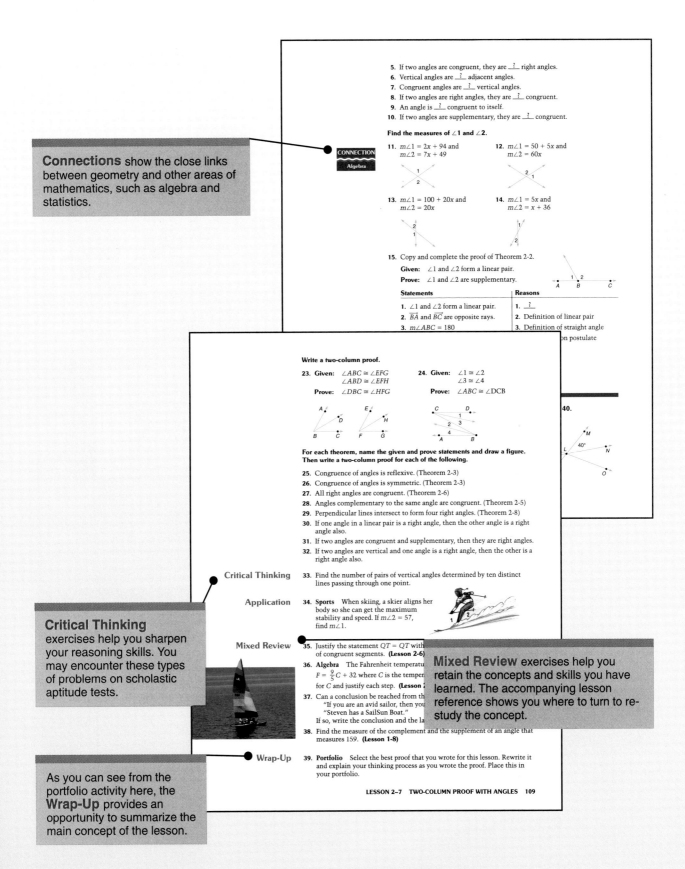

Connections show the close links between geometry and other areas of mathematics, such as algebra and statistics.

5. If two angles are congruent, they are ___?___ right angles.
6. Vertical angles are ___?___ adjacent angles.
7. Congruent angles are ___?___ vertical angles.
8. If two angles are right angles, they are ___?___ congruent.
9. An angle is ___?___ congruent to itself.
10. If two angles are supplementary, they are ___?___ congruent.

Find the measures of ∠1 and ∠2.

CONNECTION
Algebra

11. $m\angle 1 = 2x + 94$ and $m\angle 2 = 7x + 49$

12. $m\angle 1 = 50 + 5x$ and $m\angle 2 = 60x$

13. $m\angle 1 = 100 + 20x$ and $m\angle 2 = 20x$

14. $m\angle 1 = 5x$ and $m\angle 2 = x + 36$

15. Copy and complete the proof of Theorem 2-2.

Given: ∠1 and ∠2 form a linear pair.
Prove: ∠1 and ∠2 are supplementary.

Statements	Reasons
1. ∠1 and ∠2 form a linear pair.	1. ___?___
2. $\overrightarrow{BA}$ and $\overrightarrow{BC}$ are opposite rays.	2. Definition of linear pair
3. $m\angle ABC = 180$	3. Definition of straight angle
	...on postulate

Write a two-column proof.

23. Given: $\angle ABC \cong \angle EFG$
$\angle ABD \cong \angle EFH$
Prove: $\angle DBC \cong \angle HFG$

24. Given: $\angle 1 \cong \angle 2$
$\angle 3 \cong \angle 4$
Prove: $\angle ABC \cong \angle DCB$

40.

For each theorem, name the given and prove statements and draw a figure. Then write a two-column proof for each of the following.

25. Congruence of angles is reflexive. (Theorem 2-3)
26. Congruence of angles is symmetric. (Theorem 2-3)
27. All right angles are congruent. (Theorem 2-6)
28. Angles complementary to the same angle are congruent. (Theorem 2-5)
29. Perpendicular lines intersect to form four right angles. (Theorem 2-8)
30. If one angle in a linear pair is a right angle, then the other angle is a right angle also.
31. If two angles are congruent and supplementary, then they are right angles.
32. If two angles are vertical and one angle is a right angle, then the other is a right angle also.

Critical Thinking

33. Find the number of pairs of vertical angles determined by ten distinct lines passing through one point.

Application

34. **Sports** When skiing, a skier aligns her body so she can get the maximum stability and speed. If $m\angle 2 = 57$, find $m\angle 1$.

Critical Thinking exercises help you sharpen your reasoning skills. You may encounter these types of problems on scholastic aptitude tests.

Mixed Review

35. Justify the statement $QT = QT$ with ... of congruent segments. **(Lesson 2-6)**
36. **Algebra** The Fahrenheit temperatu... $F = \frac{9}{5}C + 32$ where C is the temper... for C and justify each step. **(Lesson ...**

Mixed Review exercises help you retain the concepts and skills you have learned. The accompanying lesson reference shows you where to turn to re-study the concept.

37. Can a conclusion be reached from th... "If you are an avid sailor, then you ... "Steven has a SailSun Boat." If so, write the conclusion and the la...
38. Find the measure of the complement and the supplement of an angle that measures 159. **(Lesson 1-8)**

Wrap-Up

39. **Portfolio** Select the best proof that you wrote for this lesson. Rewrite it and explain your thinking process as you wrote the proof. Place this in your portfolio.

As you can see from the portfolio activity here, the **Wrap-Up** provides an opportunity to summarize the main concept of the lesson.

LESSON 2–7 TWO-COLUMN PROOF WITH ANGLES 109

Getting into the Chapter

Every chapter begins with a two-page multicultural connection to geometry. The large full-color photograph and engaging copy will help you tie the geometry you will learn in the chapter to real people and cultures.

C H A P T E R 2

The list of **chapter objectives** lets you know what you can expect to learn in the chapter.

Reasoning and Introduction to Proof

GEOMETRY AROUND THE WORLD
United States

What rectangle filled with three circles helps keep you safe every day? Give up? The answer is a traffic light, developed in 1923 by an African-American inventor named Garrett Morgan.

Born in Kentucky in 1877, Morgan moved to Cleveland, Ohio, when he was 18. There, he found work repairing sewing machines and soon invented a belt fastener to make the machines operate more efficiently. Later, he invented a gas mask to protect fire fighters inside smoke-filled buildings. The patented device won a gold medal from the International Exposition for Sanitation and Safety. Morgan and three others wore the masks when they entered a gas-filled tunnel to save workers trapped by an explosion. During World War I, Morgan's invention protected Allied soldiers from breathing the deadly gases their enemies used in battle.

Concern for safety also motivated Morgan, at age 48, to invent a three-way automatic electric traffic light. At the time, he was said to be the only African-American in Cleveland who owned a car. Morgan patented his device and later sold the rights to market it to the General Electric Company. He died in 1963.

GEOMETRY IN ACTION

Garrett Morgan systematically went about developing his inventions. First, he identified the problem to be solved. How can sewing machines be made to operate more efficiently? How can people be protected from breathing deadly fumes? How can traffic be regulated to protect pedestrians and drivers?

What do you suppose Morgan did after he identified the problem? Write the steps you think he might have gone through to invent the traffic light.

◀ *Park Avenue in New York City*   Inset: *Garrett Morgan*

CHAPTER OBJECTIVES

In this chapter, you will:
Make conjectures.
Use the laws of logic to make conclusions.
Write proofs involving segment and angle theorems.

Geometry Around the World connects geometry to real life by describing its usefulness to inventors, artists, architects, and other professionals, and its relevance to the lives of ordinary people in many cultures.

This modern traffic light looks different from Garrett Morgan's original invention, but the purpose is the same — saving lives.

In **Geometry in Action,** you can begin to actually apply the geometry that is presented in the chapter.

69

4

Wrapping Up the Chapter

Review pages at the end of each chapter allow you to complete your mastery of the material. The vocabulary, objectives, examples, and exercises help you make sure you understand the skills and concepts presented in the chapter.

Each Chapter Summary and Review opens with a listing of **vocabulary** words that were introduced in the chapter. You can use these to check your understanding of the chapter.

VOCABULARY

Upon completing this chapter, you should be familiar with the following terms:

conclusion	76	76	if-then statement
conditional statement	76	70	inductive reasoning
conjecture	70	82	law of detachment
converse	77	83	law of syllogism
counterexample	71	77	postulate
deductive reasoning	82	98	theorem
hypothesis	76	89	two-column proof

The second part of the Chapter Summary and Review helps you review the important **skills and concepts** you developed in the chapter. You can use the objectives and examples provided in the left column to help you complete the exercises in the right column.

SKILLS AND CONCEPTS

OBJECTIVES AND EXAMPLES	REVIEW EXERCISES
Upon completing this chapter, you should be able to:	Use these exercises to review and prepare for the chapter test.
make geometric conjectures based on given information. **(Lesson 2-1)**	**Determine if the conjecture is *true* or *false* based on the given information. Explain your answer.**
To determine if a conjecture made from inductive reasoning is true or false, look at situations where the given information is true. Determine if there are situations where the given is true and the conjecture is false.	**1. Given:** A, B, and C are collinear and $AB = BC$ **Conjecture:** B is the midpoint of $\overline{AC}$. **2. Given:** $\angle 1$ and $\angle 2$ are supplementary. **Conjecture:** $\angle 1 \cong \angle 2$
write conditionals in if-then form. **(Lesson 2-2)**	**Write the conditional statement in if-then form.**
Write the statement "Adjacent angles have a common ray" in if-then form.	3. Every cloud has a silver lining.
"If angles are adjacent, then they have a common ray."	4. A rectangle has four right angles.
	5. Obsidian is a glassy rock produced by a volcano.
	6. The intersection of two planes is a line.

OBJECTIVES AND EXAMPLES

- complete proofs involving angle theorems **(Lesson 2-7)**

Given: $\angle 1 \cong \angle 2$
Prove: $\angle 3 \cong \angle 4$

Statements	Reasons
1. $\angle 1 \cong \angle 2$	1. Given
2. $\angle 1 \cong \angle 3$ $\angle 2 \cong \angle 4$	2. Vertical $\angle$s are
3. $\angle 3 \cong \angle 4$	3. Congruence of angles is transitive. (used twice)

APPLICATIONS AND CONNECTIONS

16. Advertising Write the conditional "Hard-working people deserve a night on the town at Gil's Grill" in if-then form. Identify the hypothesis and the conclusion of the conditional. Then write the converse. **(Lesson 2-2)**

17. Botany If possible, write a valid conclusion. State the law of logic that you used. **(Lesson 2-3)**
- A sponge is a sessile animal.
- A sessile animal is one that remains permanently attached to a surface all of its adult life.

The **Applications and Connections** problems help you connect the material to the world beyond your textbook.

18. Algebra Name the property of equality that justifies the statement "If $x + y = 3$ and $3 = w + v$, then $x + y = w + v$." **(Lesson 2-4)**

19. Geology The underground temperature of rocks varies with the depth below the surface. The deeper a rock is in the Earth, the hotter it is. The temperature, t, in degrees Celsius estimated by the equation $t = 35d$ where d is the depth in kilometers. Solve the formula for d and justify step. **(Lesson 2-4)**

theorems. **(Lesson 2-6)**

Theorem 2-1 states that congruence of segments is reflexive, symmetric, and transitive.

13. Given: $\overline{AM} \cong \overline{CN}$
$\overline{MB} \cong \overline{ND}$
Prove: $\overline{AB} \cong \overline{CD}$

20. Use the process of elimination to solve this problem. **(Lesson 2-5)**
Alana, Becky, and Carl each had different lunches in the school cafeteria. One had spaghetti, one had a salad, and one had macaroni and cheese. Alana did not have a sa Becky did not have spaghetti or a salad. What did each person have for lunch?

C H A P T E R 1

The Language of Geometry

GEOMETRY AROUND THE WORLD

Have you ever seen a map of Great Britain? If so, then you know its long coastline juts and jags like tricky pieces of a jigsaw puzzle. Like clouds, mountains, and other natural formations, coastlines are irregular in shape. Nature is not usually straight and smooth! Yet, mathematics—particularly geometry—prefers to deal with straight lines and smooth curves. How then can we use mathematics to make accurate models of the irregular shapes found in nature and elsewhere?

This question intrigued mathematician Benoit Mandelbrot. Before Mandelbrot began seeking answers in the mid-1970s, no way existed for creating accurate mathematical representations of nature's irregularities. Because other mathematicians either ignored or minimized the problem, Mandelbrot developed a mathematical method for describing and reconstructing what he named **fractals**—shapes that are irregular or broken. This new field of mathematics is called **fractal geometry.**

GEOMETRY IN ACTION

In geometry, **similar** forms have the same shape, even if they are different sizes. Fractals have **self-similar** shapes. This means that the smaller and smaller details of a form have the same geometrical character as the original, larger form. To verify this, use a magnifying glass to enlarge a small portion of the fractal image shown at the right. This beautiful computer-generated picture was generated using the mathematical formulas of fractal geometry.

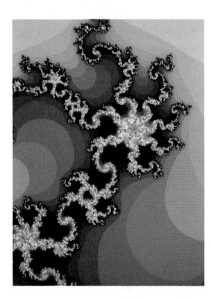

Do you see the pattern repeated in this fractal image? What would you see if you enlarged any section?

◀ *Coastline of Great Britain*   Inset photo: *Benoit Mandelbrot*

1-1 Connections from Algebra: The Coordinate Plane

Objective

After studying this lesson, you should be able to:

- graph ordered pairs on a coordinate plane.

Application

This past summer, Jaime had the opportunity to attend an Amy Grant concert at Poplar Creek. His ticket was for Seat 27, Row KK. When Jaime arrived at the concert, he used a branch of math called **coordinate geometry** to find his seat.

In this lesson, you will use what you already know about algebra and coordinate geometry to help you make a smooth transition into high school geometry.

We don't often recognize the everyday uses of mathematics in the world around us, but without math, things would not run as smoothly as they do. In Jaime's case, the number and letters on his ticket actually represent a point in a coordinate system used to identify seats. Without such a system, it would be hard to get people seated in a timely and organized way.

FYI ···

According to legend, René Descartes got the idea for coordinate geometry while watching a fly walk on a tiled ceiling.

Some of the terms you have used in your study of the coordinate system in algebra are summarized below and shown in the figure at the right.

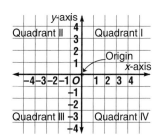

Origin: point *O*, ordered pair (0, 0)

Axes: x-axis and y-axis

Quadrants: regions labeled I, II, III, and IV

Coordinate Plane: plane containing the x-axis and y-axis

The notation A(-3, 2) can also be used to indicate that point A is named by the ordered pair (-3, 2).

Point *A* is located in quadrant II and has **coordinates** (-3, 2). These coordinates, given in an **ordered pair**, locate point *A* relative to the origin and axes. The first coordinate, -3, called the **x-coordinate**, indicates the number of units to move left or right from the origin. The second coordinate, 2, called the **y-coordinate**, indicates the number of units to move up or down from the origin.

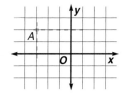

8   CHAPTER 1   THE LANGUAGE OF GEOMETRY

Example 1

APPLICATION

Cartography

On the map at the right, letters and numbers are used to form ordered pairs that name sectors on the map. Name all the sectors that Interstate Highway 75 passes through.

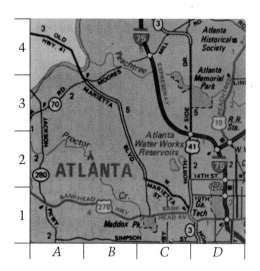

Interstate 75 first appears at the top of the map in sector (*C*, 4). It then travels down through sectors (*C*, 3) and (*C*, 2) and right through sector (*D*, 2) before traveling down again and exiting the map through sector (*D*, 1).

Locating cities on a map is very similar to locating points on the coordinate plane.

Example 2

Write the ordered pairs that name points *R*, *S*, and *T*.

Point *R*: The *x*-coordinate is -2 and the *y*-coordinate is -5. Thus, the ordered pair is (-2, -5).

Point *S*: The *x*-coordinate is 1 and the *y*-coordinate is 4. Thus, the ordered pair is (1, 4).

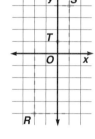

Point *T*: The ordered pair for point *T* is (0, 1). Why?

Look at *R*, *S*, and *T* in Example 2. It appears from their graphs that these points all lie on the same line. We call such points **collinear.** To say that points in the coordinate plane are collinear means that the coordinates of those points all satisfy the same linear equation. In this case, points *R*, *S*, and *T* all lie on the graph of $y = 3x + 1$ since the coordinates of each of the points satisfy (are a solution of) that equation.

Point	x	y	y = 3x + 1
R(-2, -5)	-2	-5	-5 = 3(-2) + 1 ✔
S(1, 4)	1	4	4 = 3(1) + 1 ✔
T(0, 1)	0	1	1 = 3(0) + 1 ✔

Is the point *U*(3, 8) collinear with points *R*, *S*, and *T*? Since $8 \neq 3(3) + 1$, the coordinates of *U* do not satisfy the equation $y = 3x + 1$, which means that point *U* *does not* lie on the same line as *R*, *S*, and *T*. Since point *U* is *not* collinear with points *R*, *S*, and *T*, points *R*, *S*, *T*, and *U* are called **noncollinear** points.

Example 3	How can you locate five points in a coordinate plane so that you are sure that no three of these points will be collinear?

First, plot two points in the plane. Then, draw a line through the points and plot another point not on that line.

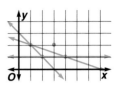

Next, draw all possible lines through pairs of these three points and plot a fourth point not on any of these lines.

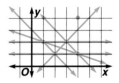

Finally, draw all possible lines through pairs of these four points and plot a fifth point not on any of these lines. These five points must be noncollinear.

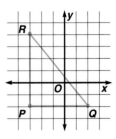

CHECKING FOR UNDERSTANDING

Communicating Mathematics

Read and study the lesson to answer these question.

1. Explain the difference between the ordered pairs (3, 5) and (5, 3).
2. Refer to the situation at the beginning of the lesson. For the concert, Jaime was seated in Seat 27, Row KK. Why do you think letters were used for the rows and numbers were used for the seats in each row?
3. How would you determine whether or not the graphs of $y = 3x + 2$ and $y = -x - 6$ intersect at the point (-2, 3)?
4. How could you show that the points representing the cities of Chicago, Denver, and Seattle on a map are noncollinear?

Guided Practice

In the figure below, triangle PQR (△PQR) is drawn in the coordinate plane. Use the figure to answer each question.

5. In which quadrant is vertex P located?
6. In which quadrant is vertex Q located?
7. Which axis intersects side $\overline{PQ}$?
8. Which vertex of △PQR has the greatest x-coordinate? What are its coordinates?
9. Which vertex of △PQR has the greatest y-coordinate? What are its coordinates?
10. Name the type of angle formed by sides $\overline{RP}$ and $\overline{PQ}$.
11. Describe how you would find the perimeter of △PQR.

EXERCISES

Practice   **Write the ordered pair for each point shown at the right.**

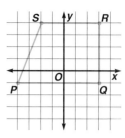

12. P **13.** Q

14. R **15.** S

Graph each point on one coordinate plane.

16. $A(5, -2)$ **17.** $B(3, 6)$ **18.** $C(-6, 0)$

19. $D(-4, 3)$ **20.** $E(-3, -3)$ **21.** $F(0, 4)$

In the figure at the right, all of the segments shown are parallel to either the *x*- or the *y*-axis. Determine the ordered pair that represents each point.

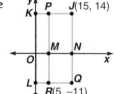

22. K **23.** L

24. M **25.** N

26. P **27.** Q

Points $A(5, 7)$ and $B(-1, 1)$ lie on the graph of $y = x + 2$. Determine whether the following points are collinear with A and B.

28. $C(0, 2)$ **29.** $D(1, -1)$ **30.** $E(-3, -2)$ **31.** $F(-3, -1)$

32. Refer to the figure at the right to answer each question.

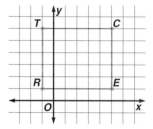

   a. What is the *x*-coordinate of any point collinear with points R and T?

   b. What is the *y*-coordinate of any point collinear with points C and T?

   c. Find the perimeter of figure $RECT$.

33. Make a table of values to determine the coordinates of three points that lie on the graph of $y = 3x + 5$. Use the headings x, $3x + 5$, y, and (x, y).

   a. Use these three points to draw the line in a coordinate plane.

   b. Make a table of values to determine three points on the graph of $y = -2x - 10$. Then draw the line in the same coordinate plane.

   c. Based on your graphs, what is the point of intersection of the lines with equations $y = 3x + 5$ and $y = -2x - 10$?

Critical Thinking   **34.** Describe the possible locations, in terms of quadrants or axes, for point $A(x, y)$ if x and y satisfy the following conditions.

   a. $xy < 0$ **b.** $xy > 0$ **c.** $xy = 0$

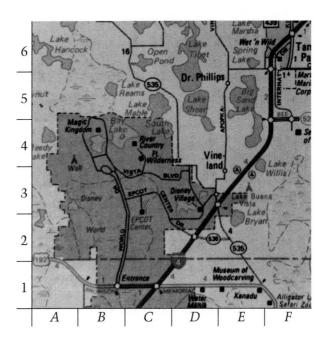

Application

35. Cartography Refer to the map above to answer each question.
  a. What part of Walt Disney World is in sector $(B, 4)$?
  b. In what sector is the city of Vineland?
  c. In what sectors is Lake Tibet?
  d. What road goes from sector $(B, 3)$ to sector $(D, 2)$?
  e. Name all the sectors that Interstate Highway 4 passes through.

Wrap-Up

36. Journal Entry Occasionally you will be asked to record some of your thoughts about the geometry you are learning in a journal. Start your journal by writing a few sentences about the importance of having a coordinate system and knowing how it works.

~ ARCHITECTURE CONNECTION ~

Walt Disney's EPCOT Center is designed to be a model of a city of the future. EPCOT, the Experimental Prototype Community of Tomorrow, is home to the world's largest geodesic sphere, Spaceship Earth. The sphere is eighteen stories high with a diameter of 180 feet. The outer skin of the sphere is made up of interlocking aluminum triangles that look like glass.

The geodesic forms that inspired the design for Spaceship Earth were developed by R. Buckminster Fuller. Mr. Fuller spent years tinkering with geometric shapes before discovering the basis for the geodesic structure. He found that buildings of this type are able to contain more space and use less building material than buildings of conventional design.

1-2 Points, Lines, and Planes

Objectives

After studying this lesson, you should be able to:
- identify and draw models of points, lines, and planes, and
- identify collinear and coplanar points and intersecting lines and planes.

Application

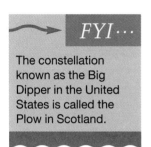

FYI···

The constellation known as the Big Dipper in the United States is called the Plow in Scotland.

Astronomers usually name the different constellations in the night sky by looking at the figure formed when segments are drawn to connect some or all the stars that make up the constellation. Thus, the same constellation might have different names in different parts of the world because of the way that its stars are connected.

Each star in constellations like the Big Dipper suggests the simplest figure studied in geometry—a **point**. In geometry, points do not have any actual size, though they sometimes represent objects, such as stars, that do have size. A point is usually named by a capital letter, and, as you have already seen, in the coordinate plane, a point can also be named by an ordered pair. *All geometric figures are made up of points.*

Suppose a segment connecting two points, *A* and *B*, is extended indefinitely in both directions. This geometric figure is called a **line**.

In plane geometry, a line means a straight line.

A line has no thickness or width, although a picture of a line does. A line is often named by a lowercase script letter. The line at the right is line *k*. If the names of two points on a line are known, such as points *A* and *B* in the figure at the right, then the line can be denoted as follows.

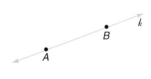

line *AB* $\overleftrightarrow{AB}$ line *BA* $\overleftrightarrow{BA}$

Arrows are drawn on each end of the line to suggest that it goes on forever.

A third basic geometric figure is a **plane**. You are already familiar with the coordinate plane from your study of algebra. Planes are also suggested by flat surfaces such as window panes and walls. Unlike these surfaces, a plane has no thickness and extends indefinitely in all directions.

The four-sided figure used to represent a plane is usually a parallelogram, even though a plane has no edges.

Planes are often represented or modeled using four-sided figures like the one at the right. A plane can be named by a capital script letter or by three *noncollinear points in the plane.* Thus, the plane at the right is plane *R* or plane *PQS*.

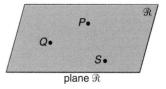

plane *R*

All planes, lines, and points are contained in space. **Space** is the set of all points.

In geometry, the terms *point*, *line*, and *plane* are considered *undefined terms* since they have only been explained using examples and descriptions. Even though they are undefined, these terms can still be used to define other geometric terms and properties.

You may recall from the previous lesson that collinear points are points that lie on the same line. Similarly, **coplanar** points are points that lie in the same plane. The points that you plotted in a coordinate plane in the previous lesson were coplanar points. *All points plotted in a coordinate plane are coplanar.*

Example 1

Refer to the figure at the right to answer each question.

a. Are points A, E, and D collinear?

Since points A, E, and D lie on $\overleftrightarrow{AD}$, they are collinear.

b. Are points A, B, C, and D coplanar?

Points A, B, and C lie in plane $\mathcal{M}$, but point D does *not* lie in plane $\mathcal{M}$. Thus, the four points are not coplanar.

c. How many planes appear in this figure?

There are 4 planes: Plane $\mathcal{M}$ (or plane ABC), plane ACD, plane ABD, and plane CBD.

Notice that segments AB, CB, and DB are dashed rather than solid. For figures in space, dashed segments and lines are used to represent parts of the three-dimensional figure that are hidden from view.

Figures play an important role in understanding geometric concepts. The drawing and labeling of figures can help you model and visualize various geometric relationships. For example, the figures and descriptions below can help you visualize some important relationships among points, lines, and planes.

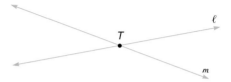

P is on m.
P is in m.
m contains P.
m passes through P.

ℓ and m intersect at T.
ℓ and m intersect in T.
ℓ and m both contain T.
T is the intersection of ℓ and m.
*The **intersection** of two figures is the set of points that are in both figures.*

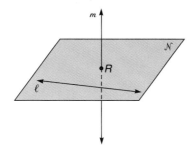

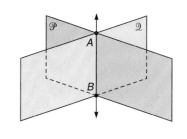

ℓ and R are in $\mathcal{N}$.

$\mathcal{N}$ contains R and ℓ.

m intersects $\mathcal{N}$ at R.

R is the intersection of m with $\mathcal{N}$.

$\overleftrightarrow{AB}$ is in $\mathcal{P}$ and is in $\mathcal{Q}$.

$\mathcal{P}$ and $\mathcal{Q}$ both contain $\overleftrightarrow{AB}$.

$\mathcal{P}$ and $\mathcal{Q}$ intersect in $\overleftrightarrow{AB}$.

$\overleftrightarrow{AB}$ is the intersection of $\mathcal{P}$ and $\mathcal{Q}$.

Example 2

Draw and label a figure in the coordinate plane showing lines $\overleftrightarrow{MN}$ and $\overleftrightarrow{PQ}$ intersecting at X and a point Y not on either $\overleftrightarrow{MN}$ or $\overleftrightarrow{PQ}$. Use $M(1, 1)$, $N(-6, 4)$, $P(0, -3)$, and $Q(5, 2)$ as given points.

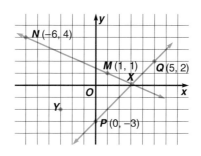

X must be the point of intersection of $\overleftrightarrow{MN}$ and $\overleftrightarrow{PQ}$. Y can be <u>any</u> point that is not on $\overleftrightarrow{MN}$ and is not on $\overleftrightarrow{PQ}$.

CHECKING FOR UNDERSTANDING

Communicating Mathematics

Read and study the lesson to answer these questions.

1. Refer to the application at the beginning of the lesson. What geometric figures do the stars in the night sky represent?

2. What are the three undefined terms presented in this lesson?

3. Draw and label line ℓ intersecting plane $\mathcal{P}$ at point D where ℓ is not in plane $\mathcal{P}$.

4. *True* or *false*: Collinear points are also coplanar.

5. *True* or *false*: Coplanar points are also collinear.

Guided Practice

Draw and label a figure for each relationship.

6. Point A lies on line MN.

7. Plane $\mathcal{M}$ contains line n and point R.

8. Lines ℓ and m intersect at point P.

9. Plane $\mathcal{N}$ and line CD intersect at point T.

Determine whether each statement is *true* or *false*.

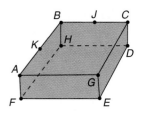

10. A, B, and C are collinear.
11. A, B, and C are coplanar.
12. E, F, G, and H are coplanar.
13. B, J, and G are coplanar.
14. A, K, C, and G are coplanar.

Refer to the figure at the right to answer each question.

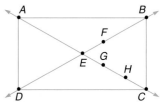

15. What is the intersection of $\overleftrightarrow{DB}$ and $\overleftrightarrow{GH}$?
16. Are F, B, and E collinear? coplanar?
17. Are F, G, and H collinear? coplanar?
18. What points are collinear with A and E?

Refer to the figure at the right to answer each question.

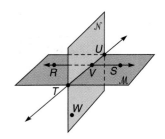

19. What is the intersection of planes $\mathcal{M}$ and $\mathcal{N}$?
20. Are R, S, V, and T coplanar?
21. What points are coplanar with S, T, and V?
22. Are there more than three points on $\overleftrightarrow{RS}$? If so, how many points are on $\overleftrightarrow{RS}$?
23. Are T, U, V, and W the only points on plane $\mathcal{N}$? If no, how many points are on plane $\mathcal{N}$?

EXERCISES

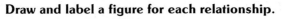

Practice

State whether each of the following is best modeled by a point, line, or plane.

24. the runway landing lights at Byrd International Airport
25. the floor in your classroom
26. the median strip on a two-lane highway
27. the contrail left by an airplane flying at high altitudes
28. each colored dot, or pixel, on a television screen
29. a page from this book

Draw and label a figure for each relationship.

30. $\overleftrightarrow{AB}$ contains point R.
31. ℓ contains point P and lies in plane $\mathcal{N}$.
32. Lines AB and ℓ both contain point X.
33. The intersection of planes $\mathcal{G}$ and $\mathcal{H}$ is $\overleftrightarrow{CD}$.
34. Lines PQ and m, and plane $\mathcal{H}$ intersect at T.

35. Planes $\mathcal{A}$, $\mathcal{B}$, and $\mathcal{C}$ intersect at P.

36. $\overleftrightarrow{RS}$ and plane $\mathcal{L}$ do not intersect.

37. Planes $\mathcal{M}$, $\mathcal{N}$, and $\mathcal{R}$ do not intersect.

38. A, B, and C lie on ℓ, but D does not lie on ℓ.

39. n contains Q and R, but does not contain P and S.

Refer to the figure at the right to answer each question. *This figure is a pyramid, which you will study in Chapter 11.*

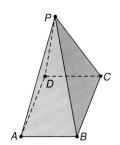

40. The flat surfaces of the pyramid are called **faces**. Name the five planes that contain the faces of the pyramid.

41. Are points A, D, P, and C coplanar?

42. Name three lines that intersect at B.

43. Name two planes that intersect in line PD.

44. Name a line and a plane that intersect in point P.

45. What do the dashed segments in the figure represent?

46. On a separate sheet of paper, repeat the steps shown below to make a drawing of a box. *This figure is also called a rectangular solid.*

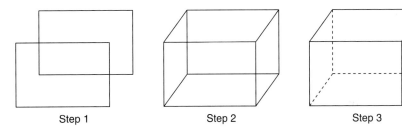

Step 1   Step 2   Step 3

Draw and label a figure for each relationship.

47. Noncollinear points P, Q, and R lie in plane $\mathcal{A}$, point S does not lie in plane $\mathcal{A}$, and P, Q, R, and S lie in plane $\mathcal{B}$.

48. The intersection of planes $\mathcal{C}$, $\mathcal{D}$, $\mathcal{E}$, and $\mathcal{F}$ is point X.

49. Planes $\mathcal{G}$ and $\mathcal{H}$ each intersect plane $\mathcal{J}$ but do not intersect each other.

50. Planes $\mathcal{K}$ and $\mathcal{L}$ intersect each other and they both intersect plane $\mathcal{M}$, but there are *no* points common to all three planes.

Critical Thinking

In your classroom, you can think of the floor and ceiling as models of *horizontal* planes and the walls as models for *vertical* planes. Answer the following questions about horizontal and vertical planes.

51. Can two horizontal planes intersect?

52. Can two vertical planes intersect?

53. Must any line that lies in a vertical plane be a vertical line?

54. Must any line that lies in a horizontal plane be a horizontal line?

Applications

55. Carpentry Mr. Johnstone is building some new chairs for his dinette table. If he wants to be sure that these chairs will not wobble, should he build chairs with three legs or with four legs? Explain.

56. Printing In order to print a color photo for this book, dots of four colors are used to break a photo down into its component parts. You can use a magnifying glass to see the dots. Each photo is separated into four primary colors as shown below.

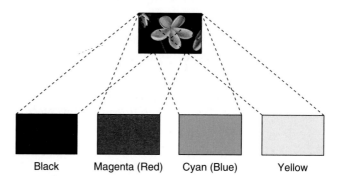

Black   Magenta (Red)   Cyan (Blue)   Yellow

The photos in this book are printed at 1200 dpi (dots per square inch). What is the maximum number of different colored dots that could be used to print a 5-inch by 8-inch rectangular color photo? *Hint: Recall that the formula for the area of a rectangle is area = length × width.*

Mixed Review

Graph each point on one coordinate plane. (Lesson 1-1)

57. $A(-4, 3)$ **58.** $B(3, -4)$ **59.** $C(-4, 0)$

Refer to the figure at the right to answer each question. (Lesson 1-1)

60. What ordered pair names W?

61. What ordered pair names Z?

62. What is the x-coordinate of any point that is collinear with X and Y?

63. What is the y-coordinate of any point that is collinear with Z and Y?

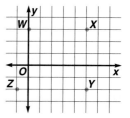

Wrap-Up

64. Name the three important geometric figures described in this lesson. Then give an example of something that is modeled by each of the figures.

Looking Ahead
Algebra Review

You will need to find square roots in the next lesson. Find each square root to the nearest hundredth.

Examples:

a. $\sqrt{1296} = \sqrt{9 \cdot 9 \cdot 4 \cdot 4}$
$= \sqrt{36^2}$
$= 36$

b. $\sqrt{530}$

Enter: 530 $\boxed{\sqrt{x}}$ 23.0217289

Therefore, $\sqrt{530}$ to the nearest hundredth is 23.02.

---

65. $\sqrt{961}$ **66.** $\sqrt{3025}$ **67.** $\sqrt{775}$ **68.** $\sqrt{6436}$

1-3 Problem-Solving Strategy: List the Possibilities

Objective

After studying this lesson, you should be able to:

- solve problems by making a list of possibilities.

Application

Mr. and Mrs. Jeremy are planning to plant a rectangular vegetable garden in their backyard. They want the garden to have an area of at least 15 square yards, but they only have 18 yards of wire mesh fence to use for fencing. What are the possible dimensions for the Jeremys' garden if they plan to use all the fencing and for the sides to have whole number lengths?

The four steps for solving any problem are listed below.

Problem-Solving

1. **Explore the problem.**
2. **Plan the solution.**
3. **Solve the problem.**
4. **Examine the solution.**

Let's use the four-step plan to find the dimensions of the Jeremys' garden.

EXPLORE   The garden is to be a rectangle with an area of at least 15 square yards. The Jeremys only have 18 yards of fencing material, so the perimeter of the rectangle will be at most 18 yards.

PLAN   You may recall that the formulas for the perimeter and area of a rectangle are as follows.

$$\text{perimeter} = 2\ell + 2w \qquad\qquad \text{area} = \ell w$$

ℓ represents the length and w represents the width.

Thus, to solve this problem, we must find all whole numbers ℓ and w such that $18 = 2\ell + 2w$ and $15 \le \ell w$.

One possible approach we can use to find the solution is to *list the possibilities.* By making an organized list or table of all the rectangles with perimeters of 18 yards and with sides of whole number length, we can determine the ones with areas greater than or equal to 15 square yards.

SOLVE First, we need to list all whole numbers ℓ and w such that $18 = 2\ell + 2w$ along with the corresponding values for ℓw. Then, we can find all the values in the list where $15 \le \ell w$.

ℓ	w	$2\ell + 2w$	ℓw	$15 \le \ell w$
1	8	18	8	
2	7	18	14	
3	6	18	18	✔
4	5	18	20	✔
5	4	18	20	✔
6	3	18	18	✔
7	2	18	14	
8	1	18	8	

Why are the values $\ell = 0$, $w = 9$ and $\ell = 9$, $w = 0$ not listed?

Thus, there are four possibilities for the dimensions of the Jeremys' garden: 3 yards by 6 yards, 4 yards by 5 yards, 5 yards by 4 yards, and 6 yards by 3 yards.

EXAMINE Our answers satisfy the conditions of the problem. A rectangular garden that is 3 yards by 6 yards, 4 yards by 5 yards, 5 yards by 4 yards, or 6 yards by 3 yards has a perimeter less than or equal to 18 yards and an area greater than or equal to 15 square yards.

Whenever you are listing possibilities, it may help you to make a table or a tree diagram to be sure that you do not omit any possibilities.

Example

On Wednesdays, the Cliff House restaurant offers three dinner specials. One is a shrimp dinner with either rice pilaf or a baked potato. The second is a lobster dinner with either rice pilaf or a vegetable medley. The third is a chicken dinner with either the rice pilaf, vegetable medley, or a baked potato. Each special dinner also comes with either soup or salad. How many different dinners specials can you order on Wednesdays?

The tree diagram on page 21 shows the various combinations.

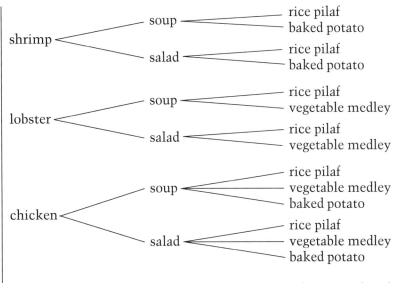

shrimp
- soup
  - rice pilaf
  - baked potato
- salad
  - rice pilaf
  - baked potato

lobster
- soup
  - rice pilaf
  - vegetable medley
- salad
  - rice pilaf
  - vegetable medley

chicken
- soup
  - rice pilaf
  - vegetable medley
  - baked potato
- salad
  - rice pilaf
  - vegetable medley
  - baked potato

There are 14 different dinner specials you can order on Wednesdays.

CHECKING FOR UNDERSTANDING

Communicating Mathematics

Read and study the lesson to answer each question.

1. When you use the strategy of listing possibilities, why is it important to make your list organized?

2. Name three different ways you can display possibilities.

3. Refer to the application at the beginning of the lesson. What does it mean to say a rectangle has a perimeter of 18 yards?

Guided Practice

Solve each problem by listing possibilities.

4. The president, vice president, secretary, and treasurer of the Debate Team are to be seated in a row of four chairs for a yearbook picture. How many different seating arrangements are possible?

5. A vending machine dispenses products that each cost 50¢. The machine will only accept quarters, dimes, and nickels. How many different combinations of coins must the machine be programmed to accept?

6. All of the points *A*, *B*, *C*, *D*, and *E* are on a circle. List all of the lines that contain exactly two of these five points.

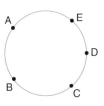

7. Sarah, Victor, Janel, and Kirby checked their coats at the door of the Colonnade Room Restaurant. The coat checker is in quite a hurry when he returns their coats. How many different ways can their coats be returned so that no one receives the right one?

EXERCISES

Solve. Use any strategy.

Strategies

Look for a pattern.
Solve a simpler problem.
Act it out.
Guess and check.
Draw a diagram.
Make a chart.
Work backward.

8. Tim, Renee, and Sandra are walking down the hall single file. Tim always tells the truth, and Sandra *never* tells the truth. The student walking in the front says, "Tim is in the middle." The student in the middle says," I am Renee." The student in the back says, "Sandra is in the middle." Name the students in order from front to back.

9. You can cut a pizza into a maximum of 7 pieces with only 3 straight cuts. What is the greatest number of pieces you can make with 6 straight cuts?

10. In the figure at the right, exactly two line segments can be removed to make four squares. Which two segments should be removed?

11. Place operation symbols and any necessary grouping symbols in the sentence below to make it correct.

$$3\ 3\ 3\ 3\ 3\ 3 = 11$$

12. Line segments are used to make up the digits of the numbers in the display of a digital clock.
    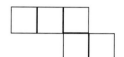
    a. Which line segment is used least often when forming the digits 0 to 9?
    b. Which segment is used most often?

13. Ted and Mary traded in their old car which averaged 22 miles per gallon. The EPA sticker on the new car stated that it should average 37 miles per gallon. If Ted and Mary drive about 12,000 miles per year and gasoline is about $1.20 per gallon, how much should they expect to save on gasoline in the first year of owning their new car?

14. An airline gives each of its flight attendants one red shirt, one white shirt, one blue shirt, a navy blazer, one pair of navy pants, and one pair of navy pin-striped pants. How many different outfits can a flight attendant wear if the blazer is optional? *(Hint: Find the outfits possible without the blazer, then add the blazer to each outfit.)*

COOPERATIVE LEARNING PROJECT

Work in groups. Each person in the group must understand the solution and be able to explain it to any person in class.

Three students are blindfolded and stand in a single-file line. Karen takes three hats from a box containing three red hats and two yellow hats and places one on each of the blindfolded students. She tells the blindfolded students how many hats of each color there were in the box and removes their blindfolds. The student in the back looks at the hats on the two students in front of him and says "I don't know what color hat I am wearing." The student in the center hears that statement, looks at the hat on the student in front of her, and says the same thing. The student in the front says "I know what color hat I am wearing." What color hat is he wearing? Explain his logic.

1-4 Finding the Measures of Segments

Objectives

After studying this lesson, you should be able to:
- find the distance between points on a number line, and
- find the distance between points in a coordinate plane.

Application

In designing their deck, Bill and Teresa Hartford decided it should be unlike any they had ever seen. However, this design created a real challenge for contractors who were asked to provide construction bids. The bids varied widely because of the way each contractor figured the measurements for the lumber needed and the lumber that would be wasted after cutting.

The idea of measurement is an important mathematical concept. Without it, many things would be impossible. For example, the contractors would not know how much lumber is needed to build the Hartfords' deck. Like Sally, you have already had quite a bit of experience measuring with rulers.

In geometry, the distance between two points is used to define the measure of a segment. We discussed segments in the previous lesson when we talked about connecting pairs of stars in a constellation. Segments can be defined by using the idea of *betweenness* of points.

In the figure at the right, point N is **between** M and P while point Q is *not* between M and P. For N to be between M and P, all three points must be collinear. **Segment MP**, written $\overline{MP}$, consists of points M and P and all points between M and P. The **measure** of $\overline{MP}$, written MP (without a bar over the letters), is the distance between M and P. Thus, the measure of a segment is the same as the distance between its two endpoints.

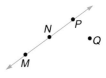

In order to quantify the measure of a segment, you must measure the segment using a device, like a ruler, that has a *unit of measure*, such as inches.

In the comic above, Sally placed her ruler so that one of Snoopy's lips was on 6 and the other was on 9. She could have placed the ruler so that one lip was on 0 and the other was on 3. In fact, two points on any line can always be paired with real numbers so that one point is paired with zero and the other is paired with a positive number. This correspondence allows you to measure the distance between any two points, and thus, the length of any segment. This is called the **ruler postulate**.

Distance can be measured using centimeters, inches, or any other convenient unit of measure.

Postulate 1-1 Ruler Postulate	The points on any line can be paired with the real numbers so that, given any two points P and Q on the line, P corresponds to zero, and Q corresponds to a positive number.

*A **postulate** is a statement that is assumed to be true. You will study postulates more in depth in Chapter 2.*

The distance between points may be measured using a *number line*.

To find the measure of $\overline{AB}$, you first need to identify the coordinates of A and B. The coordinate of A is 1 and the coordinate of B is 5. Since measure is always a positive number, you can subtract the lesser coordinate from the greater one, or find the *absolute value* of the difference. When you use absolute value, the order in which you subtract the coordinates does not matter.

Finding the distance from A to B or from B to A results in the same measure.

distance from A to B
$$|5 - 1| = |4|$$
$$= 4$$

distance from B to A
$$|1 - 5| = |-4|$$
$$= 4$$

The measure of $\overline{AB}$ is 4 or $AB = 4$.

Example 1

Find *AB*, *BC*, and *AC* on the number line shown below.

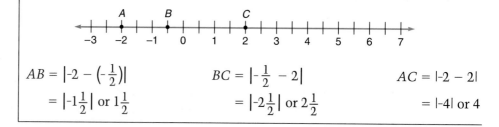

$$AB = \left|-2 - \left(-\tfrac{1}{2}\right)\right|$$
$$= \left|-1\tfrac{1}{2}\right| \text{ or } 1\tfrac{1}{2}$$

$$BC = \left|-\tfrac{1}{2} - 2\right|$$
$$= \left|-2\tfrac{1}{2}\right| \text{ or } 2\tfrac{1}{2}$$

$$AC = |-2 - 2|$$
$$= |-4| \text{ or } 4$$

In Example 1, B is between A and C and $AB + BC = AC$ since $1\tfrac{1}{2} + 2\tfrac{1}{2} = 4$. This example and others like it lead us to the following postulate.

| Postulate 1-2
Segment Addition
Postulate | **If Q is between P and R, then $PQ + QR = PR$.**
If $PQ + QR = PR$ then Q is between P and R. |

Example 2

Find the measure of $\overline{MN}$ if M is between K and N, $KM = 2x - 4$, $MN = 3x$, and $KN = 26$.

Since M is between K and N, $KM + MN = KN$.

$$KM + MN = KN$$
$$(2x - 4) + 3x = 26 \qquad KM = 2x - 4,\ MN = 3x,\ KN = 26$$
$$5x - 4 = 26 \qquad \textit{Add 4 to both sides.}$$
$$5x = 30$$
$$x = 6$$

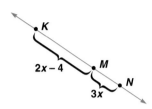

$$MN = 3x$$
$$= 3(6)\ \text{or}\ 18$$

The measure of $\overline{MN}$ is 18.

You can find the measure of segments in the coordinate plane by using the distance formula. You will derive this formula in Chapter 8.

| *The Distance Formula* | **The distance, d, between any points with coordinates (x_1, y_1) and (x_2, y_2) is given by the following formula.**
$$d = \sqrt{(x_2 - x_1)^2 + (y_2 - y_1)^2}$$
The distance between points is the same no matter which point is called (x_1, y_1). |

Example 3

Algebra

Find the length of the segment with endpoints $A(-2, 3)$ and $B(5, -3)$.

Let $(-2, 3)$ be (x_1, y_1) and $(5, -3)$ be (x_2, y_2).
$$d = \sqrt{(x_2 - x_1)^2 + (y_2 - y_1)^2} \qquad \textit{Distance formula}$$
$$= \sqrt{[5 - (-2)]^2 + (-3 - 3)^2}$$
$$= \sqrt{7^2 + (-6)^2}$$
$$= \sqrt{85}$$
$$\approx 9.22$$

The length of $\overline{AB}$ is about 9.22 units.

A compass and straightedge can be used to construct a segment with the same length as a given segment without knowing the exact length of the segment.

Construct a segment that has the same length as $\overline{XY}$.

1. Use a straightedge to draw a line on your paper.

2. Choose any point on that line. Label it P.

3. Place the compass at point X and adjust the compass setting so that the pencil is at point Y.

4. Using that setting, place the compass at point P and draw an arc that intersects the line. Label the point of intersection, Q.

5. By construction, $XY = PQ$.

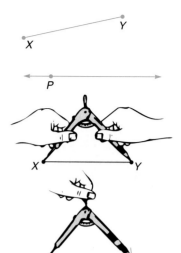

Because the measures of segments are real numbers, they can be compared. To compare the measure of $\overline{AB}$ to the measures of $\overline{CD}$, $\overline{EF}$, and $\overline{GH}$ shown below, you could set your compass width to match the measure of $\overline{AB}$ and then compare this to the measure of each segment.

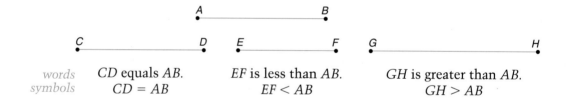

words	CD equals AB.	EF is less than AB.	GH is greater than AB.
symbols	$CD = AB$	$EF < AB$	$GH > AB$

CHECKING FOR UNDERSTANDING

Communicating Mathematics

Read and study the lesson to answer each question.

1. When using the distance formula to find the distance between points $A(18, 8)$ and $B(5, 7)$, do you have to choose 18 for x_1? Explain.

2. Refer to the comic on page 23. Explain how Sally knew that the length of Snoopy's mouth was "lip to lip, three inches."

3. Draw $\overline{AB}$ and $\overline{BC}$ such that $AB + BC \neq AC$.

4. Use a compass to compare the lengths of the following segments. List them in order from shortest to longest.

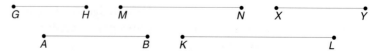

Refer to the number line at the right to find each measure.

5. AB 6. CD

7. BD 8. CB

9. DA 10. AC

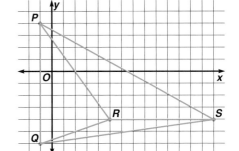

Refer to the coordinate plane at the right to find each measure. If the measure is not a whole number, round the result to the nearest hundredth.

11. PQ 12. SR

13. RP 14. PS

15. QR 16. QS

17. If B is between A and C, $AB = x$, $BC = 2x + 1$, and $AC = 22$, find the value of x and the measure of $\overline{BC}$.

18. Construct a segment that is twice as long as $\overline{MN}$, shown at the right.

EXERCISES

Practice **Refer to the number line below to find each measure.**

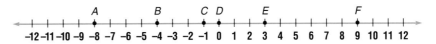

19. CD 20. BF 21. CF 22. EB

23. BA 24. FE 25. FA 26. AC

Given that J is between H and K, find each missing measure.

27. $HJ = 17$, $JK = 6$, $HK = \underline{\ ?\ }$ 28. $HJ = 4.8$, $JK = 7$, $HK = \underline{\ ?\ }$

29. $HJ = 23.7$, $JK = \underline{\ ?\ }$, $HK = 35.2$ 30. $HJ = \underline{\ ?\ }$, $JK = 2\frac{1}{2}$, $HK = 6\frac{2}{5}$

Refer to the coordinate plane at the right to find each measure. If the measure is not a whole number, round the result to the nearest hundredth.

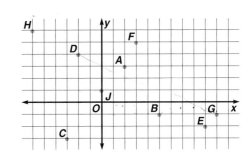

31. AB 32. CF

33. DG 34. HE

35. JF 36. AD

37. GE 38. JB

39. CH 40. HG

If B is between A and C, find the value of x and the measure of $\overline{BC}$.

41. $AB = 3$, $BC = 4x + 1$, $AC = 8$

42. $AB = x + 2$, $BC = 2x - 6$, $AC = 20$

43. $AB = 24$, $BC = 3x$, $AC = 7x - 4$

44. $AB = 3$, $BC = 2x + 5$, $AC = 11x + 2$

Given $\overline{AB}$ and $\overline{CD}$ shown at the right, construct segments with measures equal to the following measures.

45. $AB + CD$ **46.** $AB - CD$

47. $3(AB)$ **48.** $2(CD) - AB$

49. Find the perimeter of the triangle with vertices $X(2, -1)$, $Y(5, 3)$, and $Z(-3, 11)$. Round your result to the nearest hundredth.

50. Find the value of a so that the distance between points $A(4, 7)$ and $B(a, 3)$ is 5 units.

51. Modeling   Draw a figure that satisfies all of the following conditions.
(1) Points V, W, X, Y, and Z are all collinear,
(2) V is between Y and Z,
(3) X is next to V, and
(4) $WY = YX$.

Critical Thinking

52. Use the segment addition postulate to show that points $R(1, 1)$, $S(5, 4)$, and $T(-7, -5)$ are collinear.

53. Points A, B, and C are collinear. If $AB = x$, $BC = 2x + 3$, and $AC = 3x - 6$, which point is between the other two points?

Application

54. Telecommunications   In order to set long distance rates, phone companies will first superimpose an imaginary coordinate grid over the United States. Then the location of each exchange is represented by an ordered pair. The units on this grid are approximately equal to 0.316 mile. So, a distance of 3 units on the grid equals an actual distance of about 3(0.316) or 0.948 mile. Suppose the exchanges in two cities are located at (158, 562) and (387, 213). Find the actual distance between these cities, to the nearest mile.

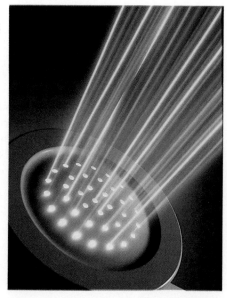

Computer 55. Use the BASIC program at the right to find the distance between each pair of points whose coordinates are given.
  a. What is represented by line 30 of the program?
  b. (5, -1), (11, 7)
  c. (0.67, -4), (3, -2)
  d. (12, -2), (-3, 5)

```
10 INPUT "ENTER COORDINATES OF
   POINT (X1, Y1)."; X1, Y1
20 INPUT "ENTER COORDINATES OF
   POINT (X2, Y2)."; X2, Y2
30 D = SQR((X2 - X1)^2 +
   (Y2 - Y1)^2)
40 PRINT "DISTANCE FROM ("; X1;
   ", "; Y1;") TO (";X2;", "
   Y2;") IS ";D;" UNITS."
50 END
```

Mixed Review   **Refer to the figure at the right to answer each question.**

56. What ordered pair names D? **(Lesson 1-1)**

57. Which points are in quadrant IV? **(Lesson 1-1)**

58. What is the point of intersection of $\overleftrightarrow{BD}$ and $\overleftrightarrow{EC}$? **(Lesson 1-2)**

59. Which points are *not* on $\overleftrightarrow{BC}$? **(Lesson 1-2)**

60. **Modeling**   Draw and label a figure to show plane $\mathcal{R}$ contains point A and line ℓ. **(Lesson 1-2)**

61. Solve by listing possibilities. **(Lesson 1-3)**
    Telecommunications   The telephone area codes in the U.S. and Canada are three-digit numbers where the first digit is 2, 3, 4, 5, 6, 7, 8, or 9, the second digit is 0 or 1, and the third digit is any digit other than 0. How many different area codes start with the digit 6?

Wrap-Up   62. Write a five-question quiz about this lesson. Be sure to include answers to your questions.

~~~~~~~~ **HISTORY CONNECTION** ~~~~~~~~

Hypatia

   Hypatia was the first woman known to have made significant contributions to the field of mathematics. She was the daughter of the Greek mathematician Theon, whose version of Euclid's *Elements* became the traditional geometry text. She taught at the famous Library of Alexandria in Egypt, considered the intellectual center of the ancient world. Hypatia wrote important commentaries on the works of the mathematician Appollonius and the scientist Ptolemy. Her interests also included the study of astronomy and philosophy.

   Because Hypatia was important and respected among scholars of the era, she became the target of criticism from fanatics who equated science to paganism. In March of A.D. 415, she was brutally murdered by an angry mob. Soon after her death, the library was destroyed, and the Dark Ages began, limiting the serious study of mathematics for the next 500 years.

1-5 Segment Relationships

Objectives

After studying this lesson, you should be able to:
- find the midpoint of a segment, and
- identify and use congruent segments.

Application

The machine used to drill the holes in record albums must place the hole in the exact center of each album. If a record is 12 inches in diameter, how far from the edge of the record should the center of the drill be placed?

In order to find the exact center of a record, you must locate the *middle point* of a line segment that is a diameter of the record. This point, called the **midpoint** of the segment, divides the diameter into two segments that have the same length.

└─── 12 in. ───┘

Definition of Midpoint

The midpoint, M, of $\overline{PQ}$ is the point between P and Q such that $PM = MQ$.

A diameter of the record album is 12 inches long. So the length of a segment from the edge of the record to the center is 6 inches. *Why?* Thus, the center of the drill should be placed 6 inches from the edge of the record.

In algebra, you may have determined the midpoint of a segment on a number line or in the coordinate plane using one of the following formulas.

Midpoint on a Number Line

The coordinate of the midpoint of a line segment whose endpoints have coordinates a and b is

$$\frac{a+b}{2}.$$

Midpoint in the Coordinate Plane

The coordinates of the midpoint of a line segment whose endpoints have coordinates (x_1, y_1) and (x_2, y_2) are

$$\left(\frac{x_1 + x_2}{2}, \frac{y_1 + y_2}{2} \right).$$

Example 1

CONNECTION

Algebra

If $M(-1, 7)$ is the midpoint of $\overline{GH}$ and the coordinates of G are $(2, 5)$, what are the coordinates of H?

Let $(2, 5)$ be (x_1, y_1) and let (x_2, y_2) be the coordinates of H.
Now use the expression for the coordinates of the midpoint to find x_2 and y_2.

$$\left(\frac{x_1 + x_2}{2}, \frac{y_1 + y_2}{2}\right) = (-1, 7)$$

The coordinates of the midpoint, M, are (-1, 7).

$$\frac{x_1 + x_2}{2} = -1 \qquad \frac{y_1 + y_2}{2} = 7$$

The x- and y-coordinates must be equal.

$$\frac{2 + x_2}{2} = -1 \qquad \frac{5 + y_2}{2} = 7$$

$$2 + x_2 = -2 \qquad 5 + y_2 = 14$$

$$x_2 = -4 \qquad y_2 = 9$$

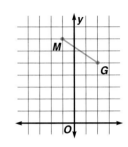

The coordinates of $H(x_2, y_2)$ are $(-4, 9)$.

Any segment, line, or plane that intersects a segment at its midpoint is called a **segment bisector**. In the figure at the right, M is the midpoint of $\overline{PQ}$. Thus, point M, $\overline{TM}$, $\overleftrightarrow{RM}$, and plane $\mathcal{N}$ are all bisectors of $\overline{PQ}$ and are said to bisect $\overline{PQ}$.

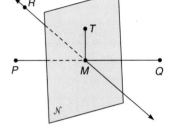

The midpoint can also be considered a segment bisector.

You can use a compass to find the midpoint of any segment, and thus, bisect the segment.

CONSTRUCTION

This construction draws a bisector that is perpendicular to the segment.

Using a compass and straightedge, bisect a segment.

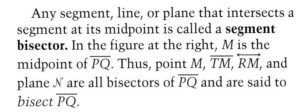

1. Use a straightedge to draw the segment you wish to bisect. Name it $\overline{XZ}$.

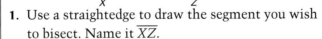

2. Place the compass at point X. Adjust the compass so that its width is greater than $\frac{1}{2}XZ$.

3. Draw arcs above and below $\overline{XZ}$.

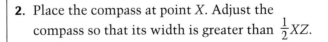

4. Using the same compass setting, place the compass at point Z. Draw arcs above and below $\overline{XZ}$ that intersect the two arcs previously drawn. Label the points of intersection A and B.

5. Use a straightedge to draw $\overleftrightarrow{AB}$ that will intersect $\overline{XZ}$. Label the point of intersection Y.

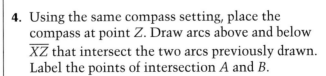

Point Y is the midpoint of $\overline{XZ}$, and $\overleftrightarrow{AB}$ is a bisector of $\overline{XZ}$. Also, $XY = YZ = \frac{1}{2}XZ$.

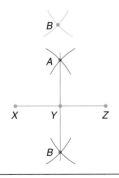

In the figure at the right, $AB = PQ$ since $\overline{AB}$ and $\overline{PQ}$ have the same lengths. Segments that are equal in length are called **congruent segments.** To indicate that $\overline{AB}$ and $\overline{PQ}$ are congruent, we write $\overline{AB} \cong \overline{PQ}$, which is read "segment AB is congruent to segment PQ." In a diagram, small "hash marks," like the ones in red at the right, are used to indicate congruent segments.

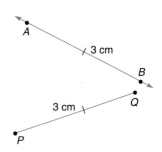

The definition of congruent segments tells us that the statements $AB = PQ$ and $\overline{AB} \cong \overline{PQ}$ are equivalent. Therefore, we will use them interchangeably. For example, if you know that M is the midpoint of $\overline{KN}$, then you could write that $\overline{KM} \cong \overline{MN}$ rather than $KM = MN$.

The concept of congruence can be applied to other figures and objects. In geometry, two objects are **congruent** if they have the exact same size *and* shape. For many of the geometric figures we will be discussing in this text, such as angles, triangles, circles, and arcs, more specific definitions for congruence will be provided.

CHECKING FOR UNDERSTANDING

Communicating Mathematics

Read and study the lesson to answer each question.

1. If $AB = BC$, must B be the midpoint of $\overline{AC}$? Explain.

2. Draw two segments that bisect each other.

3. How many midpoints does a segment have? How many bisectors? Prepare a convincing argument to share with your classmates.

INVESTIGATION

4. Suppose two points, X and Y, are drawn on a piece of paper. Show how you could find the midpoint of $\overline{XY}$ by simply folding the paper.

Guided Practice

Refer to the number line below to find the coordinate of the midpoint of each segment.

5. $\overline{BD}$ 6. $\overline{AB}$

7. $\overline{DA}$ 8. $\overline{BC}$

9. $\overline{CD}$ 10. $\overline{CA}$

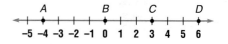

Refer to the coordinate plane at the right to find the coordinates of the midpoint of each segment.

11. $\overline{PQ}$ 12. $\overline{PR}$

13. $\overline{RQ}$ 14. $\overline{SP}$

15. $\overline{QS}$ 16. $\overline{RS}$

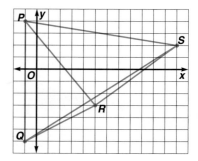

17. If E is the midpoint of $\overline{DF}$, $DE = 5x - 3$, and $EF = 3x + 5$, find the value of x and the measure of $\overline{DF}$.

18. Copy $\overline{AB}$ shown at the right. Then use a compass and straightedge to bisect the segment.

EXERCISES

Practice

Refer to the figure below to determine whether each statement is *true* or *false*.

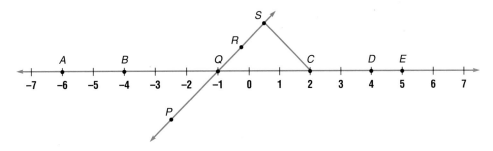

19. Q is the midpoint of $\overline{BC}$.

20. D is the midpoint of $\overline{CE}$.

21. $\overleftrightarrow{RP}$ bisects $\overline{AD}$.

22. $\overleftrightarrow{CS}$ bisects $\overline{EQ}$.

23. $\overline{AB} \cong \overline{BQ}$

24. $\overline{QB} \cong \overline{CE}$

25. $\overline{BC} \cong \overline{QE}$

26. $AQ > QE$

27. $EC > BQ$

28. $AC \le DB$

Given the coordinates of one endpoint of $\overline{AB}$ and its midpoint, M, find the coordinates of the other endpoint.

29. $A(-1, 5)$, $M(2, 5)$

30. $B(3, -2)$, $M(-3, 1)$

31. $M(-2, 2)$, $B(-6, -4)$

32. $A(-1, -3)$, $M(0.5, -6.5)$

CONNECTION

Algebra

In the figure at the right, $\overline{CX}$ bisects $\overline{AB}$ at X and $\overleftrightarrow{CD}$ bisects $\overline{XB}$ at Y. Given the following conditions, find the value of x and the measure of the indicated segment.

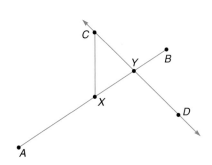

33. $AX = 2x + 11$, $XB = 4x - 5$; $\overline{AB}$

34. $AB = x + 3$, $AX = 3x - 1$; $\overline{XB}$

35. $YB = 23 - 2x$, $XY = 2x + 3$; $\overline{AB}$

36. $AX = 27 - x$, $XB = 13 - 3x$; $\overline{XY}$

37. $AB = 5x - 4$; $XY = x + 1$; $\overline{AX}$

If XY = 8x − 5, YZ = 4x + 7, and XZ = 34, answer each question.

38. For what value of x are $\overline{XY}$ and $\overline{YZ}$ congruent?

39. If $\overline{XY}$ and $\overline{YZ}$ are congruent, is Y the midpoint of $\overline{XZ}$? *(Hint: Investigate the measure of $\overline{XZ}$.)*

Copy each segment shown below. Then use a compass and straightedge to bisect each segment.

40. W   R   **41.**   B   **42.**   F   P   Y

43. Find the coordinates of points P and Q on $\overline{AB}$ which has endpoints A(2, 3) and B(8, -9). P is between A and Q and $\overline{AP} \cong \overline{PQ} \cong \overline{QB}$.

44. The coordinates of K and L are (8, 12) and (-4, 0). Find the coordinates of a point N on $\overline{KL}$ such that $KN = \frac{1}{4}KL$.

45. If S is between R and T, RT = 24, RS = x^2 + 8, and ST = 3x + 6, find the value of x and determine if S is the midpoint of $\overline{RT}$.

Critical Thinking

46. Copy △ABC shown at the right.
  a. Construct the lines that bisect and are perpendicular to each side of the triangle using the construction on page 31.
  b. What conjecture could you make about these three lines?
  c. Draw two different triangles and test your conjecture on these triangles.

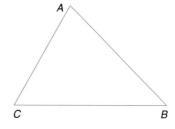

47. Make another copy of △ABC. Use the construction on page 31 to find the midpoints of $\overline{AB}$ and $\overline{AC}$. Label them M and N. Draw $\overline{MN}$. Use a compass to compare the lengths of $\overline{MN}$ and $\overline{BC}$. What conjectures could you make about $\overline{MN}$ and $\overline{BC}$?

Application

48. Carpentry Mr. Juarez needs to cut some pieces of lumber into 10-foot lengths, but he left his tape measure at home. He has a 4-foot long board and a 2-foot long board.
  a. Describe how he could use the boards to cut the lumber to the required length.
  b. What geometric properties is Mr. Juarez using in order to cut the lumber to the required length?

49. Point *B* lies on the same vertical line as *A*(-1, 2). If *AB* = 3 what are the coordinates of *B*? **(Lesson 1-1)**

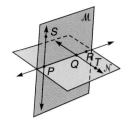

50. Refer to the figure at the right. **(Lesson 1-2)**
   a. What points lie in plane *M*?
   b. Are points *S*, *Q*, and *T* coplanar? Why or why not?

51. Points *W*, *X*, *Y*, and *Z* are collinear. How many ways can you use two letters to name the line that contains these four points? **(Lesson 1-3)**

52. Find *AB*, given *A*(-1, 4) and *B*(6, -20). **(Lesson 1-4)**

53. If *D*, *E*, and *F* are collinear, *DE* = 12, *EF* = 9, and *DF* = 3, which point is between the other two? **(Lesson 1-4)**

Wrap-Up   **54.** Write a description for the coordinates of the midpoint of a line segment in the coordinate plane using the word "average."

MID-CHAPTER REVIEW

Write the ordered pair for each point. Refer to the coordinate plane shown at the right. (Lesson 1-1)

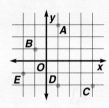

1. *A*   **2.** *B*   **3.** *C*

4. *True* or *false:* Point *E* is in Quadrant III. **(Lesson 1-1)**

5. What is the *y*-coordinate of any point collinear with *C* and *D*? **(Lesson 1-1)**

Refer to the figure at the right to answer each question. (Lesson 1-2)

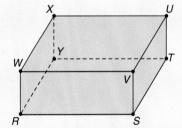

6. Name a point not on plane *RST*.

7. Name three lines that intersect at point *S*.

8. Name the intersection of planes *RWX* and *UTY*.

9. *True* or *false:* $\overleftrightarrow{YT}$ and $\overleftrightarrow{WX}$ are noncoplanar lines.

10. Wheels "*R*" Us sells bicycles, tricycles, and unicycles. They have one more bicycle than unicycle in stock. If there are 60 pedals and 80 wheels, how many bicycles, tricycles, and unicycles are there in stock? **(Lesson 1-3)**

11. If *R*, *S*, and *T* are collinear, *RS* = 3x − 6, *ST* = 2x + 9, and *RT* = 13, which point is between the other two? **(Lesson 1-4)**

12. If *B* is between *A* and *C*, *AB* = 4x − 9, *BC* = 7 − x, and *AC* = 2x + 3, find the value of *x* and the measure of $\overline{AC}$. **(Lesson 1-4)**

13. Find *MN* and the coordinates of the midpoint of $\overline{MN}$, given *M*(7, 7) and *N*(-1, 1). **(Lesson 1-5)**

14. If *DE* = 4x − 9, and *EF* = 7 + 2x, find the value of *x* so that $\overline{DE} \cong \overline{EF}$. **(Lesson 1-5)**

1-6 Rays and Angles

Objectives

After studying this lesson, you should be able to:

- identify angles and parts of angles, and
- use the angle addition postulate to find the measures of angles.

Application

Compass headings are always measured in a clockwise direction.

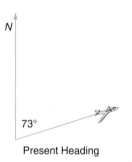

N

73°

Present Heading

Captain Cindy Berkeley, a pilot for United Airlines, is on approach for a landing at Chicago's O'Hare International Airport. Her present compass heading is 73 degrees. This heading refers to the measure of the *angle* formed by the flight path of the plane and an imaginary path in the direction due north. The tower has informed Captain Berkeley to land on runway 9. She knows that multiplying the runway number by 10 gives her the compass heading for a landing on that runway. So the compass heading for her landing must be 90°. How many degrees and in what direction must Captain Berkeley turn in order to land on runway 9?

This problem will be solved in Example 3, but first we need to review and discuss angles and how to find their measures.

In geometry, an **angle** is defined in terms of the two *rays* that form the angle. You can think of a **ray** as a segment that is extended indefinitely in *one* direction. Rays have exactly one endpoint and that point is always named first when naming the ray. Like segments, rays can also be defined using betweenness of points.

A laser beam shooting out into space would be a model for a ray.

Ray PQ, written $\overrightarrow{PQ}$, consists of the points on $\overline{PQ}$ and all points S on $\overrightarrow{PQ}$ such that Q is between P and S.

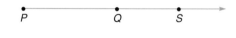

Any given point on a line determines exactly two rays, called **opposite rays.** This point is the common endpoint of the opposite rays. In the figure below, $\overrightarrow{PQ}$ and $\overrightarrow{PR}$ are opposite rays, and P is the common endpoint.

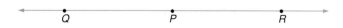

Opposite rays can be defined as a figure formed by two collinear rays with a common endpoint, since the two rays lie on the same line. Similarly, an **angle** can be defined as a figure formed by two *noncollinear* rays with a common endpoint. The two rays are called the **sides** of the angle. The common endpoint is called the **vertex**. The figure formed by opposite rays is often referred to as a **straight angle** even though it does *not* satisfy our definition for an angle.

In the figure at the right, the sides of the angle are $\overrightarrow{YX}$ and $\overrightarrow{YZ}$, and the vertex is Y. This angle could be named $\angle Y$, $\angle XYZ$, $\angle ZYX$, or $\angle 1$. When letters are used to name an angle, the letter that names the vertex is used either as the only letter or as the middle of three letters.

An angle is named by a single letter only when there is no chance of confusion. For example, it is not obvious which angle shown at the right is $\angle A$ since there are three different angles that have A as a vertex. *Can you name them?*

Whenever two or more angles have a common vertex, you need to use either three letters or a number to name each angle.

Example 1

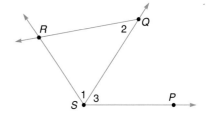

Refer to the figure at the right to answer each question.

a. What number names $\angle QSP$? *3*

b. What is the vertex of $\angle 2$? *Q*

c. What are the sides of $\angle 1$? $\overrightarrow{SQ}$ *and* $\overrightarrow{SR}$

An angle separates a plane into three distinct parts, the **interior** of the angle, the **exterior** of the angle, and the angle itself.

A point is in the interior of an angle if it does not lie on the angle itself and it lies on a segment whose endpoints are on the sides of the angle. Neither of the endpoints of this segment can be the vertex of the angle.

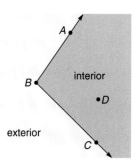

In the figure at the right, point D and all other points in the *blue* region are in the interior of $\angle B$. Any point that is not on the angle or in the interior of the angle is in the exterior of the angle. The *yellow* region is the exterior of $\angle B$.

Just as a ruler can be used to measure the length of a segment, a *protractor* can be used to find the **measure of an angle** in degrees. To find the measure of an angle, place the center point of the protractor over the vertex of the angle.

Then align the mark labeled 0 on either side of the scale with one side of the angle. This has been done for ∠XYZ shown below.

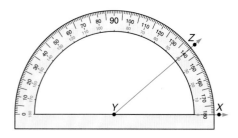

Using the inner scale of the protractor, shown in blue, you can see that ∠Y is a 40-degree (40°) angle. Thus, we say that the degree measure of ∠XYZ is 40. This can also be written as m∠XYZ = 40.

The protractor postulate guarantees that there is only one 40° angle on each side of $\overrightarrow{YX}$.

Postulate 1-3 *Protractor Postulate*	Given $\overrightarrow{AB}$ and a number *r* between 0 and 180, there is exactly one ray with endpoint *A*, extending on each side of $\overrightarrow{AB}$, such that the measure of the angle formed is *r*.

Example 2

Use a protractor to find the degree measure of each numbered angle.

$m\angle 1 = 30$ $m\angle 2 = 95$

$m\angle 3 = 18$ $m\angle 4 = 37$

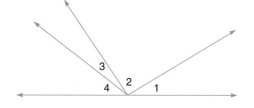

In Lesson 1-4, a measurement relationship between the lengths of segments, called the *segment addition postulate*, was introduced. A similar relationship exists between the measure of angles.

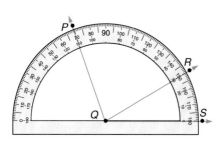

In the figure at the left, you can see that point *R* is in the interior of ∠PQS. m∠PQS = 110 and m∠RQS = 30.

The sides of ∠PQR align with the marks labeled 110 and 30 on the inner scale. So m∠PQR = 110 − 30 or 80.

Since 80 + 30 = 110, m∠PQR + m∠RQS = m∠PQS. This example and others like it, lead us to the following postulate about angle measures.

Postulate 1-4 Angle Addition Postulate	If R is in the interior of $\angle PQS$, then $m\angle PQR + m\angle RQS = m\angle PQS$. If $m\angle PQR + m\angle RQS = m\angle PQS$ then R is in the interior of $\angle PQS$.

Example 3

Aviation

Refer to the problem presented at the beginning of the lesson. Let $\overrightarrow{AP}$ represent the path of Captain Berkeley's plane, and let $\overrightarrow{AR}$ represent the path for a landing on runway 9. Determine the number of degrees that the plane must be turned to land on runway 9 by determining the measure of $\angle PAR$.

The compass heading for the path of Captain Berkeley's plane, $\overrightarrow{AP}$, is 73°. Using the formula given in the problem, we know that the compass heading for a landing on runway 9 is 9(10) or 90°. $\overrightarrow{AP}$ and $\overrightarrow{AR}$ at the right represent the paths corresponding to these compass headings. We can use the angle addition postulate to find $m\angle PAR$.

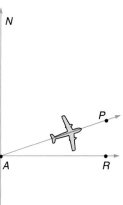

$m\angle NAP + m\angle PAR = m\angle NAR$
$73 + m\angle PAR = 90$
$m\angle PAR = 17$

Thus, Captain Berkeley must turn the plane 17° right to land on runway 9.

CHECKING FOR UNDERSTANDING

Communicating Mathematics

Read and study the lesson to answer each question.

1. Refer to the application at the beginning of the lesson. What would be the compass heading of an airplane for a landing on runway 13?

2. What is the intersection of two opposite rays?

3. Explain in your own words why a straight angle does not fit our definition of angle.

4. Draw $\overrightarrow{PQ}$ and $\overrightarrow{RS}$ so that the intersection of $\overrightarrow{PQ}$ and $\overrightarrow{RS}$ is $\overline{PR}$.

Guided Practice

Draw two angles that satisfy the following conditions.

5. The angles intersect in a single point.

6. The angles intersect in two points.

7. The angles intersect in a ray.

Algebra

8. Suppose $\overrightarrow{AB}$ and $\overrightarrow{AC}$ are opposite rays on a number line. If the coordinate of A is 0 and the coordinate of B is 2, is the coordinate of C a positive or a negative number?

Use the figure below to answer each question.

9. What are two other names for $\overrightarrow{QS}$?

10. What is the endpoint of $\overrightarrow{SP}$?

11. *True* or *false:* $\overrightarrow{RX}$ and $\overrightarrow{RT}$ are opposite rays.

12. What are the sides of $\angle 2$?

13. Name a point that lies on $\angle 3$.

14. Name all of the angles that have $\overrightarrow{RY}$ for a side.

15. Name a point in the exterior of $\angle PRY$.

16. Complete: $m\angle XRT = m\angle 2 +$ __?__.

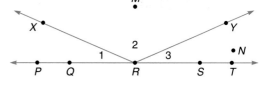

Find the measure of the following angles in the figure at the right.

17. $\angle PQA$

18. $\angle RQE$

19. $\angle PQC$

20. $\angle AQB$

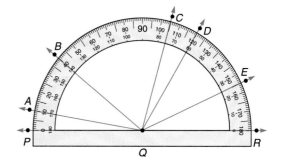

EXERCISES

Practice

Find the measure of the following angles in the figure above.

21. $\angle BQD$ 22. $\angle EQC$ 23. $\angle AQC$ 24. $\angle AQE$

Use the figure below to answer each question.

25. What is the vertex of angle 2?

26. Name a straight angle.

27. Name a point in the interior of $\angle 4$.

28. Name all the angles that have J as the vertex.

29. Do $\angle 3$ and $\angle 4$ have a common side? If so, name it.

30. Do $\angle 2$ and $\angle J$ name the same angle? Explain.

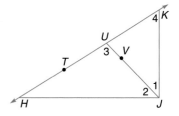

Draw two angles that satisfy the following conditions.

31. The angles intersect in three points.

32. The angles intersect in four points.

33. The angles intersect in a segment.

CONNECTION
Algebra

In the figure, $\overrightarrow{XP}$ and $\overrightarrow{XT}$ are opposite rays. Given the following conditions, find the value of x and the measure of the indicated angle.

34. $m\angle SXT = 3x - 4, m\angle RXS = 2x + 5,$
$m\angle RXT = 111; m\angle RXS$

35. $m\angle PXQ = 2x, m\angle QXT = 5x - 23;$
$m\angle QXT$

36. $m\angle QXR = x + 10, m\angle QXS = 4x - 1,$
$m\angle RXS = 91; m\angle QXS$

37. $m\angle QXR = 3x + 5, m\angle QXP = 2x - 3, m\angle RXP = x + 50; m\angle RXT$

38. $m\angle TXS = x + 4, m\angle SXR = 3x + 4, m\angle RXP = 2x + 4; m\angle PXS$

How many angles, not including straight angles, are shown in each figure?

39.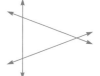

40.

41.

42.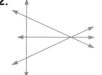

43. How many points on $\overrightarrow{PQ}$ are 2 units from P? 2 units from Q?

Critical Thinking **44.** Each figure below shows noncollinear rays with a common endpoint.

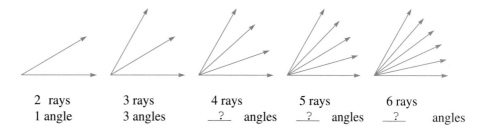

| 2 rays | 3 rays | 4 rays | 5 rays | 6 rays |
| 1 angle | 3 angles | _?_ angles | _?_ angles | _?_ angles |

a. Count the number of angles in each figure.

b. Do you see a pattern? Try to predict the number of angles that are formed by 7 rays. by 10 rays.

c. Write a formula for the number of angles formed by n noncollinear rays with a common endpoint.

Applications

45. Golf The *loft* of a golf club is the measure of the angle that the head of the club would form with a vertical ray. Golf clubs of the same type (irons or woods) are numbered so that greater numbers indicate greater amounts of loft. For example, a 9-iron has more loft than a 3-iron.

a. Explain how you think the loft of a club affects the path of a shot.

b. Draw a picture to show how the path of a shot hit with a 9-iron might differ from the path of a shot hit with a 3-iron.

46. Botany Botanists have found that the angle between the main branches of a tree and its trunk remains constant in each species. Measure the angles between several branches and the trunk of the tree in the photograph at the right. Are the angles all about the same? What is the measure of each angle?

Mixed Review

47. Find the ordered pair for the point with *x*-coordinate 2 that lies on the line $x + 3y = -1$. **(Lesson 1-1)**

48. Draw and label a figure to show that the intersection of planes $\mathcal{P}$, $\mathcal{Q}$, and $\mathcal{R}$ is line ℓ. **(Lesson 1-2)**

49. If *M* is the midpoint between *K* and *L* on a number line, the coordinate of *M* is 3, and the coordinate of *L* is -2, is the coordinate of *K* greater than 0 or less than 0? **(Lesson 1-4)**

50. Find *BC* if *B* is between *A* and *C*, $AC = 12$, $AB = 5x - 3$, and $BC = 3x - 1$. **(Lesson 1-4)**

51. Find the midpoint of $\overline{CD}$ if the coordinates of *C* are (-5, -2) and the coordinates of *D* are (3, 6). **(Lesson 1-5)**

52. If $\overline{XY}$ bisects $\overline{WZ}$ at *X*, $WX = 2x - 3$, and $XZ = 5x - 24$, what is the value of *x*? **(Lesson 1-5)**

Wrap-Up

53. Write an explanation of the angle addition postulate, different from the one presented in this lesson, that you could present to a classmate.

DEVELOPING REASONING SKILLS

Copy the figure shown at the right. Then draw line segments through the nine dots that satisfy the following conditions.

1. Use only straight line segments.

2. You may only pass through a dot one time.

3. Once you begin drawing the lines, you may not lift your pencil until you are finished.

4. You may use up to 4 line segments with three changes of direction.

Classifying Angles

Objectives

After studying this lesson, you should be able to:
- classify angles as acute, obtuse, right, and straight, and
- identify and use congruent angles and the bisector of an angle.

Application

Have you ever noticed how some diamonds "sparkle" more than others even though there appears to be no difference in their shapes? The amount of brilliance with which a diamond sparkles depends on the angle of its faces and how they catch and reflect light. Early jewelers were unaware that they could cut diamonds in certain ways to give them more brilliance. Today, thanks to the work of the Polish mathematician Tolkowsky, gem cutters know the exact angles to cut the faces of diamonds to produce many different levels of brilliance.

Gem cutters use many different tools to determine exactly where to cut a diamond so that each of its faces will be at a desired angle. They must use these tools since they cannot simply use a protractor to draw the appropriate angles on the diamond. You, however, can use a protractor to draw angles with specific measures.

Example 1

Use a protractor to draw a 65° angle.

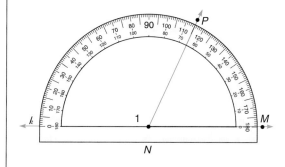

What is the measure of ∠1 in the figure above?

1. Draw line *k* with point *N* on line *k*.
2. Place the center point of the protractor on *N* and align the mark labeled 0 on the inner scale with line *k*. Draw point *M* at the 0 mark.
3. Locate and draw point *P* at the mark labeled 65 on the inner scale.
4. Draw $\overrightarrow{NP}$. Angle *MNP* is a 65° angle.

In geometry, angles are classified according to their measures.

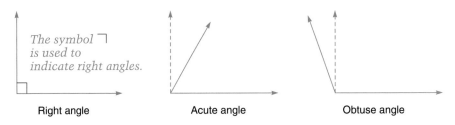

Right angle   Acute angle   Obtuse angle

The symbol ⌐ is used to indicate right angles.

A corner of a sheet of paper forms a right angle. You may use it to help you determine if an angle is right, acute, or obtuse. The edge of the paper can be used to determine straight angles.

Definition of Right Acute, and Obtuse Angles	**A right angle is an angle whose measure is 90. An acute angle is one whose measure is less than 90. An obtuse angle is one whose measure is greater than 90.**

In Lesson 1-5, you learned that congruent segments have the same length. Similarly, **congruent angles** have the same measure.

In the figure at the right, $m\angle PQR = m\angle SQV$ since each has a measure of 50. Therefore, the two angles are congruent, which you can indicate by writing $\angle PQR \cong \angle SQV$. In a diagram, small "arcs," like the ones in red shown at the right, are used to indicate congruent angles.

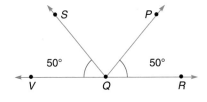

The definition of congruent angles tells us that statements such as $m\angle A = m\angle B$ and $\angle A \cong \angle B$ are equivalent. Therefore, we will use them interchangeably throughout the remainder of this book.

Example 2

CONNECTION

Algebra

Point *D* is in the interior of $\angle ABC$. Find $m\angle ABC$ if $\angle ABD \cong \angle DBC$, $m\angle ABD = 11x - 13$, and $m\angle DBC = 5x + 23$.

Since $\angle ABD \cong \angle DBC$, $m\angle ABD = m\angle DBC$.

$$m\angle ABD = m\angle DBC$$
$$11x - 13 = 5x + 23$$
$$6x - 13 = 23$$
$$6x = 36$$
$$x = 6$$

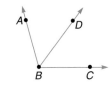

$$m\angle ABD = 11x - 13 \qquad m\angle DBC = 5x + 23$$
$$= 11(6) - 13 \qquad\qquad = 5(6) + 23$$
$$= 53 \qquad\qquad\qquad = 53$$

$$m\angle ABC = m\angle ABD + m\angle DBC \qquad \textit{Angle addition postulate}$$
$$= 53 + 53$$
$$= 106$$

A compass and straightedge can be used to construct an angle that is congruent to a given angle without knowing the degree measure of the angle.

CONSTRUCTION

Construct an angle congruent to a given angle.

1. Draw an angle like ∠A on your paper.

2. Use a straightedge to draw a ray on your paper. Label its endpoint E.

3. Put the compass at point A and draw a large arc that intersects both sides of ∠A. Label the points of intersection B and C.

4. Using the same compass setting, put the compass at point E and draw a large arc that starts above the ray and intersects the ray. Label the point of intersection F.

5. Set the point of your compass on C and adjust so that the pencil tip is on B.

6. Using that setting, place the compass at point F and draw an arc to intersect the larger arc you drew in Step 4. Label the point of intersection D.

7. Use a straightedge to draw $\overrightarrow{ED}$.

Thus, $m\angle BAC = m\angle DEF$, or $\angle BAC \cong \angle DEF$, by construction.

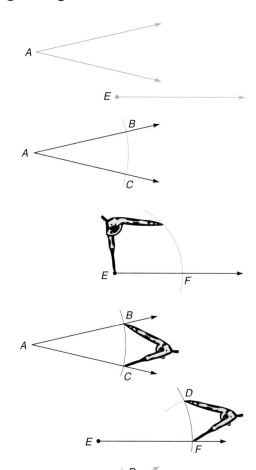

A *segment bisector* separates a segment into two congruent segments since it intersects the midpoint of the segment. Similarly, we can talk about an *angle bisector* separating an angle into two congruent angles.

For $\overrightarrow{QS}$ to be the **angle bisector** of $\angle PQR$, point S must be on the interior of $\angle PQR$ and $\angle PQS \cong \angle SQR$, as shown at the right.

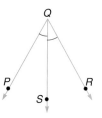

The construction on the next page shows how a compass and straightedge can be used to find the angle bisector of a given angle.

CONSTRUCTION

Construct the bisector of a given angle.

1. Draw an angle like ∠P on your paper.

2. Place the compass at point P and draw a large arc that intersects both sides of ∠P. Label the points of intersection Q and R.

3. Place the compass at point Q and draw an arc in the interior of the angle.

4. Using the same compass setting, place the compass at point R and draw a large arc that intersects the arc drawn in Step 3. Label the point of the intersection W.

5. Draw $\overrightarrow{PW}$.

$\overrightarrow{PW}$ is the bisector of ∠QPR.

$m\angle QPW = m\angle WPR$, or $\angle QPW \cong \angle WPR$

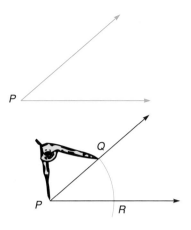

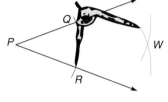

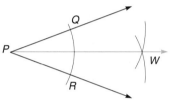

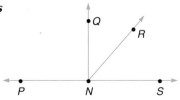

CHECKING FOR UNDERSTANDING

Communicating Mathematics

Read and study the lesson to answer each question.

1. An angle has how many different bisectors? Explain.
2. Explain why all right angles are congruent.
3. Explain why not all acute angles are congruent.
4. Draw and label a figure to show ∠AOB ≅ ∠BOC and ∠AOD ≅ ∠DOC. Be sure to indicate each pair of congruent angles.

Guided Practice

Use a protractor to draw angles having each measure.

5. 45
6. 60
7. 125
8. 100
9. 29
10. 144

State whether each angle in the figure *appears* to be acute, obtuse, right, or straight.

11. ∠SNR
12. ∠QNP
13. ∠SNP
14. ∠RNP
15. ∠SNQ
16. ∠QNR

17. If $\angle A$ is an acute angle and $m\angle A = 3x + 12$, write a compound inequality to describe all the possible values for x.

18. If $\angle M \cong \angle N$, $m\angle M = 8n - 17$, and $m\angle N = 7n - 3$, is $\angle N$ acute, obtuse, right, or straight?

EXERCISES

Practice

State whether each angle in the figure *appears* to be acute, obtuse, right, or straight.

19. $\angle MAL$
20. $\angle LBC$
21. $\angle ALC$
22. $\angle BLN$
23. $\angle ABC$
24. $\angle LMA$
25. $\angle ACN$
26. $\angle CNM$

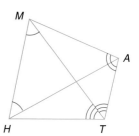

Refer to the figure to answer each question.

27. Name all pairs of congruent angles.

28. Name all segments that bisect angles and name the angles that they bisect.

29. If $m\angle AHM = 3x + 5$ and $m\angle TMH = 7x - 27$, find $m\angle AHM$.

30. If $m\angle MAH = 2x + 5$ and $m\angle TAH = 7x - 10$, find $m\angle MAT$.

31. If $m\angle HTA = 130 - x$ and $m\angle HTM = 3x + 2$, find $m\angle MTA$.

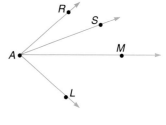

In the figure, $\overrightarrow{AM}$ bisects $\angle LAR$ and $\overrightarrow{AS}$ bisects $\angle MAR$. Given the following conditions, find the value of x and the measure of the indicated angle.

32. $m\angle MAR = 2x + 13$, $m\angle MAL = 4x - 3$; $m\angle RAL$

33. $m\angle RAL = x + 32$, $m\angle MAR = x - 31$; $m\angle LAM$

34. $m\angle RAS = 25 - 2x$, $m\angle SAM = 3x + 5$; $m\angle LAR$

35. $m\angle RAM = 31 - x$, $m\angle LAM = 17 - 3x$, $m\angle SAR$

36. $m\angle RAL = 5x - 7$, $m\angle MAS = x + 3$, $m\angle MAR$

37. Suppose $\angle PQR$ is a right angle and T is in the interior of $\angle PQR$. If $m\angle PQT$ is four times $m\angle TQR$, find $m\angle TQR$.

38. In the figure, $\angle LOJ$ is an obtuse angle. If $m\angle LOJ = 5x + 25$, write a compound inequality to describe all the possible values for x.

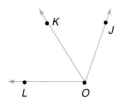

39. If $m\angle KOJ = 8x - 17$, $m\angle KOL = 3x + 28$, and $m\angle JOL = 110$, does $\overrightarrow{OK}$ bisect $\angle LOJ$?

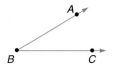

40. Construct an angle that is congruent to ∠ABC shown at the right. Then construct the bisector of that angle.

41. Suppose ∠JKM and ∠MKJ are right angles. If $m\angle JKM = 23x - 6y$ and $m\angle MKJ = 7x + 6y$, find the values of x and y.

42. If S is in the interior of ∠RPT, $m\angle RPT = 76$, $m\angle RPS = n^2 - 12$, and $m\angle SPT = 4n + 11$, find the value of n and determine if $\overrightarrow{PS}$ bisects ∠RPT.

Critical Thinking

43. As the hands of a clock move from 3:00 A.M. to 3:00 P.M., how many times will they be at right angles with each other?

44. Copy △ABC shown at the right. Then draw the rays that bisect each angle of the triangle using the construction on page 46.
  a. What conjecture could you make about these three rays?
  b. Draw two different triangles and test your conjecture on these triangles.

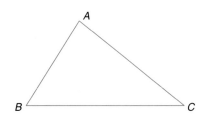

Application

45. Physics When a beam of light reflects off a plane mirror, its *angle of incidence* is congruent to its *angle of reflection*. If $m\angle IMR = 80$, find the angle of incidence and $m\angle IMN$.

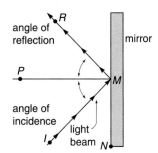

Mixed Review

46. Would the best model for the intersection of two adjacent walls and the ceiling in a room be a point, a line, or a plane? **(Lesson 1-2)**

In the figure below, $\overleftrightarrow{MS}$ and $\overleftrightarrow{TQ}$ intersect at P. Refer to this figure to answer each question.

47. If $SP = 3x + 11$, $PM = 2x + 1$, and $MS = 27$, find the value of x. **(Lesson 1-4)**

48. If $\overrightarrow{PR}$ bisects $\overline{TQ}$, $PQ = x + 7$, and $TQ = 5x - 1$, find PT. **(Lesson 1-5)**

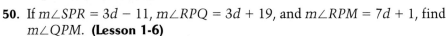

49. *True* or *false*: $m\angle TPR + m\angle RPM = m\angle TPM$. Explain. **(Lesson 1-6)**

50. If $m\angle SPR = 3d - 11$, $m\angle RPQ = 3d + 19$, and $m\angle RPM = 7d + 1$, find $m\angle QPM$. **(Lesson 1-6)**

Wrap-Up

51. Look up the words *acute* and *obtuse* in a dictionary. Compare their everyday meanings to their mathematical meanings.

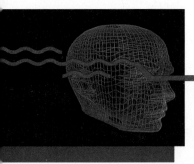

Technology
Using LOGO

BASIC
Geometric Supposer
Graphing calculators
▶ **LOGO**
Spreadsheets

LOGO is a very powerful computer programming language. You can use LOGO to examine and manipulate words, numbers, and lists. Its best-known feature is its turtle graphics. This feature allows us to create drawings with simple commands.

To use LOGO, you must first load the language into your computer. Insert the disk into drive 1 and close the door. To boot the LOGO disk, hold down the CONTROL and 🍎 keys with two fingers on your left hand and press the RESET key with a finger on your right hand. Now, release all three keys at once.

Once LOGO is loaded, type DRAW to clear the screen and locate the turtle, the small triangle used to draw, in the center of the screen. The basic LOGO commands are as follows:

FD Forward	LT Left	CS Clear screen
BK Back	RT Right	DRAW
PU Pen Up	PD Pen Down	HOME

The LOGO commands shown below have the turtle draw a rectangle and a triangle.

```
DRAW
FD 50 RT 90
FD 75 RT 90
FD 50 RT 90
FD 75 PU FD 10
PD FD 30 LT 120
FD 30 LT 120
FD 30 HT
```

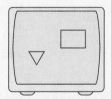

EXERCISES

1. Describe the results of the commands DRAW FD 80 BK 80.

2. Type the command HT. What happens? Now type ST.

Write the LOGO commands to draw each figure.

3.

4.

5.

6. Write the LOGO commands to draw your initials.

1-8 Pairs of Angles

Objective

After studying this lesson, you should be able to:

- identify and use adjacent angles, vertical angles, complementary angles, supplementary angles, and linear pairs of angles.

Investigation

On a piece of paper, draw two lines, $\overleftrightarrow{AB}$ and $\overleftrightarrow{PQ}$, intersecting at point X. Your figure should look similar to the one shown below. These two intersecting lines form four nonstraight angles, $\angle AXP$, $\angle PXB$, $\angle BXQ$, and $\angle QXA$. Certain pairs of these angles have special names that are used to describe the relationship between the angles.

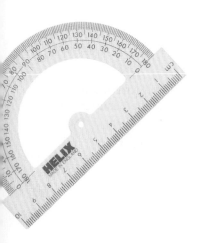

$\angle AXP$ and $\angle PXB$ are *adjacent angles*. **Adjacent angles** are angles in the same plane that have a common vertex and a common side, but no common interior points. The two congruent angles formed by an angle bisector is another example of a pair of adjacent angles. How many pairs of adjacent angles are in this figure? *4 pairs*

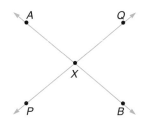

$\angle AXP$ and $\angle BXQ$ are *vertical angles*. **Vertical angles** are two nonadjacent angles formed by two intersecting lines. How many pairs of vertical angles are in this figure? *2 pairs*   *They are $\angle AXP$ and $\angle BXQ$, and $\angle AXQ$ and $\angle PXB$.*

Now, using a protractor, carefully measure each of the four angles. What do you notice about the measures of vertical angles $\angle AXP$ and $\angle BXQ$? $\angle PXB$ and $\angle QXA$? Draw another pair of intersecting lines, measure the four angles, and see if you get the same results. The results of your investigation might lead you to the following conclusion, which we will prove in Chapter 2.

Vertical angles are congruent.

Let's refer to our original figure, shown again at the right. $\angle AXP$ and $\angle PXB$ are an example of a special pair of adjacent angles called a *linear pair*. A **linear pair** of angles are adjacent angles whose noncommon sides are opposite rays. Based on the angle measurements you took previously, what can you conclude about the measures of the angles in each linear pair in this figure? Draw another pair of intersecting lines, measure the four angles, and see if you get the same results.

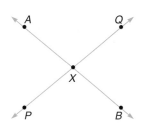

In the figure, there are four linear pairs: $\angle AXP$ and $\angle PXB$, $\angle PXB$ and $\angle BXQ$, $\angle BXQ$ and $\angle QXA$, and $\angle QXA$ and $\angle AXP$.

The results of these two examples might lead you to the following conclusion which we will also prove in Chapter 2.

The sum of the measures of the angles in a linear pair is 180.

Example 1

In the figure, $\overleftrightarrow{AB}$ and $\overrightarrow{CD}$ intersect at Z. Find the value of x and the measure of $\angle CZB$.

Since $\angle AZC$ and $\angle BZD$ are vertical angles, $m\angle AZC = m\angle BZD$.

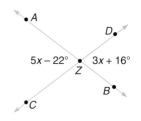

$$m\angle AZC = m\angle BZD$$
$$5x - 22 = 3x + 16$$
$$2x = 38$$
$$x = 19$$
$$m\angle AZC = 5x - 22$$
$$= 5(19) - 22$$
$$= 73$$

Since $\angle AZC$ and $\angle CZB$ are a linear pair, $m\angle AZC + m\angle CZB = 180$.

$$m\angle AZC + m\angle CZB = 180$$
$$73 + m\angle CZB = 180$$
$$m\angle CZB = 107 \qquad \text{The measure of } \angle CZB \text{ is 107.}$$

In Example 1, the sum of the measures of $\angle AZC$ and $\angle CZB$ is 180. If the sum of the measures of two angles is 180, the angles are called **supplementary angles.** When two angles are supplementary, each angle is said to be a *supplement* of the other angle.

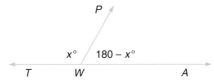

$\angle TWP$ and $\angle PWA$ are supplementary.

$\angle H$ is a supplement of $\angle M$, or $\angle M$ is a supplement of $\angle H$.

Since the sum of the measures of a linear pair of angles is 180, we can now say that *two angles that form a linear pair must be supplementary.*

Example 2

If $\angle PQS$ and $\angle SQR$ are supplementary, find $m\angle SQT$ and $m\angle TQR$.

Since $\angle PQS$ and $\angle SQR$ are supplementary, the sum of their measures is 180.

$$m\angle PQS + m\angle SQR = 180$$
$$90 + m\angle SQR = 180 \qquad \textit{Substitute 90 for } m\angle PQS.$$
$$90 + (m\angle SQT + m\angle TQR) = 180 \qquad \textit{Angle addition postulate}$$
$$90 + (x + 28) + (6x - 15) = 180 \qquad \textit{Substitute } x + 28 \textit{ for } m\angle SQT \textit{ and}$$
$$7x + 103 = 180 \qquad \textit{6x} - 15 \textit{ for } m\angle TQR.$$
$$7x = 77$$
$$x = 11$$

$$m\angle SQT = x + 28 \qquad m\angle TQR = 6x - 15$$
$$= (11) + 28 \text{ or } 39 \qquad = 6(11) - 15 \text{ or } 51$$

In Example 2, the sum of the measures of $\angle SQT$ and $\angle TQR$ is 90. If the sum of the measures of two angles is 90, the angles are called **complementary angles.** Whenever two angles are complementary, each angle is said to be the *complement* of the other angle.

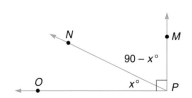

$\angle NPO$ is a complement of $\angle NPM$.
$\angle NPM$ is a complement of $\angle NPO$.

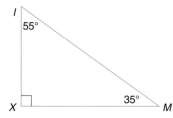

$\angle I$ and $\angle M$ are complementary.

Example 3

The measure of a supplement of an angle is 2.5 times as large as the measure of a complement of the angle. Find the measure of the angle.

EXPLORE   Let x = the measure of the angle.
Then $180 - x$ = the measure of its supplement, and
$90 - x$ = the measure of its complement.

PLAN   Write an equation that represents the relationship in this problem.

The supplement   is   2.5 times   the complement.
$$180 - x \quad = \quad 2.5 \times \quad (90 - x)$$

SOLVE
$$180 - x = 2.5(90 - x)$$
$$180 - x = 225 - 2.5x$$
$$1.5x = 45$$
$$x = 30$$

The measure of the angle is 30.

EXAMINE   Is the measure of the supplement of a 30° angle 2.5 times as large as the measure of its complement?

Measure of supplement: $180 - 30 = 150$
Measure of complement: $90 - 30 = 60$
Since $2.5(60) = 150$, the solution is correct.

CHECKING FOR UNDERSTANDING

Communicating Mathematics

Read and study the lesson to answer each question.

1. Give an explanation for why you think a linear pair of angles is called "linear."

2. *True* or *false:*   All pairs of supplementary angles are also linear pairs. Explain.

3. Complete: Adjacent angles are angles in the same plane that have a common __?__ and __?__ but no common __?__.

4. Describe in your own words the difference between complementary and supplementary angles.

Guided Practice

Find the measure of the complement and the supplement of an angle having the indicated measure.

5. 38   **6.** 63   **7.** 110   **8.** x

Refer to the figure at the right to name the following pairs of angles.

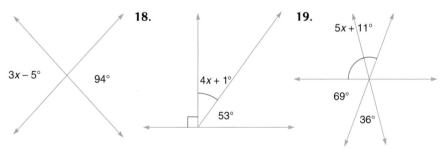

9. obtuse vertical angles

10. adjacent complementary angles

11. congruent supplementary angles

12. noncongruent supplementary angles

13. adjacent angles that do *not* form a linear pair

14. nonadjacent complementary angles

CONNECTION

Algebra

15. Suppose $\angle A$ is a complement of $\angle B$. Find the value of x, $m\angle A$, and $m\angle B$ if $m\angle A = 7x + 4$ and $m\angle B = 4x + 9$.

16. Suppose $\angle P$ is a supplement of $\angle Q$. Find the value of x, $m\angle P$, and $m\angle Q$ if $m\angle P = 6x + 4$ and $m\angle Q = 10x$.

EXERCISES

Practice   **Find the value of x.**

17.
$3x - 5°$ $94°$

18.
$4x + 1°$ $53°$

19.
$5x + 11°$
$69°$
$36°$

Identify each pair of angles as adjacent, vertical, complementary, supplementary, and/or as a linear pair.

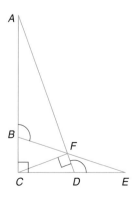

20. $\angle CFE$ and $\angle AFC$

21. $\angle BCF$ and $\angle FCD$

22. $\angle AFE$ and $\angle DFB$

23. $\angle CBF$ and $\angle ABF$

24. $\angle AFB$ and $\angle CFB$

25. $\angle CBF$ and $\angle FDE$

Find the value of *x* and *m∠ABC*.

26.   27.   28.

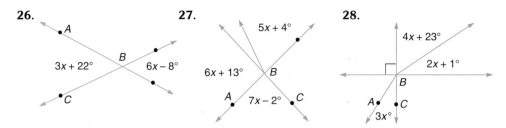

Solve.

29. Find the measures of two supplementary angles if the measure of the larger angle is 44 more than the measure of the smaller.

30. Find the measure of two complementary angles if the difference in the measures of the two angles is 12.

31. The measure of an angle is one-third the measure of its supplement. Find the measure of the angle.

32. The measure of an angle is one-fourth the measure of its complement. Find the measure of the angle.

33. The measure of an angle is 6 more than twice the measure of its complement. Find the measures of both angles.

34. The measure of an angle is 5 less than 4 times the measure of its supplement. Find the measure of both angles.

35. The measure of the supplement of an angle is 6 times the measure of its complement. Find the measures of the angle, its supplement, and its complement.

36. The measure of the supplement of an angle is 60 less than 3 times the measure of its complement. Find the measures of the angle, its supplement, and its complement.

37. Suppose ∠X and ∠Y are supplementary angles. If $m\angle X = x^2 - 9x$ and $m\angle Y = 11x + 12$, find the value of x, $m\angle X$, and $m\angle Y$.

38. Suppose ∠A and ∠B are complementary angles, and ∠C and ∠D are also complementary angles. If $m\angle A = 2x + 3$, $m\angle B = y - 2$, $m\angle C = 2x - y$, and $m\angle D = x - 1$, find the values of x and y, $m\angle A$, $m\angle B$, $m\angle C$, and $m\angle D$.

Critical Thinking

39. Determine whether the following statement is *always, sometimes,* or *never* true: *The measure of an acute angle is equal to the difference of the measure of its supplement and twice the measure of its complement.* Be sure to justify your answer.

40. In the figure at the right, ∠2 and ∠3 are complementary angles. Are ∠1 and ∠4 complementary? Why or why not?

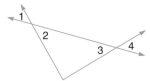

Application

41. Ski Jumping In order for ski jumpers to achieve the maximum distance on a jump, they need to make the angle between their body and the front of their skis as small as possible. This allows them to get the proper extension over the tips of their skis. If a ski jumper's body is aligned so that the angle between the body and the front of the skis is 10°, what will be the angle that the tail of the skis forms with the body? *Hint: Think about the relationship between the ski jumper's body and the skis.*

Mixed Review

Use the number line at the right to answer each question.

42. If $LN = 10$, find NS. **(Lesson 1-4)**

43. Find the coordinate of N if it is the midpoint of $\overline{LS}$. **(Lesson 1-5)**

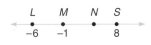

Use the figure at the right to answer each question.

44. Find the value of x if $m\angle 1 = 3x + 2$, $m\angle 2 = 4x - 1$, and $m\angle ADC = 148$. **(Lesson 1-6)**

45. Is $\overrightarrow{DB}$ the bisector of $\angle ADC$ if $m\angle 1 = 2x$, $m\angle 2 = 28$, and $m\angle ADC = 5x - 14$? Justify your answer. **(Lesson 1-7)**

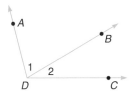

46. Distinct lines $\overleftrightarrow{AB}$, $\overleftrightarrow{CD}$, and $\overleftrightarrow{EF}$ all lie in the same plane and intersect at point X. How many distinct pairs of vertical angles are formed at the point of intersection? **(Lesson 1-8)**

Wrap-Up

47. Name the five different angle pairs discussed in this lesson and draw an example of each of these pairs.

1-9 Right Angles and Perpendicular Lines

Objectives

After studying this lesson, you should be able to:
- identify and use right angles and perpendicular lines, and
- determine what information can and cannot be assumed from a figure.

Application

When carpenters put up the studs for the wall in a new home, they must be sure that each stud is at a *right angle* to the floor of the home. To check this, they can use a device called a carpenter's square. When the right angle of the carpenter's square is placed against the angle formed by a stud and the floor, its edges should lie flush against both the stud and the floor.

The figure at the right can be used as a model for the proper positioning of the studs in a wall relative to the floor. In this figure, each of the studs is *perpendicular* to the floor.

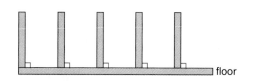

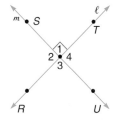

Perpendicular lines are two lines that intersect to form a right angle. In the figure at the left, lines ℓ and m are perpendicular. To indicate this, we write $\ell \perp m$, which is read "ℓ is perpendicular to m." Similarly, line segments and rays can be perpendicular to lines or other line segments and rays if they intersect to form a right angle. For example, in the figure, $\overrightarrow{RT} \perp \overrightarrow{SU}$, $\overrightarrow{US} \perp \overrightarrow{RT}$, and $\overline{TR} \perp \overline{US}$.

In the figure, $\angle 1$ is a right angle. Since $\angle 1$ and $\angle 3$ are vertical angles, what can you conclude about $\angle 3$? *It is also a right angle.*

Now, consider $\angle 2$ and $\angle 4$. Since $\angle 1$ forms a linear pair with $\angle 2$ and with $\angle 4$, what can you conclude about $\angle 2$ and $\angle 4$? *They are both right angles.*

If you draw two different lines, n and p, that are perpendicular, do you think the relationships between the four angles formed will be the same? *yes*

Based on this example, we could make the following conclusion, which will be proved in Chapter 2.

Perpendicular lines intersect to form four right angles.

Example 1

If $\overrightarrow{EB} \perp \overrightarrow{EC}$ and $\angle AEC$ and $\angle DEC$ form a linear pair, find $m\angle AEB$ and $m\angle DEC$.

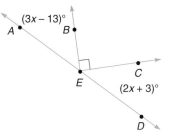

Since $\overrightarrow{EB} \perp \overrightarrow{EC}$, $\angle BEC$ is a right angle. Thus, $m\angle BEC = 90$. Since $\angle AEC$ and $\angle DEC$ form a linear pair, we know from our work in Lesson 1-8 that $m\angle AEC + m\angle DEC = 180$.

We can use the given information and the angle addition postulate to find the value of x.

$$m\angle AEC + m\angle DEC = 180$$
$$(m\angle BEC + m\angle AEB) + m\angle DEC = 180 \quad \textit{Angle addition postulate}$$
$$[90 + (3x - 13)] + (2x + 3) = 180 \quad \textit{Substitute 90 for } m\angle BEC,$$
$$5x + 80 = 180 \quad \textit{3x - 13 for } m\angle AEB, \textit{ and}$$
$$5x = 100 \quad \textit{2x + 3 for } m\angle DEC.$$
$$x = 20$$

$$
\begin{array}{ll}
m\angle AEB = 3x - 13 & \qquad m\angle DEC = 2x + 3 \\
\quad\ = 3(20) - 13 & \qquad \qquad\quad = 2(20) + 3 \\
\quad\ = 47 & \qquad \qquad\quad = 43
\end{array}
$$

A compass and straightedge can be used to construct a line perpendicular to a given line through a point on the line, *or* through a point not on the line.

Construct a line perpendicular to line ℓ and passing through point T on ℓ.

1. Place the compass at point T. Using the same compass setting, draw arcs to the right and left of T, intersecting line ℓ. Label the points of intersection D and K.

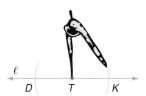

2. Open the compass to a setting greater than DT. Put the compass at point D and draw an arc above line ℓ.

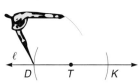

3. Using the same compass setting as in Step 2, place the compass at point K and draw an arc intersecting the arc previously drawn. Label the point of intersection S.

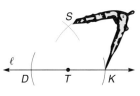

4. Use a straightedge to draw $\overleftrightarrow{ST}$.

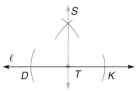

$\overleftrightarrow{ST}$ is perpendicular to ℓ at T.

CONSTRUCTION

Construct a line perpendicular to line ℓ and passing through point Q not on ℓ.

1. Place the compass at point Q. Draw an arc that intersects line $ℓ$ in two different places. Label the points of intersection R and S.

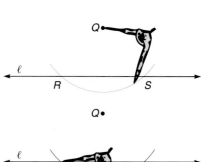

2. Open the compass to a setting greater than $\frac{1}{2}RS$. Put the compass at point R and draw an arc below line $ℓ$.

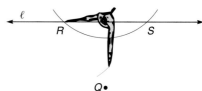

3. Using the same compass setting, place the compass at point S and draw an arc intersecting the arc drawn in Step 2. Label the point of intersection B.

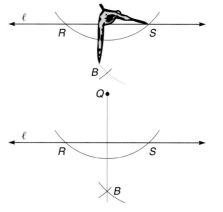

4. Use a straightedge to draw $\overleftrightarrow{QB}$.

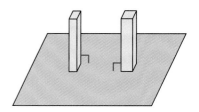

$\overleftrightarrow{QB}$ is perpendicular to $ℓ$.

In the application at the beginning of the lesson, we could have modeled the floor of the home using a plane instead of a line segment, as in the figure at the right. In this case, how could we determine whether the stud is perpendicular to the floor?

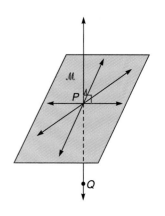

If a stud is perpendicular to the floor, then each side of the stud must be perpendicular to the portion of the floor that it intersects. Similarly, if *a line is perpendicular to a plane*, then the line must be perpendicular to every line in the plane that intersects it. Thus, $\overleftrightarrow{PQ} \perp \mathcal{M}$ and $\overleftrightarrow{PQ}$ must be perpendicular to every line in $\mathcal{M}$ that intersects it.

We can also talk about line segments and rays being perpendicular to planes. For example, in the figure, $\overline{PQ} \perp \mathcal{M}$ and $\overrightarrow{PQ} \perp \mathcal{M}$.

In this chapter, figures have been used to help describe or demonstrate different relationships among points, segments, lines, rays, and angles. Whenever you draw a figure, there are certain relationships that can be assumed from the figure and others that cannot be assumed.

Can be Assumed from Figure 1

All points shown are coplanar.

$\overleftrightarrow{CE}$, $\overrightarrow{DA}$, and $\overrightarrow{DB}$ intersect at D.

C, D, and E are collinear.

D is between C and E.

B is in the interior of $\angle ADE$.

$\angle CDE$ is a straight angle.

$\angle CDA$ and $\angle ADB$ are adjacent angles.

$\angle CDB$ and $\angle BDE$ are a linear pair.

$\angle CDA$ and $\angle ADE$ are supplementary.

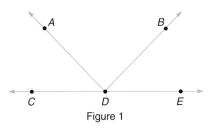

Figure 1

Cannot be Assumed from Figure 1

$\overline{CD} \cong \overline{DE}$

$\angle CDA \cong \angle EDB$

$\overrightarrow{DA} \perp \overrightarrow{DB}$

Figure 2, shown at the right, is marked so that these additional relationships are true.

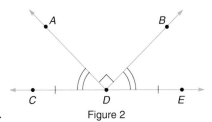

Figure 2

CHECKING FOR UNDERSTANDING

Communicating Mathematics

Read and study the lesson to answer each question.

1. What symbol do we use to indicate that two lines are perpendicular?

2. Two lines are perpendicular if they intersect to form a linear pair of angles that are also __?__.

3. If line ℓ intersects plane $\mathcal{P}$ at point T and $\ell \perp \mathcal{P}$, what must be true about any line, m, in $\mathcal{P}$ that passes through T?

4. From the figure at the right, can you assume that B is between A and C? that B is the midpoint of $\overline{AC}$? that $AB + BC = AC$?

Guided Practice

Refer to the figure at the right to answer each question.

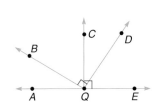

5. Which segment is perpendicular to $\overline{QA}$?

6. Which angles are complementary to $\angle BQC$?

7. Is $\angle BQC \cong \angle DQE$? Explain.

8. If $m\angle AQB = 4x - 15$ and $m\angle BQC = 2x + 9$, what is the value of x and $m\angle CQD$?

Determine whether each relationship can be assumed from the figure.

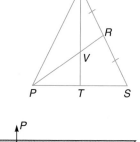

9. V is between Q and T.
10. $\overline{PR}$ bisects $\angle QPS$.
11. $PT = TS$
12. $\overline{PR}$ bisects $\overline{QS}$.
13. $\overline{QT} \perp \overline{PS}$
14. $\angle QRV$ is a right angle.
15. $\angle QVP$ and $\angle QVR$ form a linear pair.

16. In the figure at the right, $\overrightarrow{OP} \perp \mathcal{M}$ and $\overrightarrow{AB} \perp \overleftrightarrow{XY}$. How many right angles are in this figure?

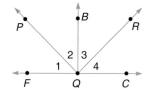

EXERCISES

Practice

Determine if the given information is enough to conclude that $\overrightarrow{QP} \perp \overrightarrow{QR}$.

17. $\angle 1$ and $\angle 3$ are complementary.
18. $\angle 1$ and $\angle 4$ are complementary.
19. $\angle 2 \cong \angle 3$
20. $\angle 1 \cong \angle 4$
21. $m\angle 1 + m\angle 4 = m\angle 2 + m\angle 3$
22. $m\angle 1 + m\angle 2 = m\angle 3 + m\angle 4$
23. $\angle 1 \cong \angle 2$ and $\angle 3 \cong \angle 4$

For each figure, find the value of x and determine if $\overline{AB} \perp \overline{CD}$.

24.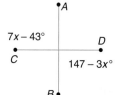
25.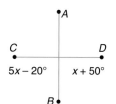
26.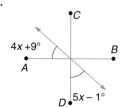

Determine whether each relationship can be assumed from the figure.

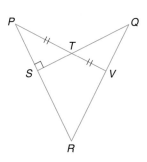

27. $\angle QVT$ is a right angle.
28. $\overline{QR} \perp \overline{PV}$
29. T is the midpoint of $\overline{PV}$.
30. $\overline{PV}$ bisects $\overline{QS}$.
31. $\angle SPT$ and $\angle VQT$ are congruent.
32. $PS = QV$
33. $PV = TV + PT$
34. $\angle STP \cong \angle VTQ$

CONSTRUCTION

35. Draw $\overrightarrow{PR}$. Construct line ℓ perpendicular to $\overleftrightarrow{PR}$ through P and line m perpendicular to $\overleftrightarrow{PR}$ through R. Now, locate point S on line m so that $RS = RP$. Finally, construct line k perpendicular to ℓ through S. What figure have you constructed?

36. $\overleftrightarrow{AB}$ and $\overleftrightarrow{CD}$ intersect at E. F lies in the interior of $\angle CEA$ and G lies in the interior of $\angle AED$. Also, $\angle CEA$ and $\angle FEG$ are right angles. Find $m\angle FEA$ if $m\angle CEF = 3x - 24$ and $m\angle AEG = 2x + 10$.

37. $\overleftrightarrow{AB}$ and $\overleftrightarrow{CD}$ intersect at Q and R is in the interior of $\angle AQC$. If $m\angle DQA = 2x - y$, $m\angle AQR = x$ and $m\angle RQC = y$, find the values of x and y so that $\overleftrightarrow{AB} \perp \overleftrightarrow{CD}$.

38. Find $m\angle BPC$ in the figure if $\overline{AC} \perp \overline{BC}$, $m\angle APC = 7x + 3$, $m\angle BPC = 16y$, $m\angle ACP = 3x + 2y$, and $m\angle BCP = 3x + 4y$.

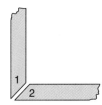

39. Find the values of x and y if $\overleftrightarrow{AB} \perp \overleftrightarrow{MN}$ and $\mathcal{L}$ contains $\overleftrightarrow{MN}$, $m\angle MTB = 2x + 6y$, $m\angle ATN = 4x + 3y$, and $m\angle BTN = \frac{8}{3}x + 5y$.

Critical Thinking

40. Given a line in a plane and a point on that line, explain why there is exactly one line in the plane perpendicular to the given line through the given point.

Application

41. **Framing** The corner of a frame for a painting is made by joining two pieces of wood so that they are perpendicular, as shown at the right. If $\angle 1$ is a 45° angle, what must be the measure of $\angle 2$?

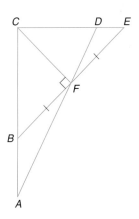

Mixed Review

Refer to the figure to answer each question.

42. If $AB = 2x - 3$, $BC = 3x + 13$, and $AC = 30$, find the value of x. **(Lesson 1-4)**

43. Name all the bisectors of $\overline{BE}$. **(Lesson 1-5)**

44. What angles have $\overline{CE}$ as a side? **(Lesson 1-6)**

45. If $m\angle BCF = 5x + 11$, $m\angle FCD = 3x + 23$, for what value of x does $\overline{CF}$ bisect $\angle ACE$? **(Lesson 1-7)**

46. If $m\angle AFB = d + 10$ and $m\angle CFD = 6d - 11$, find $m\angle DFE$. **(Lesson 1-8)**

Wrap-Up

47. Based on your homework and classroom discussions, what concepts from this lesson do you think should appear on a chapter test?

VOCABULARY

Upon completing this chapter, you should be
familiar with the following terms:

44 acute angles	**13** line	**8** quadrants
50 adjacent angles	**50** linear pair	**36** ray
36 angle	**30** midpoint	**44** right angle
45 angle bisector	**44** obtuse angle	**23** segment
9 collinear	**36** opposite rays	**31** segment bisector
52 complementary angles	**9** ordered pair	**37** straight angle
32 congruent segments	**8** origin	**51** supplementary angles
8 coordinate plane	**56** perpendicular lines	**50** vertical angles
14 coplanar	**13** plane	**8** x- and y-axes
14 intersection	**13** point	**8** x- and y-coordinates

SKILLS AND CONCEPTS

OBJECTIVES AND EXAMPLES	REVIEW EXERCISES

Upon completing this chapter, you should
be able to:

Use these exercises to review and prepare
for the chapter test.

■ graph ordered pairs
on a coordinate
plane. **(Lesson 1-1)**

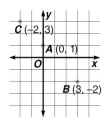

Graph $A(0, 1)$,
$B(3, -2)$, and
$C(-2, 3)$.

Graph each point.

1. $D(2, 4)$ **2.** $E(-4, -2)$

3. $F(-3, 0)$ **4.** $G(-1, 1)$

5. If $x > 0$ and $y < 0$, in which quadrant is
the point $P(x, y)$ located?

■ identify collinear and coplanar points and
intersecting lines and planes. **(Lesson 1-2)**

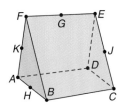

Name the
intersection
of plane JDC
and $\overleftrightarrow{FG}$.

Point E

**Refer to the figure at the left to answer
each question.**

6. Are E, A, K, and G coplanar?

7. What points do plane $BCEF$ and $\overline{ED}$
have in common?

8. What three lines intersect at D?

OBJECTIVES AND EXAMPLES	REVIEW EXERCISES

■ find the distance between points on a number line. **(Lesson 1-4)**

```
   A      B      C   D   E   G   H
 ──┼──┼──┼──┼──┼──┼──┼──┼──▶
 -20 -15 -10  -5   0   5  10  15  20
```

What is the distance between A and E?

$$AE = |-20 - 10|$$
$$= |-30|$$
$$= 30$$

Refer to the number line at the left to answer each question.

9. Find the distance between A and G.

10. Find the distance between E and B.

11. Which point(s) is 10 units from point E?

■ Find the midpoint of a segment. **(Lesson 1-5)**

For the number line above, which segments have D as the midpoint?

Since $CD = DE$ and $BD = DH$, D is the midpoint of $\overline{CE}$ and of $\overline{BH}$.

12. Refer to the number line above to find the coordinate of the midpoint of $\overline{AG}$.

13. **Algebra** Find the midpoint of $\overline{XY}$ given points $X(-2, 7)$ and $Y(8, 1)$.

14. **Algebra** If Q is the midpoint of $\overline{PR}$, $PQ = x + 6$, and $PR = 5x - 3$, find QR.

■ identify and use congruent segments. **(Lesson 1-5)**

For the number line above, which segments are congruent to $\overline{CG}$?

$CG = |15 - 0|$ or 15

Since $BD = |5 - (-10)|$ or 15, $\overline{BD} \cong \overline{CG}$.
Since $DH = |20 - 5|$ or 15, $\overline{DH} \cong \overline{CG}$.

15. If $GH = 17$ and the coordinate of J is 3, find the coordinate of K if $\overline{JK} \cong \overline{GH}$.

If $AB = 8x - 7$, $BC = 4x + 9$, and $AC = 13x - 2$, answer each question.

16. For what value of x are $\overline{AB}$ and $\overline{BC}$ congruent?

17. If A, B, and C are collinear and $\overline{AB} \cong \overline{BC}$, is B the midpoint of $\overline{AC}$?

■ use angle addition to find the measures of angles. **(Lesson 1-6)**

In the figure, if $m\angle PXR = 125$, $m\angle 1 = 40 - x$, and $m\angle 2 = 2x + 70$, find the value of x.

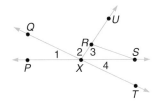

$$m\angle 1 + m\angle 2 = m\angle PXR$$
$$(40 - x) + (2x + 70) = 125$$
$$x = 15$$

Refer to the figure at the left to answer each question.

18. If $m\angle 1 = 37$, find $m\angle QXS$.

19. If $m\angle 2 = 3x - 20$, $m\angle 3 = 3x - 19$, and $m\angle QXS = 147$, find the value of x.

20. If $m\angle 3 = 77 - x$, $m\angle 4 = 2x + 7$, and $m\angle RXT = 3x$, find $m\angle 3$ and $m\angle 4$.

OBJECTIVES AND EXAMPLES	REVIEW EXERCISES

OBJECTIVES AND EXAMPLES

- identify and use congruent angles and the bisector of an angle. **(Lesson 1-7)**

  In the figure for exercises 18-20, if $m\angle 4 = m\angle 1$, $\overrightarrow{XR}$ bisects $\angle QXS$, and $\overrightarrow{XS}$ bisects $\angle TXR$, does $\overrightarrow{XQ}$ bisect $\angle PXR$?

  Since $\overrightarrow{XR}$ bisects $\angle QXS$, $m\angle 2 = m\angle 3$.
  Since $\overrightarrow{XS}$ bisects $\angle TXR$, $m\angle 3 = m\angle 4$.
  Thus, $m\angle 2 = m\angle 1$ and $\overrightarrow{XQ}$ bisects $\angle PXR$.

- identify and use adjacent angles, vertical angles, complementary angles, supplementary angles, and linear pairs of angles. **(Lesson 1-8)**

  vertical angles:
  $\angle 1$ and $\angle 5$:
  $\angle ANF$ and
  $\angle BNE$
  congruent
  adjacent angles:
  $\angle BND$ and
  $\angle DNF$

  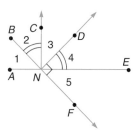

- identify and use right angles and perpendicular lines. **(Lesson 1-9)**

  In the figure above, is $\overline{AE} \perp \overline{CN}$?

  $\angle 4$ and $\angle 5$ are complementary. Since $\angle 1$ and $\angle 5$ are vertical angles, $\angle 1 \cong \angle 5$. Since $\angle 2 \cong \angle 4$, $\angle 1$ and $\angle 2$ are complementary. Thus, $\overline{AE} \perp \overline{CN}$.

REVIEW EXERCISES

Refer to the figure for exercises 18-20 to answer each question.

21. If $\angle 2 \cong \angle 3$, is any angle bisected? If so, which one?

22. If $\overline{QX}$ bisects $\angle PXR$, $m\angle 1 = 6x - 7$, and $m\angle 2 = 9x - 31$, find $m\angle PXR$.

23. If $\angle A \cong \angle B$, $m\angle A = 3n + 55$, and $m\angle B = 8n$, is $\angle B$ acute or obtuse?

Refer to the figure at the left to name the following pairs of angles.

24. adjacent complementary angles

25. linear pair of angles that includes $\angle 5$

26. supplementary angles that includes $\angle 2$

27. **Algebra** The measure of an angle is 10 less than three times the measure of its complement. Find the measure of both angles.

28. In the figure above, if $m\angle 4 = 2x - 1$ and $m\angle 5 = 3x - 4$, find $m\angle 1$.

29. In the figure above, could you conclude that $\overline{AE} \perp \overline{CN}$ if $\angle 1$ was congruent to $\angle 4$ and $\angle 2$ was *not* congruent to $\angle 4$?

30. $\angle J$ and $\angle K$ are vertical angles. Are $\angle J$ and $\angle K$ right angles if $m\angle J = 20x - 11$ and $m\angle K = 13x + 24$?

APPLICATIONS AND CONNECTIONS

31. **Statistics** The weekly geometry quiz in Mrs. Lao's class consists of 5 true-false questions. Is it possible for each of her 33 students to have a different pattern of answers? **(Lesson 1-3)**

32. **Aviation** The path of an airplane is described as 56° east of north. Draw a diagram that represents this flight path. **(Lesson 1-5)**

Draw and label a figure for each relationship.

1. Lines ℓ, *m*, and *n* all intersect at point *X*.   2. Planes $\mathcal{Q}$ and $\mathcal{R}$ do not intersect.
3. Plane $\mathcal{P}$ contains point *A* but does not contain $\overleftrightarrow{BC}$.

4. Graph *A*(4, 3), *B*(-2, -5), *C*(-3, 4), and *D*(0, -1) on a coordinate plane.

Refer to the coordinate grid to answer each question.

5. What ordered pair names point *S*?
6. What is the length of $\overline{PQ}$?
7. What are the coordinates of the midpoint of $\overline{QR}$?
8. What is the *y*-coordinate of any point collinear to *Q* and *S*?

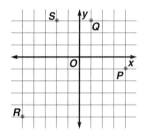

Refer to the number line to answer each question.

9. What is the measure of $\overline{AC}$?
10. What is the coordinate of the midpoint of $\overline{CF}$?
11. What segment is congruent to $\overline{BF}$?
12. What is the coordinate of *G* if *C* is between *D* and *G* and *DG* = 14?

Refer to the figure at the right to answer each question.

13. Name two points on line ℓ.
14. Name the sides of $\angle 1$.
15. Which of the numbered angles appears to be obtuse?
16. Name a pair of congruent supplementary angles.
17. Name a pair of adjacent angles that do not form a linear pair.
18. If $\angle 1 \cong \angle 5$, does $\overrightarrow{VG}$ bisect $\angle BVF$?
19. *True* or *false*: $\overrightarrow{VB} \perp \overrightarrow{VF}$
20. *True* or *false*: $\overleftrightarrow{CG}$ bisects $\overline{AE}$.
21. If *AC* = 4*x* + 1 and *CE* = 16 − *x*, find *AE*.
22. If $m\angle BVF = 7x - 1$ and $m\angle FVA = 6x + 12$, is $\overleftrightarrow{AB} \perp \overleftrightarrow{VF}$?
23. Which two angles must be complementary if $\angle AVF$ is a right angle?
24. If $m\angle 5 = 3x + 14$, $m\angle 6 = x + 30$, and $m\angle FVB = 9x - 11$, find $m\angle FVB$.

25. **Sports**   Three darts are thrown at the target shown at the right. If we assume that each of the darts lands within one of the rings, how many different point totals are possible?

Bonus   *P*(-8, 11) and *Q*(16, 1) are the endpoints of a diameter of a circle. Find the coordinates of the center of the circle and two other points on the circle.

Algebra Review

OBJECTIVES AND EXAMPLES	REVIEW EXERCISES

■ Add or subtract rational numbers.

$-14 + 3 = -11$ $-14 - 3 = -17$

$2.3 + (-3.52) = -1.22$

$\frac{3}{4} - \left(-\frac{7}{10}\right) = \frac{15}{20} + \frac{14}{20} = \frac{29}{20}$

Find each sum or difference.

1. $17 + (-9)$ 2. $-7 + (-13)$

3. $2.4 - 3.7$ 4. $-3.72 - (-8.651)$

5. $\frac{5}{4} + \left(-\frac{7}{8}\right)$ 6. $-\frac{6}{5} - \frac{11}{12}$

■ Use order of operations to evaluate expressions.

Evaluate $(2x + 3y) \div z^2 - 3$ if $x = 5$, $y = 6$, and $z = 2$.

$(2x + 3y) \div z^2 - 3$

$= (2 \cdot 5 + 3 \cdot 6) \div (2)^2 - 3$

$= (10 + 18) \div 4 - 3$

$= 28 \div 4 - 3$

$= 7 - 3$

$= 4$

Evaluate each expression if $a = 18$, $b = 3$, $c = 4$, and $d = 5$.

7. $a - b \cdot c + d$

8. $a - (b \cdot c + d)$

9. $a - b \cdot (c + d)$

10. $(a - b) \cdot c + d$

11. $(a - b) \cdot (c + d)$

■ Use the distributive property to simplify expressions.

$2x - 5xy + 6x = 2x + 6x - 5xy$

$\qquad\qquad = (2 + 6)x - 5xy$

$\qquad\qquad = 8x - 5xy$

Simplify each expression.

12. $6x + 7y + 8x - 2y$

13. $9(r - s) + 4s$

14. $3m(n - 2m) - 2n(2m - 3n)$

15. $\frac{3a^2}{4} + \frac{2ab}{3} + ab - a^2$

■ Multiply or divide rational numbers.

$-\frac{4}{5} \cdot \frac{7}{2} = -\frac{28}{10}$ $-\frac{4}{5} \div \frac{7}{2} = -\frac{4}{5} \cdot \frac{2}{7}$

$\qquad\quad = -\frac{14}{5}$ $\qquad\qquad = -\frac{8}{35}$

Simplify.

16. $-\frac{10}{7}\left(-\frac{5}{9}\right)$ 17. $-\frac{10}{7} \div \left(-\frac{5}{9}\right)$

18. $\frac{33a - 66}{-11}$

19. $-3\left(-\frac{7}{4}a + \frac{1}{6}\right) + \frac{5}{2}\left(3 - \frac{a}{2}\right)$

■ Find the absolute value of a number.

Evaluate $|n - 2|$ if $n = -3$.

$|n - 2| = |-3 - 2|$

$\qquad\quad = |-5|$

$\qquad\quad = 5$

Evaluate each expression.

20. $|4 - x|$, if $x = -2$

21. $|a| - |2b|$, if $a = -5$ and $b = 1$

22. $-|m + n|$, if $m = 3$ and $n = -12$

OBJECTIVES AND EXAMPLES	REVIEW EXERCISES

OBJECTIVES AND EXAMPLES

■ Translate verbal sentences into equations.

Write an equation for the sentence *The sum of x and the square of y is equal to twice z.*

$$x + y^2 = 2z$$

REVIEW EXERCISES

Write an equation for each sentence.

23. Twelve decreased by the square of a is equal to b.

24. The number c equals the cube of the sum of 2 and three times m.

25. The product of x and the square of y is t.

■ Solve equations by using addition or subtraction.

$$x - 3 = 5 \qquad q + 9 = 2$$
$$x - 3 + 3 = 5 + 3 \qquad q + 9 - 9 = 2 - 9$$
$$x = 8 \qquad q = -7$$

Solve each equation. Check your solution.

26. $x - 16 = 37$ 27. $z + 15 = -9$

28. $r - (-4) = 21$ 29. $m + (-5) = 17$

30. $-19 = -8 + d$ 31. $9 = 18 + d$

■ Solve equations by using multiplication or division.

$$14x = 42 \qquad \frac{m}{-6} = 8$$
$$\frac{14x}{14} = \frac{42}{14} \qquad \frac{-6m}{-6} = -6(8)$$
$$x = 3 \qquad m = -48$$

Solve each equation. Check your solution.

32. $-7r = -56$ 33. $23y = 103.5$

34. $-534 = 89a$ 35. $\frac{x}{5} = 7$

36. $-\frac{3}{4}n = 12$ 37. $\frac{-5}{3}z = -\frac{3}{2}$

Applications and Connections

38. **Foods** There are 113 calories in one cup of orange juice. Write an equation to represent the number of calories in 3 cups of orange juice. Then solve the equation.

39. **Riding** Alma rides her bicycle for three fourths of an hour every day. Find the distance she rides if she averages 13.65 miles per hour.

40. **Golfing** Jessica's golf score was 68. This was 4 less than Mark's golf score. What was Mark's score?

41. **Smart Shopping** Which is the better buy: 0.75 liters of soft drink for 89¢ or 1.25 liters of soft drink for $1.31?

C H A P T E R 2

Reasoning and Introduction to Proof

CHAPTER OBJECTIVES

In this chapter, you will:
- Make conjectures.
- Use the laws of logic to make conclusions.
- Write proofs involving segment and angle theorems.

GEOMETRY AROUND THE WORLD
United States

What rectangle filled with three circles helps keep you safe every day? Give up? The answer is a traffic light, developed in 1923 by an African-American inventor named Garrett Morgan.

Born in Kentucky in 1877, Morgan moved to Cleveland, Ohio, when he was 18. There, he found work repairing sewing machines and soon invented a belt fastener to make the machines operate more efficiently. Later, he invented a gas mask to protect fire fighters inside smoke-filled buildings. The patented device won a gold medal from the International Exposition for Sanitation and Safety. Morgan and three others wore the masks when they entered a gas-filled tunnel to save workers trapped by an explosion. During World War I, Morgan's invention protected Allied soldiers from breathing the deadly gases their enemies used in battle.

Concern for safety also motivated Morgan, at age 48, to invent a three-way automatic electric traffic light. At the time, he was said to be the only African-American in Cleveland who owned a car. Morgan patented his device and later sold the rights to market it to the General Electric Company. He died in 1963.

This modern traffic light looks different from Garrett Morgan's original invention, but the purpose is the same — saving lives.

GEOMETRY IN ACTION

Garrett Morgan systematically went about developing his inventions. First, he identified the problem to be solved. How can sewing machines be made to operate more efficiently? How can people be protected from breathing deadly fumes? How can traffic be regulated to protect pedestrians and drivers?

What do you suppose Morgan did after he identified the problem? Write the steps you think he might have gone through to invent the traffic light.

◄ *Park Avenue in New York City*   Inset: *Garrett Morgan*

2-1 Inductive Reasoning and Conjecturing

Objective

After studying this lesson, you should be able to:
- make geometric conjectures based on inductive reasoning.

Application

After making a few observations, the caveman in the comic has made a **conjecture.** A conjecture is an educated guess. Looking at several specific situations to arrive at a conjecture is called **inductive reasoning.** For centuries, mathematicians have used inductive reasoning to develop the geometry that we study and use today.

"Water boils down to nothing . . . snow boils down to nothing . . . ice boils down to nothing . . . everything boils down to nothing."

Drawing by Ed Fisher; © 1966 Saturday Review, Inc.

You have had some experiences with geometry, so you can make conjectures about geometry from given information. However, not all conjectures are true. For example, the caveman conjectured that everything boils down to nothing. But, if he boiled something like a potato, he would find that his conjecture was false.

Example 1

> Points *A*, *B*, and *C* lie on a segment. Write a conjecture about the relationship of points *A*, *B*, and *C*. Draw a figure to illustrate your conjecture.
>
> **Given:**   Points *A*, *B*, and *C* lie on a segment.
> **Conjecture:**   *A*, *B*, and *C* are collinear.
>
>

A conjecture that is based on several observations may be true, but not necessarily.

Example 2

APPLICATION

Botany

> Shana has been studying the growth of several different species of plants that she has planted in her backyard. The weather has been unusually dry for this time of the year. Plant *A* is thriving and plant *B* is shriveled and dying. Both plants receive several hours of strong sunlight each day. Make a list of conjectures that Shana can make and investigate as to why the plants are performing differently.

Some conjectures are

- Plant *A* requires less water than plant *B*;
- Plant *A* requires strong sunlight and plant *B* prefers shade;
- Plant *B* requires different soil than is available in Shana's area.

Can you name some more conjectures about Shana's plants?

Sometimes we, like the caveman, use inductive reasoning to make a conjecture and later determine that the conjecture is false. It takes only one false example to show that a conjecture is not true. The false example is called a **counterexample**.

Example 3

Given that points *A*, *B*, and *C* lie on a line segment, Jerry conjectured that *B* is between *A* and *C*. Determine if his conjecture is true or false. Explain your answer.

Given: Points *A*, *B*, and *C* lie on a segment.

Conjecture: *B* is between *A* and *C*.

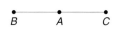

In the figure above, *A*, *B*, and *C* lie on the segment, but *A* is between *B* and *C*. Since we can find a counterexample for the conjecture, the conjecture is false.

Example 4

Write a conjecture that is based on the given information. Draw a figure to illustrate your conjecture. Explain why you think your conjecture is true.

Given: ∠*B* is supplementary to ∠*A*.
∠*B* is supplementary to ∠*C*.

Conjecture: ∠*A* ≅ ∠*C*

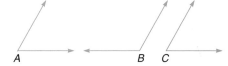

For example, suppose $m\angle B$ is 143. Then $m\angle A$ would be 37 since $180 - 143 = 37$. $m\angle C$ would also be $180 - 143$ or 37. Therefore, ∠*A* ≅ ∠*C* since they have the same measure. *This example supports the conjecture.*

Later on in this chapter, you will learn how to establish the truth of a conjecture. But for now, remember that one example or even one hundred examples do not establish the truth of a conjecture.

CHECKING FOR UNDERSTANDING

Communicating Mathematics

Read and study the lesson to answer these questions.

1. In your own words, explain the meaning of *conjecture*.

2. How can you prove that a conjecture is false?

3. Explain how the caveman's inductive reasoning led him to a false conjecture. What could he have done differently that would lead him to a different conjecture?

4. For centuries, people believed that Earth was flat. Name some other famous conjectures that proved to be false.

Guided Practice

Determine if the conjecture is *true* or *false* based on the given information. Explain your answer.

5. Given: points A, B, and C
Conjecture: A, B, and C are collinear.

6. Given: $\angle A$ and $\angle C$ are complementary angles.
  $\angle B$ and $\angle C$ are complementary angles.
Conjecture: $\angle A \cong \angle B$

Write a conjecture based on the given information. Draw a figure to illustrate your conjecture.

7. Given: $\angle A$ and $\angle B$ are right angles.

8. Given: $\overline{AB}$, $\overline{BC}$, $\overline{CD}$

9. Given: $P(0, -3)$, $R(0, 3)$, $Q(0, 0)$

10. Given: $A(-1, -3)$, $B(3, 0)$, $C(3, -3)$

EXERCISES

Practice

Determine if the conjecture is *true* or *false* based on the given information. Explain your answer and give a counterexample for any false conjecture.

11. Given: noncollinear points A, B, and C
Conjecture: $\overline{AB}$, $\overline{BC}$, and $\overline{AC}$ form a triangle.

12. Given: collinear points A, B, and C
Conjecture: $\overline{AB}$, $\overline{BC}$, and $\overline{AC}$ form a triangle.

13. Given: $\overline{AB}$, $\overline{BC}$, and $\overline{CD}$
Conjecture: A, B, C, and D are collinear.

Write a conjecture based on the given information. If appropriate, draw a figure to illustrate your conjecture.

14. $\overline{AB}$ and $\overline{CD}$ intersect at E.

15. Point Q is between R and S.

16. $\angle ABC$ and $\angle DBE$ are vertical angles.

17. $Q(5, 2)$, $P(-2, 2)$, $R(-5, 2)$

18. $A(1, 5)$, $B(1, 1)$, $C(-2, 4)$, $D(-2, 0)$

Write the equation you think should come next in each sequence. Check your answers with a calculator.

19.
$$1^2 = 1$$
$$11^2 = 121$$
$$111^2 = 12{,}321$$

20.
$$1^3 = 1^2 - 0^2$$
$$2^3 = 3^2 - 1^2$$
$$3^3 = 6^2 - 3^2$$

Write a conjecture based on the given information.

21. D is the midpoint of $\overline{AB}$, and E is the midpoint of $\overline{AC}$.

22. $WXYZ$ is a rectangle.

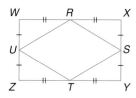

Determine if each conjecture is *true* or *false.* Explain your answers and give a counterexample for any false conjecture.

23. Given: $\overline{ST} \cong \overline{TU}$
   Conjecture: T is the midpoint of $\overline{SU}$.

24. Given: x is a real number.
   Conjecture: x^2 is a nonnegative number.

25. Given: $ABCD$ is a rectangle.
   Conjecture: $\overline{AC}$ and $\overline{BD}$ are congruent.

26. Given: $W(-2, 3)$, $X(1, 7)$, $Y(5, 4)$, and $Z(2, 0)$
   Conjecture: $WXYZ$ is a square.

Write a conjecture based on the given information. Draw a figure to illustrate your conjecture. Write a sentence or two to explain why you think your conjecture is true.

27. Points A, B, C, D, E with no three collinear

28. $\overline{AB}$, $\overline{BC}$, $\overline{CD}$, $\overline{DE}$, $\overline{EA}$ with only A, B, and C collinear

29. $A(-3, 0)$, $B(0, 3)$, $C(4, 1)$, $D(-1, -4)$

30. Draw several quadrilaterals with each pair of opposite sides parallel. Use a protractor to measure the pairs of opposite angles in each figure. Make a conjecture.

Critical Thinking

31. Number Theory The expression $n^2 - n + 41$ has a prime value for $n = 1$, $n = 2$, and $n = 3$. You might conjecture that this expression always generates a prime number for any positive integral value of n. Try different values of n to test the conjecture. Answer true if you think it is always true. Answer false and give a counterexample if you think it is false.

Applications

32. Statistics Work in pairs. Use a tape measure to measure your partner's height and the distance around your partner's head.

   a. Make a conjecture about the relationship of the two measurements.

   b. Compare your conjecture to other students' measurements.

33. **Botany** In the 1920s, some Japanese farmers observed that some rice plants were growing taller and thinner than normal rice plants and then drooping over, making them impossible to harvest.

    a. Make some conjectures about why the plants were drooping.

    b. The scientists who researched the problem discovered that fungus was growing on the drooping rice plants, while the healthy plants had no fungus. Does this observation prompt a new conjecture?

    c. How might the scientists have tested the conjecture they made?

Mixed Review

34. Determine whether you can assume that ∠ACD and ∠BCD are adjacent from the figure at the right. **(Lesson 1-8)**

35. Use a protractor to draw a 65° angle. **(Lesson 1-7)**

36. Kim and Ryan are making pizzas for the band's fall fund-raiser. A pizza can have thin or thick crust and either pepperoni, mushrooms, or green peppers. How many different pizzas can be made? **(Lesson 1-3)**

37. Is the top of your desk best modeled by a point, line, or a plane? **(Lesson 1-1)**

Wrap-Up

38. **Journal Entry** Select a problem you did from this exercise set that illustrates some of the main ideas of this section. Rewrite the problem in your journal. Thoroughly explain how the problem was done and how it relates to conjectures.

～～～～ HISTORY CONNECTION ～～～～

Most of the geometry that you are familiar with is *Euclidean Geometry*, named after the Greek mathematician Euclid. In about 300 B.C., Euclid unified the works of many schools and mathematicians into his *Elements*. It was so comprehensive a book that it made every other book on geometry obsolete. *Elements* begins by establishing fundamental concepts from which all of the remaining concepts are proven using deductive reasoning.

The process of deductive reasoning that Euclid used is not limited to mathematics. Lawyers use deductive reasoning to present their legal arguments. When Abraham Lincoln was a young lawyer trying to perfect his skills of logic, he studied Euclid's *Elements*. He would read by candlelight after the other lawyers who shared his room had gone to sleep.

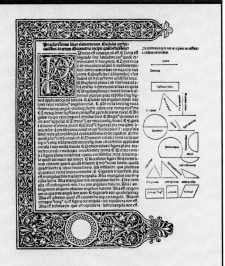

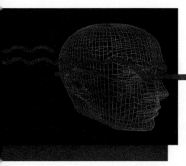

Technology
Conjectures

BASIC
▶ **Geometric Supposer**
Graphing calculators
LOGO
Spreadsheets

As you know, a conjecture is an educated guess based on observations of a particular situation. Let's use the Geometric Supposer to make some conjectures about the diagonals of quadrilaterals.

The Geometric Supposer: Quadrilaterals will allow you to draw and investigate many different quadrilaterals. A quadrilateral is a four-sided plane figure. We will begin by drawing a parallelogram, which is a special quadrilateral in which both pairs of opposite sides are parallel. To draw a parallelogram, press *N* to begin. Then choose *(1) Parallelogram* from the menu. A menu of different types of parallelograms will appear. Choose *(1) Random*. A parallelogram will appear on the screen.

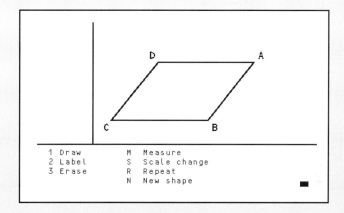

A diagonal of a quadrilateral is a segment that connects a pair of opposite vertices. So the diagonals of your parallelogram *ABCD* are $\overline{AC}$ and $\overline{BD}$. To find the lengths of the diagonals, press *(M) Measure*. Then choose *(1) Length* from the menu. Enter *AC* and *RETURN*. Record the length of $\overline{AC}$ and repeat for $\overline{BD}$.

EXERCISES

Use the Geometric Supposer to draw three examples of each different type of parallelogram, trapezoid, and kite.

1. Use a chart to record the lengths of the diagonals for each type of quadrilateral that you draw.

2. Is there a pattern in the measures of the diagonals of a certain type of quadrilateral? Explain.

2-2 If-Then Statements, Converses, and Postulates

Objectives

After studying this lesson, you should be able to:
- identify the hypothesis and conclusion of an "if-then" statement,
- write the converse of an "if-then" statement, and
- identify and use basic postulates about points, lines, and planes.

Application

President Bush's statement about the environment is an **if-then** or **conditional statement.** The portion of the sentence immediately following *if* is called the **hypothesis,** and the part immediately following *then* is called the **conclusion.**

Conditional statements are the basis of logic. In logic, the hypothesis is often represented by p and the conclusion by q. The conditional "If p then q" is written in symbols as $p \rightarrow q$ and is read "p implies q."

"If our response is to be effective, then all the nations of the world must make common cause in defense of our environment."

Example 1

Identify the hypothesis and conclusion of the conditional "If a hot dog is a foot long, then it is 12 inches long."

Hypothesis: a hot dog is a foot long

Conclusion: it is 12 inches long

Note that "if" is not used when you write the hypothesis and "then" is not used when you write the conclusion.

Sometimes a conditional statement is written without using "if" and "then." For example, Ben Franklin's saying, "A stitch in time saves nine," can be written as the conditional "If you make a stitch in time, then you will save nine."

Example 2

Write the statement "Adjacent angles have a common vertex" in if-then form.

The hypothesis is that two angles are adjacent and the conclusion is that the angles have a common vertex. So, the conditional can be written as follows:

If two angles are adjacent, then they have a common vertex.

Sometimes you must add information to a statement when you write it in if-then form. For example, in Example 2 it was necessary to know adjacent angles came in pairs in order for the if-then statement to be clear.

You can form another if-then statement by interchanging the hypothesis and conclusion of a conditional. This new statement is called the **converse** of the original conditional. The converse of "If two angles are adjacent, then they have a common vertex" is "If two angles have a common vertex, then they are adjacent angles."

If a conditional is not in if-then form, it may be easier to write it in that form before writing the converse. The converse of a true conditional is not necessarily true. *The converse of $p \rightarrow q$ is $q \rightarrow p$.*

Example 3

Write the converse of the true conditional "Vertical angles are congruent." Determine if the converse is true or false. If it is false, give a counterexample.

First write the conditional in if-then form.

The hypothesis is that two angles are vertical and the conclusion is that the angles are congruent. So, the if-then form of the conditional is as follows.

If-Then Form: If two angles are vertical, then they are congruent.

Now exchange the hypothesis and the conclusion to form the converse of the conditional.

Converse: If two angles are congruent, then they are vertical.

The converse of the conditional is false since two congruent angles are not necessarily vertical. A counterexample is shown below.

See pages 695-704 for a complete list of the postulates in this book.

Geometry is built on conditional statements called **postulates.** Postulates are principles that are accepted to be true without proof. The first two postulates describe the ways that points, lines, and planes are related.

Postulate 2-1	**Through any two points there is exactly one line.**
	If there are two points, then there is exactly one line that contains them.
Postulate 2-2	**Through any three points not on the same line there is exactly one plane.**
	If there are three points not on the same line, then there is exactly one plane that contains them.

The relationships between points, lines, and planes can be used to solve practical problems.

Example 4

APPLICATION

Construction

There are four buildings on the Woodlawn High School campus. Assuming that no three of the buildings lie in a straight line, how many sidewalks would need to be constructed so that each building is directly connected to each other building?

If we represent the buildings as points A, B, C, and D, there are four points in the plane. For every two points there is exactly one line. So for four points there are six lines that can be drawn.

In the figure, $\overline{AB}$, $\overline{BC}$, $\overline{CD}$, $\overline{AD}$, $\overline{BD}$, and $\overline{AC}$ can be drawn.

Six sidewalks need to be constructed.

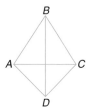

The next four postulates state more relationships of points, lines, and planes.

Postulate 2-3	**A line contains at least two points.**
Postulate 2-4	**A plane contains at least three points not on the same line.**
Postulate 2-5	**If two points lie in a plane, then the entire line containing those two points lies in that plane.**
Postulate 2-6	**If two planes intersect, then their intersection is a line.**

CHECKING FOR UNDERSTANDING

Communicating Mathematics

Read and study the lesson to answer these questions.

1. Define a conditional statement and write an example of a conditional.

2. Translate the conditional statement "Waste not want not" into if-then form.

3. Look through some magazines or newspapers for conditional statements in advertising and write them in if-then form. For example, the well-known United Negro College Fund slogan "A mind is a terrible thing to waste" can be written as "If something is a mind, then it is a terrible thing to waste."

4. How do you form the converse of a conditional statement? Write the converse of "If it is raining, then there is a rainbow."

5. Postulates are basic principles that are assumed without proof. Can you think of some postulates that apply to other fields of study?

Guided Practice

Identify the hypothesis and conclusion of each conditional statement.

6. If you work for 8 hours, then you work for one third of a day.
7. If two lines are perpendicular, then they intersect.
8. If you are sixteen years old, then you may get a driver's license.
9. If $x = 4$, then $x^2 = 16$.

Write the converse of each conditional. Determine if the converse is *true* or *false*. If it is false, give a counterexample.

10. If it is raining, then there are clouds.
11. If an angle measures 37°, then it is acute.
12. If three points are collinear, then they lie on a straight line.
13. If a native-born United States citizen is at least thirty-five years old, then he or she may serve as President.

Write each conditional statement in if-then form.

14. Congruent angles have the same measure.
15. Two planes intersect in a line.
16. What goes up must come down.
17. A recycled aluminum can is remelted and back in the store within six weeks.

In the figure, *R*, *P*, and *S* are collinear. Points *R*, *P*, *S*, and *T* lie in plane $\mathcal{M}$. Use the postulates you have learned to determine whether each statement is *true* or *false*.

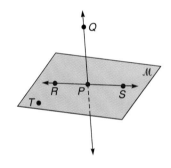

18. P, Q, and R lie in plane $\mathcal{M}$.
19. $\overline{PS}$ does not lie in plane $\mathcal{M}$.
20. P, Q, R, and S are coplanar.
21. Q, R, and T are collinear.

EXERCISES

Practice

Identify the hypothesis and conclusion of each conditional statement.

22. If it is Memorial Day, then it is a holiday.
23. If a candy bar is a Milky Way®, then it contains caramel.
24. If a container holds 32 ounces, then it holds a quart.

In the figure, *P*, *Q*, and *R* are collinear. Points *P* and *S* lie in plane $\mathcal{M}$. Points *Q* and *T* lie in plane $\mathcal{N}$. Determine whether each statement is *true* or *false*.

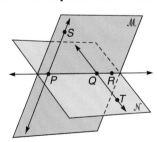

25. Q lies in plane $\mathcal{M}$.
26. P, Q, and R lie in plane $\mathcal{M}$.
27. Q does not lie in plane $\mathcal{N}$.
28. P, Q, S, and T are coplanar.
29. $\overline{QT}$ lies in plane $\mathcal{N}$.

Write each conditional statement in if-then form.

30. Right angles are congruent.

31. A car has 4 wheels.

32. A square has 4 sides.

33. Acute angles are less than 90°.

34. A triangle contains exactly 3 angles.

35. Parallel lines do not intersect.

Write the converse of each conditional. Determine if the converse is *true* or *false*. If it is false, give a counterexample.

36. If the month is January, then it has 31 days.

37. If the distance of a race is 10 kilometers, then it is about 6.2 miles.

38. Springer spaniels are dogs.

State the postulate that explains each relationship.

39. Exactly one line contains Q and P.

40. There are at least two points on line QP.

41. The entire line containing R and S lies in plane M.

42. There are at least three points in plane M.

43. The intersection of plane M and the plane determined by Q, P, and R is line RP.

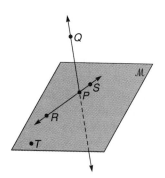

State the number of lines that can be drawn that contain the given set of points taken two at a time.

44. two points

45. three collinear points

46. three noncollinear points

47. four points, no three of which are collinear

48. five points, no three of which are collinear

49. six points, no three of which are collinear

50. *Look for a pattern* in your answers to Exercises 44 through 49. What conjecture can you make about the number of lines through seven points, no three of which are collinear?

State the number of planes that can be drawn that contain the given set of points taken three at a time.

51. three noncollinear points

52. three collinear points

53. five points, no four of which are coplanar

54. Give examples of each of the following.

   a. a conditional and its converse that are both true

   b. a conditional and its converse that are both false

   c. a true conditional whose converse is false

   d. a false statement whose converse is true

Applications

55. Biology Use a diagram to illustrate the following conditional about the animal kingdom. "If an animal is a butterfly, then it is an arthropod."

56. Advertising A billboard reads "If you want a fabulous vacation, try Georgia."

   a. Write the converse of the conditional.

   b. What do you think the advertiser wants you to conclude about vacations in Georgia?

   c. Does the advertisement say that vacations in Georgia are fabulous?

Computer

In BASIC, an IF-THEN statement is used to compare two numbers. It tells the computer what to do based on the results of the comparison. The program below uses an IF-THEN statement to find the greatest number in a list.

57. Change the DATA statement to find the greatest number in each list.

   a. 84, 70, 22, 90, 31, 68, 92, 19, 36, 75

   b. 112, 524, 923, 987, 473, 811, 476, 216, 892, 1023

```
10   FOR I = 1 TO 10
20   READ A(I)
30   NEXT I
40   LET L = 1
50   FOR K = 2 TO 10
60   IF A(K) > A(L) THEN GOTO 80
70   GOTO 90
80   LET L = K
90   NEXT K
100  PRINT A(L)
110  END
120  DATA 43, 65, 78, 95, 21,
     56, 1, 42, 101, 97
```

58. How could you change the program to find the least number in the list?

Mixed Review

59. Given that $AP = PB$, David made the conjecture that P is the midpoint of $\overline{AB}$. Do you think his conjecture is true or false? Explain. **(Lesson 2-1)**

60. Find the measures of two supplementary angles if the measure of the one angle is 36 more than the measure of the other. **(Lesson 1-8)**

61. Suppose $\angle MON$ is a right angle and L is in the interior of $\angle MON$. If $m\angle MOL$ is five times $m\angle LON$, find $m\angle LON$. **(Lesson 1-7)**

62. $M(-3, 5)$ is the midpoint of $\overline{AB}$. If the coordinates of A are $(-7, 6)$, find the coordinates of B. **(Lesson 1-5)**

63. B is between A and C, $AB = 6x - 1$, $BC = 2x + 4$, and $AC = 9x - 3$. Find AC. **(Lesson 1-4)**

64. Draw and label a figure that shows lines m and ℓ intersecting at point P. **(Lesson 1-2)**

65. Points $A(4, -3)$ and $B(-2, 9)$ lie on the line whose equation is $y = -2x + 5$. Determine whether $N(7, -2)$ is collinear to A and B. **(Lesson 1-1)**

Wrap-Up

66. Write a five-question quiz about this lesson. Be sure to include answers to your questions.

2-3 Deductive Reasoning

Objective

After studying this lesson, you should be able to:
- use the law of detachment or the law of syllogism in deductive reasoning.

Application

If you like walking, you'll love AirSports.

Jeanine enjoys walking the nature trail at the park each day. Assuming that the conditional statement in the advertisement is true, she will love AirSports.

This is an example of the use of the **law of detachment.** The law of detachment offers us a way to draw conclusions from if-then statements. It says that whenever a conditional is true and its hypothesis is true, we can conclude that its conclusion is true.

Law of Detachment	If $p \rightarrow q$ is a true conditional and p is true, then q is true.

Deductive reasoning uses a rule to make a specific conclusion. Inductive reasoning uses several examples to make a conjecture or rule.

The law of detachment and the other laws of logic can be used to provide a system for reaching logical conclusions. This system is called **deductive reasoning.** Deductive reasoning is one of the cornerstones of the study of geometry. The following example illustrates how the law of detachment can be used to draw a valid conclusion.

Example 1

"If two numbers are negative, then their product is positive" is a true conditional, and -3 and -4 are negative numbers. Use the law of detachment to state a valid conclusion.

The hypothesis is that two numbers are negative. -3 and -4 are indeed two negative numbers. Since the conditional is true and the given statement satisfies the hypothesis, the conclusion is true. So, the product of -3 and -4 is positive.

Knowing that the conditional is true and that the conclusion is satisfied will not allow us to say the hypothesis follows. Consider this counterexample. The product of 5 and 9 is 45, a positive number, but 5 and 9 are not two negative numbers.

Valid reasoning includes a true conditional paired with given information that satisfies the hypothesis. Invalid reasoning results when a true conditional is paired with information that satisfies the conclusion.

Example 2

APPLICATION

Advertising

Determine if a true conclusion can be reached from the two statements "If you want a steak that's grilled to perfection, then go to Morton's" and "Gail went to Morton's" using the law of detachment. If a valid conclusion is possible, state it.

Hypothesis: you want a steak that's grilled to perfection

Conclusion: go to Morton's

Given: Gail went to Morton's.

Morton's Restaurant wants you to conclude that Gail went there because she wanted a steak grilled to perfection. However, she may have gone there because they have good service. Since the given information satisfied the conclusion instead of the hypothesis, we cannot make a valid conclusion.

A second law in logic is the **law of syllogism.** It looks very much like the transitive property of equality that you should remember from your studies in algebra.

| *Law of Syllogism* | **If $p \rightarrow q$ and $q \rightarrow r$ are true conditionals, then $p \rightarrow r$ is also true.** |

Example 3

APPLICATION

Chemistry

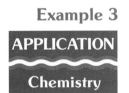

FYI···

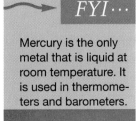

Mercury is the only metal that is liquid at room temperature. It is used in thermometers and barometers.

Determine if a valid conclusion can be reached from the two statements "If a metal is liquid at room temperature, then it is mercury" and "If a metal is mercury, then its chemical symbol is Hg" using the law of syllogism.

Let p, q, and r represent each part of the statements.

$p:$ a metal is liquid at room temperature

$q:$ it is mercury

$r:$ its chemical symbol is Hg

The given statements can be represented as $p \rightarrow q$ and $q \rightarrow r$. So, according to the law of syllogism we can conclude that $p \rightarrow r$. That is "If a metal is liquid at room temperature, then its chemical symbol is Hg."

CHECKING FOR UNDERSTANDING

Communicating Mathematics

Read and study the lesson to answer these questions.

1. Give an example of a correct use of the law of detachment.

2. What is required to use the law of detachment to reach a valid conclusion?

3. "Those who choose Tint and Trim Hair Salon have impeccable taste; and you have impeccable taste" is an example of how an advertiser can misuse the law of detachment to make you come to an invalid conclusion.
   a. What conclusion do they want you to make?
   b. Write another example of this type of incorrect use of logic.

4. Use the law of syllogism to derive a valid conclusion from these two conditionals. "If two angles are a linear pair, then they are supplementary" and "If two angles are supplementary, then their measures total 180."

Guided Practice

Determine if statement (3) follows from statements (1) and (2) by the law of detachment or the law of syllogism. If it does, state which law you used.

5. (1) If a student is enrolled at Lyons High, then the student has an ID number.
   (2) Joel Nathan is enrolled at Lyons High.
   (3) Joel Nathan has an ID number.

6. (1) If an angle measures 123, then it is obtuse.
   (2) $m\angle C = 123$.
   (3) $\angle C$ is obtuse.

7. (1) If your car needs more power, use Powerpack Motor Oil.
   (2) Marcus uses Powerpack Motor Oil.
   (3) Marcus needed more power in his car.

8. (1) If you like pizza with everything, then you'll like Jimmy's Pizza.
   (2) If you like Jimmy's Pizza, then you are a pizza connoisseur.
   (3) If you like pizza with everything, then you are a pizza connoisseur.

9. (1) If a rectangle has four congruent sides, then it is a square.
   (2) A square has diagonals that are perpendicular.
   (3) A rectangle has diagonals that are perpendicular.

Determine if a conclusion can be reached from the two given statements using the law of detachment or the law of syllogism. If a conclusion is possible, state it and the law that is used. If a conclusion does not follow, state *"No conclusion."*

10. (1) If you want the best hamburger in town, then buy a Biggie Burger.
    (2) Pat Gorman bought a Biggie Burger.

11. (1) If $\overline{AB}$ is a segment, then $\overline{AB} \cong \overline{AB}$.
    (2) $\overline{CD}$ is a segment.

12. (1) If two angles form a linear pair, then they share a common ray.
    (2) If two angles share a common ray then they are adjacent.

13. (1) Two planes intersect in a line.
    (2) Planes $\mathcal{M}$ and $\mathcal{N}$ intersect.

14. (1) Sponges belong to the phylum porifera.
    (2) Sponges are animals.

15. (1) Bobby Rahal has raced in the Indianapolis 500.
    (2) Only professional race car drivers have raced in the Indianapolis 500.

EXERCISES

Practice

Determine if statement (3) follows from statements (1) and (2) by the law of detachment or the law of syllogism. If it does, state which law was used.

16. (1) If you are not satisfied with a tape, then you can return it within a week for a full refund.
    (2) Joe is not satisfied with a tape.
    (3) Joe can return the tape within a week for a full refund.

17. (1) If fossil fuels are burned, then acid rain is produced.
    (2) If acid rain falls, wildlife suffers.
    (3) If fossil fuels are burned, then wildlife suffers.

18. (1) If x is a real number, then x^2 is nonnegative.
    (2) x^2 is nonnegative.
    (3) x is a real number.

19. (1) If an angle measures less than 90, then it is acute.
    (2) $m\angle A$ is less than 90.
    (3) $\angle A$ is acute.

20. (1) Careful bicycle riders wear helmets.
  (2) Riders who wear helmets have fewer injuries.
  (3) Careful bicycle riders have fewer injuries.

21. (1) All pilots must pass a physical examination.
  (2) Kris Thomas must pass a physical examination.
  (3) Kris Thomas is a pilot.

Determine if a conclusion can be reached from the two given statements using the law of detachment or the law of syllogism. If a conclusion is possible, state it and the law that is used. If a conclusion does not follow, state *"No conclusion."*

22. (1) If you are looking for the excitement of a new car, then you should make tracks to your Pontiac® dealer.
  (2) Ann made tracks to the Pontiac® dealer.

23. (1) If A is between B and C, then A, B, and C are collinear.
  (2) A is between B and C.

24. (1) If a quadrilateral is a rectangle, then it has four right angles.
  (2) A rectangle has diagonals that are congruent.

25. (1) If two lines intersect, then they have a point in common.
  (2) Lines p and q intersect.

26. (1) If an angle is right, then it measures 90°.
  (2) $m\angle A = 90$

27. (1) If an ordered pair for a point has 0 as its x-coordinate, then the point lies on the y-axis.
  (2) If a point lies on the y-axis, then it is not contained in any of the four quadrants.

28. (1) In an ordered pair, if the x-coordinate is negative and the y-coordinate is positive, then the point lies in Quadrant II.
  (2) Point Q lies in Quadrant II.

29. (1) If $\angle A \cong \angle B$ and $\angle B \cong \angle C$, then $\angle A \cong \angle C$.
  (2) $\angle A \cong \angle C$.

30. (1) Parallel lines do not intersect.
  (2) If lines do not intersect, then they have no points in common.

31. (1) If M is the midpoint of $\overline{AB}$, then $AM = MB$.
  (2) $FG = GH$.

32. (1) Right angles are congruent.
  (2) $\angle A \cong \angle B$

33. (1) Basalt is an igneous rock.
  (2) Igneous rocks were formed by volcanos.

Using the given statement, create a second statement and a valid conclusion that illustrates the correct use of the law of detachment. Then write a statement and a conclusion that illustrate the correct use of the law of syllogism.

34. If you're looking for a fun car to drive, then you need a Tigercub.

35. Angles that are adjacent and supplementary are right angles.

36. If the coordinates of a point satisfy the equation of a line, then the point lies on the line.

37. All physicians have graduated from medical school.

Critical Thinking

38. An advertisement states that "If you like skiing, then you'll love SnowLinks." You like to ski, but when you went to SnowLinks, you didn't like it there. Where did the logic break down?

Applications

If possible, write a valid conclusion. State the law of logic that you used.

39. Geology
- If a mineral sample is a sample of quartz, then the sample has a hardness factor of 7.
- If a mineral sample has a hardness factor greater than 5, then it can scratch glass.

40. Biology
- All species in the plant family Maalvaceae have five petals.
- A wild rose has five petals.

41. Advertising   Conditionals and logic are often used in advertising. Look through some newspapers and magazines for advertisements that use logic and evaluate the arguments.

Mixed Review

42. Write the statement "An 89 is an above average score on the geometry test" in if-then form. **(Lesson 2-2)**

43. Is an angle that measures 67° acute, obtuse, right, or straight? **(Lesson 1-7)**

44. Find the ordered pair for the midpoint of $\overline{GH}$ with endpoints $G(4, 8)$ and $H(-3, 0)$. **(Lesson 1-5)**

45. Use the distance formula to find the measure of $\overline{XY}$ with endpoints $X(5, -3)$ and $Y(0, -5)$. **(Lesson 1-4)**

46. Draw and label a figure that shows perpendicular lines ℓ and m. **(Lesson 1-2)**

Wrap-Up

47. Write and design an advertisement for one of your favorite products that uses the law of detachment to convince a person your argument is true.

2-4 Properties from Algebra and Proof

Objective After studying this lesson, you should be able to:

- ▪ use properties of equality in algebraic and geometric proofs.

Certain rules apply to every field of study. Gravity is one of the rules of science.

You are familiar with many of the rules of algebra. These rules, along with various defined operations and sets of numbers form a mathematical system. Working within the rules of the system allows you to perform algebraic operations.

Geometry is another example of a mathematical system. Since it also deals with variables, numbers, and operations, many of the rules of algebra are also used in geometry. Some of the important ones are listed in the table below.

Properties of Equality for Real Numbers	
Reflexive Property	For every number a, $a = a$.
Symmetric Property	For all numbers a and b, if $a = b$, then $b = a$.
Transitive Property	For all numbers a, b, and c, if $a = b$ and $b = c$, then $a = c$.
Addition and Subtraction Properties	For all numbers a, b, and c, if $a = b$, then $a + c = b + c$ and $a - c = b - c$.
Multiplication and Division Properties	For all numbers a, b, and c, if $a = b$, then $a \cdot c = b \cdot c$, and if $c \neq 0$, $\frac{a}{c} = \frac{b}{c}$.
Substitution Property	For all numbers a and b, if $a = b$, then a may be replaced by b in any equation or expression.
Distributive Property	For all numbers a, b, and c, $a(b + c) = ab + ac$.

You will need to recognize and use these properties in problems.

Since segment measures and angle measures are real numbers, these properties from algebra can be used to discuss their relationships. Some examples of these applications are shown below.

Property	Segments	Angles
Reflexive	$AB = AB$	$m\angle C = m\angle C$
Symmetric	If $XY = YZ$, then $YZ = XY$.	If $m\angle 1 = m\angle 2$, then $m\angle 2 = m\angle 1$.
Transitive	If $MN = NO$ and $NO = OP$, then $MN = OP$.	If $m\angle K = m\angle L$ and $m\angle L = m\angle M$, then $m\angle K = m\angle M$.

Example 1

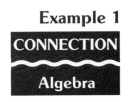

Name the property of equality that justifies each statement.

Statements

a. If $5 = x$, then $x = 5$.

b. If $\frac{1}{2}x = 9$, then $x = 18$.

c. If $AB = 2x$ and $AB = CD$, then $CD = 2x$.

d. If $2AB = 2CD$, then $AB = CD$.

Properties

a. Symmetric property

b. Multiplication property

c. Substitution and symmetric properties

d. Division property

You can use these properties as reasons for the step-by-step solution of an equation.

Example 2

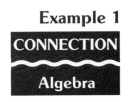

Justify each step in solving $2x - 3 = \frac{2}{3}$.

Statements	Reasons
1. $2x - 3 = \frac{2}{3}$	1. Given
2. $3(2x - 3) = 2$	2. Multiplication property
3. $6x - 9 = 2$	3. Distributive property
4. $6x = 11$	4. Addition property
5. $x = \frac{11}{6}$	5. Division property

Example 2 is a proof of the conditional "If $2x - 3 = \frac{2}{3}$, then $x = \frac{11}{6}$." The given information relates to the hypothesis of the conditional. It is the starting point of the proof. The conclusion, $x = \frac{11}{6}$, is the end of the proof. The fact that reasons (properties) are listed with the steps leading to the conclusion makes this sequence a proof.

Proofs in geometry can be organized in the same manner. The algebra properties and definitions, postulates, and other true statements can be used as reasons. Most of the time we will write proofs in geometry in **two-column** form. These are considered to be formal proofs.

Example 3

Justify the steps for the proof of the conditional "If $AB = CD$, then $AC = BD$." *Remember that AB, CD, AC, and BD represent real numbers.*

Given: $AB = CD$

Prove: $AC = BD$

A————B————C————D

Statements	Reasons
1. $AB = CD$	**1.** __?__
2. $BC = BC$	**2.** __?__
3. $AB + BC = CD + BC$	**3.** __?__
4. $AC = AB + BC$ $BD = BC + CD$	**4.** __?__
5. $AC = BD$	**5.** __?__

Reason 1: "Given" since it follows from the hypothesis

Reason 2: Reflexive property of equality

Reason 3: Addition property of equality

Reason 4: Segment addition postulate

Reason 5: Substitution property of equality *Applied twice to Step 3.*

CHECKING FOR UNDERSTANDING

Communicating Mathematics

Read and study the lesson to answer these questions.

1. Make a flash card for each of the properties of equality in this lesson. On the front of the card put the name of the property. On the back of the card, put the explanation that goes with the property. Practice your flash cards.

2. Describe the differences between the reflexive, symmetric, and transitive properties.

3. In your own words, explain the meaning of *proof.*

4. What part of a conditional is related to the *Given* statement of a proof? What part is related to the *Prove* statement?

Guided Practice

Name the property of equality that justifies each statement.

5. If $3x + 7 = 12$, then $3x = 5$.

6. If $2(x + 5) = 13$, then $2x + 10 = 13$.

7. If $5x = 7$, then $x = \frac{7}{5}$.

8. If $AB = CD$, then $AB + EF = CD + EF$.

9. If $m\angle A + m\angle B = 180$ and $m\angle B = 30$, then $m\angle A + 30 = 180$.

Copy each proof, then name the property that justifies each statement.

10. Prove that if $2x + 3 = 7$, then $x = 2$.

  Given: $2x + 3 = 7$
  Prove: $x = 2$

Statements	Reasons
a. $2x + 3 = 7$	**a.** _?_
b. $2x = 4$	**b.** _?_
c. $x = 2$	**c.** _?_

11. Prove that if $m\angle 1 = m\angle 3$, then $m\angle ABD = m\angle CBE$.

  Given: $m\angle 1 = m\angle 3$
  Prove: $m\angle ABD = m\angle CBE$

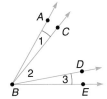

Statements	Reasons
a. $m\angle 1 = m\angle 3$	**a.** _?_
b. $m\angle 2 = m\angle 2$	**b.** _?_
c. $m\angle 1 + m\angle 2 = m\angle 3 + m\angle 2$	**c.** _?_
d. $m\angle ABD = m\angle 1 + m\angle 2$ $\ \ \ \ m\angle CBE = m\angle 3 + m\angle 2$	**d.** _?_
e. $m\angle ABD = m\angle CBE$	**e.** _?_

EXERCISES

Practice   **Name the property of equality that justifies each statement.**

12. If $x + 4 = -3$, then $x = -7$.

13. If $y = 2x + 3$ and $x = 2$, then $y = 7$.

14. If $\frac{1}{2}m\angle F = \frac{1}{2}m\angle G$, then $m\angle F = m\angle G$.

15. If $AB + BC = AC$ and $AC = EF + GH$, then $AB + BC = EF + GH$.

16. If $m\angle A = 90$ and $m\angle B = 90$, then $m\angle A = m\angle B$.

17. $m\angle A = m\angle A$.

18. If $m\angle 1 = m\angle 2$, then $m\angle 2 = m\angle 1$.

19. If $AB + BC = 12$, then $BC = 12 - AB$.

20. If $x + y = 9$ and $x - y = 12$, then $2x = 21$.

21. If $AB - CD = EF - CD$, then $AB = EF$.

Copy each proof, then name the property that justifies each statement.

22. Prove that if $\frac{2}{3}x = -8$, then $x = -12$.

Given: $\frac{2}{3}x = -8$

Prove: $x = -12$

Statements	Reasons
a. $\frac{2}{3}x = -8$	a. __?__
b. $2x = -24$	b. __?__
c. $x = -12$	c. __?__

23. Prove that if $5 = 2 - \frac{1}{2}x$, then $x = -6$.

Given: $5 = 2 - \frac{1}{2}x$

Prove: $x = -6$

Statements	Reasons
a. $5 = 2 - \frac{1}{2}x$	a. __?__
b. $2 - \frac{1}{2}x = 5$	b. __?__
c. $2(2 - \frac{1}{2}x) = 10$	c. __?__
d. $4 - x = 10$	d. __?__
e. $-x = 6$	e. __?__
f. $x = -6$	f. __?__

24. Prove that if $2x - 7 = \frac{1}{3}x - 2$, then $x = 3$.

Given: $2x - 7 = \frac{1}{3}x - 2$

Prove: $x = 3$

Statements	Reasons
a. $2x - 7 = \frac{1}{3}x - 2$	a. __?__
b. $3(2x - 7) = 3(\frac{1}{3}x - 2)$	b. __?__
c. $6x - 21 = x - 6$	c. __?__
d. $5x - 21 = -6$	d. __?__
e. $5x = 15$	e. __?__
f. $x = 3$	f. __?__

25. Prove that if $m\angle ABD = m\angle EFH$ and $m\angle 2 = m\angle 4$, then $m\angle 1 = m\angle 3$.

Given: $m\angle ABD = m\angle EFH$
$m\angle 2 = m\angle 4$

Prove: $m\angle 1 = m\angle 3$

Statements	Reasons
a. $m\angle ABD = m\angle EFH$ $m\angle 2 = m\angle 4$	**a.** _?_
b. $m\angle ABD = m\angle 1 + m\angle 2$ $m\angle EFH = m\angle 3 + m\angle 4$	**b.** _?_
c. $m\angle 1 + m\angle 2 = m\angle 3 + m\angle 4$	**c.** _?_
d. $m\angle 1 + m\angle 4 = m\angle 3 + m\angle 4$	**d.** _?_
e. $m\angle 1 = m\angle 3$	**e.** _?_

Write a complete proof for each of the following.

26. If $4 - x = 10$, then $x = -6$.

27. If $m\angle M = m\angle P$ and $m\angle N = m\angle P$, then $m\angle M = m\angle N$.

28. If $x - 1 = \dfrac{x - 10}{-2}$, then $x = 4$.

29. If $m\angle ABC = 90$, $m\angle EDC = 90$, and $m\angle 1 = m\angle 3$, then $m\angle 2 = m\angle 4$.

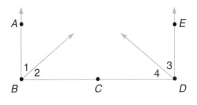

Critical Thinking

30. Choose a basic rule from a field of study or a sport and negate it. Investigate the results. For example, what would change about the game of golf if the player with the highest score wins?

Applications

31. Banking The formula for finding the amount in a bank account after adding the amount earned from simple interest is $A = p + prt$, where p is the principal, r is the annual interest rate, and t is the time in years. Solve the formula for p and justify each step.

32. Physics The distance, s, that an object travels is found by $s = \frac{1}{2}at^2 + v_o t$, where a is the acceleration, t is the time, and v_o is the initial velocity. Solve the formula for a and justify each step.

33. Language Arts Find relationships other than equality that are reflexive, symmetric, or transitive. For example, "is a relative of" is symmetric and "is taller than" is transitive.

34. If possible, write a valid conclusion. State the law of logic that you used. **(Lesson 2-3)**

- If a is a real number, then $a = a$.
- 7 is a real number.

35. Use a calculator to find the decimal equivalents of $\frac{1}{9}$, $\frac{2}{9}$, $\frac{3}{9}$, and $\frac{4}{9}$. Make a conjecture about the decimal equivalent of $\frac{5}{9}$. Use your calculator to check your answer. **(Lesson 2-1)**

36. Find the coordinates of the midpoint of the segment whose endpoints are at (9, 3), and (-3, 8). **(Lesson 1-5)**

37. Draw and label a figure that shows the relationship "N lies on line ℓ." **(Lesson 1-2)**

38. Write three multiple-choice questions that could be used as part of a quiz on this lesson.

MID-CHAPTER REVIEW

Determine if each conjecture is *true* or *false*. Explain your answers and give a counterexample for any false conjecture. (Lesson 2-1)

1. Given: A is the midpoint of $\overline{BC}$.
Conjecture: $AB = AC$

2. Given: x is a real number.
Conjecture: x^3 is a real number.

3. Given: Points A, B, and C are collinear.
Conjecture: B is between A and C.

4. Given: $\angle 1$ and $\angle 2$ are right angles.
Conjecture: $\angle 1 \cong \angle 2$

Write each conditional statement in if-then form. (Lesson 2-2)

5. Perpendicular lines form four right angles.

6. The chemical formula for ordinary table sugar is $C_{12}H_{22}O_{11}$.

7. The month of February has 28 days in a non-leap year.

Determine if a conclusion can be drawn from the two statements using the law of detachment or the law of syllogism. If a conclusion is possible, state it and the law used. If a conclusion does not follow, state *No conclusion*. (Lesson 2-3)

8. (1) If you are a music lover, then you can read music.
(2) Shelly can read music.

9. (1) If a polygon is a square, then its diagonals are congruent.
(2) $ABCD$ is a square.

10. (1) If two angles are right angles, then they are congruent.
(2) If two angles are congruent, then they have the same measure.

Write a two-column proof. (Lesson 2-4)

11. Prove that if $x = 7$, then $4x^2 = 196$.

12. Prove that if $AC = AB$, $AC = 4x + 1$, and $AB = 6x - 13$ then $x = 7$.

2-5

Problem-Solving Strategy: Process of Elimination

EXPLORE
PLAN
SOLVE
EXAMINE

Objective

After studying this lesson, you should be able to:

■ solve problems by eliminating possibilities.

Many everyday problems have several possible solutions. Deductive reasoning can be used to eliminate some of those possibilities and choose the correct solution.

Example

CONNECTION

Geology

On a recent geology test, Rena was given five different mineral samples to identify using the portion of a mineral characteristics chart shown at the right. Rena made the following observations about the samples.

1. **Sample C is softer than glass.**
2. **Samples D and E are red and Sample C is brown.**
3. **Samples B and E are harder than glass.**

Mineral	Color	Hardness
Biotite	brown or black	softer than glass
Halite	white	softer than glass
Hematite	red	softer than glass
Feldspar	white, pink, or green	harder than glass
Jasper	red	harder than glass

Make a chart showing the sample names and the minerals. Use the observations to mark Xs where possibilities are eliminated and ✔s where matches are found.

Observation 1 indicates that Sample C is softer than glass. So, the minerals that are harder than glass, feldspar and jasper, can be eliminated as possible choices for Sample C.

Sample	A	B	C	D	E
Biotite					
Halite					
Hematite					
Feldspar			X		
Jasper			X		

Observation 2 tells us that Samples D and E must be the minerals that are red, namely hematite and jasper. Sample C must be biotite.

Sample	A	B	C	D	E
Biotite			✔	X	X
Halite				X	X
Hematite					
Feldspar			X	X	X
Jasper			X		

Sample	A	B	C	D	E
Biotite		X	✔	X	X
Halite		X		X	X
Hematite		X			X
Feldspar			X	X	X
Jasper			X		

Observation 3 indicates that Sample B and Sample E must be either feldspar or jasper.

Sample	A	B	C	D	E
Biotite	X	X	✔	X	X
Halite	✔	X	X	X	X
Hematite	X	X	X	✔	X
Feldspar	X	✔	X	X	X
Jasper	X	X	X	X	✔

Each row and column must have only one ✔, so we can now complete the chart and draw our conclusions.

Sample A - halite, Sample B - feldspar, Sample C - biotite, Sample D - hematite, Sample E - jasper

CHECKING FOR UNDERSTANDING

Communicating Mathematics

Read and study the lesson to answer these questions.

1. Describe an effective way of keeping track of the possible solutions you have eliminated.

2. Would the chart used to solve the example work if there were two samples of the same mineral? Why or why not?

Guided Practice

Use the process of elimination to solve each problem.

3. Anthony, Erin, Lisa, and Brad each represent a different grade in the Briggs High School Student Council. Erin is older than Anthony, but younger than Brad. Lisa is older than Brad. Who represents each grade?

4. In Chemistry lab, Al was given four different metal samples to identify. They were gallium, a metal that melts when held in the hand; mercury, which is liquid at room temperature; lithium, which will float in water; and calcium, which bubbles slowly when placed in water. When Al was holding Sample 1, it was a liquid. Sample 2 is a solid and sank to the bottom when dropped in a beaker of water. Bubbling action occurred when Al dropped Sample 4 in the water. Identify each sample.

EXERCISES

Practice

Strategies

Look for a pattern.
Solve a simpler problem.
Act it out.
Guess and check.
Draw a diagram.
Make a chart.
Work backward.

Solve. Use any strategy.

5. As I stood in the cafeteria line, I observed that there were nine more people behind me than there were ahead of me. There were three times as many people in line as there were people ahead of me. How many people were behind me?

6. If you cut one corner off of a square, how many corners would you have left?

7. Umeko, Jim, and Gwen each participate in an extra-curricular activity and have an after-school job. They are in the Spanish Club, the Drama Club, and the marching band. One of them is a pizza delivery person, one is a math tutor, and one is a lifeguard at the community pool. Jim is tutoring the Drama Club member's brother in Algebra. The Spanish Club member cannot swim or drive a car. The lifeguard is teaching Umeko to read music. Who does what?

8. At how many different times during a 24-hour period are all of the digits on a digital clock the same?

9. Three cartons of clothes were delivered to Dunbar's. One contained jeans, one T-shirts, and one had both jeans and T-shirts. All of the boxes were mislabeled. Jack pulled a T-shirt from the box marked "jeans and T-shirts." Which box contains what?

10. The Boston Tea Party was a notable event of the eighteenth century. The sum of the digits of the year it occurred is 18. The ones digit of the year is $\sqrt[4]{81}$. The Boston Tea Party took place on December 16, _?_ .

COOPERATIVE LEARNING PROJECT

Work in groups. Each person in the group must understand the solution and be able to explain it to any person in class.

During the first five days of Michelle's summer vacation, she spent each day with a different person: Zach, Luisa, Marcus, Kelly, and her mother. During that week, Michelle spent her mornings swimming, playing chess, shopping, biking, and picnicking. Her afternoons were spent playing tennis, walking, listening to the radio, skateboarding, and painting her room. Use the strategy of eliminating possibilities to determine which day she spent with each person and what they did together in the morning and afternoon of each day.

1. Michelle and Zach spent the afternoon skateboarding.
2. It rained all day on Friday, so Michelle spent the day in the house.
3. Luisa, who cannot swim, spent Thursday with Michelle. They didn't go shopping.
4. Michelle went biking with Marcus two days before helping her mother paint her room.
5. Michelle didn't play tennis on Tuesday or on the day she went bicycling. She didn't go swimming on the day she painted her room.
6. Michelle's mother doesn't play chess.

2-6 Two-Column Proofs with Segments

Objective

After studying this lesson, you should be able to:

- complete proofs involving segment theorems.

Constructing a proof can be confusing.

DO YOU REMEMBER WHAT WE'RE TRYING TO PROVE?

But if you remember that a proof must include the following five essential steps, you will be able to construct any proof.

- State the theorem to be proved.
- List the given information.
- If possible, draw a diagram to illustrate the given information.
- State what is to be proved.
- Develop a system of deductive reasoning.

In order to use deductive reasoning to construct a valid proof, we must rely on statements that are accepted to be true. In geometry, those statements are the definitions and postulates. We also depend on the list of undefined terms.

The statements that are proved through deductive reasoning using definitions, postulates, and undefined terms are called **theorems**. Once a theorem is proved, it becomes another tool that we can use in the system. That is, proved theorems can be used in the proofs of new theorems.

The first theorem that we will look at is similar to some familiar properties from algebra.

Theorem 2-1	**Congruence of segments is reflexive, symmetric, and transitive.**

Theorem 2-1 can be written in symbols as follows.

Reflexive Property

$\overline{AB} \cong \overline{AB}$

Symmetric Property

If $\overline{AB} \cong \overline{CD}$, then $\overline{CD} \cong \overline{AB}$.

Transitive Property

If $\overline{AB} \cong \overline{CD}$ and $\overline{CD} \cong \overline{EF}$, then $\overline{AB} \cong \overline{EF}$.

The symmetric part of Theorem 2-1 is proved below. You will be asked to prove the reflexive and transitive parts in Exercises 4 and 3 respectively.

You can use the properties of algebra in geometric proofs. Notice that the symmetric property of equality is used in the proof of Theorem 2-1.

Proof of Symmetric Part of Theorem 2-1

Given: $\overline{AB} \cong \overline{CD}$

Prove: $\overline{CD} \cong \overline{AB}$

Recall that since the measures of segments are real numbers, the properties of algebra can be used to prove relationships with segment measures.

Statements	Reasons
1. $\overline{AB} \cong \overline{CD}$	1. Given
2. $AB = CD$	2. Definition of congruent segments
3. $CD = AB$	3. Symmetric property of equality
4. $\overline{CD} \cong \overline{AB}$	4. Definition of congruent segments

Since AB and CD represent real numbers, the Symmetric property of equality can be used in Step 3.

Example 1

$\overline{STEP}$ represents the segment containing points S, T, E, and P. T is between S and E and E is between T and P.

Justify each step in the proof.

Given: $\overline{STEP}$

Prove: $SP = ST + TE + EP$

Statements	Reasons
1. $\overline{STEP}$	1. $\underline{\ ?\ }$
2. $SP = ST + TP$	2. $\underline{\ ?\ }$
3. $TP = TE + EP$	3. $\underline{\ ?\ }$
4. $SP = ST + TE + EP$	4. $\underline{\ ?\ }$

Reason 1: Given

Reason 2: Segment addition postulate

Reason 3: Segment addition postulate

Reason 4: Substitution property of equality

Example 2

Write a two-column proof.

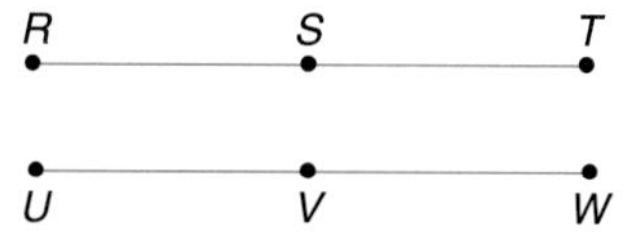

Given: $\overline{RS} \cong \overline{UV}$
$\overline{ST} \cong \overline{VW}$

Prove: $\overline{RT} \cong \overline{UW}$

Statements	Reasons
1. $\overline{RS} \cong \overline{UV}$ $\overline{ST} \cong \overline{VW}$	1. Given
2. $RS = UV$ $ST = VW$	2. Definition of congruent segments
3. $RS + ST = UV + VW$	3. Addition property of equality
4. $RS + ST = RT$ $UV + VW = UW$	4. Segment addition postulate
5. $RT = UW$	5. Substitution property of equality *Twice*
6. $\overline{RT} \cong \overline{UW}$	6. Definition of congruent segments

CHECKING FOR UNDERSTANDING

Communicating Mathematics

Read and study the lesson to answer these questions.

1. In your own words, describe what must be included in a valid proof.

2. Why can the properties of algebra be used to establish relationships between the measures of segments?

3. Copy and complete the proof of the transitive part of Theorem 2-1.

Given: $\overline{AB} \cong \overline{CD}$
$\overline{CD} \cong \overline{EF}$

Prove: $\overline{AB} \cong \overline{EF}$

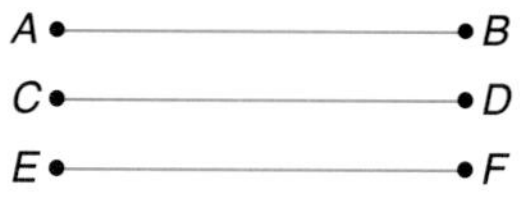

Statements	Reasons
a. $\overline{AB} \cong \overline{CD}$ $\overline{CD} \cong \overline{EF}$	a. __?__
b. $AB = CD$ $CD = EF$	b. __?__
c. $AB = EF$	c. __?__
d. $\overline{AB} \cong \overline{EF}$	d. __?__

4. Write a two-column proof of the reflexive part of Theorem 2-1.

Guided Practice

Justify each statement with a property from algebra or a property of congruent segments.

5. $\overline{PS} \cong \overline{PS}$

6. If $AB + BC = BC + CD$, then $AB = CD$.

7. If $\overline{XY} \cong \overline{OP}$, then $\overline{OP} \cong \overline{XY}$.

8. If $2MN = TS$, then $MN = \frac{1}{2}TS$.

9. If $GH = 12$ and $GH + HI = GI$, then $12 + HI = GI$.

10. If $\overline{PQ} \cong \overline{QR}$ and $\overline{QR} \cong \overline{RT}$, then $\overline{PQ} \cong \overline{RT}$.

11. If $AN - 8 = IN - 8$, then $AN = IN$.

12. If $EF = GH$ and $GH = JK$, then $EF = JK$.

Write the given and the prove statements that you would use to prove each theorem. Draw a figure if applicable.

13. If an angle is a right angle, then its measure is 90.

14. All rational numbers are real.

15. If two angles are vertical, then they are congruent.

16. If two lines are perpendicular, then they form four right angles.

17. The sum of the degree measures of the angles of a triangle is 180.

18. The diagonals of a rectangle are congruent.

19. Copy and complete the proof.

Given: $\overline{LE} \cong \overline{MR}$
$\overline{EG} \cong \overline{RA}$

Prove: $\overline{LG} \cong \overline{MA}$

Statements	Reasons
a. $\underline{\quad?\quad}$ $\underline{\quad?\quad}$	a. Given
b. $LE = MR$ $EG = RA$	b. $\underline{\quad?\quad}$
c. $LE + EG = LG$ $MR + RA = MA$	c. $\underline{\quad?\quad}$
d. $LE + EG = MR + RA$	d. Addition property of equality
e. $\underline{\quad?\quad}$	e. Substitution property of equality
f. $\overline{LG} \cong \overline{MA}$	f. $\underline{\quad?\quad}$

EXERCISES

Practice **Copy and complete each proof.**

20. Given: $DA = EL$
   Prove: $DE = AL$

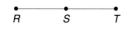

D E   A L

Statements	Reasons
a. __?__	a. Given
b. $DA = DE + EA$ $EL = EA + AL$	b. __?__
c. __?__	c. Substitution property of equality
d. $DE = AL$	d. __?__

21. Given: $\overline{AB} \cong \overline{CD}$
   M is the midpoint of $\overline{AB}$.
   N is the midpoint of $\overline{CD}$.
   Prove: $\overline{AM} \cong \overline{CN}$

A   M   B

C   N   D

Statements	Reasons
a. $\overline{AB} \cong \overline{CD}$ M is the midpoint of $\overline{AB}$. N is the midpoint of $\overline{CD}$.	a. __?__
b. $AB = CD$	b. __?__
c. __?__ __?__	c. Definition of midpoint
d. $AM + MB = AB$ $CN + ND = CD$	d. __?__
e. $AM + MB = CN + ND$	e. __?__
f. $AM + AM = CN + CN$	f. __?__
g. $2AM = 2CN$	g. Substitution property of equality
h. $AM = CN$	h. __?__
i. $\overline{AM} \cong \overline{CN}$	i. __?__

Write a two-column proof.

22. Given: $RS = ST$
   Prove: $RT = 2ST$

R   S   T

23. Given: $MP = NP$
   $PO = PL$
   Prove: $MO = NL$

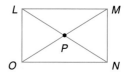

24. Given: $AC = AD$
$AB = AE$
Prove: $BC = ED$

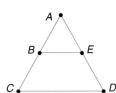

25. Given: $\overline{SA} \cong \overline{ND}$
Prove: $\overline{SN} \cong \overline{AD}$

26. Given: $\overline{BC} \cong \overline{YX}$
$\overline{AC} \cong \overline{ZX}$
Prove: $\overline{AB} \cong \overline{ZY}$

27. Given: $\overline{QT} \cong \overline{RT}$
$\overline{TS} \cong \overline{TP}$
Prove: $\overline{QS} \cong \overline{RP}$

For each statement, name the given and prove statements, and draw a figure. Then write a two-column proof.

28. If two points separate a segment AB into three congruent segments, then the measure of $\overline{AB}$ is three times the measure of one of the three shorter segments.

29. The midpoints of two segments of equal measure separate the segments into segments with equal measures.

Critical Thinking

30. If you were given that $\overline{EG} \cong \overline{KM}$, $\overline{KM} \cong \overline{JH}$, $\overline{EJ} \cong \overline{GH}$, $\overline{FI} \cong \overline{GH}$, F is the midpoint of EG, L is the midpoint of $\overline{KM}$, and I is the midpoint of $\overline{JH}$, list three statements that you could prove using the postulates, theorems, and definitions that you have learned.

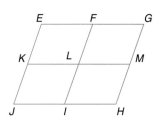

Applications

31. Law Translate the following lawyer's argument into a two-column proof. "The law states that if a driver proceeds through a red traffic light that is in proper working order, that driver is subject to a $50 fine. The defendant was seen driving through a red traffic light at the corner of Washington and Elm. The traffic computer shows no indication that the signal was down. Therefore, the defendant is guilty and subject to a $50 fine."

32. Advertising A television ad states "The Runaround is the best small truck on the market. It is priced well below any comparable truck. The Runaround is fun to drive, with room for four and plenty of cargo space. We think you'll find that it's all you need in a truck!"

a. What reasons do the advertisers give for claiming that "The Runaround is the best small truck on the market"?

b. Do you think they present a convincing argument?

c. What other concerns might a car buyer have that the advertisers have not mentioned?

Mixed Review

33. Four friends have birthdays in January, February, August, and September. Amy was not born in the winter. Emma celebrates her birthday during summer vacation. Timothy's birthday is the month after Pablo's. When is each one's birthday? **(Lesson 2-5)**

34. Name the algebraic property that justifies the statement "If $3x = 12$, then $x = 4$." **(Lesson 2-4)**

35. Translate the statement "A student must maintain a C average to be eligible to play a varsity sport" into if-then form. **(Lesson 2-2)**

36. Give a counterexample to show that "What goes up must come down" is a false conditional. **(Lesson 2-1)**

37. Angles *AND* and *NOR* are complementary. If $m\angle AND = 4m\angle NOR$, find the measures of the angles. **(Lesson 1-8)**

38. Refer to the number line to find each length. **(Lesson 1-4)**

a. *MN*
b. *NO*
c. *MO*
d. *PM*

Wrap-Up

39. Write an example to illustrate what you think is the most important concept in this lesson.

2-7 Two-Column Proof with Angles

Objective

After studying this lesson, you should be able to:
- complete proofs involving angle theorems.

Application

The Leaning Tower of Pisa, the famous bell tower in Pisa, Italy, is considered to be one of the seven wonders of the modern world. The tower has leaned ever since the ground beneath the tower began to shift after the first three stories were built. Today, the tower stands about 17 feet off the perpendicular. Its lean increases about $\frac{1}{20}$ of an inch per year.

The angle that the tower makes with the ground is about 84° on one side and 96° on the other. If you look at the Leaning Tower of Pisa as a ray and the ground as a line, then the angles the tower forms with the ground form a linear pair. Theorem 2-2 states that if two angles form a linear pair, the angles are supplementary.

Theorem 2-2 **Supplement** **Theorem**	**If two angles form a linear pair, then they are supplementary angles.**
	You will be asked to prove Theorem 2-2 in Exercise 15.

Example 1

APPLICATION

Construction

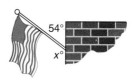

The flagpole shown at the right forms an angle of 54° with the wall. Find the measure, x, of the larger angle formed by the flagpole and the wall.

The angles form a linear pair. Using Theorem 2-2, the angles must be supplementary. Since the measures of supplementary angles have a sum of 180, we can write this equation.

$$x + 54 = 180$$
$$x + 54 - 54 = 180 - 54 \qquad \text{Subtract 54 from each side.}$$
$$x = 126$$

The larger angle measures 126.

The relationships from algebra that we found to be true for segments and the measures of segments are also true for angles and the measures of angles. The congruence relationships are stated in Theorem 2-3.

Theorem 2-3	Congruence of angles is reflexive, symmetric, and transitive.

Proof of Transitive Part of Theorem 2-3

Given: $\angle X \cong \angle Y$
$\angle Y \cong \angle Z$
Prove: $\angle X \cong \angle Z$

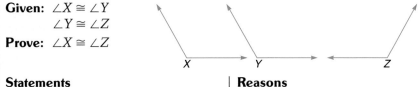

The proof of the transitive portion of Theorem 2-3 is given. You will be asked to prove the reflexive and symmetric portions in Exercises 25 and 26 respectively.

Statements	Reasons
1. $\angle X \cong \angle Y$ $\angle Y \cong \angle Z$	1. Given
2. $m\angle X = m\angle Y$ $m\angle Y = m\angle Z$	2. Definition of congruent angles
3. $m\angle X = m\angle Z$	3. Transitive property of equality
4. $\angle X \cong \angle Z$	4. Definition of congruent angles

If two angles are supplementary to the same angle, what do you think is true about the angles? Draw several examples and make a conjecture.

Theorem 2-4	Angles supplementary to the same angle or to congruent angles are congruent.

Proof of Theorem 2-4

Given: $\angle A$ and $\angle B$ are supplementary.
$\angle C$ and $\angle B$ are supplementary.
Prove: $\angle A \cong \angle C$

Statements	Reasons
1. $\angle A$ and $\angle B$ are supplementary. $\angle C$ and $\angle B$ are supplementary.	1. Given
2. $m\angle A + m\angle B = 180$ $m\angle C + m\angle B = 180$	2. Definition of supplementary
3. $m\angle A + m\angle B = m\angle C + m\angle B$	3. Substitution property of equality
4. $m\angle A = m\angle C$	4. Subtraction property of equality
5. $\angle A \cong \angle C$	5. Definition of congruent angles

You will be asked to prove this theorem in Exercise 28.

A similar theorem for complementary angles is stated below.

Theorem 2-5	Angles complementary to the same angle or to congruent angles are congruent.

Draw two right angles. What do you think is true about these angles? They are congruent. Since right angles are defined as having a measure of 90, they are all congruent by the definition of congruent angles. You will be asked to prove this in Exercise 27.

Theorem 2-6	**All right angles are congruent.**

Recall that in your investigations in Chapter 1, you discovered that vertical angles are congruent. We will prove this as a theorem now.

Theorem 2-7	**Vertical angles are congruent.**

Proof of Theorem 2-7

Given: ∠1 and ∠2 are vertical angles.
Prove: ∠1 ≅ ∠2

Statements	Reasons
1. ∠1 and ∠2 are vertical angles.	1. Given
2. ∠2 and ∠4 form a linear pair. ∠1 and ∠4 form a linear pair.	2. Definition of linear pair
3. ∠2 and ∠4 are supplementary. ∠1 and ∠4 are supplementary.	3. If 2 ∡ form a linear pair, then they are supp.
4. ∠1 ≅ ∠2	4. ∡ supp. to the same ∠ are ≅.

A theorem that follows from Theorem 2-7 is stated below. You will prove this theorem in Exercise 29.

Theorem 2-8	**Perpendicular lines intersect to form four right angles.**

CHECKING FOR UNDERSTANDING

Communicating Mathematics

Read and study the lesson to answer these questions.

1. We have just proved that vertical angles are congruent. Find some examples of vertical angles in the objects around you. For example, a chain-link fence contains many pairs of vertical angles.

2. Use a protractor to draw an example for each theorem in the lesson. Write the angle measures on the figures that you draw and explain why your drawing shows the relationship in the theorem.

Guided Practice

Complete each statement with *always*, *sometimes*, or *never*.

3. If two angles are right angles, they are __?__ adjacent.

4. If two angles are complementary, they are __?__ right angles.

5. If two angles are congruent, they are ___?___ right angles.
6. Vertical angles are ___?___ adjacent angles.
7. Congruent angles are ___?___ vertical angles.
8. If two angles are right angles, they are ___?___ congruent.
9. An angle is ___?___ congruent to itself.
10. If two angles are supplementary, they are ___?___ congruent.

Find the measures of ∠1 and ∠2.

11. $m\angle 1 = 2x + 94$ and
    $m\angle 2 = 7x + 49$

12. $m\angle 1 = 50 + 5x$ and
    $m\angle 2 = 60x$

13. $m\angle 1 = 100 + 20x$ and
    $m\angle 2 = 20x$

14. $m\angle 1 = 5x$ and
    $m\angle 2 = x + 36$

15. Copy and complete the proof of Theorem 2-2.

    Given: ∠1 and ∠2 form a linear pair.

    Prove: ∠1 and ∠2 are supplementary.

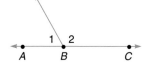

Statements	Reasons
1. ∠1 and ∠2 form a linear pair.	1. ___?___
2. $\overrightarrow{BA}$ and $\overrightarrow{BC}$ are opposite rays.	2. Definition of linear pair
3. $m\angle ABC = 180$	3. Definition of straight angle
4. ___?___	4. Angle addition postulate
5. $180 = m\angle 1 + m\angle 2$	5. ___?___
6. ∠1 and ∠2 are supplementary.	6. ___?___

EXERCISES

Practice

Complete each statement if $m\angle RLQ = 30$ and $m\angle MLN = 40$.

16. $\angle MLR \cong$ ___?___
17. $\angle QLM$ and ___?___ are supplementary.
18. $m\angle MLR =$ ___?___
19. $m\angle OLP =$ ___?___
20. $\angle NLP$ and ___?___ are supplementary.
21. $m\angle NLO =$ ___?___
22. $m\angle RLP =$ ___?___

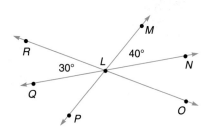

Write a two-column proof.

23. **Given:** $\angle ABC \cong \angle EFG$
        $\angle ABD \cong \angle EFH$

    Prove: $\angle DBC \cong \angle HFG$

24. **Given:** $\angle 1 \cong \angle 2$
        $\angle 3 \cong \angle 4$

    Prove: $\angle ABC \cong \angle DCB$

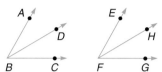
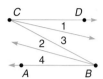

For each theorem, name the given and prove statements and draw a figure. Then write a two-column proof for each of the following.

25. Congruence of angles is reflexive. (Theorem 2-3)

26. Congruence of angles is symmetric. (Theorem 2-3)

27. All right angles are congruent. (Theorem 2-6)

28. Angles complementary to the same angle are congruent. (Theorem 2-5)

29. Perpendicular lines intersect to form four right angles. (Theorem 2-8)

30. If one angle in a linear pair is a right angle, then the other angle is a right angle also.

31. If two angles are congruent and supplementary, then they are right angles.

32. If two angles are vertical and one angle is a right angle, then the other is a right angle also.

Critical Thinking

33. Find the number of pairs of vertical angles determined by ten distinct lines passing through one point.

Application

34. **Sports**   When skiing, a skier aligns her body so she can get the maximum stability and speed. If $m\angle 2 = 57$, find $m\angle 1$.

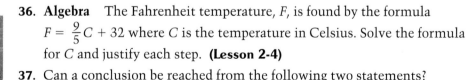

Mixed Review

35. Justify the statement $QT = QT$ with a property from algebra or a property of congruent segments. **(Lesson 2-6)**

36. **Algebra**   The Fahrenheit temperature, F, is found by the formula $F = \frac{9}{5}C + 32$ where C is the temperature in Celsius. Solve the formula for C and justify each step. **(Lesson 2-4)**

37. Can a conclusion be reached from the following two statements?
    "If you are an avid sailor, then you need a SailSun Boat."
    "Steven has a SailSun Boat."
    If so, write the conclusion and the law of logic used. **(Lesson 2-3)**

38. Find the measure of the complement and the supplement of an angle that measures 159. **(Lesson 1-8)**

Wrap-Up

39. **Portfolio**   Select the best proof that you wrote for this lesson. Rewrite it and explain your thinking process as you wrote the proof. Place this in your portfolio.

VOCABULARY

Upon completing this chapter, you should be familiar with the following terms:

conclusion	**76**	**76**	if-then statement
conditional statement	**76**	**70**	inductive reasoning
conjecture	**70**	**82**	law of detachment
converse	**77**	**83**	law of syllogism
counterexample	**71**	**77**	postulate
deductive reasoning	**82**	**98**	theorem
hypothesis	**76**	**89**	two-column proof

SKILLS AND CONCEPTS

OBJECTIVES AND EXAMPLES

Upon completing this chapter, you should be able to:

- make geometric conjectures based on given information. **(Lesson 2-1)**

To determine if a conjecture made from inductive reasoning is true or false, look at situations where the given information is true. Determine if there are situations where the given is true and the conjecture is false.

- write conditionals in if-then form. **(Lesson 2-2)**

Write the statement "Adjacent angles have a common ray" in if-then form.

"If angles are adjacent, then they have a common ray."

REVIEW EXERCISES

Use these exercises to review and prepare for the chapter test.

Determine if the conjecture is *true* or *false* based on the given information. Explain your answer.

1. **Given:** A, B, and C are collinear and $AB = BC$
   Conjecture: B is the midpoint of $\overline{AC}$.

2. **Given:** $\angle 1$ and $\angle 2$ are supplementary.
   Conjecture: $\angle 1 \cong \angle 2$

Write the conditional statement in if-then form.

3. Every cloud has a silver lining.

4. A rectangle has four right angles.

5. Obsidian is a glassy rock produced by a volcano.

6. The intersection of two planes is a line.

OBJECTIVES AND EXAMPLES	REVIEW EXERCISES

■ use the laws of logic to draw a conclusion. **(Lesson 2-3)**

The law of detachment states that if $p \rightarrow q$ is a true conditional and p is true, then q is true.

The law of syllogism states that if $p \rightarrow q$ the $q \rightarrow r$ are true conditionals, then $p \rightarrow r$ is also true.

If possible, write a conclusion. State the law of logic that you used.

7. (1) Angles that are complementary have measures with a sum of 90.
(2) $\angle A$ and $\angle B$ are complementary.

8. (1) Well-known athletes appear on Wheaties™ boxes.
(2) Michael Jordan appeared on Wheaties™ boxes.

9. (1) The sun is a star.
(2) Stars are in constant motion.

■ use properties of equality in algebraic and geometric proofs. **(Lesson 2-4)**

Prove that if $ST = UV$ then $SU = TV$.

Given: $ST = UV$
Prove: $SU = TV$

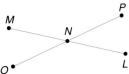

Statements	Reasons
1. $ST = UV$	1. Given
2. $ST + TU$ $= TU + UV$	2. Addition property of equality
3. $SU = ST + TU$ $TV = TU + UV$	3. Segment addition postulate
4. $SU = TV$	4. Substitution property

Write a complete proof for each of the following.

10. If $12x + 24 = 0$, then $x = -2$.

11. If $MN = PN$ and $NL = NO$, then $ML = PO$.

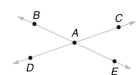

12. If $m\angle BAC + m\angle BAD = 180$, $m\angle DAE + m\angle CAE = 180$, and $m\angle BAD = m\angle CAE$, then $m\angle DAE = m\angle BAC$.

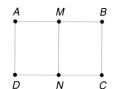

■ complete proofs involving segment theorems. **(Lesson 2-6)**

Theorem 2-1 states that congruence of segments is reflexive, symmetric, and transitive.

Write a two-column proof.

13. Given: $\overline{AM} \cong \overline{CN}$
$\overline{MB} \cong \overline{ND}$
Prove: $\overline{AB} \cong \overline{CD}$

- complete proofs involving angle theorems. **(Lesson 2-7)**

Given: $\angle 1 \cong \angle 2$
Prove: $\angle 3 \cong \angle 4$

Statements	Reasons
1. $\angle 1 \cong \angle 2$	1. Given
2. $\angle 1 \cong \angle 3$ $\quad \angle 2 \cong \angle 4$	2. Vertical $\angle$s are $\cong$.
3. $\angle 3 \cong \angle 4$	3. Congruence of angles is transitive. (used twice)

Write a two-column proof.

14. **Given:** $\angle 1$ and $\angle 3$ are supplementary.
    $\angle 3$ and $\angle 4$ form a linear pair.
    Prove: $\angle 1 \cong \angle 4$

15. **Given:** $\angle 1$ and $\angle 2$ form a linear pair
    $\angle 1 \cong \angle 2$
    Prove: $\angle 1$ and $\angle 2$ are right angles.

APPLICATIONS AND CONNECTIONS

16. **Advertising**   Write the conditional "Hard-working people deserve a night on the town at Gil's Grill" in if-then form. Identify the hypothesis and the conclusion of the conditional. Then write the converse. **(Lesson 2-2)**

17. **Botany**   If possible, write a valid conclusion. State the law of logic that you used. **(Lesson 2-3)**
    - A sponge is a sessile animal.
    - A sessile animal is one that remains permanently attached to a surface for all of its adult life.

18. **Algebra**   Name the property of equality that justifies the statement "If $x + y = 3$ and $3 = w + v$, then $x + y = w + v$." **(Lesson 2-4)**

19. **Geology**   The underground temperature of rocks varies with their depth below the surface. The deeper that a rock is in the Earth, the hotter it is. The temperature, t, in degrees Celsius is estimated by the equation $t = 35d + 20$, where d is the depth in kilometers. Solve the formula for d and justify each step. **(Lesson 2-4)**

20. Use the process of elimination to solve this problem. **(Lesson 2-5)**
    Alana, Becky, and Carl each had different lunches in the school cafeteria. One had spaghetti, one had a salad, and one had macaroni and cheese. Alana did not have a salad. Becky did not have spaghetti or a salad. What did each person have for lunch?

Determine if each conjecture is *true* or *false*. Explain your answers and give a counterexample for any false conjecture.

1. **Given:** x is a real number.
   Conjecture: $-x < 0$

2. **Given:** $\angle 1 \cong \angle 2$
   Conjecture: $\angle 2 \cong \angle 1$

3. **Given:** $\angle 1$ and $\angle 2$ form a linear pair.
   Conjecture: $m\angle 1 + m\angle 2 = 180$

4. **Given:** $3x^2 = 48$
   Conjecture: $x = 4$

Write the conditional statement in if-then form. Identify the hypothesis and the conclusion of the conditional. Then write the converse.

5. Through any two points there is exactly one line.

6. A rolling stone gathers no moss.

7. Wise investments with Petty-Bates pay off.

8. Two parallel planes do not intersect.

If possible, write a conclusion. State the law of logic that you used.

9. (1) Wise investments with Petty-Bates pay off.
   (2) Investments that pay off build for the future.

10. (1) Perpendicular lines intersect.
    (2) Lines ℓ and m are perpendicular.

11. (1) Vertical angles are congruent.
    (2) $\angle 1$ is congruent to $\angle 2$.

12. (1) All integers are real numbers.
    (2) 7 is an integer.

Name the property of equality that justifies each statement.

13. If $m\angle A = m\angle B$, then $m\angle B = m\angle A$.

14. If $x + 9 = 12$, then $x = 3$.

15. If $2ST = 4UV$, then $ST = 2UV$.

16. If $AB = 7$ and $CD = 7$, then $AB = CD$.

17. Anthony, Eric, and Karen each bought a car. One bought a Honda, one bought a Ford, and one bought a Volkswagen. Anthony did not buy a foreign car. Karen did not buy a Honda. Who bought which car?

Write a two-column proof.

18. **Given:** $\overline{AC} \cong \overline{BD}$
    Prove: $\overline{AB} \cong \overline{CD}$

19. **Given:** $m\angle 1 = m\angle 3 + m\angle 4$
    Prove: $m\angle 3 + m\angle 4 + m\angle 2 = 180$

20. Prove that if two angles are supplementary and one angle is a right angle, then the other one is a right angle also.

Bonus The *inverse* of a conditional $p \rightarrow q$ is not $p \rightarrow$ not q. Write the inverse of "All dogs are mammals."

College Entrance Exam Preview

Directions: Choose the one best answer. Write A, B, C, or D. You may use a calculator.

1. If $0.06x = 24$, then $x =$
   - (A) 600
   - (B) 400
   - (C) 0.04
   - (D) 1.44

2.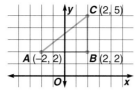

   The distance between A and C is
   - (A) 5 units
   - (B) 8 units
   - (C) 12 units
   - (D) 6 units

3. If $x < 12$ and $y < 18$, then
   - (A) $y - x = 6$
   - (B) $x < y$
   - (C) $y < x$
   - (D) $x + y < 30$

4. 40% of 120 is one third of
   - (A) 144
   - (B) 120
   - (C) 48
   - (D) 40

5.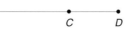

   In the figure, if $AD = 50$, $CD = 12$, and B is the midpoint of $\overline{AC}$, then $BC =$
   - (A) 18
   - (B) 38
   - (C) 19
   - (D) 12

6. The points $(4, 2)$ and $(-1, y)$ are $\sqrt{74}$ units apart. What is the value of y?
   - (A) 7
   - (B) -7
   - (C) 11
   - (D) 9

7. If $x + 5 = \frac{1}{3}(3x - 5)$, then $x =$
   - (A) 5
   - (B) 3
   - (C) any real number
   - (D) no real number

8. If a calculator costs $12.90 after a 25% discount, what is the original price of the calculator?
   - (A) $16.13
   - (B) $17.20
   - (C) $9.68
   - (D) $15.00

9. If $\angle A$ and $\angle B$ are supplementary and $\angle A$ and $\angle C$ are complementary, then
   - I. $m\angle A < 90$, $m\angle B > 90$, and $m\angle C < 90$
   - II. $m\angle A < m\angle B$ and $m\angle C < 90$
   - III. $m\angle A > 90$ and $m\angle C < m\angle B$
   - (A) I only
   - (B) II only
   - (C) III only
   - (D) I and II only

10. If two planes intersect, their intersection can be
    - I. a line.
    - II. three non-collinear points.
    - III. two intersecting lines.
    - (A) I only
    - (B) II only
    - (C) III only
    - (D) I and II only

11. The product of 12^3 and 12^8 is
    - (A) 12^5
    - (B) 12^{24}
    - (C) 12^{11}
    - (D) 24^{11}

12. The distance from city X to city Y is 200 miles. The distance from city X to city Z is 140 miles. Which of the following must be true?
    - (A) The distance from Y to Z is 60 miles.
    - (B) Seven times the distance from X to Y is ten times the distance from X to Z.
    - (C) The distance from Y to Z is 340 miles.
    - (D) The distance from Y to Z is one-third the distance from X to Z.

Solve. You may use a calculator.

13. If one half of the female students and one third of the male students at a school take Spanish classes, what part of the student body takes Spanish?

14. A gallon of water evaporates from a 9-gallon drum of 4% salt solution. What is the percentage of salt in the remaining solution?

15. In the graph below, the axes and the origin are not shown. If point P has coordinates *(4, 2)*, what are the coordinates of point Q?

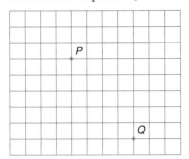

16. Westville has a population of 7200, which is decreasing at a rate of 80 people per year. Troy has a population of 5000 and is gaining 120 people per year. In how many years will the populations of Westville and Troy be the same?

17. One evening, the candy counter at the Cineplex sold 532 buckets of popcorn for $1489.50. A large bucket sells for $2.25 and a jumbo bucket sells for $3.75. How many jumbo buckets of popcorn were sold?

18. On the blue prints for a house, 2 inches represents 3 feet. If the width of a room on the plan is $6\frac{1}{2}$ inches, what is the actual width of the room?

19. In the figure below, $\angle ABC$ is a straight angle, and $\overline{DB}$ is perpendicular to $\overline{BE}$. If $\angle ABD$ measures x degrees, write an expression to represent the degree measure of $\angle CBE$.

20. Jordan answered all of the questions on a 20-question test. A score on the test is computed by adding 5 points for every correct answer and subtracting 1 point for every incorrect answer. If Jordan's score was 82, how many questions did he answer correctly?

CHAPTER 3

Parallels

CHAPTER OBJECTIVES

In this chapter you will:
- Use the properties of parallel lines.
- Prove lines parallel.
- Find and use the slope of lines.
- Recognize and use distance relationships between points, lines, and planes.

GEOMETRY AROUND THE WORLD
Japan

What do you get when you take the angles and planes of geometry off the printed page and combine them in innovative ways with traditional forms? The answer: buildings by Japanese architect Kenzo Tange that will knock your socks off!

Tange, who received a gold medal from the American Institute of Architects, is one of the greatest architects of the twentieth century. He was trained at the University of Tokyo and also studied architecture in Paris. There, he learned about the geometric building designs favored by some western architects. As a result, his buildings reflect both traditional Japanese design and the more angular forms used in the United States and other countries in the western hemisphere.

For the City Hall building in Kurashiki, Japan, Tange used a geometric design called a **parallelpiped.** A parallelpiped has six faces, each of which is a **parallelogram.** A parallelogram has four sides; its opposite sides are parallel and congruent. The Kurashiki City Hall also has wooden panels and other features used in traditional Japanese buildings. For contrast, Tange used dramatic geometric forms for its roof.

GEOMETRY IN ACTION

As you can see from the photograph, Kurashiki's City Hall is a remarkable-looking building. If you measured the slopes of the left and right inclines of the roof, you would find that the right slope is -0.36 and the left is 0.21. What method do you think was used to compute these slopes?

What geometric forms can you find in this photo of a building designed by Japanese architect Kenzo Tange?

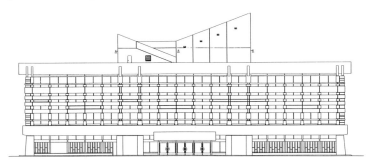

◀ *Kurashiki City Hall*   Inset photo: *Kenzo Tange*

3-1

Problem-Solving Strategy: Draw a Diagram

EXPLORE
PLAN
SOLVE
EXAMINE

Objective

After studying this lesson, you should be able to:

- solve problems by using a diagram.

Sometimes drawing a diagram of the situation described in a problem can help you find the solution. A diagram can help you choose a strategy, organize your information, or may even show you the answer to the problem.

Example 1

APPLICATION

Sports

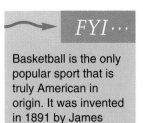

FYI···

Basketball is the only popular sport that is truly American in origin. It was invented in 1891 by James Naismith.

The varsity basketball teams of the seven high schools in the Mid-State Conference each play every other team twice this season. How many games will there be in the conference this year?

Since there are seven teams and each team plays every other team twice, it seems like there should be a total of $7 \times 6 \times 2$ or 84 games. Let's draw a diagram and see if this conjecture is true.

Draw a diagram of seven noncollinear points to represent the seven teams and label the points A through G. Use a line segment between two points to represent a game between the two teams. How many line segments are needed to join all seven dots?

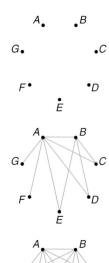

First, draw all of the segments from point A. There are six. Next, draw all of the segments from B. This adds five more.

Continue to draw line segments until all of the pairs of points are connected. There are $6 + 5 + 4 + 3 + 2 + 1$ or 21 segments. Since all the teams play each other twice, there will be 2×21 or 42 games in the Mid-State Conference this season.

Our conjecture was incorrect. There will be 42, not 84, games.

Diagrams can help you organize your information when you are writing proofs of geometric theorems. You will need to be able to draw a diagram that shows the given information for a theorem.

Example 2

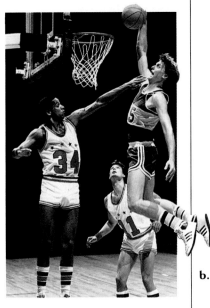

Draw and label the diagram that you would use to prove each statement. Then state the given information and the statement to be proved.

a. The bisector of an angle separates the angle into two angles whose measures are each half of the measure of the original angle.

We are given that an angle has a bisector. Draw an angle $\angle ABC$ with a bisector $\overrightarrow{BD}$.

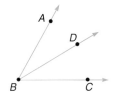

Given: $\overrightarrow{BD}$ bisects $\angle ABC$.

Prove: $m\angle ABD = \frac{1}{2}m\angle ABC$

$m\angle DBC = \frac{1}{2}m\angle ABC$

Notice that both statements must be proved to prove the statement.

b. If two adjacent angles are complementary, then their exterior sides are perpendicular.

There are two adjacent angles that are complementary. Draw adjacent angles $\angle KLM$ and $\angle MLN$.

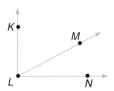

We should not mark $\overrightarrow{LK}$ and $\overrightarrow{LN}$ as perpendicular, since that is what is to be proved.

Given: $\angle KLM$ and $\angle MLN$ are complementary.

Prove: $\overrightarrow{LK} \perp \overrightarrow{LN}$

CHECKING FOR UNDERSTANDING

Communicating Mathematics

Read and study the lesson to answer these questions.

1. How can drawing a diagram help you solve a problem?

2. If two new teams joined the Mid-State Conference, how many games would be played? Explain your solution.

3. What information should be shown in a diagram that accompanies a proof?

Guided Practice

Solve. Draw a diagram.

4. Halfway through her bus trip from Savannah to Jacksonville, Gloria fell asleep. When she awoke, she still had to travel half of the distance she traveled when asleep. For what fraction of the trip was Gloria asleep?

5. Keith and Peggy are building steps to their new shed. It takes one concrete block to build one step, three blocks for two steps, and six blocks for three steps. How many concrete blocks will it take to build six steps?

Draw and label the diagram that you would use to prove each statement. Then state the given information and the statement to be proved.

6. If the exterior sides of two adjacent angles are perpendicular, then the angles are complementary.

7. If two vertical angles are supplementary, then they are right angles.

EXERCISES

Practice

Solve. Use any strategy.

8. How many times in a 12-hour period will the sum of the digits on a digital clock be greater than 16?

9. Take two numbers whose sum is one. Which is greater, the square of the greater added to the lesser or the square of the lesser added to the greater?

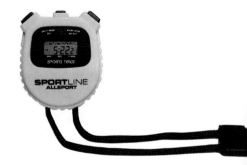

Strategies

Look for a pattern.
Solve a simpler problem.
Act it out.
Guess and check.
Draw a diagram.
Make a chart.
Work backward.

10. Draw and label the diagram that you would use to prove the statement *If two right angles are adjacent, then their exterior sides are opposite rays.* Then state the given information and the statement to be proved.

11. Find 33^2, 333^2, and 3333^2. Without calculating, what is $33,333^2$?

12. How many different acute angles can be traced using the rays in the figure at the right?

13. In how many ways can you receive change for a quarter if at least one coin is a dime?

14. In how many ways can a straight line separate a square into two identical regions?

15. Follow the steps to draw a diagram of a *hexagonal prism*. This is a solid that looks like a six-sided box.

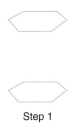

Step 1

Step 2

Step 3

Draw a six-sided top and a congruent six-sided bottom directly below the top.

Draw vertical edges.

Make the edges of the prism that would not be seen from this perspective dashed.

16. The bike path at the new park in Glenville will be 1.2 miles long, 3 feet wide, and 2 inches thick. If asphalt is sold in cubic yards, how much asphalt should the contractor order?
(1 mile = 1760 yards)

17. Two distinct circles and two distinct lines intersect each other. What is the maximum number of points of intersection where at least two of them intersect?

COOPERATIVE LEARNING PROJECT

Work in groups. Each person in the group must understand the solution and be able to explain it to any person in class.

The Parents Association of Easton High School is planning to make and sell school directories to raise money for a new science lab. They will make the directories by printing four pages at a time on the front and back of double-sized sheets of paper. The sheets will then be folded and stapled together inside a cover. If there are 200 pages in the directory, which pages should be printed on each sheet of paper so that the pages will be positioned correctly in the directory? How should the pages be numbered if there are 150 pages? Suppose there are *N* pages. Develop a formula to find the numbers on any sheet of paper for the directory.

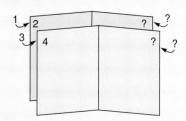

Parallels and Transversals

Objectives

After studying this lesson, you should be able to:

- describe the relationships between two lines and between two planes, and
- identify the relationships between pairs of angles formed by pairs of lines and transversals.

Application

Bill Randell is an air-traffic controller. One of his responsibilities is to assign airplanes their cruising altitudes as they head toward their destinations. For safety, airplanes heading eastbound are assigned an altitude level that is an odd number of thousands of feet above the ground and airplanes headed westbound are assigned an altitude level that is an even number of thousands of feet above the ground. The altitude levels can be thought of as **parallel planes**.

While the airplanes are cruising, the paths of two airplanes flying at the same altitude are in the same plane, or **coplanar**. Suppose they are flying some distance apart but going in the same direction. They can be represented by two lines in a plane that never meet. Two lines in a plane that never meet are called **parallel lines**.

In geometry, the symbol ∥ means *is parallel to*. In the figure below, the two lines are parallel.

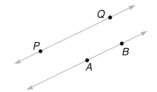

symbols: $\overleftrightarrow{AB} \parallel \overleftrightarrow{PQ}$

words: line AB is parallel to line PQ.

The term parallel and the notation ∥ are also used for segments, rays, and planes. The symbol ∦ means "is not parallel to."

Parts of lines are parallel if the lines that contain them are parallel. For example in the figure above, $\overline{AB} \parallel \overleftrightarrow{QP}$.

Arrows are used in diagrams to indicate that lines are parallel. In the figure at the right, the arrows on the segments indicate that $\overline{MN} \parallel \overline{OP}$.

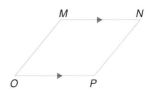

Two planes can intersect or be parallel just like two lines. Recall that the altitude levels of the airplanes are parallel planes. You learned in Chapter 1 that if two planes intersect, they intersect in a line. The figure below is a rectangular prism. Its faces are contained in parallel and intersecting planes.

Remember that plane ABC refers to the plane containing points A, B, and C.

parallel planes:

plane *ADR* ∥ plane *BCS*

plane *RSC* ∥ plane *ABT*

plane *RST* ∥ plane *ADC*

planes intersecting plane *ABC*:

plane *ABT*, in $\overleftrightarrow{AB}$

plane *ADR*, in $\overleftrightarrow{AD}$

plane *RSC*, in $\overleftrightarrow{CD}$

plane *BCS*, in $\overleftrightarrow{BC}$

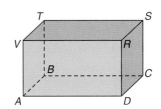

Using Mr. Randell's system, eastbound and westbound airplanes will never be in danger of colliding. The closest that an eastbound airplane can come to a westbound airplane is 1000 feet. Suppose you are in an airplane that is flying northeast at an altitude of 27,000 feet and you see a second airplane flying due west at an altitude of 24,000. The paths of your airplane and the second airplane will never meet, but they are not in the same plane. Lines like those represented by these paths are called **skew lines**.

Skew Lines

Two lines are skew if they do not intersect and are not in the same plane.

Example 1

Identify each pair as intersecting, parallel, or skew.

a. $\overline{BA}$ and $\overline{GH}$ parallel

b. $\overline{EH}$ and $\overline{CD}$ skew

c. plane *EAB* and plane *GCB*   intersecting

d. $\overline{HG}$ and plane *EAB*   parallel

e. plane *HGC* and $\overline{BC}$ intersecting

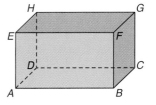

The runways at Mitchell Field are shown at the right. Notice that runway *t* intersects runways ℓ and *m*. A line that intersects two or more lines in a plane at different points is called a **transversal**.

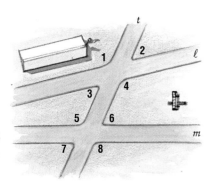

When the transversal *t* intersects lines ℓ and *m*, it forms eight angles with these lines. Several of the angles and pairs of angles are given special names.

Interior Angles	$\angle 3$, $\angle 4$, $\angle 5$, $\angle 6$
Alternate Interior Angles	$\angle 3$ and $\angle 6$, $\angle 4$ and $\angle 5$
Consecutive Interior Angles	$\angle 3$ and $\angle 5$, $\angle 4$ and $\angle 6$
Exterior Angles	$\angle 1$, $\angle 2$, $\angle 7$, $\angle 8$
Alternate Exterior Angles	$\angle 1$ and $\angle 8$, $\angle 2$ and $\angle 7$
Corresponding Angles	$\angle 1$ and $\angle 5$, $\angle 2$ and $\angle 6$, $\angle 3$ and $\angle 7$, $\angle 4$ and $\angle 8$

Example 2

Identify each pair of angles as *alternate interior, consecutive interior, alternate exterior,* or *corresponding angles.*

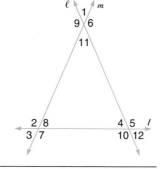

a. $\angle 7$ and $\angle 11$ corresponding

b. $\angle 2$ and $\angle 9$ consecutive interior

c. $\angle 7$ and $\angle 12$ corresponding

d. $\angle 8$ and $\angle 10$ alternate interior

e. $\angle 4$ and $\angle 8$ consecutive interior

f. $\angle 2$ and $\angle 12$ alternate exterior

CHECKING FOR UNDERSTANDING

Communicating Mathematics

Read and study the lesson to answer these questions.

1. What symbol is used to indicate that two lines are parallel? Draw and label two parallel lines and indicate they are parallel using the symbols.

2. Give a real-world example or model of two skew lines.

3. Draw two lines. Label them *q* and *p*. Then draw a transversal *t* that intersects the two lines. Label a pair of corresponding angles, $\angle 1$ and $\angle 2$. Discuss your strategy for identifying corresponding angles.

Guided Practice

Classify each situation as a model of intersecting, parallel, or skew lines.

4. airplane flight paths

5. train tracks

6. lines on writing paper

7. skis on skier

8. plaid fabric

9. airport runway

Determine whether each statement is *true* or *false*. If false, explain why.

10. A line that intersects two skew lines is a transversal.

11. Two lines are parallel if they do not intersect.

12. Skew lines are parallel.

13. If a line intersects two parallel lines, then it is a transversal.

Determine whether each statement is *true* or *false*.

14. ∠6 and ∠11 are alternate interior angles.

15. ∠4 and ∠6 are vertical angles.

16. ∠5 and ∠8 are consecutive interior angles.

17. ∠4 and ∠9 are alternate exterior angles.

18. ∠10 and ∠11 are alternate interior angles.

19. ∠7 and ∠11 are corresponding angles.

20. ∠14 and ∠7 are alternate exterior angles.

21. ∠3 and ∠8 are corresponding angles.

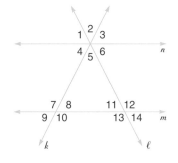

EXERCISES

Practice

State the transversal that forms each pair of angles. Then identify the special angle pair name for the angles.

22. ∠2 and ∠4

23. ∠1 and ∠3

24. ∠*TVR* and ∠*VTS*

25. ∠*SRV* and ∠*RVT*

26. ∠11 and ∠8

27. ∠1 and ∠2

28. ∠7 and ∠4

29. ∠7 and ∠12

30. ∠3 and ∠5

31. ∠11 and ∠4

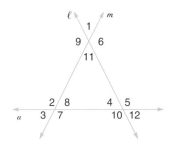

Identify each as a model of intersecting or parallel planes.

32. the pieces of glass in a double-paned window

33. the sides of a box

34. a floor and a ceiling

35. the sides of a pup tent

36. the floor and the top of a table

37. the sides of a roof

Draw a figure to illustrate each situation.

38. two skew lines

39. two lines parallel to a third line

40. two parallel planes with a line perpendicular to both planes

41. two intersecting planes with a line parallel to both planes

42. two lines perpendicular to a plane

43. two parallel planes with a line parallel to both planes

The three-dimensional figure shown at the right is called a right hexagonal prism.

44. Identify all segments that appear to be skew to $\overline{XY}$.

45. Which segments seem parallel to $\overline{ST}$?

46. Which segments seem parallel to $\overline{VW}$?

47. Identify all planes that appear parallel to the plane STU.

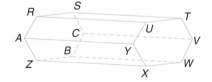

48. If line ℓ is parallel to line m and line m is parallel to plane $\mathcal{N}$, is $\ell \parallel \mathcal{N}$? If yes, describe a real-life model that demonstrates this concept. If no, draw a counterexample.

49. If plane A is parallel to plane B and plane B is parallel to plane C, then plane A is parallel to plane C. Explain what this means and state a model in your school that demonstrates this property.

Critical Thinking

50. Choose two parallel lines on a piece of lined notebook paper and draw a transversal through them.

   a. Use a protractor to measure a pair of corresponding angles.

   b. Make a conjecture about the relationship between the measures of the corresponding angles you measured.

Application

51. **Air-Traffic Control** Above 30,000 feet, eastbound airplanes are assigned cruising altitudes of 33,000 feet, 37,000 feet, 41,000 feet, and so on. The westbound airplanes cruise altitudes of 31,000 feet, 35,000 feet, 39,000 feet, and so on.

   a. What will now be the closest vertical distance two airplanes passing over each other will encounter?

   b. What are some of the advantages of this system of assigning cruising altitudes?

Mixed Review

52. Draw the diagram that you would use to prove the theorem *If the exterior sides of two adjacent angles are opposite rays, then the angles are supplementary.* Then state the given and the statement to be proved. **(Lesson 3-1)**

53. Write a two-column proof. **(Lesson 2-7)**
   Given: ∠1 ≅ ∠2
   Prove: ∠1 ≅ ∠3

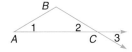

54. Name the property of equality that justifies the statement *If m∠A = m∠B, then m∠B = m∠A.* **(Lesson 2-4)**

55. Write the converse of the conditional *If two lines are parallel, then they lie in the same plane and do not intersect.* **(Lesson 2-2)**

56. Find the coordinates of the midpoint of the segment whose endpoints are *A*(10, -4) and *B*(-6, 0). **(Lesson 1-5)**

57. *T* is between *R* and *S*. If *TS* = 7 and *RS* = 20, find *RT*. **(Lesson 1-4)**

Wrap-Up

58. Draw two lines and a transversal and label the angles formed. Name the interior angles, the exterior angles, a pair of alternate interior angles, a pair of alternate exterior angles, a pair of corresponding angles, and a pair of consecutive interior angles.

HISTORY CONNECTION

The Maori people of South America and New Zealand and the Polynesians of the Pacific Islands were navigating the Pacific Ocean long before the invention of the compass and the sextant. They knew how to navigate by the stars and amazingly they had discovered the geometry and physics behind the wave patterns in the waters around islands.

The Maoris and the Polynesians understood that parallel waves are reflected around islands in patterns. They taught their children to read the patterns of the waves with a *mattang*. Once they learned the patterns, a navigator could feel the motion of the water around his boat to locate nearby islands.

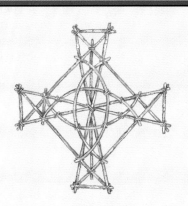

3-3 Using Parallel Lines

Objective
After studying this lesson, you should be able to:
- use the properties of parallel lines to determine angle measures.

In 300 B.C. parallel lines were defined as lines that "ran along beside each other." The B.C. cartoon is an illustration of an artist's use of a perspective drawing; that is, a drawing that looks like things look to our eyes.

To Peter, the parallel lines appear to meet at a vanishing point in the distance. However parallel lines do not meet in Euclidean geometry. For mathematicians, trying to prove that there is exactly one line parallel to a given line through a point not on the given line has been a historical debate.

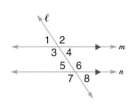

There are several postulates and theorems that help you gain insight into properties of parallel lines. If you investigate the relationships between the corresponding angles formed by two parallel lines cut by a transversal, you will observe that the angles are congruent. This property is accepted as a postulate.

Postulate 3-1 Corresponding Angles Postulate	If two parallel lines are cut by a transversal, then each pair of corresponding angles is congruent.

This postulate, combined with linear pair and vertical angle properties, helps to establish several angle relationships.

Example 1

A road crosses a set of railroad tracks. If the measure of ∠6 is 110, find m∠3.

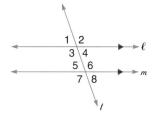

Since ∠2 and ∠6 are corresponding angles, m∠2 = m∠6. m∠2 = m∠3 because they are vertical angles. Therefore, m∠6 = m∠3 by the transitive property of equality. So, m∠3 = 110.

Notice that in Example 1, ∠4 and ∠5 form a pair of alternate interior angles. This is an application of another of the special relationships between the angles formed by two parallel lines and a transversal. These relationships are summarized in Theorems 3-1 through 3-3. You will be asked to prove Theorems 3-1 and 3-2 in Exercises 46 and 47, respectively.

Theorem 3-1 **Alternate Interior Angle Theorem**	**If two parallel lines are cut by a transversal, then each pair of alternate interior angles is congruent.**
Theorem 3-2 **Consecutive Interior Angle Theorem**	**If two parallel lines are cut by a transversal, then each pair of consecutive interior angles is supplementary.**
Theorem 3-3 **Alternate Exterior Angle Theorem**	**If two parallel lines are cut by a transversal, then each pair of alternate exterior angles is congruent.**

The proof of Theorem 3-3 that is given below is called a **paragraph proof**. The statements and reasons are written informally in a paragraph. But the steps in a paragraph proof are the same as those in a two-column proof.

Proof of Theorem 3-3

Given: $p \parallel q$
ℓ is a transversal of p and q.

Prove: ∠1 ≅ ∠8; ∠2 ≅ ∠7

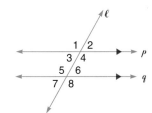

Paragraph Proof:
We are given that $p \parallel q$. If two parallel lines are cut by a transversal, corresponding angles are congruent. So, ∠1 ≅ ∠5 and ∠2 ≅ ∠6. ∠5 ≅ ∠8 and ∠6 ≅ ∠7 because vertical angles are congruent. Therefore, ∠1 ≅ ∠8 and ∠2 ≅ ∠7 since congruence of angles is transitive.

Example 2

Find the values of x, y, and z.

Since $\overleftrightarrow{AG} \parallel \overleftrightarrow{BH}$, $\angle CEF \cong \angle EFH$ by the alternate interior angle theorem.

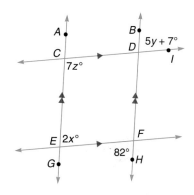

$$m\angle CEF = m\angle EFH$$
$$2x = 82 \qquad \text{\small m∠CEF = 2x, m∠EFH = 82.}$$
$$x = 41$$

Since $\overleftrightarrow{CD} \parallel \overleftrightarrow{EF}$, $\angle DCE$ and $\angle CEF$ are supplementary by the consecutive interior angle theorem.

$$m\angle DCE + m\angle CEF = 180$$
$$7z + 2x = 180 \qquad \text{\small m∠DCE = 7z, m∠CEF = 2x.}$$
$$7z + 2(41) = 180 \qquad \text{\small Substitute 41 for x.}$$
$$7z = 98$$
$$z = 14$$

Since $\overleftrightarrow{CD} \parallel \overleftrightarrow{EF}$, $\angle BDI \cong \angle EFH$ by the alternate exterior angle theorem.

$$m\angle BDI = m\angle EFH$$
$$5y + 7 = 82$$
$$5y = 75$$
$$y = 15$$

Therefore, $x = 41$, $y = 15$, and $z = 14$.

There is a special relationship that occurs when one of two parallel lines is cut by a perpendicular line. You will prove this theorem in Exercise 45.

Theorem 3-4 *Perpendicular* *Transversal Theorem*	**In a plane, if a line is perpendicular to one of two parallel lines, then it is perpendicular to the other.**

CHECKING FOR UNDERSTANDING

Communicating Mathematics

Read and study the lesson to answer these questions.

1. Explain why $\angle 4$ and $\angle 6$ must be supplementary.

2. If you know that $m\angle 1 = 70$, explain two different strategies you could use to find $m\angle 5$.

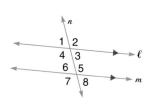

3. Explain what the arrowheads on the lines in both diagrams at the right indicate.

4. If $\overleftrightarrow{BD} \parallel \overleftrightarrow{AT}$ then $\angle 1 \cong \angle 3$ and $\angle 6 \cong \angle 4$. Explain why this is true.

5. If $\overrightarrow{WP} \parallel \overrightarrow{KR}$, then $\angle 1 \cong \angle 2$. Explain why this is true.

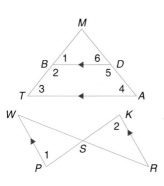

Guided Practice List the conclusions that can be drawn from each figure.

6.

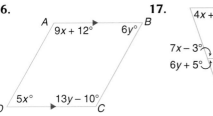

7.

8.

Given $\ell \parallel m$, $m\angle 1 = 98$, and $m\angle 2 = 40$, find the measure of each angle.

9. $m\angle 3$ **10.** $m\angle 4$

11. $m\angle 5$ **12.** $m\angle 6$

13. $m\angle 7$ **14.** $m\angle 8$

15. $m\angle 9$

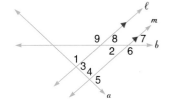

Find the values of x and y.

16.   **17.**   **18.**

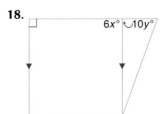

EXERCISES

Practice In the figure, $n \parallel p$, $q \perp p$, $\overline{CD} \parallel \overline{AB}$, and $m\angle 1 = 125$. Find the measure of each angle.

19. $\angle 2$ **20.** $\angle 3$

21. $\angle 4$ **22.** $\angle 5$

23. $\angle 6$ **24.** $\angle 7$

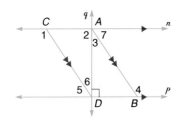

Find the values of x and y.

25.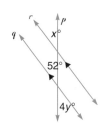

26.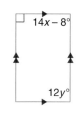

27.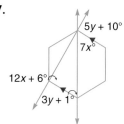

In the figure, $m\angle 2 = 62$, $m\angle 1 = 41$, $\overline{XS} \parallel \overline{YT}$, and $\overline{SY} \parallel \overline{TZ}$. Find the measure of each angle.

28. $\angle 4$

29. $\angle 3$

30. $\angle 5$

31. $\angle 6$

32. $\angle 7$

33. $\angle 8$

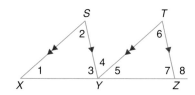

Find the values of x, y, and z.

34.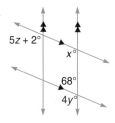

35.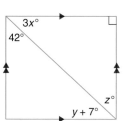

36.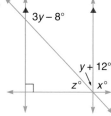

37.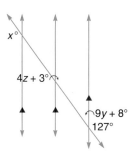

38.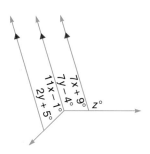

In the figure, $\overline{AC} \parallel \overline{BD}$, $\overline{CB} \parallel \overline{DE}$, $m\angle 1 = 35$, $m\angle 4 = 20$, and $\overline{DE}$ bisects $\angle BDF$. Find the measure of each angle.

39. $\angle 2$

40. $\angle 3$

41. $\angle 5$

42. $\angle 6$

43. $\angle 7$

44. $\angle 8$

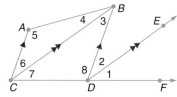

45. Copy and complete the proof of Theorem 3-4.

Given: $m \perp \ell$
          $\ell \parallel p$

Prove: $m \perp p$

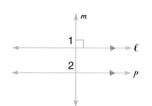

Statements	Reasons
a. $m \perp \ell$ $\ell \parallel p$	**a.** _?_
b. $\angle 1$ is a right angle.	**b.** _?_
c. _?_	**c.** Definition of right angle
d. $\angle 1 \cong \angle 2$	**d.** _?_
e. _?_	**e.** Substitution property of equality
f. $\angle 2$ is a right angle.	**f.** _?_
g. _?_	**g.** Definition of perpendicular lines

46. Write a two-column proof of Theorem 3-1.

47. Write a paragraph proof of Theorem 3-2.

48. Find measures x and y in the figure at the right, if $\overline{AF} \parallel \overline{BC}$ and $\overline{AE} \parallel \overline{CD}$.

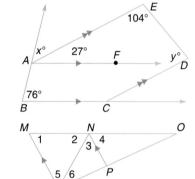

49. Given: $\overline{MQ} \parallel \overline{NP}$
          $\angle 1 \cong \angle 5$
  Prove: $\angle 4 \cong \angle 3$.

50. If $\overline{MQ} \parallel \overline{NP}$, $\angle 1 \cong \angle 5$, and $m\angle 2 = 34$, find the $m\angle 3$.

Critical Thinking

51. Planes BCD and AEF are parallel. Name an angle congruent to each given angle.

  a. $\angle EAD$

  b. $\angle CAF$

  c. $\angle ACF$

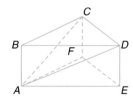

Application

52. Interior Decorating The walls in houses are not perfectly vertical, but wallpaper should be hung vertically to make the pattern look nice. So to hang wallpaper, a true vertical line must be established. The paperhanger uses a plumb bob, which is a piece of string with a weight at the bottom to make the vertical line for the first piece of wallpaper. How can she be sure that all of the other pieces of wallpaper are vertical if she doesn't use the plumb bob again?

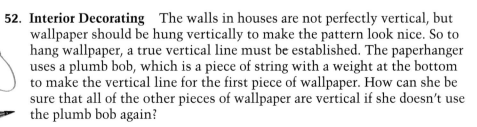

53. Are the ceiling and the floor of your classroom a model of parallel or intersecting planes? **(Lesson 3-2)**

54. Write a two-column proof. **(Lesson 2-6)**

Given: $\overline{AB} \cong \overline{FE}$
$\overline{BC} \cong \overline{ED}$

Prove: $\overline{AC} \cong \overline{FD}$

55. If possible, write a conclusion from the two statements *If two lines are parallel, then they never meet* and *Lines p and m are parallel.* State the law of logic that you used. **(Lesson 2-3)**

56. The measure of an angle is $9x + 14$ and the measure of its supplement is $12x + 19$. Find the value of x. **(Lesson 1-8)**

57. Point T is 6 units from point S on a number line. If the coordinate of point S is 3, what are the possible coordinates for point T? **(Lesson 1-4)**

58. If $x < 0$ and $y < 0$, in which quadrant is the point $Q(x, y)$ located? **(Lesson 1-1)**

Wrap-Up
59. This lesson presented four major conclusions you can make if you know that two parallel lines are cut by a transversal. List those conclusions and draw a diagram that describes these results.

~~~~~ MID-CHAPTER REVIEW ~~~~~

1. Draw and label the diagram that you would use to prove the theorem *If two parallel lines are cut by a transversal so that consecutive interior angles are congruent, then the transversal is perpendicular to the parallels.* Then state the given information and the statement to be proved. **(Lesson 3-1)**

The three-dimensional figure shown at the right is called a right-triangular prism. (Lesson 3-2)

2. Which segment appears parallel to $\overline{AC}$?

3. Which segments appear to be skew to $\overline{BD}$?

4. Name the plane that appears parallel to plane ABC.

5. Which plane appears parallel to plane BCF?

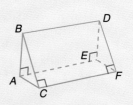

In the figure, $\ell \parallel m$, and $m\angle 2 = 115$. Find the measure of each angle. (Lesson 3-3)

6. $\angle 1$

7. $\angle 3$

8. $\angle 4$

9. $\angle 5$

10. $\angle 6$

11. $\angle 7$

12. $\angle 8$

Proving Lines Parallel

Objectives

After studying this lesson, you should be able to:

- recognize angle conditions that produce parallel lines, and
- prove two lines are parallel based on given angle relationships.

Application

Have you ever noticed that the yardage markers on a football field are parallel? How would the game be affected if they weren't? The grounds crew must be very careful to position the lines correctly so that the game can run smoothly. They use the properties of parallel lines to ensure a fair game.

INVESTIGATION

Construct a line parallel to a given line through a point *not* on the line.

1. Use a straightedge to draw line ℓ and locate point P not on line ℓ.

2. Now draw a line through P that intersects ℓ. Label the point of intersection X and label angle 1 as shown.

FYI···

Originally, football games didn't have halves and quarters. The players just took a break after each score.

3. Construct an angle congruent to angle 1 using P as a vertex and one side on $\overleftrightarrow{PX}$. Draw a line through P to form an angle congruent to $\angle 1$. Label the line n and the angle 2.

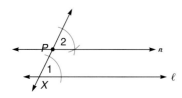

Use a ruler to measure the perpendicular distance between ℓ and n in several different places. Make a conjecture about ℓ and n. *They are parallel.*

What kind of angles are $\angle 1$ and $\angle 2$? *corresponding*

The investigation illustrates a postulate that helps us to prove that two lines are parallel. Notice that this postulate is the converse of Postulate 3-1.

Postulate 3-2	**If two lines in a plane are cut by a transversal so that corresponding angles are congruent, then the lines are parallel.**

The investigation also shows us that there is at least one line through P parallel to ℓ. The following postulate states that there is *exactly* one line parallel to a line through a given point not on the line.

Postulate 3-3 **Parallel Postulate**	If there is a line and a point not on the line, then there exists exactly one line through the point that is parallel to the given line.

There are sets of conditions other than Postulate 3-2 that prove that two lines are parallel. One of them is stated in Theorem 3-5.

Theorem 3-5	If two lines in a plane are cut by a transversal so that a pair of alternate interior angles is congruent, then the two lines are parallel.

Proof of Theorem 3-5

Given: $\angle 1 \cong \angle 2$

Prove: $p \parallel q$

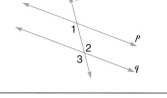

Statements	Reasons
1. $\angle 1 \cong \angle 2$	**1.** Given
2. $\angle 2 \cong \angle 3$	**2.** Vertical angles are congruent.
3. $\angle 1 \cong \angle 3$	**3.** Congruence of angles is transitive.
4. $p \parallel q$	**4.** If 2 lines in a plane are cut by a transversal and corr. ∠s are ≅, the lines are ∥.

Theorems 3-6, 3-7, and 3-8 state three more ways to prove that two lines are parallel. You will be asked to prove these theorems in Exercises 37, 30, and 38 respectively.

Theorem 3-6	If two lines in a plane are cut by a transversal so that a pair of consecutive interior angles is supplementary, then the lines are parallel.
Theorem 3-7	If two lines in a plane are cut by a transversal so that a pair of alternate exterior angles is congruent, then the lines are parallel.
Theorem 3-8	In a plane, if two lines are perpendicular to the same line, then they are parallel.

Notice that Theorems 3-5 through 3-8 are the converses of Theorems 3-1 through 3-4.

Example 1

CONNECTION

Algebra

Find the value of x so that $\ell \parallel m$.

If two lines in a plane are cut by a transversal so that a pair of consecutive interior angles is supplementary, then the lines are parallel. So if $m\angle ABC + m\angle BCD = 180$, then $\ell \parallel m$.

$$(3x - 8) + (5x + 4) = 180$$
$$8x - 4 = 180$$
$$8x = 184$$
$$x = 23$$

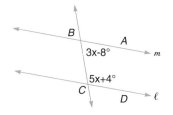

You can use the relationships between angles to determine if lines are parallel.

Example 2

If $\angle 1 \cong \angle 2$ and $\angle 3 \cong \angle 4$, which lines are parallel? Explain.

$\overrightarrow{NQ}$ is a transversal for $\overleftrightarrow{MN}$ and $\overleftrightarrow{QP}$. $\angle 1$ and $\angle 2$ are congruent alternate interior angles. So since the alternate interior angles are congruent, $\overleftrightarrow{MN} \parallel \overleftrightarrow{QP}$.

$\overleftrightarrow{QP}$ is a transversal for $\overleftrightarrow{NP}$ and $\overleftrightarrow{MQ}$, and $\angle 3$ and $\angle 4$ are corresponding angles, so $\overleftrightarrow{NP} \parallel \overleftrightarrow{MQ}$.

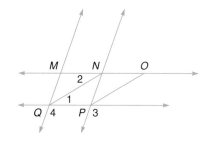

CHECKING FOR UNDERSTANDING

Communicating Mathematics

Read and study the lesson to answer these questions.

1. How do you think the grounds crew positions the yardage lines on a football field so that all of the lines are parallel? Explain.

2. Name some places besides football fields where it is important that lines be parallel.

3. For the figure at the right, justify the statement $\overline{AM} \parallel \overline{HT}$.

4. Your friend claims that $\overline{AT} \parallel \overline{MH}$. Explain why that can't be true.

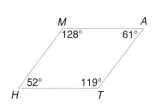

5. Copy the figure at the right and draw $\overleftrightarrow{AH}$. Is $\angle MAH \cong \angle AHT$? Explain.

Guided Practice   Find the value of *x* so that $\ell \parallel m$.

6.

7.

8.

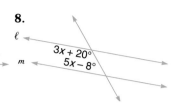

9.

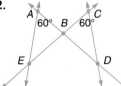

10.

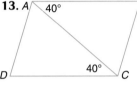

11.

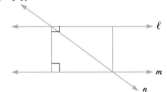

State which segments, if any, are parallel. State the postulate or theorem that justifies your answer.

12.

13.

14.

Determine if each statement is *true* or *false*. If false, explain why.

15. Two lines that are perpendicular to a third line must be parallel.

16. In a plane, two lines that are parallel to a third line must be parallel.

17. Two lines are parallel if the alternate interior angles formed by a transversal are supplementary.

18. Through a point not on a line, there exists exactly one line parallel to the given line.

EXERCISES

Practice   Given the following information, determine which lines, if any, are parallel. Justify your answer.

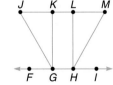

19. $\angle HLK \cong \angle GKJ$ **20.** $\angle IHL \cong \angle HLK$

21. $\angle FGJ \cong \angle KJG$

22. $m\angle GJK + m\angle HLK = 180$

23. $\overline{HL} \perp \overline{GH},\ \overline{GK} \perp \overline{JM}$

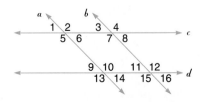

24. $\angle 14 \cong \angle 11$ **25.** $\angle 1 \cong \angle 9$

26. $\angle 10 \cong \angle 15$

27. $m\angle 16 + m\angle 15 = 180$

28. $m\angle 6 + m\angle 10 = 180$

29. $m\angle 7 + m\angle 10 = 180$

30. Copy and complete the proof of Theorem 3-7.

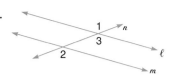

Given: $\angle 2 \cong \angle 1$

Prove: $\ell \parallel m$

Statements	Reasons
a. $\underline{\ ?\ }$	**a.** Given
b. $\angle 1 \cong \angle 3$	**b.** $\underline{\ ?\ }$
c. $\underline{\ ?\ }$	**c.** Congruence of angles is transitive.
d. $\ell \parallel m$	**d.** $\underline{\ ?\ }$

Find the values of x and y that make the blue lines parallel and the red lines parallel.

31.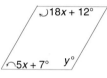

32.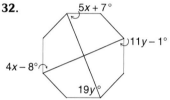

33. You are given that two of the numbered angles are supplementary. Using this information, you can prove that $p \parallel q$. List the pairs of angles that you could be given as supplementary.

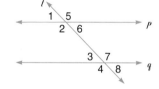

34. If $\ell \nparallel m$, can $x = 12$? Justify your answer.

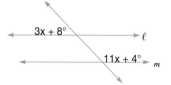

35. Use the information in the figure to determine which lines are parallel. State the theorems that justify your conclusions.

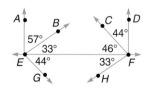

36. Find $m\angle 1$ for the figure at the right.
(Hint: Draw a line through X parallel to ℓ and m.)

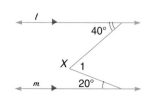

37. Copy and complete the proof of Theorem 3-6.

Given: ∠2 and ∠3 are supplementary.

Prove: ℓ ∥ m

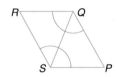

Statements	Reasons
a. ∠2 and ∠3 are supplementary.	**a.** ?
b. ?	**b.** Definition of linear pair
c. ∠1 and ∠2 are supplementary.	**c.** ?
d. ?	**d.** If 2 ∠s are supp. to the same ∠ they are ≅.
e. ℓ ∥ m.	**e.** ?

38. Write a two-column proof of Theorem 3-8.

Write a two-column proof.

39. Given: ∠RQP ≅ ∠PSR
 ∠SRQ and ∠PSR are supplementary.

Prove: $\overline{QP}$ ∥ $\overline{RS}$

40. Given: $\overline{JK}$ ⊥ $\overline{KM}$
 ∠1 ≅ ∠2

Prove: $\overline{LM}$ ⊥ $\overline{KM}$

41. Draw and label the diagram that you would use in a proof of the theorem *If two lines are cut by a transversal so a pair of corresponding angles are congruent, then the lines that bisect those angles are parallel.*

a. State the given information and the statement to be proved.

b. Write the proof.

Critical Thinking

42. Explain why △ABC as it is shown below cannot exist.

Application

43. Construction A carpenter uses a special instrument to draw parallel line segments. Darlene wants to make two parallel cuts at an angle of 40° through points *D* and *P*. Explain why these lines will be parallel.

Mixed Review

44. Classify the statement *If a line intersects two parallel lines, then it is a transversal* as *true* or *false*. **(Lesson 3-3)**

45. Are the spokes on a wheel a model of parallel, intersecting, or skew lines? **(Lesson 3-2)**

46. The formula for finding the total surface area of a cylinder is $A = 2\pi r^2 + 2\pi rh$, where r is the radius of the base and h is the height. Solve the formula for h and justify each step. **(Lesson 2-4)**

47. Write *A cloud is composed of millions of water droplets* in if-then form. **(Lesson 2-2)**

48. $\angle M$ and $\angle N$ are vertical angles. Are $\angle M$ and $\angle N$ right angles if $m\angle M = 4x + 14$ and $m\angle N = 6x - 24$? **(Lesson 1-9)**

49. Jenny forgot the combination to her gym locker. She remembers that the numbers are 18, 37, and 12, but doesn't remember the order. How many different combinations are possible? **(Lesson 1-3)**

Wrap-Up

50. Summarize the five ways that you can prove two lines parallel.

FINE ARTS CONNECTION

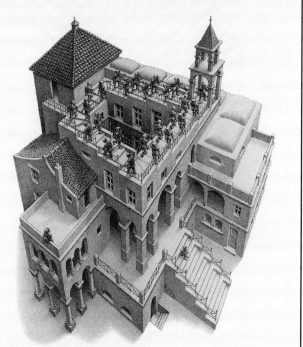

Dutch graphic artist M.C. Escher created some of the most interesting drawings of all time. Born in the Netherlands, he spent a number of years studying art and traveling in Europe. Mr. Escher is famous for his ability to use realistic detail to create bizarre optical effects. Two of his major themes are pattern and visual paradox.

Escher's *Ascending and Descending* first appears to be a picture of a building. But look closely at the stairs! They are impossible in the real world, a visual paradox. But then again, if the stairs are impossible, how did Escher draw them?

©1960 M.C. Escher/Cordon Art-Baarn, Holland

Slopes of Lines

Objectives

After studying this lesson, you should be able to:
- find the slope of a line, and
- use slope to identify parallel and perpendicular lines.

Connection

So far in this chapter we have investigated the properties of parallel and perpendicular lines based on geometric relationships. In the seventeenth century philosopher and mathematician Rene Descartes introduced coordinates to the study of geometry. This allowed discussions about lines and shapes to become discussions about numbers and equations. Thus, Descartes literally began the connection of geometry to algebra.

Descartes identified parallel and perpendicular lines in the coordinate plane by investigating the numerical values of the slopes of these lines. As you recall from algebra, one way to find the slope of a line is to examine the vertical and horizontal change between two points on a line. The slope is the ratio of these two changes.

Definition of Slope

The slope of a line containing two points with coordinates (x_1, y_1) and (x_2, y_2) is given by the formula

$$m = \frac{y_2 - y_1}{x_2 - x_1}, \text{ where } x_1 \neq x_2.$$

The slope of a vertical line, where $x_1 = x_2$, is undefined.

Example 1

Determine the slope of each line.

a.

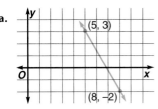

(5, 3)

(8, –2)

$$m = \frac{y_2 - y_1}{x_2 - x_1}$$

$$= \frac{3 - (-2)}{5 - 8} \text{ or } -\frac{5}{3}$$

Lines with negative slope fall as you go along them from left to right.

b.

(3,4)

(–1,1)

$$m = \frac{y_2 - y_1}{x_2 - x_1}$$

$$= \frac{4 - 1}{3 - (-1)} \text{ or } \frac{3}{4}$$

Lines with positive slope rise as you go along them from left to right.

c.

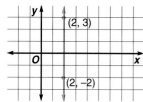

$$m = \frac{y_2 - y_1}{x_2 - x_1}$$

$$= \frac{2 - 2}{5 - 1} \text{ or } 0$$

Lines with a slope of 0 are horizontal.

d.

$$m = \frac{y_2 - y_1}{x_2 - x_1}$$

$$= \frac{3 - (-2)}{2 - 2} \quad \text{undefined}$$

Lines with an undefined slope are vertical.

The graphs of lines p, q, and l are shown at the right. Notice that lines p and q are parallel and l is perpendicular to p and q. Let's investigate the slopes of these lines.

slope of p slope of q slope of l

$$m = \frac{1 - (-1)}{0 - 3} \quad m = \frac{-2 - (-4)}{0 - 3} \quad m = \frac{2 - (-1)}{0 - (-2)}$$

$$= -\frac{2}{3} \qquad\qquad = -\frac{2}{3} \qquad\qquad = \frac{3}{2}$$

Lines p and q are parallel, and their slopes are the same. Line l is perpendicular to lines p and q, and its slope is the negative reciprocal of the slopes of p and q. These results suggest two important algebraic properties of parallel and perpendicular lines.

Postulate 3-4	**Two nonvertical lines have the same slope if and only if they are parallel.**
Postulate 3-5	**Two nonvertical lines are perpendicular if and only if the product of their slopes is -1.**

Note that Postulates 3-4 and 3-5 are written in "if and only if" form. If a conditional and its converse are true, it can be written in "if and only if" form.

Example 2

Draw a line passing through $A(3, 2)$ that is parallel to line ℓ.

First, find the slope of ℓ. $m = \frac{6 - 2}{1 - (-2)}$ or $\frac{4}{3}$

Next, find a point P so that the slope of $\overline{AP}$ is $\frac{4}{3}$. Since slope is $\frac{rise}{run}$, start at point A. "Rise" 4 units then "run" 3 units. Locate point P. Draw $\overleftrightarrow{PA}$.

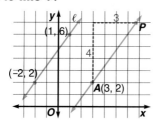

CONNECTION

Algebra

Example 3

Find the value of x so the line through $(x, 6)$ and $(4, -3)$ is perpendicular to the line that passes through $(1, 6)$ and $(7, -2)$.

First, find the slope of the line that passes through $(1, 6)$ and $(7, -2)$.

$$m = \frac{y_2 - y_1}{x_2 - x_1} \qquad \text{\textit{Definition of slope}}$$

$$= \frac{6 - (-2)}{1 - 7} \qquad \text{\textit{Let } } (x_1, y_1) = (7, -2) \text{ \textit{and} } (x_2, y_2) = (1, 6).$$

$$= -\frac{4}{3}$$

The product of the slopes of the two perpendicular lines is -1. Since $-\frac{4}{3} \cdot \frac{3}{4} = -1$, the slope of the line through $(x, 6)$ and $(4, -3)$ is $\frac{3}{4}$.

Now use the formula for the slope of a line to find the value of x.

$$m = \frac{y_2 - y_1}{x_2 - x_1}$$

$$\frac{3}{4} = \frac{6 - (-3)}{x - 4} \qquad \text{\textit{Slope} } = \frac{3}{4}; (x_1, y_1) = (4, -3); (x_2, y_2) = (x, 6)$$

$$\frac{3}{4} = \frac{9}{x - 4}$$

$$3(x - 4) = 36 \qquad \text{\textit{Cross multiply.}}$$

$$3x - 12 = 36$$

$$3x = 48$$

$$x = 16$$

The line through $(16, 6)$ and $(4, -3)$ is perpendicular to the line that passes through $(1, 6)$ and $(7, -2)$.

CHECKING FOR UNDERSTANDING

Communicating Mathematics

Read and study the lesson to answer these questions.

1. Describe a line whose slope is 0 and a line whose slope is 5.

2. Why does the algebraic definition of parallel lines exclude vertical lines? Are vertical lines parallel to each other? Explain.

3. What type of line is perpendicular to a vertical line?

4. Find the slope for the line containing $A(0, -3)$ and $B(16, 5)$. State the slope of any line perpendicular to $\overleftrightarrow{AB}$.

5. On the same coordinate system, draw a line that fits each description.
   a. line a with positive slope
   b. line b with zero slope
   c. line c with negative slope
   d. line d with an undefined slope

Find the slope of the line passing through the given points. Then describe each line as you move from left to right as *rising, falling, horizontal,* or *vertical.*

6. (1, 1), (3, 1) **7.** (-1, 0), (3, -2) **8.** (5, 0), (0, 4)

9. (3, 4), (1, 2) **10.** (-2, 4), (-1, 3) **11.** (8, 2), (8, -5)

Use slope to identify each pair of lines as *parallel, perpendicular,* or *neither.*

12.   **13.**   **14.**

State the slope of a line parallel to a line passing through each pair of points. Then state the slope of a line perpendicular to the line passing through each pair of points.

15. (-1, -1), (3, 2) **16.** (4, 2), (3, -1) **17.** (1, 5), (3, 2)

18. (-7, 5), (1, 1) **19.** (-2, -3), (1, 8) **20.** (-2, -2), (1, 6)

EXERCISES

Determine the slope of each line named below.

21. a

22. b

23. c

24. d

25. any line perpendicular to b

26. any line parallel to a

27. any line perpendicular to c

28. any line parallel to d

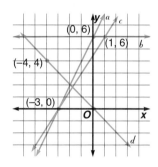

Graph the line that satisfies each description.

29. slope = 0, passes through $P(2, 6)$

30. slope = -4, passes through $P(-2, 1)$

31. undefined slope, passes through $P(2, 0)$

32. slope = $\frac{2}{5}$, passing through $P(0, -2)$

33. passes through $P(2, 1)$ and is parallel to $\overleftrightarrow{AB}$ with $A(-2, 5)$ and $B(1, 8)$

34. passes through $P(4, 1)$ and is perpendicular to $\overleftrightarrow{CD}$ with $C(0, 3)$ and $D(-3, 0)$

Use slope to write an argument to show that each group of points is collinear.

35. A(6, 2), B(-6, 6), C(3, 3)   **36.** M(4, 1), N(-2, -11), O(1, -5)

Determine if $\overleftrightarrow{AB} \parallel \overleftrightarrow{CD}$. Justify your answer.

37. A(-2, -7), B(1, 2), C(0, -1), D(3, 8)

38. A(2, 1), B(3, 2), C(4, 3), D(1, 4)

Determine the value of r so that a line through the points with the given coordinates has the given slope. Draw a sketch of each situation.

39. $(r, 2)$, $(4, -6)$; slope $= -\dfrac{8}{3}$

40. $(5, r)$, $(2, 3)$; slope $= 2$

41. $(r, 6)$, $(8, 4)$; slope $= \dfrac{1}{2}$

42. $(6, r)$, $(9, 2)$; slope $= -\dfrac{1}{3}$

43. A parallelogram is a four-sided polygon whose opposite sides are parallel. Given A(1, 1), B(6, 2), and C(2, 4), find the coordinates of a point D so that A, B, C, and D form a parallelogram.

44. The vertices of a figure ABCD are A(-5, -3), B(5, 3), C(7, 9), and D(-3, 3).
  a. Show that the opposite sides of ABCD are parallel.
  b. Show that the opposite sides of ABCD are congruent. *(Hint: Use the distance formula.)*
  c. What type of figure is ABCD?

45. A line contains the points (9, 1) and (5, 5). Write a convincing argument that the line intersects the y-axis at (0, 10).

Critical Thinking

46. The graph at the right shows the speed of a car at different points in a trip. Describe the movement of the car for each line segment, $\overline{AB}$, $\overline{BC}$, $\overline{CD}$, $\overline{DE}$, $\overline{EF}$, $\overline{FG}$, and $\overline{GH}$. Make a conjecture as to what would cause each type of movement.

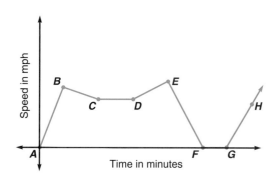

47. Demographics The population of Troy was 150,000 in 1980 and 225,000 in 1990.
  a. What is the rate of change for the population of Troy? That is, how much does the population change in one year?
  b. How does the rate of change relate to the slope of a line?
  c. What do you think the population of Troy will be in the year 2000?

48. Aviation An airplane passing over Richmond at an elevation of 33,000 feet begins its descent to land at Washington, D.C., 107 miles away. How many feet should the airplane descend per mile to land in Washington?

49. Travel The western entrance to the Eisenhower Tunnel in Colorado is at an elevation of 11,160 feet. The tunnel has a downward slope of 0.00895 toward the east and its horizontal distance is 8941 feet long.
  a. Draw and label a diagram of the tunnel.
  b. Find the elevation of the eastern end of the tunnel.

Computer

50. The BASIC program at the right finds the slope between two points.
  a. What does line 50 in the program do?
  b. Use the BASIC program to check your answers to Exercises 6-11.

```
10 PRINT "ENTER THE
     COORDINATES OF POINT A."
20 INPUT X1, Y1
30 PRINT "ENTER THE
     COORDINATES OF POINT B."
40 INPUT X2, Y2
50 M = (Y2-Y1)/(X2-X1)
60 PRINT "THE SLOPE BETWEEN
     POINTS A AND B IS "; M;"."
```

Mixed Review

51. Write a two-column proof. **(Lesson 3-3)**

Given: $\angle 1 \cong \angle 2$
$\overline{PQ} \perp \overline{QR}$
Prove: $\overline{ST} \perp \overline{PQ}$

52. List three of the conclusions that can be drawn from the figure. **(Lesson 3-2)**

53. Write a complete proof of *If 5x − 7 = x + 1, then x = 2.* **(Lesson 2-4)**

54. Erin observed that 2^2 is greater than 2 and 5^2 is greater than 5, and made a conjecture that *The square of any real number is greater than the number.* Give a counterexample to this conjecture. **(Lesson 2-1)**

55. Find the midpoint of the segment whose endpoints have coordinates (8, 11) and (-4, 7). **(Lesson 1-5)**

56. Graph point *A*(-4, 7) on a coordinate plane. **(Lesson 1-1)**

Wrap-Up

57. Explain three different methods you can use to prove that two lines are parallel. At least one method should be algebraic.

Parallels and Distance

Objective

After studying this lesson, you should be able to:

- recognize and use distance relationships between points, lines, and planes.

Application

A team of botanists are traveling to begin a research project at the wildlife preserve on Cumberland Island, Georgia. They are loading their supplies into a boat and wish to take the shortest route possible to the island. What is the shortest route?

The shortest segment from a point to a line is the perpendicular segment from the point to the line. This fact is used to define the **distance from a point to a line** and can help the team of botanists choose the shortest route to the island.

Definition of the Distance Between a Point and a Line	**The distance from a line to a point not on the line is the length of the segment perpendicular to the line from the point.** *The measure of the distance between a line and a point on the line is zero.*

FYI...

Botanists found that the angle between a tree's main branches and its trunk is constant. The same angle appears between the small and large veins of the tree's leaves.

So, for the shortest route to the island, the botanists should use a path that is perpendicular to the shoreline.

You can also find the distance between two parallel lines. Consider the lines with equations $y = 2x + 6$ and $y = 2x - 2$. These lines are parallel since their slopes are the same.

According to the definition, two parallel lines do not intersect. An alternate definition says that two lines in a plane are parallel if and only if they are everywhere equidistant. Equidistant means that the distance between the two lines measured along a line perpendicular to the lines is always the same.

Remember that the distance is measured on a perpendicular segment.

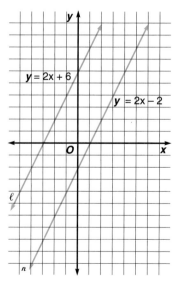

<table>
<tr><td>**Definition of the Distance Between Parallel Lines**</td><td>**The distance between two parallel lines is the distance between one of the lines and any point on the other line.**</td></tr>
</table>

So to find the distance between the two parallel lines ℓ and r we need to choose a point on one of the lines. As shown in the following construction, we choose $A(-3, 0)$ on line ℓ and then construct the perpendicular. The length of this perpendicular segment will represent the distance between the lines.

We could also construct a line perpendicular to r using a point on r and the construction you learned in Chapter 1.

CONSTRUCTION

Construct a line perpendicular to line r through point A(-3, 0), not on r.

1. Place the compass point at point A. Make the setting wide enough so that when an arc is drawn, it intersects r in two places. Label these points of intersection B and C.

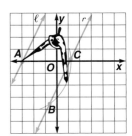

2. Using the same compass setting, put the compass at point B and draw an arc below line ℓ. *Any compass setting greater than $\frac{1}{2} BC$ will work.*

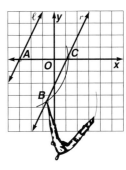

3. Then put the compass at point C and draw an arc to intersect the one drawn in step 2. Be sure to use the same compass setting. Label the point of intersection D.

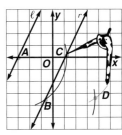

4. Draw $\overleftrightarrow{AD}$. $\overleftrightarrow{AD} \perp \ell$

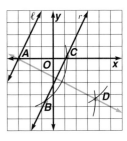

Notice that the construction of the line segment from point A perpendicular to line r intersects line r at approximately $(0, -2)$. To find the distance between the two parallel lines, we can use the distance formula we learned in Lesson 1-4.

$$\text{distance} = \sqrt{(x_2 - x_1)^2 + (y_2 - y_1)^2}$$

$$= \sqrt{(-3 - 0)^2 + (0 - (-2))^2}$$

The coordinates of point A are (-3, 0) and the coordinates of point E are (0, -2).

$$= \sqrt{13}$$

The distance between ℓ and r is $\sqrt{13}$ or about 3.6 units.

Example 1

APPLICATION

Planning

The plan for the new Westerville city park is shown below. Which line segment would you use to find each distance if $\overline{AB} \perp \overline{BC}$ and $\overline{BE} \perp \overline{AC}$?

a. the distance between the gazebo and the path between the footbridge and the shelterhouse

$\overline{AB}$, because $\overline{AB} \perp \overline{BC}$.

b. the distance from the entrance to the path between the gazebo and the shelterhouse

$\overline{DE}$, since $\overline{DE} \perp \overline{AC}$.

c. the distance from the shelterhouse to the path between the footbridge and the gazebo

$\overline{BC}$, since $\overline{BC} \perp \overline{AB}$.

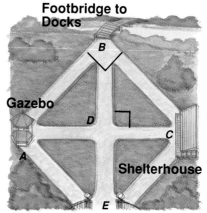

Footbridge to Docks

Gazebo

Shelterhouse

Entrance

You can use the properties of parallel lines to discover properties of other geometric figures.

Example 2

Write a two-column proof.

Given: Quadrilateral $WXYZ$,
$\overline{XW} \parallel \overline{YZ}, \overline{XW} \perp \overline{WZ}, \overline{XY} \perp \overline{YZ}$

Prove: $\overline{XY} \parallel \overline{WZ}$

Statements	Reasons
1. $\overline{XW} \parallel \overline{YZ}$ $\overline{XW} \perp \overline{WZ}$ $\overline{XY} \perp \overline{YZ}$	**1.** Given
2. $\overline{WZ} \perp \overline{YZ}$	**2.** In a plane, if a line is $\perp$ to one of 2 $\parallel$ lines, then it is $\perp$ to the other.
3. $\overline{XY} \parallel \overline{WZ}$	**3.** In a plane, if 2 lines are $\perp$ to the same line, then the lines are $\parallel$.

CHECKING FOR UNDERSTANDING

Communicating Mathematics

Read and study the lesson to answer these questions.

1. If two lines are parallel, explain how you can construct a segment that represents the distance between the two lines.

2. Can you find the distance between two coplanar lines that are not parallel? Explain why or why not.

3. Use graph paper and work with a partner to find the approximate distance between the lines whose equations are $y = 3x - 1$ and $y = 3x + 9$.

4. Think of everyday situations when you might need to find the distance from one point to another point, the distance from a point to a line, and the distance from a line to a line. For example, finding the shortest route from your campsite to the river is an example of the distance from a point to a line.

Guided Practice

Copy each diagram and construct the segment that represents the distance indicated.

5. A to $\overleftrightarrow{XY}$

6. ℓ to P

7. a to b

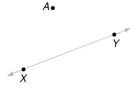

Determine if each statement is *true* or *false*. If the statement is false, explain why.

8. If two lines are everywhere equidistant, then the lines are parallel.

9. The distance between a line on a plane and the plane is 0.

10. The distance between two parallel lines is the length of any segment that connects points on the two lines.

11. If AB is the distance between two lines and A is on one line and B is on the other, then $\overline{AB}$ is perpendicular to both lines.

Use a ruler and the definition of parallel lines to determine whether the lines shown in color are parallel. Write *yes* or *no*. Explain your answer.

12.

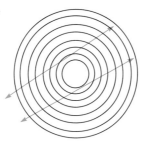

13.

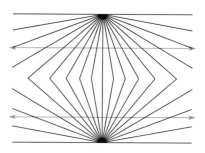

EXERCISES

Practice

Copy each figure and draw the segment that represents the distance indicated.

14. C to $\overline{FE}$

15. C to $\overleftrightarrow{DE}$

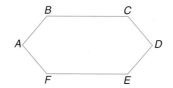

16. C to $\overline{AB}$

17. N to $\overline{MP}$

18. R to $\overleftrightarrow{TS}$

Using graph paper, graph each equation and plot the given point. Then construct a perpendicular segment and find the distance from the point to the line.

19. $x + 3y = 6$, $(-5, -3)$

20. $2x - y = 3$, $(2, 6)$

21. $y = -4x$, $(-2, 8)$

22. $y = 5$, $(-2, 4)$

23. $3x + 4y = 1$, $(2, 5)$

24. $x = 1$, $(4, 5)$

Draw a figure to illustrate each of the following. Then write any possible conclusions that can be made from the given information.

25. $\overline{AB}$ is perpendicular to $\overline{CD}$ at B.

26. $\overline{MN}$ is perpendicular to $\overline{NO}$.

27. ℓ is perpendicular to m and n is perpendicular to m.

28. quadrilateral $QUAD$ with $\overline{QU}$ parallel to $\overline{AD}$

In the figure below, $\overline{PS} \perp \overline{SQ}$, $\overline{PQ} \perp \overline{QR}$, and $\overline{QR} \perp \overline{SR}$. Name the segment whose length represents the distance between the following points and lines.

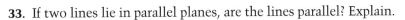

29. P to $\overline{SQ}$

30. R to $\overline{PQ}$

31. Q to $\overline{SR}$

32. S to $\overline{QR}$

33. If two lines lie in parallel planes, are the lines parallel? Explain.

34. Explain how you could find the distance between two parallel planes.

35. If a line is perpendicular to one of two parallel planes, is it perpendicular to the other? Explain.

Critical Thinking

36. An advanced algebra book states that the distance from a point $P(x_1, y_1)$ to a line ℓ with equation $Ax + By + C = 0$ can be found by the formula $d = \dfrac{|Ax_1 + By_1 + C|}{\sqrt{A^2 + B^2}}$. Use the formula to find the distance between the points and lines given in Exercises 19-24. Do you think the formula is valid? Explain.

Applications

37. Transportation An accident has occurred on Central Avenue and an ambulance has been dispatched from Polyclinic Hospital.

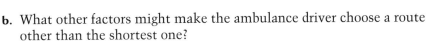

a. What is the shortest route the ambulance could take to the accident? Justify your answer.

b. What other factors might make the ambulance driver choose a route other than the shortest one?

38. Interior Design Rosa is putting a chair railing on her dining room walls. In order to ensure that the rail is parallel to the baseboards, she measures and marks 36 inches up from the baseboard in several places along the walls. If she installs the rail at these markings, how does Rosa know that the rail will be parallel to the baseboards?

39. Transportation The map below shows a part of downtown Minneapolis-St. Paul. On it, 1 centimeter represents 0.6 kilometers.

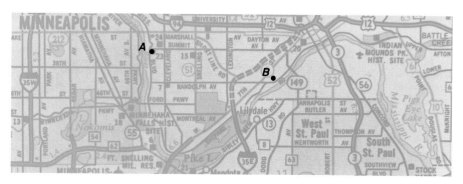

a. Estimate the distance that a bird would fly to get from Saint Thomas College at point *A* to point *B*.

b. Estimate the distance a new college student would travel in his car to get from Saint Thomas College to point *B*. Assume that he takes the shortest possible route.

c. About how far would the captain of a riverboat travel to get from point *A* to point *B*?

d. Compare the distances each one traveled. Who traveled the farthest? Who traveled the shortest? Explain.

Mixed Review

40. Find the slope of the line that passes through the points (3, 0) and (8, -2). **(Lesson 3-5)**

41. Draw a figure to illustrate two lines that are perpendicular to a third line, but are not parallel to each other. **(Lesson 3-1)**

42. Determine if the statement *Parallelism is transitive* is *true* or *false*. **(Lesson 3-1)**

43. Find the value of *x*. **(Lesson 2-7)**

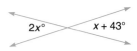

44. The sign in front of the Screaming Eagle Rollercoaster says *If you are over 48 inches tall, then you may ride the Screaming Eagle.* Jamal is 54 inches tall. Can he ride the Screaming Eagle? Which law of logic leads you to this conclusion? **(Lesson 2-3)**

45. State the hypothesis and the conclusion of the statement *If two lines are parallel then they are everywhere equidistant.* **(Lesson 2-2)**

46. Aviation The path of an airplane is described as 22° west of south. Draw a diagram that represents this flight path. **(Lesson 1-5)**

Wrap-Up

47. Describe how to find the distance between two points, the distance between a point and a line, and the distance between two lines.

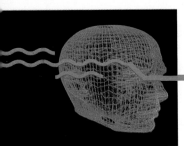

Technology

BASIC
Geometric Supposer
▶ **Graphing calculators**
LOGO
Spreadsheets

Finding the Distance Between a Point and a Line

You can use a graphing calculator to find the distance between a point and a line. Let's find the distance between the line whose equation is $y = 2x - 5$ and $(7, -1)$. The equation of the line perpendicular to the line with equation $y = 2x - 5$ line through $(7, -1)$ is $y = -\frac{1}{2}x + \frac{5}{2}$.

Clear the graphics screen. Press SHIFT Cls then EXE for a Casio fx-7000G. Press Y= , then use the arrow keys and the CLEAR key to select and clear any equations from the Y = list for a TI-81.

Now graph the lines.

Casio

ENTER: GRAPH 2 ALPHA X (−) 5
: GRAPH (−) 0.5 ALPHA
X + 2.5 EXE

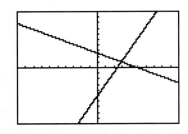

TI-81

ENTER: Y= 2 XT (−) 5 ENTER
(−) 0.5 XT + 2.5 GRAPH

Press SHIFT TRACE on the Casio or TRACE on the TI-81. Then move the cursor to the point of intersection of the two lines and approximate its coordinates.

Now use the distance formula to find the distance between the intersection point and the given point.

$$\text{distance} = \sqrt{(7 - 3)^2 + (-1 - 1)^2}$$
$$= \sqrt{16 + 4}$$
$$= 2\sqrt{5} \quad \text{or about 4.5}$$

EXERCISES

Use a graphing calculator to find the distance between each given line and point on a perpendicular line.

1. $y = x + 7$;
$(2, 7)$ on $y = -x + 9$

2. $y = \frac{2}{5}x + 3$
$(-8, -6)$ on $y = -\frac{5}{2}x - 26$

3. $y = 3x - 13$
$(1, 0)$ on $y = -\frac{1}{3}x + \frac{1}{3}$

VOCABULARY

Upon completing this chapter, you should be familiar with the following terms:

coplanar	**122**	**142**	slope
parallel	**122**	**123**	skew lines
parallel planes	**122**	**124**	transversal

SKILLS AND CONCEPTS

OBJECTIVES AND EXAMPLES	REVIEW EXERCISES

Upon completing this chapter, you should be able to:

- describe the relationships between two lines and two planes. **(Lesson 3-2)**

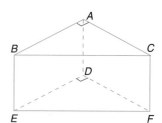

Plane *ABC* and plane *DEF* are parallel.

Segments $\overline{BC}$ and $\overline{EF}$ are parallel.

Segments $\overline{AB}$ and $\overline{DF}$ are skew.

Use these exercises to review and prepare for the chapter test.

Use the figure below to find an example of each of the following.

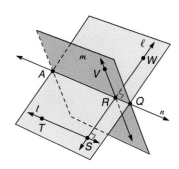

1. two parallel lines

2. two skew lines

3. two parallel lines and a transversal

4. two intersecting planes

5. two noncoplanar lines

OBJECTIVES AND EXAMPLES	REVIEW EXERCISES

■ use the properties of parallel lines to determine angle measures. **(Lesson 3-3)**

List the conclusions that can be drawn if $\ell \parallel m$.

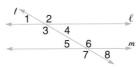

Corresponding angles: $\angle 1 \cong \angle 5$, $\angle 2 \cong \angle 6$, $\angle 3 \cong \angle 7$, and $\angle 4 \cong \angle 8$

Alternate interior angles: $\angle 3 \cong \angle 6$, and $\angle 4 \cong \angle 5$

Consecutive interior angles: $\angle 3$ and $\angle 5$, and $\angle 4$ and $\angle 6$ are supplementary.

Alternate exterior angles: $\angle 1 \cong \angle 8$, and $\angle 2 \cong \angle 7$

Use the figure below to answer each question.

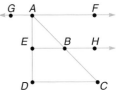

6. If $\overline{AF} \parallel \overline{EB}$, which angles are congruent?

7. If $\overline{AF} \parallel \overline{DC}$, which angles are congruent?

8. If $\overline{AF} \parallel \overline{DC}$, which angle is supplementary to $\angle CAG$?

9. If $\overline{EB} \parallel \overline{DC}$, which angle is supplementary to $\angle BCD$?

■ prove that two lines are parallel. **(Lesson 3-4)**

Ways to show two lines parallel:

1. Alternate interior angles are congruent.
2. Corresponding angles are congruent.
3. Alternate exterior angles are congruent.
4. Consecutive interior angles are supplementary.
5. Two lines are perpendicular to a third line.

Use the figure below to answer each question.

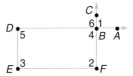

10. Given $\angle 1$ and $\angle 2$ are supplementary, which lines are parallel and why?

11. Given $\angle 5 \cong \angle 6$, which lines are parallel and why?

12. Given $\angle 6 \cong \angle 2$, which lines are parallel and why?

■ find and use the slopes of lines. **(Lesson 3-5)**

Find the slope of the lines parallel to and perpendicular to a line through $(4, -2)$ and $(5, 3)$.

Find the slope of line through $(4, -2)$ and $(5, 3)$.

$$m = \frac{-2 - 3}{4 - 5} \text{ or } 5$$

A line parallel has slope 5, and a line perpendicular has slope $-\frac{1}{5}$.

Find the slope of the line through the given points.

13. $(0, 4)$, $(-1, -2)$ 14. $(2, 0)$, $(0, -6)$

15. $(11, 2)$, $(5, 4)$ 16. $(-1, -5)$, $(3, -7)$

Find the slope of the lines parallel and perpendicular to the line through each pair of points.

17. $(-3, 7)$, $(4, -2)$ 18. $(0, 6)$, $(3, 6)$

19. $(7, -2)$, $(1, -3)$ 20. $(9, -2)$, $(-1, 4)$

■ use distance relationships between points, lines, and planes. **(Lesson 3-6)**

The distance between a point and a line is the length of the segment perpendicular to the line from the point.

The distance between parallel lines is the distance between one of the lines and any point on the other line.

In the figure below, $\overline{PS} \perp \overline{SQ}$, $\overline{PQ} \perp \overline{QR}$, $\overline{PM} \perp \overline{RM}$, and $\overline{QR} \perp \overline{SR}$. Name the segment whose length represents the distance between the following points and lines.

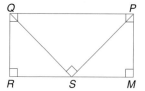

21. from P to $\overline{SQ}$

22. from R to $\overline{PQ}$

23. from $\overline{RM}$ to $\overline{QP}$

24. from $\overline{RQ}$ to $\overline{MP}$

APPLICATIONS AND CONNECTIONS

25. Education   The students in a gym class are standing in a circle. When they count off, the students with numbers 5 and 23 are standing exactly opposite one another. Assuming the students are evenly spaced around the circle, how many students are in the class? **(Lesson 3-1)**

26. Nature Studies   A park ranger estimates that there are 6000 deer in the Blendon Woods Park. He also estimates that one year ago there were 6100 deer in the park. **(Lesson 3-5)**

  a. What is the rate of change for the number of deer in Blendon Woods Park?

  b. At the same rate, how many deer will there be in the park in 10 years?

27. Travel   At 10:00 A.M., Liz had completed 195 miles of her cross-country trip. By 2:00 P.M., she had traveled a total of 455 miles. Use slope to determine Liz's rate of travel. **(Lesson 3-5)**

28. Construction   A ramp was installed to give handicapped people access to the new public library. The top of the ramp is three feet higher than the bottom. The lower end of the ramp is 36 feet from the door of the library. **(Lesson 3-5)**

  a. Draw a diagram of the ramp.

  b. Find the slope of the ramp.

In the figure, $\ell \parallel m$. Determine whether each statement is *true* or *false*. Justify your answer.

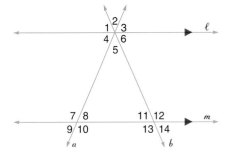

1. $\angle 1$ and $\angle 14$ are alternate exterior angles.
2. $\angle 5$ and $\angle 11$ are consecutive interior angles.
3. $\angle 2$ and $\angle 6$ are vertical angles.
4. $\angle 6$ and $\angle 12$ are supplementary angles.
5. $\angle 3 \cong \angle 8$
6. $\angle 12 \cong \angle 13$
7. $m\angle 7 + m\angle 10 = 180$
8. $m\angle 4 + m\angle 5 + m\angle 11 = 180$

Use the figure at the right to answer each question.

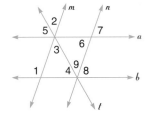

9. Given $\angle 5 \cong \angle 4$, which lines are parallel and why?
10. Given $\angle 3 \cong \angle 9$, which lines are parallel and why?
11. Given $m\angle 4 + m\angle 9 + m\angle 6 = 180$, which lines are parallel and why?

Find the slope of the lines parallel to and perpendicular to a line through the given points.

12. $(4, 5), (-2, 5)$
13. $(11, 8), (5, -3)$
14. $(-3, 1), (7, -1)$

Name the segment that represents the distance between the following points and lines.

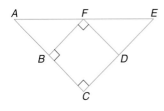

15. from F to $\overline{CE}$
16. from F to $\overline{AC}$
17. from A to $\overline{CE}$
18. from D to $\overline{BC}$

19. Write a two-column proof.

    Given: $\angle 1 \cong \angle 2$
    $\overline{ST} \parallel \overline{PR}$

    Prove: $\angle P \cong \angle R$

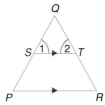

20. **Business** The Carpet Experts cleaning team can clean carpet in a room that is 10 feet by 10 feet in 20 minutes. Draw a diagram and find how long it would take them to do a walk-in closet that is 5 feet by 5 feet.

Bonus In the figure at the right, $a \parallel b, \ell \parallel m, m \parallel n$, and $m\angle 1 = 40$. Find $m\angle 2$.

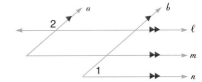

Algebra Review

OBJECTIVES AND EXAMPLES	REVIEW EXERCISES

■ Solve equations involving more than one operation.

$$2x + 15 = 19$$
$$2x + 15 - 15 = 19 - 15$$
$$\frac{2x}{2} = \frac{4}{2}$$
$$x = 2$$

Solve each equation. Check the solution.

1. $3x - 8 = 22$ **2.** $-4y + 5 = 35$

3. $0.5n + 2 = -7$ **4.** $-6 = 3.1t + 6.4$

5. $\frac{x}{-3} + 2 = -21$ **6.** $\frac{8 - 5r}{6} = 3$

■ Solve proportions.

$$\frac{x}{3} = \frac{x-5}{2}$$
$$x \cdot 2 = 3 \cdot (x - 5) \quad \textit{Cross multiply.}$$
$$2x = 3x - 15$$
$$-x = -15$$
$$x = 15$$

Solve each proportion.

7. $\frac{n}{45} = \frac{6}{15}$ **8.** $\frac{35}{55} = \frac{x}{11}$

9. $\frac{4}{8} = \frac{11}{t}$ **10.** $\frac{5}{6} = \frac{a-2}{4}$

11. $\frac{y+4}{y-1} = \frac{4}{3}$ **12.** $\frac{z+7}{6} = \frac{z-3}{7}$

■ Solve percent problems.

What percent of 75 is 9?
$$\frac{9}{75} = \frac{r}{100}$$
$$100 \left(\frac{9}{75} \right) = r$$
$$12 = r \quad \text{9 is 12\% of 75.}$$

Solve.

13. What number is 60% of 80?

14. Twenty-one is 35% of what number?

15. Eighty-four is what percent of 96?

16. What number is 0.3% of 62.7?

■ Graph inequalities on number lines.

Graph the solution set of $x \geq -3$.

Graph the solution set of each inequality on a number line.

17. $x < -2$

18. $x \neq -4$

■ Solve inequalities by using addition or subtraction.

$$3x - 2 < 4x$$
$$3x - 3x - 2 < 4x - 3x$$
$$-2 < x$$

The solution set is $\{x \mid x > -2\}$.

Solve each inequality. Check the solution.

19. $n - 4 < 9$ **20.** $r + 8 \leq -3$

21. $a - 2.6 \geq -8.1$ **22.** $5z - 6 > 4z$

23. $3x \leq 2x + 7$ **24.** $y + \frac{7}{8} > \frac{13}{24}$

- Multiply and divide monomials.

$$(2ab^2)(3a^2b^3) = (2 \cdot 3)(a \cdot a^2)(b^2 \cdot b^3)$$
$$= 6a^3b^5$$

$$\frac{2x^6 y}{8x^2y^2} = \frac{2}{8} \cdot \frac{x^6}{x^2} \cdot \frac{y}{y^2}$$
$$= \frac{x^4}{4y}$$

Simplify. Assume that no denominator is equal to zero.

25. $y^3 \cdot y^4 \cdot y$ 26. $(3mn)(-4m^2n^3)$

27. $(-4a^2x)(-5a^3x^4)$ 28. $\frac{42b^7}{14b^4}$

29. $\frac{y^3xw}{-yxw^2}$ 30. $\frac{-16a^3b^2x^4y}{-48a^4bxy^3}$

- Find the degree of a polynomial.

Find the degree of $2xy^3 + x^2y$.

degree of $2xy^3$: $1 + 3$ or 4

degree of x^2y: $2 + 1$ or 3

Thus, the degree of $2xy^3 + x^2y$ is 4.

Find the degree of each polynomial.

31. $2^3n^2 + 17n^2t^2$

32. $4xy + 7rs^4 + 9x^2z^2$

33. $-6y - 2y^3 + 4 - 8y^2$

- Find the greatest common factor (GCF) for a set of monomials.

Find the GCF of $15x^2y$ and $45xy^2$.

$15x^2y = ③ \cdot ⑤ \cdot x \cdot ⓧ \cdot ⓨ$
$45xy^2 = ③ \cdot 3 \cdot ⑤ \cdot ⓧ \cdot ⓨ \cdot y$

The GCF is $3 \cdot 5 \cdot x \cdot y$ or $15xy$.

Find the GCF of the given monomials.

34. $15ab, -5a^2b^2$

35. $16mrt, 30m^2r$

36. $20n^3q, 24n^2p^2$

37. $2x^2y^3z^4, 8xy^2z^3, 5x^2yz^3$

Applications and Connections

38. **Number Theory** Find two consecutive integers such that twice the greater integer increased by the lesser integer is 50.

39. **Travel** Susan drove 3.35 hours at a rate of 50 miles per hour. To the nearest tenth, how long would it take her to drive the same distance at a rate of 45 miles per hour?

40. **Entertainment** The television program with the largest audience to date was the final episode of "M*A*S*H." Of the 162 million people watching television that evening, 77% saw the program. How many people were watching?

41. **Savings** Each week, Mei deposits $68.25 of her paycheck into a savings account for college. She began the summer with $420.75 and now has $898.50 in the account. For how many weeks has Mei made deposits if she received no money in interest during this time?

42. **Statistics** Namid had an average of 74 on four history tests. What score does he have to get on the 100-point final exam if it counts double and he wants to have an average of 80 or better?

CHAPTER 4

Congruent Triangles

GEOMETRY AROUND THE WORLD
France

If you have ever watched an artist at work, you know the canvas is gradually filled with geometric shapes that, together, compose the finishing painting. It's the way an artist combines and balances these shapes that determines his or her style and gives paintings a different look and "feel."

Although you may not realize it, your eyes respond in different ways to the shapes within a painting. Diagonal lines provide a painting with energy. A circular shape within a painting will pull your eyes toward its center. Often, the arrangement of objects within a painting creates a pleasing geometric shape for your eyes to travel around.

One of the masters of geometry-based still-life paintings was a French 18th-century artist named Jean Baptiste Simeon Chardin. Chardin sometimes spent days arranging objects he planned to paint so their lines and form pleased him. Only then did he set to work with his palette and brushes.

GEOMETRY IN ACTION

Look closely at Chardin's painting "Laid Out Table" shown at the left. Can you see any triangles in the arrangement of objects in the painting?

With your finger, trace the shape of the triangle formed by these flowers. What type of triangle is it?

◀ *"Laid Out Table"*   Inset photo: *Jean Baptiste Simeon Chardin*

163

4-1 Classifying Triangles

Objectives

After studying this lesson, you should be able to:
- identify the parts of a triangle, and
- classify triangles.

Application

A side of the cab, the boom, and the support cable of a crane form a triangle. A triangle is a three-sided polygon. A polygon is a closed figure in a plane that is made up of segments called **sides** that intersect *only* at their endpoints, called **vertices**. When the crane reaches out to allow the vertical cable to pick up a boulder, the angle between the side of the cab and the boom is obtuse. The triangle formed is called an **obtuse triangle**.

FYI···

The 893-ton Rosenkranz crane has a height of 663 feet and can lift 33 tons to a height of 525 feet.

Triangle ABC, written $\triangle ABC$, has the following parts.

The vertices of the triangle can be named in any order.

sides: $\overline{AB}, \overline{BC}, \overline{CA}$

vertices: A, B, C

angles: $\angle BAC$ or $\angle A$, $\angle ABC$ or $\angle B$,
$\angle BCA$ or $\angle C$

The side opposite $\angle A$ is $\overline{BC}$. The angle opposite $\overline{AB}$ is $\angle C$. A similar statement can be made about the remaining side and angle.

One way of classifying triangles is by their angles. All triangles have at least two acute angles, but the third angle can be acute, right, or obtuse. A triangle can be classified using the third angle. We called the triangle formed by the parts of the crane an obtuse triangle because the third angle was obtuse.

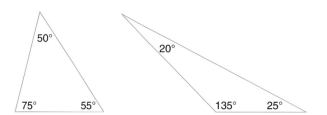

In an **acute triangle,** all the angles are acute.

In an **obtuse triangle,** one angle is obtuse.

In a **right triangle,** one angle is a right angle.

When all of the angles of a triangle are congruent, the triangle is **equiangular**.

Some parts of a right triangle have special names. In right triangle RST, $\overline{RT}$, the side opposite the right angle, is called the **hypotenuse**. The other two sides, $\overline{RS}$ and $\overline{ST}$, are called the **legs**.

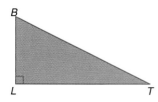

Example 1

> **The quilt piece shown at the right is a right triangle. If the vertices _B_, _L_, and _T_ are labeled as shown, name the angles, the right angle, the hypotenuse, the legs, the side opposite ∠_B_, and the angle opposite $\overline{BL}$.**
>
>
>
> The angles are $\angle B$, $\angle T$, and $\angle L$. The right angle is $\angle L$. The hypotenuse is the side opposite the right angle, $\overline{BT}$. The legs are $\overline{LB}$ and $\overline{LT}$. The side opposite $\angle B$ is $\overline{LT}$. The angle opposite $\overline{BL}$ is $\angle T$.

Triangles can also be classified according to the number of congruent sides. The slashes on the sides of a triangle mean those sides are congruent.

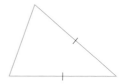

No two sides of a **scalene triangle** are congruent.

At least two sides of an **isosceles triangle** are congruent.

All the sides of an **equilateral triangle** are congruent.

Like the right triangle, the parts of an isosceles triangle have special names. The congruent sides are called **legs**. The angle formed by the legs is the **vertex angle**, and the other two angles are **base angles**. The **base** is the side opposite the vertex angle.

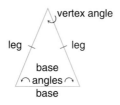

Example 2

> **Triangle _ABC_ is an isosceles triangle. ∠_A_ is the vertex angle, _AB_ = 4_x_ − 14 and _AC_ = _x_ + 10. Find the length of the legs.**
>
> If $\angle A$ is the vertex angle, then $\overline{BC}$ is the base and $\overline{AB}$ and $\overline{AC}$ are the legs. So, $AB = AC$.
>
>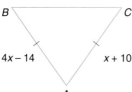
>
> Solve the following equation.
>
> $$AB = AC$$
> $$4x - 14 = x + 10 \qquad \textit{Substitution property of equality}$$
> $$3x = 24 \qquad \textit{Addition property of equality}$$
> $$x = 8 \qquad \textit{Division property of equality}$$
>
> If $x = 8$, then $AB = 4(8) - 14$ or 18, and $AC = (8) + 10$ or 18. The legs of isosceles $\triangle ABC$ are 18 units long.

Triangles can be graphed on the coordinate plane.

Example 3

CONNECTION

Algebra

Given △MNP with vertices M(2, -4), N(-3, 1), and P(1, 6), use the distance formula to prove △MNP is scalene.

According to the distance formula, the distance between (x_1, y_1) and (x_2, y_2) is $\sqrt{(x_2 - x_1)^2 + (y_2 - y_1)^2}$ units. *The distance formula is given on page 25.*

$$MN = \sqrt{(2 - (-3))^2 + (-4 - 1)^2}$$
$$= \sqrt{25 + 25}$$
$$= \sqrt{50} \text{ or } 5\sqrt{2}$$

$$NP = \sqrt{(-3 - 1)^2 + (1 - 6)^2}$$
$$= \sqrt{16 + 25}$$
$$= \sqrt{41}$$

$$MP = \sqrt{(2 - 1)^2 + (-4 - 6)^2}$$
$$= \sqrt{1 + 100}$$
$$= \sqrt{101}$$

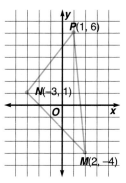

Since no two sides have the same length, the triangle is scalene.

CHECKING FOR UNDERSTANDING

Communicating Mathematics

Read and study the lesson to answer these questions.

1. Name some everyday items that are shaped like triangles. Classify each triangle by angles and by sides.

2. Draw an isosceles right triangle and label the hypotenuse and legs.

3. Can a triangle be both isosceles and scalene? Explain why or why not.

4. Draw and label a scalene obtuse triangle. Identify the side opposite the obtuse angle.

5. Draw an equilateral triangle and describe the lengths of the sides.

Guided Practice

Look around your classroom, your house, or your neighborhood to find an example of each.

6. scalene triangle
7. right scalene triangle
8. isosceles triangle
9. obtuse isosceles triangle
10. equilateral triangle
11. acute scalene triangle

In figure _ACDE_, ∠_E_ and ∠_ADC_ are right angles and the congruent parts are indicated.

12. Name the right triangle(s).
13. Name the isosceles triangle(s).
14. Which triangle(s) is obtuse?
15. Which triangle(s) is equilateral?
16. Which segment(s) can be called the hypotenuse?
17. Which segment(s) is opposite ∠_C?_

18. Can an isosceles triangle be equilateral? Explain your answer.
19. Find the perimeter of equilateral triangle _JLK_ if $JL = x + 3$, and $KJ = 2x - 5$.

EXERCISES

Practice

Draw and label △_ALT_ using the given conditions. If possible, classify each triangle by its angles and by its sides.

20. $m\angle A < 90$, $\overline{AL}$ is the hypotenuse.
21. $AL = LT$; $m\angle L = 90$
22. $m\angle A > 90$; $AL < LT$
23. $AL < LT < AT$
24. $AL = LT = AT$
25. ∠_A_ is obtuse; △_ALT_ is isosceles

Triangle _ROM_ is isosceles with the congruent sides as marked. Name each of the following.

26. sides
27. angles
28. vertex angle
29. base angles
30. side opposite ∠_R_
31. congruent sides
32. angle opposite $\overline{OR}$

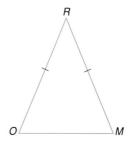

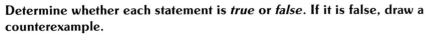

Determine whether each statement is _true_ or _false_. If it is false, draw a counterexample.

33. All equilateral triangles are isosceles.

34. A scalene triangle can be obtuse.

35. A scalene triangle can be acute.

36. A right triangle is never isosceles.

37. All obtuse triangles are scalene.

Use a ruler and protractor to draw triangles with the given parts. Classify each triangle in terms of its angles and in terms of its sides.

38. $\triangle OAT$, $OA = 8$ cm, $m\angle A = 60$, $AT = 4$ cm

39. $\triangle WHT$, $m\angle H = 60$, $WH = HT = 4$ cm

40. $\triangle RYE$, $m\angle Y = 90$, $RY = 4$ cm, $m\angle YRE = 60$

41. $\triangle CAR$, $CA = CR = RA = 6$ cm

42. $\triangle COR$, $m\angle O = 120$, $CO = 4$ cm, $RO = 4$ cm

43. Which of the triangles described in Exercises 38-42 seem to have the same shape? Which seem to have the same shape and size?

44. Find the measures of the legs of the isosceles triangle RLP if $RL = 4x - 5$, $RP = 2x + 11$, and $LP = x$. The perimeter of $\triangle RLP$ is 62 cm.

45. Given vertices $A(6, 4)$, $L(-2, 4)$, and $F(2, 7)$, describe $\triangle ALF$ in terms of its angles and sides. Explain your reasoning.

46. Given $\overline{BC} \parallel \overline{DE}$ and $\overline{AD} \perp \overline{DE}$, prove that $\triangle ABC$ is a right triangle.

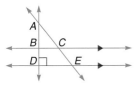

47. Given $m\angle NMO = 20$, prove that $\triangle LMN$ is an obtuse triangle.

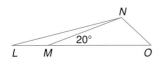

48. $\triangle RST$ is isosceles. Find the perimeter.

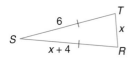

49. $\triangle DEF$ is isosceles with a perimeter between 23 and 32 units. Which angle is the vertex angle? Explain your answer.

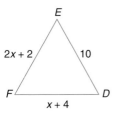

50. Describe the figure formed by connecting the points $R(7, -1)$, $S(9, 4)$, and $T(13, 14)$.

51. Pyramid *ABCDE* is a solid with square base *BCDE* and faces that are all triangles. If vertex *A* is directly over the center of the square, describe the shape of the four faces. Explain your reasoning.

Critical Thinking

52. Using four coplanar points as vertices, no three of which are collinear, what is the maximum number of right triangles that can be drawn?

53. Using four coplanar points as vertices, no three of which are collinear, what is the maximum number of equilateral triangles that can be drawn?

Applications

54. Public Works A fire hydrant is to be located on the highway in such a way that it is as close as possible to two buildings each 1 block from the highway and 1 block apart.
   a. What kind of triangle will be formed using the buildings and the hydrant as vertices?
   b. Why will this triangle satisfy the conditions necessary to locate the fire hydrant?

55. Consider the pattern formed by the dots.

The numbers used to describe each array of dots are called **triangular numbers**. The third triangular number is 6.
   a. Draw the array for the fourth triangular number.
   b. How many dots will be in the array for the eighth triangular number?

Mixed Review

56. Are two parallel planes everywhere equidistant? Explain. **(Lesson 3-6)**

57. Algebra State the slope of the line passing through the points (3, 9), and (-7, 8). **(Lesson 3-6)**

58. Determine if $\overleftrightarrow{RS} \parallel \overleftrightarrow{LM}$ given that $m\angle 1 = 42$ and $m\angle 5 = 48$. Justify your answer. **(Lesson 3-4)**

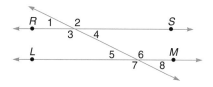

59. Draw a figure to illustrate two intersecting lines with a plane parallel to both lines. **(Lesson 3-2)**

60. Algebra Name the algebraic property that justifies the statement "If $a = x$ and $a = y$, then $x = y$." **(Lesson 2-4)**

61. Algebra If $\angle 1 \cong \angle 2$, $m\angle 1 = 4x - 7$, and $m\angle 2 = 2x + 5$, find $m\angle 1$. **(Lesson 1-7)**

Wrap-Up

62. In your own words, write three definitions that were presented in this lesson and illustrate each with a labeled diagram.

Angle Measures in Triangles

Objectives

After studying this lesson, you should be able to:
- apply the angle sum theorem, and
- apply the exterior angle theorem.

Is there any relationship among the angles in a triangle? The students in Ms. Braun's geometry class each drew a triangle and measured the angles. When the results were recorded on the chalkboard, they calculated the sum of the angles to see if there was a pattern. The results looked like those displayed in the stem-and-leaf plot below.

Stem	Leaf
17	6 6 7
	8 8 8 9 9 9
18	0 0 0 0 0 1 1 1
	2 2 2 3 4

18|4 = 184

18|4 means that one student found the sum of the angles in his or her triangle to be 184. Make a conjecture about the sum of the measures of the angles in a triangle.

Test your conjecture by drawing any triangle *ABC* and cutting it out. Fold it along a line parallel to $\overline{AC}$ so vertex *B* is on $\overline{AC}$. Then fold the triangle so that *A* and *C* are on point *B'* as shown.

Note that $\angle AB'D$ and $\angle 3$ form a linear pair. Since the angles of a linear pair are supplementary, $m\angle AB'D + m\angle 3 = 180$. By the angle addition postulate, $m\angle 1 + m\angle 2 = m\angle AB'D$. Therefore, by substitution, $m\angle 1 + m\angle 2 + m\angle 3 = 180$.

This leads to the following theorem.

Theorem 4-1 *Angle Sum Theorem*	**The sum of the measures of the angles of a triangle is 180.**

In order to prove the angle sum theorem, we will need to draw an **auxiliary line**. An auxiliary line is a line or line segment added to a diagram to help in a proof. These are shown as dashed lines in the diagram. Be sure that it is possible to draw any auxiliary lines that you use.

Proof of the
Angle Sum Theorem

Given: $\triangle PQR$

Prove: $m\angle 1 + m\angle 2 + m\angle 3 = 180$

Proof:

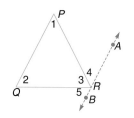

Statements	Reasons
1. $\triangle PQR$	1. Given
2. Draw $\overleftrightarrow{AB}$ through R parallel to $\overrightarrow{PQ}$.	2. Parallel postulate
3. $\angle 4$ and $\angle PRB$ form a linear pair.	3. Definition of linear pair
4. $\angle 4$ and $\angle PRB$ are supplementary.	4. If 2 △s form a linear pair, they are supp.
5. $m\angle 4 + m\angle PRB = 180$	5. Definition of supplementary
6. $m\angle 5 + m\angle 3 = m\angle PRB$	6. Angle addition postulate
7. $m\angle 4 + m\angle 5 + m\angle 3 = 180$	7. Substitution property of equality
8. $m\angle 1 = m\angle 4$ $m\angle 2 = m\angle 5$	8. If 2 ∥ lines are cut by a transversal, alt. int. △s are ≅.
9. $m\angle 1 + m\angle 2 + m\angle 3 = 180$	9. Substitution property of equality

If you know the measures of two angles of a triangle, you can find the measure of the third.

Example 1

APPLICATION

Construction

The roof support at the right is shaped like a triangle. Two angles each have a measure of 25. Find the measure of the third angle.

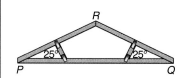

If we label the vertices of the triangle P, Q, and R, then $m\angle P = 25$ and $m\angle Q = 25$. Since the sum of the angles measures is 180, we can write the equation below.

$$m\angle P + m\angle Q + m\angle R = 180 \quad \text{\textit{Angle sum theorem}}$$
$$25 + 25 + m\angle R = 180 \quad \text{\textit{Substitution property of equality}}$$
$$m\angle R = 130 \quad \text{\textit{Subtraction property of equality}}$$

The measure of the third angle is 130.

The Angle Sum Theorem leads to a useful theorem about the angles in two triangles. You will prove this theorem in Exercise 42.

Theorem 4-2 Third Angle Theorem	If two angles of one triangle are congruent to two angles of a second triangle, then the third angles of the triangles are congruent.

Suppose you were picking strawberries and started walking from a certain point on a north-south path. You walked at an angle of 60° northeast for 800 feet, turned directly south and walked 400 feet, and then turned 90° clockwise and walked back to the same place where you started.

A diagram of your path is shown at the right. The north/south lines are parallel, forming congruent alternate interior angles. By the supplement theorem, the angle formed when you turned measures 120°.

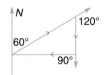

remote interior angles

exterior angle

The 120° and 90° angles formed when you turned are called **exterior angles** of the triangle. An exterior angle is formed by one side of a triangle and another side extended. The interior angles of the triangle not adjacent to a given exterior angle are called **remote interior angles** of the triangle. In the figure at the left, $\angle BCD$ is an exterior angle with $\angle A$ and $\angle B$ as its remote interior angles.

The measure of an exterior angle and its remote interior angles are related. Look at your path again. Do you have a conjecture about the relationship? Let's investigate.

$m\angle 1 + m\angle 2 + m\angle 3 = 180$	*Angle sum theorem*
$m\angle 1 + m\angle 4 = 180$	*Supplement theorem*
$m\angle 1 + m\angle 2 + m\angle 3 = m\angle 1 + m\angle 4$	*Substitution property of equality*
$m\angle 2 + m\angle 3 = m\angle 4$	*Subtraction property of equality*

The measure of an exterior angle of a triangle is equal to the sum of the measures of its remote interior angles. This is called the Exterior Angle Theorem. You will prove this theorem in Exercise 43.

Theorem 4-3 Exterior Angle Theorem	The measure of an exterior angle of a triangle is equal to the sum of the measures of the two remote interior angles.

Example 2

Find the measure of each numbered angle in the figure if $\overline{RK} \parallel \overline{SL}$ and $\overline{RS} \perp \overline{SL}$.

By the exterior angle theorem,
$m\angle KRL + m\angle 4 = m\angle MKR$.

$$m\angle KRL + m\angle 4 = m\angle MKR$$
$$40 + m\angle 4 = 115$$
$$m\angle 4 = 75$$

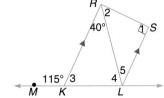

By the angle sum theorem, $m\angle KRL + m\angle 3 + m\angle 4 = 180$.

$$m\angle KRL + m\angle 3 + m\angle 4 = 180$$
$$40 + m\angle 3 + 75 = 180$$
$$m\angle 3 = 65$$

Since $\overline{RK} \parallel \overline{SL}$, and $\angle 5$ and $\angle KRL$ are alternate interior angles, $m\angle 5 = m\angle KRL$.

$$m\angle 5 = 40$$

$\overline{RS} \perp \overline{SL}$ and perpendicular lines form right angles. Therefore, $m\angle 1 = 90$.

$$m\angle 1 + m\angle 5 + m\angle 2 = 180 \qquad \textit{Angle sum theorem}$$
$$90 + 40 + m\angle 2 = 180$$
$$m\angle 2 = 50$$

Therefore, $m\angle 1 = 90$, $m\angle 2 = 50$, $m\angle 3 = 65$, $m\angle 4 = 75$, and $m\angle 5 = 40$.

A statement that can easily be proven using a theorem is often called a **corollary** of that theorem. A corollary, just like a theorem, can be used as a reason in a proof. You will be asked to prove Corollary 4-2 in Exercise 44.

Corollary 4-1	**The acute angles of a right triangle are complementary.**
Corollary 4-2	**There can be at most one right or obtuse angle in a triangle.**

Proof of Corollary 4-1

Given: $\triangle RST$, $\angle R$ is a right angle.

Prove: $\angle S$ and $\angle T$ are complementary.

Proof:

Statements	Reasons
1. $\angle R$ is a right angle.	**1.** Given
2. $m\angle R + m\angle S + m\angle T = 180$	**2.** Angle sum theorem
3. $m\angle R = 90$	**3.** Definition of right angle
4. $90 + m\angle S + m\angle T = 180$	**4.** Substitution property of equality
5. $m\angle S + m\angle T = 90$	**5.** Subtraction property of equality
6. $\angle S$ and $\angle T$ are complementary.	**6.** Definition of complementary angles

CHECKING FOR UNDERSTANDING

Communicating Mathematics

Read and study the lesson to answer these questions.

1. What is the sum of the measures of the angles in △GHI? Compare the sums of the angle measures for the three triangles.

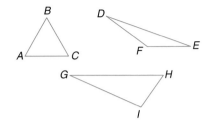

2. If $m\angle I = 90$, what must be true about $\angle G$ and $\angle H$? Explain.

3. If the measures of $\angle A$, $\angle B$, and $\angle C$ are equal, what is the measure of each angle? Justify your answer.

4. Draw a triangle and label an exterior angle at each vertex.

Guided Practice

Find the measure of each angle in the figure at the right.

5. $m\angle 1$　　　　6. $m\angle 2$

7. $m\angle 3$　　　　8. $m\angle 4$

9. $m\angle 5$

Find the value of x.

10.

11.

12.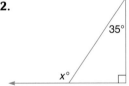

13. Name the exterior angles and their corresponding remote interior angles in the figure at the right.

14. Write an argument to explain why a triangle can or cannot have two right angles.

15. Draw a triangle that has exactly two congruent angles. Can you draw another triangle that has two congruent angles, but is a different shape?

List two conclusions that you can make about the angles in each drawing.

16.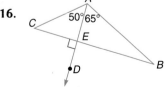

17.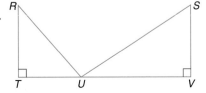

EXERCISES

Practice

Find the value of *x*.

18.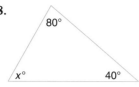

19.

20.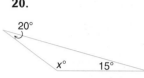

21.

22.

23.

Find the measure of each angle.

24. $m\angle 1$ **25.** $m\angle 2$

26. $m\angle 3$ **27.** $m\angle 4$

28. $m\angle 5$ **29.** $m\angle 6$

30. $m\angle 7$ **31.** $m\angle 8$

32. $m\angle 9$

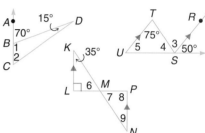

In the figure, $\overline{AB} \perp \overline{BC}$, $\overline{CD} \perp \overline{BC}$, $m\angle BEC = 150$, and $\angle ABE \cong \angle DCE$. Find the measure of each angle.

33. $\angle AEB$ **34.** $\angle EBC$

35. $\angle CED$ **36.** $\angle ECD$

37. $\angle ECB$ **38.** $\angle ABE$

List three conclusions that you can make about the measures of the angles in each drawing.

39. $\overline{AB} \parallel \overline{DC}$
   $m\angle ATC = 140$
   $m\angle C = 20$

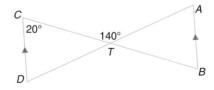

40. $\overline{ML} \perp \overline{LN}$
   $\overline{LP} \perp \overline{MN}$
   $m\angle N = 32$

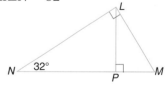

41. Prove that if a triangle is equiangular, the measure of each angle is 60.

42. Prove the third angle theorem (Theorem 4-2).

43. Prove the exterior angle theorem (Theorem 4-3).

44. Prove that there can be at most one right or obtuse angle in a triangle. (Corollary 4-2)

45. Given: $\overline{LT} \perp \overline{TS}$
$\overline{ST} \perp \overline{SR}$
Prove: $\angle TLR \cong \angle LRS$

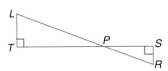

46. Given: $\angle RUW \cong \angle VSR$
Prove: $\angle V \cong \angle W$

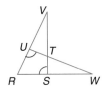

47. If $\overline{AB}$ bisects $\angle CAD$, $\overline{CD} \perp \overline{AC}$ and $\overline{BD} \perp \overline{AD}$, which numbered angles must be congruent?

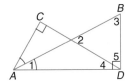

48. Triangle ABC is scalene, with two of its angles trisected as shown. If $m\angle A = 30$, find $m\angle M$ and $m\angle T$.

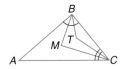

49. What is the sum of the interior angles of the quadrilateral in the figure? *(Hint: Think in terms of triangles.)*

Critical Thinking

50. If you draw a triangle on a globe of Earth, what conjectures could you make about the angles in the triangle? Does the Angle Sum Theorem work? *(Hint: Look at the latitude and longitude lines on a globe.)*

51. If you could draw a triangle on the inside of a large balloon, what would the triangle look like? What conjectures could you make about the angles of this triangle?

Application

52. Surveying Surveyors use a method called *triangulation* to map a region. This method locates points by means of a network of triangles. Find the remaining angle measures in the survey shown at the right.

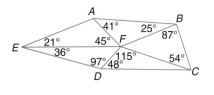

Mixed Review

53. Can a scalene triangle be a right triangle also? **(Lesson 4-1)**

54. Algebra The slope of line m is 9. What is the slope of any line perpendicular to m? **(Lesson 3-5)**

55. Name the congruence property that justifies the statement "$\angle A \cong \angle A$." **(Lesson 2-7)**

56. The measure of an angle is one-third the measure of its supplement. Find the measure of the angle. **(Lesson 1-8)**

57. Find the value of a so that the distance between $(5, 6)$ and $(a, 10)$ is 5 units. **(Lesson 1-4)**

Wrap-Up

58. State the main theorems in this lesson. Give examples of how they apply to an obtuse triangle, an acute triangle, or a right triangle.

4-3 Congruent Triangles

Objectives

After studying this lesson, you should be able to:
- identify congruent triangles, and
- name and label corresponding parts of congruent triangles.

Application

In 1913, Henry Ford began producing automobiles using an assembly line. When products are mass produced, each piece must be interchangeable, so they must have the same size and shape. Each piece is an exact copy of the others, and any piece can be made to coincide with all of the others. Remember from Chapter 1 that figures with the same size and same shape are *congruent*.

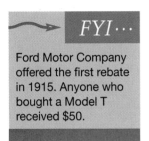

FYI...

Ford Motor Company offered the first rebate in 1915. Anyone who bought a Model T received $50.

Because the triangle is the simplest of the polygons, it seems reasonable to begin a study of congruent polygons by investigating congruent triangles.

INVESTIGATION

Draw a triangle and cut it out. Use it as a pattern to draw a second triangle and cut that triangle out. If one triangle is placed on top of the other, the two coincide or match exactly. This means that each part of the first triangle matches exactly the corresponding part of the second triangle. You have made a pair of congruent triangles.

Note that the order of the letters in the congruence statement indicates the correspondence of the vertices.

If $\triangle ABC$ is congruent to $\triangle RST$ ($\triangle ABC \cong \triangle RST$), the vertex labeled A corresponds to the vertex labeled R, vertex B corresponds to S, and vertex C corresponds to T. This correspondence can be described in terms of angles and sides as follows.

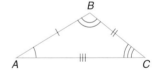

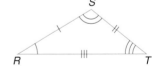

$\angle A$ corresponds to $\angle R$. $\overline{AB}$ corresponds to $\overline{RS}$.

$\angle B$ corresponds to $\angle S$. $\overline{BC}$ corresponds to $\overline{ST}$.

$\angle C$ corresponds to $\angle T$. $\overline{AC}$ corresponds to $\overline{RT}$.

Since the two triangles match exactly, the corresponding parts are congruent.

Definition of Congruent Triangles (CPCTC)	**Two triangles are congruent if and only if their corresponding parts are congruent.** *The abbreviation CPCTC means Corresponding Parts of Congruent Triangles are Congruent.*

Example 1

A triangular wedge is used to anchor the seat belts of a car.

a. **Draw two identical wedges and label the vertices *P*, *R*, and *S* on one part and *K*, *L*, and *M* on the other so that △*PRS* ≅ △*KLM*. Then mark the congruent parts.**

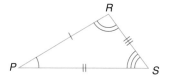

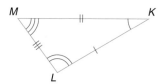

b. **What angle in △*PRS* is congruent to ∠*K* in △*KLM*?**

∠*P* is congruent to ∠*K*.

c. **Which side of △*KLM* is congruent to $\overline{PS}$ in △*PRS*?**

$\overline{KM}$ is congruent to $\overline{PS}$.

Congruence of triangles, like congruence of segments and angles, is reflexive, symmetric, and transitive. This is stated in Theorem 4-4. The proof of the transitive part of this theorem is shown below. You will be asked to prove the reflexive and symmetric parts of the theorem in Exercises 35 and 24, respectively.

Theorem 4-4	**Congruence of triangles is reflexive, symmetric, and transitive.**

Proof of Theorem 4-4 (Transitive Part)

Given: △*ABC* ≅ △*DEF*
△*DEF* ≅ △*GHI*

Prove: △*ABC* ≅ △*GHI*

Paragraph Proof:

We are given that △*ABC* ≅ △*DEF*. By the definition of congruent triangles, the corresponding parts of the triangles are congruent. So, ∠*A* ≅ ∠*D*, ∠*B* ≅ ∠*E*, ∠*C* ≅ ∠*F*, $\overline{AB}$ ≅ $\overline{DE}$, $\overline{BC}$ ≅ $\overline{EF}$, and $\overline{AC}$ ≅ $\overline{DF}$. It is also given that △*DEF* ≅ △*GHI*, so by the definition of congruent triangles, ∠*D* ≅ ∠*G*, ∠*E* ≅ ∠*H*, ∠*F* ≅ ∠*I*, $\overline{DE}$ ≅ $\overline{GH}$, $\overline{EF}$ ≅ $\overline{HI}$, and $\overline{DF}$ ≅ $\overline{GI}$. Since congruence of angles is transitive, ∠*A* ≅ ∠*G*, ∠*B* ≅ ∠*H*, and ∠*C* ≅ ∠*I*. Congruence of segments is transitive, so $\overline{AB}$ ≅ $\overline{GH}$, $\overline{BC}$ ≅ $\overline{HI}$, and $\overline{AC}$ ≅ $\overline{GI}$. Therefore, △*ABC* ≅ △*GHI* by the definition of congruent triangles.

CHECKING FOR UNDERSTANDING

Communicating Mathematics

Read and study the lesson to answer these questions.

1. Draw triangles *TLA* and *RSB*. Mark the corresponding parts for $\triangle TLA \cong \triangle RSB$.

2. Describe how you would tell if two triangles were congruent.

3. If two triangles are congruent, what conclusions can you make? Give an example to illustrate your answer.

Guided Practice

Complete each congruence statement.

4. $\triangle ARM \cong$ __?__

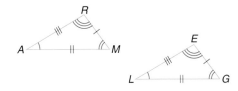

5. $\triangle SPT \cong$ __?__

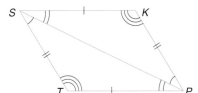

Copy the figures and use the given information to determine which parts are congruent. Determine if the given triangles listed in the *Prove* statement are congruent.

6. **Given:** *N* is the midpoint of $\overline{AB}$ and $\overline{CD}$.
$\overline{AD} \parallel \overline{BC}$
$\overline{AD} \cong \overline{BC}$

   Prove: $\triangle AND \cong \triangle BNC$

7. **Given:** *W* is the midpoint of $\overline{XZ}$.
$\triangle XYZ$ is isosceles and right.
$\overline{YW} \perp \overline{XZ}$
$\angle X \cong \angle Z$

   Prove: $\triangle WXY \cong \triangle WZY$

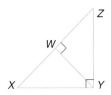

8. Draw two triangles with the same perimeter that are not congruent.

EXERCISES

Practice

9. The corresponding parts of the two triangles are congruent as marked in the figure. This can be written as $\triangle ABC \cong$ __?__ .

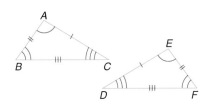

Draw triangles *KAT* and *BRO*. Label the corresponding parts for △*KAT* ≅ △*BRO*. Use the figures to complete each statement.

10. ∠*A* ≅ ___?___

11. ∠*B* ≅ ___?___

12. $\overline{OR}$ ≅ ___?___

13. $\overline{KA}$ ≅ ___?___

Write a congruence statement for the congruent triangles in each diagram.

14.  **15.** 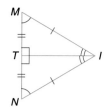 **16.**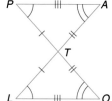

Explain why the following pairs of triangles are not congruent.

17. 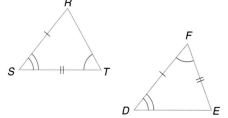 **18.**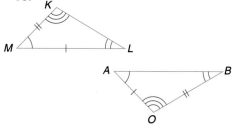

Complete each congruence statement.

19. △*CDB* ≅ ___?___

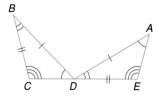

20. △*PTL* ≅ ___?___

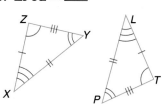

21. Given the following, identify the congruent triangles in the figure.

$\overline{AR} \perp \overline{RB}$

$\overline{AS} \perp \overline{SB}$

$\overline{AS} \cong \overline{SB}$

$\overline{AR} \cong \overline{AS}$

$\overline{RB} \cong \overline{SB}$

∠*RAB* ≅ ∠*SAB*

∠*RBA* ≅ ∠*SBA*

$\overline{AB}$ is the perpendicular bisector of $\overline{RS}$.

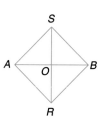

22. Given $\triangle CAT \cong \triangle DOG$, $CA = 14$, $AT = 18$, $TC = 21$, and $DG = 2x + 7$, find the value of x.

23. Given $\triangle BLU \cong \triangle RED$, $m\angle L = 57$, $m\angle R = 64$, and $m\angle U = 5x + 4$, find the value of x.

24. Justify each step in the proof of the symmetric part of Theorem 4-4.

Given: $\triangle LMN \cong \triangle OPQ$

Prove: $\triangle OPQ \cong \triangle LMN$

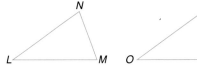

Statements	Reasons
a. $\triangle LMN \cong \triangle OPQ$	**a.** $\underline{\ ?\ }$
b. $\angle L \cong \angle O$ $\angle M \cong \angle P$ $\angle N \cong \angle Q$ $\overline{LM} \cong \overline{OP}$ $\overline{MN} \cong \overline{PQ}$ $\overline{LN} \cong \overline{OQ}$	**b.** $\underline{\ ?\ }$
c. $\angle O \cong \angle L$ $\angle P \cong \angle M$ $\angle Q \cong \angle N$	**c.** $\underline{\ ?\ }$
d. $\overline{OP} \cong \overline{LM}$ $\overline{PQ} \cong \overline{MN}$ $\overline{OQ} \cong \overline{LN}$	**d.** $\underline{\ ?\ }$
e. $\triangle OPQ \cong \triangle LMN$	**e.** $\underline{\ ?\ }$

If $\triangle BCD \cong \triangle ECA$, determine if each of the statements is *true* or *not necessarily true*. Explain your answers.

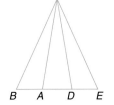

25. $\overline{BD} \cong \overline{AE}$

26. $\overline{AB} \cong \overline{DE}$

27. $\overline{BC} \cong \overline{AC}$

28. $\overline{AC} \cong \overline{DC}$

29. $\angle CBA \cong \angle CED$

30. $\angle BCA \cong \angle ACE$

31. Draw two triangles that have equal areas and are congruent.

32. Draw two triangles that have equal perimeters and are congruent.

33. Draw two triangles that have equal areas but are not congruent.

34. Given: $\overline{AB} \parallel \overline{RT}$
$\overline{AR} \perp \overline{AB}$
$\overline{BT} \perp \overline{RT}$
$\overline{AB} \cong \overline{RT}$
$\overline{AR} \cong \overline{TB}$

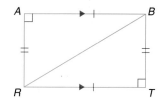

Prove: $\triangle ABR \cong \triangle TRB$

35. Prove that congruence of triangles is reflexive. (Theorem 4-4)

36. $\triangle ABC$ is equilateral and equiangular. $\triangle TUV$ has three congruent sides, $m\angle T = 60$, and $m\angle U = 60$. Are the two triangles congruent? Why or why not?

37. If $\triangle MNO \cong \triangle ONM$, prove $\triangle MNO$ is isosceles.

38. If $\triangle RST \cong \triangle TSR$ and $\triangle RST \cong \triangle RTS$, prove that $\triangle RST$ is equilateral and equiangular.

Critical Thinking

39. Draw two triangles that have five pairs of congruent parts, but are not congruent.

Applications

40. Describe the congruence involved in each of the following situations.
  a. Crafts   a pottery maker making a set of dishes
  b. Arts   a quilt maker cutting pieces for a quilt
  c. Entertainment   a Pittsburgh Pirates pennant
  d. Finance   a five-dollar bill and a one-dollar bill

FYI ···

Robert Mangold's interest in geometric art began while he was working as a guard at the Museum of Modern Art in New York City.

41. Art   Draw a sketch of the painting by Robert Mangold shown at the right. Then label the intersections of lines and name the triangles that appear to be congruent.

Mixed Review

42. The measures of two interior angles of a triangle are 54 and 79. What is the measure of the exterior angle opposite these angles? **(Lesson 4-2)**

43. Algebra   The measures of the angles of a triangle are $8x + 1$, $3x - 6$, and $4x - 10$. What are the measures of the angles? **(Lesson 4-2)**

44. Is the statement "A right triangle can be scalene." *true* or *false?* Explain. **(Lesson 4-1)**

45. Find the slope of the line passing through the points with coordinates $(7, -3)$ and $(6, -1)$. **(Lesson 3-5)**

46. Angles L and S are vertical. If $m\angle L = 3x + 7$ and $m\angle S = 43$, find the value of x. **(Lesson 1-9)**

Wrap-Up

47. Make up a five-question quiz about this section. Be sure to give the answers to your questions.

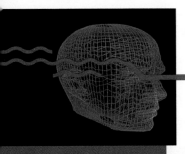

Technology

Congruent Triangles

BASIC
Geometric Supposer
▶ **Graphing calculators**
LOGO
Spreadsheets

A graphing calculator can be used to plot points and draw the line segments that connect them. We will use this feature to graph triangles.

Example

Graph △ABC whose vertices are A(0, 2), B(6, 2) and C(5, 4) and △A'B'C' whose vertices are A'(7, 5), B'(13, 5), and C'(12, 7). Do the two triangles appear to be congruent?

To draw a triangle on the Casio fx-7000G, plot each point and the line segment between each pair of points.

Enter: [SHIFT] [PLOT] 0 [SHIFT] , 2 [EXE] [SHIFT] [PLOT] 6
[SHIFT] , 2 [EXE] [SHIFT] [LINE] [EXE] [SHIFT] [PLOT] 5
[SHIFT] , 4 [EXE] [SHIFT] [LINE] [EXE] [SHIFT] [PLOT] 0
[SHIFT] , 2 [EXE] [SHIFT] [LINE] [EXE]

To draw a triangle on the TI-81, we will begin by using the line feature in the draw menu. Given two points, a line will be drawn between them.

Enter: [2nd] [DRAW] 2 0 [ALPHA] , 2 [ALPHA] , 6 [ALPHA] ,
2 [ENTER]

Now complete the triangle. Press [2nd] [DRAW] 2 and use the arrow keys to move the cursor to one endpoint of the segment. Press [ENTER]. Then move the cursor to the third vertex. Watch the coordinates given at the bottom of the screen to approximate the position of the point. Press [ENTER] twice. Finally, move the cursor to the other endpoint of the segment and press [ENTER] to complete the triangle.

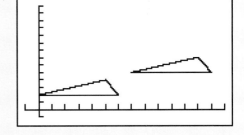

Repeat the steps to draw the second triangle. It appears that △ABC ≅ △A'B'C'.

EXERCISES

Graph △ABC and △A'B'C'. Do the two triangles appear to be congruent?

1. A(-8, 1), B(-8, -6), C(-5, -6); A'(3, 5), B'(3, -2), C'(6, 5)
2. A(-3, 8), B(-3, 0), C(-5, 4); A'(4, -3), B'(4, -10), C'(2, -6)

4-4   Tests for Congruent Triangles

Objective   After studying this lesson, you should be able to:

■ use SAS, SSS and ASA postulates to test for triangle congruence.

Application   Is it always necessary to show that all of the corresponding parts of two triangles are congruent to be sure that the two triangles are congruent? For example, if you are designing supports for the beams in a roof, must you measure all three sides and all three angles to ensure that the supports are all identical?

Suppose you are given that the lengths of the sides of a triangular support are 3 meters, 5 meters, and 6 meters. How many different braces could you make?

One way to think about this is to construct a model of the triangle with the given dimensions. How many different triangles could you construct? Let's construct a triangle with sides of lengths 3 centimeters, 5 centimeters, and 6 centimeters by following the steps below.

1. On any line ℓ, select a point A.
2. Construct $\overline{AB}$ on ℓ such that $AB = 6$ cm.
3. Using A as the center, draw an arc with radius 5 cm.
4. Using B as the center, draw an arc with radius 3 cm.
5. Let C be the point of intersection of the two arcs.
6. Draw $\overline{AC}$ and $\overline{BC}$.

The figures shown below are not actual size.

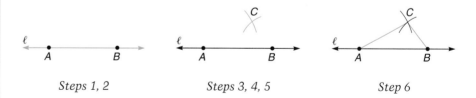

| Steps 1, 2 | Steps 3, 4, 5 | Step 6 |

Try using the same procedure with $AB = 5$ and then with $AB = 3$, and compare all of the triangles. How many different triangles is it possible to construct with sides of the given measures? It seems as if knowing that three sides are congruent is sufficient to guarantee the triangles are congruent. The side-side-side, or SSS, postulate states this fact.

SSS Postulate
Side-Side-Side

If the sides of one triangle are congruent to the sides of a second triangle, then the triangles are congruent.

---

184 CHAPTER 4 CONGRUENT TRIANGLES

The SSS postulate can be used to prove triangles congruent.

Example 1

Given $\triangle ABC$ **with vertices** $A(0, 5)$, $B(2, 0)$, **and** $C(0, 0)$ **and** $\triangle RST$ **with vertices** $R(5, 8)$, $S(5, 3)$, **and** $T(3, 3)$, **show that** $\triangle ACB \cong \triangle RST$.

Use the distance formula to show that the corresponding sides are congruent.

$$AC = \sqrt{(0 - 0)^2 + (5 - 0)^2}$$
$$= \sqrt{25} \text{ or } 5$$

$$RS = \sqrt{(5 - 5)^2 + (8 - 3)^2}$$
$$= \sqrt{25} \text{ or } 5$$

$$AB = \sqrt{(0 - 2)^2 + (5 - 0)^2}$$
$$= \sqrt{29}$$

$$RT = \sqrt{(5 - 3)^2 + (8 - 3)^2}$$
$$= \sqrt{29}$$

$$CB = \sqrt{(0 - 2)^2 + (0 - 0)^2}$$
$$= \sqrt{4} \text{ or } 2$$

$$ST = \sqrt{(5 - 3)^2 + (3 - 3)^2}$$
$$= \sqrt{4} \text{ or } 2$$

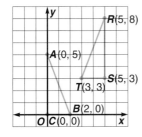

All the pairs of corresponding sides are congruent, so $\triangle ACB \cong \triangle RST$ by SSS.

Will any other combinations of corresponding and congruent sides and angles determine a unique triangle? Suppose you were given the measures of two sides and the angle that they form, which is called the **included angle**. How many different triangles would you be able to make?

Again, we can investigate by constructing a triangle, given an angle A **and sides of lengths** $\overline{AB}$ **and** $\overline{AC}$. **Let's use** $m\angle A = 60$, $AB = 4$ **cm, and** $AC = 6$ **cm.**

1. Use a protractor to draw a 60° angle so that one side of the angle is on line ℓ. Label the vertex A.
2. Using A as the center, draw an arc with radius 4 cm. Label the point of intersection with the ray B.
3. Using A as the center, draw an arc with radius 6 cm. Label the point of intersection with ℓ C.
4. Draw segment BC.

The figures shown are not actual size.

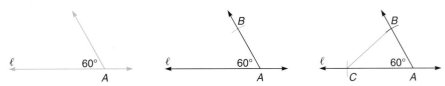

Now try to construct different triangles with the given dimensions. Are they congruent to $\triangle ABC$?

The investigation leads to the following postulate.

SAS Postulate *Side-Angle-Side*	**If two sides and the included angle of one triangle are congruent to two sides and an included angle of another triangle, then the triangles are congruent.**

The following proof uses the SAS postulate.

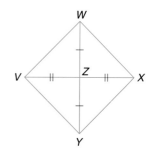

Example 2

Write a two-column proof.

Given: $\overline{WZ} \cong \overline{YZ}$
$\overline{VZ} \cong \overline{ZX}$

Prove: $\triangle VZW \cong \triangle XZY$

Proof:

Statements	Reasons
1. $\overline{WZ} \cong \overline{YZ}$ $\overline{VZ} \cong \overline{ZX}$	1. Given *(side)* *(side)*
2. $\angle WZV \cong \angle YZX$	2. Vertical ⩬ are ≅. *(included angle)*
3. $\triangle VZW \cong \triangle XZY$	3. SAS

If a triangle is constructed using two given angles and the included side, the triangle will be unique. This suggests a third postulate to determine congruent triangles.

ASA Postulate *Angle-Side-Angle*	**If two angles and the included side of one triangle are congruent to two angles and the included side of another triangle, the triangles are congruent.**

Some proofs ask you to show that a pair of corresponding parts of two triangles are congruent. Often you can do this by first proving that the two triangles are congruent. Then use the definition of congruent triangles to show that the corresponding parts are congruent.

Example 3

Write a two-column proof.

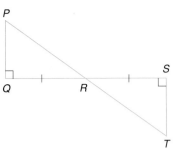

Given: $\angle Q$ and $\angle S$ are right angles.
$\overline{QR} \cong \overline{SR}$

Prove: $\angle P \cong \angle T$

Proof:

Statements	Reasons
1. $\overline{QR} \cong \overline{SR}$	1. Given *(included side)*
2. $\angle PRQ \cong \angle TRS$	2. Vertical ⩳ are ≅. *(angle)*
3. $\angle Q$ and $\angle S$ are right angles.	3. Given
4. $\angle Q \cong \angle S$	4. All rt. ⩳ are ≅. *(angle)*
5. $\triangle PRQ \cong \triangle TRS$	5. ASA
6. $\angle P \cong \angle T$	6. CPCTC

CHECKING FOR UNDERSTANDING

Communicating Mathematics

Read and study the lesson to answer these questions.

1. Refer to $\triangle ALM$ and $\triangle PRT$ at the right.

   a. Name one additional pair of corresponding parts that need to be congruent in order to prove that $\triangle ALM \cong \triangle PTR$.

   b. What postulate would you use to prove the triangles are congruent?

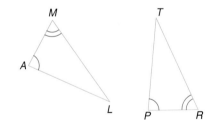

2. Refer to $\triangle TUW$ and $\triangle QOS$ at the right.

   a. Name one additional pair of corresponding parts that need to be congruent in order to prove that $\triangle TUW \cong \triangle QOS$ by SAS.

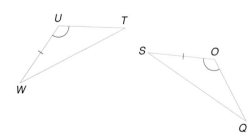

   b. Name one additional pair of corresponding parts that need to be congruent in order to prove that $\triangle TUW \cong \triangle QOS$ by ASA.

3. Will two triangles be congruent if only two pairs of corresponding parts are congruent? Explain your reasoning.

Determine whether each pair of triangles are congruent. If they are congruent, indicate the postulate that can be used to prove their congruence.

4.

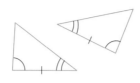

5.

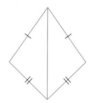

6.

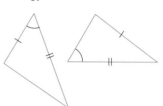

7. Justify each step in the proof.

Given: $\overline{AM} \parallel \overline{CR}$

B is the midpoint of $\overline{AR}$.

Prove: $\triangle ABM \cong \triangle RBC$

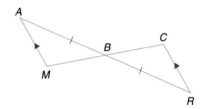

Proof:

Statements	Reasons
a. B is the midpoint of $\overline{AR}$.	**a.** ___?___
b. $\overline{AB} \cong \overline{BR}$	**b.** ___?___
c. $\overline{AM} \parallel \overline{CR}$	**c.** ___?___
d. $\angle A \cong \angle R$	**d.** ___?___
e. $\angle ABM \cong \angle RBC$	**e.** ___?___
f. $\triangle ABM \cong \triangle RBC$	**f.** ___?___

Copy each figure and mark all congruent parts. Indicate the postulate that can be used to prove their congruence.

8. $\overline{RL} \perp \overline{ST}$

$\overline{ST}$ bisects $\angle RSL$

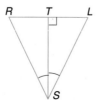

9. $\overline{AD} \parallel \overline{GR}$

$\overline{AD} \cong \overline{GR}$

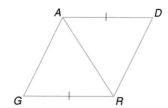

10. $\overline{AB} \cong \overline{BC}$

$\overline{AD} \cong \overline{CD}$

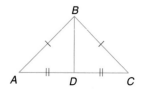

EXERCISES

Practice

Given figure *ABCD* with $\overline{DB} \perp \overline{DC}$ and $\overline{DB} \perp \overline{AB}$, indicate whether each statement is *true, could be true,* or appears to be *false.*

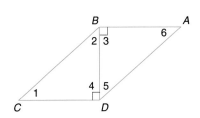

11. $\overline{DC} \parallel \overline{AB}$

12. ∠4 and ∠5 are a linear pair.

13. $\overline{AD} \cong \overline{BC}$

14. $\overline{AD} \cong \overline{AD}$

15. $\triangle ABD \cong \triangle CDB$

16. Triangle *DCB* is isosceles.

17. ∠1 ≅ ∠2

18. ∠2 ≅ ∠3

Determine which postulate can be used to prove the triangles congruent. If it is not possible to prove them congruent, write *not possible.*

19.

20.

21.

Copy each figure and mark all congruent parts. Then complete the prove statement and identify the postulate that can be used to prove the triangles congruent.

22.

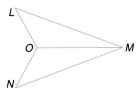

Given: ∠*LMO* ≅ ∠*NMO*
∠*LOM* ≅ ∠*NOM*

Prove: $\triangle MOL \cong$ ___?___

23.

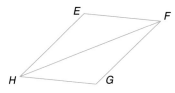

Given: $\overline{EF} \cong \overline{GH}$
$\overline{EH} \cong \overline{GF}$

Prove: $\triangle EFH \cong$ ___?___

24. *ABCD* is a quadrilateral, $\overline{BD}$ is a diagonal, $\overline{AD} \cong \overline{BC}$ and $\overline{AB} \cong \overline{DC}$.

    a. What conclusions can be drawn about $\triangle ABD$ and $\triangle CDB$? Explain.

    b. What conclusions can be drawn about ∠*A* and ∠*C*? Explain.

Write a two-column proof.

25. Given: $\angle A \cong \angle D$
$\overline{AO} \cong \overline{OD}$

Prove: $\triangle AOB \cong \triangle DOC$

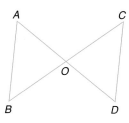

26. Given: $\overline{AD}$ bisects $\overline{BC}$.
$\overline{BC}$ bisects $\overline{AD}$.

Prove: $\triangle AOB \cong \triangle DOC$

27. Given: $\overline{MO} \cong \overline{PO}$
$\overline{NO}$ bisects $\overline{MP}$.

Prove: $\triangle MNO \cong \triangle PNO$

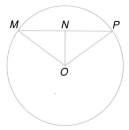

28. Given: $\overline{NO}$ bisects $\angle POM$.
$\overline{NO} \perp \overline{MP}$

Prove: $\triangle MNO \cong \triangle PNO$

29. Graph $\triangle ABC$ with vertices $A(-3, 1)$, $B(-8, 5)$, $C(-1, 8)$ and $\triangle DEF$ with vertices $D(0, 1)$, $E(4, 6)$ and $F(7, -1)$. Use the distance formula to show that $\triangle ABC \cong \triangle DEF$.

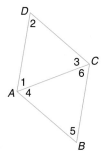

30. Given: $\angle 1 \cong \angle 6$
$\angle 3 \cong \angle 4$

Prove: $\overline{AD} \cong \overline{CB}$

31. Given: $\angle 3 \cong \angle 4$
$\overline{DC} \cong \overline{BA}$

Prove: $\angle 1 \cong \angle 6$

Critical Thinking

32. In the figure $\triangle ACD$, $\overline{DB} \cong \overline{DC}$. List all of the congruent parts of $\triangle ADB$ and $\triangle ADC$. Is $\triangle ADB \cong \triangle ADC$? What does this tell you about SSA relation between corresponding parts of two triangles?

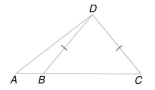

Applications

33. Construction   A truss is a triangular-based supportive frame used in construction. One pattern for a roof truss is shown at the right. The vertices of the triangles have been labeled. Name the triangle that appears to be congruent to each of the following triangles.

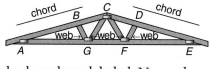

a. $\triangle BCG$ **b.** $\triangle AGC$ **c.** $\triangle ABG$

d. $\triangle GCF$ **e.** $\triangle ACF$ **f.** $\triangle CFE$

34. Suppose $\triangle BIG \cong \triangle TOP$. List all the pairs of corresponding parts that are congruent because corresponding parts of congruent triangles are congruent. **(Lesson 4-3)**

35. The legs of an isosceles triangle are $6x - 6$ units and $x + 9$ units long. Find the length of the legs. **(Lesson 4-1)**

36. State the hypothesis and the conclusion of the statement "If you want a great pizza go to Katie's." **(Lesson 2-2)**

37. $\overrightarrow{QT}$ and $\overrightarrow{QS}$ are opposite rays. Describe $\angle TQS$. **(Lesson 1-6)**

Wrap-Up   **38.** Write a summary of the ways to prove triangles congruent that you learned in this lesson.

~~~~~ MID-CHAPTER REVIEW ~~~~~

Determine whether each following statement is *true* or *false*. If it is false, draw a counterexample. (Lesson 4-1)

 1. A right triangle is equilateral.

 2. A scalene triangle can be a right triangle.

 3. An obtuse triangle can be isosceles.

 4. An isosceles triangle can be equilateral.

Decide whether it is possible to have a triangle with the following characteristics. (Lesson 4-2)

 5. two acute angles and one right angle

 6. no acute angles

 7. two congruent angles

 8. three noncongruent angles

Draw triangles $\triangle HAT$ and $\triangle TOP$. Label the corresponding parts if $\triangle HAT \cong \triangle TOP$. Use the figures to complete each statement. (Lesson 4-3)

 9. $\angle A \cong$ ___?___

 10. $\overline{HT} \cong$ ___?___

 11. $\angle P \cong$ ___?___

Write a two-column proof. (Lesson 4-4)

12. Given: $\overline{QP} \cong \overline{ST}$
   $\angle P$ and $\angle T$ are right angles.
   R is the midpoint of $\overline{PT}$.

   Prove: $\overline{QR} \cong \overline{SR}$

4-5 Another Test for Congruent Triangles

Objective

After studying this lesson, you should be able to:
- use AAS theorem to test for triangle congruence.

Application

FYI···

Quilting became popular in the United States in colonial days because cloth was scarce. In the 1800s when cloth became readily available, quilting grew to be an art form that is still popular today.

Karen is making a quilt that is to be constructed from congruent triangles. How few pairs of corresponding parts of two triangular pieces could she check and still be sure that the pieces are congruent? From the previous lesson, we know that three given sides determine a unique triangle, as well as two sides and an included angle, or two angles and the included side. Karen could also be assured that all of her triangles were congruent if one side and any two pairs of corresponding angles were congruent. This can be proved as a theorem.

Theorem 4-5 AAS Angle-Angle-Side	**If two angles and a nonincluded side of one triangle are congruent to the corresponding two angles and side of a second triangle, the two triangles are congruent.**

Proof of Theorem 4-5

Given: $\angle P \cong \angle A$
$\angle Q \cong \angle B$
$\overline{QR} \cong \overline{BC}$

Prove: $\triangle PQR \cong \triangle ABC$

Proof:

Statements	Reasons
1. $\angle P \cong \angle A$ $\angle Q \cong \angle B$ $\overline{QR} \cong \overline{BC}$	1. Given *angle* *side*
2. $\angle R \cong \angle C$	2. If 2 △s in a △ are ≅ to 2 △s in another △ the third △s are ≅ also. *angle*
3. $\triangle PQR \cong \triangle ABC$	3. *ASA*

Could Karen be assured that two of her triangular quilt pieces were congruent if she determined that two pairs of corresponding sides and a pair of corresponding nonincluded angles were congruent (SSA)? Remember that a *counterexample* is enough to show that a statement is not true. The figures at the right show two triangles that meet the conditions for SSA, but clearly are not congruent. Therefore, SSA is not a test for congruence.

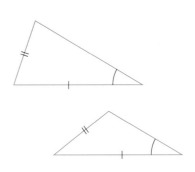

Example 1

Write a paragraph proof.

Given: ∠Q and ∠S are right angles.
∠1 ≅ ∠3

Prove: $\overline{QP} \cong \overline{SR}$

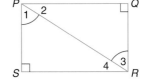

Plan:

$\overline{QP}$ and $\overline{SR}$ are in △PQR and △RSP, respectively. If those two triangles are congruent, the two sides will be congruent by CPCTC.

Paragraph Proof:
We are given that ∠1 ≅ ∠3. *angle* It is also given that ∠Q and ∠S are right angles. Since all right angles are congruent, ∠Q ≅ ∠S. *angle* $\overline{PR} \cong \overline{PR}$ since congruence of segments is reflexive. *side* Therefore, △PQR ≅ △RSP by AAS. So, $\overline{QP} \cong \overline{SR}$ by the definition of congruent triangles (CPCTC).

Sometimes the triangles we would like to prove congruent are overlapping. Then it is helpful to draw the two figures separately or to use different colors to distinguish the parts of each triangle.

Example 2

Write a two-column proof to verify the statement *Segments from the vertices of the base angles of an isosceles triangle to the midpoints of the opposite sides are congruent.*

Given: $\overline{AC} \cong \overline{AB}$
D is the midpoint of $\overline{AC}$.
E is the midpoint of $\overline{AB}$.

Prove: $\overline{DB} \cong \overline{EC}$

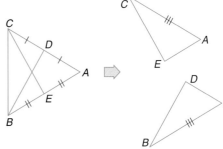

Plan:

$\overline{DB}$ is a side of △DBA, and $\overline{EC}$ is a side of △ECA. Prove these two triangles congruent and then use CPCTC to prove that $\overline{DB}$ is congruent to $\overline{EC}$.

Proof:

Statements	Reasons
1. $\overline{AC} \cong \overline{AB}$	1. Given *side*
2. $AC = AB$	2. Definition of congruence
3. $AC = AD + DC$ $AB = AE + EB$	3. Segment addition postulate
4. $AD + DC = AE + EB$	4. Substitution property of equality
5. D is the midpoint of $\overline{AC}$. E is the midpoint of $\overline{AB}$.	5. Given
6. $AD = DC$ $AE = EB$	6. Definition of midpoint
7. $2AD = 2AE$	7. Substitution property of equality
8. $AD = AE$	8. Division property of equality
9. $\overline{AD} \cong \overline{AE}$	9. Definition of congruent segments *side*
10. $\angle A \cong \angle A$	10. Congruence of angles is reflexive. *angle*
11. $\triangle DBA \cong \triangle ECA$	11. SAS
12. $\overline{DB} \cong \overline{EC}$	12. CPCTC

Therefore, the segments from the base angles of an isosceles triangle to the midpoints of the opposite sides are congruent.

CHECKING FOR UNDERSTANDING

Communicating Mathematics

Read and study the lesson to answer these questions.

1. Make a list of the ways to prove triangles congruent. How are these ways the same? How are they different?

2. Decide which pair(s) of the following triangles is congruent. Justify your answer.

   a. **b.** **c.** **d.** **e.** **f.**

   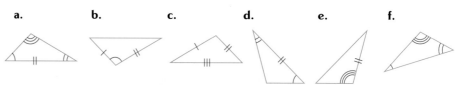

3. List three ways to show that two angles are congruent.

4. In $\triangle DOA$, name the sides that include $\angle 10$, $\angle 11$, and $\angle 12$.

5. In $\triangle DOC$, name the angles that include $\overline{DC}$, $\overline{DO}$, and $\overline{CO}$.

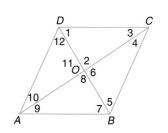

Indicate the additional pairs of corresponding parts that would have to be proved congruent in order to use the given postulate or theorem to prove the triangles congruent.

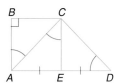

6. $\triangle ACE \cong \triangle DCE$ by SSS

7. $\triangle ACB \cong \triangle CAE$ by SAS

8. $\triangle ACB \cong \triangle CAE$ by AAS

9. $\triangle ACB \cong \triangle CAE$ by ASA

10. $\triangle ACB \cong \triangle DCE$ by AAS

Prove each conclusion if possible. If not possible, state the additional information that could be given to prove the conclusion.

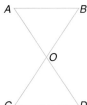

11. $\overline{AB} \cong \overline{CD}$
$\overline{AB} \parallel \overline{CD}$
Therefore, $\triangle AOB \cong \triangle DOC$.

12. $\overline{AB} \parallel \overline{CD}$
Therefore, $\triangle AOB \cong \triangle DOC$.

Determine which triangles in *ABCD* are congruent under the given conditions. Justify your answers.

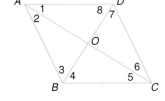

13. $\angle 8 \cong \angle 4$
$\overline{AD} \cong \overline{BC}$

14. $\overline{AB} \parallel \overline{DC}$
$\overline{AB} \cong \overline{DC}$

15. O is the midpoint of both $\overline{DB}$ and $\overline{AC}$.

16. If $\overline{PR}$ and $\overline{QS}$ bisect each other, write a paragraph proof to show that $\triangle PQT \cong \triangle RST$.

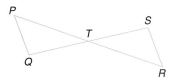

EXERCISES

Draw and label triangles *DEF* and *RST*. Indicate the additional pairs of corresponding parts that would have to be proved congruent in order to use the given postulate or theorem to prove the triangles congruent.

17. $\angle D \cong \angle R$ and $\overline{DE} \cong \overline{RS}$ by ASA

18. $\angle E \cong \angle S$ and $\overline{EF} \cong \overline{ST}$ by AAS

19. $\angle F \cong \angle T$ and $\angle D \cong \angle R$ by AAS

20. $\angle E \cong \angle S$ and $\angle F \cong \angle T$ by ASA

Determine which triangles are congruent under the given conditions. Justify your answers.

21.

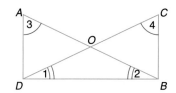

22.

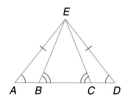

23. The steps in the following proof are *not* in logical order. Rearrange them in a correct sequence and give the reasons.

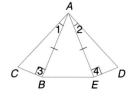

Given: $\overline{AB} \perp \overline{BC}$, $\overline{AE} \perp \overline{DE}$,
$\angle 1 \cong \angle 2$, $\overline{AB} \cong \overline{AE}$

Prove: $\overline{AC} \cong \overline{AD}$

Statements	Reasons
1. $\angle 3$ is a right angle.	**1.** __?__
2. $\angle 1 \cong \angle 2$	**2.** __?__
3. $\overline{AB} \perp \overline{BC}$	**3.** __?__
4. $\overline{AC} \cong \overline{AD}$	**4.** __?__
5. $\overline{AE} \perp \overline{DE}$	**5.** __?__
6. $\triangle ABC \cong \triangle AED$	**6.** __?__
7. $\overline{AB} \cong \overline{AE}$	**7.** __?__
8. $\angle 3 \cong \angle 4$	**8.** __?__
9. $\angle 4$ is a right angle.	**9.** __?__

Determine if each conclusion is valid based on the information given in the figure. Justify your answer.

24.

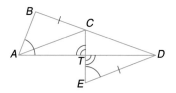

$\triangle ABC \cong \triangle DTE$

25.

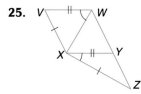

$\triangle VWX \cong \triangle YXZ$

Write a paragraph proof for each.

26. Given: $\angle 1 \cong \angle 2$
$\angle L \cong \angle M$

Prove: $\overline{EM} \cong \overline{AL}$

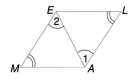

27. Given: $\overline{ST} \cong \overline{QN}$
$\angle S \cong \angle Q$
$\overline{PS} \cong \overline{PQ}$

Prove: $\triangle TPN$ is isosceles.

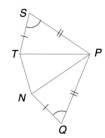

Write a two-column proof for each.

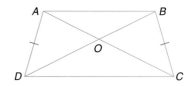

28. Given: $\overline{BC} \cong \overline{AD}$
$\overline{BD} \cong \overline{AC}$
Prove: $\angle BAC \cong \angle ABD$

29. Prove that a segment drawn from the vertex angle of an isosceles triangle through the midpoint of the opposite side divides the triangle into two congruent triangles.

30. Prove that the bisector of the vertex angle of an isosceles triangle is perpendicular to the base of the triangle.

Critical Thinking

31. Monty says that two triangles can be proved congruent by using AAA (angle-angle-angle). Missy disagrees. Who do you agree with? Justify your answer completely.

Applications

32. **History** It is said that Thales determined the distance from the shore to enemy Greek ships during an early war by sighting the angle to the ship from a point P on the shore, walking a distance to point Q, and then sighting the angle to the ship from that point. He then reproduced the angles on the other side of line PQ and continued these lines until they intersected.

a. How did he determine the distance to the ship in this way?

b. Why does it work?

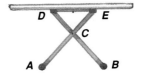

33. **Home Economics** Explain why the plane of an ironing board is parallel to the plane of the floor if the legs of the board bisect each other.

Mixed Review

34. Write a two-column proof. **(Lesson 4-4)**
Given: $\overline{PR} \cong \overline{TR}$
$\angle 1 \cong \angle 2$
$\angle P$ and $\angle T$ are right angles.
Prove: $\overline{QR} \cong \overline{SR}$

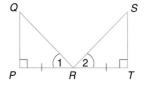

35. If $\triangle CAK \cong \triangle PIE$, which angle in $\triangle CAK$ is congruent to $\angle IPE$ in $\triangle PIE$? **(Lesson 4-3)**

36. Name five ways to prove that two lines are parallel. **(Lesson 3-4)**

37. **Algebra** Find the length and the midpoint of the segment with endpoints having coordinates (7, 4) and (-3, 8). **(Lesson 1-4)**

Wrap-Up

38. Write a paragraph to describe the difference between the AAS theorem and ASA postulate. Give examples to show how each one is used.

4-6

Problem-Solving Strategy: Identify Subgoals

Objective

After studying this lesson, you should be able to:

■ solve problems by identifying and achieving subgoals.

Solving a problem is like walking a mile; it takes many small steps. Being able to identify the steps, or subgoals, that need to be accomplished to solve a problem makes solving the problem simpler.

Example 1

What fraction of the five-digit whole numbers are divisible by 1, 2, 3, 4, and 5?

We could write out all of the five-digit numbers, determine which ones are divisible by 1, 2, 3, 4, and 5, and then find the fraction of the five-digit numbers that these numbers represent. But, that approach would be quite time consuming.

First subgoal   To solve the problem efficiently, identify what kind of numbers have 1, 2, 3, 4, and 5 as factors. Numbers that have the least common multiple of 1, 2, 3, 4, and 5 as a factor also have 1, 2, 3, 4, and 5 as factors. Since the least common multiple of 1, 2, 3, 4, and 5 is 60, the five-digit numbers that have 60 as a factor are the ones we wish to identify.

Second subgoal   Use a calculator to find how many five-digit whole numbers are divisible by 60.

The numbers are:

10,020,	10,080,	10,140,	...,	99,960.
60×167,	60×168,	60×169,	...,	60×1666

So, there is a multiple of 60 for each number from 167 to 1666 inclusive. Therefore, there are 1500 five-digit numbers that are divisible by 1, 2, 3, 4, and 5.

Final subgoal   Find the fraction of the five-digit numbers that are divisible by 1, 2, 3, 4, and 5. Since there are 90,000 five-digit numbers, the fraction is $\frac{1500}{90,000}$ or $\frac{1}{60}$.

In geometric proofs, we often have to take several steps to complete the proof. Setting subgoals can be helpful in taking those steps.

Example 2

Write a two-column proof.

Given: $\overline{KT} \perp \overline{IE}$
$\angle IKS \cong \angle ITS$

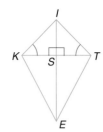

Prove: $\angle KEI \cong \angle TEI$

Plan:

Let's set some subgoals before we start the proof. Reasoning backward, we can show that $\angle KEI \cong \angle TEI$ if $\triangle KEI \cong \triangle TEI$. To prove these triangles congruent, we need to show that $\angle KIE \cong \angle TIE$ and $\overline{KI} \cong \overline{TI}$. This can be proved by showing that they are corresponding parts of congruent triangles *IKS* and *ITS*. This can be proved from the given information.

Our subgoals are:

1. Prove $\triangle IKS \cong \triangle ITS$.
2. Use CPCTC to show that $\angle KIE \cong \angle TIE$ and $\overline{KI} \cong \overline{TI}$.
3. Prove $\triangle KEI \cong \triangle TEI$.
4. Use CPCTC to show that $\angle KEI \cong \angle TEI$.

Now, use the subgoals to write the proof.

Proof:

Statements	Reasons
1. $\overline{KT} \perp \overline{IE}$	1. Given
2. $\angle ISK$ and $\angle IST$ are right angles.	2. $\perp$ lines form four rt $\angle$s.
3. $\angle ISK \cong \angle IST$	3. All rt. $\angle$s are $\cong$.
4. $\angle IKS \cong \angle ITS$	4. Given
5. $\overline{IS} \cong \overline{IS}$	5. Congruence of segments is reflexive.
6. $\triangle IKS \cong \triangle ITS$	6. AAS
7. $\angle KIE \cong \angle TIE$ $\overline{KI} \cong \overline{TI}$	7. CPCTC
8. $\overline{IE} \cong \overline{IE}$	8. Congruence of segments is reflexive.
9. $\triangle KIE \cong \triangle TIE$	9. SAS
10. $\angle KEI \cong \angle TEI$	10. CPCTC

CHECKING FOR UNDERSTANDING

Communicating Mathematics

Read and study the lesson to answer these questions.

1. Why is it helpful to identify subgoals when trying to solve a problem?

2. When do you anticipate using the strategy of identifying subgoals in your study of geometry?

3. What fraction of the five-digit whole numbers are divisible by 1, 2, 3, 4, 5, 6, and 7?

Guided Practice

Solve. Use the strategy of identifying subgoals.

4. How many whole numbers less than 1000 have digits whose sum is 8?

5. Find the sum of the reciprocals of all the factors of 48.

6. Find the least positive integer having remainders of 2, 3, and 2 when divided by 3, 5, and 7 respectively.

7. Identify the subgoals you would need to accomplish to complete the proof.

Given: $\overline{AG} \cong \overline{ED}$
$\overline{BC} \cong \overline{FC}$
$\overline{BG} \perp \overline{AE}$
$\overline{FD} \perp \overline{AE}$

Prove: $\angle A \cong \angle E$

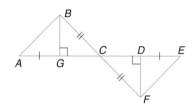

EXERCISES

Practice

Solve. Use any strategy.

8. An antique dealer bought an antique chair for $400 and sold it for $500 the following week. A month later, the dealer bought the chair back for $600 and then sold it to another dealer for $700. How much profit did the antique dealer make?

9. Find the least four-digit perfect square whose digits are all even.

10. James invited his friends Jill and Enrico over for a cookout. His small grill is only big enough to cook two hamburgers at a time. If each hamburger must cook for 5 minutes on each side, what is the least amount of time James could take to cook all three hamburgers?

11. Identify the subgoals you would need to accomplish to complete the proof. Then complete the proof.

Given: $\overleftrightarrow{AB} \perp$ plane BCD.
$\overline{DB} \cong \overline{CB}$

Prove: $\angle DAB \cong \angle CAB$

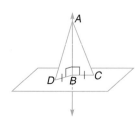

Strategies

Look for a pattern.
Solve a simpler problem.
Act it out.
Guess and check.
Draw a diagram.
Make a chart.
Work backward.

12. The date May 19, 1995, is represented by 5/19/95, and is called a *product date* since $5 \times 19 = 95$. Find all of the product dates in the 1990s.

13. All my pets are dogs except two, all are cats except two, and all are hamsters except two. How many pets do I have?

14. Delete or change some '+' symbols to '−' in the equation below to make it true. *(Hint: The spacing between numbers can be changed to make greater numbers.)*

$$1 + 2 + 3 + 4 + 5 + 6 + 7 + 8 + 9 = 100$$

15. In how many ways can you roll two dice to obtain a sum divisible by 3?

16. Five couples who had known each other for years were reminiscing about predictions they had made. Art, Naren, Will, Jared, and Anthony predicted who would marry Justine, Vivian, Cynthia, Kim, and Sarah. Art said that Will would marry Justine. Naren said that Anthony would marry Vivian. Will thought that Art would not marry Kim or Sarah and Anthony was sure that Naren would not marry Sarah.

    a. What marriages would have taken place if all the predictions were correct?

    b. It turned out that no one's prediction was right! One of the women is Anthony's sister. If we know who she is, we can tell who is really married to whom. What marriages did take place and who is Anthony's sister?

~~~ COOPERATIVE LEARNING PROJECT ~~~

Work in groups. Each person in the group must understand the solution and be able to explain it to any person in class.

When a billiard ball hits the cushion of a billiard table, the angle at which it rebounds has the same measure as the angle at which it hit the cushion. The billiard table shown at the right has sides that are in the ratio of 5 to 7. Copy the table and use a protractor and straightedge to draw the path of a billiard ball that is hit at a 45° angle from one corner of the table. The ball will rebound several times before ending up in one of the corners. How many times did the ball rebound before reaching a corner? Draw several other billiard tables with sides in different ratios and repeat the process. Write a formula for finding the number of rebounds on a billiard table with sides a and b units long where a and b are whole numbers with no common factors other than 1.

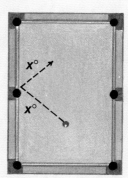

4-7 Isosceles Triangles

Objective

After studying this lesson, you should be able to:

- use properties of isosceles and equilateral triangles.

Application

Jerome Taylor has been contracted by the Harper family to draw the blueprints for their A-frame house. He must find the measurements for the angles formed where the roof meets the ground. Mr. Taylor knows that the two sides of the roof are to meet at an angle of 40° at the top of the house. Based on this information, at what angle should the roof meet the ground for this house? *You will solve this problem in Example 1.*

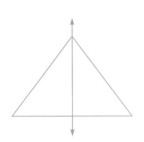

The front of an A-frame house is shaped like an isosceles triangle. Recall that an isosceles triangle has at least two sides congruent. Another important property of an isosceles triangle is that it has *line symmetry*. That is, if you fold an isosceles triangle along the bisector of its vertex angle, the two base angles match exactly. This observation leads us to the following theorem.

Theorem 4-6 Isosceles Triangle Theorem	If two sides of a triangle are congruent, then the angles opposite those sides are congruent.

Proof of Theorem 4-6

Given: $\triangle PQR$
$\overline{PQ} \cong \overline{RQ}$

Prove: $\angle P \cong \angle R$

Proof:

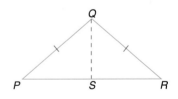

Statements	Reasons
1. Let S be the midpoint of $\overline{PR}$.	1. Every segment has exactly one midpoint.
2. Draw auxiliary segment QS.	2. Through any 2 pts. there is 1 line.
3. $\overline{PS} \cong \overline{RS}$	3. Definition of midpoint
4. $\overline{QS} \cong \overline{QS}$	4. Congruence of segments is reflexive.
5. $\overline{PQ} \cong \overline{RQ}$	5. Given
6. $\triangle PQS \cong \triangle RQS$	6. SSS
7. $\angle P \cong \angle R$	7. CPCTC

Example 1

Find the angle that the roof of the Harper's house should make with the ground.

The base of the isosceles triangle lies on the ground, and the two sides of the roof are the legs. So the angles formed by the roof and ground are base angles and must be congruent, by the Isosceles Triangle Theorem. Also, the vertex angle of this triangle measures 40°.

We can use the Angle Sum Theorem to determine the value of x. Let x represent the measure of each base angle.

$$x + x + 40 = 180$$
$$2x = 140$$
$$x = 70$$

The roof of the A-frame house should meet the ground at a 70° angle.

Since isosceles triangles have many applications to construction and other fields, it is important that we be able to find the measures of their angles.

Example 2

In isosceles triangle *RST*, ∠*R* is the vertex angle. If $m\angle S = 7x - 17$ and $m\angle T = 3x + 35$, find the measure of each angle of the triangle.

Since $\triangle RST$ is isosceles, the base angles are congruent.

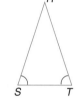

$$m\angle S = m\angle T$$
$$7x - 17 = 3x + 35 \qquad \textit{Substitution property of equality}$$
$$4x = 52$$
$$x = 13$$

$$m\angle S = 7x - 17 \qquad m\angle T = 3x + 35 \qquad \textit{Calculating both angle}$$
$$= 7(13) - 17 \qquad \quad = 3(13) + 35 \qquad \textit{measures verifies that}$$
$$= 74 \qquad \qquad \quad = 74 \qquad \qquad \textit{m}\angle S = m\angle T.$$

$$m\angle R = 180 - m\angle S - m\angle T$$
$$= 180 - 74 - 74 \text{ or } 32$$

Thus, the measures of ∠*R*, ∠*S*, and ∠*T* are 32, 74, and 74, respectively.

An auxiliary ray is used in the proof of the following theorem, which is the converse of the Isosceles Triangle Theorem.

Theorem 4-7	**If two angles of a triangle are congruent, then the sides opposite those angles are congruent.**

Example 3	**Write a plan for a two-column proof for Theorem 4-7.**

You will be asked to complete the proof in Exercise 39.

Draw an auxiliary ray that is the bisector of $\angle ABC$ and let D be the point where the bisector intersects $\overline{AC}$. Show that $\triangle ABD \cong \triangle CBD$ by AAS using the given information and the auxiliary ray. Then, $\overline{AB} \cong \overline{CB}$ by CPCTC.

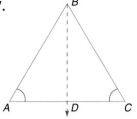

The Isosceles Triangle Theorem leads us to some interesting corollaries. You will be asked to prove these in Exercises 38 and 37 respectively.

Corollary 4-3	**A triangle is equilateral if and only if it is equiangular.**
Corollary 4-4	**Each angle of an equilateral triangle measures 60°.**

CHECKING FOR UNDERSTANDING

Communicating Mathematics

Read and study the lesson to answer these questions.

1. Describe the special properties of an isosceles triangle.
2. If two angles of $\triangle ABC$ are congruent, what kind of triangle is ABC?
3. Does an equilateral triangle have a line of symmetry? Explain.
4. Draw and label an isosceles triangle including the congruent parts. Identify the base, the legs, the vertex angle, and the base angles.

Guided Practice

For each exercise, draw $\triangle ABC$ with point D on $\overline{BC}$ such that the following conditions are satisfied.

5. $\overline{AD}$ bisects $\angle BAC$, but $\overline{BD}$ and $\overline{DC}$ are *not* congruent.
6. $\overline{AD} \perp \overline{BC}$, but $\overline{BD}$ and $\overline{DC}$ are *not* congruent.
7. $\overline{BD} \cong \overline{DC}$, but $\overline{AD}$ is *not* perpendicular to $\overline{BC}$.
8. $\overline{BD} \cong \overline{DC}$, but $\overline{AD}$ does *not* bisect $\angle BAC$.

9. If $\overline{XW}$ is the perpendicular bisector of $\overline{YZ}$ in $\triangle XYZ$ and W is on $\overline{YZ}$, what type of triangle is $\triangle XYZ$?
10. If $\overline{XW}$ bisects both $\angle YXZ$ and $\overline{YZ}$ in $\triangle XYZ$, must $\triangle XYZ$ be isosceles?

Given the following information, name two angles that must be congruent.

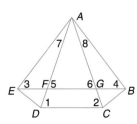

11. $\overline{EA} \cong \overline{EB}$

12. $\overline{AD} \cong \overline{AC}$

13. $\overline{AF} \cong \overline{AG}$

14. $\overline{FA} \cong \overline{FE}$

204   CHAPTER 4   CONGRUENT TRIANGLES

Find the value of *x*.

15.

16.

17.
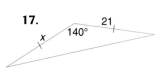

18. Graph $\triangle RST$ with vertices $R(4, 2)$, $S(8, 2)$, and $T(6, 6)$. Prove that $\triangle RST$ is isosceles.

19. Write a two-column proof.

    Given: $\overline{AB} \cong \overline{BC}$

    Prove: $\angle 3 \cong \angle 5$

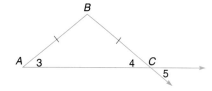

EXERCISES

Practice

Find the value of *x*.

20.

21.
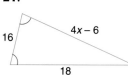

22.
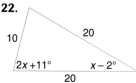

23.
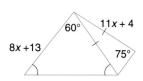

24.
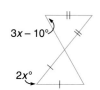

25.
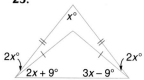

State the conclusions that can be drawn from the given information. Justify your answers.

26. $\angle 1$ and $\angle 3$ are complementary.
$\angle 2$ and $\angle 4$ are complementary.
$\angle 1 \cong \angle 2$

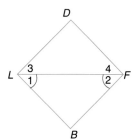

27. $\overline{TB} \cong \overline{TR}$
$\angle 1 \cong \angle 2$

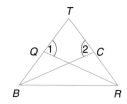

Write a two-column proof for each.

28. **Given:** $\overline{AB} \cong \overline{BC}$
Prove: $\angle 3 \cong \angle 4$

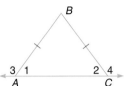

29. **Given:** $\overline{PS} \cong \overline{QR}$
$\angle 3 \cong \angle 4$
Prove: $\angle 1 \cong \angle 2$

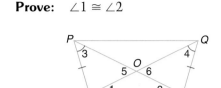

30. **Given:** $\overline{ZT} \cong \overline{ZR}$
$\overline{TX} \cong \overline{RY}$
Prove: $\angle 5 \cong \angle 7$

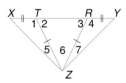

31. **Given:** $\triangle ABC$ is isosceles.
$\overline{DE} \parallel \overline{AB}$, $\overline{AC} \cong \overline{BC}$
Prove: $\triangle DEC$ is isosceles.

32. **Given:** $\angle 5 \cong \angle 6$
$\overline{GJ} \perp \overline{FH}$
Prove: $\triangle FHJ$ is isosceles.

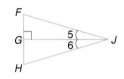

33. **Given:** $\overline{AB} \cong \overline{AC}$
$\overline{BX}$ bisects $\angle ABC$
$\overline{CX}$ bisects $\angle ACB$
Prove: $\overline{BX} \cong \overline{CX}$

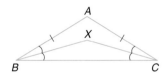

34. Find each angle measure if $m\angle 1 = 30$.

35. Find each angle measure if $m\angle 8 = 60$.

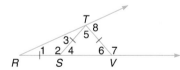

36. $\triangle RST$ is an isosceles triangle with two vertices $R(3, 8)$ and $S(-2, 1)$. If $\angle R$ is the vertex angle, could $(8, 1)$ be the coordinates of T? Explain.

Write a two-column proof for each.

37. Each angle of an equilateral triangle measures 60°. (Corollary 4-4)

38. A triangle is equilateral if and only if it is equiangular. (Corollary 4-3)
Hint: Since the statement contains if and only if, the proof must show that an equilateral triangle is equiangular and that an equiangular triangle is equilateral.

39. If two angles of a triangle are congruent, then the sides opposite those angles are congruent. (Theorem 4-7).

Critical Thinking   **40.** Draw an isosceles triangle. Then, draw line segments connecting the midpoints of the sides to form a new triangle. What type of triangle is formed? Explain.

Applications

41. Hang Gliding The sail of a certain type of hang glider consists of two congruent isosceles triangles joined along the keel so that a 90° angle is formed at the nose of the sail. In order to construct such a sail, what must be the measure of $\angle BCD$?

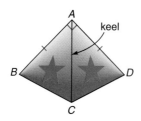
keel

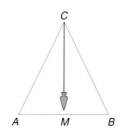

42. Carpentry Before the invention of the bubble level, carpenters used a device called a *plumb level* to verify that a surface was level. This level consisted of a frame in the shape of an isosceles triangle with the midpoint of the base (point M in the figure at the left) marked. A plumb line was suspended from the vertex angle. To use this instrument, the carpenter would hold it upright with the base resting on the surface to be leveled. If the surface was level, over what point on the base do you think the plumb line would hang? Explain.

Computer

43. Use the BASIC program to find the measures of the base angles of an isosceles triangle which has a vertex angle of the given measure.

```
10 INPUT "ENTER THE MEASURE OF THE VERTEX ANGLE OF AN
   ISOSCELES TRIANGLE."; A
20 B = (180 – A)/2
30 "THE BASE ANGLES OF THE ISOSCELES TRIANGLE ARE EACH ";
   B; " DEGREES."
40 END
```

a. 26   **b.** 120   **c.** 32

d. 78   **e.** 85   **f.** 101

Mixed Review

44. Identify the subgoals you would need to accomplish to complete the proof. **(Lesson 4-6)**

Given: $\overline{BL} \cong \overline{BE}$
$\overline{LS} \cong \overline{ES}$

Prove: $\angle LUS \cong \angle EUS$

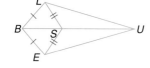

45. List combinations of congruent corresponding parts that could be used to prove that $\triangle TIN \cong \triangle CAN$ by AAS. **(Lesson 4-5)**

46. Algebra The measures of the angles of $\triangle MAP$ are $x + 14$, $3x + 1$, and $6x - 5$. Find the measures of the angles. **(Lesson 4-2)**

47. List all of the conclusions that can be drawn from the figure. **(Lesson 3-2)**

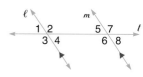

48. Determine if a valid conclusion can be made from these two statements.
- If a student scores 92 or above on the Geometry test, he or she will receive an A.
- Don scored 94 on the Geometry test.

State the law of logic you used. **(Lesson 2-3)**

Wrap-Up

49. Write a five-question quiz that covers the major concepts of this lesson.

CHAPTER 4 | SUMMARY AND REVIEW

VOCABULARY

After completing this chapter, you should be
familiar with the following terms:

acute triangle	**164**	**165**	legs
base	**165**	**202**	line symmetry
base angles	**165**	**164**	obtuse triangle
corollary	**173**	**172**	remote interior angles
equiangular	**164**	**164**	right triangle
equilateral	**165**	**165**	scalene triangle
exterior angle	**172**	**164**	sides
hypotenuse	**165**	**165**	vertex angle
included angle	**185**	**164**	vertices
isosceles triangle	**165**		

SKILLS AND CONCEPTS

OBJECTIVES AND EXAMPLES	REVIEW EXERCISES

Upon completing this chapter, you should
be able to:

- identify parts of triangles and classify
  triangles by their parts. **(Lesson 4-1)**

Types of Triangles

Classification by Angles	
acute	three acute angles
obtuse	one obtuse angle
right	one right angle
equiangular	three congruent angles

Classification by Sides	
scalene	no two sides congruent
isosceles	at least two sides congruent
equilateral	three sides congruent

Use these exercises to review and prepare
for the chapter test.

**In figure *ABCDE*, ∠*BAE* and ∠*BDE* are
right angles and the congruent parts are
indicated.**

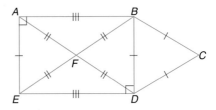

1. Name the right triangle(s).
2. Which triangle is equilateral?
3. Which segment is opposite ∠*C*?
4. Name the acute triangle(s).
5. Which triangles are isosceles?
6. Name the obtuse triangle(s).

OBJECTIVES AND EXAMPLES	REVIEW EXERCISES

■ use the angle sum and exterior angle theorems. **(Lesson 4-2)**

Find the values of *x* and *y*.

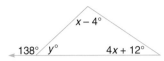

By the exterior angle theorem,
$$(x - 4) + (4x + 12) = 138$$
$$5x = 130$$
$$x = 26$$

By the angle sum theorem,
$$(x - 4) + (4x + 12) + y = 180$$
$$(26 - 4) + (4(26) + 12) + y = 180$$
$$138 + y = 180$$
$$y = 42$$

In Δ*MNO*, $\overline{MO} \perp \overline{ON}$, $\overline{OP} \perp \overline{MN}$, and $m\angle N = 37$. Find the measure of each angle.

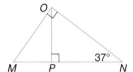

7. $\angle OMN$
8. $\angle MON$
9. $\angle PON$
10. $\angle MOP$

Use the information from the figure to find each measure.

11. *x*   12. *w*
13. *v*   14. *r*
15. *s*   16. *t*
17. *y*   18. *z*

■ identify congruent triangles and name corresponding parts of congruent triangles. **(Lesson 4-3)**

Definition of Congruent Triangles

Two triangles are congruent if and only if their corresponding parts are congruent.

Draw triangles Δ*GHI* and Δ*JKL*. Label the corresponding parts if Δ*GHI* ≅ Δ*JKL*. Use the figures to complete each statement.

19. $\angle L \cong$ __?__   20. $\overline{GI} \cong$ __?__
21. $\overline{KJ} \cong$ __?__   22. $\angle LKJ \cong$ __?__
23. $\overline{LJ} \cong$ __?__   24. $\angle HIG \cong$ __?__

■ use SAS, SSS, and ASA postulates to test for triangle congruence. **(Lesson 4-4)**

Given: $\overline{LN}$ and $\overline{OP}$ bisect each other at *M*.

Prove: $\angle O \cong \angle P$

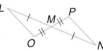

Statements	Reasons
1. $\overline{LN}$ and $\overline{OP}$ bisect each other at *M*.	1. Given
2. $\overline{LM} \cong \overline{MN}$ $\overline{PM} \cong \overline{MO}$	2. Definition of bisector
3. $\angle LMO \cong \angle NMP$	3. Vertical ⓐ are ≅.
4. Δ*LMO* ≅ Δ*NMP*	4. SAS
5. $\angle O \cong \angle P$	5. CPCTC

Write a two-column proof for each.

25. **Given:** *E* is the midpoint of $\overline{AC}$.
    $\angle 1 \cong \angle 2$

    Prove: $\angle 3 \cong \angle 4$

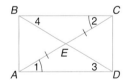

26. **Given:** Δ*WXZ* is isosceles.
    Y is the midpoint of $\overline{XZ}$.

    Prove: $\overline{YW}$ bisects $\angle XWZ$.

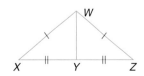

OBJECTIVES AND EXAMPLES	REVIEW EXERCISES

- use AAS theorem to test for triangle congruence. **(Lesson 4-5)**

  AAS

  If two angles and a nonincluded side of one triangle are congruent to the corresponding two angles and side of a second triangle, the two triangles are congruent.

Write a paragraph proof for each.

27. **Given:** $\overline{KL} \cong \overline{ML}$
    $\angle J \cong \angle N$
    $\angle 1 \cong \angle 2$
    Prove: $\overline{JK} \cong \overline{NM}$

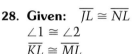
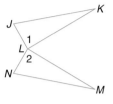

28. **Given:** $\overline{JL} \cong \overline{NL}$
    $\angle 1 \cong \angle 2$
    $\overline{KL} \cong \overline{ML}$
    Prove: $\angle K \cong \angle M$

- use the properties of isosceles and equilateral triangles. **(Lesson 4-7)**

  In isosceles triangle FGH, $\angle H$ is the vertex angle. If $m\angle F = 5x - 6$ and $m\angle G = 3x + 14$, find the measure of each angle of the triangle.

  Since $\triangle FGH$ is isosceles and $\angle H$ is the vertex angle, $m\angle F = m\angle G$.

  $$5x - 6 = 3x + 14 \qquad m\angle F = 5x - 6$$
  $$2x = 20 \qquad\qquad = 5(10) - 6$$
  $$x = 10 \qquad\qquad = 44$$

  $$m\angle F + m\angle G + m\angle H = 180$$
  $$44 + 44 + m\angle H = 180$$
  $$m\angle H = 92$$

  The measures of the angles are 44, 44, and 92.

Find the value of x.

29.

30.

Write a two-column proof.

31. **Given:** $\angle A \cong \angle D$
    $\overline{AB} \cong \overline{DC}$
    E is the midpoint of $\overline{AD}$.
    Prove: $\angle 3 \cong \angle 4$

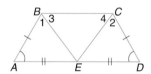

APPLICATIONS AND CONNECTIONS

32. **Sports** The sail for a sailboat is a right triangle. If the angle at the top of the sail measures 54°, what is the measure of the angle at the bottom? **(Lesson 4-2)**

33. **Number Theory** Find a three-digit number that is a perfect square and a perfect cube. **(Lesson 4-6)**

34. **Navigation** The captain of a ship uses an instrument called a *pelorus* to note the angle between the ship's path and the line from the ship to a lighthouse. The captain finds the distance that the ship travels and the change in the measure of the angle with the lighthouse as the ship progresses. When the angle with the lighthouse is twice that of the original angle, the captain knows that the ship is as far from the lighthouse as the ship has traveled since the lighthouse was first sighted. Why? **(Lesson 4-7)**

In figure PQRST, ∠R is right and the congruent parts are indicated.

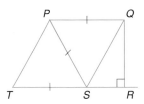

1. Name the isosceles triangle(s).
2. Which triangle is right?
3. Which segment of ΔPTS is opposite∠T?
4. Name the triangle(s) that appear to be scalene.

In the figure, $\overline{GH} \cong \overline{GL}$, $\overline{GI} \cong \overline{GK}$, $\overline{GJ} \perp \overline{HL}$ m∠3 = 30, and m∠4 = 20. Find each measure.

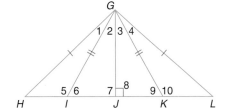

5. m∠8
6. m∠7
7. m∠9
8. m∠6
9. m∠2
10. m∠5
11. m∠10
12. m∠L
13. m∠H
14. m∠1

Name the additional pairs of corresponding parts that would have to be proved congruent in order to use the given postulate or theorem to prove the triangles congruent.

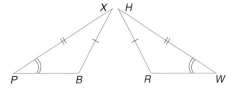

15. SSS
16. SAS
17. AAS
18. ASA

19. Find the greatest four-digit whole number with exactly three factors.

20. **Given:** $\overline{AC} \perp \overline{BD}$
    $\angle B \cong \angle D$
    Prove: C is the midpoint of $\overline{BD}$.

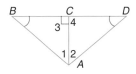

Bonus

History When a march by Napoleon's army was blocked by a stream, a young soldier found the distance across the stream by sighting the angle to the opposite edge of the stream with the brim of his hat. He then turned and, without raising or lowering his head, sighted the point in line with the brim of his hat. He paced off the distance to this point and reported the distance across the stream. How did the soldier know the distance? Why did his method work?

College Entrance Exam Preview

Directions: Choose the one best answer. Write A, B, C, or D. You may use a calculator.

1.

If $\ell \parallel m$ in the figure above, which of the following must be equal to 180?

   I. $m\angle 3 + m\angle 5$
   II. $m\angle 4 + m\angle 6$
   III. $m\angle 1 + m\angle 7$
   IV. $m\angle 2 + m\angle 8$
   V. $m\angle 7 + m\angle 8$

(A) I and II only

(B) III and IV only

(C) V only

(D) I, II, III, and IV only

2. The sum of an odd number and an even number is

(A) sometimes an odd number.

(B) always an even number.

(C) always an odd number.

(D) always divisible by 3, 5, or 7.

3. If $|6x - 8| = 10$, then $x =$

(A) 3   (B) 3 or $-\frac{1}{3}$

(C) $\frac{1}{3}$ or -3   (D) 0

4.

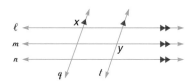

If $\ell \parallel m \parallel n$ and $q \parallel \ell$, then

(A) $m\angle x = m\angle y$

(B) $m\angle x > m\angle y$

(C) $m\angle x < m\angle y$

(D) cannot determine the relationship between $m\angle x$ and $m\angle y$

5. A basket holds 65 apples. Thirteen of the apples are rotten. What percentage of the apples are good?

(A) 20%   (B) 52%

(C) 95%   (D) 80%

6. What is the slope of the line that passes through the points (-4, 8) and (3, -7)?

(A) $-\frac{15}{7}$ (B) -2

(C) -1   (D) $-\frac{7}{15}$

7. What is the average of $a + 5$, $2a - 4$, and $3a + 8$?

(A) $2a$ (B) $3a + 3$

(C) $2a + 3$ (C) $6a + 9$

8. Three vertices of a parallelogram are at (2, 1), (-1, -3), and (6, 4). The fourth vertex is at

(A) (0, 3)   (B) (3, 0)

(C) (-1, 4)   (D) (6, 1)

9. For all $x \neq 0$, $\dfrac{x^6 + x^6 + x^6}{x^3} =$

(A) x^6 (B) $3x^2$

(C) $x^3 + 2x^6$ (D) none of these

10. Tickets to the spring play were $2 for students and $4 for adults. A total of 250 tickets were sold. If s is the number of student tickets sold, which of the following is a formula for the total sales in dollars?

(A) $2s + 1000$

(B) $2s + 4(250 - s)$

(C) $4s + 2(250 + s)$

(D) $4s + 500$

Solve. You may use a calculator.

11. In the figure below, lines ℓ and m are parallel. List three angles, all having the same measure.

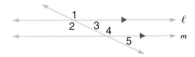

12. One hundred high school students were asked how many courses they were taking that school year. The results are recorded in the table below. What is the average number of courses that a student in this group is taking?

Number of Courses	5	6	7	8
Number of Students	9	19	52	20

13. Twenty-two students have chartered a bus to travel to the championship basketball game and will split the cost equally. After 17 of the students have turned in their share, $131.75 has been collected. What is the cost of chartering the bus?

14. A box contains 75 red, blue, and green pens. If 24% are red and 17 pens are blue, what fraction of the pens are green?

TEST-TAKING TIP

Managing Your Time

College entrance exams are timed tests, so good use of time is essential. Pace yourself as you work and try not to spend too much time on any one problem. Develop a game plan before you take the test. Three possible plans are

- Answer all questions in order, taking as much time as needed for each one.

- Answer all of the easy questions first. Mark more difficult problems with a star and then go back to finish them after all of the easy ones are finished. This plan may allow you to answer many of the questions in a short period of time and spend more time on the difficult ones.

- Answer all of the difficult questions first, then go back and do the easy ones. This approach may allow you to complete the difficult questions when your mind is fresh and leave the easy problems for the end of the test time when you may be tired or rushed.

Whatever plan you use to attack the test, organize your work and write legibly as you solve problems. This will allow you to check your work easily if you have time remaining after you finish all of the problems.

15. Mark can type 45 words per minute, and there is an average of 450 words per page. At this rate, how many hours would it take Mark to type a paper x pages long?

C H A P T E R 5

Applying Congruent Triangles

CHAPTER OBJECTIVES

In this chapter, you will:
- Identify and use the special segments in triangles.
- Prove right triangles congruent.
- Recognize and apply relationships between the sides and angles in a triangle.

GEOMETRY AROUND THE WORLD
United States

Like the geometric forms used in art, the lines and angles of beautiful buildings delight us. Besides being attractive, well-designed homes and public buildings must also be structurally sound. Making sure that buildings are sturdy and livable as well as striking is the job of technical architects such as Norma Merrick Sklarek.

Sklarek, a graduate of Columbia University School of Architecture, was the first African American woman registered as an architect in the United States. For over 30 years, she has been ensuring that a variety of buildings all over the world are safely wired, energy-efficient, and capable of withstanding decades of time.

In some cities, such as San Francisco and Tokyo, Sklarek has also had to make sure that the buildings she helps design will not collapse during earthquakes. As you have probably guessed, the practical skills of technical architects are needed whenever new buildings are erected. Sklarek has worked on a variety of projects, including office and apartment buildings, hotels, hospitals, and other public buildings, such as the U.S. Embassy in Tokyo.

Have you ever seen a triangular-shaped building? Which rooms would be the most angular? Which rooms would have the fewest angles?

GEOMETRY IN ACTION

Because of their distinct shape, triangles make what architects describe as "strong architectural statements." Basically, this means there's no way to mistake a triangle's sharp angles for any other form. Measure the angles and the sides of two small triangles in the design for a triangular-shaped building shown at the right. Are all the angles and sides the same? Are their corresponding parts congruent?

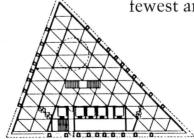

◀ *U.S. Embassy in Tokyo*   Inset: *Norma Merrick Sklarek*

215

5-1 Special Segments in Triangles

Objective

After studying this lesson, you should be able to:

- identify and use medians, altitudes, angle bisectors, and perpendicular bisectors in a triangle.

Application

Almost everywhere you look there are triangles. Triangles are used to create beautiful patterns in stained-glass windows and quilts, and also in designing bridges and buildings. The Eiffel Tower is constructed with triangles because, of all the polygons, they are the most rigid.

Triangles have four types of special segments.

One kind of special segment of a triangle is called a **median.** A median connects a vertex of a triangle to the midpoint of the opposite side. For $\triangle ABC$ at the right, $\overline{AX}$, $\overline{BY}$, and $\overline{CZ}$ are the medians. Every triangle has three medians.

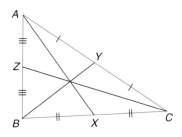

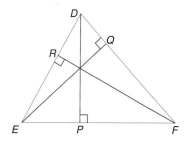

A second kind of special segment is an **altitude**. An altitude has one endpoint at a vertex of a triangle and the other on the line that contains the side opposite that vertex so that the segment is perpendicular to this line. For acute triangle DEF at the left, $\overline{DP}$, $\overline{EQ}$, and $\overline{FR}$ are the altitudes. Every triangle has three altitudes.

The altitudes for a right and an obtuse triangle are shown below. The legs of a right triangle are two of the altitudes. For an obtuse triangle, two of the altitudes are outside of the triangle.

One endpoint of an altitude may be a point not on the triangle.

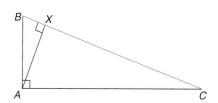

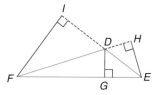

In right triangle ABC, $\overline{CA}$ is the altitude from C and $\overline{BA}$ is the altitude from B.

In obtuse triangle DEF, $\overline{DG}$, $\overline{EH}$ and $\overline{FI}$ are the altitudes. Notice that $\overline{EH}$ and $\overline{FI}$ are outside of $\triangle DEF$.

Example 1

$\triangle KCT$ has vertices $K(12, -6)$, $C(9, 2)$, and $T(2, 4)$. Determine the coordinates of point M on $\overline{KT}$ so that $\overline{CM}$ is a median of $\triangle KCT$.

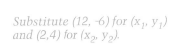

According to the definition of median, $\overline{CM}$ will be a median of $\triangle KCT$ if M is the midpoint of $\overline{KT}$.

You may want to review midpoints in Lesson 1-5 and slopes of perpendicular lines in Lesson 3-5.

$\left(\dfrac{x_1 + x_2}{2}, \dfrac{y_1 + y_2}{2}\right) = \left(\dfrac{12 + 2}{2}, \dfrac{-6 + 4}{2}\right)$ *Substitute $(12, -6)$ for (x_1, y_1) and $(2, 4)$ for (x_2, y_2).*

$= (7, -1)$

The coordinates of M on $\overline{KT}$ are $(7, -1)$.

Determine if $\overline{CM}$ is an altitude of $\triangle KCT$.

For $\overline{CM}$ to be an altitude of $\triangle KCT$, $\overleftrightarrow{CM}$ must be perpendicular to $\overleftrightarrow{KT}$. This means that the product of the slopes of $\overleftrightarrow{CM}$ and $\overleftrightarrow{KT}$ must be -1.

$$\text{slope} = \frac{y_2 - y_1}{x_2 - x_1}$$

$$\text{slope of } \overleftrightarrow{CM} = \frac{2 - (-1)}{9 - 7} \qquad \text{slope of } \overleftrightarrow{KT} = \frac{-6 - 4}{12 - 2}$$

$$= \frac{3}{2} \qquad\qquad\qquad = -\frac{10}{10} \text{ or } -1$$

$$\text{product of slopes} = \frac{3}{2} \cdot (-1) \text{ or } -\frac{3}{2}$$

Since the product of these slopes is not -1, $\overline{CM}$ is not perpendicular to $\overleftrightarrow{KT}$. Thus, $\overline{CM}$ is not an altitude of $\triangle KCT$.

The third of the four special segments of a triangle is an **angle bisector.** An angle bisector is a segment that bisects an angle of the triangle and has one endpoint at a vertex of the triangle and the other endpoint at another point on the triangle.

For $\triangle PQR$ at the right, $\overline{PX}$, $\overline{QY}$, and $\overline{RZ}$ are the angle bisectors. Since there are three angles in every triangle, every triangle has three angle bisectors.

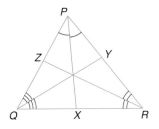

The following example illustrates how the special segments of triangles can be used to prove special properties of triangles.

Example 2

Prove the statement *If an angle bisector of a triangle is also an altitude, then the triangle is isosceles.*

Given: $\overline{BD}$ is an angle bisector of $\triangle ABC$.
$\overline{BD}$ is an altitude of $\triangle ABC$.

Prove: $\triangle ABC$ is isosceles.

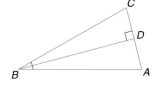

Proof:

Statements	Reasons
1. $\overline{BD}$ is an angle bisector of $\triangle ABC$.	**1.** Given
2. $\angle ABD \cong \angle CBD$	**2.** Definition of bisector
3. $\overline{BD}$ is an altitude of $\triangle ABC$.	**3.** Given
4. $\overline{BD} \perp \overleftrightarrow{AC}$	**4.** Definition of altitude
5. $\angle ADB$ and $\angle CDB$ are right angles.	**5.** $\perp$ lines form four rt. $\angle$s.
6. $\angle ADB \cong \angle CDB$	**6.** All rt. $\angle$s are $\cong$.
7. $\overline{BD} \cong \overline{BD}$	**7.** Congruence of segments is reflexive.
8. $\triangle ADB \cong \triangle CDB$	**8.** ASA
9. $\overline{CB} \cong \overline{AB}$	**9.** CPCTC
10. $\triangle ABC$ is isosceles.	**10.** Definition of isosceles triangle

Now let's investigate the fourth type of special segment.

INVESTIGATION

Cut out any large acute triangle. Fold the triangle so that one vertex falls on a second vertex as shown below.

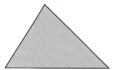 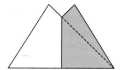

Unfold the triangle. Repeat the process with the other two pairs of vertices. What do you notice about the three folds? Each is perpendicular to a side and they all intersect in one point.

In the investigation, the folds in the triangle were the **perpendicular bisectors** of the sides of the triangle. A line or a line segment that passes through the midpoint of a side of a triangle *and* is perpendicular to that side is the *perpendicular bisector* of the side of triangle.

Perpendicular bisectors of segments have some special properties. These properties are listed in Theorems 5-1 and 5-2. You will be asked to prove these theorems in Exercises 28 and 29. Theorem 5-3 involves the bisector of an angle and will be proved in Exercise 30.

Theorem 5-1	**A point on the perpendicular bisector of a segment is equidistant from the endpoints of the segment.**
Theorem 5-2	**A point equidistant from the endpoints of a segment lies on the perpendicular bisector of the segment.**
Theorem 5-3	**A point on the bisector of an angle is equidistant from the sides of the angle.**
Theorem 5-4	**A point in the interior of or on an angle and equidistant from the sides of an angle lies on the bisector of the angle.**

CHECKING FOR UNDERSTANDING

Communicating Mathematics

Read and study the lesson to answer these questions.

1. What is the difference between an angle bisector and a perpendicular bisector of a triangle?

2. Could a median be a perpendicular bisector of a side of a triangle? If so, in what type of triangle would this occur? Justify your answer.

3. How are Theorems 5-1 and 5-2 related?

Guided Practice

Complete. Refer to the figure at the right.

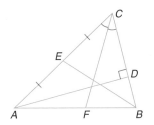

4. $\overline{EB}$ is a __?__ of $\triangle ABC$.

5. __?__ is an altitude of $\triangle ABC$.

6. If $\overline{CF}$ is both an angle bisector and an altitude of $\triangle ABC$, then $\triangle ABC$ is __?__.

Draw and label a figure to illustrate each situation.

7. $\overline{PS}$ is a median of $\triangle PQR$ and S is between Q and R.

8. $\overline{QT}$ is an angle bisector of $\triangle PQR$ and T is between P and R.

9. $\overline{RU}$ is an altitude of $\triangle PQR$ and Q is between U and P.

10. $\overline{AC}$ and $\overline{BC}$ are altitudes of $\triangle ABC$.

11. $\overline{DX}$ is an altitude of $\triangle DEF$ and F is between E and X.

12. Answer each question if $R(3, 3)$, $S(-1, 6)$, and $T(1, 8)$ are the vertices of $\triangle RST$, and $\overline{RX}$ is a median of $\triangle RST$ with X on $\overline{ST}$.

    a. What are the coordinates of X?

    b. What is the length of $\overline{RX}$?

    c. What is the slope of $\overleftrightarrow{RX}$?

    d. Is $\overline{RX}$ an altitude of $\triangle RST$? Explain.

13. Write a two-column proof.

    Given: $\overline{AB} \cong \overline{CB}$
                    $\overline{BD}$ is a median of $\triangle ABC$.

    Prove: $\overline{BD}$ is an altitude of $\triangle ABC$.

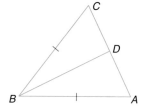

EXERCISES

Draw and label a figure to illustrate each situation.

14. $\overline{PT}$ and $\overline{RS}$ are medians of $\triangle PQR$ and intersect at V.

15. $\overline{AD}$ is a median and an altitude of $\triangle ABC$.

16. $\overline{XM}$ is a median and an angle bisector of $\triangle XYZ$, and $\overline{ZX}$ is an altitude.

17. $\triangle DEF$ is a right triangle with right angle at F. $\overline{FG}$ is a median of $\triangle DEF$ and $\overline{GH}$ is the perpendicular bisector of $\overline{DE}$.

18. For parts a–d, draw a large acute triangle that is not equilateral. Then complete the indicated construction and answer each question.

    a. Construct the three medians of the triangle. What do you notice?

    b. Construct the three altitudes of the triangle. What do you notice?

    c. Construct the three angle bisectors of the triangle. What do you notice?

    d. Construct the three perpendicular bisectors of the sides of the triangle. What do you notice?

    e. Cut out each of the triangles from parts a-d. Then place each triangle on the head of the pin at the point where the constructed lines intersect. Describe what you observe.

    f. What changes would occur if the constructions in parts a-d were done on a right triangle or an obtuse triangle, instead of an acute triangle?

State whether each sentence is *always, sometimes,* or *never* true.

19. The three medians of a triangle intersect at a point inside the triangle.

20. The three angle bisectors of a triangle intersect at a point outside the triangle.

21. The three altitudes of a triangle intersect at a vertex of the triangle.

22. The three perpendicular bisectors of a triangle intersect at a point outside the triangle.

23. Answer each question if $A(1, 6)$, $B(13, 2)$, and $C(-7, 12)$ are the vertices of $\triangle ABC$.

    a. What are the coordinates of the midpoint of $\overline{AB}$?

    b. What is the slope of the perpendicular bisector of $\overline{AB}$?

    c. The coordinates of point N on $\overleftrightarrow{BC}$ are $\left(\frac{9}{5}, \frac{38}{5}\right)$. Is $\overline{NA}$ an altitude of $\triangle ABC$?

24. Find the value of x if $\overline{AD}$ is an altitude of $\triangle ABC$.

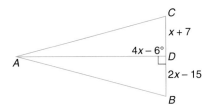

25. Find the value of x if $\overline{PS}$ is a median of $\triangle PQR$.

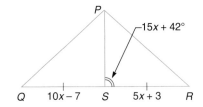

26. $\overline{YV}$ is an angle bisector of $\triangle XYZ$. Determine $m\angle ZYV$ and $m\angle XYZ$ if $m\angle XYZ = 8x - 6$ and $m\angle XYV = 2x + 7$.

Draw and label a figure for each statement. List the information that is given and the statement to be proved in terms of your figure. Then write a two-column proof.

27. If a median of a triangle is also an altitude, then the triangle is isosceles.

28. A point on the perpendicular bisector of a segment is equidistant from the endpoints of the segment. (Theorem 5-1)

29. A point equidistant from the endpoints of a segment lies on the perpendicular bisector of the segment. (Theorem 5-2)

30. A point on the bisector of an angle is equidistant from the sides of the angle. (Theorem 5-3)

31. If a triangle is isosceles, then the altitude to its base is also a median.

32. Corresponding medians of congruent triangles are congruent.

33. Corresponding angle bisectors of congruent triangles are congruent.

34. The medians drawn to the congruent sides of an isosceles triangle are congruent.

Critical Thinking

35. What can you conclude about the perpendicular bisector of a side of a triangle and the altitude drawn to that same side of the triangle?

36. What can you conclude about the point of intersection of the perpendicular bisectors of the legs of a right triangle?

Application

37. Construction The SunFun Amusement Park is placing *Photo Opportunity* stations with statues of cartoon characters around the park. Copy the diagram of the park and locate the stations described.

a. One station will be placed at the edge of Fountain Lake equidistant from the Iron Snake and The Trip to the Moon. Show the point where it will be located as *A*.

b. A second station will be placed by the side of the Magic Carousel equidistant from Excitement Alley and SunFun Street. Show the point where it will be located as *B*.

c. The point equidistant from the Gateway to Fun, Henry Ford's Carworks, and Billy the Kid's Hideout is the location of the third station. Mark this point as *C*.

Mixed Review

Use the figure at the right for Exercises 38-40.

38. Find the value of *x*. **(Lesson 4-7)**

39. If $m\angle NBC = 34$, find $m\angle ANB$. **(Lesson 4-2)**

40. Find the value of *y* if $AN = 4y - 7$ and $NC = 9 + 2y$. **(Lesson 1-5)**

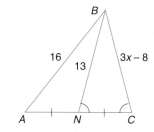

41. Without graphing, determine if $\overleftrightarrow{PQ} \parallel \overleftrightarrow{RS}$ given points $P(0, 0)$, $Q(3, -2)$, $R(1, -7)$, and $S(-5, -3)$. Explain. **(Lesson 3-4)**

42. If possible, write a conclusion. State the law of logic that you use. **(Lesson 2-3)**

(1) $\overline{AX}$ is an altitude of $\triangle ABC$ if $\overline{AX} \perp \overline{BC}$ and *X* is on $\overline{BC}$.

(2) $\overline{AX} \perp \overline{BC}$ and *X* is on $\overline{BC}$.

Wrap-Up

43. Journal Entry Draw a figure that shows a median, altitude, and angle bisector of a triangle and a perpendicular bisector of a side of the triangle. Then describe how they differ from each other.

5-2

Right Triangles

Objective

After studying this lesson, you should be able to:

■ recognize and use tests for congruence of right triangles.

FYI···

The America's Cup was originally called the Hundred-Guinea Cup. It was renamed for the *America*, the first yacht from the United States to win the cup.

The America's Cup race is the most famous international yachting competition. In 1987, *Stars and Stripes* brought the America's Cup back to the United States after it had been lost to Australia in 1983. The mainsail for *Stars and Stripes*, like the mainsails for most modern sailboats, is shaped like a right triangle. The spare mainsail must be congruent to the original mainsail. *Why?* What do you need to know about the sails to determine if they are congruent?

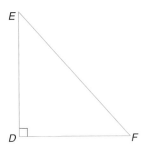

Suppose right triangle *DEF*, shown at the left, represents one of the mainsails. Since all right angles are congruent and all right triangles have a right angle, any other right triangle has at least one angle congruent to an angle in △*DEF*. If the corresponding legs are congruent, then the triangles will be congruent by SAS.

Theorem 5-5 **LL**	**If the legs of one right triangle are congruent to the corresponding legs of another right triangle, then the triangles are congruent.**

You will be asked to prove Theorem 5-5 in Exercise 27.

Suppose you know that the hypotenuse and an acute angle of a right triangular sail are congruent to the corresponding parts of the mainsail. Must these two sails be congruent? Since we know that the right angles are congruent, knowing that another angle and the hypotenuse are also congruent tells us that two corresponding angles and a corresponding nonincluded side are congruent. So, the triangles must be congruent by AAS.

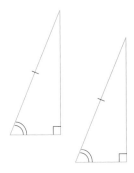

Theorem 5-6 *HA*	If the hypotenuse and an acute angle of one right triangle are congruent to the hypotenuse and corresponding acute angle of another right triangle, then the two triangles are congruent.

Proof of
Theorem 5-6

Given: $\triangle ABC$ and $\triangle XYZ$ are right triangles.
$\angle A$ and $\angle X$ are right angles.
$\overline{BC} \cong \overline{YZ}$
$\angle B \cong \angle Y$

Prove: $\triangle ABC \cong \triangle XYZ$

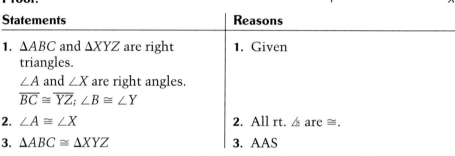

Proof:

Statements	Reasons
1. $\triangle ABC$ and $\triangle XYZ$ are right triangles. $\angle A$ and $\angle X$ are right angles. $\overline{BC} \cong \overline{YZ}$; $\angle B \cong \angle Y$	1. Given
2. $\angle A \cong \angle X$	2. All rt. ∠s are ≅.
3. $\triangle ABC \cong \triangle XYZ$	3. AAS

You can use the HA Theorem in many proofs involving right triangles.

Example 1

A truss for a roof is in the shape of an isosceles triangle with an altitude from the vertex angle to the base. Prove that the two right triangles formed by the altitude are congruent.

Given: $\overline{WY}$ is an altitude of $\triangle XYZ$.
$\triangle XYZ$ is an isosceles triangle.

Prove: $\triangle XWY \cong \triangle ZWY$

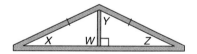

Proof:

Statements	Reasons
1. $\overline{WY}$ is an altitude of $\triangle XYZ$.	1. Given
2. $\overline{WY} \perp \overline{XZ}$	2. Definition of altitude
3. $\angle XWY$ and $\angle ZWY$ are right angles.	3. ⊥ lines form four rt. ∠s.
4. $\triangle XWY$ and $\triangle ZWY$ are right triangles.	4. Definition of right triangle
5. $\triangle XYZ$ is an isosceles triangle.	5. Given
6. $\overline{XY} \cong \overline{ZY}$	6. Definition of isosceles triangle
7. $\angle X \cong \angle Z$	7. If 2 sides of a $\triangle$ are ≅ the ∠s opp. the sides are ≅.
8. $\triangle XWY \cong \triangle ZWY$	8. HA

You have already seen that right triangles are congruent if corresponding legs are congruent (LL) or if the hypotenuses and corresponding acute angles are congruent (HA). As a third possibility, suppose you know that a leg and an acute angle of the *Stars and Stripes* mainsail are congruent to the corresponding leg and acute angle of the spare sail. Must these two sails be congruent? To see, we must consider two different cases.

Case 1
The leg is included between the acute angle and the right angle.

Case 2
The leg is *not* included between the acute angle and the right angle.

In both cases, the triangles are congruent. For Case 1, the triangles are congruent by ASA. For Case 2, the triangles are congruent by AAS. This observation suggests Theorem 5-7. *You will prove Theorem 5-7 in Exercise 28.*

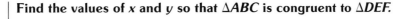

Theorem 5-7 *LA*	**If one leg and an acute angle of one right triangle are congruent to the corresponding leg and acute angle of another right triangle, then the triangles are congruent.**

Example 2

Find the values of *x* and *y* so that △*ABC* is congruent to △*DEF*.

△*ABC* will be congruent to △*DEF* by
LA if $\overline{AB} \cong \overline{DE}$ and $\angle A \cong \angle D$.

$$AB = DE \qquad m\angle D = m\angle A$$
$$15 - 2x = 5 \qquad 9y - 32 = 67$$
$$-2x = -10 \qquad 9y = 99$$
$$x = 5 \qquad y = 11$$

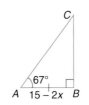 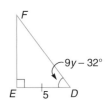

If *x* = 5, then $\overline{AB} \cong \overline{DE}$; and if *y* = 11, then $\angle A \cong \angle D$. Therefore, △*ABC* ≅ △*DEF*.

For a final application of congruent right triangles, let's look at pitching a tent. If the tent pole is perpendicular to the ground and the ground is level, then the tent pole, ground, and tent rope pulled taut from the tent pole form a right triangle. If two tent ropes of the same length are each pulled taut from tent poles of the same height, then the stakes at the end of each rope will be placed in the same spot on the ground relative to the tent poles. This situation suggests the following postulate.

Postulate 5-1 HL	If the hypotenuse and a leg of one right triangle are congruent to the hypotenuse and corresponding leg of another right triangle, then the triangles are congruent.

CHECKING FOR UNDERSTANDING

Communicating Mathematics

Read and study the lesson to answer each question.

1. Write a sentence to explain why the tests for congruency of right triangles have only two requirements while the tests for other triangles have three.

2. Which of the tests for congruent triangles would you use to justify the Leg-Leg test for congruent right triangles?

3. Explain to a classmate why two cases must be considered when proving Theorem 5-7 (LA).

Guided Practice

Name the theorem or postulate used to determine whether each pair of triangles is congruent. If there is not enough information write *none*.

4.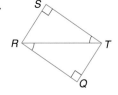

5.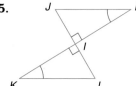

6.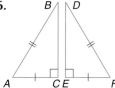

7.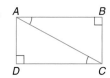

8.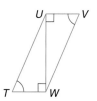

9.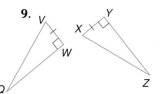

Find the value of *x* for each figure.

10.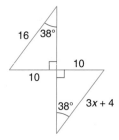

11.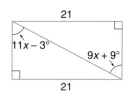

12.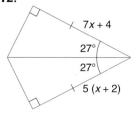

13. Use the figure at the right to write a two-column proof.

> **Given:** $\angle Q$ and $\angle S$ are right angles.
>
> $\angle 1 \cong \angle 2$
>
> **Prove:** $\triangle PQR \cong \triangle RSP$

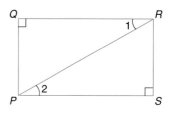

EXERCISES

Practice

For each situation, determine whether $\triangle ABC \cong \triangle XYZ$. Justify your answer.

14. Given: $\overline{BC} \cong \overline{YZ}$ and $\angle A \cong \angle X$

15. Given: $\angle A \cong \angle X$ and $\angle B \cong \angle Y$

16. Given: $\overline{AC} \cong \overline{XZ}$ and $\overline{BC} \cong \overline{YZ}$

17. Given: $\overline{AB} \cong \overline{XY}$ and $\angle B \cong \angle Y$

18. Given: $\overline{AB} \cong \overline{XY}$ and $\overline{BC} \cong \overline{YZ}$

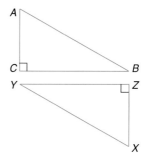

For each figure, find the values of x and y so that $\triangle DEF \cong \triangle PQR$ by the indicated theorem or postulate.

19. HL

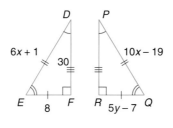

20. HA

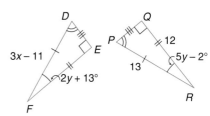

21. LA

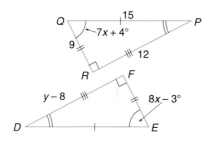

22. LL

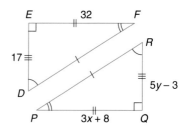

Use the figure at the right to write a two-column proof.

23. **Given:** $\overline{QP} \cong \overline{SR}$
    $\angle Q$ and $\angle S$ are right angles.
    Prove: $\angle 1 \cong \angle 2$

24. **Given:** $\angle Q$ and $\angle S$ are right angles.
    $\overline{QR} \parallel \overline{PS}$
    Prove: $\overline{PQ} \cong \overline{RS}$

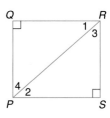

Use the figure at the right to write a two-column proof.

25. **Given:** $\triangle ABY$ and $\triangle CBY$ are right triangles.
    $\overline{AB} \cong \overline{CB}$
    $\overline{YX} \perp \overline{AC}$
    Prove: $\overline{AX} \cong \overline{CX}$

26. **Given:** $\overline{YX}$ is an altitude of $\triangle AYC$.
    $\angle AYX \cong \angle CYX$
    $\overline{YB} \perp$ Plane $\mathcal{M}$
    Prove: $\angle AYB \cong \angle CYB$

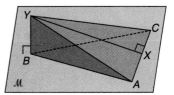

Draw and label a figure for each statement. List the information that is given and the statement to be proved in terms of your figure. Then write a two-column proof.

27. If the legs of one right triangle are congruent to the corresponding legs of another right triangle, then the triangles are congruent. (Theorem 5-5)

28. If one leg and an acute angle of one right triangle are congruent to the corresponding leg and acute angle of another right triangle, then the triangles are congruent. (Theorem 5-7)

29. Corresponding altitudes of congruent triangles are congruent.

30. The two segments that have the midpoint of each leg of an isosceles triangle as one endpoint and are perpendicular to the base of the triangle are congruent.

Critical Thinking

31. Using the figure at the right, write two paragraph proofs, using two different pairs of congruent right triangles.

    Given: $\overline{AC} \perp \overline{BE}$
    $\overline{CE} \perp \overline{AD}$
    $\overline{AC} \cong \overline{CE}$
    Prove: $\overline{BE} \cong \overline{AD}$

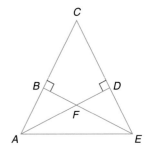

Application

32. Manufacturing Donald Owens is having a new foresail made for his boat. The old foresail is in the shape of a right triangle with the dimensions shown at the right. If the sailmaker knows that the new foresail is shaped like a right triangle, list all of the different possible sets of information that Mr. Owens could give the sailmaker so that she could verify that the new foresail is the same size and shape as the old foresail.

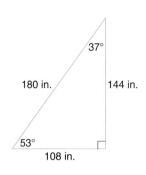

33. Construction Four wood braces of equal length are to be used to support a deck near where the deck meets the wall of a house. One end of each brace will be attached to the bottom of the deck at the same distance from the wall. Explain why the other end of each brace will be attached to the wall at the same distance from the bottom of the deck.

Mixed Review

Refer to the figure for Exercises 34-36.

34. If X is the midpoint of $\overline{AD}$, name the additional parts of $\triangle AXC$ and $\triangle DXB$ that would have to be congruent to prove that $\triangle AXC \cong \triangle DXB$ by ASA. **(Lesson 4-5)**

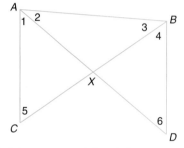

35. If $\overline{AC} \parallel \overline{BD}$, which pairs of numbered angles must be congruent? **(Lesson 3-2)**

36. If $m\angle 3 = 3d + 11$, $m\angle 4 = 5d - 2$, and $m\angle ABD = 97$, find the value of d. **(Lesson 1-6)**

37. Write the statement *A median of a triangle bisects one side of the triangle* in if-then form. **(Lesson 2-2)**

Wrap-Up

38. Write a short paragraph describing each of the tests for congruence of right triangles presented in this lesson.

HISTORY CONNECTION

The oldest pictures of sailboats that have been found are on Egyptian vases made around 3200 B.C. Since Egypt had no wood strong enough for large ships, their boats were made of green papyrus stalks tied in bundles. The Nile River was perfect for shipping cargo between northern and southern Egypt since the water flows north and the wind blows from north to south. The sailors could ride the current to travel north and raise the sails to use the wind to travel south.

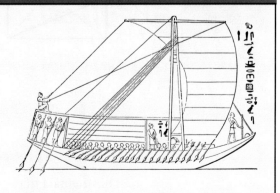

5-3 Problem-Solving Strategy: Work Backward

Objective

After studying this lesson, you should be able to:

■ solve problems by working backward.

Application

Mr. Dean spent $1491.21 on a personal computer from Wholesale Electronics. The computer was on sale for 25% off. In addition, Mr. Dean received 10% off of the sale price for paying in cash instead of financing his purchase. Sales tax of 5.2% was added after the discounts. Mr. Dean plans to insure his computer for its original price. For what amount should he insure the computer?

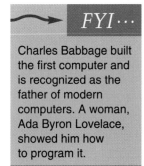

FYI···

Charles Babbage built the first computer and is recognized as the father of modern computers. A woman, Ada Byron Lovelace, showed him how to program it.

We know the discounted price, so we can work backward to find the original price.

Sales Tax The last operation in computing the price was to add sales tax.

Let x = the price of the computer before the tax was added. Then $0.052x$ = the amount of sales tax on the computer. *5.2% = 0.052*

price before sales tax	plus	sales tax	equals	price after sales tax
x	+	$0.052x$	=	1491.21
		$1.052x$	=	1491.21
		x	=	1417.50

Discount for Cash Mr. Dean's 10% off for paying cash was a discount *before* the sales tax.

Let y = the price before the 10% discount. Then $0.1y$ = the amount of the 10% discount. *10% = 0.1*

price before discount	minus	discount	equals	price before tax and after discount
y	−	$0.1y$	=	1417.50
		$0.9y$	=	1417.50
		y	=	1575

Sale Discount The sale price was 25% off the original price.

Let z = the original price. Then $0.25z$ = the amount of the 25% discount. *25% = 0.25*

original price	minus	discount	equals	price before tax and after sale discount
z	−	$0.25z$	=	1575
		$0.75z$	=	1575
		z	=	2100

The original price of the computer was $2100. Mr. Dean should insure his personal computer for $2100.

CHECKING FOR UNDERSTANDING

Communicating Mathematics

Read and study the lesson to answer each question.

1. Describe how you could use *guess and check* to find the original price of the personal computer.

2. Describe how to check the answer to the application problem.

3. Explain how you might use working backward when writing a proof.

Guided Practice

TAMIKO

Solve each problem by working backward.

4. On a game show, all the contestants begin with the same number of points. They are awarded points for questions answered correctly and lose points for questions answered incorrectly. Tamiko answered six 20-point questions correctly. Then she answered a 50-point question and an 80-point question incorrectly. In her final round question, Tamiko tripled her score and won with a score of 270 points. How many points did each player have at the beginning of the game?

5. Decrease a number by 52, then multiply the result by 12, then add 20 and divide by 4. If the final result is 32, with what number did you start?

6. Tom collected baseball cards for a few years before he decided to give them all away and try a new hobby. First he gave half the cards plus one extra card to Leila. Then, he gave half of what was left plus one extra card to Karl. Then, he gave half of what was left plus one extra card to Jarrod. Finally, Tom gave the remaining 74 cards to his sister, Charlene. How many baseball cards did Tom have in his collection?

EXERCISES

Practice

Solve. Use any strategy.

Strategies
Look for a pattern.
Solve a simpler problem.
Act it out.
Guess and check.
Draw a diagram.
Make a chart.
Work backward.

7. Chicken Express sells chicken wings in boxes of 15 or 25 pieces. Ken and Julie bought seven boxes and got 125 wings for a party. How many boxes of 15 did they buy?

8. The cans of soda in a soda machine cost 65¢. If the machine will accept quarters, dimes, and nickels, how many different combinations of coins must the machine be programmed to accept?

9. Enrico cashed his paycheck this morning. He put half of the paycheck in the savings account for his college tuition. He then paid $20 for a concert ticket. He spent one eighth of the remaining money on pizza with friends. If Enrico had $42 left, how much was his paycheck?

10. Continue the pattern. Explain your reasoning.

$$1^3 = 1^2 - 0^2$$
$$2^3 = 3^2 - 1^2$$
$$3^3 = 6^2 - 3^2$$
$$4^3 = \underline{\ ?\ }$$
$$5^3 = \underline{\ ?\ }$$

11. Each letter in the following equation represents a single digit. Find the digits that make the equation correct.

$$ON + ON + ON + ON = GO$$

12. Shelly's little sister changed the 5 on one of Shelly's dice to a 3. She changed the 6 on the other die to a 4. Shelly rolled both the dice.
  a. How many different sums could Shelly have rolled?
  b. What is the total Shelly most likely rolled?

13. When the length of each side of a square is decreased by 5 units, the area of the square is $\frac{4}{9}$ of the area of the original square. What is the area of the original square?

14. Montgomery's Department Store is having an electronics sale. A VCR is on sale for $50 less than the list price. The store employees receive an additional 15% discount. The sales tax added to the price is 6%. If a store employee paid $198.22 for the VCR, what is the regular price?

∼ COOPERATIVE LEARNING PROJECT ∼

Work in groups. Each person in the group must understand the solution and be able to explain it to any person in class.

Lewis Carroll (1832-1898), the author of *Alice in Wonderland*, was very interested in logic, mathematics, and word games. One of the word games that he invented, Doublets, became very popular in his day. Many competitions in the game were sponsored by the magazine *Vanity Fair*. The object of the game is to take two different words of the same length and transform one word into the other by a series of intermediate words that differ by only one letter. Proper names are not allowed and each word should be common enough to find in a dictionary. An example is turning CAT to DOG.

CAT
COT   *Step 1: Change A to O. The new word is COT.*
DOT   *Step 2: Change C to D. The new word is DOT.*
DOG   *Step 3: Change T to G. The new word, DOG, is the desired word.*

Change TOP to HAT, make a SEED GROW, and turn WILD to TAME. What is the least number of steps it takes to change one word to another?

5-4 Indirect Proof and Inequalities

Objectives

After studying this lesson, you should be able to:
- use indirect reasoning and indirect proof to reach a conclusion, and
- recognize and apply the properties of inequalities to the measures of segments and angles.

Application

During criminal trials, defendants sometimes try to use an *alibi* to prove that they could not have committed the crime for which they are accused. An alibi is a claim that the defendant was somewhere else when the crime was committed. If the jury believes the alibi, then the defendant is proved innocent. The use of an alibi is an example of a form of **indirect reasoning.**

Direct reasoning is just an application of the Law of Detachment. You may wish to review this law in Lesson 2-3.

Up to this point, the proofs you have encountered have used direct reasoning. With direct reasoning, you start with a true hypothesis and prove that the conclusion is true. With indirect reasoning, you *assume* that the conclusion is false and then show that this assumption leads to a contradiction of the hypothesis or some other accepted fact, like a postulate, theorem, or corollary. That is, since your assumption has been proved false, the conclusion must be true.

The following steps summarize the process of indirect reasoning. You can use these steps when doing an **indirect proof.**

Steps for Writing an Indirect Proof

1. **Assume that the conclusion is false.**
2. **Show that the assumption leads to a contradiction of the hypothesis or some other fact, such as a postulate, theorem, or corollary.**
3. **Point out that the assumption must be false, and therefore, the conclusion must be true.**

Example 1

State the assumption you would make to start an indirect proof of each statement. *Do not write the proofs.*
a. $\overline{XW}$ is an altitude of $\triangle XYZ$.
b. $\triangle MNO$ is not a right triangle.
c. $m\angle A > m\angle B$

To write an indirect proof, we assume that the conclusion is false. So negate each statement.
a. $\overline{XW}$ is not an altitude of $\triangle XYZ$.
b. $\triangle MNO$ is a right triangle.
c. $m\angle A \le m\angle B$ *Remember that if $m\angle A$ is not greater than $m\angle B$, then it could be less than or equal to $m\angle B$.*

In Lesson 4-2, you learned about a relationship between exterior angles of a triangle and their remote interior angles. The following theorem is a result of this relationship and can be proven using an indirect proof.

Theorem 5-8 *Exterior Angle* *Inequality Theorem*	**If an angle is an exterior angle of a triangle, then its measure is greater than the measure of either of its corresponding remote interior angles.**

Indirect Proof of Theorem 5-8

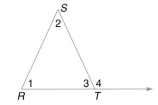

Given: $\angle 4$ is an exterior angle of $\triangle RST$.

Prove: $m\angle 4 > m\angle 1$
$m\angle 4 > m\angle 2$

Indirect Proof:

Step 1: Make the assumption that $m\angle 4 \leq m\angle 1$ or $m\angle 4 \leq m\angle 2$.

Step 2: We will only show that the assumption $m\angle 4 \leq m\angle 1$ leads to a contradiction, since the argument for $m\angle 4 \leq m\angle 2$ uses the same reasoning.

$m\angle 4 \leq m\angle 1$, means that either $m\angle 4 = m\angle 1$ or $m\angle 4 < m \angle 1$. So, we need to consider both cases.

Case 1: If $m\angle 4 = m\angle 1$, then, since $m\angle 1 + m\angle 2 = m\angle 4$ by the Exterior Angle Theorem, we have $m\angle 1 + m\angle 2 = m\angle 1$ by substitution. Then $m\angle 2 = 0$, which contradicts the fact that the measure of an angle is greater than 0.

Case 2: If $m\angle 4 < m\angle 1$, then since $m\angle 1 + m\angle 2 = m\angle 4$ by the Exterior Angle Theorem, we have $m\angle 1 + m\angle 2 < m\angle 1$ by substitution. If $m\angle 1 + m\angle 2 < m\angle 1$, then $m\angle 2 < 0$, which contradicts the fact that the measure of an angle is greater than 0.

Step 3: In both cases, the assumption leads to the contradiction of a known fact. Therefore, the assumption that $m\angle 4 \leq m\angle 1$ must be false, which means that $m\angle 4 > m\angle 1$ must be true. Likewise, $m\angle 4 > m\angle 2$.

In the proof of Theorem 5-8, certain properties of inequalities that you encountered in algebra were used. For example, in Step 1 of the proof it was stated that if $m\angle 4 \leq m\angle 1$, then $m\angle 4 = m\angle 1$ or $m\angle 4 < m\angle 1$. This statement is an application of the *Comparison or Trichotomy Property*, which states that for any two numbers a and b, either $a > b$, $a = b$, or $a < b$. The chart at the top of the next page gives a list of properties of inequalities that you studied in algebra.

Properties of Inequality for Real Numbers	
For all numbers a, b, and c,	
Comparison Property	$a < b$, $a = b$, or $a > b$.
Transitive Property	**1.** If $a < b$ and $b < c$, then $a < c$. **2.** If $a > b$ and $b > c$, then $a > c$.
Addition and Subtraction Properties	**1.** If $a > b$, then $a + c > b + c$ and $a - c > b - c$. **2.** If $a < b$, then $a + c < b + c$ and $a - c < b - c$.
Multiplication and Division Properties	**1.** If $c > 0$ and $a < b$, then $ac < bc$ and $\frac{a}{c} < \frac{b}{c}$. **2.** If $c > 0$ and $a > b$, then $ac > bc$ and $\frac{a}{c} > \frac{b}{c}$. **3.** If $c < 0$ and $a < b$, then $ac > bc$ and $\frac{a}{c} > \frac{b}{c}$. **4.** If $c < 0$ and $a > b$, then $ac < bc$ and $\frac{a}{c} < \frac{b}{c}$.

The following statement can be used to define the inequality relationship between two numbers.

Definition of Inequality	**For any real numbers a and b, $a > b$ if there is a positive number c such that $a = b + c$.**

CHECKING FOR UNDERSTANDING

Communicating Mathematics

Read and study the lesson to answer each question.

1. Explain why the use of an alibi during a trial is an example of indirect reasoning.

2. In Lesson 2-4, the properties of equality for real numbers were presented. Compare these properties to the properties of inequality presented in this lesson. Describe the similarities and differences between them.

3. Given that $a > b$ and $c > d$, Lynn concluded that $a + b > c + d$.
   a. Find values of a, b, c, and d to show that Lynn's conclusion is not always true.
   b. Write a true inequality that results from the given information.

Guided Practice

State the assumption you would make to start an indirect proof of each statement.

4. Points M, N, and P are collinear.

5. Triangle ABC is acute.

6. The disk is defective.

7. If two parallel lines are cut by a transversal, then alternate exterior angles are congruent.

8. The angle bisector of the vertex angle of an isosceles triangle is also an altitude of the triangle.

Use the figure at the right. Complete each statement with either < or >.

9. $m\angle UNK \underline{\ ?\ } m\angle 1$

10. $m\angle 5 \underline{\ ?\ } m\angle 4$

11. $m\angle 1 \underline{\ ?\ } m\angle 7$

12. If $m\angle 1 < m\angle 3$, then
    $m\angle 5 \underline{\ ?\ } m\angle 1$.

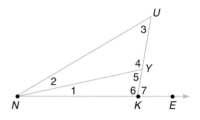

Name the property of inequality that justifies each statement.

13. If $-4x < 20$, then $x > -5$.

14. If $AB + CD > EF + CD$, then $AB > EF$.

15. If $m\angle 1 < m\angle 2$ and $m\angle 3 < m\angle 1$, then $m\angle 3 < m\angle 2$.

16. If $XY \neq 2x - 25$, then $XY < 2x - 25$ or $XY > 2x - 25$.

17. Write an indirect proof.
    Given: $\overline{PQ} \cong \overline{PR}$
    $\angle 1 \not\cong \angle 2$
    Prove: $\overline{PZ}$ is *not* a median of $\triangle PQR$.

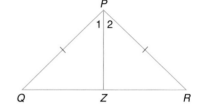

EXERCISES

Practice

18. Complete the indirect proof in paragraph form by completing each sentence.

    Through a point <u>not</u> on a given line, there is exactly one line perpendicular to the given line.

    Given: P is a point not on line ℓ.

    Prove: $\overrightarrow{PQ}$ is the only line through P perpendicular to ℓ.

    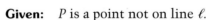

    a. Make the assumption that $\underline{\ ?\ }$.
    b. Both $\overrightarrow{PQ}$ and $\overrightarrow{PR}$ are perpendicular to ℓ because $\underline{\ ?\ }$.
    c. Since $\underline{\ ?\ }$, $\angle 1$ and $\angle 2$ are right angles.
    d. By $\underline{\ ?\ }$, $m\angle 1 = 90$ and $m\angle 2 = 90$.
    e. $m\angle 1 + m\angle 2 + m\angle QPR = 180$ by $\underline{\ ?\ }$.
    f. By $\underline{\ ?\ }$, $90 + 90 + m\angle QPR = 180$.
    g. Therefore by $\underline{\ ?\ }$, $m\angle QPR = 0$.
    h. But, $m\angle QPR > 0$ since $\underline{\ ?\ }$.
    i. Therefore, our assumption is incorrect, so $\underline{\ ?\ }$.

Use the figure to complete each statement with either $<$ or $>$.

19. $m\angle 6 \underline{\ ?\ } m\angle 4$

20. $m\angle 11 \underline{\ ?\ } m\angle 8$

21. $m\angle 9 \underline{\ ?\ } m\angle 1$

22. If $m\angle 10 = m\angle 13$, $m\angle 10 \underline{\ ?\ } m\angle 6$

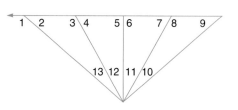

Write a two-column proof.

23. **Given:** $\triangle KNL$
$\overline{NM} \cong \overline{OM}$

   Prove: $m\angle 1 > m\angle 2$

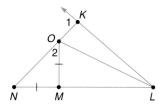

24. **Given:** $\triangle SQR$
$\overline{SP} \cong \overline{QP}$

   Prove: $m\angle\text{SQR} > m\angle 2$

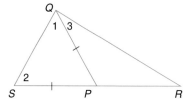

25. **Given:** $\triangle ABC$
$m\angle ABC = m\angle BCA$

   Prove: $x < y$

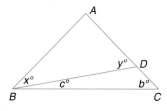

26. **Given:** $\triangle AEC$
$\triangle CFB$

   Prove: $m\angle 4 < m\angle 1$

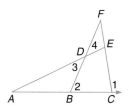

Write an indirect proof.

27. **Given:** $\angle 2 \not\cong \angle 1$

   Prove: ℓ is not parallel to m.

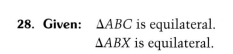

28. **Given:** $\triangle ABC$ is equilateral.
$\triangle ABX$ is equilateral.
$\triangle ACX$ is *not* equilateral.

   Prove: $\triangle BCX$ is *not* equilateral.

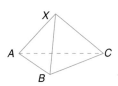

29. If two lines intersect, then they intersect in no more than one point.

30. If two sides of a triangle are *not* congruent, then angles opposite those sides are *not* congruent.

31. If no two altitudes of a triangle are congruent, then the triangle is scalene.

32. If $\overleftrightarrow{AB}$ and $\overleftrightarrow{PQ}$ are skew lines, then $\overleftrightarrow{AP}$ and $\overleftrightarrow{BQ}$ are skew lines.

33. Write a two-column proof.

Given: X is in the interior of $\triangle PQR$.

Prove: $m\angle X > m\angle Q$

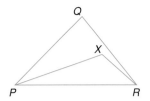

Critical Thinking

34. Use indirect reasoning and a chart to solve this problem:
A guard's prisoner is given the choice of opening one of two doors. Each door leads either to freedom or to the dungeon. A sign on the door on the right reads *This door leads to freedom and the other door leads to the dungeon.* The door on the left has a sign that reads *One of these doors leads to freedom and the other leads to the dungeon.* The guard tells the prisoner that one of the signs is true and the other is false. Which door should the prisoner choose? Why?

Application

35. Entertainment As a dinner party game, Mr. Block had his murder staged. Use the clues from the following scenario to determine who "killed" Mr. Block. Explain your reasoning.

Four people were attending an exclusive dinner party at the island estate of Mr. Block. After dinner, Inspector Photos and Ms. Tebbe went for a walk on the grounds. Upon returning to the house, they found that Mr. Block had been killed, shot in the back with an arrow. Mr. Sopher, who is legally blind without his glasses, told the inspector that he had been in the lounge smoking a pipe when he heard Mr. Block scream. Mrs. Bloom, with her left arm in a sling and right hand wrapped in bandages due to a recent accident, said she had been upstairs looking at paintings when she heard a scream. The security guard who is stationed in a building near the dock informed Inspector Photos that no one other than the four guests had been on the island that day.

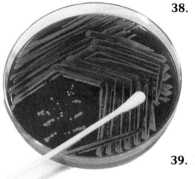

36. If R, S, and T are collinear, $RS = 13$, $ST = 2x + 7$, and $RT = 3x + 8$, which point is between the other two? Justify your answer. **(Lesson 1-4)**

37. Name the properties of equality that justify the statement *If $2x + 6 = 8$, then $x + 3 = 4$.* **(Lesson 2-4)**

38. Find the value of x. Explain your reasoning. **(Lesson 5-2)**

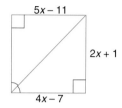

39. Biology A bacteria population triples in number each day. If there are 2,187,000 bacteria on the seventh day, how many bacteria were there on the first day? **(Lesson 5-3)**

Wrap-Up

40. Write an explanation of the steps used for an indirect proof that you could use to teach another student about indirect reasoning.

MID-CHAPTER REVIEW

Refer to the figure at the right to answer each question.

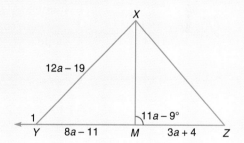

1. Find the value of a if $\overline{XM}$ is a median of $\triangle XYZ$. **(Lesson 5-1)**

2. Find the value of a if $\overline{XM}$ is an altitude of $\triangle XYZ$. **(Lesson 5-1)**

3. If $\overline{XM}$ is the perpendicular bisector of $\overline{YZ}$, is $\triangle XYM \cong \triangle XZM$? Justify your answer. **(Lesson 5-2)**

4. If $\overline{XM}$ is an altitude of $\triangle XYZ$, what additional information do you need to prove $\triangle XYM \cong \triangle XZM$ by HL. **(Lesson 5-2)**

5. Complete with either $<$ or $>$: $m\angle Z \underline{\ ?\ } m\angle 1$. Justify your answer. **(Lesson 5-4)**

6. If $YM < YX$, find all possible values of a. **(Lesson 5-4)**

7. Write an indirect proof for the statement *If two lines not in the same plane do not intersect, then the lines are skew.* **(Lesson 5-4)**

8. Eric, Sally, and Jon are playing a card game. They have a rule that when a player loses a hand, he or she must subtract enough points from his score to double each of the other players' scores. First Eric loses a hand, then Jon, then Sally. If each player now has 8 points, who lost the most points? **(Lesson 5-3)**

Inequalities for Sides and Angles of a Triangle

Objective

After studying this lesson, you should be able to:

- recognize and use relationships between sides and angles in a triangle.

INVESTIGATION

You know from Chapter 4 that in a triangle if two sides are congruent then the angles opposite those sides are congruent. What is the relationship between two angles of a triangle if the sides opposite those angles are not congruent? Draw several different triangles and measure each side and angle. Describe the measure of the angle opposite the longest side of the triangle.

The investigation leads us to Theorem 5-9.

Theorem 5-9	**If one side of a triangle is longer than another side, then the angle opposite the longer side is greater than the angle opposite the shorter side.**

The converse of Theorem 5-9 is also true.

Theorem 5-10	**If one angle of a triangle is greater than another angle, then the side opposite the greater angle is longer than the side opposite the lesser angle.**

A paragraph proof for Theorem 5-9 is given below. You will be asked to write an indirect proof for Theorem 5-10 in Exercise 28.

Proof of Theorem 5-9

Given: $\triangle PQR$

$PQ > RQ$

Prove: $m\angle QRP > m\angle P$

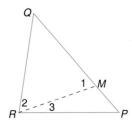

Paragraph Proof:

Draw auxiliary segment RM so that M is between P and Q and $\overline{QM} \cong \overline{QR}$.

Thus, $\triangle QMR$ is isosceles, and $m\angle 1 = m\angle 2$ since if two sides of a triangle are congruent the angles opposite those sides are congruent.

Notice that $\angle 1$ is an exterior angle of $\triangle PRM$. Since the measure of an exterior angle is greater than the measure of either of its corresponding remote interior angles, $m\angle P < m\angle 1$.

Since $m\angle 2 + m\angle 3 = m\angle QRP$ by the angle addition postulate, $m\angle 2 < m\angle QRP$ by the definition of inequality. Thus, $m\angle 1 < m\angle QRP$ by the substitution property of equality.

Finally, we can apply the transitive property of inequality to the inequalities $m\angle P < m\angle 1$ and $m\angle 1 < m\angle QRP$ to get $m\angle P < m\angle QRP$.

You can use algebra to help solve problems involving triangles.

Example 1

CONNECTION

Algebra

In $\triangle PQR$, $m\angle P = 6x + 4$, $m\angle Q = 7x + 12$, and $m\angle R = 6x - 7$. List the sides of $\triangle PQR$ in order from the longest to the shortest.

In order to compare the lengths of the sides of $\triangle PQR$, you must determine the measures of the angles in $\triangle PQR$. Since the sum of the measures of the angles of a triangle is 180, $m\angle P + m\angle Q + m\angle R = 180$.

$$m\angle P + m\angle Q + m\angle R = 180$$
$$(6x + 4) + (7x + 12) + (6x - 7) = 180$$
$$19x + 9 = 180$$
$$19x = 171$$
$$x = 9$$

$$m\angle P = 6x + 4 \qquad m\angle Q = 7x + 12 \qquad m\angle R = 6x - 7$$
$$= 6(9) + 4 \qquad\qquad = 7(9) + 12 \qquad\qquad = 6(9) - 7$$
$$= 58 \qquad\qquad\quad = 75 \qquad\qquad\quad = 47$$

Since $m\angle R < m\angle P < m\angle Q$, by Theorem 5-10, the side opposite $\angle R$ is shorter than the side opposite $\angle P$, and the side opposite $\angle P$ is shorter than the side opposite $\angle Q$. Thus, in $\triangle PQR$, $\overline{PQ}$ is shorter than $\overline{QR}$ is shorter than $\overline{PR}$. So the sides in order from longest to shortest are $\overline{PR}$, $\overline{QR}$, $\overline{PQ}$.

Check the answer by drawing $\triangle PQR$ with angles of 58°, 75°, and 47° and measuring the sides.

In the figure at the right, P is a point not on line t, and Q is the point on t such that $\overline{PQ} \perp t$. In Lesson 3-6, we defined the distance between P and t as the length of $\overline{PQ}$. At that time, it was stated, without proof, that $\overline{PQ}$ was the shortest segment from P to t. Assuming this is true, if R is any point, other than Q, on t, $\overline{PR}$ would have to be longer than $\overline{PQ}$. *Why?*

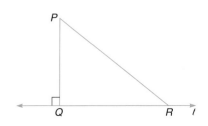

We will now restate our assumption about the shortest segment from a point to a line as a theorem and show that it is true as a direct result of Theorem 5-10.

Theorem 5-11	**The perpendicular segment from a point to a line is the shortest segment from the point to the line.**

Proof of Theorem 5-11

Given: $\overline{PQ} \perp \ell$
$\overline{PR}$ is any segment from P to ℓ that is different from $\overline{PQ}$.

Prove: $PR > PQ$

Proof:

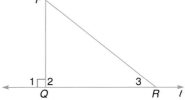

Statements	Reasons
1. $\overline{PQ} \perp \ell$	1. Given
2. $\angle 1$ and $\angle 2$ are right angles.	2. $\perp$ lines form four rt. $\angle$s.
3. $\angle 1 \cong \angle 2$	3. All rt. $\angle$s are $\cong$.
4. $m\angle 1 = m\angle 2$	4. Definition of congruent angles
5. $m\angle 1 > m\angle 3$	5. If an $\angle$ is an ext. $\angle$ of a Δ, then its measure is greater than the measure of either of its corr. remote int. $\angle$s.
6. $m\angle 2 > m\angle 3$	6. Substitution
7. $PR > PQ$	7. If one $\angle$ of a Δ is greater than another $\angle$, then the side opp. the greater $\angle$ is longer than the side opp. the lesser $\angle$.

The proof of Corollary 5-1 follows directly from Theorem 5-11. You will be asked to prove the corollary in Exercise 29.

Corollary 5-1	**The perpendicular segment from a point to a plane is the shortest segment from the point to the plane.**

CHECKING FOR UNDERSTANDING

Communicating Mathematics

Read and study the lesson to answer each question.

1. A triangle has angles measuring about 30°, 55°, and 95° and sides measuring about 4 cm, 5 cm, and 2.5 cm. Draw the triangle and label the measures of all sides and angles. Explain how to construct the triangle.

2. Justify the statement *In an obtuse triangle, the longest side is opposite the obtuse angle.*

3. Write the conclusions you can draw about $\triangle PAL$, if

   a. $PA = PL$

   b. $m\angle A > m\angle L$

   c. $PA < PL$.

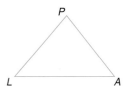

Guided Practice

For each triangle, list the angles in order from greatest to least.

4.

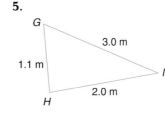

5.

6.

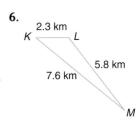

For each triangle, list the sides in order from longest to shortest.

7.

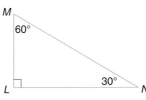

8.

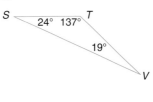

9.

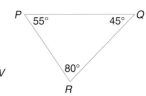

Given the angles indicated in the figure, answer each question.

10. Which side of $\triangle PAT$ is the longest?

11. Which side of $\triangle HPT$ is the shortest?

12. How do the two sides in Exercises 10 and 11 compare?

13. What is the longest side of the figure *PATH?*

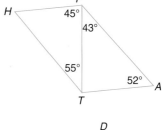

14. Write a two-column proof.

   Given: $\triangle DEF$
   $\angle D$ is a right angle.

   Prove: $EF > ED$

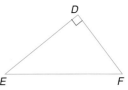

EXERCISES

Practice

15. Name the least and greatest angles in △*ABC*.

16. Name the least and greatest angles in △*BCD*.

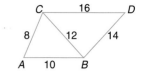

17. Find the shortest segment in the figure. This figure is not drawn to scale.

18. How many of the segments in the figure are longer than $\overline{PS}$?

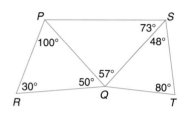

For each triangle in the figure, list the angles in order from least to greatest.

19. △*ADE*   20. △*ABC*

21. △*ABD*   22. △*ACD*

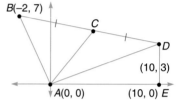

List the sides of △*PQR* in order from longest to shortest if the angles of △*PQR* have the indicated measures.

23. $m\angle P = 7x + 8$, $m\angle Q = 8x - 10$, $m\angle R = 7x + 6$

24. $m\angle P = 3x + 44$, $m\angle Q = 68 - 3x$, $m\angle R = x + 61$

Write a two-column proof.

25. **Given:** $QR > QP$
    $\overline{PR} \cong \overline{PQ}$

    Prove: $m\angle P > m\angle Q$

26. **Given:** $\overline{AC} \cong \overline{AE}$
    $\overline{AE} \cong \overline{KE}$

    Prove: $m\angle 1 > m\angle 2$

27. **Given:** $TE > AE$
    $m\angle P > m\angle PAE$

    Prove: $TE > PE$

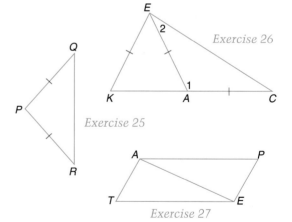

Exercise 26

Exercise 25

Exercise 27

28. Write an indirect proof for Theorem 5-10: If one angle of a triangle is greater than another angle, then the side opposite the greater angle is longer than the side opposite the lesser angle.

29. Write a paragraph proof for Corollary 5-1: The perpendicular segment from a point to a plane is the shortest segment from the point to the plane.

Critical Thinking

30. If $AB > BC > AC$ in $\triangle ABC$ and $\overline{AM}$, $\overline{BN}$, and $\overline{CO}$ are the medians of the triangle, list $\overline{AM}$, $\overline{BN}$, and $\overline{CO}$ in order from longest to shortest.

Application

31. Genetics The line graphed at the right represents the relation obtained by predicting the height of a son given the height of his father. Points A, B, and C each represent a particular father and son. For which pair is the prediction the worst? Justify your claim.

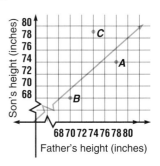

Mixed Review

32. If B is between A and C, $AC = 33$, and $AB = 21 - 4x$, find all possible values of x. **(Lesson 5-4)**

33. Given $\triangle ANT \cong \triangle TOE$, $AN = 15$, $NT = 19$, $AT = 27$, and $EO = 4x - 1$, find the value of x. **(Lesson 4-3)**

34. Draw a figure to illustrate two parallel lines with a plane perpendicular to both lines. **(Lesson 3-1)**

35. If $\angle A \cong \angle B$, $m\angle A = 7d - 1$, and $m\angle B = 11d - 53$, is $\angle B$ acute, right, or obtuse? **(Lesson 1-7)**

Wrap-Up

36. Write a summary of the characteristics of triangles that you have learned in this lesson.

HISTORY CONNECTION

The geometry that we use and study today has its roots in surveying. The word *geometry* literally means to measure the earth. Although the purpose of surveying has remained the same, to accurately describe land, the instruments of surveying have changed over the centuries.

In A.D. 60, an Egyptian named Heron of Alexandria invented the *dioptra* for surveying and making astronomical observations. A dioptra could measure horizontal and vertical angles.

The Romans used a *groma* to measure squares and rectangles for laying out city blocks, streets, and aqueducts. A groma consists of two crosspieces mounted to swivel on a pole. Each arm of the crosspieces has a plumb-bob.

Today's surveyors use *transits* and *theodolites*. These instruments measure both angles and distances. To find a distance, surveyors measure the time it takes for light to travel from one point to another and back. Since they know the speed of light, they can find the distance the light traveled. Modern transits and theodolites use lasers to transmit the light.

5-6 The Triangle Inequality

Objective

After studying this lesson, you should be able to:
- apply the triangle inequality theorem.

Application

Jane's art class has been asked to make a variety of triangular frames for mathematical mobiles for an art show. The students are to make these frames by bending pieces of wire 100 centimeters long at two points and soldering the ends together.

Will it make any difference where along the wire Jane makes her two bends in order to create the triangular frames? Can these two points be chosen purely at random? What are the factors Jane needs to take into account before bending the wire?

INVESTIGATION

Try simulating Jane's wire bending by taking a straw and bending it in two distinct points and then putting a pin through the ends to fasten them together. Try several different combinations of bends similar to those suggested below. Will the process always form a triangle? Why or why not?

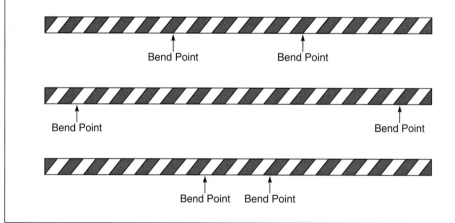

What conclusion can you draw about the lengths of the sides of triangles? One possible conclusion is stated in the following theorem.

Theorem 5-12 *Triangle Inequality Theorem*	**The sum of the lengths of any two sides of a triangle is greater than the length of the third side.**

You will be asked to prove the Triangle Inequality Theorem in Exercise 37.

The Triangle Inequality Theorem shows that some sets of line segments cannot be used to form a triangle, because their lengths do not satisfy the inequality.

Example 1

Mrs. Bailey, Jane's art teacher, gave Jane four pieces of copper tubing to use to make a triangular base for her mobiles. The lengths of the pieces of tubing are 2 meters, 4.5 meters, 5.8 meters, and 10.2 meters. How many different triangles could Jane make?

There are four possible ways for Jane to choose three pieces of tubing for the triangular base. She can choose the following combinations:

a. 2 m, 4.5 m, and 5.8 m
b. 2 m, 4.5 m, and 10.2 m
c. 2 m, 5.8 m, and 10.2 m
d. 4.5 m, 5.8 m, and 10.2 m

To determine if each of the four combinations can be used to form a triangle, you must test that all combinations for the lengths of the three sides satisfy the triangle inequality.

a. 2 m, 4.5 m, 5.8 m

Is 2 + 4.5 > 5.8?   *yes*
Is 2 + 5.8 > 4.5?   *yes*
Is 4.5 + 5.8 > 2?   *yes*

These three pieces can be used to form a triangle.

b. 2 m, 4.5 m, 10.2 m

Is 2 + 4.5 > 10.2?   *no*

These three pieces cannot be used to form a triangle. *Why don't you need to check the other two combinations for the sides?*

c. 2 m, 5.8 m, 10.2 m

Is 2 + 5.8 > 10.2?   *no*

These three pieces cannot be used to form a triangle.

d. 4.5 m, 5.8 m, 10.2 m

Is 4.5 + 5.8 > 10.2?   *yes*
Is 4.5 + 10.2 > 5.8?   *yes*
Is 5.8 + 10.2 > 4.5?   *yes*

These three pieces can be used to form a triangle.

Jane can make mobiles with either of two different sets of tubing, Set A or Set D.

If the lengths of two sides of a triangle are known, then it is possible to determine the range of possible lengths for the third side.

Example 2

The lengths of two sides of a triangle are 6 cm and 10 cm. What are the possible lengths for the third side of this triangle?

Let x = the length of the third side.

By the Triangle Inequality Theorem, each of these inequalities must be true.

$$x + 6 > 10 \qquad x + 10 > 6$$
$$x > 4 \qquad\qquad x > -4$$

$$10 + 6 > x$$
$$16 > x \quad x < 16$$

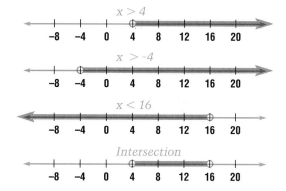

The length of the third side must fall in the range included in all three inequalities. The graphs show us that the length of the third side must be between 4 cm and 16 cm.

The possible measures can be expressed as follows: $\{x \mid 4 < x < 16\}$.

CHECKING FOR UNDERSTANDING

Communicating Mathematics

Read and study the lesson to answer each question.

1. State the Triangle Inequality Theorem in your own words.

2. Give an example of a set of numbers that can be the lengths of the sides of a triangle and a set of numbers that cannot be. Use drawings or sketches to justify your reasoning.

3. An isosceles triangle has a base that is 8 inches long.
   a. Are there any restrictions on how short or how long the legs of this triangle can be?
   b. Draw diagrams to help explain your answer.

4. Two sides of a triangle are 5 cm and 13 cm long. Must 5 cm be the length of the shortest side of the triangle? Must 13 cm be the length of the longest side? Explain.

Guided Practice

Determine whether it is possible to draw a triangle with sides of the given measures. Write *yes* or *no*. If yes, then draw the triangle.

5. 1, 2, 5

6. 11, 10, 17

7. 2.4, 6.8, 4.5

The measures of two sides of a triangle are given. Between what two numbers must the measure of the third side fall?

8. 12 and 15

9. 4 and 13

10. 21 and 17

11. Is it possible for the points $A(0, 0)$, $B(3, 2)$, and $C(-6, -4)$ to be the vertices of a triangle? Explain your answer.

12. Write a two-column proof.

    Given: $\angle B \cong \angle ACB$

    Prove: $AD + AB > CD$

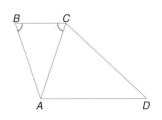

EXERCISES

Practice

Determine whether it is possible to draw a triangle with sides of the given measures. Write *yes* or *no.*

13. 12, 11, 17   **14.** 1, 2, 3   **15.** 4.7, 9, 4.1

16. 2.5, 6, 6.5   **17.** 12, 2.2, 14.3   **18.** 2.3, 12, 12.2

19. 9, 40, 41   **20.** 5, 100, 100   **21.** 204, 7, 215

Two sides of a triangle are **18** and **21** centimeters in length. Determine whether each measurement can be the length of the third side.

22. 10 cm   **23.** 40 cm   **24.** 7 cm

25. 21 cm   **26.** 3 cm   **27.** 57 cm

Is it possible to have a triangle with the given vertices? Write *yes* or *no.* Explain your answer.

28. $A(4, -3)$, $B(0, 0)$, $C(-4, 3)$ **29.** $D(-2, 1)$, $E(2, -1)$, $F(-6, 3)$

30. $G(-2, 4)$, $H(-6, 5)$, $I(-3, -3)$ **31.** $J(3, -3)$, $K(8, 2)$, $L(5, 5)$

32. Answer each question, given that Victor has five straws with lengths of 3 cm, 4 cm, 5 cm, 6 cm, and 12 cm.

    a. How many different triangles can Victor make with the straws?

    b. How many different triangles can Victor make that have a perimeter that is divisible by 3?

    c. There are 10 different combinations of straws. What is the probability that if Victor chooses three straws at random he will be able to make a triangle? *(number of choices that will make a triangle) ÷ (total number of ways to choose 3 straws from a group of 5)*

If the sides of a triangle have the following lengths, find all possible values for *x.*

33. $(3x + 2)$ cm, $(8x - 10)$ cm, $(5x + 8)$ cm

34. $2x$ ft, $(15 - x)$ ft, $(4x - 6)$ ft

35. How many different triangles are possible if the measures of the three sides must be selected, without repetition, from the measures 1, 2, 3, and 4?

Write a two-column proof for each.

36. Given: $MN = MQ$

    Prove: $OP + ON > PQ$

37. Given: $\overline{AD}$, $\overline{BE}$, and $\overline{CF}$ are altitudes of $\triangle ABC$.
   Prove: $AB + BC + AC > AD + BE + CF$

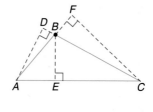

38. Write a two-column proof for the Triangle Inequality Theorem. (Theorem 5-12)
   Given: $\triangle PQR$
   Prove: $PQ + PR > RQ$
   Hint: Draw auxiliary segment PS so that P is between S and R and $\overline{PQ} \cong \overline{PS}$.

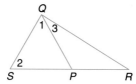

Critical Thinking

39. State and prove a theorem that compares the measure of the longest side of a quadrilateral with the measures of the other three sides.

Graphing Calculator

40. Choose a partner. Each partner should enter the code [INT] [( ] 100 [×] [RAND] [ )] in a TI-81 graphing calculator or the code [INT] [( ] 100 [×] [RAN#] [ )] in a Casio graphing calculator. Then each should press [ENTER] or [EXE] three times and record the random values from the display. The first person who obtains five sets of random values that are possible measures for the sides of a triangle wins the game.

Applications

41. Construction A metal rod that is 9 inches long is to be cut into three pieces and welded together to form a triangular brace for a stair step.
   a. If the pieces all have an integral length, name all of the possible combinations of lengths.
   b. Is the triangle with the shortest side isosceles, scalene, or equilateral?

42. Gardening Mr. and Mrs. Zellar have six old railroad ties that they would like to use to border two different triangular flower beds. They have two ties that are 3 feet long, two ties that are 5 feet long, one tie that is 7 feet long, and one tie that is 4 feet long. Can these ties be used to border two flower beds without having to cut them? If so, what are the possible dimensions of each flower bed?

Mixed Review

43. What is the longest segment of $\triangle XYZ$ if $m\angle X = 4n + 61$, $m\angle Y = 67 - 3n$, and $m\angle Z = n + 74$? **(Lesson 5-5)**

44. If $\triangle NQD$ has vertices $N(2, -1)$, $Q(-4, -1)$, and $D(-1, 3)$, describe $\triangle NQD$ in terms of its angles and sides. **(Lesson 4-1)**

45. Find the slope of the line that passes through points $(6, -11)$ and $(4, 9)$. **(Lesson 3-5)**

46. For the following theorem, name the given and the prove statements and draw a figure. Then write a two-column proof. **(Lesson 2-6)**
   If $\overline{PQ}$ bisects $\overline{AB}$ at point M, then $\overline{AM} \cong \overline{MB}$.

47. Algebra The measure of an angle is 5 more than four times the measure of its complement. Find the measure of both angles. **(Lesson 1-8)**

Wrap-Up

48. Write a sentence to describe a situation where the triangle inequality theorem may be used in daily living.

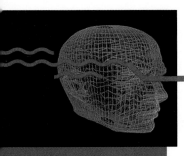

Technology

▶ **BASIC**
Geometric Supposer
Graphing calculators
LOGO
Spreadsheets

The Triangle Inequality

You have learned that the sum of the measures of any two sides of a triangle must be greater than the measure of the third side. You can determine whether any three numbers can be the measures of the sides of a triangle by addition. However, this process can be long and tiresome if you have several sets of numbers to check. A computer can help you complete the task in very little time.

The BASIC program below can be used to do the arithmetic necessary to test any three numbers to see if they may be the measures of the sides of a triangle. You simply enter the numbers you wish to check.

```
10 PRINT "THE TRIANGLE TESTER"
20 PRINT
30 PRINT "WHAT THREE MEASURES WOULD YOU LIKE TO TEST?"
40 PRINT "MAKE SURE THAT YOU ENTER THE MEASURES OF THE
   SIDES IN ORDER FROM LEAST TO GREATEST."
50 INPUT A, B, C
60 PRINT
70 IF A + B > C THEN 100
80 PRINT "A, B, AND C CANNOT BE THE MEASURES OF THE
   SIDES OF A TRIANGLE."
90 GOTO 110
100 PRINT "A, B, AND C CAN BE THE MEASURES OF THE SIDES OF
   A TRIANGLE."
110 END
```

EXERCISES

Use the BASIC program to determine whether the numbers given could be the measures of the sides of a triangle.

1. 4, 6, 11

2. 3.5, 7.75, 5.25

3. 5.776, 11.845, 5.803

4. 389, 227, 101

5. What does line 70 of the program do?

6. The program only checks one inequality instead of three. Why?

Inequalities Involving Two Triangles

Objective

After studying this lesson, you should be able to:
- use the SAS Inequality and SSS Inequality in proofs and to solve problems.

Application

Over the winter, Gina Salazar follows a program to stretch and strengthen different muscles in preparation for track season in the spring. One exercise to strengthen the muscles around her ankles is shown below. For this exercise, she moves one foot against the tension supplied by an elastic band.

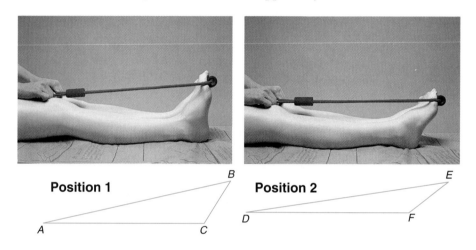

Position 1   **Position 2**

The sides that represent Gina's leg and foot have the same length in each triangle. Thus, $\overline{AC} \cong \overline{DF}$ and $\overline{BC} \cong \overline{EF}$. Because of the way Gina moves her foot, you know the measure of the angle between her leg and foot is greater in position 2 than in position 1. Also, the elastic band is longer in position 2 than in position 1. Thus, $m\angle F > m\angle C$ and $DE > AB$.

In general, it appears that as the measure of the angle between Gina's leg and foot increases, the length of the elastic band increases. Similarly, if the length of the elastic band is increased, the angle between the leg and foot increases to maintain the tension. This suggests the following two theorems.

Theorem 5-13 **SAS Inequality** **(Hinge Theorem)**	**If two sides of one triangle are congruent to two sides of another triangle, and the included angle in one triangle is greater than the included angle in the other, then the third side of the first triangle is longer than the third side in the second triangle.**
Theorem 5-14 **SSS Inequality**	**If two sides of one triangle are congruent to two sides of another triangle and the third side in one triangle is longer than the third side in the other, then the angle between the pair of congruent sides in the first triangle is greater than the corresponding included angle in the second triangle.**

An indirect proof for the SSS Inequality is given below. You will be asked to write a paragraph proof for the SAS Inequality in Exercise 27.

Indirect Proof of SSS Inequality (Theorem 5-14)

Given: $\overline{AC} \cong \overline{DF}$
$\overline{BC} \cong \overline{EF}$
$DE > AB$

Prove: $m\angle F > m\angle C$

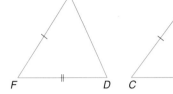

Proof:

Step 1: Assume $m\angle F \leq m\angle C$.

Step 2: If $m\angle F \leq m\angle C$, then either $m\angle F = m\angle C$ or $m\angle F < m\angle C$.
Case 1: If $m\angle F = m\angle C$, then $\triangle ABC \cong \triangle DEF$ by SAS, and $AB = DE$. *Why?*   *CPCTC*
Case 2: If $m\angle F < m\angle C$, then $AB > DE$ by SAS Inequality.

Step 3: In both cases, our assumption lead to a contradiction of the hypothesis that $DE > AB$. Therefore, the assumption that $m\angle F \leq m\angle C$ must be false, and the conclusion, $m\angle F > m\angle C$, must be true.

Examples 1 and 2 illustrate different geometric and algebraic applications of the SAS Inequality and the SSS Inequality.

Example 1

Write a two-column proof.

Given: $\overline{EA}$ is a median of $\triangle KEC$.
$m\angle 1 > m\angle 2$

Prove: $m\angle K > m\angle C$

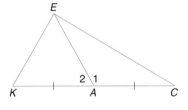

Proof:

Statements	Reasons
1. $\overline{EA}$ is a median of $\triangle KEC$.	1. Given
2. A is the midpoint of $\overline{KC}$.	2. Definition of median
3. $\overline{KA} \cong \overline{AC}$	3. Definition of midpoint
4. $\overline{EA} \cong \overline{EA}$	4. Congruence of segments is reflexive.
5. $m\angle 1 > m\angle 2$	5. Given
6. $EC > EK$	6. SAS Inequality
7. $m\angle K > m\angle C$	7. If one side of a $\triangle$ is longer than another side, then the $\angle$ opp. the longer side is greater than the $\angle$ opp. the shorter side.

Example 2

In the figure, $\angle 1 \cong \angle 2$, $\overline{PT} \cong \overline{QR}$, and $PQ > SR$. Write two inequalities to describe the possible values for x.

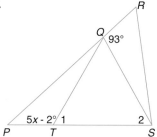

If two angles in a triangle are congruent, then the sides opposite those angles are congruent. So, since $\angle 1 \cong \angle 2$, $\overline{QS} \cong \overline{QT}$.

Since $\overline{PT} \cong \overline{QR}$, $\overline{QT} \cong \overline{QS}$, and $PQ > SR$, you know $m\angle PTQ > m\angle RQS$ by the SSS Inequality.

Since $\angle PTQ$ is an angle of a triangle, you know that $m\angle PTQ < 180$.

Use these two inequalities to describe the possible values for x.

$m\angle PTQ > m\angle RQS$	$m\angle PTQ < 180$
$5x - 2 > 93$	$5x - 2 < 180$
$5x > 95$	$5x < 182$
$x > 19$	$x < 36.4$

The inequalities $x > 19$ and $x < 36.4$, or $19 < x < 36.4$, describe the possible values of x.

CHECKING FOR UNDERSTANDING

Communicating Mathematics

Read and study the lesson to answer each question.

1. Draw two isosceles triangles with legs 2 centimeters long. Draw one triangle with a vertex angle of 40° and the other with a vertex angle of 100°.

   a. Measure the base of each triangle.

   b. Which triangle has a longer base?

   c. What theorem does this investigation demonstrate?

2. The SAS Inequality Theorem is subtitled the Hinge Theorem. Explain why you think it is given this name.

3. Use an everyday object to illustrate the SAS Inequality. For example, as the angle between a door and its frame increases, the edge of the door gets farther from the edge of the frame.

Refer to the figure at the right to write an equation or inequality relating each pair of angle measures.

4. $m\angle ALK$, $m\angle ALN$

5. $m\angle ALK$, $m\angle NLO$

6. $m\angle OLK$, $m\angle NLO$

7. $m\angle KLO$, $m\angle ALN$

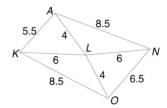

CONNECTION
Algebra

In $\triangle ABC$, M is the midpoint of $\overline{AB}$. If $m\angle 1 = 5x + 20$ and $m\angle 2 = 8x - 100$, determine which measure is greater. Justify your answers.

8. BC or AC

9. $m\angle B$ or $m\angle A$

10. $m\angle 1$ or $m\angle 2$

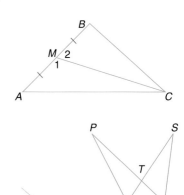

11. Write a two-column proof.
 Given: $\overline{PQ} \cong \overline{SQ}$
 Prove: $PR > SR$

EXERCISES

Practice

Write an inequality relating each pair of measures.

12. $m\angle ADC$, $m\angle ADB$

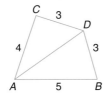

13. AB, AC

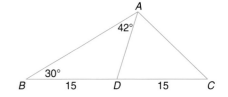

14. PT, RS

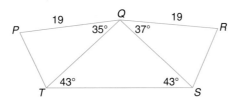

15. $m\angle 1$, $m\angle 2$

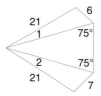

16. ZR, XR

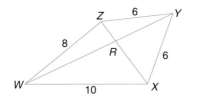

17. $m\angle DFE$, $m\angle DFG$

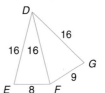

For each figure, write an inequality or pair of inequalities to describe the possible values of *x*.

18.
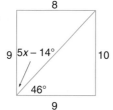

19.
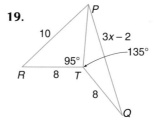

20.
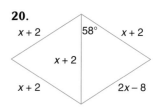

Write a two-column proof.

21. **Given:** $\overline{PQ} \cong \overline{RS}$
$QR < PS$

Prove: $m\angle 3 < m\angle 1$

22. **Given:** $\overline{PR} \cong \overline{PQ}$
$SQ > SR$

Prove: $m\angle 1 < m\angle 2$

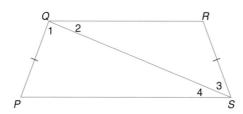
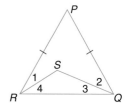

23. **Given:** $\triangle TER$
$\overline{TR} \cong \overline{EU}$

Prove: $TE > RU$

24. **Given:** $\overline{TU} \cong \overline{US}$
$\overline{US} \cong \overline{SV}$
$m\angle SVU > m\angle USV$

Prove: $ST > UV$

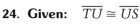
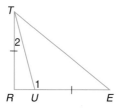

25. **Given:** $\overline{ED} \cong \overline{DF}$
$m\angle 1 > m\angle 2$
D is the midpoint of $\overline{CB}$.
$\overline{AE} \cong \overline{AF}$

Prove: $AC > AB$

26. **Given:** $m\angle DBC = m\angle DCB$
$m\angle ADB < m\angle ADC$

Prove: $m\angle ACB < m\angle ABC$

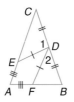
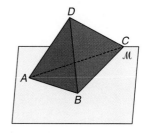

27. Write a paragraph proof for the SAS Inequality. (Theorem 5-13)

Given: $\overline{AC} \cong \overline{DF}$

$\overline{BC} \cong \overline{EF}$

$m\angle F > m\angle C$

Prove: $DE > AB$

Hint: Draw auxiliary ray FZ such that $m\angle DFZ = m\angle C$ and $ZF = BC$. Then, consider two cases: Z lies on $\overline{DE}$, and Z does not lie on $\overline{DE}$.

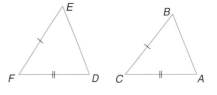

Critical Thinking

28. In the figure at the right, plane $\mathcal{P}$ bisects $\overline{XZ}$ at Y and $WZ > WX$. What can you conclude about the relation between $\overline{XZ}$ and plane $\mathcal{P}$?

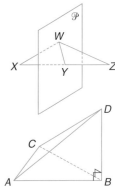

29. In the figure at the right, $\triangle ABC$ is equilateral. Name the angle that is congruent to $\angle DCA$. Justify your answer.

Application

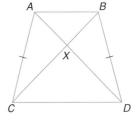

30. Biology In the 1970s, R. McNeill Alexander created a formula to estimate the speed (velocity) of an animal. This formula is

$$v = \frac{0.78s^{1.67}}{h^{1.17}},$$

where v is the speed of the animal in meters per second, s is the length of the animal's stride in meters, and h is the height of the animal's hip in meters.

a. Determine the velocity of two animals that each have a hip height of 1.08 meters and that have strides of 2.26 meters and 2.40 meters.

b. Draw a mathematical model of the triangles formed by the two animals in Part A if one point of the triangle represents the position of each animal's hip and the other two points represent the beginning and end of each animal's stride. Then discuss how this model is related to either the SAS Inequality or the SSS Inequality.

Mixed Review

31. Algebra Is it possible to have a triangle with vertices $A(1, -1)$, $B(7, 7)$, $C(2, -5)$? Explain. **(Lesson 5-6)**

32. Write a two-column proof. **(Lesson 4-6)**

Given: $\overline{AC} \cong \overline{BD}$

$\overline{AD} \cong \overline{BC}$

Prove: $\triangle AXC \cong \triangle BXD$

33. If $\overline{AB} \parallel \overline{CD}$, $m\angle BAC = 3n + 32$, and $m\angle ACD = 14n - 5$, find the value of n. **(Lesson 3-3)**

34. $\angle A$ and $\angle B$ are complementary. If $m\angle A = 6x + 6$ and $m\angle B = 11x - 1$, find $m\angle A$ and $m\angle B$. **(Lesson 1-8)**

Wrap-Up

35. Journal Entry Write a sentence or two about the SAS Inequality and the SSS Inequality in your journal.

VOCABULARY

Upon completing this chapter, you should be
familiar with the following terms:

altitude of a triangle **216**	**233** indirect reasoning
angle bisector of a triangle **217**	**216** median of a triangle
indirect proof **233**	**218** perpendicular bisector

SKILLS AND CONCEPTS

OBJECTIVES AND EXAMPLES

Upon completing this chapter, you should
be able to:

- identify and use medians, altitudes, angle
  bisectors, and perpendicular bisectors of a
  triangle. **(Lesson 5-1)**

If $\overline{BG}$ is an altitude of
$\triangle BDC$, then $\overline{BG} \perp \overline{DC}$.

If $\overline{AE}$ is a median of
$\triangle ABD$, then $\overline{DE} \cong \overline{BE}$.

If $\overline{EF}$ is an angle bisector
of $\triangle AED$, then $\angle 1 \cong \angle 2$.

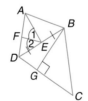

REVIEW EXERCISES

Use these exercises to review and prepare
for the chapter test.

Refer to the figure to answer each question.

1. $\overline{BG}$ is an altitude and an angle bisector
   of $\triangle BCD$. If $m\angle DBG = 33$, find the
   measure of the three angles of $\triangle BCD$.

2. Find the value of x and $m\angle 2$ if $\overline{AE}$ is an
   altitude of $\triangle ABD$, $m\angle 1 = 3x + 11$, and
   $m\angle 2 = 7x + 9$.

3. If $\overline{AE}$ is a median of $\triangle ABD$,
   $DE = 3x - 14$, $EB = 2x - 1$, and
   $m\angle AED = 7x + 1$, is $\overline{AE}$ also an
   altitude of $\triangle ABD$? Explain.

- recognize and use
  tests for congruence
  of right triangles.
  (Lesson 5-2)

Find the values of
x and y.

Since $\triangle ABC \cong \triangle DEF$ by HA, $AB = DE$
and $BC = EF$.

$$AB = DE \qquad BC = EF$$
$$3x + 7 = 19 \qquad 2y - 11 = 15$$
$$3x = 12 \qquad 2y = 26$$
$$x = 4 \qquad y = 13$$

Find the values of x and y.

4.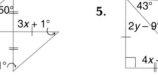

5.

6.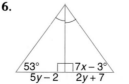

7.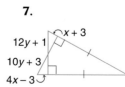

| **OBJECTIVES AND EXAMPLES** | **REVIEW EXERCISES** |

- use indirect reasoning and indirect proof to reach a conclusion. **(Lesson 5-4)**

Steps for Writing an Indirect Proof

1. Assume that the conclusion is false.

2. Show that the assumption leads to a contradiction of the hypothesis or some other fact, such as a postulate, theorem, or corollary.

3. Conclude that the assumption must be false, and therefore the conclusion must be true.

Write an indirect proof.

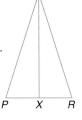

8. **Given:** $\overline{QP} \cong \overline{QR}$
   $\overline{QX}$ does not bisect $\angle PQR$.
   Prove: $\overline{QX}$ is not a median of $\triangle PQR$.

9. **Given:** $\triangle QXP \cong \triangle QXR$
   Prove: $\overline{QX}$ is an altitude of $\triangle PQR$.

- recognize and apply the properties of inequalities to the measures of segments and angles. **(Lesson 5-4)**

Find the value of a if $AC = AB$, $AX = 16$, and $AC = 3a - 2$.

$AC = AB$
$AC = AX + XB$
$AC > AX$
$3a - 2 > 16$
$3a > 18$
$a > 6$

Complete each statement with < or >. Use the figure at the left.

10. $m\angle 3 \underline{\ ?\ } m\angle ACB$

11. $m\angle 4 \underline{\ ?\ } m\angle 10$

12. $m\angle 6 \underline{\ ?\ } m\angle 11$

13. $m\angle 5 \underline{\ ?\ } m\angle YXB$

14. If $m\angle 7 < m\angle 4$, then $m\angle 1 \underline{\ ?\ } m\angle 8$

- recognize and use relationships between sides and angles in a triangle. **(Lesson 5-5)**

Find the longest segment in the figure. The figure is not drawn to scale.

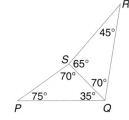

The longest side of $\triangle PQS$ is the side opposite the 75° angle, $\overline{SQ}$. This side is also the shortest side of $\triangle QRS$ since it is opposite the 45° angle. Thus, the longest side is the side opposite the 70° angle in $\triangle QRS$, $\overline{RS}$.

15. Name the shortest side in the figure at the left.

16. List the sides of $\triangle ABC$ in order from shortest to longest.

17. List the angles of $\triangle BCD$ in order from least to greatest.

18. Write a two-column proof.
    Given: $FG < FH$
    Prove: $m\angle 1 > m\angle 2$

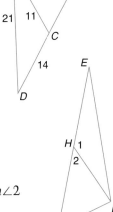

■ apply the Triangle Inequality Theorem.
(Lesson 5-6)

The measures of two sides of a triangle are 7 and 9. Between what two numbers is the measure of the third side?

Let x be the measure of the third side. Then by the Triangle Inequality Theorem, the following inequalities must be true.

$x + 7 > 9$	$x + 9 > 7$	$7 + 9 > x$
$x > 2$	$x > -2$	$16 > x$

Thus, the measure of the third side must be between 2 and 16.

The measures of two sides of a triangle are given. Between what two numbers is the measure of the third side?

19. 5 and 11

20. 24 and 7

Is it possible to have a triangle with the given vertices?

21. $A(-5, 12)$, $B(4, -3)$, $C(0, 0)$

22. $D(-3, 4)$, $E(3, -5)$, $F(-1, 1)$

23. How many different triangles are possible if the measures of the three sides must be selected from the measures 2, 3, 4, and 5?

■ use the SAS Inequality and SSS Inequality in proofs and to solve problems.
(Lesson 5-7)

Write an inequality relating AB and AC.

$m\angle AXC = 75 + 18$
$\qquad = 93$

$m\angle AXB = 180 - 93$
$\qquad = 87$

Since $m\angle AXC > m\angle AXB$, by SAS Inequality, $AB < AC$.

Refer to the figure to write an inequality relating each pair of measures.

24. $m\angle PNQ$, $m\angle QNR$

25. $m\angle PNS$, $m\angle RNS$

26. $m\angle NPQ$, $m\angle NRS$

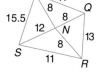

27. Write a two-column proof.

Given: $AD = BC$
Prove: $AC > DB$

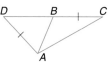

APPLICATIONS AND CONNECTIONS

28. Solve by working backward. An ice sculpture is melting at a rate of one-half its weight every one hour. After 8 hours, the sculpture weighs $\frac{5}{16}$ of a pound. How much did the sculpture weigh to begin with? **(Lesson 5-3)**

29. **Algebra** Find the coordinates of the point on line ℓ, whose equation is $y = 2x - 5$, that is the endpoint of the shortest segment from the origin to ℓ. **(Lesson 5-5)**

30. Pine City is 6 kilometers from Susanton, 9 kilometers from Blockburg, and 13 km from Leshville. Also, Susanton, Blockburg, and Leshville do not lie on a straight line. Bonnie begins in Blockburg and drives directly to Susanton, then to Leshville, and then back to Blockburg. What are the maximum and minimum distances she could have traveled on this trip? **(Lesson 5-7)**

CHAPTER 5 | TEST

In $\triangle AHW$, $m\angle A = 64$ and $m\angle AWH = 36$. If $\overline{WP}$ is an angle bisector and $\overline{HQ}$ is an altitude, find each measure.

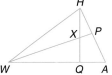

1. $m\angle AQH$
2. $m\angle AHQ$
3. $m\angle APW$
4. $m\angle HXW$

5. If $\overline{WP}$ is a median, $AP = 3y + 11$ and $PH = 7y - 5$, find AH.

Complete each statement with $<$ or $>$. Use the figure at the right.

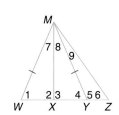

6. $m\angle 2 \underline{\ ?\ } m\angle 9$
7. $m\angle 6 \underline{\ ?\ } m\angle 1$
8. If $m\angle 8 < m\angle 7$, then $WX \underline{\ ?\ } XY$.
9. If $MX < MZ$, then $m\angle 4 \underline{\ ?\ } m\angle 5$.

Complete each statement. Use the figure at the right.

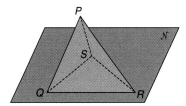

10. If $PS > PQ$, then $m\angle PQS \underline{\ ?\ } m\angle PSQ$.
11. If $m\angle SQR > m\angle SRQ$, then $\underline{\ ?\ } > \underline{\ ?\ }$.
12. If $\overline{PQ} \cong \overline{PR}$ and $SQ > SR$, then $\underline{\ ?\ } \underline{\ ?\ } m\angle SPQ$.
13. If $\overline{QR} \cong \overline{SR}$, $QS < PQ$, and $m\angle PRQ < m\angle PRS$, then the longest side of $\triangle PQS$ is $\underline{\ ?\ }$.
14. If $PS = 31$ and $SQ = 13$, then PQ is between what two numbers?

15. Find the longest segment in $\triangle ABC$ if $m\angle A = 5x + 31$, $m\angle B = 74 - 3x$, and $m\angle C = 4x + 9$.

16. If the sides of a triangle have lengths of $3x + 8$ m, $5x + 2$ m and $8x - 10$ m, find all possible values for x.

17. If M is the midpoint of $\overline{DE}$, $m\angle 1 > m\angle 2$, $DF = 13x - 5$, and $EF = 7x + 25$, find all possible values of x.

18. Write an indirect proof.
    Given: $\overline{FM}$ is a median of $\triangle DEF$.
    $m\angle 1 > m\angle 2$
    Prove: $DF \neq EF$

Write a two-column proof.

19. **Given:** $NO = QP$
    $PN > OQ$
    Prove: $MP > MO$

20. **Given:** $\overline{AD} \perp \overline{DC}$,
    $\overline{AB} \perp \overline{BC}$
    $AB = DC$
    Prove: $\overline{DC} \perp \overline{BC}$

Bonus Write a two-column proof for the following statement:
If M is any point in the interior of $\triangle PQR$, then $PR + QR > PM + QM$.

Algebra Review

OBJECTIVES AND EXAMPLES	REVIEW EXERCISES

Solve equations with variables on both sides.

$$3c + 22 = 8c - 3$$
$$3c - 3c + 22 = 8c - 3c - 3$$
$$22 + 3 = 5c - 3 + 3$$
$$\frac{25}{5} = \frac{5c}{5}$$
$$5 = c$$

Solve each equation. Check the solution.

1. $5a - 5 = 7a - 19$
2. $\frac{2}{3}x + 5 = \frac{1}{2}x + 4$
3. $5(4 - n) = 2n - 1$
4. $2(2y - 3) = -9(y - 6) + y$

Solve problems involving percent of increase or decrease.

A shirt's price was decreased from $25 to $20. Find the percent of decrease.

$$\frac{r}{100} = \frac{25 - 20}{25}$$
$$r = 100\left(\frac{5}{25}\right) \text{ or } 20$$

The decrease was 20%.

Solve.

5. A skirt's price was increased from $20 to $25. Find the percent of increase.
6. The price of a half-gallon of ice cream plus 5% tax is $3.15. What is the original price of the ice cream?
7. A pair of jeans sells for $36 after a 25% discount. What is the original price of the jeans?

Solve problems involving direct variation.

If y varies directly as x, and $x = 15$ when $y = 1.5$, find x when $y = 9$.

$$\frac{x}{9} = \frac{15}{1.5} \qquad \frac{x_1}{x_2} = \frac{y_1}{y_2}$$
$$x = 9\left(\frac{15}{1.5}\right) \text{ or } 90$$

Solve. Assume that y varies directly as x.

8. If $y = 15$ when $x = 5$, find y when $x = 7$.
9. If $y = 35$ when $x = 175$, find y when $x = 75$.
10. If $y = 1.2$ when $x = 21$, find x when $y = 21$.

Solve inequalities using multiplication or division.

$$-\frac{2m}{3} \leq 10$$
$$-\frac{3}{2}\left(-\frac{2m}{3}\right) \geq -\frac{3}{2}(10)$$
$$m \geq -15$$

The direction of the inequality must be reversed because each side is multiplied by a negative number.

The solution set is $\{m | m \geq -15\}$.

Solve each inequality. Check the solution.

11. $6x \leq -24$
12. $-7y \geq -91$
13. $-0.8t < -0.96$
14. $\frac{4}{3}a < 16$
15. $\frac{2}{3}k \geq \frac{2}{15}$
16. $\frac{4}{7}z > -\frac{2}{5}$

OBJECTIVES AND EXAMPLES

■ Simplify expressions involving powers of monomials or negative exponents.

$(2x^2y^3)^3 = 2^3(x^2)^3(y^3)^3$
$\qquad = 8x^6y^9$

$\dfrac{3a^{-2}}{4a^6} = \dfrac{3}{4}(a^{-2-6}) = \dfrac{3}{4a^8}$

REVIEW EXERCISES

Simplify. Assume no denominator is equal to zero.

17. $(4a^2b)^3$

18. $(-3xy)^2(4x)^3$

19. $(-2c^{-2}d)^4(-3cd^2)^3$

20. $\dfrac{(3a^3b^{-1}c^2)^2}{18a^2b^3c^4}$

■ Express numbers in scientific notation.

$3{,}600{,}000 = 3.6 \times 10^6$

$0.0021 = 2.1 \times 10^{-3}$

Express each number in scientific notation.

21. 240,000

22. 4,880,000,000

23. 0.000314

24. 0.00000187

■ Add or subtract polynomials.

$(4x^2 - 3x + 7) + (2x^2 + 4x)$
$\quad = (4x^2 + 2x^2) + (-3x + 4x) + 7$
$\quad = 6x^2 + x + 7$

$(7r^2 + 9r) - (12r^2 - 4r - 3)$
$\quad = (7r^2 - 12r^2) + [9r - (-4r)] - (-3)$
$\quad = -5r^2 + 13r + 3$

Find each sum or difference.

25. $(2x^2 - 5x + 7) - (3x^3 + x^2 + 2)$

26. $(x^2 - 6xy + 7y^2) + (3x^2 + xy - y^2)$

27. $(11m^2n^2 + 4mn - 6) + (5m^2n^2 - 6mn + 17)$

28. $(7a^2 + 4) - (3a^2 + 2a - 6)$

■ Factor quadratic trinomials.

$a^2 - 3a - 4 = (a + 1)(a - 4)$

$4x^2 - 4xy - 15y^2$
$\quad = 4x^2 + (-10 + 6)xy - 15y^2$
$\quad = (4x^2 - 10xy) + (6xy - 15y^2)$
$\quad = 2x(2x - 5y) + 3y(2x - 5y)$
$\quad = (2x + 3y)(2x - 5y)$

Factor each trinomial.

29. $y^2 + 7y + 12$

30. $b^2 + 5b - 6$

31. $a^2 - 10ab + 9b^2$

32. $2r^2 - 3r - 20$

33. $6x^2 - 5x - 6$

34. $56m^2 - 93mn + 27n^2$

Applications and Connections

35. **Cartography** The scale on a map is 2 cm to 5 km. Kern and Dent are 16 km apart. How far apart are they on the map?

36. **Geometry** For what values of d is an angle with measure $3.6d$ an acute angle?

37. **Sales** Juanita bought sixteen 12-packs of soft drinks for the picnic. Some cost $2.79 each, and the rest cost $2.99 each. If she spent $46.04 on soft drinks, how many of each did she buy?

38. **Travel** Peter drove to work at 40 miles per hour and arrived one minute late. If he had driven at 45 miles per hour, he would have arrived one minute early. How far does Peter drive to work?

CHAPTER 6

Quadrilaterals

GEOMETRY AROUND THE WORLD
Russia

Do you like traditional paintings that depict things exactly as they appear in real life? Or are you drawn to more offbeat works in which the artist's idea of reality is different from what you see every day?

If you like the offbeat, you'll enjoy puzzling over the dramatic paintings of Russian artist Marc Chagall. Born in 1887, in the small Russian city of Pestkowatil, Chagall was a poor student who daydreamed in class. Only during drawing lessons and geometry class did he sit up and take notice. Perhaps that's why Chagall's paintings contain so many distinct geometric shapes!

His family was poor, but they managed to send Chagall to an art school in the city of St. Petersburg. He later studied at another Russian art school before moving to Paris in 1913 to pursue his career as a painter.

GEOMETRY IN ACTION

Chagall's style is marked by bold colors, strong geometric shapes, and dreamlike images of floating bodies. All these are clearly evident in "Paris Through the Window," painted by Chagall when he arrived there in 1913. Chagall wears two faces in this painting—one looking out toward Paris; the other back toward Russia. How many different geometric shapes do you see in this painting? How many **quadrilaterals** make up the window?

Do you see the quadrilaterals in the trouser legs of Chagall's 1924 painting "Green Violinist?" What shapes are in the man's coat?

◀ *"Paris Through the Window"*   Inset: *Marc Chagall*

Parallelograms

Objectives

After studying this lesson, you should be able to:

- recognize and define a parallelogram, and
- recognize, use, and prove the properties of a parallelogram.

Application

The beautiful pattern in this wooden tile is made up of **polygons**, each one with the same general shape. A polygon is a figure made up of coplanar segments, called **sides** which intersect at points called **vertices**. The sides each intersect exactly two other sides, one at each endpoint, and no two sides with a common endpoint are collinear.

The polygons in the wooden tile are **quadrilaterals**, or four-sided polygons. Below are some examples and nonexamples of quadrilaterals.

> *FYI···*
>
> The art of making furniture and other items with inlaid wood is called marquetry. Marquetry first became popular in France in the late sixteenth century.

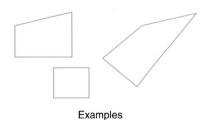

Examples   Nonexamples

We will often refer to the **diagonals** in a quadrilateral. The diagonals of a figure are the segments which connect any two nonconsecutive vertices. In quadrilateral *EFGH* at the right, the dashed line segments, $\overline{EG}$ and $\overline{FH}$, are the diagonals.

There is something else special about the quadrilaterals used in the wooden tile design. Study the pattern carefully. What do you think it is?

The special quadrilaterals you see in the wooden tile pattern are called **parallelograms**. A parallelogram is a quadrilateral with both pairs of opposite sides parallel. The quadrilateral at the right is called parallelogram *ABCD*. The pairs of opposite sides are $\overline{AB}$ and $\overline{DC}$, and $\overline{AD}$ and $\overline{BC}$. The pairs of opposite angles are $\angle A$ and $\angle C$, and $\angle B$ and $\angle D$.

Parallelogram ABCD is written $\square ABCD$.

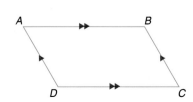

What conjectures can you make about the sides of a parallelogram? Can you develop a convincing argument or proof to show that your conjectures are correct?

Theorem 6-1	**Opposite sides of a parallelogram are congruent.**

Proof of Theorem 6-1

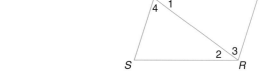

Given: $\square PQRS$

Prove: $\overline{PQ} \cong \overline{RS}$
$\overline{QR} \cong \overline{SP}$

Paragraph Proof:

Draw an auxiliary segment PR and label angles 1, 2, 3, and 4 as shown. Since the opposite sides of a parallelogram are parallel, $\angle 1 \cong \angle 2$, and $\angle 3 \cong \angle 4$ because they are alternate interior angles. Since congruence of segments is reflexive, $\overline{PR} \cong \overline{PR}$. So $\triangle QPR \cong \triangle SRP$ by ASA. $\overline{PQ} \cong \overline{RS}$ and $\overline{QR} \cong \overline{SP}$ by CPCTC.

The angles of a parallelogram have a special relationship also.

Theorem 6-2	**Opposite angles of a parallelogram are congruent.**
Theorem 6-3	**Consecutive angles in a parallelogram are supplementary.**
	You will be asked to prove Theorems 6-2 and 6-3 in Exercises 41 and 42.

Example 1

CONNECTION

Algebra

DUCK **is a parallelogram. Find the values of** *w, x, y,* **and** *z.*

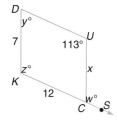

Since the opposite sides of a parallelogram are congruent, $x = 7$.

The opposite angles are congruent, so $z = 113$.

The consecutive angles of a parallelogram are supplementary, so $y + 113 = 180$. Therefore, $y = 67$.

$\angle KCU$ and $\angle UCS$ form a linear pair, so $m\angle KCU + m\angle UCS = 180$. The opposite angles of a parallelogram are congruent, so $m\angle KCU = 67$. Therefore, $w = 113$.

If the diagonals of a parallelogram are drawn, what appears to be true about them? Our next theorem states the relationship.

Theorem 6-4	**The diagonals of a parallelogram bisect each other.**

Example 2 | **Write a plan for the proof of Theorem 6-4.**

Given: ▱*EAST*

Prove: $\overline{ES}$ bisects $\overline{AT}$.
$\overline{AT}$ bisects $\overline{ES}$.

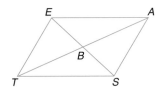

Plan for Proof:

First, use the alternate interior angle theorem, and Theorem 6-1 to prove that $\triangle EBA \cong \triangle SBT$ or that $\triangle EBT \cong \triangle SBA$ by ASA. Next, show that B is the midpoint of both $\overline{ES}$ and $\overline{AT}$. Then conclude that $\overline{ES}$ bisects $\overline{AT}$ and $\overline{AT}$ bisects $\overline{ES}$, by the definition of a segment bisector. *You will be asked to complete the proof in Exercise 43.*

CHECKING FOR UNDERSTANDING

Communicating Mathematics

Read and study the lesson to answer these questions.

1. Describe a quadrilateral and sketch an example and a nonexample.

2. Look around your classroom and name at least five different quadrilaterals you are able to find.

3. By definition, what is the special characteristic a quadrilateral must have in order to be a parallelogram?

4. If quadrilateral *FRAC* is a parallelogram, what congruence statements can you make?

Guided Practice

Complete each statement about parallelogram *ABCD* at the right. Then name the theorem or definition that justifies your answer.

5. $\overline{AB} \parallel$ ___?___

6. $\overline{DA} \cong$ ___?___

7. $\triangle ADC \cong$ ___?___

8. $\angle CDA \cong$ ___?___

9. $\overline{DE} \cong$ ___?___

10. $\angle BAC \cong$ ___?___

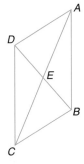

11. Is quadrilateral *ABCD* a parallelogram if its vertices have coordinates $A(1, 1)$, $B(3, 6)$, $C(8, 8)$, and $D(6, 3)$? Justify your answer.

12. If quadrilateral *SLAM* is a parallelogram and $m\angle S = 92$, what are the measures of angles *L*, *A*, and *M*?

13. Prove that a diagonal and the sides of a parallelogram form two congruent triangles.

14. In ▱*ABCD*, $AB = 2x + 5$, $CD = y + 1$, $AD = y + 5$, and $BC = 3x - 4$. Find the measures of the sides.

DUNK is a parallelogram. Name a theorem or definition that justifies each statement.

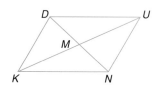

15. $\overline{DU} \parallel \overline{KN}$

16. M is the midpoint of $\overline{KU}$.

17. $m\angle DUN = m\angle NKD$

18. $\angle KDU$ is a supplement of $\angle DKN$.

19. If $DU = 3x + 6$, $UN = 8y - 4$, $KN = 8x - 4$, and $KD = 2y + 14$, find the perimeter of $DUNK$.

EXERCISES

Practice

EFGH is a parallelogram. Determine whether each statement must be true. If it must be true, state the theorem or definition that justifies the statement.

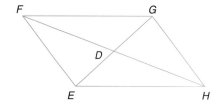

20. $\overline{FE} \parallel \overline{GH}$

21. $\triangle FDE \cong \triangle HDG$

22. $\angle FGH \cong \angle FEH$

23. $\overline{FD} \cong \overline{DG}$

24. $\triangle FHE \cong \triangle GHE$

25. $DE = \frac{1}{2} EG$

If each quadrilateral is a parallelogram, find the value of x, y, and z.

26. 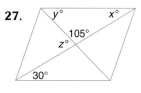   **27.** 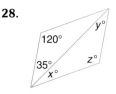   **28.**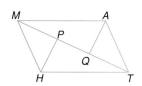

29. Given parallelogram $PQRS$ with $m\angle P = y$ and $m\angle Q = 4y + 20$, find the measures of $\angle R$ and $\angle S$.

30. In parallelogram $ABCD$, $m\angle C = x + 75$ and $m\angle D = 3x - 199$. Find the measure of each angle.

31. Find all the possible ordered pairs for the fourth vertex of a parallelogram with vertices at $J(1, 1)$, $U(3, 4)$, and $N(7, 1)$.

32. Given: Parallelogram $MATH$
$\overline{MP} \cong \overline{TQ}$

     Prove: $\overline{PH} \parallel \overline{AQ}$

Is each quadrilateral a parallelogram? Justify your answer.

33.

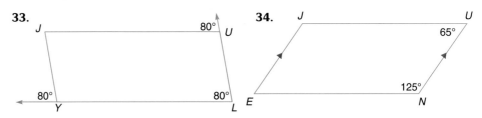

34.

Explain why it is impossible for each figure to be a parallelogram.

35.

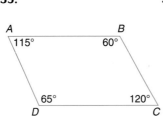

36.

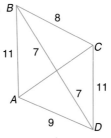

37.

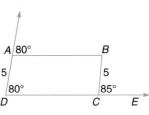

38. If $NCTM$ is a parallelogram, $m\angle N = 12x + 10y + 5$, $m\angle C = 9x$, and $m\angle T = 6x + 15y$, find $m\angle M$.

39. $NCSM$ is a parallelogram with diagonals $\overline{NS}$ and $\overline{MC}$ that intersect at point P. If $NP = 4a + 20$, $NS = 13a$, $PC = a + b$, and $PM = 2b - 2$, find CM.

40. If $PQSW$ and $RTUV$ are parallelograms, find $m\angle SXR$.

41. Write a two-column proof of Theorem 6-2.

42. Write a paragraph proof of Theorem 6-3.

43. Write a two-column proof of Theorem 6-4.

Write a two-column proof.

44. Given: $PQST$ is a parallelogram.
$\overline{RP}$ bisects $\angle QPT$.
$\overline{VS}$ bisects $\angle QST$.

      Prove: $\overline{RP} \parallel \overline{VS}$

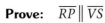

45. Given: $PQST$ is a parallelogram.
$\overline{RP}$ bisects $\angle QPT$.
$\overline{VS}$ bisects $\angle QST$.

      Prove: $\overline{RP} \cong \overline{VS}$

Construction

46. Construct a parallelogram with one angle congruent to the given angle and sides congruent to the given segments.

Critical Thinking

47. Draw some figures to determine if the statement *If the midpoints of the sides of any quadrilateral are connected in clockwise order, they form a parallelogram* is *true* or *false.* Justify your conclusion.

Applications

48. **Physics**   Vectors are represented by line segments that have a certain length and direction. In physics, vectors are used to indicate force or motion. If two forces, *A* and *B*, are acting on an object, then the net force, or resultant vector, *R*, can be found by drawing a parallelogram. The resultant is represented by the diagonal that begins at the common endpoint of the vectors. Copy vectors *X* and *Y* and find the resultant.

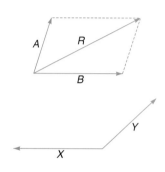

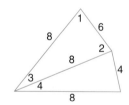

Mixed Review

49. Complete the statements with $<$, $>$, or $=$. **(Lesson 5-7)**

$m\angle 1 \underline{\ ?\ } m\angle 2$

$m\angle 3 \underline{\ ?\ } m\angle 4$

50. Name the postulates and theorems that can be used to prove two right triangles congruent. Explain each. **(Lesson 5-2)**

51. If $\triangle DOG \cong \triangle CAT$, what segment in $\triangle CAT$ is congruent to $\overline{GO}$ in $\triangle DOG$? **(Lesson 4-3)**

52. If $m\angle R = 90$ in $\triangle RST$, describe $\triangle RST$ as acute, obtuse, or right. Could $\triangle RST$ be isosceles? Could $\triangle RST$ be equiangular? **(Lesson 4-1)**

53. Find the slope of the line that passes through (-4, 8) and (3, 0). **(Lesson 3-5)**

54. Are the edge lines on a straight highway a model of lines that are intersecting, parallel, or skew? **(Lesson 3-1)**

55. Write the conditional *A parallelogram is a quadrilateral with opposite sides parallel* in if-then form. **(Lesson 2-2)**

Wrap-Up

56. Summarize all the conclusions that you can draw if you know that a quadrilateral is a parallelogram.

6-2 Problem-Solving Strategy: Look for a Pattern

Objective

After studying this lesson, you should be able to:

■ solve a problem by looking for a pattern and using the pattern to find the missing information.

Application

A critical thinking problem on Wendy's geometry quiz asked her to find the number of angles formed by 10 distinct rays with a common endpoint. The figure she drew looked a little confusing, so Wendy drew some figures with fewer rays and made a table to record how many angles they formed.

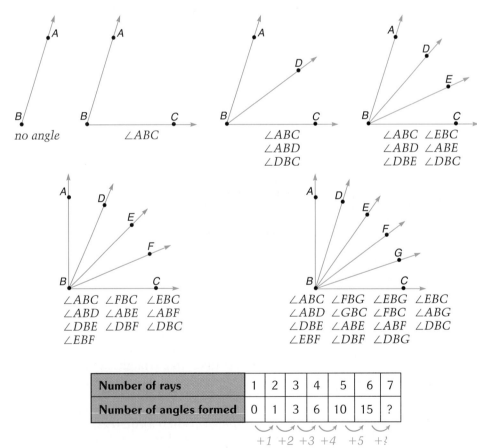

Number of rays	1	2	3	4	5	6	7
Number of angles formed	0	1	3	6	10	15	?

+1 +2 +3 +4 +5 +?

After drawing a figure with 6 rays, Wendy noticed a pattern. She conjectured that a figure with 10 rays must form 45 angles. Do you agree?

Example

CONNECTION

Algebra

What is the remainder when 3^{200} is divided by 5?

Make a table of the remainders when successive powers of 3 are divided by 5 and look for a pattern. Use the $\boxed{y^x}$ key on your calculator to find the powers of 3.

n	3^n	Remainder of $3^n \div 5$
1	$3^1 = 3$	3
2	$3^2 = 9$	4
3	$3^3 = 27$	2
4	$3^4 = 81$	1
5	$3^5 = 243$	3
6	$3^6 = 729$	4
7	$3^7 = 3187$	2
8	$3^8 = 6561$	1
9	$3^9 = 19,683$	3

The remainders of $3^n \div 5$ repeat in a pattern of four:
3, 4, 2, 1, 3, 4, 2, 1, ...

Every fourth power has a remainder of 1 when divided by 5. So since 200 is divisible by 4, $3^{200} \div 5$ has a remainder of 1.

CHECKING FOR UNDERSTANDING

Communicating Mathematics

Read and study the lesson to answer these questions.

1. How many angles are formed by 12 rays with a common endpoint?
2. What is the remainder when $3^{250} \div 5$?
3. Do you see a pattern in the ones' digits of successive powers of 3?

Guided Practice

Find the next number in each pattern.

4. 1, 2, 4, 8, 16, 32, ___?___
5. 1, 2, 4, 7, 11, 16, 22, ___?___
6. 6, 2, 4, 6, 2, 4, 6, 2, 4, 6, ___?___

EXERCISES

Solve. Use any strategy.

7. Shina and Jeff have invited seven other couples over for a picnic. They will rent square cardtables to seat their friends for dinner. One of the cardtables can seat four people. If the tables are placed end to end to form one long table, how many must Shina and Jeff rent to seat everyone?

Strategies

Look for a pattern.
Solve a simpler problem.
Act it out.
Guess and check.
Draw a diagram.
Make a chart.
Work backward.

8. The Moores used 4 miles of fencing to enclose their square 640-acre cornfield. Their neighbors, the Escaladas, bought 2 miles of fencing to enclose their square horse pasture. How many acres is the Escaladas' pasture?

9. Find the number of pairs of vertical angles determined by eight distinct lines passing through a point.

10. At Grandma's Bakery, muffins are sold in boxes of 4, 6, or 13. Jena can buy 8 muffins by choosing two boxes of 4 muffins, but she can't buy 9 muffins with any combination of boxes. Find all of the numbers of muffins less than 25 that Jena cannot buy.

11. How many triangles are there in the figure shown at the right? *(Hint: There are more than 12 triangles.)*

12. Nathan beat Tim by 5 yards in a 100-yard dash. The next time, Nathan evened-up the race by starting 5 yards behind Tim. If they each ran at the same rate as in the previous race, who won the second race? Explain.

13. Use your calculator to find the values of 11^2, 111^2 and 1111^2. Without calculating, use a pattern to determine the value of $11,111^2$.

～～～ COOPERATIVE LEARNING PROJECT ～～～

Work in groups. Each person in the group must understand the solution and be able to explain it to any person in class.

Choose a three-digit number in which the digit in the hundreds place is greater than the digit in the ones place. Reverse the digits to form a different three-digit number and subtract the new number from the original. If necessary, add leading zeros to the difference to make it a three-digit number. Reverse the digits of the difference and add that number to the difference. What is the result? Try a few more examples. Is the result always the same? What is the result if the hundreds digit equals the ones digit?

Tests for Parallelograms

Objective

After studying this lesson, you should be able to:

■ recognize and apply the conditions that ensure that a quadrilateral is a parallelogram.

Application

The artist who designed this stained glass window used parallelograms and color effectively to create an unusual effect. In order for the pattern to work, it was essential that the basic components were all parallelograms. The artist had to know how to ensure that a quadrilateral is a parallelogram. There are several tests, other than the definition for a parallelogram, that she might have used.

Theorem 6-5

If both pairs of opposite sides of a quadrilateral are congruent, then the quadrilateral is a parallelogram.

Proof of Theorem 6-5

Given: $\overline{MA} \cong \overline{DE}$
$\overline{ME} \cong \overline{DA}$

Prove: *MADE* is a parallelogram.

Proof:

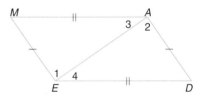

Statements	Reasons
1. $\overline{MA} \cong \overline{DE}$ $\overline{ME} \cong \overline{DA}$	1. Given
2. $\overline{AE} \cong \overline{EA}$	2. Congruence of segments is reflexive.
3. $\triangle MAE \cong \triangle DEA$	3. SSS
4. $\angle 1 \cong \angle 2$ $\angle 3 \cong \angle 4$	4. CPCTC
5. $\overline{ME} \parallel \overline{DA}$ $\overline{MA} \parallel \overline{DE}$	5. If 2 lines are cut by a transversal and alt. int. ∠s are ≅, then the lines are ∥.
6. *MADE* is a parallelogram.	6. Definition of parallelogram

The following theorem provides another test to determine if a quadrilateral is a parallelogram. You will be asked to prove this theorem in Exercise 17.

Theorem 6-6	If one pair of opposite sides of a quadrilateral are both parallel and congruent, then the quadrilateral is a parallelogram.

We can use the distance and slope formulas to determine if the conditions of Theorem 6-6 are satisfied.

Example 1

The coordinates of the vertices of quadrilateral *ABCD* are *A*(-1, 3), *B*(2, 1), *C*(9, 2), and *D*(6, 4). Determine if quadrilateral *ABCD* is a parallelogram.

Find BC and AD to determine if this pair of opposite sides are congruent.

$BC = \sqrt{(2-9)^2 + (1-2)^2}$ or $\sqrt{50}$
$AD = \sqrt{(-1-6)^2 + (3-4)^2}$ or $\sqrt{50}$

Since $BC = AD$, $\overline{BC} \cong \overline{AD}$.

Find the slopes of $\overline{BC}$ and $\overline{AD}$ to determine if these opposite sides are parallel.

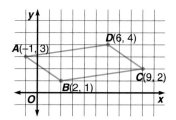

slope of $\overline{BC} = \frac{1-2}{2-9}$ or $\frac{1}{7}$

slope of $\overline{AD} = \frac{3-4}{-1-6}$ or $\frac{1}{7}$

Since the slopes $\overline{BC}$ and $\overline{AD}$ are equal, the sides are parallel. Since one pair of sides is both congruent and parallel, *ABCD* is a parallelogram by Theorem 6-6.

There are two more tests for parallelograms. You will be asked to prove Theorem 6-7 in Exercise 41.

Theorem 6-7	If the diagonals of a quadrilateral bisect each other, then the quadrilateral is a parallelogram.
Theorem 6-8	If both pairs of opposite angles in a quadrilateral are congruent, then the quadrilateral is a parallelogram.

Here is a summary of the tests to show that a quadrilateral is a parallelogram.

Show that:
1. Both pairs of opposite sides are parallel. (Definition)
2. Both pairs of opposite sides are congruent. (Theorem 6-5)
3. A pair of opposite sides are both parallel and congruent. (Theorem 6-6)
4. Diagonals bisect each other. (Theorem 6-7)
5. Both pairs of opposite angles are congruent. (Theorem 6-8)

CHECKING FOR UNDERSTANDING

Communicating Mathematics

Read and study the lesson to answer these questions.

1. Which of the tests for a parallelogram do you think the artist used to ensure that the quadrilaterals in the stained glass window design were parallelograms? Why do you think the artist chose this test?

Determine if each conditional is *true* or *false*. If it is false, draw a counterexample.

2. If the opposite angles in a quadrilateral are congruent, then the quadrilateral is a parallelogram.

3. If two sides of a quadrilateral are congruent, then the quadrilateral is a parallelogram.

4. If a diagonal and the sides of quadrilateral *WXYZ* form two congruent triangles, then *WXYZ* is a parallelogram.

5. If a pair of consecutive angles in a quadrilateral are congruent, then the quadrilateral is a parallelogram.

Guided Practice

Determine if each quadrilateral must be a parallelogram. Justify your answer.

6.

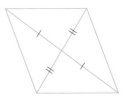

7.

8.

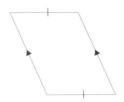

9.

What values must x and y have in order for each quadrilateral to be a parallelogram?

10.

11.

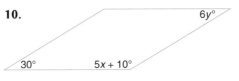

Quadrilateral *JKLM* has vertices *J*(3, -2), *K*(8, -2), *L*(7, -4), and *M*(3, -6).

12. Show that *JKLM* is not a parallelogram.

13. Show that the quadrilateral formed by joining consecutive midpoints of the sides of *JKLM* is a parallelogram.

Write a two-column proof.

14. **Given:** $\triangle AEU \cong \triangle OUE$

   Prove: *AEOU* is a parallelogram.

15. **Given:** $\square PQRS$

   $\overline{XS} \cong \overline{QY}$

   Prove: *PYRX* is a parallelogram.

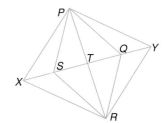

16. **Given:** $\overline{UN} \parallel \overline{KE}$

   $\angle YUK \cong \angle REN$

   Prove: *UNEK* is a parallelogram.

17. Write a two-column proof of Theorem 6-6.

EXERCISES

Practice Determine if each quadrilateral must be a parallelogram. Justify your answer.

18.

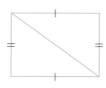

19.

20.

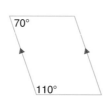

21.

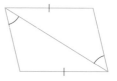

Use parallelogram *ABCD* and the given information to find each value.

22. $m\angle ABC = 137$. Find $m\angle DAB$.

23. $AC = 5x - 12$ and $AT = 14$. Find x.

24. $AB = 6$, $BC = 9$, and
$m\angle ABC = 80$. Find CD.

25. $BC = 4x + 7$ and $AD = 8x - 5$.
Find x.

26. $BT = 3x + 1$ and $BD = 4x + 8$.
Find x.

27. $m\angle BCD = 3x + 14$ and $m\angle ADC = x + 10$. Find $m\angle ADC$.

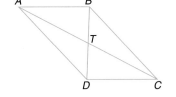

What values must *x* and *y* have in order that the quadrilateral is a parallelogram?

28.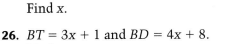

29.

Determine if each statement is *true* or *false*. If false, find a counterexample.

30. A quadrilateral is a parallelogram if it has two pairs of congruent sides.

31. A quadrilateral is a parallelogram if it has one pair of congruent sides and one pair of parallel sides.

Determine whether *ABCD* is a parallelogram given each set of vertices. Explain.

32. $A(8, 10)$, $B(16, 17)$, $C(16, 11)$, $D(8, 4)$

33. $A(8, 6)$, $B(6, 0)$, $C(4, 2)$, $D(7, 3)$

34. $A(-4, 0)$, $B(6, 0)$, $C(5, 4)$, $D(0, 4)$

35. $A(-2, 6)$, $B(2, 8)$, $C(3, 8)$, $D(-1, 3)$

36. Draw a quadrilateral that has two pairs of congruent sides, but is not a parallelogram.

37. Draw a quadrilateral that is not a parallelogram and has one pair of parallel sides and one pair of congruent sides.

Write a two-column proof.

38. **Given:** $\square NCTM$
$\overline{NA} \cong \overline{ST}$

Prove: $ACSM$ is a parallelogram.

39. **Given:** $\triangle TWA$ is equilateral.
$TBWA$ is a parallelogram.
$TWAI$ is a parallelogram.

Prove: $\triangle IBM$ is equilateral.

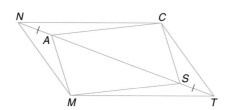

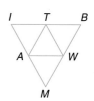

40. **Given:** $\triangle PQR \cong \triangle STV$
$\overline{PR} \parallel \overline{VS}$

Prove: $PRSV$ is a parallelogram.

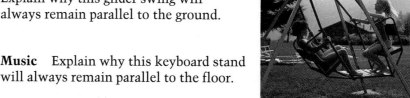

41. Write a paragraph proof of Theorem 6-7.

Critical Thinking

42. Explain why this glider swing will always remain parallel to the ground.

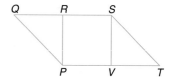

Applications

43. **Music** Explain why this keyboard stand will always remain parallel to the floor.

44. **Construction** Mr. Gallagher is building an open staircase for a new home. He has measured several three-foot support bars to hold up the handrail. If he nails the support bars to the handrail and to the edge of the staircase so that the bars are six inches apart, will the handrail be parallel to the staircase? Explain.

Mixed Review

45. Find the next number in the pattern 2, 5, 10, 17, 26, ___?___. **(Lesson 6-2)**

46. The opposite angles of a parallelogram have measures of $9x + 12$ and $15x$. Find the measures of the angles. **(Lesson 6-1)**

47. Could 30, 35, and 66 be the measures of the sides of a triangle? Explain. **(Lesson 5-6)**

48. The angles in a triangle have measures of $7x - 8$, $3x + 3$, and $18x - 11$. Is the triangle acute, obtuse, or right? Explain. **(Lesson 4-1)**

49. The measure of an acute angle is $18y$. What are the possible values of y? **(Lesson 1-7)**

Wrap-Up

50. **Journal Entry** Write a sentence in your journal about each test that you can use to prove that a quadrilateral is a parallelogram.

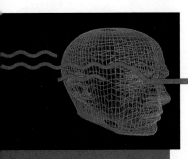

Technology
Parallelograms

▶ **BASIC**
Geometric Supposer
Graphing calculators
LOGO
Spreadsheets

A BASIC computer program can be written to use one of the tests for determining if a quadrilateral is a parallelogram. The program below uses the definition of a parallelogram to determine if the coordinates of the points the user enters name the vertices of a parallelogram.

```
 10 REM THIS PROGRAM DETERMINES WHETHER POINTS ARE
    THE VERTICES OF A PARALLELOGRAM
 20 INPUT "ENTER THE COORDINATES OF POINT A.";
    X1, Y1
 30 INPUT "ENTER THE COORDINATES OF POINT B.";
    X2, Y2
 40 INPUT "ENTER THE COORDINATES OF POINT C.";
    X3, Y3
 50 INPUT "ENTER THE COORDINATES OF POINT D.";
    X4, Y4
 60 IF (X2 - X1) = 0 THEN GOTO 80
 70 M1 = (Y2 - Y1)/(X2 - X1)
 80 IF (X4 - X3) = 0 THEN GOTO 110
 90 M2 = (Y4 - Y3)/(X4 - X3)
100 IF M1 <> M2 THEN GOTO 200
110 IF (X3 - X2) = 0 AND (X4 - X1) = 0 AND (X2 - X1)
    = 0 THEN GOTO 200
120 IF (X3 - X2) = 0 AND (X4 - X1) = 0
    THEN GOTO 180
130 IF (X3 - X2) = 0 THEN GOTO 200
140 M3 = (Y3 - Y2)/(X3 - X2)
150 IF (X4 - X1) = 0 THEN GOTO 200
160 M4 = (Y4 - Y1)/(X4 - X1)
170 IF M3 <> M4 THEN GOTO 200
180 PRINT "ABCD IS A PARALLELOGRAM."
190 GOTO 210
200 PRINT "ABCD IS NOT A PARALLELOGRAM."
210 END
```

EXERCISES

Use the BASIC program to determine if each set of points are the vertices of a parallelogram.

1. $A(9, 7)$, $B(11, 13)$, $C(6, 12)$, $D(4, 6)$

2. $A(-4, 1)$, $B(-3, 2)$, $C(8, 5)$, $D(6, 3)$

3. How could you change the program to determine whether a quadrilateral is a parallelogram using the theorem that says that a quadrilateral with a pair of opposite sides that are congruent and parallel is a parallelogram?

6-4 Rectangles

Objectives

After studying this lesson, you should be able to:
- recognize the properties of rectangles, and
- use properties of rectangles in proofs.

Application

Have you ever thought about the importance of rectangles in the sports you play or enjoy watching? Without them, many of our most popular sports would be very different or maybe not even exist at all. The photographs on this page show sports that involve rectangles. How many of them do you enjoy?

A **rectangle** is a quadrilateral with four right angles. It follows that because both pairs of opposite angles are congruent, a rectangle is a parallelogram. Since a rectangle is a parallelogram, it has all the properties of a parallelogram.

The diagonals of a rectangle also have a special relationship.

Theorem 6-9	If a parallelogram is a rectangle then its diagonals are congruent.

Proof of Theorem 6-9

Given: $ABCD$ is a rectangle with diagonals $\overline{AC}$ and $\overline{BD}$.

Prove: $\overline{AC} \cong \overline{BD}$

Proof:

Statements	Reasons
1. $ABCD$ is a rectangle.	1. Given
2. $\overline{DC} \cong \overline{DC}$	2. Congruence of segments is reflexive.
3. $\overline{AD} \cong \overline{BC}$	3. Opp. sides of a $\square$ are $\cong$.
4. $\angle ADC$ and $\angle BCD$ are right angles.	4. Def. of rectangle
5. $\angle ADC \cong \angle BCD$	5. All rt. $\angle$s are $\cong$.
6. $\triangle ADC \cong \triangle BCD$	6. SAS
7. $\overline{AC} \cong \overline{BD}$	7. CPCTC

The converse of Theorem 6-9 is often used in the building trades to ensure that an angle is a right angle.

Example 1

APPLICATION

Building

The Owens family is constructing a deck in their backyard. Mrs. Owens has laid out stakes where the corners of the deck will be as shown below. She has made sure that the opposite sides are congruent. If the diagonals are congruent, can Mrs. Owens be sure that the deck will be a rectangle?

Label the vertices *M, N, O,* and *P.* Then $\overline{MN} \cong \overline{OP}$, $\overline{MP} \cong \overline{ON}$, and $\overline{MO} \cong \overline{NP}$. We need to prove that *MNOP* is a rectangle.

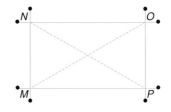

Given: $\overline{MN} \cong \overline{OP}, \overline{MP} \cong \overline{ON}, \overline{MO} \cong \overline{NP}$

Prove: *MNOP* is a rectangle.

Paragraph proof:

Since congruence of segments is reflexive, $\overline{MP} \cong \overline{PM}$. So, $\triangle MNP \cong \triangle POM$ by SSS. Corresponding parts of congruent triangles are congruent, so $\angle PMN \cong \angle MPO$. *MNOP* is a parallelogram since the opposite sides are congruent. Therefore, $\angle PMN$ and $\angle MPO$ are supplementary since consecutive angles of a parallelogram are supplementary. So, $m\angle PMN + m\angle MPO = 180$ and $m\angle PMN = m\angle MPO$.

By the substitution property of equality, $2m\angle PMN = 180$ so $m\angle PMN = 90$. Therefore, $\angle PMN$ and $\angle MPO$ are right angles. Since the opposite angles of a parallelogram are congruent, $\angle PON$ and $\angle MNO$ are right angles also. *MNOP* is a rectangle.

Yes, Mrs. Owens can be sure that the layout for the deck is a rectangle if the diagonals are congruent. *This example proves the converse of Theorem 6-9.*

We can construct a rectangle using right angles and the definition of a parallelogram.

CONSTRUCTION

Construct a rectangle with a length of 5 cm and a width of 4 cm.

1. Use a straightedge to draw line ℓ. Label a point *J* on ℓ. With your compass set at 5 cm, place the point at *J* and locate point *K* on ℓ so that $JK = 5$. Now construct lines perpendicular to ℓ through *J* and through *K*. Label them *m* and *n*.

2. Set your compass at 4 cm. Place the compass point at *J* and mark off a segment on *m*. Using the same compass setting, place the compass at *K* and mark a segment on *n*. Label these points *O* and *P*.

3. Draw $\overline{OP}$.

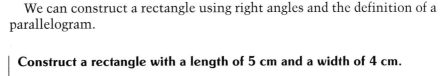

Quadrilateral *JKPO* is a parallelogram and all angles are right angles. Therefore, quadrilateral *JKPO* is a rectangle with a length of 5 cm and a width of 4 cm.

Here is a summary of the properties of a rectangle.

1. Opposite sides are congruent.
2. Opposite angles are congruent.
3. Consecutive angles are supplementary.
4. Diagonals bisect each other.
5. All four angles are right angles.
6. Diagonals are congruent.
7. Opposite sides are parallel.

CHECKING FOR UNDERSTANDING

Communicating Mathematics

Read and study the lesson to answer these questions.

1. Look for examples of rectangles in the objects around you. Could the objects you see be another shape and still be effective? For example, would a circular door be as useful as a rectangular door?

2. If quadrilateral *GHIJ* is a rectangle, what can you say about its diagonals?

3. Are there any properties of a parallelogram that do not hold true for rectangles? Explain.

4. Name two properties that hold for all rectangles, but not necessarily for all parallelograms.

Guided Practice

CONNECTION
Algebra

If quadrilateral *JKLM* is a rectangle, find the value of *x*.

5. $LP = 3x + 7$
   $MK = 26$

6. $LJ = 4x - 12$
   $KM = 7x - 36$

7. $KP = x^2$
   $PJ = 7x - 10$

CONSTRUCTION

8. Construct a rectangle with sides congruent to the given segments.

9. Is it true that, "If two sides of a quadrilateral are perpendicular, then the quadrilateral is a rectangle?" If not, draw a counterexample.

10. Show that quadrilateral *WXYZ* with vertices $W(1, 1)$, $X(5, 5)$, $Y(8, 2)$, and $Z(4, -2)$ is a rectangle.

11. Write a two-column proof.

Given: $\square WXYZ$
$\angle 1$ and $\angle 2$ are complementary.

Prove: $WXYZ$ is a rectangle.

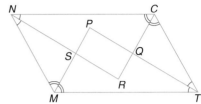

12. If quadrilateral $NCTM$ is a parallelogram, explain why quadrilateral $PQRS$ must be a rectangle.

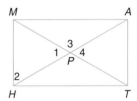

EXERCISES

Practice

Use rectangle *MATH* and the given information to solve each problem.

13. $MP = 6$, find HA.

14. $MH = 8$, find AT.

15. $HP = 3x$ and $PT = 18$, find x.

16. $m\angle 1 = 55$, find $m\angle 2$.

17. $m\angle 3 = 110$, find $m\angle 4$.

Draw a counterexample to show that each statement below is false.

18. If a quadrilateral has congruent diagonals, it is a rectangle.

19. If a quadrilateral has opposite sides congruent, it is a rectangle.

20. If a quadrilateral has diagonals that bisect each other, it is a rectangle.

21. Graph $J(2, -3)$, $K(-3, 1)$, $L(1, 6)$, and $M(6, 2)$.

    a. Describe two ways of determining if $JKLM$ is a rectangle.

    b. Is $JKLM$ a rectangle? Justify your answer.

CONNECTION

Algebra

Find the values of *x* and *y* in rectangle *PQRS*.

22. $PT = 3x - y$
$ST = x + y$
$TQ = 5$

23. $PS = y$
$QR = x + 7$
$PQ = y - 2x$
$SR = x + 1$

24. $PT = x + y$
$ST = 2y - 7$
$PR = -3x$

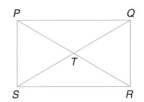

Use rectangle *MNRS* and the given information to solve each problem.

25. If $m\angle 1 = 32$, find the $m\angle 2$, $m\angle 3$, and $m\angle 4$.

26. If $ST = 14.25$, find MR.

27. If $m\angle MTN = 116$, find $m\angle 1$ and $m\angle 4$.

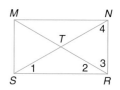

Determine whether *ABCD* is a rectangle. Explain.

28. $A(12, 2)$, $B(12, 8)$, $C(-3, 8)$, $D(-3, 2)$

29. $A(0, -3)$, $B(4, 8)$, $C(7, -4)$, $D(11, 7)$

30. $A(4, 0)$, $B(6, -3)$, $C(8, 4)$, $D(10, 1)$

31. $A(-5, 8)$, $B(6, 9)$, $C(7, -2)$, $D(-4, -3)$

The faces of the solid shown below are rectangles and all meet at right angles. Does each set of vertices form a rectangle?

32. A, H, G, and B

33. A, C, E, and G

34. E, D, C, and F

35. B, D, H, and A

36. E, H, F, and B

37. F, C, G, and B

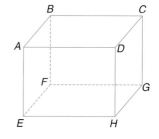

38. Prove that if one angle of a parallelogram is a right angle, then the parallelogram is a rectangle.

Critical Thinking

39. The ordered pairs $(2, 2)$, $(16, 4)$, and $(8, -4)$ are the coordinates of three of the vertices of a parallelogram.
   a. Find all possible ordered pairs for the fourth vertex.
   b. Do any of these ordered pairs result in a parallelogram that is a rectangle?

Applications

40. **Construction**   A cement contractor is getting ready to pour a footer for a new home. The outside dimensions of the basement walls are to be 42 feet by 56 feet. She places stakes and strings to mark the outside walls with the corners at J, K, L, and M. To make sure that quadrilateral *JKLM* is a rectangle, the contractor measures $\overline{MK}$ and $\overline{JL}$. If $MK <$ JL, describe how she should move stakes F and G to make quadrilateral *JKLM* a rectangle. Explain your answer.

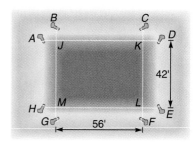

41. **Building Materials**   Explain why the rectangle is used as the shape for bricks used in construction of walls in homes and buildings.

42. Sports Investigate the playing field of each sport. Is it a rectangle? If so, what are its dimensions?

    a. racquetball   **b.** wrestling   **c.** polo

    d. volleyball   **e.** tennis   **f.** bowling

    g. soccer   **h.** baseball   **i.** jai alai

Mixed Review

43. Determine whether quadrilateral $ABCD$ with vertices $A(9, 4)$, $B(0, -2)$, $C(-4, 6)$, and $D(5, 6)$ is a parallelogram. **(Lesson 6-3)**

44. Quadrilateral $WEST$ is a parallelogram. If $WE = 3x + 7$, $ES = 7y - 1$, $ST = 6x - 2$, and $TW = 2y + 9$, find the perimeter of $WEST$. **(Lesson 6-1)**

45. Find the slope of the line that passes through $(9, 3)$ and $(8, -4)$. **(Lesson 3-5)**

46. Write *A rectangle is a parallelogram* in if-then form. **(Lesson 2-2)**

47. The West High School band is selling pizzas. Pizzas can be plain cheese or topped with any combination of pepperoni, green peppers, mushrooms, and onions. How many different pizzas are possible? **(Lesson 1-3)**

Wrap-Up

48. Explain how rectangles and parallelograms are the same and how they are different.

～～～～ MID-CHAPTER REVIEW ～～～～

Complete each statement about parallelogram *EFGH*. Then name the theorem or definition that justifies your answer. (Lesson 6-1)

1. $\overline{GH} \parallel$ ___?___

2. $\angle GHE \cong$ ___?___

3. $\overline{FD} \cong$ ___?___

4. $\triangle FDE \cong$ ___?___

5. $\overline{EH} \cong$ ___?___

6. $\overline{GF} \parallel$ ___?___

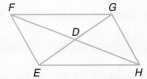

Find the next number in each pattern. (Lesson 6-2)

7. 0, 1, 3, 6, 10, 15, ___?___

8. 1, 8, 27, 64, ___?___

9. -1, 2, -4, 8, -16, ___?___

10. 6, $\frac{9}{2}$, 3, $\frac{3}{2}$, 0, ___?___

State the definition or theorem that proves that quadrilateral *ABCD* is a parallelogram using the given information. (Lesson 6-3)

11. $\overline{AB} \parallel \overline{DC}$ and $\overline{BC} \parallel \overline{AD}$

12. $\overline{BC} \cong \overline{AD}$ and $\overline{BC} \parallel \overline{AD}$

13. $\angle ABC \cong \angle ADC$ and $\angle BAD \cong \angle BCD$

14. $\overline{OC} \cong \overline{OA}$ and $\overline{DO} \cong \overline{BO}$

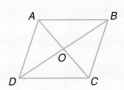

15. Write a two-column proof. **(Lesson 6-4)**

    Given: rectangle $JKLM$

                $\overline{KF} \cong \overline{MH}$

                $\overline{JE} \cong \overline{LG}$

    Prove: $EFGH$ is a parallelogram.

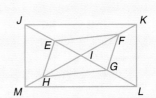

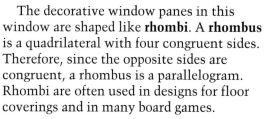

6-5 | Squares and Rhombi

Objectives

After studying this lesson, you should be able to:

 ▪ recognize the properties of squares and rhombi, and
 ▪ use properties of squares and rhombi in proofs.

Application

Rhombi is the plural of rhombus.

   The decorative window panes in this window are shaped like **rhombi**. A **rhombus** is a quadrilateral with four congruent sides. Therefore, since the opposite sides are congruent, a rhombus is a parallelogram. Rhombi are often used in designs for floor coverings and in many board games.

   A rhombus has all of the properties of a parallelogram. The diagonals of a rhombus have two special relationships that are described in the following theorems.

Theorem 6-10	**The diagonals of a rhombus are perpendicular.**
	You will be asked to prove this theorem in Exercise 36.
Theorem 6-11	**Each diagonal of a rhombus bisects a pair of opposite angles.**

Proof of Theorem 6-11

Given: *ABCD* is a rhombus.

Prove: Each diagonal bisects a pair of opposite angles.

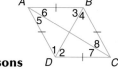

Statements	Reasons
1. *ABCD* is a rhombus.	1. Given
2. *ABCD* is a parallelogram.	2. Definition of rhombus
3. ∠*ABC* ≅ ∠*ADC* ∠*BAD* ≅ ∠*BCD*	3. Opp. ∡ of a ▱ are ≅.
4. $\overline{AB} \cong \overline{BC} \cong \overline{CD} \cong \overline{DA}$	4. Definition of rhombus
5. △*ABC* ≅ △*ADC*	5. SAS
6. ∠5 ≅ ∠6 ∠7 ≅ ∠8	6. CPCTC
7. △*BAD* ≅ △*BCD*	7. SAS
8. ∠1 ≅ ∠2 ∠3 ≅ ∠4	8. CPCTC
9. Each diagonal bisects a pair of opposite angles.	9. Definition of angle bisector

The characteristics of a rhombus can be helpful in solving problems.

Example 1

Use rhombus *RSTV* with *SV* = 42 to determine whether each statement is *true* or *false*. Justify your answers.

a. *RT* = 42   generally false; The diagonals of a rhombus are not congruent unless it is a rectangle.

b. *PS* = 21   true; The diagonals of a parallelogram bisect each other.

c. $\overline{RT} \perp \overline{SV}$ true; The diagonals of a rhombus are perpendicular.

You can construct a rhombus with a compass and straightedge.

CONSTRUCTION

Construct a rhombus.

1. Draw $\overline{AD}$. Set the compass to match the length of $\overline{AD}$. You will use this compass setting for all arcs drawn.

2. Place the compass at point *A* and draw an arc above $\overline{AD}$. Choose any point on that arc and label it *B*.

3. Place the compass at point *B* and draw an arc to the right of *B*.

4. Then place the compass at point *D* and draw an arc to intersect the arc drawn from point *B*. Label the point of intersection *C*.

5. Use a straightedge to draw $\overline{AB}$, $\overline{BC}$, and $\overline{CD}$.

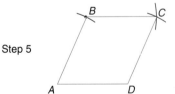

Quadrilateral *ABCD* is a rhombus since all of the sides are congruent.

When a quadrilateral is both a rhombus and a rectangle, it is a **square**. A square is a quadrilateral with four right angles and four congruent sides. You can construct a square in the same way that you constructed a rhombus, but a square must have a right angle.

CHECKING FOR UNDERSTANDING

Communicating Mathematics

Read and study the lesson to answer these questions.

1. A rhombus can be defined as any quadrilateral with four congruent sides. Explain why this definition is equivalent to *A rhombus is a parallelogram with four congruent sides.*

2. Compare a parallelogram and a rhombus. How are they the same? How are they different?

3. Explain the relationship between squares and rhombi. Is every rhombus a square? Is every square a rhombus?

Guided Practice

Determine whether *ABCD* is a parallelogram, rectangle, rhombus, or square. List all that apply.

4. $A(0, 1)$, $B(2, 0)$, $C(3, 2)$, $D(1, 3)$

5. $A(-1, 0)$, $B(1, 0)$, $C(3, 5)$, $D(1, 5)$

6. $A(-3, 2)$, $B(-3, 8)$, $C(12, 8)$, $D(12, 2)$

7. $A(8, 11)$, $B(2, 3)$, $C(3, 7)$, $D(9, 15)$

8. $A(-1, 8)$, $B(-6, -2)$, $C(5, 0)$, $D(10, 10)$

9. Copy and complete the following table. Determine if each quadrilateral has the given property. Write *yes* or *no*.

	Property	Parallelogram	Rectangle	Rhombus	Square
a.	The diagonals bisect each other.	_?_	_?_	_?_	_?_
b.	The diagonals are congruent.	_?_	_?_	_?_	_?_
c.	Each diagonal bisects a pair of opposite angles.	_?_	_?_	_?_	_?_
d.	The diagonals are perpendicular.	_?_	_?_	_?_	_?_

Use rhombus *BCDE* and the given information to solve each problem.

10. If $m\angle EBC = 132.6$, find $m\angle EBD$.

11. If $m\angle BDC = 25.9$, find $m\angle EDC$.

12. If $m\angle BEC = 2x + 10$ and $m\angle CED = 5x - 20$, find x.

13. If $m\angle CBD = 2x + 24$ and $m\angle EBD = x^2$, find x.

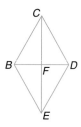

EXERCISES

Name all the quadrilaterals—parallelogram, rectangle, rhombus, or square—that have each property.

14. All angles are congruent.   **15.** The opposite sides are parallel.

16. All sides are congruent.   **17.** The opposite sides are congruent.

18. It is equiangular and equilateral.

Use rhombus _BEAC_ with _BA_ = 26 to determine whether each statement is _true_ or _false_. Justify your answers.

19. $CE = 26$ **20.** $HA = 13$

21. $\overline{BA} \perp \overline{EC}$ **22.** $\triangle BHE \cong \triangle AHC$

23. $m\angle BEH = m\angle EBH$

24. $\angle CBE$ and $\angle BCA$ are supplementary.

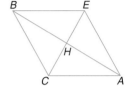

Determine whether _EFGH_ is a parallelogram, rectangle, rhombus, or square. List all that apply.

25. $E(0, 1)$, $F(2, 0)$, $G(4, 4)$, $H(2, 5)$

26. $E(0, 0)$, $F(4, -3)$, $G(8, 0)$, $H(4, 3)$

27. $E(2, -3)$, $F(-3, 1)$, $G(1, 6)$, $H(6, 2)$

28. $E(0, -4)$, $F(-4, 0)$, $G(0, 4)$, $H(4, 0)$

Use rhombus _IJKL_ and the given information to solve each problem.

29. If $m\angle 3 = 62$, find $m\angle 1$, $m\angle 4$, and $m\angle 6$.

30. If $m\angle 3 = 2x + 30$ and $m\angle 4 = 3x - 1$, find x.

31. If $m\angle 3 = 4(x + 1)$ and $m\angle 5 = 2(x + 1)$, find x.

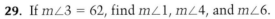

32. If $WXYZ$ is a square, find $m\angle ZXY$.

33. $PQMN$ is a parallelogram. If $PN = 7x - 10$ and $PQ = 5x + 6$, for what value of x is $PQMN$ a rhombus?

34. $ABXY$ is a parallelogram. If $AB = 5x + 24$ and $BX = x^2$, for what values of x is $ABXY$ a rhombus?

A kite is a quadrilateral with exactly two pairs of consecutive sides congruent and no two opposite sides congruent.

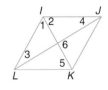

35. What conjecture can you make about the diagonals of a kite? Prove your conjecture.

36. Write a two-column proof of Theorem 6-10.

37. Write a two-column proof.

> **Given:** $ABCD$ is a rhombus.
> $\overline{AF} \cong \overline{BG}$
> $\overline{BG} \cong \overline{CH}$
> $\overline{CH} \cong \overline{DE}$
> $\overline{DE} \cong \overline{AF}$
>
> **Prove:** $EFGH$ is a parallelogram.

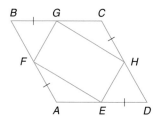

Critical Thinking

38. A line of symmetry for a figure is a line that can be drawn through the figure so that it can be folded along the line and the two halves match exactly. Draw a rhombus. How many lines of symmetry can you find?

Applications

39. Mechanics If you change the measures of the angles in a rhombus and the sides remain the same length, the sides will still be parallel. This property is useful in mechanical objects such as the car jack shown at the right. As the crank is turned, the measures of the angles at A and C become greater and those at B and D become lesser.

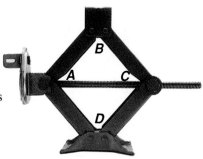

a. Describe what happens to the diagonals of the rhombus as the crank is turned.

b. What property of a rhombus ensures that the car will remain parallel to the ground as it lifts? Explain.

c. Could a parallelogram that is not a rhombus be used in this car jack? Why or why not?

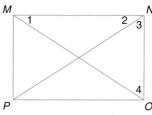

Mixed Review

40. In rectangle $MNOP$, $m\angle 1 = 32$. Find $m\angle 2$, $m\angle 3$, and $m\angle 4$. **(Lesson 6-4)**

41. $ABCD$ is a rectangle. If $AB = 2x + 5$, $CD = y + 1$, $AD = y + 5$, and $BC = 3x - 4$, find the measures of the sides. **(Lesson 6-1)**

42. If the sides of a triangle have measures of $3x + 2$, $8x + 10$, and $5x + 8$, find all possible values of x. **(Lesson 5-6)**

43. Determine whether the figure contains a pair of congruent triangles. Justify your answer. **(Lesson 5-2)**

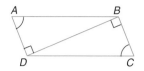

44. What algebraic property allows us to say that if $m\angle 1 = m\angle 2$ and $m\angle 2 = m\angle 3$, then $m\angle 1 = m\angle 3$? **(Lesson 2-4)**

45. Is the top of your desk best modeled by a point, a line, or a plane? **(Lesson 1-2)**

Wrap-Up

46. Copy the diagram at the right. Label the regions quadrilaterals, parallelograms, rhombi, rectangles, and squares to show the relationships among the figures.

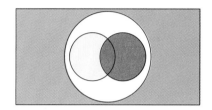

HISTORY CONNECTION

In India, geometry is used to expand the Hindu temples in a way that reflects a person's growth. During each stage of development, a body grows but remains similar to the earlier form. The temples are built beginning with the altar and then each stage of expansion adds to the structure.

When the temples are expanded, the expansion is a *gnomon*. A gnomon is any figure which when added to another figure leaves the new figure **similar** to the original. Similar figures have the same shape and a different size. For example, if a temple begins with a square altar, each consecutive step would add a shape that makes the temple into a larger square.

Three different ways of expanding a triangle with a gnomon.

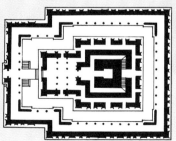

The floor plan of a Hindu temple.

6-6 Trapezoids

Objectives

After studying this lesson, you should be able to:

- recognize the properties of trapezoids, and
- use the properties of trapezoids in proofs and other problems.

Application

Frank Lloyd Wright was one of the most influential American architects. He created the prairie style of homes to harmonize with the landscape. This style emphasizes horizontal lines and natural materials because Wright thought a building should "grow" from its site. The faces of the roof of the prairie style home shown at the right are shaped like **trapezoids**.

A trapezoid is a quadrilateral with exactly one pair of parallel sides. The parallel sides are called **bases**, and the nonparallel sides are called **legs**. $\angle A$ and $\angle D$ and $\angle B$ and $\angle C$ are pairs of **base angles**.

A trapezoid is an **isosceles trapezoid** if its legs are congruent. You can fold an isosceles trapezoid so that the legs coincide. What do you think is true about the base angles?

Theorem 6-12

Both pairs of base angles of an isosceles trapezoid are congruent.

Proof of Theorem 6-12

Given: $TRAP$ is an isosceles trapezoid.
$\overline{RA} \parallel \overline{TP}$
$\overline{TR} \cong \overline{PA}$

Prove: $\angle T \cong \angle P$
$\angle TRA \cong \angle PAR$

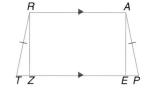

Paragraph proof:

Draw auxiliary segments so that $\overline{RZ} \perp \overline{TP}$ and $\overline{AE} \perp \overline{TP}$. Since $\overline{RA} \parallel \overline{TP}$ and parallel lines are everywhere equidistant, $\overline{RZ} \cong \overline{AE}$. Perpendicular lines form right angles, so $\angle RZT$ and $\angle AEP$ are right angles. $\triangle RZT$ and $\triangle AEP$ are right triangles by definition. Therefore, $\triangle RZT \cong \triangle AEP$ by HL. $\angle T \cong \angle P$ by CPCTC.

Since $\angle ARZ$ and $\angle RAE$ are right angles and all right angles are congruent, $\angle ARZ \cong \angle RAE$. $\angle TRZ \cong \angle PAE$ by CPCTC. So $\angle TRA \cong \angle PAR$ by angle addition.

Isosceles trapezoids have another unique property.

Theorem 6-13	**The diagonals of an isosceles trapezoid are congruent.**
	You will be asked to prove this theorem in Exercise 28.

The **median** of a trapezoid is the segment that joins the midpoints of the legs. The median has a special relationship to the bases. Draw a trapezoid and measure the bases and the median. What is their relationship?

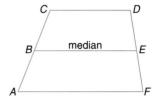

Theorem 6-14	**The median of a trapezoid is parallel to the bases and its measure is one half the sum of the measures of the bases.**

You can use the lengths of the bases to find the length of the median.

Example 1

CONNECTION

Algebra

Find the length of the median $\overline{QR}$ of trapezoid *MNOP* with vertices *M*(2, -9), *N*(-1, 1), *O*(3, 8), and *P*(10, 5).

First, graph the trapezoid to find the bases.

$\overline{NO}$ and $\overline{MP}$ are the bases. Now use the distance formula to find *NO* and *MP*.

$$NO = \sqrt{(-1-3)^2 + (1-8)^2}$$
$$= \sqrt{65}$$

$$MP = \sqrt{(2-10)^2 + (-9-5)^2}$$
$$= \sqrt{260}$$

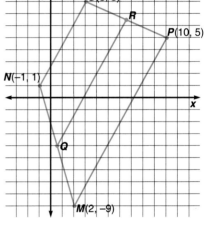

According to Theorem 6-14, the length of the median is half of the sum of the lengths of the bases, so $QR = \frac{1}{2}(NO + MP)$.

$$QR = \frac{1}{2}(NO + MP)$$
$$= \frac{1}{2}\left(\sqrt{65} + \sqrt{260}\right) \qquad \textit{Use your calculator.}$$
$$\approx 12.09$$

The length of the median is approximately 12 units.
Check by finding the coordinates of the midpoints and then applying the distance formula.

The length of the median can be used to find the lengths of the bases.

Example 2

Given trapezoid *ABCD* with median $\overline{EF}$, find the value of *x*.

EF is half the sum of *AB* and *DC*, so we can write the following equation.

$EF = \frac{1}{2}(AB + DC)$

$21 = \frac{1}{2}(3x + 4 + 5x - 2)$ *Substitution*

$21 = \frac{1}{2}(8x + 2)$

$21 = 4x + 1$

$20 = 4x$

$5 = x$

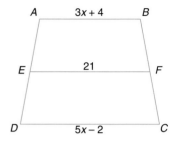

The value of *x* is 5.

Check: Is the sum of the measures of the bases twice the measure of the median?

CHECKING FOR UNDERSTANDING

Communicating Mathematics

Read and study the lesson to answer these questions.

1. Draw a trapezoid with each set of characteristics.
   a. both bases are shorter than the legs
   b. contains a right angle
   c. has two obtuse angles

2. Draw an isosceles trapezoid and label the legs and the bases.

3. If the measure of the median of a trapezoid is 15, what could the measures of the bases be? Explain.

Guided Practice

***ABCD* is an isosceles trapezoid with bases $\overline{AD}$ and $\overline{BC}$. Use the figure and the given information to solve each problem.**

4. If *BA* = 9, find *CD*.

5. If *AC* = 21, find *BD*.

6. If *AC* = 4*y* − 5 and *BD* = 2*y* + 3, find *AC* and *BD*.

7. If *m*∠*BAD* = 123, find *m*∠*CBA*.

8. If *m*∠*ADC* = 105, find *m*∠*DAB*.

9. Find the length of the median of a trapezoid with vertices at (1, 0), (3, −1), (6, 2), and (7, 6).

Find the value of x.

10.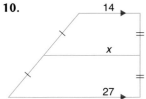

11.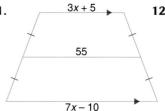

12.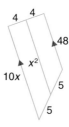

EXERCISES

Practice

If possible, draw a trapezoid that has the following characteristics. If the trapezoid cannot be drawn, explain why.

13. three congruent sides

14. congruent bases

15. a leg longer than both bases

16. bisecting diagonals

17. two right angles

18. four acute angles

19. one pair of opposite angles congruent

$PQRS$ is an isosceles trapezoid with bases $\overline{PS}$ and $\overline{QR}$. Use the figure and the given information to solve each problem.

20. If $PS = 20$ and $QR = 14$, find TV.

21. If $QR = 14.3$ and $TV = 23.2$, find PS.

22. If $TV = x + 7$ and $PS + QR = 5x + 2$, find x.

23. If $m\angle RVT = 57$, find $m\angle QTV$.

24. If $m\angle VTP = a$, find $m\angle TPS$ in terms of a.

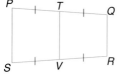

25. If the measure of the median of an isosceles trapezoid is 4.5, what are the possible integral measures for the bases?

26. $\overline{UR}$ is the median of a trapezoid with bases $\overline{ON}$ and $\overline{TS}$. If the coordinates of the points are $U(1, 3)$, $R(8, 3)$, $O(0, 0)$, and $N(8, 0)$, find the coordinates of T and S.

27. What type of quadrilateral is $PQRS$ if its vertices are $P(1, 3)$, $Q(-3, -1)$, $R(-2, -8)$, and $S(8, 2)$? Justify your answer.

28. Write a paragraph proof of Theorem 6-13.

Write a two-column proof.

29. **Given:** Trapezoid $RSPT$ is isosceles.

    Prove: $\triangle RSQ$ is isosceles.

30. **Given:** $\triangle SQR$ is isosceles.
    $\triangle PQT$ is isosceles.
    $\overline{TP} \parallel \overline{RS}$

    Prove: $RSPT$ is an isosceles trapezoid.

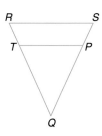

31. In the figure at the right, P is a point not in plane $\mathcal{A}$ and $\triangle XYZ$ is an equilateral triangle in plane $\mathcal{A}$. Plane $\mathcal{B}$ is parallel to plane $\mathcal{A}$ and intersects $\overline{PX}$, $\overline{PY}$, and $\overline{PZ}$ in points J, I, and K respectively.

    a. How many trapezoids are formed? Name them.

    b. If the trapezoids are isosceles, are they congruent? Justify your answer.

    c. Where could P be if none of the trapezoids are congruent?

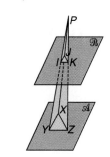

Critical Thinking

32. The figure at the right is a trapezoid with bases $\overline{SR}$ and $\overline{PQ}$. $PS = QS$, $m\angle SRQ = 120$, and $m\angle RQS = 20$. Find the measure of $\angle PSQ$.

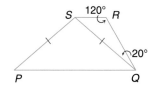

Applications

33. **Agriculture** Farmers use a gravity grain box to transport grain to a storage bin or grain elevator. The side and end views of a gravity grain box are shown below.

    a. Copy each view and label the vertices of the quadrilaterals. Describe each quadrilateral.

    b. Are any of the trapezoids on the grain box congruent?

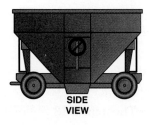

SIDE VIEW

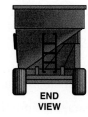

END VIEW

34. **Real Estate** A land developer bought a parcel of land for building new homes. A scale drawing of the lots into which the land will be divided is shown at the right. $\overline{BG}$ and $\overline{CF}$ are the medians of trapezoids $ADEH$ and $BDEG$ respectively. If $c = 81$, find a, b, d, and e.

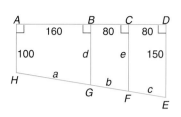

Computer

35. The BASIC computer program given below will find the measure of the median of a trapezoid $ABCD$ with legs $\overline{AB}$ and $\overline{CD}$.

```
10  INPUT "ENTER THE COORDINATES OF A"; X1, Y1
20  INPUT "ENTER THE COORDINATES OF B"; X2, Y2
30  INPUT "ENTER THE COORDINATES OF C"; X3, Y3
40  INPUT "ENTER THE COORDINATES OF D"; X4, Y4
50  X5 = (X1 + X2)/2
60  Y5 = (Y1 + Y2)/2
70  X6 = (X3 + X4)/2
80  Y6 = (Y3 + Y4)/2
90  M = SQR ((X5 - X6)^2 + (Y5 - Y6)^2)
100 PRINT "THE MEASURE OF THE MEDIAN IS "; M; "."
110 END
```

Use the program to find the measure of the median of the trapezoid with the given vertices. Round your answers to the nearest hundredth.

  a. $A(3, 5)$, $B(-1, 3)$, $C(7, 5)$, $D(7, 6)$

  b. $A(-1, 7)$, $B(8, 8)$, $C(2, 10)$, $D(-4, 8)$

Mixed Review

36. Determine if the statement *If a quadrilateral has all four sides congruent, then it is a square* is *true* or *false*. If false, give a counter-example. **(Lesson 6-5)**

37. Find the value of x if $\ell \parallel m$. **(Lesson 3-3)**

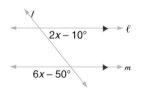

38. Determine if statement (3) follows from statements (1) and (2) by the law of detachment or the law of syllogism. If it does, state which law was used. **(Lesson 2-3)**
  (1) If a quadrilateral has exactly two opposite sides parallel, then it is a trapezoid.
  (2) $ABCD$ has exactly two opposite sides parallel.
  (3) $ABCD$ is a trapezoid.

39. If $\angle B \cong \angle L$, $m\angle B = 7x + 29$, and $m\angle L = 9x - 1$, is $\angle B$ acute, right, or obtuse? **(Lesson 1-7)**

Wrap-Up

40. **Journal Entry** Write a few sentences about the differences and similarities between trapezoids and parallelograms in your journal.

VOCABULARY

Upon completing this chapter, you should be
familiar with the following terms:

base	**294**	**266**	quadrilateral
base angle	**294**	**282**	rectangle
diagonal	**266**	**288**	rhombus
isosceles trapezoid	**294**	**266**	sides
leg	**294**	**289**	square
median	**295**	**294**	trapezoid
polygon	**266**	**266**	vertices
parallelogram	**266**		

SKILLS AND CONCEPTS

OBJECTIVES AND EXAMPLES	**REVIEW EXERCISES**

Upon completing this chapter, you should
be able to:

Use these exercises to review and prepare
for the chapter test.

■ recognize, use, and prove the properties of
a parallelogram. **(Lesson 6-1)**

In $\square MNOP$, if $PQ = 8x - 16$ and
$QN = 2x + 8$, find PN.

Since the diagonals of a parallelogram
bisect each other, $PQ = QN$.

$$PQ = QN$$
$$8x - 16 = 2x + 8$$
$$6x = 24$$
$$x = 4$$

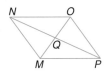

Now substitute 4 for x to find either PQ
or QN.

$$PQ = 8(4) - 16 \text{ or } 16$$

$$PN = 2(PQ) \text{ or } 32$$

**Complete each statement about
parallelogram ABCD below. Then name the
theorem or postulate that justifies your
answer.**

1. $\overline{BE} \cong$ ___?___
2. $\overline{AB} \cong$ ___?___
3. $\angle ADC \cong$ ___?___
4. $\overline{BC} \parallel$ ___?___
5. $\triangle BCD \cong$ ___?___
6. $\angle 1 \cong$ ___?___
7. $\overline{CE} \cong$ ___?___
8. $\angle ADC$ and ___?___ are supplementary.

OBJECTIVES AND EXAMPLES	REVIEW EXERCISES

- recognize the conditions that determine if a quadrilateral is a parallelogram. **(Lesson 6-3)**

If $m\angle DFG = 4y + 9$ and $m\angle FDH = 3y + 24$, find the value of y in order for $DFGH$ to be a parallelogram.

If the consecutive interior angles of a quadrilateral are supplementary, then it is a parallelogram. So, $m\angle DFG + m\angle FDH = 180$.

$$4y + 9 + 3y + 24 = 180$$
$$7y = 147$$
$$y = 21$$

Write a two-column proof.

9. **Given:** $\square PRSV$
   $\triangle PQR \cong \triangle STV$
   Prove: Quadrilateral $PQST$ is a parallelogram.

10. **Given:** $\square PQST$
    $\overline{QR} \cong \overline{TV}$
    Prove: Quadrilateral $PRSV$ is a parallelogram.

- recognize and use the properties of rectangles. **(Lesson 6-4)**

Properties of a rectangle
- Opposite sides are congruent.
- Opposite angles are congruent.
- Consecutive angles are supplementary.
- Diagonals bisect each other.
- All four angles are right angles.
- Diagonals are congruent.

If *ABCD* is a rectangle, find the value of *x*.

11. $DB = 5x - 4$
    $AC = 6x - 10$

12. $m\angle DAC = 12x + 1$
    $m\angle CAB = 6x - 1$

13. $EB = 8x + 4$
    $AC = 24x - 8$

14. $AB = x^2$
    $DC = 3x - 2$

15. $m\angle BCD = 10x^2$
    $m\angle CDA = 9x^2 + 9$

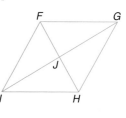

- recognize and use the properties of squares and rhombi. **(Lesson 6-5)**

$RHOM$ is a rhombus. If $m\angle 1 = 33$, find $m\angle 2$, $m\angle 3$, and $m\angle 4$.

The diagonals of a rhombus bisect a pair of opposite angles, so $\angle 1 \cong \angle 2$. $m\angle 2 = 33$

The diagonals are perpendicular. So $\angle 3$ is a right angle. $m\angle 3 = 90$

In $\triangle HRB$, $m\angle 2 + m\angle 3 + m\angle 4 = 180$. Using substitution, $33 + 90 + m\angle 4 = 180$, so $m\angle 4 = 57$.

Use rhombus *FGHI* and the given information to solve each problem.

16. If $FJ = 6$, find FH.

17. If $m\angle FGJ = 23$, find $m\angle FGH$.

18. Find $m\angle IJF$.

19. If $HG = 4x - 1$ and $FG = 20 + x$, find x.

Identify each quadrilateral as a parallelogram, rectangle, rhombus, or square. List all that apply.

20. $K(2, 3)$, $L(7, 3)$, $M(5, 0)$, $N(0, 0)$

21. $P(0, 9)$, $Q(-2, 1)$, $R(2, 0)$, $S(4, 8)$

22. $T(0, 0)$, $U(6, 6)$, $V(12, 0)$, $W(6, -6)$

■ recognize and use the properties of trapezoids. **(Lesson 6-6)**

Find the measures of the numbered angles.

$m\angle 1 = 70 \qquad m\angle 2 = 60$
$m\angle 3 = 35 \qquad m\angle 4 = 60$
$m\angle 5 = 50 \qquad m\angle 6 = 35$

RSTW is a trapezoid with bases $\overline{RS}$ and $\overline{TW}$. Use the figure and the given information to solve each problem.

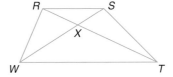

23. If $m\angle SRX = 35$, find $m\angle RTW$.
24. If $m\angle RSW = 45$, find $m\angle TWS$.
25. If $m\angle RTW = 47$, find $m\angle SRT$.
26. If $RS = 23$ and $TW = 19$, find the measure of the median of *RSTW*.
27. If $m\angle XTW = 23$ and $m\angle SXR = 127$, find $m\angle XWT$.

APPLICATIONS AND CONNECTIONS

28. Business   Jason is stacking cans of juice in a pyramid at the end of a grocery aisle. The base of the pyramid is a single row of cans and each row has one can less than the one below it. If each can is 12 centimeters tall, what is the minimum number of cans Jason could use to build a pyramid 120 centimeters tall? **(Lesson 6-2)**

29. Art   Tami is constructing a frame for her art project. She has two 12-inch pieces of framing material and two 16-inch pieces. If Tami cuts the ends of each piece at a 45° angle and places the congruent pieces on the opposite sides, will the frame be a rectangle? Explain. **(Lesson 6-4)**

30. Construction   A diagram of a bridge support is shown below. The support is an isosceles trapezoid and the cross-pieces, $\overline{AG}$ and $\overline{BG}$ are congruent. Find $m\angle ADF$. **(Lesson 6-6)**

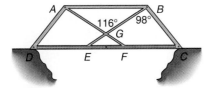

Complete each statement about parallelogram *DHGF* at the right. Then name the theorem or postulate that justifies your answer.

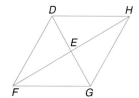

1. $\overline{DE} \cong$ __?__

2. $\angle FDH \cong$ __?__

3. $\overline{FD} \parallel$ __?__

4. $\triangle FDG \cong$ __?__

5. If *DHGF* is a rhombus, then $\overline{DG} \perp$ __?__.

6. If *DHGF* is a rhombus, then $\overline{DG}$ bisects __?__ and __?__.

7. The sequence 1, 1, 2, 3, 5, 8, . . . is called the *Fibonacci sequence*. Find the next four numbers in the pattern.

Determine whether *ABCD* is a parallelogram. Explain.

8. $A(-2, 6)$, $B(2, 11)$, $C(3, 8)$, $D(-1, 3)$

9. $A(-3, 7)$, $B(3, 2)$, $C(0, -1)$, $D(-6, 3)$

10. $A(7, -3)$, $B(4, -2)$, $C(6, 4)$, $D(12, 2)$

11. $A(11, 3)$, $B(4, 2)$, $C(1, -2)$, $D(8, -1)$

Determine if each statement is *true* or *false*. If false, draw a counterexample.

12. If a quadrilateral has four right angles, then it is a rectangle.

13. If the diagonals of a quadrilateral are perpendicular, then it is a rhombus.

14. If a quadrilateral has all four sides congruent, then it is a square.

15. If a quadrilateral has opposite sides congruent and one right angle, then it is a rectangle.

***PQRS* is an isosceles trapezoid with bases $\overline{QR}$ and $\overline{PS}$. Use the figure and the given information to solve each problem.**

16. If $PS = 32$ and $TV = 26$, find QR.

17. If $PT = 18$, find VR.

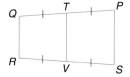

18. If $m\angle QTV = 79$, find $m\angle TVS$.

19. If $QR = 9x$ and $PS = 13x$, find TV.

20. If $QR = x - 3$, $PS = 2x + 4$, and $TV = 3x - 10$, find QR, PS, and TV.

Bonus

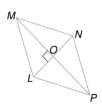

Given: $\square LMNP$
$\overline{MP} \perp \overline{LN}$

Prove: *LMNP* is a rhombus.

College Entrance Exam Preview

The questions on these pages involve comparing two quantities, one in Column A and one in Column B. In certain questions, information related to one or both quantities is centered above them.

Directions:
Write A if the quantity in Column A is greater. Write B if the quantity in Column B is greater. Write C if the quantities are equal. Write D if there is not enough information to determine the relationship.

	Column A	Column B
1.	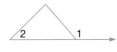	
	$m\angle 2$	$m\angle 1$
2.	$0 < x < 1$	
	$2x$	x^2
3.	a, b, and c are positive integers.	
	$\dfrac{a + b + c}{3}$	the average of a, b, and c
4.	$y \neq 3$	
	$2y + 1$	$\dfrac{y^2 - 3y}{\frac{1}{2}(y - 3)}$
5.	In $\triangle ABC$, $m\angle A > m\angle B$ and $m\angle A > m\angle C$.	
	AC	AB
6.	$\ell \parallel m$	

	$m\angle 2 + m\angle 3 - m\angle 1$	$m\angle 2$

	Column A	Column B
7.	The sum of the integers from 1 to 20	210
8.	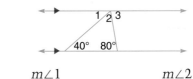	
	$m\angle 1$	$m\angle 2$
9.	$a > 0$	
	$\dfrac{1}{a}$	a
10.	$x > y$	
	$|x|$	$|y|$
11.		
	$m\angle 1$	$m\angle 2$
12.	$r > 0$ $s < 0$	
	$r + s$	$r - s$
13.	$a^\star = a^3 - 1$	
	$5^\star$	$(\text{-}5)^\star$
14.	10^{10}	$10^{11} - 10^{10}$
15.	The largest prime factor of 858	The largest prime factor of 2310
16.	$3x - 4y = -2$ $4x + 2y = 12$	
	x	y

Solve each of the following. You may use a calculator.

17. Express the quotient $4{,}000{,}000 \div 6400$ in decimal and scientific notation.

18. Line p is parallel to line l, and lines ℓ and m intersect at point P on line l. If $\angle 1$ measures $80°$ and $\angle 2$ measures $30°$, find the degree measure of $\angle 3$.

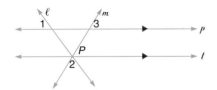

19. Write the equation of the line that passes through the points $(9, 7)$ and $(-3, 5)$ in standard form.

20. A person's optimal heart rate for exercising, h, is determined by subtracting their age, a, from 220 and then taking 75% of the result. Write a formula to represent this relation.

21. How many units apart are the points $(-11, 2)$ and $(6, 4)$ in the coordinate plane?

22. Use the formula $d = st$ where d represents distance in miles, s represents speed in miles per hour, and t represents time in hours, to find the speed of a race car if it travels 5 miles in 2 minutes and 30 seconds.

TEST-TAKING TIP

Quantitative Comparisons

Even though you don't know if you are dealing with an equality or an inequality in a quantitative comparison, you can use the rules for inequalities to simplify the comparison. In algebra, you learned that you can add or subtract a quantity from both sides of an inequality without changing the direction of the inequality sign. You can also multiply or divide by positive values and maintain the direction of the inequality. Study the example below.

Column A	Column B

x is a positive integer, and $0 < k < 1$.

Column A	Column B
x	$\dfrac{x}{k}$

Since k is a positive number, you can multiply the value in each column by k.

Column A	Column B
kx	x

Then you can divide each value by x.

Column A	Column B
k	1

Since the instructions say that $0 < k < 1$, the quantity in Column B must be greater.

Remember that multiplying or dividing by a negative number will reverse an inequality sign. *Never assume* that variables represent positive numbers if that is not stated in the instructions.

23. What are the values of x for which $\dfrac{x(x + 1)}{(x - 4)(x + 3)}$ is undefined?

C H A P T E R   7

Similarity

GEOMETRY AROUND THE WORLD
Germany

Have you ever wondered how artists paint three-dimensional scenes—pictures with height, width, and depth—on flat, two-dimensional canvases? They use a geometric concept called linear perspective, in which the lines of objects come together in a way that gives the viewer the illusion of depth.

One of the masters of linear perspective drawing was a German painter and wood engraver named Albrecht Dürer. Dürer, who lived from 1471 to 1528, traveled to Italy to study the visual geometry of Leonardo da Vinci. He then developed a technique to obtain linear perspective that he worked out on his canvas before he began painting. Carefully, he drew a grid of "parallel" lines receding to a vanishing point. He then created two additional points of alignment called lateral points. To these, Dürer added other geometric points and lines to guide him in providing depth to his paintings.

GEOMETRY IN ACTION

Look carefully at the copy of one of Dürer's engravings, completed in 1514. Notice how the parallel lines of the ceiling, table, and window ledge all recede to a single vanishing point just above the chair on the right. This technique is illustrated by the diagram below. What dimension did Dürer add to the engraving using this technique?

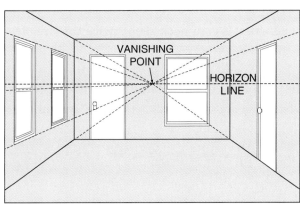

◀ *"St. Jerome in His Cell"*   Inset: *Albrecht Dürer*

Do you see the depth in this work by Albrecht Dürer? How do you think he achieved this effect?

307

7-1   Properties of Proportions

Objective   After studying this lesson, you should be able to:
■ recognize and use ratios and proportions.

Application   According to the 1990 United States census, there are about 32 people for every 10 families. This information can be expressed as the ratio 32 to 10 or $\frac{32}{10}$ or $\frac{3.2}{1}$. In other words, there were about 3.2 people per family.

A **ratio** is a comparison of two quantities. The ratio of a to b can be expressed as $\frac{a}{b}$ where b is not zero. The ratio can also be written as $a:b$.

Example 1

APPLICATION

Biology

Many parts of the human body have a common ratio. The students in a geometry class measured the sizes of their necks and their wrists in centimeters. The measures are displayed in the table. What is a good estimate for the ratio of the neck to the wrist?

Neck	30	33.5	31	33	34.5	35	34	32
Wrist	15	17.5	15.5	15	14	16	15	14

Neck	31	40	35	39.5	33	37.5	31	33
Wrist	15.5	17	18	16.5	15	17.5	15.5	15

Neck	34.5	42	37	42.5	35.5	39	36.5	36.5
Wrist	14	18	18	21	14.5	18	16	14.5

If you add all the neck sizes for the class and divide by 24, the average neck size is about 35.3 centimeters. The average wrist size is about 16.1 centimeters. Therefore the average ratio is $\frac{35.3}{16.1}$ or about 2.19. The size of their necks, on the average, is about twice the size of their wrists.

Check your neck and wrist. Is your neck about twice the size of your wrist?

Refer to the application about families. Suppose a TV report on the census stated there were 160 people for every 50 families. Is this different from the first ratio of $\frac{32}{10}$? When they are simplified, both ratios are equivalent to $\frac{3.2}{1}$.

$$\frac{32}{10} = \frac{3.2}{1} \qquad \frac{160}{50} = \frac{3.2}{1}$$

An equation stating that two ratios are equal is a **proportion**. So, $\frac{32}{10} = \frac{160}{50}$ is a proportion.

Every proportion has two **cross products**. In the proportion $\frac{32}{10} = \frac{160}{50}$, the cross products are 32 times 50 and 160 times 10. The cross products of a true proportion are equal. 32 and 50 are called the **extremes**, and 160 and 10 are called the **means**.

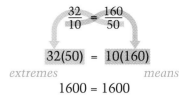

$$\frac{32}{10} = \frac{160}{50}$$

$$\underset{extremes}{32(50)} = \underset{means}{10(160)}$$

$$1600 = 1600$$

Consider the general case.

$$\frac{a}{b} = \frac{c}{d}$$

$$(bd)\frac{a}{b} = (bd)\frac{c}{d} \qquad \textit{Multiply each side by bd.}$$

$$da = bc \qquad \textit{Simplify.}$$

Equality of Cross Products	**For any numbers a and c and any nonzero numbers b and d,** $\frac{a}{b} = \frac{c}{d}$ **if and only if $ad = bc$.** *The product of the means equals the product of the extremes.*

Example 2

APPLICATION

Travel

Countries have different monetary systems. The chart at the right shows the rate of exchange for some foreign currency in terms of U.S. dollars. If a tour book costs 450 Taiwanese dollars, how much would it cost in U.S. dollars?

Foreign Currency	Rate of Exchange per U.S. Dollar
Spanish peseta	102.70
Swedish krona	5.9680
Swiss frank	1.4470
Taiwanese dollar	25.74
German mark	1.6310

Since 25.74 Taiwanese dollars can be exchanged for 1 U.S. dollar, the ratio of Taiwanese dollars to 1 U.S. dollar is $\frac{25.74}{1}$. The ratio of 450 Taiwanese dollars to x U.S. dollars would be written $\frac{450}{x}$, where x is the number of U.S. dollars. These ratios are equivalent and can be written as a proportion.

$$\frac{25.74}{1} = \frac{450}{x}$$

$$25.74x = 450 \qquad \textit{Find the cross product.}$$

$$\frac{25.74x}{25.74} = \frac{450}{25.74} \qquad \textit{Divide each side by 25.74.}$$

$$x = 17.48 \qquad \textit{Use your calculator.}$$

The book costs about $17.48 in U.S. currency.

Ratios can also be used to compare three or more numbers. The expression $a:b:c$ means that the ratio of the first two numbers is $a:b$, the ratio of the last two numbers is $b:c$, and the ratio of the first and last numbers is $a:c$.

Example 3

The ratio of the measures of the angles of a triangle is 3:5:7. What is the measure of each angle in the triangle?

Let 3x, 5x, and 7x represent the measures of the angles of the triangle.

$3x + 5x + 7x = 180$ *The sum of the angle measures in a triangle is 180.*

$15x = 180$

$x = 12$ *Divide each side by 15.*

The measures of the angles are 3(12) or 36, 5(12) or 60, and 7(12) or 84.

CHECKING FOR UNDERSTANDING

Communicating Mathematics

If there are 81 boys and 72 girls in the sophomore class, the ratio of boys to girls is $\frac{81}{72}$. Read and study the lesson to answer these questions about the proportion $\frac{81}{72} = \frac{1.125}{1}$.

1. Name the two ratios in this proportion.

2. Name the means.

3. Name the extremes.

4. Show that the product of the means is equal to the product of the extremes.

5. Explain what $\frac{1.125}{1}$ represents.

Guided Practice

Use the 1991 final standings for the American League West to find the ratios for Exercises 6-9. Express each ratio as a decimal rounded to three decimal places.

6. games won to games lost for Seattle

7. games won to games played for Chicago

8. games won to games lost for Minnesota

9. games won by Kansas City to games won by Texas

Team	Wins	Losses
Minnesota	92	63
Chicago	84	71
Texas	82	73
Oakland	82	74
Kansas City	80	76
Seattle	78	77
California	77	79

Given a = 3, b = 2, c = 6, and d = 4, determine if each pair of ratios forms a true proportion.

10. $\frac{b}{a} \stackrel{?}{=} \frac{d}{c}$

11. $\frac{a}{c} \stackrel{?}{=} \frac{b}{d}$

12. $\frac{c}{b} \stackrel{?}{=} \frac{d}{a}$

13. $\frac{(a + b)}{b} \stackrel{?}{=} \frac{(c + d)}{d}$

14. $\frac{d}{b} \stackrel{?}{=} \frac{c}{a}$

Find the value of x using cross products.

15. $\frac{x}{3} = \frac{15}{10}$

16. $\frac{5}{17} = \frac{2x}{51}$

17. $\frac{x + 1}{x} = \frac{7}{2}$

18. The ratio of AB to AC is equivalent to the ratio of FT to FE. If $AB = 8$, $BC = 6$, and $FT = 2$, what is FE?

EXERCISES

Practice

Gears on bicycles are called sprocket wheels. To determine gear ratios on bicycles, you must find the ratio of the number of rear sprocket teeth to the number of front sprocket teeth. Find each ratio. Express your answer as a decimal rounded to two places.

19. 12 rear sprocket teeth
    24 front sprocket teeth

20. 15 rear sprocket teeth
    55 front sprocket teeth

21. 13 rear sprocket teeth
    52 front sprocket teeth

22. 20 rear sprocket teeth
    30 front sprocket teeth

Solve each proportion using cross products.

23. $\frac{11}{24} = \frac{x}{24}$

24. $\frac{5}{8} = \frac{20}{x}$

25. $\frac{b}{3.24} = \frac{1}{8}$

26. $\frac{4}{n} = \frac{7}{8}$

27. $\frac{m + 3}{12} = \frac{5}{4}$

28. $\frac{1}{3} = \frac{t}{8 - t}$

$\overline{AD}$ **is a median of** $\triangle ABC$ **shown at the right.**

29. Find the ratio of BD to DC.

30. Find the ratio of DC to BC.

31. If $\triangle ABC$ is an equilateral triangle, find the ratio of $m\angle ABD$ to $m\angle ADC$.

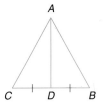

32. In the figure at the right, $\frac{RS}{SP} = \frac{RT}{TQ}$.
Use proportions to complete the table.

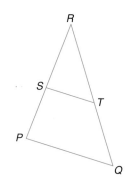

	RS	SP	RP	RT	TQ	RQ
a.	2	3	_?_	4	_?_	_?_
b.	_?_	45	_?_	32	40	72
c.	_?_	_?_	3	16	_?_	24
d.	_?_	_?_	36	36	48	_?_

Proportions can be used to change a fraction to a percent. For example to change $\frac{5}{6}$ to a percent, you solve the proportion $\frac{5}{6} = \frac{n}{100}$. Use a proportion to change each fraction to a percent. Round your answers to the nearest tenth.

33. $\frac{3}{8}$ **34.** $\frac{5}{12}$ **35.** $\frac{13}{4}$

Use equality of cross products to show each of the following is true. Assume that b and d are not zero.

36. $\frac{a}{b} = \frac{c}{d}$ if $\frac{a+b}{b} = \frac{c+d}{d}$.

37. $\frac{a}{b} = \frac{c}{d}$ if $\frac{a-b}{b} = \frac{c-d}{d}$.

38. $\frac{a}{b} = \frac{c}{d}$ if $\frac{a}{b} = \frac{a+c}{b+d}$ or $\frac{c}{d} = \frac{a+c}{b+d}$.

An expression of the form $x = y = z$ means $x = y$, $y = z$, and $x = z$. Solve for x and y.

39. $8 = \frac{10}{x} = \frac{y}{5}$

40. $\frac{x-1}{x} = \frac{3}{4} = \frac{y}{y+1}$

41. Write $xy = 22$ as a proportion.

42. The ratio of the measures of two angles of an isosceles triangle is 1 to 2. What are the possible measures of the angles of the triangle?

Critical Thinking

43. Suppose the measures of the sides of two rectangles are proportional.
   a. Will their perimeters have an equivalent ratio? Explain.
   b. Will their areas have an equivalent ratio? Explain.

Applications

44. Carpentry The pitch of a roof is the ratio of the rise (change in height) to the run (change in width). If a roof has a rise of 3.5 feet and a run of 10.5 feet, what is its pitch?

45. **Travel**   Use the table on page 309 to determine how many Swedish kronas a traveler would receive for $500 in U.S. currency.

46. **Banking**   One way to determine the strength of a bank is to calculate its capital-to-assets ratio as a percent. A strong bank should have a ratio of 4% or more. The Pilgrim National Bank has a capital of 2.3 billion dollars and assets of 52.6 billion dollars. Is it a strong bank? Explain.

47. **Sports**   On a bike, the ratio of the number of rear sprocket teeth to the number of front sprocket teeth is equivalent to the number of rear sprocket wheel revolutions to the number of pedal revolutions. If there are 24 rear sprocket teeth and 54 front sprocket teeth, how many revolutions of the rear sprocket wheel will occur for 3 revolutions of the pedal?

Mixed Review

48. Can a rectangle be a rhombus? Explain. **(Lesson 6-5)**

49. Can 19, 31, and 55 be the measures of the sides of a triangle? Explain. **(Lesson 5-6)**

50. Draw and label an obtuse isosceles triangle. **(Lesson 4-1)**

51. Write *A trapezoid has exactly two opposite sides parallel* in if-then form. **(Lesson 2-2)**

52. Draw opposite rays $\overrightarrow{QS}$ and $\overrightarrow{QR}$. **(Lesson 1-6)**

Wrap-Up

53. **Portfolio**   Choose the application problem from this lesson that you find most interesting. Copy the problem and its solution and place this in your portfolio.

～～～ ASTRONOMY CONNECTION ～～～

The fields of astronomy and geometry are closely related. Among other things, astronomers use geometry to calculate the speeds of celestial bodies. Vera Rubin, one of only 75 women to be nominated to the National Academy of Sciences, studies the speeds of faraway galaxies.

Ms. Rubin has been studying spiral galaxies for several years at the Department of Terrestrial Magnetism in Washington, D.C. She has discovered that in spiral galaxies, the stars in the outer arms of the galaxy do not slow down like the outer planets of our solar system do. Traveling at the speeds they do, these fast-moving outer stars should fly off into space. Ms. Rubin has hypothesized that there must be some type of matter between the stars that pulls them together. She calls this *dark matter*. Research on dark matter is still being conducted. But the studies that have been completed on 200 different spiral galaxies confirm Ms. Rubin's theory.

Applications of Proportions

Objective

After studying this lesson, you should be able to:

■ apply and use the properties of proportions.

Application

Suppose that you are selecting a rectangle design for the wallpaper in your room. Which of the four rectangular design shapes at the right appeals to you most?

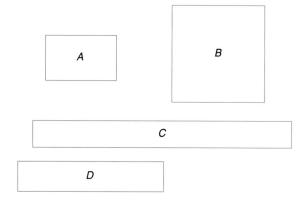

Measure the length and width of the rectangle you selected and calculate the ratio of the length to the width.

If you chose rectangle *A*, the rectangle considered most pleasing by many artists and architects, the ratio of length to width is about 1.618. This ratio is known as the *golden ratio.* Rectangles in which the ratio of their length to their width is 1.618 are called *golden rectangles.* Because of their pleasing shape, they can be found in ancient and modern architecture and in nature. Notice the number of golden rectangles in the picture of the Parthenon.

Because the ratio of length to width is the same for all golden rectangles, the measures of their corresponding sides are proportional. Rectangles *ABCD* and *PQRS* are golden rectangles.

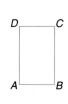
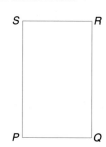

$$\frac{BC}{AB} = \frac{1.618}{1} \qquad \frac{QR}{PQ} = \frac{1.618}{1}$$

$$\frac{BC}{AB} = \frac{QR}{PQ} \qquad \textit{Transitive property}$$

$$AB \cdot QR = BC \cdot PQ \qquad \textit{Cross products}$$

$$\frac{AB \cdot QR}{QR \cdot PQ} = \frac{BC \cdot PQ}{QR \cdot PQ} \qquad \textit{Division property of equality}$$

$$\frac{AB}{PQ} = \frac{BC}{QR}$$

Proportions are also used with special ratios called **rates**. A rate is the ratio of two measurements that may have different types of units. For example, a race car at the Indianapolis 500 can travel 966 feet in 3 seconds and has a rate of speed of 966 feet/3 seconds or 322 feet per second.

Example 1

APPLICATION

Sports

In 1990 Arie Luyendyk set a track record for the Indianapolis 500. He averaged about 15.499 miles in 5 minutes. At this rate, how far could he travel in 1 hour?

This problem can be solved by writing and solving a proportion. Let d represent the distance he could travel in one hour. *Remember that 1 hour = 60 minutes.*

$$\frac{15.499 \text{ mi}}{5 \text{ min}} = \frac{d \text{ mi}}{60 \text{ min}}$$

$15.499 \cdot 60 = 5d$ *Cross products*

$929.94 = 5d$

$185.988 = d$

At this rate, Luyendyk could travel 185.988 miles in 1 hour.

Proportions and maps can be used to find distances between two locations.

Example 2

APPLICATION

Travel

On a map of Florida, three fourths of an inch represents 15 miles. If it is approximately 10 inches from Tampa to Jacksonville on the map, what is the actual distance in miles?

Let d represent the distance in miles from Tampa to Jacksonville.

$$\frac{0.75 \text{ in.}}{10 \text{ in.}} = \frac{15 \text{ mi}}{d \text{ mi}}$$

$0.75d = 15 \cdot 10$ *Cross products*

$0.75d = 150$

$d = 200$ *Division property of equality*

The distance from Tampa to Jacksonville is about 200 miles.

CHECKING FOR UNDERSTANDING

Communicating Mathematics

Read and study the lesson to answer each question.

1. Explain how you would determine whether the rectangle at the right is a golden rectangle.

2. **a.** The marriage rate for 1895 was 8.9. If the rate was per 1000 people, explain what this rate means.

   b. The marriage rate in 1989 was 9.7. What is the difference between the two rates?

   c. For every 10,000 people, how many would you expect to be married in 1989?

3. A map is scaled so that 1 centimeter represents 15 kilometers. Two towns are 7.9 centimeters apart on the map. Explain how to set up a proportion to determine the distance between the two towns.

Guided Practice

Determine which proportion can be used to solve the problem. Do not solve.

4. Find the cost of 120 pencils if 100 pencils cost $16.10.

   a. $\frac{100}{16.10} = \frac{x}{120}$ **b.** $\frac{16.10}{120} = \frac{100}{x}$ **c.** $\frac{100}{16.10} = \frac{120}{x}$ **d.** $\frac{16.10}{100} = \frac{120}{x}$

5. How much in U.S. currency would you receive for $75 in Canadian currency if the exchange rate is one Canadian dollar equals $0.88 in U.S. currency?

   a. $\frac{88}{100} = \frac{x}{75}$ **b.** $\frac{100}{88} = \frac{x}{75}$ **c.** $\frac{88}{100} = \frac{75}{x}$ **d.** $\frac{100}{x} = \frac{75}{88}$

Suppose the measures of the corresponding sides of the polygons *ABCD* and *PQRS* are proportional.

6. If $AB = 2$, $AD = 5$, and $PQ = 3$, find PS.

7. If $RS = 4.5$, $CD = 6.23$, and $BC = 7.0$, find QR.

8. If $CD = 21.7$, $DA = SP + 4$, and $SR = 14.0$, find SP.

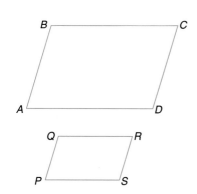

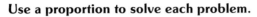

Determine if the given quantities are proportional.

9. the ratio of the number of sides of a rectangle to the number of its diagonals and the ratio of the number of sides of a pentagon to the number of its diagonals

10. the ratio of the number of sides of a triangle to the number of its exterior angles and the ratio of the number of sides of a hexagon to the number of its exterior angles

Use a proportion to solve each problem.

11. **Consumer Math** The average American drinks about 4 soft drinks every 3 days. About how many soft drinks will the average person drink in a year?

12. **Sports** A designated hitter made 8 hits in 9 games. If she continues hitting at that rate, how many hits will she make in 108 games?

EXERCISES

Practice

Use the number line at the right to determine if the given ratios are equal.

13. $\frac{CD}{CT}$ and $\frac{AC}{AT}$

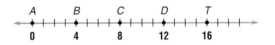

14. $\frac{BD}{BT}$ and $\frac{AD}{AT}$

15. $\frac{AB}{AT}$ and $\frac{BC}{BT}$

Suppose the measures of corresponding sides of the polygons at the right are proportional.

16. If $AB = 4$, $AD = 8$, and $PQ = 6$, find PS.

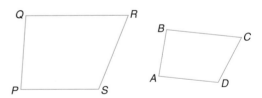

17. If $RS = 4.5$, $CD = 6.3$, and $BC = 7$, find QR.

18. If $CD = 21.8$, $DA = 43.6$, and $SR = 33$, find SP.

Use a proportion to solve each problem.

19. According to the National Safety Council, the death rate in the United States from motor vehicle accidents in 1989 was 18.9 per 100,000 people.

   a. In 1989 Los Angeles had a population of about 3,400,000. Approximately how many people would you estimate died from motor vehicle accidents in Los Angeles that year?

   b. Based on the United States population of about 249,600,000 in 1989, about how many people would you estimate died from motor vehicle accidents in the United States in 1989?

20. Geography The population and area of the five largest metropolitan areas in the world are given in the chart below. The population density of a city or country is the ratio of its population to its area.

Metropolitan Area	Population (thousands)	Area (square miles)
Tokyo, Japan	26,952	1089
Mexico City, Mexico	20,207	522
Sao Paulo, Brazil	18,052	451
Seoul, South Korea	16,268	342
New York City, United States	14,622	1274

a. Find the population density of each metropolitan area. Express each ratio as a decimal rounded to one decimal place.

b. Which area has the greatest population density? What does this mean?

Suppose the measures of corresponding sides of polygons *ABCD* and *PQRS* are proportional. Round your answer to the nearest tenth.

21. If $QR = x + 3$, $PQ = 4$, $BC = x + 5$, and $AB = 5$, find BC and QR.

22. If $DA = x + 2$, $AB = x - 3$, $PS = 5$, and $PQ = 3$, find AB and DA.

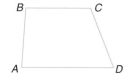

Use a proportion to solve each problem. Round your answer to the nearest tenth.

23. A 6-foot tree casts a 3.25-foot shadow. How tall is a tree that casts a 10-foot shadow at the same time of day?

24. Joyce Grauser has a drawing that is 20 centimeters long and 14 centimeters wide. She reduces the picture on a copying machine so that the copy is 10 centimeters wide.
   a. How long is the copy?
   b. What is the percentage of reduction?

25. The measures of the corresponding sides of *PDRS* and *BATL* are proportional. Given the vertices $P(2, 1)$, $D(4, 1)$, $R(4, 5)$, $S(2, 5)$, $B(-2, -1)$, and $L(-2, 5)$, find the possible coordinates of A and T.

26. Draw a star inside a regular hexagon as illustrated in the diagram at the right. Find AB, BD, AD, AE, AC, and BE by measuring the segments. Find the ratios $\dfrac{BD}{AB}$, $\dfrac{AD}{AB}$, $\dfrac{AE}{AB}$, $\dfrac{AC}{AB}$, and $\dfrac{BD}{BE}$. Which ones represent the golden ratio?

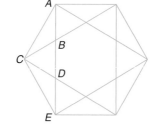

27. So far this semester, Lisa has 76 out of 90 possible points in her history class. If she gets a 100% on the next test and that raises her grade in the course to a 90%, how many points were possible on the test?

28. The perimeter of a triangle is 72 inches and the ratio of the measures of the sides is 3:4:5. Find the measure of the sides.

Critical Thinking

29. Statistics   The headlines in a newspaper read "U.S. Crime Rate up 1.4% in 1989." Use the information at the right to determine if the headlines told the whole story. Explain your answer.

Year	Population	Total Crimes
1988	245,807,000	13,923,100
1989	248,239,000	14,251,400

Applications

30. Music   The vibrations of the strings on a guitar produce the musical sounds. When the number of vibrations per second has a ratio of 2:1, the sounds are one octave apart. The sound from the higher octave has twice the number of vibrations as the sound from the lower octave. The number of vibrations per second is the frequency and is measured in a unit called Hertz. If the lowest string on a guitar has a frequency of 82.5 Hertz, find the frequency of a sound one octave higher.

31. Electronics   One basic stereo system consists of a CD component, a receiver, and two speakers. To determine how much money you should spend on the system, one rule might be to use a ratio of 1:3:2. In other words, the receiver should cost three times as much as the CD component, and the speakers should cost twice as much as the CD component. If you have saved $1000, how would you allocate your money to buy a stereo system?

32. Astronomy   Amos plans to make a scale model of Earth, the moon, and the sun. Suppose he makes the diameter of Earth 2 centimeters.

	Diameter	Distance from Earth
Earth	8000 miles	0 miles
Moon	2200 miles	240,000 miles
Sun	864,000 miles	93,000,000 miles

   a. How far away in centimeters should he place the moon from Earth?

   b. How long should the diameter of the moon be?

   c. How far away should he make the sun from Earth?

   d. How long should the diameter of the sun be?

33. Art   Joan Frank is a potter making a rectangular clay plaque 25 inches wide and 36 inches long. The plaque shrinks uniformly in the kiln to a 30-inch length. What is the width after the plaque shrinks?

34. **Baking** Tim's brownie recipe called for 2 squares of chocolate to $\frac{3}{4}$ cup flour. He wants to make as much brownie mix as possible, so he uses all $6\frac{1}{2}$ squares of chocolate that are left in the package. How much flour should he use?

35. **Consumer Math** Henri Rici paid $8000 for a car. After one year its value had decreased by $2000. By what percent had the car depreciated in value?

36. **Travel** The scale on a map is 1 centimeter to 57 kilometers. Two cities in North Dakota, Fargo and Bismark, are 4.7 cm apart on the map. What is the actual distance between these cities?

Computer

37. The *Fibonacci sequence* is also related to the golden ratio. The BASIC computer program below will find the first twenty terms of the Fibonacci sequence.

    a. Write the first twenty terms of the Fibonacci sequence.

    b. Describe the relationship between the terms of the Fibonacci sequence.

```
10 DIM X(20)
20 X(1) = 1
30 X(2) = 1
40 PRINT "THE FIRST TWENTY
   TERMS OF THE FIBONACCI
   SEQUENCE ARE ";
   X(1); ", "; X(2);
50 FOR N = 3 TO 20
60 X(N) = X(N − 1) +
   X(N − 2)
70 PRINT ", "; X(N);
80 NEXT N
90 PRINT "."
100 END
```

    c. Find the ratio between the pairs of consecutive terms of the sequence. What do you notice about these ratios?

Mixed Review

38. Solve the proportion $\frac{t}{18} = \frac{5}{6}$ using cross products. **(Lesson 7-1)**

39. The vertex angle of an isosceles triangle measures 104°. What are the measures of the base angles? **(Lesson 4-7)**

40. Draw parallel lines ℓ and m and transversal t and label a pair of corresponding angles 1 and 2. **(Lesson 3-1)**

41. Graph the point (7, 5) on a coordinate plane. **(Lesson 1-1)**

Wrap-Up

42. Give three examples of how ratios, rates, and proportions are used in your environment.

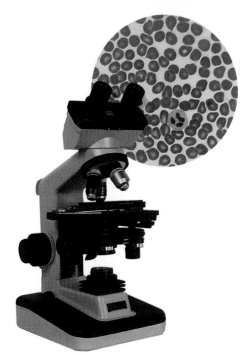

7-3 Similar Polygons

Objectives

After studying this lesson, you should be able to:

- identify similar figures, and
- solve problems involving similar figures.

Application

Research scientists use microscopes in their laboratories to study characteristics of blood cells that are not apparent to the naked eye. The microscope provides a larger view of the actual cells. When figures are the same shape but not necessarily the same size, they are called similar figures.

If the corresponding angles of two polygons are congruent and the measures of the corresponding sides are proportional, then the two polygons have the same shape. The parallelograms below are similar.

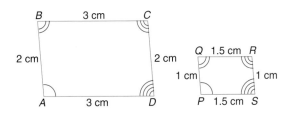

The symbol ~ means *is similar to*. We write $\square ABCD \sim \square PQRS$, which means *parallelogram ABCD is similar to parallelogram PQRS*. Just as in congruence, the order of the letters indicates the vertices that correspond. We can make the following statements about parallelograms *ABCD* and *PQRS*.

$$\angle A \cong \angle P \qquad \angle B \cong \angle Q \qquad \angle C \cong \angle R \qquad \angle D \cong \angle S$$

$$\frac{AB}{PQ} = \frac{BC}{QR} = \frac{CD}{RS} = \frac{DA}{SP} = \frac{2}{1}$$

The ratio of the lengths of two corresponding sides of two similar polygons is called the **scale factor**. The scale factor of $\square ABCD$ to $\square PQRS$ is 2.

Definition of Similar Polygons	**Two polygons are similar if and only if their corresponding angles are congruent and the measures of their corresponding sides are proportional.**

The properties of proportions can be used to solve problems involving similar polygons.

Example 1

Quadrilateral *PQRS* is similar to quadrilateral *ABCD*.

a. Find the value of *x*.

The corresponding sides are proportional, so we can write a proportion to find the value of *x*.

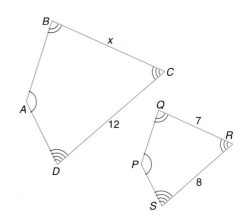

$$\frac{BC}{QR} = \frac{DC}{SR}$$

$$\frac{x}{7} = \frac{12}{8}$$

$8x = 84$ *Cross products*

$x = 10.5$

The value of *x* is 10.5.

b. Find the scale factor of quadrilateral *PQRS* to quadrilateral *ABCD*.

The scale factor is the ratio of the lengths of two corresponding sides. $\overline{SR}$ and $\overline{DC}$ are two corresponding sides.

$$\text{scale factor} = \frac{SR}{DC}$$

$$= \frac{8}{12} \text{ or } \frac{2}{3}$$

The scale factor of quadrilateral *PQRS* to quadrilateral *ABCD* is $\frac{2}{3}$.

Example 2

Contractors refer to blueprints when constructing a house. The blueprint at the right is similar to the floor plan of a constructed house. One inch on the blueprint represents 16 feet in the actual house. Using the properties of similar figures, determine the dimensions of the kitchen.

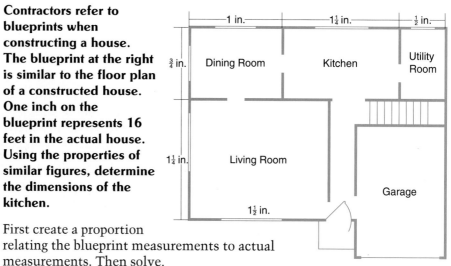

First create a proportion relating the blueprint measurements to actual measurements. Then solve.

$$\frac{blueprint\ measurement}{actual\ measurement} = \frac{blueprint\ kitchen\ measurement}{actual\ kitchen\ measurement}$$

$$\frac{1}{16} = \frac{1.25}{\ell} \qquad 1\frac{1}{4} = 1.25 \qquad\qquad \frac{1}{16} = \frac{0.75}{w} \qquad \frac{3}{4} = 0.75$$

$$\ell = 20 \qquad\qquad\qquad\qquad\qquad w = 12$$

The kitchen is 20 feet long and 12 feet wide.

CHECKING FOR UNDERSTANDING

Communicating Mathematics

Read and study the lesson to answer each question.

1. Describe the difference between congruence and similarity.

2. Must two congruent figures be similar? Explain.

3. Must two similar figures be congruent? Explain.

4. Are two equiangular polygons necessarily similar? Explain.

Guided Practice

List the conditions necessary for the pairs of figures to be similar.

5. $\triangle ABC \sim \triangle DEF$

6. polygon $RSTVWX \sim$ polygon $LMNOPQ$

Determine whether each pair of figures is similar. Justify your answer.

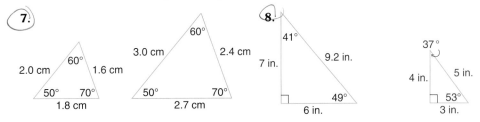

7.

8.

Is each pair of figures described necessarily similar? Explain.

9. any two right triangles

10. any two congruent quadrilaterals

11. any two rectangles

12. any two parallelograms

Given quadrilateral *LEFT* ~ quadrilateral *CORK*, find each of the following.

13. scale factor of *LEFT* to *CORK*

14. *KR*

15. *RO*

16. *CO*

17. **a.** perimeter of *LEFT*
    b. perimeter of *CORK*
    c. ratio of the perimeters of *LEFT* and *CORK*

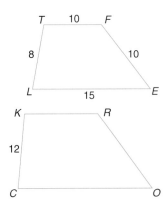

EXERCISES

Practice Two similar polygons are shown. Find the values of *x* and *y*.

18.

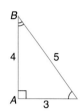

19.

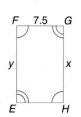

Draw and label a pair of polygons for each. If it is impossible to draw two such figures, write *impossible*.

20. two similar isosceles triangles

21. two rectangles that are not similar

22. two equilateral triangles that are not similar

23. two hexagons that are similar

24. two rhombi that are not similar

25. two similar trapezoids

The legs of a right triangle are 6.2 inches and 7 inches long. The shorter leg of a similar triangle is 17.6 inches long.

26. Find the length of the other leg of the second triangle.

27. Find the ratio of the measures of the hypotenuses.

28. Given $\triangle DEC \sim \triangle DAB$, $m\angle A = 25$, $m\angle DCE = 80$, $AE = 6$, $AD = 10$, and $DC = 5$, find each of the following.

a. $m\angle 1$ b. $m\angle 2$

c. $m\angle 3$ d. DB

e. What conclusions can be made about $\overline{EC}$ and $\overline{AB}$?

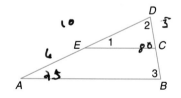

29. Use grid paper and draw enlargements of each figure. Make each segment twice the length of the original.

I.

II.

III.

IV.

30. Copy and complete each chart below, using the information from Exercise 29.

a.

	I	II	III	IV
Perimeter of Original	?	?	?	?
Perimeter of Enlargement	?	?	?	?

b.

	I	II	III	IV
Area of Original	?	?	?	?
Area of Enlargement	?	?	?	?

c. What conclusions can you draw from Chart **a**?

d. What conclusions can you draw from Chart **b**?

e. What would happen to the perimeters of the enlargements if the segment measures had been tripled?

f. What would happen to the area of the enlargements if the segment measures had been tripled?

g. Are the figures and their enlargements similar?

31. In order to reduce a 5-inch by 7-inch photograph, Tina uses a copy machine that reduces a document by 10%. Suppose she reduces the photo, then reduces the copy, and then reduces that copy, and so on. How many times would she have to use the copy machine in order to fit the photo into a 4-inch square?

Plot the given points on graph paper. Draw *RSTV* and $\overline{LB}$. Find points *M* and *N* such that *RSTV* is similar to *LBMN*.

32. $R(-1,0)$, $S(1,0)$, $T(2,3)$, $V(-2,3)$, $L(-3,0)$, $B(3,0)$

33. $R(0,4)$, $S(0,0)$, $T(8,0)$, $V(4,6)$, $L(2,0)$, $B(0,0)$

Make a scale drawing of each using the given scale.

34. A soccer field is 91 meters by 46 meters. scale: 1 mm = 1 m

35. A basketball court is 84 feet by 50 feet. scale: $\frac{1}{8}$ in. = 2 ft

36. A football field is 100 yards by 160 yards. scale: $\frac{1}{4}$ in. = 20 ft

37. A tennis court is 36 feet by 78 feet. scale: $\frac{1}{8}$ in. = 1 ft

Critical Thinking

38. Two rectangular solids are similar with ratios between the corresponding sides 3:1.

a. If you were to build a model of each using straws, would the larger solid take three times as many straws? Explain.

b. If you were to paint each solid, would the larger solid take three times as much paint? Explain.

c. Would the larger solid hold exactly three times as much water? Explain.

Applications

39. Engineering Janis Shaull designs automobiles using scale drawings. Suppose one length on an automobile is 20 centimeters and corresponds to 2 centimeters on the drawing. Find the length on the automobile that corresponds to 9 centimeters on the drawing.

40. Art The council members of Millersville wish to honor the town's founder Clara Miller by placing a statue of Ms. Miller in the town square. They want the statue to be 8 feet tall although Ms. Miller herself was only 5 feet 2 inches tall. If her arms were 2 feet long, how long should the sculptor make the arms on the statue?

41. Publishing Randall is working on the school yearbook. He must reduce a photo that is 4 inches wide by 5 inches long to fit in a space 3 inches wide. How long will the reduced photo be?

42. Education Overhead projectors project images from a transparency onto a screen so that they can be seen by many people. The scale factor of the projected image of a transparency to the transparency is 8. If the transparency is $8\frac{1}{2}$ inches by 11 inches, what are the dimensions of its image?

Mixed Review

43. The ratio of seniors to juniors on a football team is 2:3. If there are 21 juniors, how many seniors are on the team? **(Lesson 7-2)**

44. What kind of quadrilateral has exactly one pair of parallel sides? **(Lesson 6-6)**

45. If two sides of a triangle have measures of 5 and 8, what must be true about the measure of the third side? **(Lesson 5-6)**

46. Name three different ways in which two lines can be positioned in space relative to each other. **(Lesson 3-1)**

47. Rewrite the following conditional statement in if-then form. *I will wear a sweater when it is cool.* **(Lesson 2-2)**

48. An angle separates a plane into three parts. Name the parts. **(Lesson 1-6)**

Wrap-Up

49. Describe how you can tell if two polygons are similar.

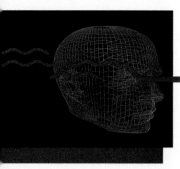

Technology

BASIC
▶ **Geometric Supposer**
Graphing calculators
LOGO
Spreadsheets

Similarity

The Geometric Supposer is a powerful tool for investigating geometric relationships. Let's use it to look at some relationships in triangles.

Begin by loading the Geometric Supposer: Triangles and drawing an acute or obtuse triangle. Your triangle will be labeled $\triangle ABC$. Choose a random point on $\overline{AB}$ by choosing *(2) Label* from the main menu, then choosing *(4) Random Point* from the Label menu. The random point will be labeled D.

Now draw a segment through D that is parallel to $\overline{BC}$. To do this choose *(1) Draw* on the main menu, then choose *(5) Parallel* from the Draw menu. Enter D as the point you wish the line to pass through and $\overline{BC}$ as the line to which it will be parallel. Define the length to intersect segments AC and AB. This parallel segment will be labeled $\overline{EF}$. *Points E and D are the same point.*

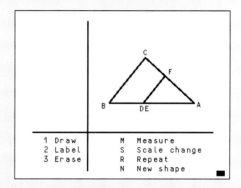

EXERCISES

Use the triangle you drew on the Geometric Supposer to answer each question.

1. Find the measure of each angle using the measure option.
   a. $\angle CBA$
   b. $\angle BCA$
   c. $\angle BAC$
   d. $\angle FEA$
   e. $\angle EFA$
   f. $\angle EAF$

2. Find the measure of each segment using the measure option.
   a. $\overline{AB}$
   b. $\overline{AC}$
   c. $\overline{BC}$
   d. $\overline{AE}$
   e. $\overline{AF}$
   f. $\overline{EF}$

3. Find each ratio.
   a. $\dfrac{AB}{AE}$
   b. $\dfrac{AC}{AF}$
   c. $\dfrac{BC}{EF}$

4. Is $\triangle ABC$ similar to $\triangle AEF$? Justify your answer.

5. Draw several other triangles on the Geometric Supposer and repeat Exercises 1-4. Can you make a conjecture?

Similar Triangles

Objectives

After studying this lesson, you should be able to:

- identify similar triangles, and
- use similar triangles to solve problems.

Application

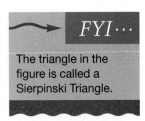

FYI···

The triangle in the figure is called a Sierpinski Triangle.

How many similar triangles are in the figure at the right? The figure is a fractal pattern. Fractals take a pattern and repeat that pattern several times. The fractal pattern on the right is of level four. There are four repeats of the large main pattern, each one smaller than the one before. Note that this pattern has many similar triangles in it. When a figure has this kind of self replication, it is called self similar.

You can learn more about fractals on pages 680–691.

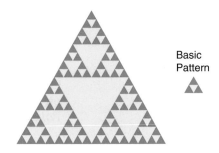

Basic Pattern

There are 200 similar triangles.

How do you know when two triangles are similar? Is it necessary to prove all of the conditions of the definition to determine whether two triangles are similar? In Chapter 4 you learned several tests to determine whether two triangles are congruent. There are also tests to determine whether two triangles are similar.

INVESTIGATION

Draw $\triangle ABC$ with $m\angle C = 45$, $AC = 5$ cm and $m\angle A = 75$. Then measure $\overline{BA}$ and calculate the ratio of the two sides, $\frac{CA}{BA}$. Draw a second triangle $\triangle DEF$ with $m\angle F = 45$, $DF = 8$ cm, and $m\angle D = 75$ and calculate the ratio of the sides $\frac{FD}{ED}$.

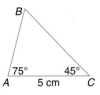

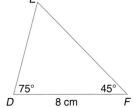

Each ratio is about 1.22. Calculate $\frac{BA}{BC}$ and $\frac{ED}{EF}$. Are these ratios the same? How about $\frac{CA}{BC}$ and $\frac{FD}{EF}$? Draw a few more triangles with the same angle measures and different side measures. Do all of the triangles appear to be similar?

The investigation leads to the following postulate.

| Postulate 7-1 AA Similarity | If two angles of one triangle are congruent to two angles of another triangle, then the triangles are similar. |

Example 1

In the figure, $\overline{ST} \parallel \overline{PR}$, $QS = 3$, $SP = 1$, and $TR = 1.2$. Show that $QT = 3.6$.

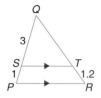

Since $\overline{ST} \parallel \overline{PR}$, $\angle QST \cong \angle QPR$ and $\angle QTS \cong \angle QRP$.

By AA Similarity, $\triangle SQT \sim \triangle PQR$. Using the definition of similar polygons, $\frac{QT}{QR} = \frac{QS}{QP}$. By the Segment Addition Postulate, $QP = QS + SP$ and $QR = QT + TR$. Substituting these values into the proportion results in the following proportion that can be solved for QT.

$$\frac{QT}{QT + TR} = \frac{QS}{QS + SP}$$

$$\frac{QT}{QT + 1.2} = \frac{3}{3 + 1} \qquad \textit{Substitution property of equality}$$

$$(QT)(3 + 1) = 3(QT + 1.2) \qquad \textit{Cross products}$$

$$4QT = 3QT + 3.6 \qquad \textit{Distributive property}$$

$$QT = 3.6 \qquad \textit{Subtraction property}$$

It is also possible to prove triangles similar by testing the measures of corresponding sides for proportionality.

| Theorem 7-1 SSS Similarity | If the measures of the corresponding sides of two triangles are proportional, then the triangles are similar. |

Proof of Theorem 7-1

Given: $\frac{PQ}{AB} = \frac{QR}{BC} = \frac{RP}{CA}$

Prove: $\triangle BAC \sim \triangle QPR$

Locate D on $\overline{AB}$ so that $\overline{DB} \cong \overline{PQ}$ and draw $\overline{DE}$ so that $\overline{DE} \parallel \overline{AC}$.

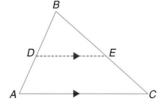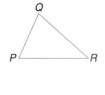

Paragraph Proof:

Since $\overline{DB} \cong \overline{PQ}$, the given proportion will become $\frac{DB}{AB} = \frac{QR}{BC} = \frac{RP}{CA}$. Since $\overline{DE} \parallel \overline{AC}$, $\angle BDE \cong \angle A$ and $\angle BED \cong \angle C$. By AA Similarity, $\triangle BDE \sim \triangle BAC$. By the definition of similar polygons, $\frac{DB}{AB} = \frac{BE}{BC} = \frac{ED}{CA}$. Using the two proportions and substitution, $\frac{QR}{BC} = \frac{BE}{BC}$ and $\frac{RP}{CA} = \frac{ED}{CA}$. This means that $QR = BE$ and $RP = ED$ or $\overline{QR} \cong \overline{BE}$ and $\overline{RP} \cong \overline{ED}$. With these congruences and $\overline{DB} \cong \overline{PQ}$, $\triangle BDE \cong \triangle QPR$ by SSS. By CPCTC, $\angle B \cong \angle Q$ and $\angle BDE \cong \angle P$. But $\angle BDE \cong \angle A$, so $\angle A \cong \angle P$. By AA Similarity, $\triangle BAC \sim \triangle QPR$.

The next theorem describes another test for similarity of triangles. You will be asked to prove this theorem in Exercise 27.

Theorem 7-2 SAS Similarity	**If the measures of two sides of a triangle are proportional to the measures of two corresponding sides of another triangle and the included angles are congruent, then the triangles are similar.**

You can use SAS similarity to determine if two triangles are similar.

Example 2

Angle *RVU* is trisected by $\overline{VS}$ and $\overline{VT}$. If *VR* = 20, *VS* = 25, *VU* = 15, and *VT* = 12, determine which triangles in the figure are similar.

$\angle RVS \cong \angle SVT \cong \angle TVU$ since $\angle RVU$ is trisected. If the corresponding sides which include the angle are proportional, then the triangles are similar. Find ratios of corresponding sides to see which ones are equivalent, thus forming a proportion.

$$\frac{VR}{VS} = \frac{20}{25} = \frac{4}{5}$$

$$\frac{VT}{VS} = \frac{12}{25}$$

$$\frac{VT}{VU} = \frac{12}{15} = \frac{4}{5}$$

Therefore, $\frac{VR}{VS} = \frac{VT}{VU}$ and $\Delta RSV \sim \Delta TUV$ by SAS Similarity.

Example 3

A hypsometer as shown at the right can be used to measure the height of a tree. Look through the straw to the top of the tree. Note where the weighted string crosses the scale. Suppose Al Henke used the readings shown. His eye was 167 cm from the ground and he was 15 m, or 1500 cm, from the tree. Find the height of the tree.

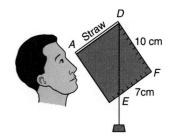

Let *x* represent the measure of the distance from eye level to the tip of the tree. If we can show that ΔDEF and ΔABC are similar, we can set up a proportion of corresponding sides and solve for *x*.

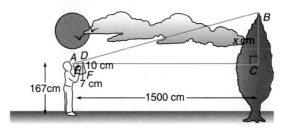

Assume that $\angle DFE$ and $\angle BCA$ are right angles. $\overline{DE} \parallel \overline{BC}$ and $\overline{AD} \parallel \overline{EF}$.

$\angle FED \cong \angle ADE$	*They are alternate interior angles.*
$\angle ADE \cong \angle ABC$	*They are corresponding angles.*
$\angle FED \cong \angle ABC$	*Congruence of angles is transitive.*
$\angle DFE \cong \angle BCA$	*All right angles are congruent.*
$\triangle DEF \sim \triangle ABC$	*AA Similarity*

Therefore, $\dfrac{BC}{EF} = \dfrac{AC}{DF}$ *Measures of corresponding sides of similar triangles are proportional.*

$$\frac{x}{7} = \frac{1500}{10}$$
$$10x = 10{,}500$$
$$x = 1050$$

The tree is 1050 + 167 or 1217 centimeters tall. *The tree is about 12 meters tall.*

Like congruence of triangles, similarity of triangles is reflexive, symmetric, and transitive. You will prove this theorem in Exercise 29.

Theorem 7-3	**Similarity of triangles is reflexive, symmetric, and transitive.**

CHECKING FOR UNDERSTANDING

Communicating Mathematics

Read and study the lesson to answer each question.

1. Describe three ways to determine whether two triangles are similar.

2. List the tests to prove triangles are congruent and compare them to the tests to prove triangles are similar. How are the tests alike? How are they different?

3. The triangles in the figure are similar with corresponding sides of lengths a and d, c and f, and b and e. Explain why you can write the proportion $\dfrac{a}{c} = \dfrac{d}{f}$.

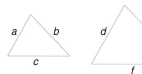

Guided Practice

4. If $\overline{LM} \parallel \overline{PT}$, name all of the congruent angles in the figure.

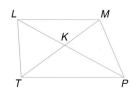

5. If $\triangle RAT \sim \triangle OFT$, name the proportional parts of the triangles.

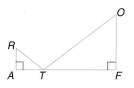

Determine whether each pair of triangles is similar using the given information. Explain your answer.

6.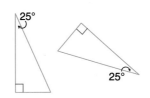

7.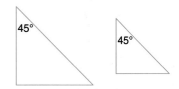

8.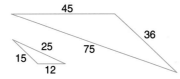

9.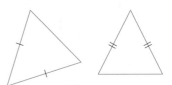

10.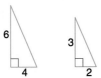

11.

Determine if each pair of triangles is similar. If similar, state the reason and find the missing measures.

12.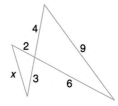

13.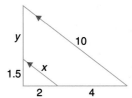

14. One angle in a right triangle measures 25°. A second right triangle has a 65° angle. Are the two triangles similar? Explain.

EXERCISES

Practice Determine whether each pair of triangles is similar using the given information. Explain your answer.

15.

16.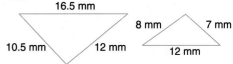

17.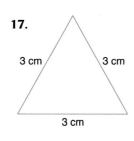

18.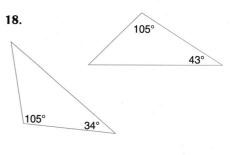
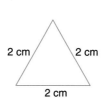

Find the value of *x*.

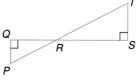

19. $QR = x + 4$
$RS = 2x + 3$
$QP = 3$
$TS = 5$

20. $TS = 6$
$QP = 4$
$RS = x + 1$
$QR = 3x - 4$

Identify the similar triangles in each figure. Explain your answer.

21.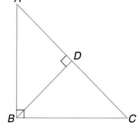

22.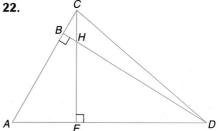

Identify the similar triangles in each figure. Explain why they are similar and find the missing measures.

23. If $\overline{BE} \parallel \overline{CD}$, find *CD*, *AC*, and *BC*.

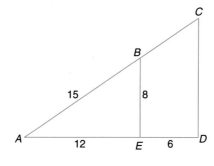

24. If *VRST* is a parallelogram, find *TC*, *SB*, and *SC*.

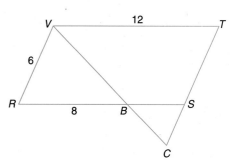

Write a two-column proof for each.

25. Given: ∠D is a right angle.
         $\overline{BE} \perp \overline{AC}$
  Prove: $\triangle ADC \sim \triangle ABE$

26. Given: $\overline{QS} \parallel \overline{PT}$
     Prove: $\triangle QRS \sim \triangle TRP$

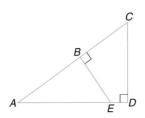

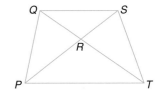

27. Prove that if the measures of two sides of a triangle are proportional to the measures of two corresponding sides of another triangle and the included angles are congruent, the triangles are similar. (Theorem 7-2)

28. Prove that if the measures of the legs of two right triangles are proportional, the triangles are similar.

29. Prove Theorem 7-3.

30. Graph $\triangle ABC$ and $\triangle TRC$ with vertices $A(-2, 7)$, $B(-2, -8)$, $C(4, 4)$, $R(0, -4)$, and $T(0, 6)$.
  a. Prove that $\triangle ABC \sim \triangle TRC$.
  b. Find the ratio of the perimeters of the triangles.

Critical Thinking

31. What are the minimum conditions necessary in order to have similar parallelograms?

Applications

32. Surveying Leslie uses a carpenter's square to find the distance across a river. She puts the square on top of a pole which is high enough to sight along $\overline{BC}$ to point P across the river. Then she sights along $\overline{BE}$ to point Q. If $QA = 2$ feet and $BA = 6$ feet, find AP.

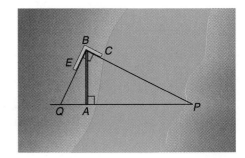

33. Landscaping Lamar Presley is planning to landscape his yard. First he needs to calculate the height of a palm tree in the backyard. He sights the top of the tree in a mirror that is 6.0 meters from the tree. It is on the ground and faces up. Lamar is 0.9 meters from the mirror and his eyes are 1.8 meters from the ground. How tall is the tree? *Hint: The angle between the ground and the line of sight is congruent to the angle between the ground and the line from the mirror to the tree.*

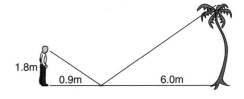

34. **Cartography**   The pantograph is an instrument that was originally used to enlarge or reduce maps. The bars are attached to form parallelogram *DCBA*. Point *P* is fixed to the drawing board and a pencil is attached at *T*. The artist traces the original picture at *B*. Suppose *PA* is 4, *CD* is 8, *CT* is 9, and *AD* is 18.

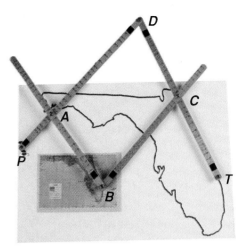

 a. Identify the similar triangles and give a reason for their similarity.

 b. What is the ratio of *PB* to *BT*?

Mixed Review

Determine whether each of the following is *true* or *false*.

35. If the corresponding sides of two polygons are proportional, the polygons are similar. **(Lesson 7-3)**

36. Consecutive sides of a rhombus are congruent. **(Lesson 6-5)**

37. A median of a triangle always lies in the interior of the triangle. **(Lesson 5-2)**

38. AAS is a triangle congruence test. **(Lesson 4-5)**

39. An isosceles triangle cannot have a right angle. **(Lesson 4-1)**

40. Congruence of angles is transitive. **(Lesson 1-8)**

Wrap-Up

41. **Journal Entry**   Write a short summary of the ways you can prove two triangles are similar in your journal. Give an example of each.

～～～ MID-CHAPTER REVIEW ～～～

1. Jackie's soccer team has won 12 games and lost 4. Find the ratio of wins to games played. **(Lesson 7-1)**

2. How many U.S. dollars would you receive for 3000 Italian lira if the exchange rate is 1350 lira for each U.S. dollar? **(Lesson 7-2)**

3. If Teresa's car gets 29 miles per gallon, how many gallons of gas will she need to drive from Chicago to Bozeman, Montana, and back, a round trip of 2700 miles? **(Lesson 7-2)**

4. Find the value of *x* if $\frac{x}{5} = \frac{x-1}{3}$. **(Lesson 7-2)**

5. Define similar polygons. **(Lesson 7-3)**

6. One angle in a right triangle is 37°. A second right triangle has an angle that is 53°. Are the two triangles similar? Explain. **(Lesson 7-4)**

7. Shelia claims that all right triangles are similar. David disagrees. Who is correct? Explain. **(Lesson 7-4)**

7-5 Proportional Parts

Objectives

After studying this lesson, you should be able to:
- use proportional parts of triangles to solve problems, and
- divide a segment into congruent parts.

Application

    Why are triangles often used in buildings? One reason is the rigid nature of a triangle. Triangles are used to brace boards in roofs, to support telephone wires, and to provide the structure of bridges. In the picture at the right, each horizontal support is the base of a triangle. Each support is parallel to the others. The following theorem states that since the supports are parallel, the triangles formed are similar.

Theorem 7-4 *Triangle* *Proportionality*	**If a line is parallel to one side of a triangle and intersects the other two sides in two distinct points, then it separates these sides into segments of proportional lengths.**

Proof of Theorem 7-4

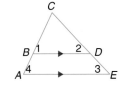

Given: $\overline{BD} \parallel \overline{AE}$

Prove: $\dfrac{BA}{CB} = \dfrac{DE}{CD}$

Paragraph Proof:

Since $\overline{BD} \parallel \overline{AE}$, $\angle 4 \cong \angle 1$ and $\angle 3 \cong \angle 2$. Then by AA Similarity, $\triangle ACE \sim \triangle BCD$. From the definition of similar polygons, $\dfrac{CA}{CB} = \dfrac{CE}{CD}$. Since B is between A and C, $CA = BA + CB$, and since D is between C and E, $CE = DE + CD$. Substituting, we get

$$\frac{BA + CB}{CB} = \frac{DE + CD}{CD}$$

$$\frac{BA}{CB} + \frac{CB}{CB} = \frac{DE}{CD} + \frac{CD}{CD}$$

$$\frac{BA}{CB} + 1 = \frac{DE}{CD} + 1 \qquad \textit{Substitution property of equality}$$

$$\frac{BA}{CB} = \frac{DE}{CD} \qquad \textit{Subtraction property of equality}$$

    Likewise, proportional parts of a triangle can be used to prove the converse of Theorem 7-4. You will be asked to prove this theorem in Exercise 34.

336 CHAPTER 7 SIMILARITY

<table>
<tr><td>Theorem 7-5</td><td>If a line intersects two sides of a triangle and separates the sides into corresponding segments of proportional lengths, then the line is parallel to the third side.</td></tr>
</table>

Example 1

In the figure $CA = 15$, $CE = 3$, $DA = 8$, and $BA = 10$. Determine if $\overline{ED} \parallel \overline{CB}$.

From the segment addition postulate, $CA = CE + EA$ and $BA = BD + DA$. Now substitute the known measures.

$$CA = CE + EA \qquad BA = BD + DA$$
$$15 = 3 + EA \qquad 10 = BD + 8$$
$$12 = EA \qquad 2 = BD$$

To show $\overline{ED} \parallel \overline{CB}$, we must show that $\dfrac{EA}{CE} = \dfrac{DA}{BD}$. From the information above, $\dfrac{EA}{CE} = \dfrac{12}{3}$ and $\dfrac{DA}{BD} = \dfrac{8}{2}$. Since $\dfrac{12}{3} = \dfrac{8}{2}$, $\dfrac{EA}{CE} = \dfrac{DA}{BD}$. Therefore, by Theorem 7-5, $\overline{ED} \parallel \overline{CB}$.

The theorems about the proportionality of triangles can be used to solve practical problems.

Example 2

APPLICATION

History

The sector compass was an instrument used in the seventeenth and eighteenth centuries. It was perfected by Galileo and was used to solve a variety of problems. A sector compass consisted of two arms fastened at one end by a pivot joint. A scale was marked on each arm as shown in the diagram. To draw a segment two fifths of the length of a given segment, the 100-marks are placed at the endpoints of the given line. A segment drawn between the two 40-marks is two fifths of the length of the given segment. Why?

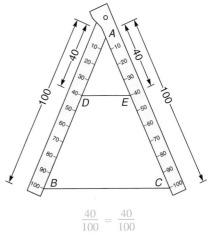

$$\frac{40}{100} = \frac{40}{100}$$

$\triangle ABC \sim \triangle ADE$	*SAS Similarity*
$\dfrac{AD}{AB} = \dfrac{DE}{BC}$	*Definition of similar polygons*
$\dfrac{40}{100} = \dfrac{DE}{BC}$	*Substitution property of equality*
$\dfrac{2}{5} = \dfrac{DE}{BC}$	
$\dfrac{2}{5} BC = DE$	*Multiplication property of equality*

The proof of the following theorem is based on Theorems 7-4 and 7-5. You will be asked to complete the proof of this theorem in Exercise 35.

Theorem 7-6	A segment whose endpoints are the midpoints of two sides of a triangle is parallel to the third side of the triangle and its length is one-half the length of the third side.

In $\triangle ACE$, suppose that B is the midpoint of $\overline{AC}$ and D is the midpoint of $\overline{CE}$. By Theorem 7-6, $\overline{BD} \parallel \overline{AE}$ and $BD = \frac{1}{2}AE$. This relationship can also be expressed $2BD = AE$.

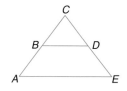

Three or more parallel lines separate transversals into proportional parts as stated in the next two corollaries.

Corollary 7-1	If three or more parallel lines intersect two transversals, then they cut off the transversals proportionally.
Corollary 7-2	If three or more parallel lines cut off congruent segments on one transversal then they cut off congruent segments on every transversal.

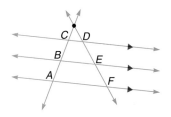

In the figure at the left, $\overleftrightarrow{CD} \parallel \overleftrightarrow{BE} \parallel \overleftrightarrow{AF}$. The transversals $\overleftrightarrow{AC}$ and $\overleftrightarrow{FD}$ have been separated into proportional segments. Sample proportions are listed below.

$$\frac{CB}{BA} = \frac{DE}{EF}; \frac{CA}{DF} = \frac{BA}{EF}; \frac{AC}{BC} = \frac{FD}{ED}$$

It is possible to separate a segment into two congruent parts by constructing the perpendicular bisector of a segment. However, a segment cannot be separated into three congruent parts by constructing perpendicular bisectors. To do this, the method illustrated below is used.

CONSTRUCTION

A similar method can be used for any given number of congruent parts.

Separate a segment into three congruent parts.

1. Copy $\overline{AB}$ and draw $\overrightarrow{AM}$.

2. With a compass point at A, mark off an arc on $\overrightarrow{AM}$ at X. Then construct $\overline{XY}$ and $\overline{YZ}$ so that $\overline{AX} \cong \overline{XY} \cong \overline{YZ}$.

3. Draw $\overline{ZB}$. Then construct lines through Y and X that are parallel to $\overline{ZB}$. Call P and Q the intersection points on $\overline{AB}$.

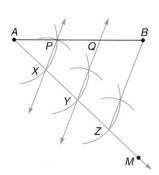

Because parallel lines cut off congruent segments on transversals, $\overline{AP} \cong \overline{PQ} \cong \overline{QB}$.

CHECKING FOR UNDERSTANDING

Communicating Mathematics

Read and study the lesson to answer each question.

1. Explain what it means to separate two segments proportionally.

2. Draw and label two segments that are separated proportionally. Then write several different true proportions.

3. In Chapter 3, you learned how to prove lines parallel by using special angle relationships. What other method can be used to prove lines are parallel?

A segment is drawn from one side to another in a triangle.

4. Under what conditions will it separate the side proportionally?

5. When will the measure of that segment be half the length of the third side?

Guided Practice

In the figure $\overline{BD} \parallel \overline{AE}$. Determine whether each statement is *true* or *false*. If it is false, explain why.

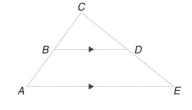

6. $\dfrac{BC}{ED} = \dfrac{AB}{CD}$

7. $\dfrac{AB}{BC} = \dfrac{DE}{CD}$

8. $\dfrac{CB}{CD} = \dfrac{CA}{CE}$

9. $\dfrac{BA}{DE} = \dfrac{CA}{CE}$

In $\triangle BAT$, determine whether $\overline{ML} \parallel \overline{BA}$ under the given conditions.

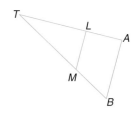

10. $\dfrac{TM}{MB} = \dfrac{TL}{LA}$

11. $\dfrac{BA}{MB} = \dfrac{ML}{LA}$

12. $\dfrac{TB}{MB} = \dfrac{TA}{LA}$

13. $\dfrac{TM}{ML} = \dfrac{TB}{BA}$

Find the value of *x*.

14.   15.   16.

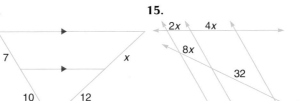

17. Draw a segment 10 centimeters long. Separate the segment into three congruent parts.

EXERCISES

Practice   In the figure at the right, $\overleftrightarrow{YA} \parallel \overleftrightarrow{OE} \parallel \overleftrightarrow{BR}$.
Complete each statement.

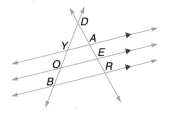

18. $\dfrac{YO}{OB} = \dfrac{AE}{?}$ **19.** $\dfrac{YB}{OB} = \dfrac{?}{ER}$

20. $\dfrac{?}{AE} = \dfrac{YB}{YO}$ **21.** $\dfrac{DY}{YO} = \dfrac{DA}{?}$

22. $\dfrac{DR}{?} = \dfrac{DB}{YB}$ **23.** $\dfrac{?}{AE} = \dfrac{DO}{YO}$

Find the value of x and y.

24.

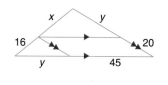

25.

26.

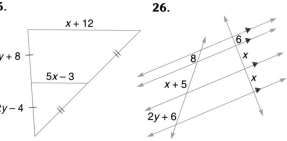

Using the figure at the right, determine the
value of x that would make $\overline{PQ} \parallel \overline{DF}$ under
each set of conditions.

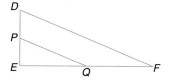

27. $EQ = 3$ **28.** $DE = 12$
$\quad\ \ DP = 12$ $\qquad\ \ PE = 7$
$\quad\ \ QF = 8$ $\qquad\ \ EQ = x + 3$
$\quad\ \ PE = x + 2$ $\qquad QF = x - 3$

Using the figure at the right, determine the
value of x under each set of conditions.

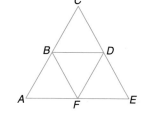

29. $\overline{BD} \parallel \overline{AE}$ **30.** $\overline{AC} \parallel \overline{DF}$
$\quad\ \ AB = 6$ $\qquad\ \ DC = 7$
$\quad\ \ DE = 8$ $\qquad\ \ DE = 5$
$\quad\ \ DC = 4$ $\qquad\ \ FA = 8$
$\quad\ \ BC = x$ $\qquad\ \ FE = x$

31. If B, D, and F are the midpoints of sides $\overline{CA}$, $\overline{CE}$, and $\overline{AE}$ respectively,
$BD = 7$, $BF = 12$, and $DF = 16$, find the perimeter of $\triangle AEC$. What is the
ratio of the perimeter of $\triangle BDF$ to the perimeter of $\triangle AEC$?

32. If B, D, and F are the midpoints of sides $\overline{CA}$, $\overline{CE}$, and $\overline{AE}$ respectively,
$BD = 8$, $CA = 10$, and $DE = 4$, find DF, AE, and BF.

33. Draw a segment that is 9 centimeters long. By construction, separate the
segment into four congruent parts.

Write a two-column proof for each.

34. If a line intersects two sides of a triangle and separates the sides into corresponding segments of proportional lengths, then the line is parallel to the third side. (Theorem 7-5)

35. A segment whose endpoints are the midpoints of two sides of a triangle is parallel to the third side of the triangle and its length is one-half the length of the third side. (Theorem 7-6)

36. Draw a segment. Then separate the segment into segments whose ratios are 2 to 3.

37. Given $A(2, 3)$ and $B(8,12)$, find P such that P separates $\overline{AB}$ into two parts with a ratio of 2 to 1.

38. In $\triangle ABC$, $\overline{MN}$ divides sides $\overline{AC}$ and $\overline{AB}$ proportionally. If the coordinates are $A(3, 7)$, $M(0, 10)$, and $N(8, 22)$ and if $\frac{AM}{MC} = \frac{3}{1}$, find the coordinates of B and C.

Critical Thinking

39. Draw any quadrilateral $RSTV$ and connect the midpoints, A, B, C, and D, of each side in order. Determine what kind of figure $ABCD$ will be. Use the information from this lesson to prove your claim.

Applications

40. **Tourism** There are 44 steps to get to the first level of a monument if you enter at the west door and 33 steps to get to the second level. If you enter at the east door, there are 52 steps to get to the first level.
   a. Draw a sketch of the monument.
   b. Which side must be steeper? How do you know?
   c. How many steps are there to get to the second level on the east side?

41. **Real Estate** In Forest Park, the home lots are laid out as shown at the right. What is the individual frontage of each lot on Piano Drive if the total frontage on the drive for the five lots is known to be 432 feet?

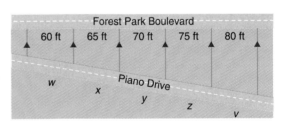

42. **History** How could the sector compass be used to find 35% of 10?

Mixed Review

43. Two triangles are similar and their corresponding sides are in a ratio of 3:5. What is the measure of a side in the second triangle that corresponds to a side that measures 15 centimeters in the first triangle? **(Lesson 7-4)**

44. In $\triangle PQR$, $PQ > PR > QR$. List the angles in $\triangle PQR$ in order from greatest to least. **(Lesson 5-6)**

45. The angles in a triangle have measures $7x - 1$, $18x + 2$, and $5x + 10$. Is the triangle acute, obtuse, or right? **(Lesson 4-1)**

46. Find the slope of the line that passes through the points with coordinates $(-22, 14)$ and $(8, -21)$. **(Lesson 3-5)**

47. Which property of algebra justifies the statement *If $5x - 1 = 8$, then $5x = 9$?* **(Lesson 2-4)**

Wrap-Up

48. Write a five-question quiz for this lesson.

7-6 Parts of Similar Triangles

Objective

After studying this lesson, you should be able to:

- recognize and use the proportional relationships of corresponding perimeters, altitudes, angle bisectors, and medians of similar triangles.

Application

Two triangular jogging paths are laid out in a park as in the figure. One route is for those who like to jog a shorter path. A second path, similar in shape, is for those who like to jog a longer path. The dimensions of the inner path are 400 meters, 500 meters, and 300 meters. The longest side of the outer path is 1000 meters. Will a jogger on the outer path run twice as far as one on the inner path?

> *FYI* ···
>
> To keep your feet warm while you jog, wear a hat. Eighty percent of your body heat escapes through your head.

The jogging paths are similar. Thus, there is a common ratio r such that $r = \dfrac{a}{e} = \dfrac{c}{g} = \dfrac{b}{f}$.

So, $re = a$, $rg = c$, and $rf = b$.

Let P_1 represent the perimeter of the shorter path and P_2 represent the perimeter of the longer path.

$P_1 = a + b + c$

$P_1 = re + rg + rf$ *Substitution property of equality*

$P_1 = r(e + g + f)$ *Distributive property*

$P_1 = rP_2$ $P_2 = e + g + f$

$\dfrac{P_1}{P_2} = r$

Therefore, $\dfrac{P_1}{P_2} = \dfrac{a}{e} = \dfrac{c}{g} = \dfrac{b}{f}$.

This proves that the perimeters of two similar triangles are proportional to the measures of the corresponding sides. This is stated in Theorem 7-7 below.

Theorem 7-7 *Proportional Perimeters*	**If two triangles are similar, then the perimeters are proportional to the measures of corresponding sides.**

Example 1

In the figure $\triangle RST \sim \triangle WVU$. If $UV = 500$, $VW = 400$, $UW = 300$, and $ST = 1000$, find the perimeter of $\triangle RST$.

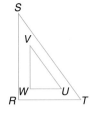

Let x represent the perimeter of $\triangle RST$.
The perimeter of $\triangle UVW = 500 + 400 + 300$ or 1200 units.

$$\frac{\text{perimeter of } \triangle UVW}{x} = \frac{VU}{ST} \qquad \textit{Proportional perimeters}$$

$$\frac{1200}{x} = \frac{500}{1000} \qquad \textit{Substitution property of equality}$$

$$x = 2400$$

The perimeter of $\triangle RST$ is 2400 units.

When two triangles are similar, the measures of their corresponding sides are proportional. What about the measures of their corresponding altitudes, medians, and angle bisectors? The following theorem states one relationship.

Theorem 7-8

If two triangles are similar, then the measures of the corresponding altitudes are proportional to the measures of the corresponding sides.

Proof of Theorem 7-8

Given: $\triangle ABC \sim \triangle PQR$
$\overline{BD}$ is an altitude of $\triangle ABC$.
$\overline{QS}$ is an altitude of $\triangle PQR$.

Prove: $\dfrac{BD}{QS} = \dfrac{BA}{QP}$

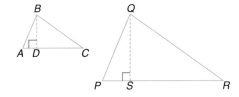

Paragraph Proof:

Since $\triangle ABC \sim \triangle PQR$ and since $\angle A$ and $\angle P$ are corresponding angles, $\angle A \cong \angle P$. Since $\overline{BD}$ is an altitude of $\triangle ABC$, $\overline{BD} \perp \overline{AC}$ and $\angle BDA$ is a right angle. Likewise, since $\overline{QS}$ is an altitude of $\triangle PQR$, $\overline{QS} \perp \overline{PR}$ and $\angle QSP$ is a right angle. So $\angle BDA \cong \angle QSP$ and $\triangle ABD \sim \triangle PQS$ by AA Similarity. Thus, $\dfrac{BD}{QS} = \dfrac{BA}{QP}$.

Likewise, the measures of corresponding angle bisectors are proportional. Using the figure below, if $\triangle RST \sim \triangle EFG$, then $\angle S \cong \angle F$ and $\angle SRT \cong \angle FEG$. If $\overline{RV}$ and $\overline{EH}$ bisect $\angle SRT$ and $\angle FEG$, then $\angle SRV \cong \angle FEH$. By AA similarity, $\triangle RSV \sim \triangle EFH$ and $\dfrac{RV}{EH} = \dfrac{RS}{EF}$.

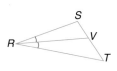
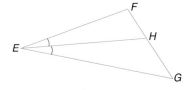

This leads us to the following theorem.

<table>
<tr><td>Theorem 7-9</td><td>If two triangles are similar, then the measures of the corresponding angle bisectors of the triangles are proportional to the measures of the corresponding sides.</td></tr>
</table>

The following theorem states the relationship between the medians of two similar triangles. You will prove this theorem in Exercise 29.

<table>
<tr><td>Theorem 7-10</td><td>If two triangles are similar, then the measures of the corresponding medians are proportional to the measures of the corresponding sides.</td></tr>
</table>

The theorem about the relationships of special segments in similar triangles can be used to solve practical problems.

Example 2

As a project for her photography class, Carol Page made a camera from a box. Suppose the film is 1.8 centimeters from the lens and the person being photographed is 360 centimeters from the camera. Find the height of the image on the film if the height of the person is 180 centimeters.

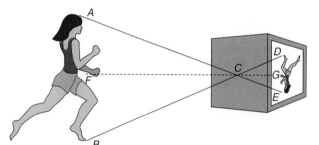

The dashed lines are the altitudes of $\triangle ABC$ and $\triangle EDC$.

Let h = the height of the image on the film. Assume that $\overline{AB} \parallel \overline{DE}$.

Since they are alternate interior angles, $\angle BAC \cong \angle DEC$ and $\angle ABC \cong \angle EDC$. Therefore, $\triangle ABC \sim \triangle EDC$ by AA Similarity.

The measures of the corresponding altitudes of similar triangles are proportional to the measures of corresponding sides.

$$\frac{AB}{FC} = \frac{ED}{GC}$$

$$\frac{180}{360} = \frac{h}{1.8} \qquad \textit{Substitution property of equality}$$

$$360h = 324$$

$$h = 0.9$$

The height of the image on the film is 0.9 centimeters.

Angle bisectors of a triangle have an additional property involving ratios.

Theorem 7-11 Angle Bisector Theorem	An angle bisector in a triangle separates the opposite side into segments that have the same ratio as the other two sides.

$\overline{CD}$ is the bisector of $\angle ACB$ of $\triangle ABC$. According to Theorem 7-11, $\frac{AD}{DB} = \frac{AC}{BC}$.

To prove this, construct a line through point A parallel to $\overrightarrow{DC}$ and meeting $\overleftrightarrow{BC}$ at E. You can prove that $\frac{AD}{DB} = \frac{EC}{BC}$ and $CA = CE$. You will complete this proof in Exercise 30.

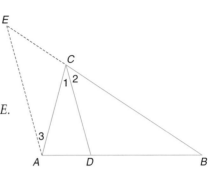

CHECKING FOR UNDERSTANDING

Communicating Mathematics

Read and study the lesson to answer each question.

1. Draw two similar obtuse triangles and draw the medians to the longest sides. Write a statement about the medians.

2. Draw two similar right triangles and draw the altitude to each hypotenuse. Write a statement about the altitudes.

3. Tom uses the figure at the right and Theorem 7-8 to conclude $\frac{BD}{BC} = \frac{AE}{AC}$. Is he right? Why or why not?

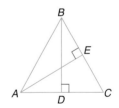

Guided Practice

In the figures at the right, $\triangle EFG \sim \triangle QRS$, $\overline{GH}$ bisects $\angle EGF$, and $\overline{ST}$ bisects $\angle QSR$. Determine whether each statement is *true* or *false*.

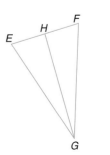

4. $\frac{HG}{TS} = \frac{EF}{QR}$ 5. $\frac{TS}{HG} = \frac{RQ}{FG}$

6. $\frac{TS}{HG} = \frac{SQ}{GE}$ 7. $\frac{FG}{RS} = \frac{HG}{TS}$

Find the value of x.

8.

9.

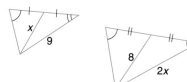

10. Suppose $\triangle PQR \sim \triangle ABC$. If $PR = 1.8$, $AC = 1.2$, and the perimeter of $\triangle ABC$ is 3.4, find the perimeter of $\triangle PQR$.

EXERCISES

Practice

In the figure at the right $\triangle ABC \sim \triangle PQR$, $\overline{BD}$ is an altitude of $\triangle ABC$, and $\overline{QS}$ is an altitude of $\triangle PQR$. Determine whether each statement is *true* or *false*. Justify your answer.

11. $\dfrac{BD}{QS} = \dfrac{AB}{PQ}$

12. $\dfrac{AD}{PS} = \dfrac{QR}{BC}$

13. $\dfrac{QP}{AB} = \dfrac{BD}{QS}$

14. $\dfrac{QR}{BC} = \dfrac{QS}{BD}$

15. $\dfrac{BD}{QS} = \dfrac{AC}{PR}$

16. $\dfrac{AB}{BD} = \dfrac{PQ}{QS}$

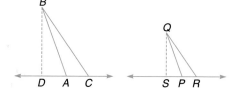

In the figure at the right, $\triangle ABC \sim \triangle DEF$, $\overline{AR} \cong \overline{RC}$, and $\overline{DS} \cong \overline{SF}$. Find the value of x.

17. $AC = 20$
$DF = 12$
$ES = 5$
$BR = x$

18. $BC = x + 2$
$BR = x - 5$
$ES = 6$
$EF = 16$

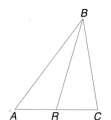

Find the value of x.

19.

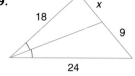

20.

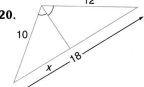

21.

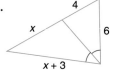

22. In the figure at the right, $\triangle WXY \sim \triangle JKL$, $\overline{XZ}$ is a median of $\triangle WXY$, and $\overline{KM}$ is a median for $\triangle JKL$. If $XZ = 4$, $WZ = 3$, $JL = x + 2$, and $KM = 2x - 5$, find JM.

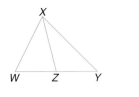

23. In the figure at the right, $\triangle ABC \sim \triangle DEF$, $\overline{AX}$ and $\overline{DY}$ are altitudes. Using the measures given, find DY.

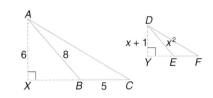

24. In the figure at the right, $\triangle STU \sim \triangle WZY$. If the perimeter of $\triangle STU$ is 30 units, find the value of x.

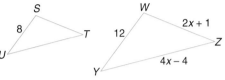

25. In the figure at the right, $\overline{UR}$ bisects $\angle VUS$, and $\overline{VS} \parallel \overline{WR}$. If $US = 5$, $WR = 15$, $UV = 10$, and $VS = 12$, find VT and UW.

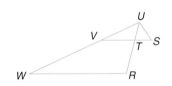

26. Suppose $\triangle LMN \sim \triangle XYZ$. If the perimeter of $\triangle LMN$ is 7.6 units, $LM = 3.0$, $MN = 2.8$, and $XY = 2.7$, find XZ.

27. If the bisector of an angle of a triangle bisects the opposite side, what is the ratio of the measures of the other two sides of the triangle?

28. The measures of the sides of a triangle are 10, 14, and 16. Find the measures of the segments formed where the bisector of the largest angle meets the opposite side.

Write a paragraph proof for each.

29. If two triangles are similar, then the measures of the corresponding medians are proportional to the measures of the corresponding sides. (Theorem 7-10)

30. An angle bisector in a triangle separates the opposite side into segments that have the same ratio as the other two sides. (Theorem 7-11)

31. In the figure at the right, $\overline{SV}$ bisects $\angle RST$, $\overline{RA} \cong \overline{RV}$, and $\overline{BT} \cong \overline{VT}$. Make and prove a conjecture about quadrilateral $ABTR$.

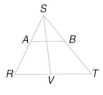

Critical Thinking

32. In the figure at the right, $\triangle ABC$ is a right triangle. $\overline{CD}$ is an altitude to the hypotenuse $\overline{AB}$. Make and prove a conjecture about the relationship of x, y, and z.

33. **Photography**   Ron is having his senior portrait taken. Suppose Ron is 300 centimeters from a camera lens and the film is 1.3 centimeters from the lens. If Ron is 180 centimeters tall, how tall is his image on the film?

34. **Photography**   Suppose the film is 1.3 centimeters from a camera lens and can have an image no more than 4.5 centimeters tall. If a person in front of the lens is 180 centimeters tall, how far from the lens can he be for a full length picture?

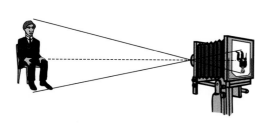

35. **Measurement**   Sally estimates Steve's height by holding a ruler in front of herself and lining up the top of the ruler with his head and the bottom of the ruler with his feet. She is 12 feet away from Steve when the ruler is lined up. If the ruler is 1 foot long and she is holding it 2 feet from her eyes, how tall is Steve?

In the figure at the right $\overline{DF} \parallel \overline{BC}$, $AD = 3$, $DB = 6$, $AF = 4$, and $BC = 15$. Find each measure. (Lesson 7-5)

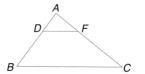

36. FC

37. DF

For each exercise state as many conclusions as you can using the figure and the given information.

38. **Given:** $\overline{MP} \cong \overline{MQ}$
    $\angle 1 \cong \angle N$
    (Lesson 4-7)

39. **Given:** $\overline{PR} \perp \overline{RS}$
    $\angle 2 \cong \angle 3$
    $\angle 1 \cong \angle 4$
    (Lesson 4-2)

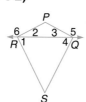

40. **Given:** $\angle F$ and $\angle 2$ and
    $\angle C$ and $\angle 1$ are
    complementary.
    (Lesson 3-2)

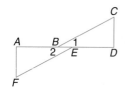

41. **Given:** $\overline{AB} \perp \overline{AD}$
    $\overline{BC} \perp \overline{CD}$
    $\angle 2 \cong \angle 4$
    (Lesson 1-8)

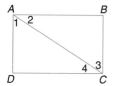

42. Write an application problem using the concepts you learned in this lesson. Be sure to include the answer to your problem.

7-7 Problem-Solving Strategy: Solve a Simpler Problem

Objective

After studying this lesson, you should be able to:

- solve problems by first solving a simpler related problem.

When you are working on a complicated or unfamiliar problem, it is often helpful to solve a similar problem first. Set aside the original problem to solve a simpler, similar problem. Then, use the strategy that worked on that simpler problem to attack the original problem.

Example 1

Find the sum of the series $1 + 3 + 5 + \ldots + 995 + 997 + 999$.

Adding all of these numbers directly, even with a calculator, would be time consuming.

Let's look at a simpler problem.

$$S = 1 + 3 + 5 + 7 + 9 + 11$$

Notice that pairs of addends have a sum of 12.

$$S = 1 + 3 + 5 + 7 + 9 + 11$$

$$\begin{array}{c} 12 \\ 12 \\ 12 \end{array}$$

There are three such pairs. So, the sum is 3(12) or 36. *Check by adding the numbers with a calculator.*

Now, try this strategy on the original problem.

$$S = 1 + 3 + 5 + \ldots + 995 + 997 + 999$$

$$\begin{array}{c} 1000 \\ 1000 \\ 1000 \end{array}$$

Each sum is 1000, and since there are 500 odd numbers between 1 and 999 inclusive, there are 250 pairs. So, the sum is 250(1000) or 250,000.

You can use the problem-solving strategy of solving a simpler problem to solve geometry problems as well.

Example 2

How many diagonals can be drawn for a polygon with 12 sides?

Let's draw several polygons with less than 12 sides. We will use a table to record the number of sides and the number of diagonals for each polygon.

Number of Sides	3	4	5	6	7	8	9	10	11	12
Number of Diagonals	0	2	5	9	14					

+2 +3 +4 +5

The pattern indicates that there are 54 diagonals in a polygon with 12 sides.

CHECKING FOR UNDERSTANDING

Communicating Mathematics

Read and study the lesson to answer these questions.

1. Describe how to use the problem-solving strategy to solve a simpler problem in your own words.

2. When is the strategy of solving a simpler problem useful?

3. What is the sum of the whole numbers from 1 to 200 inclusive?

Guided Practice

Solve each problem by solving a simpler problem.

4. How many line segments would you need to draw to connect 75 points if each point had to be connected to every other point?

5. In your sock drawer, you have 12 electric blue socks, 22 midnight blue socks, 18 sky blue socks, 6 periwinkle blue socks, 14 navy blue socks, 7 baby blue socks, 8 sapphire blue socks, and one aqua blue sock. How many socks must you pull from the drawer to ensure that you have a pair that matches? Explain your answer.

EXERCISES

Solve. Use any strategy.

6. Kara has 23 coins in her change purse. The change is worth one dollar. What combinations of coins might Kara have?

Strategies

Look for a pattern.
Solve a simpler problem.
Act it out.
Guess and check.
Draw a diagram.
Make a chart.
Work backward.

7. Use the following clues to discover the year of the completion of the first transcontinental railroad in the United States.
   - The hundreds and the tens digits are even.
   - The ones digits is one greater than the hundreds digit.
   - The sum of the digits is 24.

8. The number 10 has four divisors, namely 1, 2, 5, and 10. Find the least whole number with exactly five divisors.

9. In the Fuji Japanese Restaurant, guests are served at large tables that surround the grill where the food is prepared. Bowls of rice are to be shared by two people. Three people share a bowl of vegetables, and four people share a bowl of an entree. If there are twenty-six bowls on a table, how many guests are seated at the table?

10. Two pots of soup are placed at the end of the salad bar at the Western Ranger Steakhouse. While Wendy was checking to see which pot was vegetable soup and which was chili, she accidentally put a ladle full of chili into the vegetable soup pot. After she realized her mistake, Wendy mixed up the soup in the vegetable soup pot and poured a ladle full of the mixture into the chili pot. Does the vegetable soup pot have more or less chili in it than the chili pot has vegetable soup?

11. Find the sum of the series $\frac{1}{2} + \left(\frac{1}{3} + \frac{2}{3}\right) + \left(\frac{1}{4} + \frac{2}{4} + \frac{3}{4}\right) + \left(\frac{1}{5} + \frac{2}{5} + \frac{3}{5} + \frac{4}{5}\right) + \ldots + \left(\frac{1}{100} + \frac{2}{100} + \frac{3}{100} + \ldots + \frac{99}{100}\right)$.

12. The Healthstyle Athletic Club is having a special offer of $75 off the regular price for a one-year membership. Marc's family will receive an additional 10% discount after the $75 because his mother's employer has a corporate membership. Adding unlimited tennis privileges to the membership raises the price 5% to $463.05. What is the regular price of a one-year membership at the club?

COOPERATIVE LEARNING PROJECT

Work in groups. Each person must understand the solution and be able to explain it to any person in the class.

Jenny and Michelle were getting ready to go to a costume party dressed as Tweedledee and Tweedledum. Their costumes were identical, so Michelle wanted to pull four socks of the same color from the pile of socks in their closet. Since they were in a hurry, she grabbed enough socks to guarantee that there would be four of the same color. If the pile of socks contained red, yellow, and blue socks, how many did Michelle grab? If there had been n sisters needing the same color of socks, how many would Michelle need to grab?

CHAPTER 7 | SUMMARY AND REVIEW

VOCABULARY

Upon completing this chapter, you should be
familiar with the following terms:

cross products **309**   **308** ratio
proportion **308**   **321** similar polygons
rate **315**

SKILLS AND CONCEPTS

OBJECTIVES AND EXAMPLES	REVIEW EXERCISES

Upon completing this chapter, you should
be able to:

Use these exercises to review and prepare
for the chapter test.

■ recognize and use ratios and proportions.
(Lesson 7-1)

To solve the equation $\frac{x}{4} = \frac{27}{6}$, find the
cross products.

$$\frac{x}{4} = \frac{27}{6}$$
$$6x = 4 \cdot 27$$
$$6x = 108$$
$$x = 18$$

**Write the cross products for each proportion
and solve.**

1. $\frac{5}{x} = \frac{2}{3}$ **2.** $\frac{x}{9} = \frac{7}{15}$

3. $\frac{1}{x} = \frac{5}{x+5}$ **4.** $\frac{n+4}{3} = \frac{5n-3}{8}$

Determine if each statement is *true* or *false*.

5. If $\frac{a}{b} = \frac{c}{d}$, then $\frac{a}{b} = \frac{d}{c}$.

6. If $\frac{a}{b} = \frac{c}{d}$, then $\frac{a+b}{b} = \frac{c+d}{d}$.

■ apply and use the properties of
proportions. **(Lesson 7-2)**

A telephone pole casts a 36-foot shadow.
Nearby a 6-foot man casts an 8-foot
shadow. To find the height of the
telephone pole, write a proportion.

$$\frac{x}{36} = \frac{6}{8}$$
$$8x = 6 \cdot 36$$
$$8x = 216$$
$$x = 27 \qquad \text{The pole is 27 feet tall.}$$

**The measures of corresponding sides of
△*ABC* and △*XYZ* below are proportional.**

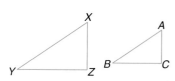

7. If $AB = 7$, $AC = 6$, and $XZ = 8$, find XY.

8. If $BC = 7$, $AC = 6$, and $XZ = 14$, find
YZ.

OBJECTIVES AND EXAMPLES	REVIEW EXERCISES

■ use proportions to solve problems involving similar figures. **(Lesson 7-3)**

$\square QRST \sim \square MNOP$

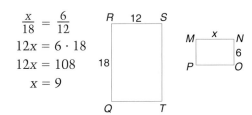

$$\frac{x}{18} = \frac{6}{12}$$
$$12x = 6 \cdot 18$$
$$12x = 108$$
$$x = 9$$

Determine whether each statement is *true* or *false*.

9. All congruent triangles are similar.

10. All similar triangles are congruent.

In the figure below $\triangle ABC \sim \triangle GFE$.

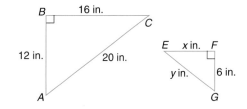

11. Find x.   12. Find y.

■ identify similar triangles. **(Lesson 7-4)**

There are three ways to prove that two triangles are similar.

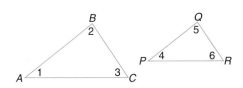

AA Similarity	$\angle 1 \cong \angle 4, \angle 3 \cong \angle 6$
SSS Similarity	$\frac{AB}{PQ} = \frac{BC}{QR} = \frac{CA}{RP}$
SAS Similarity	$\frac{AB}{PQ} = \frac{AC}{PR}, \angle 1 \cong \angle 4$

Write a two-column proof for each.

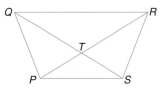

13. **Given:** $\frac{RP}{QS} = \frac{RS}{QP}$
    $\overline{QR} \parallel \overline{PS}$
    isosceles trapezoid $PQRS$
    Prove: $\triangle PQR \sim \triangle SRQ$

14. **Given:** $\overline{QR} \parallel \overline{PS}$
    Prove: $\frac{QT}{TS} = \frac{TR}{PT}$

■ use proportional parts of triangles to solve problems. **(Lesson 7-5)**

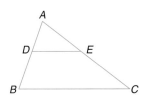

If $\overline{DE} \parallel \overline{BC}$, then $\frac{AD}{DB} = \frac{AE}{EC}$.

Use the figure and the given information to find the value of x.

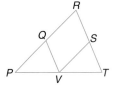

15. $\overline{SV} \parallel \overline{PR}$
    $TS = 5 + x$
    $TV = 8 + x$
    $VP = 4$
    $SR = 3$

16. $\overline{RT} \parallel \overline{QV}$
    $TV = 7.29$
    $PV = x$
    $PQ = 9$
    $QR = 27$

■ use the proportional relationships of corresponding perimeters, altitudes, angle bisectors, and medians of similar triangles. **(Lesson 7-6)**

$\triangle ABC \sim \triangle PQR$

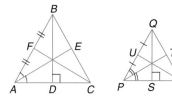

Perimeter	$\dfrac{AB + BC + CA}{PQ + QR + RP} = \dfrac{CA}{RP}$
Altitude	$\dfrac{BD}{QS} = \dfrac{BA}{QP}$
Angle bisector	$\dfrac{AE}{PT} = \dfrac{CA}{RP}$
Median	$\dfrac{CF}{RU} = \dfrac{CA}{RP}$

Find each of the following.

17. If $\triangle ABC \sim \triangle XYZ$, find YN.

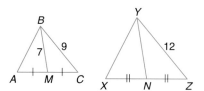

18. If $\triangle STV \sim \triangle PQM$, find the perimeter of $\triangle PQM$.

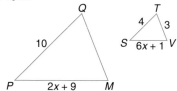

APPLICATIONS AND CONNECTIONS

19. Probability If the probability of rolling a sum of 7 with 2 dice is $\frac{1}{6}$, how many sums of 7 would you expect to get if you rolled the dice 174 times? **(Lesson 7-2)**

20. Travel A map is scaled so that 1 centimeter represents 15 kilometers. How far apart are two towns if they are 7.9 centimeters apart on the map? **(Lesson 7-2)**

21. Drafting A proportional divider is a drafting instrument which is used to enlarge or reduce a drawing. A screw at T keeps $\overline{AT} \cong \overline{TC}$ and $\overline{DT} \cong \overline{TB}$. $\overline{AB}$ and $\overline{CD}$ can be set so they are divided proportionally at T in any desired ratio. How would you adjust the dividers in order to enlarge a design in the ratio 7 to 4? **(Lesson 7-4)**

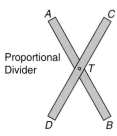

Proportional Divider

22. Hobbies A twin-jet airplane especially suited for medium-range flights has a length of 78 meters and a wingspan of 90 meters. If a scale model is made with a wingspan of 36 centimeters, find its length. **(Lesson 7-3)**

23. Solve this problem by solving a simpler problem. Find the units digit of 2^{125}. **(Lesson 7-7)**

1. If $\frac{a}{b} = \frac{x}{y}$, which one of the following statements is *not* true?

a. $ay = bx$ **b.** $\frac{a+x}{b+y} = \frac{a}{b} + \frac{x}{y}$ **c.** $\frac{b}{a} = \frac{y}{x}$ **d.** $\frac{y}{b} = \frac{x}{a}$

Solve each proportion.

2. $\frac{x}{28} = \frac{60}{16}$ **3.** $\frac{21}{1-x} = \frac{7}{x}$ **4.** $\frac{14}{21} = \frac{18}{x}$

5. A recipe for preparing material to be dyed calls for four parts alum to one part washing soda. How much washing soda should be used for 150 grams of alum?

Determine whether each statement is *true* or *false*.

6. All equilateral triangles are similar.   **7.** All regular hexagons are similar.

8. All isosceles triangles are similar.   **9.** All rhombuses are similar.

Determine if each pair of triangles is similar. Write *yes* or *no*. If they are similar, give a reason.

10. 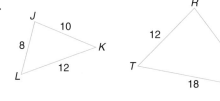   **11.**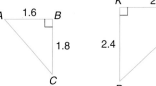

Use the figure at the right and the given information to find the value of *x*.

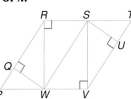

12. $\triangle RWP \sim \triangle VST$
$PW = x$ $SU = 5$
$QW = 1$ $ST = x + 5$

13. $\triangle WRS \sim \triangle VUS$
$SW = 3$ $SV = 2$
$RS = 1\frac{1}{3}$ $SU = x + \frac{2}{3}$

Complete each of the following.

14. Given: $\triangle ABC \sim \triangle RSP$
       D is the midpoint of $\overline{AC}$.
       Q is the midpoint of $\overline{PR}$.
   Prove: $\triangle SPQ \sim \triangle BCD$

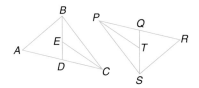

15. Use similar triangles to find the distance across the river.

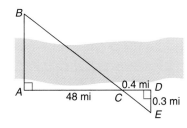

Bonus   The measures of the sides of one triangle are 5, 7, and 10. The measures of the sides of another triangle are 21, 15, and 30. Find the ratio of the areas of the triangles.

Algebra Review

- Multiply a polynomial by a monomial.

$x^2(x + 2) + 3(x^3 + 4x^2)$
$= (x^3 + 2x^2) + (3x^3 + 12x^2)$
$= (x^3 + 3x^3) + (2x^2 + 12x^2)$
$= 4x^3 + 14x^2$

Simplify.

1. $7xy(x^2 + 4xy - 8y^2)$
2. $x(3x - 5) + 7(x^2 - 2x + 9)$
3. $4x^2(x + 8) - 3x(2x^2 - 8x + 3)$

- Multiply a polynomial by a polynomial.

$(3x - 2)(x + 2)$
$= 3x(x) + 3x(2) + (-2)(x) + (-2)(2)$
$= 3x^2 + 6x - 2x - 4$
$= 3x^2 + 4x - 4$

Find each product.

4. $(r - 3)(r + 7)$ 5. $(x + 5)(3x - 2)$
6. $(4n + 3)(3n - 4)$ 7. $(2x + 9y)(3x - y)$
8. $(a - 4)(a^2 + 5a - 7)$

- Factor perfect square trinomials and binomials that are the difference of squares.

$16a^2 - 24a + 9$
$= (4a)^2 - 2(4a)(3) + (3)^2$
$= (4a - 3)^2$

$25 - 4y^2 = (5)^2 - (2y)^2$
$= (5 - 2y)(5 + 2y)$

Factor each polynomial.

9. $x^2 + 18x + 81$ 10. $32n^2 - 80n + 50$
11. $\dfrac{n^2}{4} - \dfrac{9}{16}$ 12. $3x^3 - 192x$
13. $16p^2 - 81r^4$
14. $54b^3 - 72b^2g + 24bg^2$

- Simplify rational expressions.

$\dfrac{x + y}{x^2 + 3xy + 2y^2} = \dfrac{x + y}{(x + y)(x + 2y)}$
$= \dfrac{1}{x + 2y}$

Simplify. State the excluded values of the variables.

15. $\dfrac{3x^2y}{12xy^3z}$ 16. $\dfrac{z^2 - 3z}{z - 3}$
17. $\dfrac{a^2 - 25}{a^2 + 3a - 10}$ 18. $\dfrac{x^2 + 10xy + 21y^2}{x^3 + x^2y - 42xy^2}$

- Identify the domain, range, and inverse of a relation.

State the domain, range, and inverse of $\{(3, 5), (2, 6), (-5, 6)\}$.

Domain: $\{3, 2, -5\}$ Range: $\{5, 6\}$

Inverse: $\{(5, 3), (6, 2), (6, -5)\}$

State the domain, range, and inverse of each relation.

19. $\{(-2, -1), (-1, 0), (0, 2)\}$
20. $\{(-3, 5), (-3, 6), (4, 5), (4, 6)\}$
21. $\{(4, 1), (4, -2), (4, 7), (4, -1)\}$

OBJECTIVES AND EXAMPLES

- Solve inequalities involving more than one operation.

$$14c \geq 6c + 48$$
$$14c - 6c \geq 6c - 6c + 48$$
$$\frac{8c}{8} \geq \frac{48}{8}$$
$$c \geq 6$$

The solution set is $\{c \,|\, c \geq 6\}$.

- Solve problems involving inverse variation.

If y varies inversely as x, and $y = 24$ when $x = 30$, find x when $y = 10$.

$$30 \cdot 24 = x \cdot 10 \quad x_1 y_1 = x_2 y_2$$
$$720 = 10x$$
$$72 = x$$

- Solve equations for a specified variable.

Solve the equation $\frac{x+1}{a} = b$ for x.

$$a\left(\frac{x+1}{a}\right) = a(b)$$
$$x + 1 - 1 = ab - 1$$
$$x = ab - 1$$

REVIEW EXERCISES

Solve each inequality. Check the solution.

22. $7x - 12 < 30$

23. $4y - 11 \geq 8y + 7$

24. $4(n - 1) < 7n + 8$

25. $\frac{3}{10}(4 - d) \leq -\frac{4}{5}\left(\frac{1d}{5} + 2\right)$

Solve. Assume that y varies inversely as x.

26. If $y = 28$ when $x = 42$, find y when $x = 56$.

27. If $y = 35$ when $x = 175$, find y when $x = 75$.

28. If $y = 2.7$ when $x = 8.1$, find x when $y = 3.6$

Solve for x.

29. $\frac{x+y}{c} = d$

30. $5(2a + x) = 3b$

31. $\frac{ax - 3}{2} = 7b - 6$

32. $\frac{2x - a}{3} = \frac{a + 3b}{4}$

Applications and Connections

33. Employment Ahmed's wages vary directly as the number of days he works. If his wages for 5 days are $26, how much would they be for 12 days?

34. Geometry The measure of the area of a rectangle is $25x^2 - 9$. Find the measure of its perimeter.

35. Photography To get a rectangular photo to fit into a rectangular frame, Leo had to trim a 1-inch strip from each side of the photo. In all, he trimmed off 46 square inches. If the photo is 3 inches longer than it is wide, what were its original dimensions?

36. Banking Sara budgets between $3 and $4 a month to spend on bank checking charges. Her bank charges $1.75 a month plus $0.08 per check. How many checks can Sara write each month and still meet her budget?

CHAPTER 8

Right Triangles and Trigonometry

CHAPTER OBJECTIVES

In this chapter, you will:
- Find the geometric mean between two numbers.
- Use the Pythagorean Theorem and its converse.
- Recognize and use trigonometric relationships from right triangles.
- Solve triangles and problems using the law of sines and the law of cosines.

GEOMETRY AROUND THE WORLD
Mexico

Unlike paintings, which have two dimensions, sculptures have three: length, width, and depth. Most of us are satisfied to simply look at paintings. But, because they have depth and are made of interesting materials, we usually want to touch sculptures.

Some sculptures are so tiny they can fit in your hand. Others, like "The Sculptured Space," an outdoor, public sculpture in Satellite City, Mexico, are enormous. "The Sculptured Space" was designed by a team of six sculptors headed by Helen Escobedo.

It is an environmental sculpture formed by a natural earth material called lava, produced by erupting volcanoes. Escobedo and her team of sculptors surrounded existing petrified lava with 64 concrete triangular prisms arranged in a circle. Each prism is 4 meters high, and the entire sculpture is 120 meters in diameter. There is no other sculpture like it!

GEOMETRY IN ACTION

Geometry is one of the oldest branches of mathematics. The word is derived from the Greek words *geo* meaning "earth," and *metria* meaning "measurement." You can use geometry to find missing measures for one prism in "The Sculptured Space." The shape of one prism is shown below. After you study Lesson 8-2, you will be able to find the length of the base.

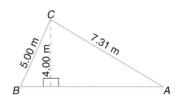

◀ *"The Sculptured Space"*

Using the people for scale, how tall does each prism in "The Sculptured Space" appear to be?

359

8-1 The Geometric Mean

Objectives

After studying this lesson, you should be able to:

- find the geometric mean between a pair of numbers, and
- solve problems using relationships between parts of a right triangle and the altitude to its hypotenuse.

Application

The shell of the chambered nautilus offers an example of right triangle relationships in nature. The segments shown allow us to approximate the spiral. The relationship among the segments in the spiral is a **geometric mean**.

The geometric mean between two positive numbers a and b is the positive number x where $\frac{a}{x} = \frac{x}{b}$. By cross multiplying, we see that $x^2 = ab$ or $x = \sqrt{ab}$.

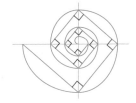

Example 1

Find the geometric mean between 12 and 18.

Let x represent the geometric mean.

$$\frac{12}{x} = \frac{x}{18} \qquad \textit{Definition of geometric mean}$$
$$x^2 = 216 \qquad \textit{Cross multiply.}$$
$$x = \sqrt{216} \qquad \textit{Take the square root of each side.}$$
$$x \approx 14.7 \qquad \textit{Use a calculator to find an approximation.}$$

The geometric mean between 12 and 18 is about 14.7.

The geometric mean is useful when studying relationships between the sides of a right triangle and its altitude. When the altitude to the hypotenuse of a right triangle is drawn, two similar triangles are formed. In the figure at the right, altitude $\overline{QS}$ separates $\triangle PQR$ into two smaller triangles that are similar to each other and also to $\triangle PQR$. This can be proved as a theorem.

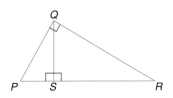

Theorem 8-1

If the altitude is drawn from the vertex of the right angle of a right triangle to its hypotenuse, then the two triangles formed are similar to the given triangle and to each other.

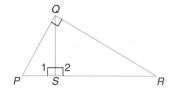

Proof of Theorem 8-1

Given: $\angle PQR$ is a right angle.
$\overline{QS}$ is an altitude of $\triangle PQR$.

Prove: $\triangle PSQ \sim \triangle PQR$
$\triangle PQR \sim \triangle QSR$
$\triangle PSQ \sim \triangle QSR$

Proof:

Statements	Reasons
1. $\angle PQR$ is a right angle. $\overline{QS}$ is an altitude of $\triangle PQR$.	1. Given
2. $\overline{QS} \perp \overline{RP}$	2. Definition of altitude
3. $\angle 1$ and $\angle 2$ are right angles.	3. $\perp$ lines form 4 rt. $\angle$s.
4. $\angle 1 \cong \angle PQR$ $\angle 2 \cong \angle PQR$	4. All rt. $\angle$s are $\cong$.
5. $\angle P \cong \angle P$ $\angle R \cong \angle R$	5. Congruence of angles is reflexive.
6. $\triangle PSQ \sim \triangle PQR$ $\triangle PQR \sim \triangle QSR$	6. AA Similarity
7. $\triangle PSQ \sim \triangle QSR$	7. Similarity of triangles is transitive.

In the figure above, $\triangle PSQ \sim \triangle QSR$. So, $\frac{PS}{QS} = \frac{QS}{SR}$ because corresponding sides of similar triangles are proportional. Thus, QS is the geometric mean between PS and SR. This is stated in the theorem below.

Theorem 8-2

The measure of the altitude drawn from the vertex of the right angle of a right triangle to its hypotenuse is the geometric mean between the measures of the two segments of the hypotenuse.

You will be asked to prove Theorem 8-2 in Exercise 33.

The geometric mean can be used to estimate hard-to-measure distances.

Example 2

APPLICATION
Measurement

To find the height of the tree in her backyard, Lori held a book near her eye so that the top and bottom of the tree were in line with the edges of the cover. If Lori's eye is 5 feet off the ground and she is standing approximately 14 feet from the tree, how tall is the tree? Assume that the tree is perpendicular to the ground and that the edges of the cover of the book are at right angles.

Draw a diagram of the situation. $\overline{EL}$ is the altitude drawn from the right angle of $\triangle ETB$.

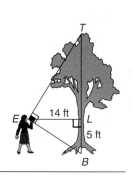

Using Theorem 8-2, we can write the equation below.

$$\frac{TL}{EL} = \frac{EL}{LB}$$

$$\frac{TL}{14} = \frac{14}{5} \qquad \textit{EL = 14, LB = 5}$$

$$5TL = 196 \qquad \textit{Cross multiply.}$$

$$TL = 39.2$$

The tree is approximately $39.2 + 5$ or 44.2 feet tall.

The altitude to the hypotenuse of a right triangle determines another relationship between segments.

Theorem 8-3	If the altitude is drawn to the hypotenuse of a right triangle, then the measure of a leg of the triangle is the geometric mean between the measures of the hypotenuse and the segment of the hypotenuse adjacent to that leg.
	You will prove this theorem in Exercise 34.

Example 3

Find the values of x and y.

According to Theorem 8-3, we can write the following proportions.

$$\frac{AD}{AB} = \frac{AB}{AC} \qquad \frac{DC}{BC} = \frac{BC}{AC}$$

$$\frac{3}{x} = \frac{x}{12} \qquad\qquad \frac{9}{y} = \frac{y}{12} \qquad \textit{AB = x, AD = 3, DC = 9, AC = 12, BC = y}$$

$$x^2 = 36 \qquad\qquad y^2 = 108 \qquad \textit{Cross multiply.}$$

$$x = 6 \qquad\qquad y = \sqrt{108} \text{ or about } 10.4$$

CHECKING FOR UNDERSTANDING

Communicating Mathematics

Read and study the lesson to answer these questions.

1. Describe how you would find the geometric mean between 6 and 10.

2. In the figure at the right, z is the geometric mean between __?__ and __?__.

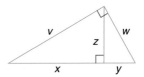

3. In the figure at the right, __?__ is the geometric mean between $(x + y)$ and x.

Guided Practice

Find the geometric mean between each pair of numbers.

4. 4 and 25

5. 15 and 3

6. 2 and 10

7. 7 and 22

8. $\frac{1}{2}$ and $\frac{2}{3}$

9. 8 and 5

10. Name three pairs of similar triangles in the figure at the right.

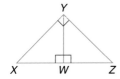

Find the values of x and y.

11.

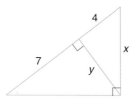

12.

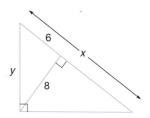

EXERCISES

Practice

Find the geometric mean between each pair of numbers.

13. 3 and 5

14. 4 and 6

15. $\frac{1}{4}$ and 9

16. 4 and $\frac{1}{9}$

17. $\frac{3}{8}$ and $\frac{8}{3}$

18. $\frac{2}{3}$ and $\frac{1}{3}$

Use the figure below and the given information to solve each problem.

19. $AD = 5$ and $DC = 9$. Find BD.

20. Find BD if $DC = 12$ and $AD = 3$.

21. If $AD = 3$ and $DC = 10$, find BD.

22. Find AB if $AC = 8$ and $AD = 3$.

23. $DA = 4$ and $DC = 4$. Find BA.

24. If $AD = 3$ and $DC = 4$, find BC.

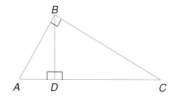

Find the values of x and y.

25.

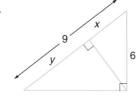

26.

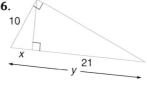

27.

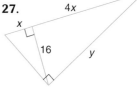

28.

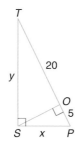

29.

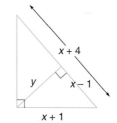

30.

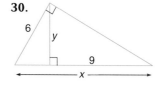

31. $VQ = 6$, $QR = 4$, $\angle PVR$ is right, and $\overline{VQ}$ is an altitude of $\triangle PVR$. Find PQ, PR, PV, and VR.

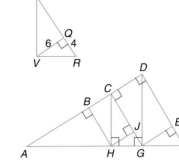

32. If $AF = 15$, $AD = 12$, and $DF = 9$ in the figure at the right, find AG, GF, DG, EF, CD, CG, and BC.

33. Write a paragraph proof of Theorem 8-2.

34. Write a two-column proof of Theorem 8-3.

Critical Thinking **35. Algebra** The arithmetic mean between two positive numbers a and b is $\frac{a+b}{2}$. Show algebraically that the arithmetic mean between two numbers is greater than the geometric mean. *(Hint: Use the problem-solving strategy of working backward.)*

Application **36. Science** The shape of the shell of the chambered nautilus can be modeled by a geometric mean. Consider the sequence of segments $\overline{OA}$, $\overline{OB}$, $\overline{OC}$, $\overline{OD}$, $\overline{OE}$, $\overline{OF}$, $\overline{OG}$, $\overline{OH}$, $\overline{OI}$, and $\overline{OJ}$. The length of each of these segments is the geometric mean between the lengths of the preceding segment and the succeeding segment. Explain. *(Hint: Consider $\triangle FGH$.)*

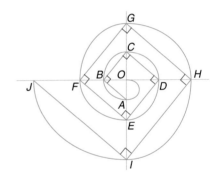

Mixed Review **37.** Kyle must finish John Steinbeck's *The Pearl* and write a report by Friday. He is starting to read page 71 and the book ends on page 197. How many pages does Kyle have yet to read? **(Lesson 7-7)**

38. The measures of the angles in $\triangle AND$ are in the ratio 4:8:12. What are the measures of the angles? **(Lesson 4-2)**

39. Name the property that allows us to say that $7(9 + 13) = 7(9) + 7(13)$. **(Lesson 2-4)**

40. Can an obtuse angle have a complement? Explain. **(Lesson 1-8)**

41. If A, B, and C are collinear and B is between A and C, how are AB and BC related to AC? **(Lesson 1-5)**

Wrap-Up **42. Journal Entry** Write two sentences about what you learned about geometric mean in your journal.

The Pythagorean Theorem

Objective

After studying this lesson, you should be able to:

- use the Pythagorean Theorem and its converse.

Application

   Marcus and his father are building a garage behind their house. A section of the side wall is six feet wide and eight feet high. Marcus is using the **Pythagorean Theorem** to find the length for the brace for this section of wall. *You will solve this problem in Example 1.*

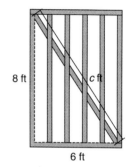

8 ft   c ft

6 ft

   The concepts behind the Pythagorean Theorem have been studied and used for thousands of years. A Chinese manuscript from about 1000 B.C. that illustrates the theorem has been found. Despite this fact, the theorem is named for Pythagoras, a Greek mathematician from the sixth century who is said to have been the first to write a proof of the theorem. Many proofs of the Pythagorean Theorem exist today. United States President James Garfield presented his own proof of the theorem in 1876.

Theorem 8-4 **Pythagorean** **Theorem**	**In a right triangle, the sum of the squares of the measures of the legs equals the square of the measure of the hypotenuse.** *If c is the measure of the hypotenuse, and a and b are the measures of the legs, then $a^2 + b^2 = c^2$.*

Proof of Pythagorean Theorem

Given: right $\triangle ABC$

Prove: $a^2 + b^2 = c^2$

Proof:

Let h be the measure of the altitude drawn from C to $\overline{AB}$. Theorem 8-3 states that two geometric means now exist.

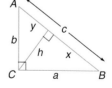

$$\frac{c}{a} = \frac{a}{x} \quad \text{and} \quad \frac{c}{b} = \frac{b}{y}$$
$$a^2 = cx \quad \text{and} \quad b^2 = cy \qquad \textit{Cross multiply.}$$

Add the equations.

$$a^2 + b^2 = cx + cy$$
$$a^2 + b^2 = c(x + y) \qquad \textit{Factor.}$$
$$a^2 + b^2 = c^2 \qquad \textit{Since c = x + y, substitute c for (x + y).}$$

Example 1

APPLICATION

Construction

Find the length of the brace Marcus needs for the section of wall that is 6 feet wide and 8 feet high.

Draw a diagram of the situation.

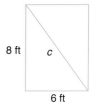

The legs of the right triangle measure 6 feet and 8 feet. We wish to find the length of the hypotenuse. Use the Pythagorean Theorem.

$$a^2 + b^2 = c^2 \qquad \textit{Pythagorean Theorem}$$
$$6^2 + 8^2 = c^2 \qquad \textit{a = 6 and b = 8}$$
$$36 + 64 = c^2$$
$$100 = c^2$$
$$10 = c \qquad \textit{Take the square root of each side. c cannot be -10}$$
$$\qquad\qquad \textit{since it represents a measure.}$$

The brace needs to be 10 feet long.

Suppose the square of the measure of the longest side of a triangle *does not* equal the sum of the squares of the measures of the other two sides. Then the triangle is *not* a right triangle. This principle, the converse of the Pythagorean Theorem, is used by carpenters when *squaring up* corners of buildings.

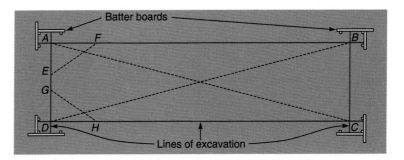

In the figure above, rectangle *ABCD* represents the lines of excavation for the foundation of a house. Since $3^2 + 4^2 = 5^2$, 3, 4, and 5 are measures of the sides of a right triangle. Therefore, from point *D* carpenters can measure 3 feet to point *G* and 4 feet to point *H*. If $\overline{GH}$ measures 5 feet, then they are assured that the corner forms a right angle. The corner is said to be *squared up*.

Theorem 8-5 *Converse of the* *Pythagorean Theorem*	**If the sum of the squares of the measures of two sides of a triangle equals the square of the measure of the longest side, then the triangle is a right triangle.** *You will prove this theorem in Exercise 37.*

Together, the Pythagorean Theorem and its converse are useful tools for solving right triangle problems.

Example 2

Determine whether a triangle with sides of 11.5 meters, 16.1 meters, and 20.7 meters is a right triangle.

The measure of the longest side of the triangle is 20.7 meters. Use the converse of the Pythagorean Theorem. Then find the value of each side of the equation.

$$16.1^2 + 11.5^2 \stackrel{?}{=} 20.7^2$$
$$259.21 + 132.25 \stackrel{?}{=} 428.49 \qquad \text{\textit{Use a calculator.}}$$
$$391.46 \neq 428.49$$

Since $391.46 \neq 428.49$, the triangle cannot be a right triangle.

CHECKING FOR UNDERSTANDING

Communicating Mathematics

Read and study the lesson to answer these questions.

1. State the Pythagorean Theorem in your own words. When does the Pythagorean Theorem hold true?

2. Look through architecture magazines in your school or public library. Find photographs of structures that appear to contain right triangles. Choose one and measure the sides of the triangle. Do they indicate that the triangle is a right triangle?

3. Write an equation that can be solved to find the length of the hypotenuse of a right triangle whose legs are 9 meters and 22 meters long. Solve your equation.

Guided Practice

Determine whether a triangle with sides having the given measures is a right triangle.

4. 5, 10, 12
5. 0.27, 0.36, 0.45
6. 1, 2, 3
7. 9, 40, 41
8. 10, 13, 17
9. 25, 60, 65

Use the Pythagorean Theorem to find each missing measure.

10.
   y in., 13 in., 12 in.

11.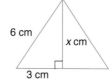
   6 cm, x cm, 3 cm

12.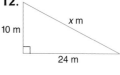
   x m, 10 m, 24 m

13. The measures of the sides of a right triangle are $x + 9$, $x + 2$, and $x + 10$. Find the value of x.

14. A picket fence is to have a gate 42 inches wide. The gate is 54 inches high. Find the length of a diagonal brace for the gate to the nearest inch.

15. A pleasure boat on Lake Erie sails 3 miles due north, 4 miles due east, and then 5 miles due south. To the nearest tenth of a mile, how far is the boat from its starting point?

EXERCISES

Practice Determine whether a triangle with sides having the given measures is a right triangle.

16. 12, 16, 20   **17.** 1.6, 3.0, 3.4   **18.** 3.87, 4.47, 5.91

19. 6, 8, 10   **20.** 25, 20, 15   **21.** 18, 34, 39

Find the value of *x*. Round your answer to the nearest tenth.

22. 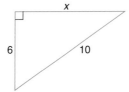   **23.** 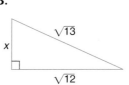   **24.**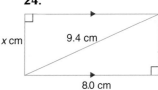

25. 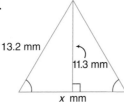   **26.** 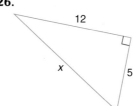   **27.**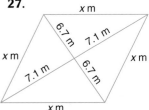

28. In a right triangle, the measures of the legs are 8 and $x + 7$, and the measure of the hypotenuse is $x + 10$. Find the value of *x*.

29. The diagonals of a rhombus measure 30 cm and 16 cm. Use the properties of a rhombus and the Pythagorean Theorem to find the perimeter of the rhombus.

30. Use the triangle at the right to find $a + b + c$.

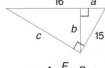

31. *ABCD* is an isosceles trapezoid. If $AB = 8$, $AC = 34$, and $EF = 30$, find the perimeter of *ABCD*.

32. A diagonal of a rhombus is 48 cm long, and a side of the rhombus is 26 cm long. Find the length of the other diagonal.

33. The rafters of a roof truss are perpendicular to each other. One rafter is 24 feet long and the other is 32 feet long, *not* counting the overhang. Find the rise of the roof.

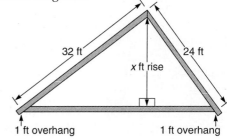

34. Draw a right triangle with vertices $A(0, a)$, $C(0, 0)$, and $B(b, 0)$ on a coordinate plane. Use the Pythagorean Theorem to derive the formula for the distance between A and B.

35. In a carton with rectangular sides, all pairs of intersecting edges are perpendicular. What is the length of the longest fishing rod that will fit inside the carton shown at the right?

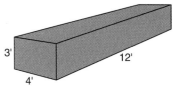

36. Explain why the drawing at the right is an illustration of the Pythagorean Theorem.

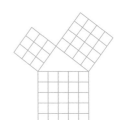

37. Write a paragraph proof of the converse of the Pythagorean Theorem.

Critical Thinking

38. Draw three different acute triangles and three different obtuse triangles. Label the measure of the longest side of each triangle c and the measures of the other two sides a and b. Measure each side in centimeters.
a. Calculate c^2 and the sum of a^2 and b^2 for each triangle.
b. Compare your results and make a conjecture from your experiments.

Applications

39. Sports   June is making a ramp to try out her car for the pinewood derby. The ramp support forms a right angle. The base is 12 feet long and the height is 5 feet. What length of plywood does she need to complete the ramp?

40. Construction   A stair stringer is a board that supports stairs. Suppose a set of stairs is to rise 8 feet over a length of 15 feet. Find the length of the stair stringer to the nearest foot.

41. History   James Garfield is the only United States president to have published a mathematical proof. Research his proof in a book on the history of mathematics. What did Garfield use as the basis of his proof?

Computer

A **Pythagorean triple** is a group of three whole numbers that satisfy the equation $a^2 + b^2 = c^2$, where c is the greatest number. The BASIC program below uses a procedure for finding Pythagorean triples that was developed by Euclid around 320 B.C.

```
 10 FOR X = 2 TO 6
 20 FOR Y = 1 TO 5
 30 IF X <= Y THEN GOTO 80
 40 A = INT(X^2 - Y^2 + 0.5)
 50 B = 2 * X * Y
 60 C = INT(X^2 + Y^2 + 0.5)
 65 IF A > B THEN PRINT B; " "; A; " "; C: GOTO 80
 70 PRINT A; " "; B; " "; C
 80 NEXT Y
 90 NEXT X
100 END
```

Run the BASIC program to generate a list of Pythagorean triples. Use the list to answer each question.

42. Notice that both (3, 4, 5) and (6, 8, 10) are Pythagorean triples. Since (6, 8, 10) = (3 · 2, 4 · 2, 5 · 2), these two triples are said to be part of a *family*. Since the numbers 3, 4, and 5 have no common factors except 1, (3, 4, 5) is called a *primitive triple*. List all of the members of the (3, 4, 5) family that are generated by the BASIC program.

43. A geometry student made the conjecture that if three whole numbers are a Pythagorean triple, then their product is divisible by 60.
  a. Does this conjecture hold true for each triple that is printed by the BASIC program?
  b. Change the values in lines 10 and 20 of the program to generate more triples. Does the conjecture hold true for these triples?
  c. Do you think the conjecture is true for all Pythagorean triples? Justify your answer.
  d. Are all sets of three whole numbers whose product is 60 Pythagorean triples? Explain.

Mixed Review

44. Find the geometric mean between 9 and 15. **(Lesson 8-1)**

45. The bases of a trapezoid are 8 meters and 22 meters long. Find the length of the median of the trapezoid. **(Lesson 6-6)**

46. Which lines are parallel if $\angle 2 \cong \angle 4$ and $\angle 3 \cong \angle 5$? Justify your answer. **(Lesson 3-3)**

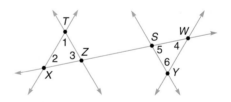

47. The measures of the legs of an isosceles triangle are $8x - 9$ and $6x - 1$. Find the value of x. **(Lesson 4-7)**

48. **Zoology** If possible, write a valid conclusion. State the law of logic that you used. **(Lesson 2-3)**

- A *monotreme* is a mammal that lays eggs.
- A duck-billed platypus is a monotreme.

Wrap-Up

49. Research Find three proofs of the Pythagorean Theorem by investigating the topic in your library.

Special Right Triangles

Objective	After studying this lesson, you should be able to:

- use the properties of 45°-45°-90° and 30°-60°-90° triangles.

Application

Baseball has been called the "great American pastime." It has been popular since the early 1800s. A professional baseball diamond is shaped like a square and has baselines that are 90 feet long. If the first baseman forces a runner out, how far will he have to throw the ball to make a double play at third? We can find the answer to this question by finding the length of a diagonal of a square with sides 90 feet long.

FYI ···

The Cleveland Indians were named in honor of Louis Sockolexis, the first Native American to play professional baseball.

The diagonal of a square and the sides of the square form two isosceles right triangles, or 45°-45°-90° triangles. We can use the Pythagorean Theorem to find the length of the diagonal.

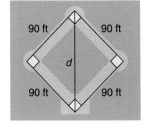

$$d^2 = 90^2 + 90^2 \quad \textit{Pythagorean Theorem}$$
$$d^2 = 8100 + 8100$$
$$d = \sqrt{16{,}200} \text{ or about } 127.3$$

So, the first baseman would have to throw the ball about 127 feet to third base.

Let's look at the general case of a square to learn more about a 45°-45°-90° triangle.

$$d^2 = s^2 + s^2 \quad \textit{Pythagorean Theorem}$$
$$d^2 = 2s^2$$
$$d = s\sqrt{2}$$

This proves the following theorem.

Theorem 8-6	**In a 45°-45°-90° triangle, the hypotenuse is $\sqrt{2}$ times as long as a leg.**

Example 1

Find the value of x.

a.

Since the two base angles are congruent, the triangle is isosceles. Therefore, $x = 8$.

b.

According to Theorem 8-6, the hypotenuse is $\sqrt{2}$ times as long as a leg. So, $x = 8\sqrt{2}$ or about 11.3.

There is a special relationship in a $30°$-$60°$-$90°$ triangle also.

If we draw an altitude from any vertex of an equilateral triangle, the triangle is separated into two congruent $30°$-$60°$-$90°$ triangles. Using the Pythagorean Theorem, it is possible to derive a formula relating the lengths of the sides to each other.

Let s = the measure of a side.
Let a = the measure of the altitude.

$\triangle PTQ$ and $\triangle RTQ$ are $30°$-$60°$-$90°$ triangles.

$$s^2 = a^2 + \left(\frac{s}{2}\right)^2 \qquad \textit{Pythagorean Theorem}$$

$$a^2 = s^2 - \left(\frac{s}{2}\right)^2 \qquad \textit{Solve for a.}$$

$$a^2 = \frac{3s^2}{4}$$

$$a = \frac{s\sqrt{3}}{2}$$

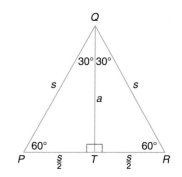

So in the $30°$-$60°$-$90°$ triangle, the measures of the sides are $\frac{s}{2}$, $\frac{s\sqrt{3}}{2}$, and s. This proves the following theorem.

Theorem 8-7

In a $30°$-$60°$-$90°$ triangle, the hypotenuse is twice as long as the shorter leg and the longer leg is $\sqrt{3}$ times as long as the shorter leg.

Example 2

APPLICATION
Architecture

A geodesic dome is made up of faces shaped like equilateral triangles. If a side of a triangle on a dome is 7 feet long, find the length of an altitude.

$\triangle ABC$ is one of the triangles. Draw $\overline{AD}$ perpendicular to $\overline{BC}$. $\triangle ADB$ is a $30°$-$60°$-$90°$ triangle.

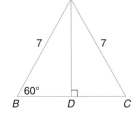

$AB = 2(BD)$ *The hypotenuse is twice the shorter leg.*

$7 = 2(BD)$ *Substitution*

$3.5 = BD$ *Division property*

$AD = \sqrt{3}\,(BD)$ *The longer leg is $\sqrt{3}$ times the shorter leg.*

$AD = \sqrt{3}\,(3.5)$ *Substitution*

The altitude is $\sqrt{3}\,(3.5)$ or about 6.1 feet.   *Check this result using the Pythagorean Theorem.*

CHECKING FOR UNDERSTANDING

Communicating Mathematics

Read and study the lesson to answer these questions.

1. If the measure of a leg of a $45°$-$45°$-$90°$ triangle is l, then the measure of the hypotenuse is __?__.

2. The measure of the shorter leg of a $30°$-$60°$-$90°$ triangle is s. The measure of the longer leg is __?__, and the measure of the hypotenuse is __?__.

3. The measure of the long leg of a $30°$-$60°$-$90°$ triangle is 14. Write expressions for the measures of the shorter leg and the hypotenuse.

Guided Practice

Find x and y.

4.

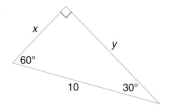

5.

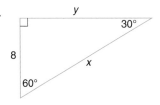

The length of a side of a square is given. Find the length of a diagonal of each square.

6. 1 ft   7. 31.2 m   8. $4\frac{2}{3}$ yd

The length of a side of an equilateral triangle is given. Find the length of an altitude of each triangle.

9. 4 ft   **10.** 27.4 m   **11.** $\frac{2}{3}$ yd

12. The perimeter of an equilateral triangle is 24 units. Find the length of an altitude of the triangle.

EXERCISES

Practice   **Find the value of x.**

13.
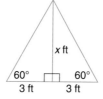

14.
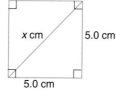

15.
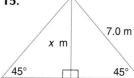

16.
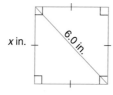

17.
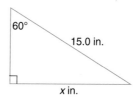

18.
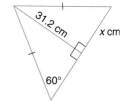

19. In the triangle at the right, $AC = 10$, $AB = 3\sqrt{3}$, and $m\angle A = 30$. Find BC.

20. Find the perimeter of $\triangle PQR$ shown at the right.

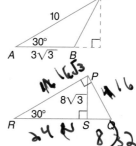

21. The sum of the squares of the measures of all sides of a rectangle is 1458. Find the measure of a diagonal of the rectangle.

22. An altitude of an equilateral triangle is 5.2 meters long. Find the perimeter of the triangle rounded to the nearest tenth.

23. The diagonals of a rectangle are 12 units long and intersect at an angle of 60°. Find the perimeter of the rectangle.

24. In $\triangle RST$ shown at the right, $RT = 10\sqrt{3}$. Find RS and VS.

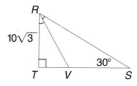

25. Find the length of the altitude of an equilateral triangle whose perimeter is 42 centimeters.

Use the figure at the right to find each measure.

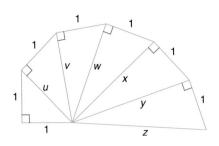

26. u **27.** v

28. w **29.** x

30. y **31.** z

Suppose each edge of the cube measures s units.

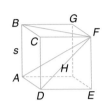

32. Develop a formula for the distance from A to F in terms of s.

33. Find $m\angle BFD$.

34. Prove Theorem 8-6.

Critical Thinking

35. A stop sign is to be made from a square piece of sheet metal. The largest stop sign that can be cut from the square has sides that are each 1 foot long. How long are the sides of the square?

Applications

36. Auto Repair   A mechanic working on a car needs to remove a hexagonal nut to remove the carburetor. One side of the nut is 6 millimeters long. How wide should the opening of the wrench be to fit the nut? *Hint: A hexagon is made up of six equilateral triangles.*

Mixed Review

37. Painting   A painter leans a ten-foot ladder against the house she is to paint. The foot of the ladder is 3 feet from the house. How far above the ground does the ladder touch the house? **(Lesson 8-2)**

38. $PA = 5$, $AQ = 3$, $PR = 12$, and $BR = 4.5$. Is $\triangle BPA \sim \triangle RPQ$? Justify your answer. **(Lesson 7-5)**

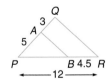

39. The base of an isosceles triangle is 18 inches long. If the legs are $3y + 21$ and $10y$ inches long, find the perimeter of the triangle. **(Lesson 4-7)**

40. Write a paragraph proof. **(Lesson 3-3)**

Given: $m\angle BAC = 90$
$m\angle ABC = 30$
$m\angle EDC = 60$

Prove: ℓ is parallel to m.

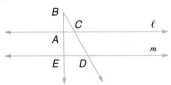

Wrap-Up

41. Write two sentences to describe the relationships for the special right triangles you learned in this lesson.

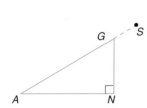

8-4 | Trigonometry

Objectives

After studying this lesson, you should be able to:

- express trigonometric ratios as fractions or decimals,
- recognize trigonometric relationships from right triangles, and
- use a calculator to find values of trigonometric ratios or measures of angles.

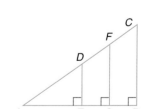

Application

When the Egyptians first used a sundial around 1500 B.C., they were using **trigonometry**. Trigonometry means triangle measurement.

The figure at the left is a model of how a sundial works. As the sun, S, shines on a fixed staff, represented by $\overline{GN}$, it casts a shadow. $\overline{AN}$ represents the shadow. Since GN is a constant, the length of $\overline{AN}$ varies with the measure of $\angle A$. The Egyptians understood that $\frac{GN}{AN}$ is a function of the measure of $\angle A$. We will define this function as the tangent of $\angle A$.

A ratio of the lengths of two sides of a right triangle is called a **trigonometric ratio**. The three most common ratios are **sine**, **cosine**, and **tangent**. Their abbreviations are *sin*, *cos*, and *tan* respectively.

$\sin A = \dfrac{opposite}{hypotenuse}$

$\cos A = \dfrac{adjacent}{hypotenuse}$

$\tan A = \dfrac{opposite}{adjacent}$

$\sin A = \dfrac{\text{side opposite } \angle A}{\text{hypotenuse}} = \dfrac{a}{c}$

$\cos A = \dfrac{\text{side adjacent to } \angle A}{\text{hypotenuse}} = \dfrac{b}{c}$

$\tan A = \dfrac{\text{side opposite } \angle A}{\text{side adjacent to } \angle A} = \dfrac{a}{b}$

Trigonometric ratios are related to the acute angles of a right triangle, not the right angle. The value of a trigonometric ratio depends *only* on the measure of the angle. It does not depend on the size of the triangle.

Consider the three overlapping triangles at the right. They are all right triangles that share a common angle, $\angle A$. So the triangles are similar by AA. If we use $\triangle ADE$, $\sin A = \dfrac{DE}{AD}$, in $\triangle AFG$, $\sin A = \dfrac{FG}{AF}$, and in $\triangle ACB$, $\sin A = \dfrac{CB}{AC}$. Because the triangles are similar, the ratios are equal. That is, $\dfrac{DE}{AD} = \dfrac{FG}{AF} = \dfrac{CB}{AC}$. *Recall that in similar triangles, corresponding sides are proportional.*

Example 1

Find sin *M*, cos *M*, tan *M*, sin *N*, cos *N*, and tan *N*. Express each ratio as a fraction and as a decimal.

$\sin M = \dfrac{ON}{MN} = \dfrac{8}{10}$ or 0.800

$\cos M = \dfrac{MO}{MN} = \dfrac{6}{10}$ or 0.600

$\tan M = \dfrac{ON}{MO} = \dfrac{8}{6}$ or $1.\overline{3}$

$\sin N = \dfrac{OM}{MN} = \dfrac{6}{10}$ or 0.600

$\cos N = \dfrac{ON}{MN} = \dfrac{8}{10}$ or 0.800

$\tan N = \dfrac{MO}{ON} = \dfrac{6}{8}$ or 0.750

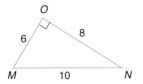

Applications of trigonometry will often require that you find a decimal approximation of a trigonometric ratio. You can use your scientific calculator to obtain approximations of trigonometric ratios. *If you do not have a scientific calculator, use the table on page 694.*

Example 2

Find each value using your calculator. *Set your calculator in degree mode.*

a. sin 75°

ENTER: 75 [SIN] 0.9659258

sin 75° ≈ 0.9659.

b. cos 65°

ENTER: 65 [COS] 0.4226182

cos 65° ≈ 0.4226.

There are many practical applications of trigonometry. You can use the ratios to find a missing measure of a right triangle.

Example 3

APPLICATION

Navigation

The *Princess III* is sailing to the Hawaiian Islands. As the ship crosses the equator, its instruments indicate that it is headed on a course that forms a 13° angle with the equator. How far will the ship be from the equator after it has traveled 90 miles on this course?

Draw a diagram of the situation.

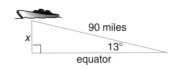

The distance from the equator can be found using the sine ratio.

$\sin 13° = \dfrac{x}{90}$ $sin = \dfrac{opposite}{hypotenuse}$

$90 \sin 13° = x$

Use your calculator to find the distance.

ENTER: 90 [×] 13 [SIN] [=] 20.245595

The *Princess III* will be about 20.2 miles from the equator after it travels 90 miles on this course.

Trigonometric ratios can be used to find the measures of the acute angles in a right triangle when you know the measures of two sides of the triangle. You must determine which ratio involves the two sides of the triangle whose measures you know. Then, use the inverse capabilities of your calculator to find the measure of the angle.

Example 4

Find the approximate measure of ∠*E.*

We know the measure of the leg adjacent to ∠*E* and the measure of the hypotenuse. The cosine ratio relates these two measures.

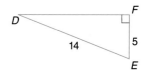

$$\cos E = \frac{EF}{DE} \qquad cos = \frac{adjacent}{hypotenuse}$$

$$= \frac{5}{14}$$

Now use a calculator to find the measure of ∠*E* to the nearest degree.

ENTER: 5 ÷ 14 = INV COS 69.075168   *The* INV *key followed by* COS *finds the angle for which the cosine function has the value* $\frac{5}{14}$.

$m\angle E \approx 69$

Example 5

APPLICATION

Aviation

Don Bates is flying a plane from Memphis to Little Rock. One minute after takeoff, his altitude is 2.8 miles and he has traveled 7.5 ground miles from the airport. If he has been climbing steadily since takeoff, what is the measure of the angle that Mr. Bates' plane makes with the ground?

We can use the tangent ratio to find the measure of the angle.

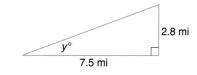

$$\tan y = \frac{2.8}{7.5} \qquad tan = \frac{opposite}{adjacent}$$

$$\tan y \approx 0.3733333 \quad \text{\textit{Use a calculator.}}$$

$$y \approx 20.47228$$

The angle that the plane's path makes with the ground measures about 20°.

CHECKING FOR UNDERSTANDING

Communicating Mathematics

Read and study the lesson to answer these questions.

1. What does trigonometry mean?

2. In the triangle below, __?__ is opposite ∠*A* and __?__ is adjacent to ∠*A*.

3. Use the triangle at the right to explain the relationship between sin *A* and cos *B*.

4. How are tan *A* and tan *B* related?

Guided Practice

Find the indicated trigonometric ratio as a fraction and as a decimal rounded to the nearest thousandth.

5. $\sin A$ **6.** $\cos A$

7. $\tan A$ **8.** $\sin B$

9. $\cos B$ **10.** $\tan B$

State the trigonometric ratio that corresponds to each value and angle given.

11. $\frac{3}{4}; \angle Q$ **12.** $\frac{4}{5}; \angle Q$

13. $\frac{4}{5}; \angle P$ **14.** $\frac{4}{3}; \angle P$

Use your calculator to find the value of each ratio to the nearest thousandth.

15. $\sin 10°$ **16.** $\cos 36°$ **17.** $\tan 38°$

Use your calculator to find the measure of each angle to the nearest degree.

18. $\sin A = 0.105$ **19.** $\tan S = 0.702$ **20.** $\cos C = 0.898$

21. A jet takes off and rises at an angle of 19° with the ground until it hits 28,000 feet. How much ground distance is covered in miles? *(5280 feet = 1 mile)*

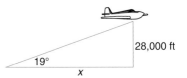

EXERCISES

Practice

Find the indicated trigonometric ratio as a fraction and as a decimal rounded to the nearest thousandth.

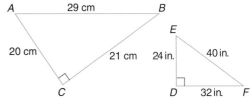

22. $\sin A$ **23.** $\sin B$

24. $\tan A$ **25.** $\cos B$

26. $\cos E$ **27.** $\tan F$

State the trigonometric ratio that corresponds to each value and angle given.

28. $\frac{\sqrt{3}}{3}; \angle C$ **29.** $\frac{1}{2}; \angle T$

30. $\frac{\sqrt{3}}{2}; \angle T$ **31.** $\sqrt{3}; \angle T$

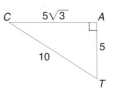

Draw a 30°-60°-90° triangle and label each side with an appropriate value. Find each ratio. State your answers as fractions.

32. $\sin 60°$ **33.** $\cos 60°$ **34.** $\tan 60°$

35. $\sin 30°$ **36.** $\cos 30°$ **37.** $\tan 30°$

Find the value of x. Round measures of segments to the nearest tenth and angle measures to the nearest degree.

38.   **39.** 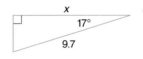   **40.**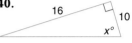

Find the values of x and y. Round measures of segments to the nearest tenth and angle measures to the nearest degree.

41.    **42.** 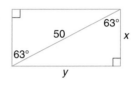   **43.**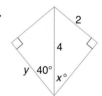

44. 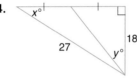   **45.** 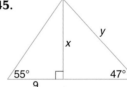   **46.**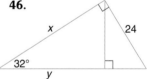

Solve. Make a drawing.

47. A guy wire is attached to a 100-foot tower that is perpendicular to the ground. The wire makes an angle of 55° with the ground. What is the length of the wire?

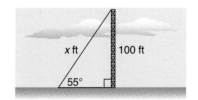

48. Each side of a rhombus is 30 units long. One diagonal makes a 25° angle with a side. What is the length of each diagonal to the nearest tenth of a unit?

49. An isosceles trapezoid has an altitude 25 inches long. One of the base angles measures 35°. Find the length of the legs of the trapezoid.

50. In $\triangle XYZ$, $\angle Z$ is a right angle. If $\sin X = \frac{3}{4}$, find $\tan Y$.

Critical Thinking

51. Use a calculator to find the sine, cosine, and tangent of several acute angle measures to make conjectures.

  a. When is $\cos A > \sin A$?

  b. When is $\cos A = \sin A$?

  c. Describe the value of $\tan A$ when $m\angle A > 45$.

  d. Describe the range of values for $\cos A$ and $\sin A$.

Applications

52. Navigation A ship travels east from Port Lincoln 24 miles before turning north. When the ship becomes disabled and radios for help, the rescue boat needs to know the fastest route to the ship. The rescue boat navigator finds that the shortest route from Port Lincoln to the ship is 48 miles long. At what angle off of due east should the rescue boat travel to take the shortest route to the ship?

53. Aviation A jet airplane begins a steady climb of 15° and flies for two miles. What was its change in altitude in feet? *(5280 ft = 1 mile)*

54. Electronics The power in watts that is absorbed by an AC circuit is given by $P = IV \cos \theta$, where I is the current in amps, V is the voltage, and the Greek letter θ (theta) is the measure of the phase angle. Use your calculator to find the power absorbed by a circuit if its current is 2 amps, its voltage is 120 volts, and its phase angle measures 67°. Round your answer to the nearest thousandth.

Mixed Review

55. A 30°-60°-90° triangle has a shorter leg that is 8 units long. Find the lengths of the hypotenuse and the longer leg. **(Lesson 8-3)**

56. Find the values of x and y. **(Lesson 8-1)**

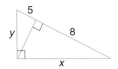

57. Determine whether $\overline{CD} \parallel \overline{BE}$. Explain your reasoning. **(Lesson 7-5)**

$AB = 9.6$
$AC = 15.0$
$AE = 6.8$
$AD = 10.0$

58. The bases of an isosceles trapezoid measure 10 inches and 22 inches. Find the length of the median of the trapezoid. **(Lesson 6-6)**

59. Determine whether the statement "The diagonals of a rectangle bisect the opposite angles" is *true* or *false*. Justify your answer. **(Lesson 6-4)**

60. Write a paragraph proof. **(Lesson 5-7)**

Given: $\overline{GA} \cong \overline{AI}$
$GL < IL$

Prove: $m\angle 1 < m\angle 2$

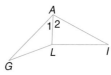

61. Find the value of x. **(Lesson 3-2)**

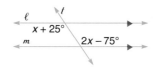

62. Two vertical angles have measures of $7x + 12$ and $4x + 42$. Find x. **(Lesson 2-7)**

Wrap-Up **63.** Draw triangle XYZ with right angle X, and label the vertices. Write a ratio for sin Y, cos Y, and tan Y.

MID-CHAPTER REVIEW

Use right triangle *GHJ* and the given information to solve each problem. (Lesson 8-1)

1. Find HK if $GK = 8$ and $KJ = 14$.

2. If $GJ = 15$ and $GK = 9$, find GH.

3. $GK = 4$ and $KJ = 7$. Find HJ.

4. Find HG if $KG = 8$ and $KJ = 8$.

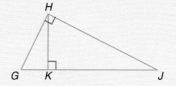

Determine whether a triangle with sides having the given measures is a right triangle. (Lesson 8-2)

5. 12, 16, 20
6. 2.2, 2.4, 3.3
7. 1, $\sqrt{2}, \sqrt{3}$
8. $\sqrt{3}, \sqrt{4}, \sqrt{5}$
9. 1.6, 3.0, 3.4
10. 15, 20, 25

11. The length of an altitude of an equilateral triangle is $12\sqrt{3}$ units. Find the perimeter of the triangle. **(Lesson 8-3)**

12. Find the length of a diagonal of a square if a side is 8 centimeters long. **(Lesson 8-3)**

13. A 16-foot ladder is leaning against a house. It touches the bottom of a window that is 12 feet 6 inches above the ground. What is the measure of the angle that the ladder forms with the ground? Round your answer to the nearest degree. **(Lesson 8-4)**

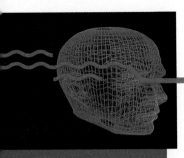

Technology

Tangent Ratio

BASIC
Geometric Supposer
▶ **Graphing calculators**
LOGO
Spreadsheets

The tangent ratio of an angle is the slope of a line. Consider the graph of $\overleftrightarrow{PQ}$ below.

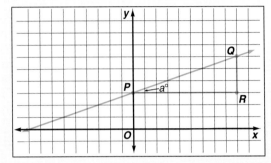

An angle with measure a is formed at P by a horizontal line and the portion of line PQ that is to the right of a vertical line, in this case the y-axis. You can find the measure of the angle using the tangent ratio for $\triangle PQR$.

$$\tan a = \frac{QR}{PR}$$

$$\tan a = \frac{3}{9} \qquad QR = 3,\ PR = 9$$

$$a \approx 18.4°$$

The slope of a line is given by the formula $m = \frac{y_2 - y_1}{x_2 - x_1}$, where (x_1, y_1) and (x_2, y_2) are the coordinates of two points on the line. Find the slope of $\overleftrightarrow{PQ}$ given the points $P(0, 3)$ and $Q(9, 6)$.

$$m = \frac{y_2 - y_1}{x_2 - x_1}$$

$$= \frac{6 - 3}{9 - 0} \text{ or } \frac{3}{9} \qquad (x_1, y_1) = P(0, 3);\ (x_2, y_2) = Q(9, 6)$$

Notice that $\tan a = m$. The vertical change, $y_2 - y_1$, is the same as the length of the side opposite, QR, and the horizontal change, $x_2 - x_1$, is the same as the length of the side adjacent, PR. So, the slope of $\overleftrightarrow{PQ}$ is the same as the tangent of the angle formed by $\overleftrightarrow{PQ}$ and a horizontal line.

EXERCISES

Graph each equation on your graphing calculator. Then find the slope of the line and the measure of the angle the line forms with a horizontal to the nearest degree.

1. $y = 8x + 4$

2. $4x - 6y = 9$

3. $y = \frac{2}{5}x$

Application: Using Trigonometry

Objectives

After studying this lesson, you should be able to:
- recognize angles of depression or elevation, and
- use trigonometry to solve problems.

Application

The airport meteorologists keep close tabs on the weather to help ensure that airplanes can fly safely. One of the things that they watch is the cloud ceiling. The cloud ceiling is the lowest altitude at which solid cloud is present. If the cloud ceiling is below a certain level, usually about 61 meters, airplanes are not allowed to take off or land.

One way that meteorologists can find the cloud ceiling at night is to shine a searchlight that is located a fixed distance from their office vertically onto the clouds. Then they measure the **angle of elevation** to the spot of light on the cloud. The angle of elevation to the spot is the angle formed by the line of sight to the spot and a horizontal segment. Using this information and some trigonometry, the cloud ceiling can be determined.

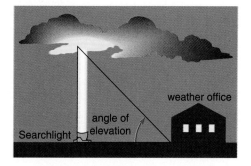

Example 1

A searchlight located 200 meters from a weather office is turned on. If the angle of elevation to the spot of light on the clouds is 35°, how high is the cloud ceiling?

Draw a diagram.

Let c represent the cloud ceiling.

$\tan 35° = \dfrac{c}{200}$ *$\tan = \dfrac{opposite}{adjacent}$*

$200 \tan 35° = c$ *Multiply each side by 200.*

Use a calculator to find c.

ENTER: 200 ⨯ 35 [TAN] [=] *140.04151*

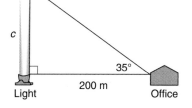

The ceiling is about 140 meters. *With this ceiling, airplanes can take off and land.*

A person on a tower or on a cliff must look down to see an object below. This person's line of sight forms an **angle of depression** with a horizontal line.

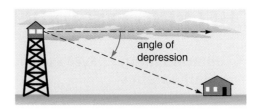

angle of depression

Example 2

APPLICATION

Fire Fighting

A fire is sighted from a fire tower in Wayne National Forest. The ranger found that the angle of depression to the fire is 22°. If the tower is 75 meters tall, how far is the fire from the base of the tower?

Let d represent the distance from the fire to the base of the fire tower.

The segments with measure d are parallel, and the line of sight is a transversal. So, by alternate interior angles, the angle between the ground and the line of sight measures 22°.

$$\tan 22° = \frac{75}{d}$$

$$d = \frac{75}{\tan 22°}$$

$$d \approx \frac{75}{0.4040}$$

$$d \approx 185.63$$

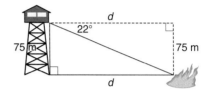

The fire is about 186 meters from the base of the tower.

Angles of depression or elevation to two different objects can be used to find the distance between those objects.

Example 3

APPLICATION

Aerospace

The lunar lander, *Eagle*, traveled aboard *Apollo 11* and descended to the surface of the moon on July 20, 1969. Before sending *Eagle* to the surface of the moon, *Apollo 11* orbited the moon three miles above the surface. At one point in the orbit, the onboard guidance system measured the angles of depression to the near and far edges of a large crater. The angles measured 25° and 18°. Find the distance across the crater.

Draw a diagram.

Let f be the distance to the far edge of the crater and n be the distance to the near edge.

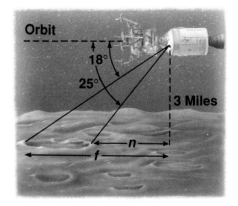

$$\tan 18° = \frac{3}{f} \qquad \tan 25° = \frac{3}{n}$$

$$f = \frac{3}{\tan 18°} \qquad\qquad n = \frac{3}{\tan 25°}$$

$$f \approx \frac{3}{0.3249} \qquad\qquad n \approx \frac{3}{0.4663}$$

$$f \approx 9.2 \qquad\qquad n \approx 6.4$$

The distance across the crater is $9.2 - 6.4$ or about 2.8 miles.

CHECKING FOR UNDERSTANDING

Communicating Mathematics

Read and study the lesson to answer these questions.

1. Make a drawing and write a few sentences to explain what is meant by an angle of elevation.

2. Give an example of how one might use an angle of depression to measure something that you cannot measure directly.

3. How do you decide whether to use sin, cos, or tan when you are finding the measure of an acute angle in a right triangle? Explain.

Guided Practice

Name the angles of elevation and depression in each figure.

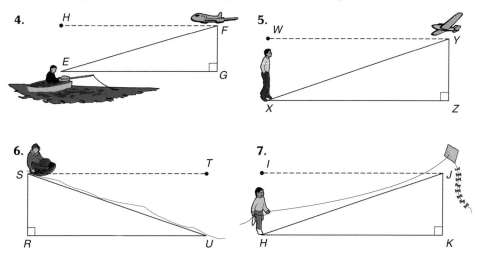

State an equation that would enable you to solve each problem. Then solve. Round answers to the nearest tenth.

8. Given $m\angle P = 15$ and $PQ = 37$, find QR.

9. Given $m\angle P = 47$ and $QR = 10$, find PQ.

10. Given $PR = 13.4$ and $m\angle P = 16$, find PQ.

11. Given $m\angle Q = 72$ and $PR = 13$, find QR.

12. Given $QR = 33.6$ and $m\angle Q = 74$, find PR.

13. Given $PR = 43.7$ and $m\angle P = 24$, find PQ.

14. A surveyor is standing 100 meters from a bridge. She determines that the angle of elevation to the top of the bridge is 35°. The surveyor's eye level is 1.45 meters above the ground. Find the height of the bridge. Round your answer to the nearest hundredth.

15. A ladder leaning against the side of a house forms an angle of 65° with the ground. The foot of the ladder is 8 feet from the building. Find the length of the ladder to the nearest foot.

EXERCISES

Practice

Use the figures below to find each measure. Round your answers to the nearest whole number.

16. $m\angle Y$ 17. $m\angle X$

18. $m\angle Z$ 19. XY

20. $m\angle U$ 21. UV

22. WV 23. $m\angle W$

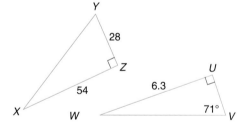

Solve each problem. Round measures of segments to the nearest hundredth and measures of angles to the nearest degree.

24. A surveyor is 100 meters from a building. He finds that the angle of elevation to the top of the building is 23°. If the surveyor's eye level is 1.55 meters above the ground, find the height of the building.

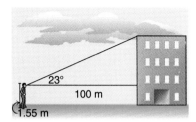

25. To secure a 500-meter radio tower against high winds, guy wires are attached to a ring 5 meters from the top of the tower. The wires form a 15° angle with the tower. Find the distance from the tower to the guy wire anchor in the ground.

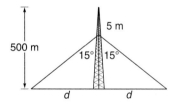

26. Mary is flying a kite on a 50-meter string. The string is making a 50° angle with the ground. How high above the ground is the kite?

27. Bill Owens is an architect designing a new parking garage for the city. The floors of the garage are to be 10 feet apart. The exit ramps between each pair of floors are to be 75 feet long. What is the measurement of the angle of elevation of each ramp?

28. From the top of a lighthouse, the angle of depression to a buoy is 25°. If the top of the lighthouse is 150 feet above sea level, find the distance from the buoy to the foot of the lighthouse.

29. At a certain time of day, the angle of elevation of the sun is 44°. Find the length of a shadow cast by a building 30 meters high.

30. A trolley car track rises vertically 40 feet over a horizontal distance of 630 feet. What is the angle of elevation of the track?

31. Danica is in the observation area of Sears Tower in Chicago overlooking Lake Michigan. She sights two sailboats going due east from the tower. The angles of depression to the two boats are 42 degrees and 29 degrees. If the observation deck is 1,353 feet high, how far apart are the boats?

32. Joe is standing on top of Marina Towers looking at the Leo Burnett World Headquarters building across the Chicago River. It is 880 feet between buildings. Joe finds the angle of elevation to the top of the Burnett building to be 8 degrees and the angle of depression to the ground level to be 20 degrees. How tall is the Burnett building to the nearest foot?

Critical Thinking

33. A fly and an ant are in one corner of a rectangular box. The end of the box is 4 inches by 6 inches and the diagonal across the bottom of the box makes an angle of 21.8° with the longer edge of the box. There is food in the corner opposite the insects.

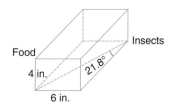

  a. What is the shortest distance the fly must fly to get to the food?

  b. What is the shortest distance the ant must crawl to get to the food?

Applications

34. Travel   A ship sails due north from its home port for 90 kilometers. It then turns east for 40 kilometers before turning north again to sail for 70 kilometers. How far is the ship from its home port?

35. Physics   A pendulum 50 centimeters long is moved 40° from the vertical. How far did the tip of the pendulum rise?

36. Meteorology   Two weather observation stations are 7 miles apart. From Station 1, the angle of elevation to a weather balloon between the stations is 35°. From Station 2, the angle of elevation to the balloon is 54°. Find the altitude of the balloon to the nearest tenth of a mile. *(Hint: Find the distance from Station 2 to the point directly below the balloon.)*

Mixed Review

37. Find cos *A*, sin *A*, and tan *A*. State each as a fraction and a decimal rounded to the nearest thousandth. **(Lesson 8-4)**

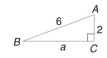

38. A Georgia map is drawn so that 1 centimeter represents 20 miles. If Savannah and Atlanta are 12.7 centimeters apart on the map, how far apart are the cities? **(Lesson 7-1)**

39. *RSTV* is a rhombus with *PT* = 7.6. Find *RT*. **(Lesson 6-5)**

40. Write a two-column proof. **(Lesson 5-1)**
   Given: $\overline{CI} \cong \overline{MI}$
          $\overline{IT}$ is a median of $\triangle CIM$.
   Prove: $\angle CIT \cong \angle MIT$

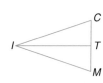

Wrap-Up

41. Write your own application problem that involves trigonometry. Provide the solution with the problem.

8-6 Law of Sines

Objective

After studying this lesson, you should be able to:
- use the law of sines to solve triangles.

Application

A triangular flower bed has one side 8 feet long and one side 11 feet long. The angle opposite the 11-foot side measures 87 °. What are the measures of the other side and angles? *You will solve this problem in Example 2.*

You can use trigonometric functions to solve problems like this that involve triangles that are *not* right triangles. One of the ways is by using the **law of sines**.

Law of Sines

Let $\triangle ABC$ be any triangle with a, b, and c representing the measures of sides opposite angles with measures A, B, and C respectively. Then,

$$\frac{\sin A}{a} = \frac{\sin B}{b} = \frac{\sin C}{c}.$$

Plan for Proof of Law of Sines

$\triangle ABC$ is a triangle with an altitude from C that intersects $\overline{AB}$ at D. Let h represent the measure of $\overline{CD}$. Since $\triangle ACD$ and $\triangle BCD$ are right triangles, we can find $\sin A$ and $\sin B$.

$$\sin A = \frac{h}{b} \qquad \sin B = \frac{h}{a}$$

$$b \sin A = h \qquad a \sin B = h$$

$$b \sin A = a \sin B \qquad \textit{Substitution}$$

$$\frac{\sin A}{a} = \frac{\sin B}{b} \qquad \textit{Divide each side by ab.}$$

The proof can be completed by using a similar technique with another altitude to show that $\frac{\sin A}{a} = \frac{\sin B}{b} = \frac{\sin C}{c}$.

Finding the measures of all the angles and sides of a triangle is called **solving the triangle**. The law of sines can be used to solve a triangle in the following cases.

1. You are given the measures of two angles and any side of a triangle.

2. You are given the measures of two sides and an angle opposite one of these sides of the triangle.

Example 1 asks you to solve a triangle given the measures of two angles and one side of the triangle.

Example 1

Solve $\triangle RPQ$ if $m\angle R = 50$, $m\angle P = 67$, and $r = 10$.

First find $m\angle Q$.

$$m\angle P + m\angle Q + m\angle R = 180 \qquad \textit{The sum of the angles}$$
$$67 + m\angle Q + 50 = 180 \qquad \textit{in a triangle is 180.}$$
$$m\angle Q = 63$$

Next, use the law of sines.

$$\frac{\sin P}{p} = \frac{\sin Q}{q} = \frac{\sin R}{r} \qquad \textit{Law of sines}$$
$$\frac{\sin 67°}{p} = \frac{\sin 63°}{q} = \frac{\sin 50°}{10} \qquad \textit{Substitution}$$

Write two proportions, each involving only one variable.

$$\frac{\sin 67°}{p} = \frac{\sin 50°}{10}$$
$$10 \sin 67° = p \sin 50° \qquad \textit{Cross products.}$$
$$\frac{10 \sin 67°}{\sin 50°} = p$$
$$12.0 \approx p$$

$$\frac{\sin 63°}{q} = \frac{\sin 50°}{10}$$
$$10 \sin 63° = q \sin 50°$$
$$\frac{10 \sin 63°}{\sin 50°} = q$$
$$11.6 \approx q$$

Therefore, $m\angle Q = 63$, $p \approx 12.0$, and $q \approx 11.6$.

Example 2 asks you to solve a triangle given the measures of two sides and an angle opposite one of those sides.

Example 2

A triangular flower bed has one side 8 feet long and one side 11 feet long. The angle opposite the 11-foot side measures $87°$. Find the measures of the other side and angles.

Draw a diagram.

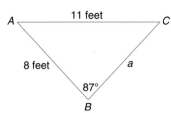

Use the law of sines.

$$\frac{\sin A}{a} = \frac{\sin 87°}{11} = \frac{\sin C}{8}$$

We can solve for $m\angle C$.

$$\frac{\sin 87°}{11} = \frac{\sin C}{8}$$
$$8 \sin 87° = 11 \sin C \qquad \textit{Cross products.}$$
$$\frac{8 \sin 87°}{11} = \sin C$$

Use your calculator to find $m\angle C$.

ENTER: 8 ⊠ 87 ⟨SIN⟩ ⟨÷⟩ 11 ⟨=⟩ ⟨INV⟩ ⟨SIN⟩ 46.575101

$m\angle C \approx 47$

Since the sum of the measures of the angles in a triangle is 180, we can find $m\angle A$.

$$m\angle A + m\angle B + m\angle C = 180$$
$$m\angle A + 87 + 47 \approx 180$$
$$m\angle A \approx 46$$

Now solve for a.

$$\frac{\sin 46°}{a} = \frac{\sin 87°}{11}$$
$$11 \sin 46° = a \sin 87°$$
$$\frac{11 \sin 46°}{\sin 87°} = a$$

Use your calculator to find a.

ENTER: 11 ⊗ 46 [SIN] ÷ 87 [SIN] [=] 7.9235968

$a \approx 8$

The missing angle measures are 47° and 46°, and the missing side measures 8 feet.

CHECKING FOR UNDERSTANDING

Communicating Mathematics

Read and study the lesson to answer these questions.

1. Draw $\triangle KLM$ and state the relationships known from the law of sines.
2. What kind of triangles can be solved using the law of sines?
3. There are two different cases when it is appropriate to use the law of sines. What are they?

Guided Practice

Use the given information about $\triangle ABC$ to write an equation that could be used to find each unknown value. Draw $\triangle ABC$ and mark it with the given information.

4. If $b = 4.7$, $m\angle A = 22$, and $m\angle B = 49$, find a.
5. If $b = 10$, $a = 14$, and $m\angle A = 50$, find $m\angle B$.
6. If $m\angle A = 40$, $m\angle B = 60$, and $a = 20$, find b.
7. If $c = 12$, $b = 16$, and $m\angle B = 42$, find $m\angle C$.

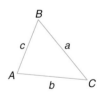

Solve each triangle *ABC*. Round measures of sides to the nearest tenth and angle measures to the nearest degree.

8. $m\angle B = 70, m\angle C = 58, a = 84$ **9.** $a = 17, b = 10, m\angle A = 106$

10. $a = 8, m\angle A = 49, m\angle B = 57$ **11.** $m\angle A = 40, m\angle B = 60, c = 20$

12. An isosceles triangle has a base of 22 centimeters and a vertex angle of 36°. Find its perimeter.

EXERCISES

Practice

Use the given information about $\triangle ABC$ to write an equation that could be used to find each value. Then find the value.

13. If $b = 2.8, m\angle A = 53$, and $m\angle B = 61$, find a.

14. If $b = 36, c = 12$, and $m\angle B = 98$, find $m\angle C$.

15. If $m\angle A = 70, m\angle B = 23$, and $c = 2.2$, find a.

16. If $m\angle C = 55, a = 9$, and $c = 11$, find $m\angle A$.

Solve each triangle *ABC*. Round measures of sides to the nearest tenth and measures of angles to the nearest degree.

17. $m\angle C = 70, c = 8, m\angle A = 30$

18. $a = 10, c = 25, m\angle C = 124$

19. $m\angle A = 29, m\angle B = 62, c = 11.5$

20. $m\angle C = 35, a = 7.5, c = 24$

21. $m\angle B = 36, m\angle C = 119, b = 8$

22. $m\angle B = 47, m\angle C = 73, a = 0.9$

23. $b = 20, c = 9.2, m\angle B = 103$

24. $a = 12, b = 14, m\angle B = 95$

25. A house is built on a triangular plot of land. Two sides of the plot are 160 feet long and they meet at an angle of 85°. If a fence is to be built around the property, how much fencing material is needed?

26. The longest side of a triangle is 34 feet. The measures of two angles of the triangle are 40 and 65. Find the lengths of the other two sides.

27. A ship is sighted at sea from two observation points on the coastline that are 30 miles apart. The angle formed by the coastline and the line between the ship and the first observation point measures 34°. The angle formed by the coastline and the line between the ship and the second observation point measures 45.6°. How far is the ship from the first observation point?

28. The 35-foot flagpole in front of the Stevenson High School stands on a uniformly sloped mound. When the angle of elevation of the sun is 37.2°, the shadow of the pole ends at the base of the mound. If the mound rises at an angle of 6.7°, find the length of the shadow.

29. Does the law of sines hold true for the acute angles of right triangles? Justify your answer.

30. **Aviation** Two airplanes leave Port Columbus International Airport at the same time. Each plane flies at a speed of 310 miles per hour. One flies in the direction 60° east of north. The other flies in the direction 40° east of south.
    a. Draw a diagram of the situation.
    b. How far apart are the planes after 3 hours?

31. **Surveying** The support for a powerline that stands on top of a hill is 60 feet tall. A surveyor stands at a point on the hill and observes that the angle of elevation to the top of the support measures 42° and to the base of the support measures 18°. How far is the surveyor from the base of the support?

32. **Surveying** A surveyor is 100 meters from the base of a dam. The angle of elevation to the top of the dam measures 26°. The surveyor's eye-level is 1.73 meters above the ground. Find the height of the dam to the nearest hundredth of a meter. **(Lesson 8-5)**

33. In the figure below, $\triangle XYZ \sim \triangle XWV$. The perimeter of $\triangle XYZ$ is 34 inches, and the perimeter of $\triangle XWV$ is 14 inches. If $WV = 4$, find YZ. **(Lesson 7-6)**

CONSTRUCTION

34. Use a compass and straightedge to construct a square with sides 6 centimeters long. **(Lesson 6-5)**

35. Write a two-column proof. **(Lesson 4-6)**

    Given: $\overline{AB} \perp \overline{BD}$
    $\overline{DE} \perp \overline{DB}$
    $\overline{DB}$ bisects $\overline{AE}$.

    Prove: $\angle A \cong \angle E$

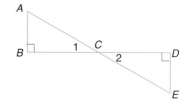

36. Draw a triangle to represent each of the two cases where it is appropriate to use the law of sines to solve a triangle. For each case, set up the proportions you would use to solve the triangle.

Law of Cosines

Objective

After studying this lesson, you should be able to:

- solve triangles and problems using the law of cosines.

Application

stride angle

Hope runs for the Grandview High School cross country team and hopes to participate in an upcoming city marathon. Recently she read of a study of runners' strides in an issue of *Runners' World*. According to the study, marathon runners run most efficiently with a *stride angle* of about 100°. The stride angle is the largest angle formed by the leading and trailing legs. Hope decided to determine if she could be running more efficiently.

A coaching staff usually uses sophisticated video equipment and computer graphics to measure a runner's stride angle. But since none of this equipment was available, Hope decided to use trigonometry to make an estimate.

First, Hope measured her legs and found that they are each 34 inches long. Next, she needed to find the length of one pace. A pace is the distance between the place where one foot is placed and where the other foot comes down. To find her pace, Hope counted the number of steps she took while running 100 yards. She took 72 steps, so each pace was $\frac{100}{72}$ yards or about 50 inches.

Hope knows the lengths of the three sides of the triangle formed by her legs and one pace. Now she can use the **law of cosines** to find the measure of her stride angle.

The law of cosines allows us to solve a triangle when the law of sines cannot be used.

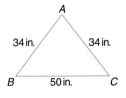

Law of Cosines

Let $\triangle ABC$ be any triangle with a, b, and c representing the measures of sides opposite angles with measures A, B, and C respectively. Then, the following equations hold true.

$$a^2 = b^2 + c^2 - 2bc \cos A$$
$$b^2 = a^2 + c^2 - 2ac \cos B$$
$$c^2 = a^2 + b^2 - 2ab \cos C$$

To find the measure of Hope's stride angle, let $b = 34$, $c = 34$, and $a = 50$. We need to find $m\angle A$.

$$a^2 = b^2 + c^2 - 2bc \cos A$$
$$50^2 = 34^2 + 34^2 - 2(34)(34) \cos A \qquad \textit{Substitute}$$
$$2500 = 1156 + 1156 - 2312 \cos A$$
$$188 = \text{-}2312 \cos A$$
$$\text{-}0.0813 = \cos A$$
$$94.7 \approx A \qquad \textit{Use your calculator.}$$

Hope's stride angle is about 95°. So, she should take longer strides to run more efficiently. *How much longer should her stride be?*

The law of cosines can be used to solve a triangle in the following cases.

1. To find the measure of the third side of any triangle if the measures of the two sides and the included angle are given.

2. To find the measure of an angle of a triangle if the measures of the three sides are given.

Example 1

Solve $\triangle ABC$ where $m\angle A = 35$, $b = 16$, and $c = 19$. Round the measures of sides to the nearest hundredth and the measures of angles to the nearest tenth of a degree.

Draw and label the triangle.

Now, determine a using the law of cosines.

$$a^2 = b^2 + c^2 - 2bc \cos A \qquad \textit{Law of cosines}$$
$$a^2 = 16^2 + 19^2 - 2(16)(19) \cos 35° \qquad \textit{A = 35, b = 16, c = 19}$$
$$a^2 \approx 118.96$$
$$a \approx 10.91$$

Next use the law of sines to determine a second angle.

$$\frac{\sin A}{a} = \frac{\sin B}{b} \qquad \textit{Law of sines}$$
$$\frac{\sin 35°}{10.91} = \frac{\sin B}{16} \qquad \textit{A = 35, a} \approx \textit{10.91, b = 16}$$
$$16 \sin 35° = 10.91 \sin B \qquad \textit{Cross products.}$$
$$\sin B \approx 0.8412$$
$$B \approx 57.3$$

Finally, determine the measure of the third angle.

$$m\angle A + m\angle B + m\angle C = 180$$
$$35 + 57.3 + m\angle C \approx 180 \qquad \textit{m}\angle\textit{A = 35, m}\angle\textit{B} \approx \textit{57.3}$$
$$m\angle C \approx 87.7$$

So, $a \approx 10.91$, $m\angle B \approx 57.3$, and $m\angle C \approx 87.7$.

Example 2

APPLICATION

Aviation

Kristin Burrows is flying from Orlando to Miami, a distance of about 220 miles. In order to avoid bad weather, she starts her flight 12 degrees off course and flies on this course for 75 miles. How far is she from Miami?

First, draw a diagram that represents the situation.

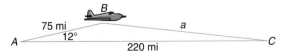

Since you know two sides and the included angle of the triangle, use the law of cosines.

$$a^2 = b^2 + c^2 - 2bc \cos A$$
$$a^2 = 220^2 + 75^2 - 2(220)(75) \cos 12° \qquad m\angle A = 12, b = 220, \text{ and } c = 75$$
$$a^2 \approx 21746.13$$
$$a \approx 147.47$$

Ms. Burrows is about 147 miles from Miami.

CHECKING FOR UNDERSTANDING

Communicating Mathematics

Read and study the lesson to answer these questions.

1. Describe the two cases when it is appropriate to use the law of cosines.

2. Create a problem with a minimum of given information for which you would use the law of cosines. Solve the problem.

3. Compare and contrast the cases where you would use the law of sines, the law of cosines, and the trigonometric ratios, sine, cosine, and tangent, to solve a triangle.

Guided Practice

Determine whether the law of sines or the law of cosines should be used first to solve each triangle. Then solve each triangle. Round measures of sides to the nearest tenth and measures of angles to the nearest degree.

4.

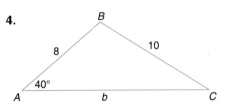

5.

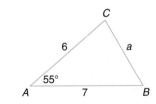

Sketch each triangle described and determine whether the law of sines or the law of cosines should be used first to solve the triangle. Then solve each triangle. Round measures of sides to the nearest tenth and measures of angles to the nearest degree.

6. $a = 14, c = 21, m\angle B = 60$ **7.** $a = 14, b = 15, c = 16$

8. $m\angle A = 51, a = 40, c = 35$ **9.** $a = 5, b = 6, c = 7$

10. $a = 140, b = 185, m\angle B = 66$ **11.** $a = 21.5, b = 13, m\angle C = 78$

EXERCISES

Practice

Determine whether the law of sines or the law of cosines should be used first to solve each triangle described below. Then solve each triangle. Round the measures of sides to the nearest tenth and the measures of angles to the nearest degree.

12. $m\angle A = 40, m\angle C = 70, c = 4$ **13.** $a = 11, b = 10.5, m\angle C = 35$

14. $a = 11, b = 17, m\angle B = 42$ **15.** $m\angle A = 56, m\angle C = 26, c = 12.2$

Solve each triangle described below.

16. $a = 51, b = 61, m\angle B = 19$ **17.** $a = 5, b = 12, c = 13$

18. $a = 20, c = 24, m\angle B = 47$ **19.** $m\angle A = 40, m\angle B = 59, c = 14$

20. $a = 345, b = 648, c = 442$ **21.** $m\angle A = 29, b = 5, c = 4.9$

22. $a = 8, m\angle A = 17, m\angle B = 71$ **23.** $c = 10.30, a = 21.50, b = 16.71$

24. $m\angle A = 29, b = 7, c = 14.1$ **25.** $a = 8, b = 24, c = 18$

26. Two sides of a triangular plot of land have lengths of 400 feet and 600 feet. The angle formed by those sides measures 46.3°. Find the perimeter of the plot to the nearest foot.

27. The measures of the sides of a triangle are 6.8, 8.4, and 4.9. Find the measure of the smallest angle to the nearest degree.

28. The sides of a parallelogram are 55 cm and 71 cm long. Find the length of each diagonal if the larger angle measures 106°.

29. Circle Q has a radius of 15 centimeters. Two radii, $\overline{QA}$ and $\overline{QB}$, form an angle of 123°. Draw a diagram of the situation and find the length of $\overline{AB}$ to the nearest centimeter.

30. The sides of a triangle are 50 meters, 70 meters, and 85 meters long. Find the measure of the angle opposite the longest side to the nearest degree.

31. A 40-foot television antenna stands on top of a building. From a point on the ground, the angles of elevation to the top and bottom of the antenna measure 56° and 42° respectively. How tall is the building?

32. A ship at sea is 70 miles from one radio transmitter and 130 miles from another. The angle formed by the signals measures 130°. How far apart are the transmitters?

Critical Thinking

33. In $\triangle ABC$, $m\angle A = 50$, $m\angle B = 70$, and $m\angle C = 60$. Is it possible to find the measures of the sides of $\triangle ABC$ using either the law of sines, the law of cosines, or the trigonometric ratios? If it is possible, then find the sides. If it is not possible, then explain why not.

34. Justify the $a^2 = b^2 + c^2 - 2bc \cos A$ portion of the law of cosines by using the Pythagorean Theorem for $\triangle BDC$. *(Hint: $\cos A = \frac{x}{c}$, so $x = c \cos A$.)*

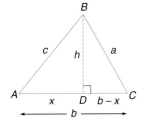

Applications

35. Navigation Two ships, the *Western Princess* and the *Hoggatt Bay*, left Savannah at 10:00 A.M. The *Princess* traveled 30° north of east at a speed of 24 knots. The *Hoggatt Bay* traveled 15° east of south at a speed of 18 knots. How far apart will the ships be at noon? *(1 knot = 1 nautical mile per hour)*

36. Aviation Eli Cooley flew his plane 900 kilometers north before turning 15° clockwise. He flew 1150 kilometers in that direction and then landed. How far is Mr. Cooley from his starting point?

37. Sports In golf, a *slice* is a shot to the right of its intended path (for a right-handed player) and a *hook* is off to the left. Peg's drive from the third tee is a 180-yard slice 12° from the path straight to the cup. If the tee is 240 yards from the cup, how far does Peg's ball lie from the cup?

Mixed Review

38. Two angles of a triangle measure 40° and 56°. If the longest side of the triangle is 38 cm long, find the length of the shortest side. Round your answer to the nearest hundredth. **(Lesson 8-6)**

39. Find the geometric mean between 8 and 18. **(Lesson 8-1)**

40. In square $LMNP$, $LN = 3x - 2$ and $MP = 2x + 3$. Find LN. **(Lesson 6-5)**

41. Is it possible to have a triangle with vertices at $K(5, 8)$, $L(0, -4)$, and $M(-1, 1)$? Explain your answer. **(Lesson 5-6)**

42. Write a paragraph proof. **(Lesson 4-7)**

Given: $\overline{AB} \cong \overline{BC}$

Prove: $\angle 3 \cong \angle 4$

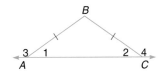

Wrap-Up

43. Write three questions that could be used for a quiz on this lesson. Be sure to provide the answers to your questions.

8-8

Problem-Solving Strategy: Decision-Making

Objective

After studying this lesson, you should be able to:

- choose the appropriate strategy for solving a problem.

Most problems can be solved in more than one way. Choosing the most efficient way is often not obvious. When deciding which problem-solving strategy to use, ask yourself these questions.

- Is the problem similar to one I have solved before? If so, what strategy worked best on that problem?

- What problem-solving strategies do I know? Does one of these seem appropriate?

- Does the problem itself suggest a strategy? Does the given information or the unknown imply which strategy will work best? For example, does it ask for several answers? Then listing the possibilities is probably a good strategy to try.

Application

Simon cashed his paycheck and deposited half of the money in his savings account. He gave his sister the $20 he had borrowed and then spent $5 on lunch. After spending $\frac{1}{5}$ of the remaining money on a movie ticket, Simon had $12 left. How much was Simon's paycheck?

Look over the list of problem-solving strategies on the next page. Many of these strategies may work. We could solve a simpler problem, act it out, guess and check, or work backward. We could also write equations for each step and solve for the amount of Simon's paycheck.

Looking at the problem itself, we see that we are given the final amount of money and are asked the beginning amount. Based on this information, working backward seems to be the most appropriate strategy.

Simon had $12 after spending $\frac{1}{5}$ of the remaining money on a movie ticket, so $12 is $\frac{4}{5}$ of that amount. Before buying the movie ticket, Simon had $15 since $\frac{4}{5}$ of 15 is 12.

There was $15 left after Simon paid his sister and had lunch. So, he had 15 + 20 + 5 or $40 before then.

Half of Simon's paycheck went into his savings account and the other half, $40, remained. So, Simon's paycheck was 2 · $40 or $80.

CHECKING FOR UNDERSTANDING

Communicating Mathematics

Read and study the lesson to answer these questions.

1. Is there always only one way to solve a problem?

2. How can you decide which problem-solving strategy is the best one to try?

Guided Practice

List the problem-solving strategies that you could use to solve each problem. Then choose the best strategy and solve.

3. The pyramid of cans shown at the right has four layers with a total of 20 cans. How many cans are there in a similar pyramid of ten layers?

4. A quarter remains still while a second quarter is rolled around it without slipping. How many times does the second coin rotate around its own axis?

EXERCISES

Solve. Use any strategy.

Strategies

Look for a pattern.
Solve a simpler problem.
Act it out.
Guess and check.
Draw a diagram.
Make a chart.
Work backward.

5. Find the next term in the sequence 0, 2, 10, 42, 170, 682. Describe the pattern.

6. What is the least number of coins you can have and be able to pay for any purchase that is less than $1?

7. You have a seven-minute and a three-minute egg timer. How could you time the boiling of spaghetti for eleven minutes?

8. Find the next number in the sequence. 3, 6, 11, 18, 27. Describe the pattern.

9. Mrs. Sterling's will states that three fourths of her estate will be left to her son, and $24,000 will go to her niece. Half of the remainder will go to her church. One third of the money remaining after the church donation will go to her alma mater and the remaining $4000 will pay Mrs. Sterling's attorney. How much is Mrs. Sterling's estate worth?

10. Complete the multiplication problem at the right. Each digit 1-9 is used exactly once.

$$\begin{array}{r} 2\ \blacksquare\ \blacksquare \\ \times\ \underline{\ 1\ \blacksquare} \\ 5\ \blacksquare\ \blacksquare\ \blacksquare \end{array}$$

11. The area of the bottom of a Health One Cereal box is 95 cm². The area of a side is 140 cm² and the area of the front is 532 cm². What is the volume of the box in cubic centimeters?

12. A square results when the length of a rectangle is reduced by 20% and the width is increased by 20%. How does the area of the rectangle compare to the area of the square? Express your answer as a percentage.

13. There are between 50 and 100 stores in the Valley View Mall. Exactly 20 percent of the stores are shoe stores, and exactly one-seventh of the stores are toy stores. Determine how many stores there are in the mall. Explain your solution.

14. The average height of the players of the Palatine High School boys basketball team is 6 feet 3 inches. The average height of the varsity players is 6 feet 8 inches, and the average height of the junior-varsity players is 6 feet. What fraction of the basketball team are varsity players?

∼ COOPERATIVE LEARNING PROJECT ∼

Work in groups. Each person in the group must understand the solution and be able to explain it to any person in class.

If the length of a side of a square is s, then the area of the square is $A = s \cdot s$ or s^2. In the diagram at the right, the area of square H is 64 square units and the area of square F is 49 square units. Find the areas of the other seven squares.

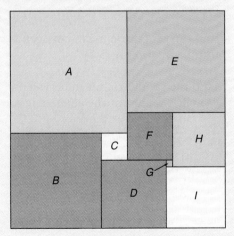

VOCABULARY

Upon completing this chapter, you should be
familiar with the following terms:

angle of depression	**385**	**365**	Pythagorean Theorem
angle of elevation	**384**	**376**	sine
cosine	**376**	**376**	tangent
geometric mean	**360**	**376**	trigonometric ratio
law of cosines	**394**	**376**	trigonometry
law of sines	**389**		

SKILLS AND CONCEPTS

OBJECTIVES AND EXAMPLES	REVIEW EXERCISES

Upon completing this chapter, you should
be able to:

Use these exercises to review and prepare
for the chapter test.

- find the geometric mean between two
  numbers. **(Lesson 8-1)**

Find the geometric mean between 28 and
44. Let x represent the geometric mean.

$\dfrac{28}{x} = \dfrac{x}{44}$ *Definition of geometric mean*

$x^2 = 1232$ *Cross products*

$x = \sqrt{1232}$

$x \approx 35.1$

The geometric mean between 28 and 44 is
$\sqrt{1232}$ or about 35.1.

**Find the geometric mean for each pair of
numbers.**

1. 12 and 27   **2.** 9 and 27

3. 60 and 52   **4.** 20 and 75

5. 1 and 4   **6.** 13 and 39

7. 99 and 121   **8.** 8 and 6

9. m and n **10.** $4p$ and $16p$

OBJECTIVES AND EXAMPLES	REVIEW EXERCISES

- solve problems involving relationships between parts of a right triangle and an altitude. **(Lesson 8-1)**

If $\triangle ABC$ is a right triangle with altitude $\overline{BD}$, then the following relationships hold true.

$$\frac{AD}{BD} = \frac{BD}{DC}$$

$$\frac{AC}{BC} = \frac{BC}{DC}$$

$$\frac{AC}{AB} = \frac{AB}{AD}$$

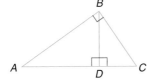

Use right triangle *KLM* and the given information to solve each problem.

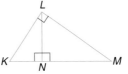

11. Find LN if $KN = 4$ and $NM = 6$.

12. Find KL if $KM = 18$ and $KN = 4$.

13. Find LM if $KN = 7$ and $NM = 3$.

14. Find LK if $KN = \frac{1}{3}$ and $NM = \frac{1}{4}$.

15. Find KM if $LM = 19$ and $NM = 14$.

16. Find KN if $LK = 0.6$ and $KM = 1.5$.

- use the Pythagorean Theorem and its converse. **(Lesson 8-2)**

A right triangle has a hypotenuse 61 inches long and a leg 11 inches long. Find the length of the other leg.

$$a^2 + b^2 = c^2$$
$$a^2 + 11^2 = 61^2$$
$$a^2 + 121 = 3721$$
$$a^2 = 3600$$
$$a = 60$$

The other leg is 60 inches long.

Find the measure of the hypotenuse of a right triangle with legs of the given measure. Round your answers to the nearest tenth.

17. 7.1, 6.7

18. 9.4, 8.0

19. 8.0, 15.0

20. 94, 88

Determine whether a triangle with sides having the given measures is a right triangle.

21. 9, 21, 23

22. 4, 7.5, 8.5

23. 17, 144, 145

24. 19, 24, 30

- use the properties of 30°-60°-90° and 45°-45°-90° triangles. **(Lesson 8-3)**

In a 45°-45°-90° triangle, the hypotenuse is $\sqrt{2}$ times as long as a leg.

In a 30°-60°-90° triangle, the hypotenuse is twice as long as the shorter leg and the longer leg is $\sqrt{3}$ times as long as the shorter leg.

Find the value of *x*.

25.

26.

27.

28.

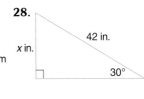

OBJECTIVES AND EXAMPLES	REVIEW EXERCISES

■ find and use trigonometric ratios.
(Lesson 8-4)

$\sin A = \dfrac{a}{c}$

$\cos A = \dfrac{b}{c}$

$\tan A = \dfrac{a}{b}$

Find the indicated trigonometric ratio as a fraction and as a decimal rounded to the nearest thousandth.

29. $\sin Q$

30. $\tan Q$

31. $\cos R$

32. $\tan R$

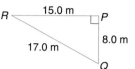

■ use the law of sines to solve triangles.
(Lesson 8-6)

According to the law of sines,

$$\dfrac{\sin A}{a} = \dfrac{\sin B}{b} = \dfrac{\sin C}{c}$$

Use the law of sines to solve each △ABC.

33. $m\angle B = 46$, $m\angle C = 83$, $b = 65$

34. $a = 80$, $b = 10$, $m\angle A = 65$

35. $a = 4.2$, $b = 6.8$, $m\angle B = 22$

■ use the law of cosines to solve triangles.
(Lesson 8-7)

According to the law of cosines,

$a^2 = b^2 + c^2 - 2bc \cos A$
$b^2 = a^2 + c^2 - 2ac \cos B$
$c^2 = a^2 + b^2 - 2ab \cos C$

Use the law of cosines to solve each △ABC.

36. $m\angle C = 55$, $a = 8$, $b = 12$

37. $a = 44$, $c = 32$, $m\angle B = 44$

38. $m\angle C = 78$, $a = 4.5$, $b = 4.9$

39. $a = 6$, $b = 9$, $c = 8$

Applications and Connections

40. **Recreation**   Dana is flying a kite whose string is making a 70° angle with the ground. The kite string is 65 meters long. How far is the kite above the ground? **(Lesson 8-5)**

41. **Transportation**   A railroad track rises 30 feet for every 400 feet of track. What is the measure of the angle the track makes with the horizontal? **(Lesson 8-5)**

42. **Navigation**   The top of a lighthouse is 120 meters above sea level. The angle of depression from the top of the lighthouse to a ship is 23°. How far is the ship from the foot of the lighthouse? **(Lesson 8-5)**

43. **Aviation**   Jim Paul flew his airplane 1000 kilometers north before turning 20° clockwise and flying another 700 kilometers. How far is Mr. Paul from his starting point? **(Lesson 8-7)**

44. **Algebra**   Which problem-solving strategies might you use to find the remainder for $5^{100} \div 7$? Find the value of the remainder for $5^{100} \div 7$. **(Lesson 8-8)**

Find the geometric mean for each pair of numbers.

1. 3 and 12   **2.** 5 and 4   **3.** 28 and 56

Use the figure below and the given information to solve each problem. Round your answers to the nearest tenth.

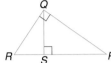

4. Find QS if $PS = 8$ and $SR = 5$.
5. Find QP if $SP = 9.5$ and $SR = 3$.

Find the measure of the hypotenuse of a right triangle with legs of the given measures. Round your answers to the nearest tenth.

6. 39, 80   **7.** 1.5, 11.2   **8.** 6.9, 7.2   **9.** 14.7, 18.1

Find the value of *x*.

10.

11.

12.

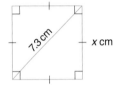

Find the indicated trigonometric ratio as a fraction and as a decimal rounded to the nearest thousandth.

13. sin A **14.** tan B
15. cos A **16.** tan A

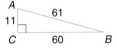

17. Fire Fighting A fire fighter's 36-foot ladder leans against a building. The top of the ladder touches the building 28 feet above the ground. What is the measure of the angle the ladder forms with the ground?

18. The longest side of a triangle is 30 centimeters long. Two of the angles have measures of 45 and 79. Find the measures of the other two sides and the remaining angle.

19. Gemology A jeweler is making a sapphire earring in the shape of an isosceles triangle with sides 22, 22, and 28 millimeters long. What is the measure of the vertex angle?

20. Algebra Which problem-solving strategy might you use to find the fraction of the odd whole numbers less than 100 that are perfect squares? Find the fraction.

Bonus Find sin A, cos A, and tan A.

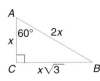

College Entrance Exam Preview

Directions: Choose the one best answer. Write A, B, C, or D. You may use a calculator.

1. In $\triangle ABC$, $AB = 8$ and $BC = 12$. Which one of the following cannot be the measure of $\overline{AC}$?

(A) 5   (B) 8
(C) 12   (D) 20

2. $\dfrac{1}{x} - \dfrac{2}{y} =$

(A) $-\dfrac{1}{xy}$ (B) $\dfrac{y - 2x}{xy}$
(C) $\dfrac{-1}{x - y}$ (D) $\dfrac{1}{y - x}$

3.

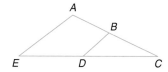

If $\overline{AE} \parallel \overline{BD}$, $AB = 10$, $BC = x$, $ED = x + 3$, and $DC = x + 6$, then $x =$

(A) 6   (B) 7
(C) 12   (D) 15

4. The graph of $2x - y - 8 = 0$ crosses the x-axis at $x =$

(A) 4   (B) 8
(C) -2   (D) -4

5. If $x - \dfrac{2}{x - 3} = \dfrac{x - 1}{3 - x}$, then $x =$

(A) 3 or -1   (B) 3 or -3
(C) -1   (D) 3

6. Find all a such that $|-a| = 7$.

(A) 7   (B) -7
(C) 7 and -7   (D) no such a exists

7. If $\triangle ACB \cong \triangle STU$, then
   I. $\overline{AB} \cong \overline{SU}$
   II. $\angle C \cong \angle S$
   III. $\overline{AC} \cong \overline{TS}$

(A) I only

(B) II only

(C) III only

(D) I and III only

8.

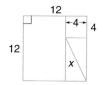

Which inequality gives the best approximation for x?

(A) $8 < x < 9$ (B) $9 < x < 10$
(C) $10 < x < 11$ (D) $11 < x < 12$

9. The measures of the sides of four triangles are given. Which one is not a right triangle?

(A) 5, 12, 13   (B) 8, 15, 17
(C) 12, 15, 18   (D) 9, 40, 41

10. The ninth term of the sequence 5, 6, 8, 11, 15, 20, 26, 33, . . . is

(A) 60   (B) 51
(C) 48   (D) 41

11. @ is defined so that $x @ y = x^2 + xy$. The value of $8 @ 2$ is

(A) 20   (B) 80
(C) 48   (D) -12

12. The ratio of $\dfrac{1}{3}$ to $\dfrac{5}{12}$ is

(A) 4 to 5   (B) 1 to 4
(C) 5 to 4   (D) 5 to 36

Solve each of the following. You may use a calculator.

13. Kelsey has taken four tests in English class this semester. Her scores were 82, 81, 79, and 87. A student must have a test average of 85 to receive a B in the class. What must Kelsey score on the next test to receive a B?

14. In the figure below, $\overline{AB}$ and $\overline{CD}$ are parallel line segments. Find the area of trapezoid *ABCD*.

15. Right triangles *ABC* and *DEF* are similar. If $BC = 9$, $AC = 21$, and $EF = 24$, find *DF*.

16. Mr. Pearson can wash his car in 30 minutes. His daughter Jan can do the job in 20 minutes. How long will it take them to wash the car if they work together?

17. $^\star x$ is defined as $^\star x = x^2 + 5x$. Find the value of $^\star 5 - {^\star}(-1)$.

18. One tenth of the water in a fish bowl evaporates in the first day after it is filled. The second day, one-twelfth of the remaining water evaporates. What fraction of the original amount of water remains after the second day?

19. The corner is cut from a rectangular piece of cardboard as shown below. Find the area of the remaining cardboard.

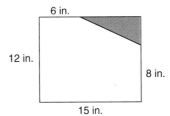

20. Two cars leave from the same place at 8:45 A.M. If both cars travel in the same direction, one car travels at 55 mph, and the other travels at 62 mph, how far apart are they at 11:15 A.M.?

CHAPTER 9

Circles

GEOMETRY AROUND THE WORLD
Kenya

What shape is your home or apartment? If you're like most Americans, its shape is basically square or rectangular. The sharp angles that characterize our buildings aren't the only home designs around, however. If you were a member of the Kikuyu, Zulu, or any of several other African tribes, your house would be round.

Because you probably haven't seen one, a circular home may sound strange. However, in areas where building materials are scarce, as they are in many parts of Africa, the circle design makes perfect sense. Why? Of all the closed geometric shapes, a circle encompasses the greatest area within a given perimeter. For this reason, U.S. architects often use a variation of a circular design called a geodesic dome to cover sports stadiums and other large public spaces.

Can you guess what geometric pattern was used to lay out the design of many African villages? A circle! Many old Zulu towns were built in a circle of 1400 round structures four or five buildings deep.

GEOMETRY IN ACTION

Like the Zulu, the Kikuyu, who live in the foothills of Mount Kenya, build comfortable, sturdy, round homes with thatched roofs. Before erecting their homes, the Kikuyu mark a circle 14 feet in diameter on the ground. Then they dig 19 holes for roof-supporting posts equal distances apart, with a wider space allowed for the door.

Turn to Lesson 9-1 of this chapter and use what you learn to write the equation for the circle of a Kikuyu home. Assume that the center of the circle is at (0, 0).

Because their homes are round, Kikuyu children who misbehave never have to stand in the corner!

◀ *Kikuyu village*   Inset: *Member of Zulu tribe*

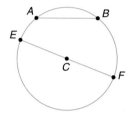

9-1 Parts of Circles

Objectives After studying this lesson, you should be able to:

- name parts of circles,
- determine relationships between lines and circles, and
- write an equation of a circle in the coordinate plane.

Application To irrigate the fields shown in the photograph, water is sprayed from pipes that rotate about central points. Circular patterns result. All the points along the edge of a field are the same distance from the center. Each edge forms a **circle.**

A circle is the set of all points in a plane that are a given distance from a given point in the plane called the **center.** A circle is named by its center. The circle above is called circle C. This is symbolized $\odot C$.

A **chord** of a circle is a segment that has its endpoints on the circle. $\overline{AB}$ is a chord of $\odot C$.

FYI ···

Around 1700 B.C., early Peruvians reclaimed and farmed an area of desert. They used an extensive irrigation system to reroute the Chillon River.

A **diameter** of a circle is a chord that contains the center of the circle. $\overline{EF}$ is a diameter of $\odot C$. A **radius** of a circle is a segment with one endpoint at the center of the circle and the other endpoint on the circle. $\overline{EC}$ and $\overline{CF}$ are radii of $\odot C$. *Radii is the plural of radius.*

It follows from the definition of a circle that all the radii of a circle are congruent. Also, all diameters of a circle are congruent.

You should recall that the measure of the diameter, d, is twice the measure of the radius, r. Formulas that relate these measures are

$$d = 2r \text{ and } r = \tfrac{1}{2}d \text{ or } r = \tfrac{d}{2}.$$

Example 1

APPLICATION

Agriculture

If the spray of water in the photograph at the top of the page can reach 90 feet, what is the diameter of each circle in the photograph?

$$d = 2r$$
$$d = 2 \cdot 90 \qquad \textit{Substitute 90 for r.}$$
$$d = 180$$

The diameter of each circle is 180 feet.

Example 2

Find the radius and the diameter of the circle shown at the right.

$\odot C$ has center at (3, 4). The circle contains points $A(3, 10)$, $B(3, -2)$, $L(-3, 4)$, and $M(9, 4)$. Chord $\overline{LM}$ is a diameter since it contains the center C. $\overline{LM}$ has a length of 12 units. *Why?*

Therefore, the diameter is 12 units long, and the radius must be 6 units long.

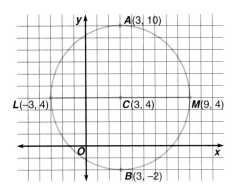

You can use the distance formula to write an equation for a circle given its center and radius. The circle at the right has its center at $(-1, 4)$ and a radius of 5 units. Let $P(x, y)$ be any point on C. Then $\overline{CP}$ is a radius of the circle. The distance between $P(x, y)$ and $C(-1, 4)$ is 5 units.

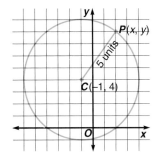

$$PC = 5$$
$$\sqrt{(x - (-1))^2 + (y - 4)^2} = 5 \qquad \textit{Distance formula}$$
$$(x + 1)^2 + (y - 4)^2 = 25 \qquad \textit{Square each side.}$$

Thus, an equation for the circle with center at $(-1, 4)$ and a radius of 5 units is $(x + 1)^2 + (y - 4)^2 = 25$.

Standard Equation of a Circle	**In general, an equation for a circle with center at (h, k) and a radius of r units is $(x - h)^2 + (y - k)^2 = r^2$.**

Example 3

Write an equation for a circle with center at (2, -3) and a diameter of 10 units.

Since $d = 10$, it follows that $r = 5$.

$$(x - h)^2 + (y - k)^2 = r^2 \qquad \textit{General equation of a circle}$$
$$(x - 2)^2 + (y - (-3))^2 = 5^2 \qquad \textit{(h, k) = (2, -3); r = 5}$$
$$(x - 2)^2 + (y + 3)^2 = 25$$

An equation for the circle is $(x - 2)^2 + (y + 3)^2 = 25$.

A circle separates a plane into three parts. The parts are the **interior**, the **exterior**, and the **circle** itself.

Suppose point *I* is in the interior of circle *P*. The measure of the segment joining *I* to *P* is less than the measure of the radius; that is, $IP < r$.

Suppose point *E* is in the exterior of circle *P*. The measure of the segment joining *E* to *P* is greater than the measure of the radius; that is, $EP > r$.

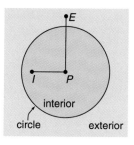

Example 4

The equation of $\odot Q$ is $x^2 + (y + 2)^2 = 16$.

a. Find the radius and the coordinates of the center, *Q*.

The equation for a circle with center at (h, k) and a radius *r* units long is $(x - h)^2 + (y - k)^2 = r^2$. The equation for $\odot Q$ can be expressed as $(x - 0)^2 + (y - (-2))^2 = 4^2$. So, the radius is 4 units long and the coordinates of *Q* are (0, -2).

b. Use the distance formula to determine if *T*(4, 3) is on $\odot Q$, in its interior, or in its exterior.

$$
\begin{aligned}
QT &= \sqrt{(4 - 0)^2 + (3 - (-2))^2} \\
&= \sqrt{16 + 25} \\
&= \sqrt{41} \\
&\approx 6.4
\end{aligned}
$$

Since the radius is 4 and $QT \approx 6.4$, $QT > r$. Therefore, *T* lies in the exterior of $\odot Q$.

c. Sketch the graph of $\odot Q$ and *T* to verify the location of point *T*.

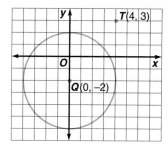

T is in the exterior of $\odot Q$.

A line in the plane of a circle can intersect the circle in one of two ways. A line can intersect a circle in exactly one point. Such a line is called a **tangent** to the circle. In the figure at the right, line ℓ intersects $\odot C$ in exactly one point, *T*, and is a tangent to $\odot C$.

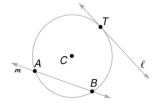

A line can also intersect a circle in exactly two points. Such a line is called a **secant** of the circle. A secant of a circle contains a chord of the circle. In the figure, line *m* contains the chord $\overline{AB}$ and is a secant of $\odot C$.

A line that is tangent to two circles in the same plane is called a **common tangent** of the two circles. There are two types of common tangents.

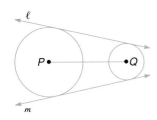

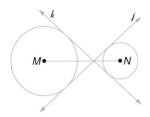

A common tangent that does not intersect the segment whose endpoints are the centers of the circles is a **common external tangent.** In the figure above, lines ℓ and m are common external tangents to $\odot P$ and $\odot Q$.

A common tangent that intersects the segment whose endpoints are the centers of the circles is a **common internal tangent.** In the figure above, lines j and k are common internal tangents to $\odot M$ and $\odot N$.

CHECKING FOR UNDERSTANDING

Communicating Mathematics

Read and study the lesson to answer these questions.

1. In $\odot A$, is A a part of the circle? Explain your answer.
2. A compass can be used to draw circles. Explain why this works.
3. Compare tangents and secants. How are they alike and/or different?
4. Consider $\odot K$, with equation $(x + 5)^2 + y^2 = 6$.
   a. Is the radius 36 or $\sqrt{6}$? Explain your answer.
   b. What are the coordinates of K?

Guided Practice

Refer to the figure at the right.

5. Name the center of $\odot P$.
6. Name three radii of the circle.
7. Name a diameter.
8. Name a chord.
9. Name a tangent.
10. Name a secant.
11. Name two points in the interior of the circle.
12. Name two points in the exterior of the circle.
13. Name five points that lie on the circle.
14. If $PC = 6$, find DB.

Determine the coordinates of the center and the measure of the radius for each circle whose equation is given.

15. $(x + 2)^2 + (y + 7)^2 = 81$

16. $(x + 5)^2 + (y - 7)^2 = 100$

EXERCISES

Practice Determine whether each statement is *true* or *false*.

17. A diameter of a circle is the longest chord of the circle.

18. A radius of a circle is a chord of the circle.

19. A chord of a circle is a secant of the circle.

20. A secant of a circle is always a diameter of the circle.

21. Two radii of a circle always form a diameter of the circle.

22. A radius of a circle is tangent to the circle.

If *r* is the measure of the radius and *d* is the measure of the diameter, find each missing measure.

23. $r = 3.8$, $d =$ ___?___   **24.** $d = 3.5$, $r =$ ___?___   **25.** $r = \frac{x}{2}$, $d =$ ___?___

The coordinates of the center and the measure of the radius of a circle are given. Write an equation of the circle.

26. $(0, 0)$, 5   **27.** $(0, 0)$, 7   **28.** $(3, 4)$, 6   **29.** $(0, 0)$, $\sqrt{14}$

Sketch the graph of the circle whose equation is given. Label the center, *C*, and the measure of the radius, *r*, on each graph.

30. $(x - 7)^2 + (y + 5)^2 = 4$ **31.** $(x + 3)^2 + (y + 6)^2 = 49$

⊙*P* has a radius of 5 units, and ⊙*T* has a radius of 3 units. Complete each statement.

32. If $QR = 1$, then $RT =$ ___?___.

33. If $PT = 7$ and $\overline{SP}$ and $\overline{ST}$ are drawn, the perimeter of $\triangle PST =$ ___?___.

34. $\overline{AR}$ is a ___?___ of ⊙*P*.

35. $QT = \frac{1}{2}$ ___?___.

36. If $QR = 1.5$, then $AQ =$ ___?___.

37. If $QR = 1$, then $AB =$ ___?___.

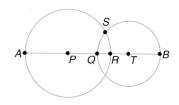

D is in the interior of ⊙*C*, A is in the exterior of ⊙*C*, and B is on ⊙*C*. Replace each ● with $<$, $>$, or $=$ to make a true statement.

38. CD ● CB **39.** CA ● CD **40.** CA ● CB

Given ⊙*T* with equation $(x - 3)^2 + (y + 4)^2 = 25$, determine if each point is on the circle, in its interior, or in its exterior.

41. $M(-2, -3)$ **42.** $A(-1, 2)$ **43.** $H(6, 0)$ **44.** $T(3, -4)$

Draw two different circles that have common tangents as described.

45. one common internal tangent and two common external tangents

46. no common internal tangents and two common external tangents

47. no common internal tangents and one common external tangent

Write an equation for each circle described.

48. The diameter is 12 units long and the center is at (-4, -7).

49. A diameter has its endpoints at (2, 7) and (-6, 15).

Critical Thinking

50. For any whole number, n, draw $2n$ radii equally spaced in a circle. Let any secant cut the sections formed by the radii into nonoverlapping regions. What will be the maximum number of such regions formed within the circle? Express your answer in terms of n. *(Hint: Use the problem-solving strategy of drawing a diagram.)*

Applications

51. Crafts   Heath is making a stained glass window. In order to cut out a circle of glass, he uses the tool at the right. Point P is the pivot point, and point A is the tip of the cutter. If he wants a circle with diameter of $10\frac{1}{2}$ inches, what length should he make $\overline{PA}$?

52. Smart Shopping   Some passenger vans use P 215/75 R14 tires. Have you ever wondered what these numbers indicate?

> P stands for passenger .
> 215 represents a tire thickness of 215 mm.
> 75 is 75% of the 215, which is the width of the tread.
> R means it's a radial tire.
> 14 is the diameter of the wheel rim in inches.

Suppose a car has tires labeled P175/65 R15. Find the width of the tread in millimeters and the radius of the wheel rim in inches.

Mixed Review

53. How many odd perfect squares are between 0 and 1000? **(Lesson 8-8)**

Determine whether each statement is *true* or *false*.

54. The diagonals of an isosceles trapezoid are congruent. **(Lesson 6-6)**

55. The diagonals of a parallelogram are congruent. **(Lesson 6-1)**

56. If two sides of a triangle are congruent, then the angles opposite those sides are congruent. **(Lesson 4-7)**

57. If two parallel lines are cut by a transversal, then two consecutive interior angles are congruent. **(Lesson 3-3)**

58. If a conditional is true, then its converse is true. **(Lesson 2-2)**

59. If two angles are supplementary, then one of the two angles is acute. **(Lesson 1-9)**

Wrap-Up

60. Journal Entry   Write a summary of the main concepts of this lesson in your journal.

Angles and Arcs

Objectives

After studying this lesson, you should be able to:

- recognize major or minor arcs or semicircles, and
- find the measures of arcs and central angles.

Application

Circle graphs are often used to compare parts of a whole. The student newspaper staff of Stevenson High School reports that of the 16 girls on the girls' varsity softball team, about 6% are freshmen, 19% are sophomores, 31% are juniors, and 44% are seniors. They wish to display these results using a circle graph. However they are not sure how to accurately measure and draw the graph. Luckily, one of the staff artists is taking geometry. He remembers that each part of the circle graph is defined by an angle whose vertex is the center of the circle. Such an angle is called a **central angle.** *You will construct the circle graph in Example 1.*

FYI ···

The Hi Ho Brakettes, a women's fast-pitch softball team from Stratford, Connecticut, won 20 United States National Championships between 1958 and 1988.

A central angle separates a circle into **arcs.** An arc is an unbroken part of a circle. For example, in the figure below, ∠APB is a central angle of ⊙P. Points A and B and all points of the circle interior to ∠APB form a **minor arc** called arc AB. This is written $\overarc{AB}$. Points A and B and all points of the circle exterior to ∠APB form a **major arc** called $\overarc{ACB}$.

words	arc *AB*	arc *ACB*	*words*
symbols	$\overarc{AB}$	$\overarc{ACB}$	*symbols*

$\overarc{AB}$ and $\overarc{BA}$ name the same minor arc.   *Three letters are needed to name a major arc. Why?*

A and B are the endpoints of $\overarc{AB}$ and $\overarc{ACB}$.

The endpoints of a segment containing the diameter of a circle separate the circle into two arcs called **semicircles.**

$\overarc{XRY}$ and $\overarc{XSY}$ are semicircles.

Arcs are measured by their corresponding central angles.

The measure of arc *LO* is 90.   *words*

$$m\overarc{LO} = 90 \qquad \text{symbols}$$

Recall that a complete rotation about a given point measures 360. Therefore, the measure of $\overarc{LNO}$ is $360 - m\overarc{LO}$, or 270.

Definition of Arc Measure	The measure of a minor arc is the measure of its central angle. The measure of a major arc is 360 minus the measure of its central angle. The measure of a semicircle is 180.

Example 1

CONNECTION

Statistics

Draw a circle graph of the data given in the lesson introduction.

In a circle graph, the artist knows that the sum of the measures of the central angles should be 360. So, the central angle for each class should be the appropriate percentage of 360.

Freshmen: 6% of 360 = 21.6
Sophomores: 19% of 360 = 68.4
Juniors: 31% of 360 = 111.6
Seniors: 44% of 360 = 158.4

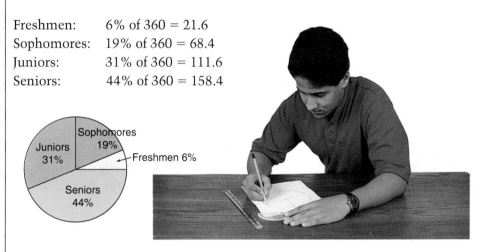

Use a protractor to measure and draw each central angle.

Adjacent arcs are arcs of a circle that have exactly one point in common. As with adjacent angles, the measures of adjacent arcs can be added to find the measure of the arc formed by the adjacent arcs.

In $\odot C$ at the right, $\overset{\frown}{PQ}$ and $\overset{\frown}{QR}$ are adjacent arcs and $\overset{\frown}{PQR}$ is formed by $\overset{\frown}{PQ}$ and $\overset{\frown}{QR}$.

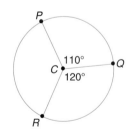

$$m\overset{\frown}{PQ} + m\overset{\frown}{QR} = m\overset{\frown}{PQR}$$
$$110 + 120 = m\overset{\frown}{PQR} \qquad m\overset{\frown}{PQ} = 110,$$
$$230 = m\overset{\frown}{PQR} \qquad m\overset{\frown}{QR} = 120$$

This result suggests the following postulate.

Postulate 9-1 Arc Addition Postulate	The measure of an arc formed by two adjacent arcs is the sum of the measures of the two arcs. That is, if Q is a point on $\overset{\frown}{PR}$, then $m\overset{\frown}{PQ} + m\overset{\frown}{QR} = m\overset{\frown}{PQR}$.

Example 2

In ⊙*P*, *m*∠*APB* = 30 and $\overline{AC}$ is a diameter. Find *m*$\widehat{AB}$, *m*$\widehat{ACB}$, *m*$\widehat{BC}$, and *m*$\widehat{BAC}$.

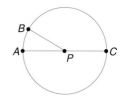

m$\widehat{AB}$　Since ∠*APB* is a central angle and *m*∠*APB* = 30, it follows that *m*$\widehat{AB}$ = 30.

m$\widehat{ACB}$　Since $\widehat{ACB}$ is the major arc for ∠*APB*, it follows that *m*$\widehat{ACB}$ = 360 − *m*$\widehat{AB}$. Since *m*$\widehat{AB}$ = 30, *m*$\widehat{ACB}$ = 360 − 30 or 330.

m$\widehat{BC}$　By the arc addition postulate, *m*$\widehat{AB}$ + *m*$\widehat{BC}$ = *m*$\widehat{ABC}$. Since *m*$\widehat{AB}$ = 30, and $\widehat{ABC}$ is a semicircle, the following holds.

$$30 + m\widehat{BC} = 180$$
$$m\widehat{BC} = 150$$

m$\widehat{BAC}$　Since $\widehat{BAC}$ is a major arc for ∠*BPC* and *m*∠*BPC* = 150, *m*$\widehat{BAC}$ = 360 − 150 or 210.

All circles have the same shape, but not all circles have the same size. **Concentric circles** are circles that lie in the same plane, have the same center, and have radii of different lengths. An example of this concept is a circular target. The edges of the rings of the target illustrate concentric circles.

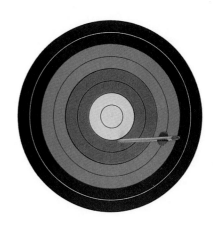

As the rings of the target show, circles with radii of different lengths are not congruent. Two circles are congruent if their radii are congruent. Two arcs of one circle are congruent if they have the same measure.

CHECKING FOR UNDERSTANDING

Communicating Mathematics

In ⊙*R* at the right, $\overline{JC}$ is a diameter. Read and study the lesson to answer these questions.

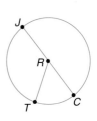

1. Name a minor arc.

2. Name a semicircle. Explain how you know that it is a semicircle.

3. Name a central angle.

4. If *m*$\widehat{JT}$ = 123, explain how to find *m*$\widehat{JCT}$. Then find *m*$\widehat{JCT}$.

5. Is the measure of a minor arc greater or less than 180? Explain.

Guided Practice

In ⊙M, m∠BMC = 40, m∠CMD = 90, and $\overline{AC}$ and $\overline{BE}$ are diameters. Determine whether each arc is a minor arc, a major arc, or a semicircle. Then find the measure of each arc.

6. $\overset{\frown}{AB}$

7. $\overset{\frown}{ECA}$

8. $\overset{\frown}{BAE}$

9. $\overset{\frown}{BDE}$

10. $\overset{\frown}{DCE}$

11. $\overset{\frown}{CBD}$

12. $\overset{\frown}{DAB}$

13. $\overset{\frown}{AE}$

14. $\overset{\frown}{BC}$

15. $\overset{\frown}{BD}$

16. $\overset{\frown}{BDC}$

17. $\overset{\frown}{AD}$

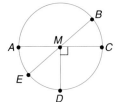

EXERCISES

Practice

In ⊙P, m∠WPX = 28, m∠ZPY = 38, and $\overline{WZ}$ and $\overline{XV}$ are diameters. Find each measure.

18. $m\overset{\frown}{YZ}$

19. $m\overset{\frown}{WX}$

20. $m\angle VPZ$

21. $m\overset{\frown}{XWY}$

22. $m\overset{\frown}{VZ}$

23. $m\overset{\frown}{VWX}$

24. $m\overset{\frown}{ZVW}$

25. $m\angle VPW$

26. $m\overset{\frown}{WYZ}$

27. $m\overset{\frown}{ZXW}$

28. $m\angle XPY$

29. $m\overset{\frown}{XY}$

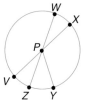

In ⊙C, m∠BCY = 2x, m∠BCQ = 4x + 15, m∠QCX = 2x + 5, and $\overline{XY}$ and $\overline{AB}$ are diameters. Find each value or measure.

30. x

31. $m\overset{\frown}{BY}$

32. $m\overset{\frown}{BQ}$

33. $m\overset{\frown}{QA}$

34. $m\overset{\frown}{QX}$

35. $m\overset{\frown}{YQ}$

36. $m\angle YCQ$

37. $m\angle QCA$

38. $m\overset{\frown}{BX}$

39. $m\angle BCX$

40. $m\overset{\frown}{XA}$

41. $m\overset{\frown}{XYA}$

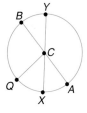

Determine whether each statement is *true* or *false*.

42. If $m\overset{\frown}{AB}$ = 32 and $m\overset{\frown}{XY}$ = 32, then $\overline{AB} \cong \overline{XY}$.

43. If $\overset{\frown}{AB} \cong \overset{\frown}{XY}$ and $m\overset{\frown}{AB}$ = 32, then $m\overset{\frown}{XY}$ = 64.

44. Two congruent circles have congruent radii.

45. All radii of a circle are congruent radii.

46. Two concentric circles always have congruent radii.

47. If two circles have the same center, they are congruent.

48. If two central angles are congruent, then their corresponding minor arcs are congruent.

49. If two minor arcs are congruent, then their corresponding central angles are congruent.

In the figure, A is the center of two concentric circles with radii $\overline{AQ}$ and $\overline{AR}$, $m\angle SAR = 32$, and $m\angle RAW = 112$. Find each measure.

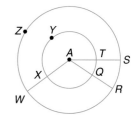

50. $m\overset{\frown}{SR}$ 51. $m\overset{\frown}{TX}$ 52. $m\overset{\frown}{SW}$

53. $m\overset{\frown}{TQ}$ 54. $m\overset{\frown}{XQ}$ 55. $m\overset{\frown}{WR}$

56. $m\overset{\frown}{TYX}$ 57. $m\overset{\frown}{SZW}$

58. Can two arcs have the same measure but not be congruent? Explain.

59. If $m\overset{\frown}{PQ} = 120$, find $m\angle Q$.

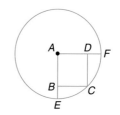

60. If $AE = 1$, find the measure of each side of square $ABCD$.

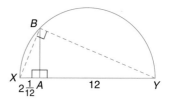

61. If B is a point on a semicircle XBY with $\overline{AB} \perp \overline{XY}$ at point A, $XA = 2\frac{1}{12}$, and $AY = 12$, find BY.

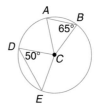

Critical Thinking   62. Find $m\angle ECB + m\angle ACD$.

Applications   63. **Food**   Wayne cuts a pie into 6 congruent pieces. What is the measure of the central angle of each piece?

64. **Engineering**   Allison Hsu designs bicycles. She designs a wheel with 30 evenly spaced spokes. Suppose the spokes are numbered consecutively from 1 through 30. Find the measure of the central angle formed by spokes 1 and 14.

65. **Teaching**   Mr. Yant is a physical education teacher. He directs the students to form a circle and count off. If the students are evenly spaced around the circle and student number 12 is directly across from student number 35, how many students are in the circle?

66. **Statistics**   The cafeteria staff surveyed the students of Middletown High School to determine which type of soup is the students' favorite. 8% prefer bean soup, 32% prefer chicken soup, 10% prefer tomato soup, and 50% prefer vegetable soup. Determine the central angle measures needed to accurately construct a circle graph using the data. Then draw an appropriate graph.

67. Write an equation for the circle with center at (1, 2) and radius of 3 units. **(Lesson 9-1)**

68. Find the value of x if the triangles below are similar. **(Lesson 7-6)**

69. Find the next number in the sequence 11, 9, 7, 5, 3, 1, -1. Describe the pattern. **(Lesson 6-2)**

70. A median of $\triangle ABC$ separates side $\overline{BC}$ of the triangle into segments $\overline{BD}$ and $\overline{DC}$. If $CD = x + 7$ and $BD = 2x - 1$, what is the value of x? **(Lesson 5-1)**

71. Describe the difference between obtuse and acute triangles. **(Lesson 4-1)**

72. Write "Concentric circles have the same center" in if-then form. **(Lesson 2-2)**

73. Find the distance between A(-11, 6) and B(-3, 7). **(Lesson 1-4)**

Wrap-Up

74. Write three questions that could be used as a quiz over this lesson.

∼ HISTORY CONNECTION ∼

Around 4500 B.C., the people who lived in Iberia, the peninsula on which Spain and Portugal are now located, built large circular tombs. These tombs, called *dolmens*, are constructed of giant stones each weighing up to 30 tons! It would take about 170 people to lift a stone this size.

The interior chambers of the dolmens that have been excavated reveal engraved and painted decorations around the walls and contain tools and pottery. Archaeologists have learned a great deal about our ancestors from these structures. But they have not yet learned why the dolmens were constructed.

9-3 Arcs and Chords

Objective

After studying this lesson, you should be able to:

- recognize and use relationships between arcs, chords, and diameters.

Application

The seats of a small Ferris wheel in the children's section of an amusement park move in a circular pattern. The diameter of this circle is 12 feet. The seats are connected with 6-foot steel bars. What is the length of the support bars that connect the center of the Ferris wheel with the midpoint of the bars? The geometric concepts in this lesson will help you answer this question in Example 1.

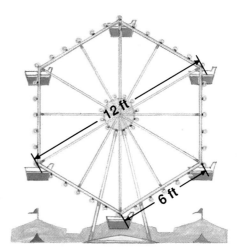

When a minor arc and a chord have the same endpoints, we call the arc the **arc of the chord.** For example, in the figure at the right, $\overset{\frown}{PQ}$ is the arc of $\overline{PQ}$.

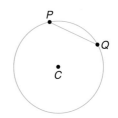

A diameter that is perpendicular to a chord has a special relationship to the chord and its arc.

Theorem 9-1

In a circle, if a diameter is perpendicular to a chord, then it bisects the chord and its arc.

Proof of Theorem 9-1

Given: $\odot P$

$\overline{AB} \perp \overline{TK}$

Prove: $\overline{AR} \cong \overline{BR}$

$\overset{\frown}{AK} \cong \overset{\frown}{BK}$

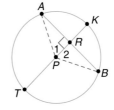

Paragraph Proof:

Draw auxiliary radii $\overline{PA}$ and $\overline{PB}$. Since all radii of a circle are congruent, $\overline{PA} \cong \overline{PB}$. $\overline{PR} \cong \overline{PR}$ because congruence of segments is reflexive. Therefore, $\triangle ARP \cong \triangle BRP$ by HL. Thus, $\overline{AR} \cong \overline{BR}$ and $\angle 1 \cong \angle 2$ by CPCTC. Thus, $\overset{\frown}{AK} \cong \overset{\frown}{BK}$ by the definition of congruent arcs.

Example 1

Find the length of the support bar, $\overline{FC}$, in the Ferris wheel.

Consider just one section of the Ferris wheel. $\overline{FA} \cong \overline{FB}$ because they are radii of the same circle. Since C is the midpoint of $\overline{AB}$, $\overline{BC} \cong \overline{AC}$. $\overline{FC} \cong \overline{FC}$ since congruence of segments is reflexive. So, $\triangle FBC \cong \triangle FAC$ by SSS. Thus, $\angle FCB \cong \angle FCA$. $\angle FCB$ and $\angle FCA$ are also supplementary, so they are right angles. We can use the Pythagorean Theorem to find x.

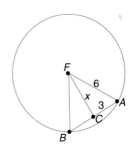

Since the diameter of the circle is 12 feet long, $FA = 6$. The steel bar represented by $\overline{BA}$ is 6 feet long. By Theorem 9-1, $AC = 3$.

$$(FC)^2 + (AC)^2 = (FA)^2 \quad \text{\textit{Pythagorean Theorem}}$$
$$x^2 + 3^2 = 6^2 \quad \text{\textit{Substitute x for FC, 3 for AC, and 6 for FA.}}$$
$$x^2 + 9 = 36$$
$$x = \sqrt{27} \text{ or about 5.2}$$

The length of the support bar is about 5.2 feet.

In a circle, a chord that is the perpendicular bisector of another chord is a diameter. The justification for the following construction is based on this relationship.

Locate the center of a given circle.

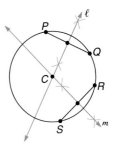

1. Draw any circle and two nonparallel chords. Label the chords $\overline{PQ}$ and $\overline{RS}$.

2. Construct the perpendicular bisectors for each chord. Call them ℓ and m. Call C the intersection of ℓ and m. Then C is the center of the circle because ℓ and m contain diameters of the circle.

Chords and their arcs are related in the following way.

Theorem 9-2	In a circle or in congruent circles, two minor arcs are congruent if and only if their corresponding chords are congruent.

The proof of this theorem is based on congruent triangles. You will be asked to complete the proof in Exercise 36.

**Proof of part of
Theorem 9-2**

**Prove that if two arcs of a circle are congruent,
then their corresponding chords are congruent.**

Given: $\odot C$
$\quad\quad\quad \overset{\frown}{AB} \cong \overset{\frown}{PQ}$

Prove: $\overline{AB} \cong \overline{PQ}$

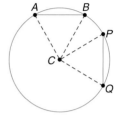

Paragraph Proof:

Draw radii $\overline{AC}, \overline{BC}, \overline{PC},$ and $\overline{QC}$ and segments $\overline{AB}$ and $\overline{PQ}$. Since all radii of a circle are congruent, $\overline{AC} \cong \overline{PC}$ and $\overline{BC} \cong \overline{QC}$. $\overset{\frown}{AB} \cong \overset{\frown}{PQ}$ and the measure of an arc is the measure of its central angle, so $\angle ACB \cong \angle PCQ$. Therefore, $\triangle ACB \cong \triangle PCQ$ by SAS and $\overline{AB} \cong \overline{PQ}$ by CPCTC.

Example 2

Use $\odot Q$ below to find the value of x.

Draw radii $\overline{QL}$ and $\overline{QY}$. Since $\overline{QC} \perp \overline{JL}$, $CL = 8$. *Why?*

$\triangle CLQ$ is a right triangle, so use the Pythagorean Theorem to find QL.

$$(QC)^2 + (CL)^2 = (QL)^2 \quad \text{\textit{Pythagorean Theorem}}$$
$$6^2 + 8^2 = (QL)^2 \quad \text{\textit{Substitute 6 for QC}}$$
$$36 + 64 = (QL)^2 \quad \text{\textit{and 8 for CL.}}$$
$$100 = (QL)^2$$
$$10 = QL$$

If $QL = 10$, then $QY = 10$. *Why?*

Use the Pythagorean Theorem with right triangle QBY to solve for x.

$$x^2 + (BY)^2 = (QY)^2 \quad \text{\textit{Pythagorean Theorem}}$$
$$x^2 + 8^2 = 10^2 \quad \text{\textit{Substitute 8 for BY}}$$
$$x^2 + 64 = 100 \quad \text{\textit{and 10 for QY.}}$$
$$x^2 = 36$$
$$x = 6$$

Notice that chords $\overline{JL}$ and $\overline{GY}$ in Example 2 are congruent. Notice also that $\overline{QC} \cong \overline{QB}$; that is, the chords are the same distance from the center Q. This leads to the next theorem, which you will be asked to prove in Exercises 37 and 38.

Theorem 9-3

In a circle or in congruent circles, two chords are congruent if and only if they are equidistant from the center.

CHECKING FOR UNDERSTANDING

Communicating Mathematics

Read and study the lesson to answer these questions about ⊙*P.*

1. Explain why Δ*PAB* is isosceles.
2. Explain what is meant by the term *arc of the chord*. Use the figure to give an example of an arc of a chord.
3. If $\overline{PM} \perp \overline{AB}$, state as many conclusions as you can.
4. If $\overline{AB} \cong \overline{CD}$, state as many conclusions as you can.

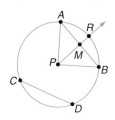

Guided Practice

Name the theorem that justifies each statement.

5. If $\overline{BC} \cong \overline{AD}$, then $\overparen{BC} \cong \overparen{AD}$.
6. If $\overparen{BC} \cong \overparen{AD}$, then $\overline{BC} \cong \overline{AD}$.
7. If $\overline{BC} \cong \overline{AD}$, then $PQ = PR$.
8. If $\overline{PF} \perp \overline{AD}$, then $\overparen{AF} \cong \overparen{FD}$.
9. If $\overline{PQ} \perp \overline{BC}$, then $\overline{BQ} \cong \overline{QC}$.
10. If $PQ = PR$, then $\overline{CB} \cong \overline{DA}$.
11. If $\overline{FX} \perp \overline{AD}$, then $\overparen{AX} \cong \overparen{XD}$.

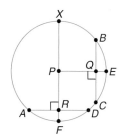

In each figure, *O* is the center of the circle. Find each measure.

12. *AC*

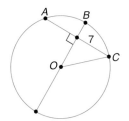

13. $m\overparen{JK}$

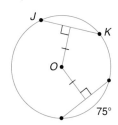

14. *ON*

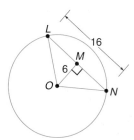

EXERCISES

Practice

In ⊙*A,* $\overline{SY} \perp \overline{QT}$ and $\overline{YS}$ and $\overline{ZR}$ are diameters.

15. Name a segment congruent to $\overline{VT}$.
16. Name the midpoint of $\overparen{QT}$.
17. Name the midpoint of $\overline{TQ}$.
18. Name an arc congruent to $\overparen{ST}$.
19. Name an arc congruent to $\overparen{QY}$.
20. Name a segment congruent to $\overline{RZ}$.
21. Which segment is longer, $\overline{WA}$ or $\overline{VA}$?
22. Which segment is shorter, $\overline{QT}$ or $\overline{YS}$?
23. If *W* is the midpoint of $\overline{QV}$, is *R* the midpoint of $\overparen{QS}$?

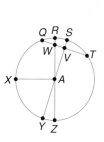

In each figure, O is the center of the circle. Find each measure.

24. $m\overset{\frown}{ST}$

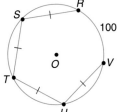

25. DF

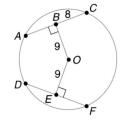

26. JL

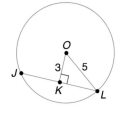

27. If $\overline{AB} \cong \overline{CD}$, is $\overset{\frown}{AB}$ congruent to $\overset{\frown}{CD}$? Explain your answer.

28. If $\overline{AB} \cong \overline{XY}$, is $\overset{\frown}{AB}$ congruent to $\overset{\frown}{XY}$? Explain your answer.

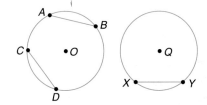

29. Suppose a chord of a circle is 10 inches long and is 12 inches from the center of the circle. Find the length of the radius.

30. Suppose a chord of a circle is 18 centimeters long and is 12 centimeters from the center of the circle. Find the length of the radius.

31. Suppose the diameter of a circle is 20 centimeters long and a chord is 16 centimeters long. Find the distance between the chord and the center of the circle.

32. Suppose the diameter of a circle is 10 inches long and a chord is 6 inches long. Find the distance between the chord and the center of the circle.

33. Draw a circle and two noncongruent chords. Which chord is closer to the center, the shorter chord or the longer chord?

34. Draw a circle by tracing around a glass or other circular object. Use a compass and straightedge to find the center of the circle.

CONSTRUCTION

35. $\overline{MN} \cong \overline{PQ}$, $MN = 7x + 13$, and $PQ = 10x - 8$. Find PS.

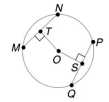

36. In a circle, if two chords are congruent, then their corresponding minor arcs are congruent. (Theorem 9-2) Draw a diagram and write a paragraph proof.

37. In a circle, if two chords are equidistant from the center, then they are congruent. (Theorem 9-3) Draw a diagram and write a paragraph proof.

38. In a circle, if two chords are congruent, then they are equidistant from the center. (Theorem 9-3) Draw a diagram and write a paragraph proof.

39. Find the length of a chord that is the perpendicular bisector of a radius of length 20 units in a circle.

40. Circles O and P, with radii 20 and 34 units, respectively, intersect at points A and B. If the length of $\overline{AB}$ is 32 units, find OP.

Critical Thinking

41. Given two circles that intersect in two points, what is the minimum number of chords you would need to construct the centers of both circles? Explain your answer.

Applications

42. Food Thelma is barbecuing chicken. The grill on her barbecue is in the shape of a circle with a diameter of 54 centimeters. The horizontal wires are supported by 2 wires that are 12 centimeters apart as shown in the figure. If the grill is symmetrical, what is the length of each support wire?

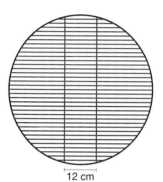

12 cm

43. Crafts Jonah plans to decoupage a picture on a circular piece of wood. He wants the picture to be able to stand freely on a shelf so he cuts along a chord of the circle to form a flat bottom. If the radius of the circle is 4 inches and the chord is 5 inches long, find the height of the final product.

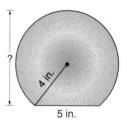

5 in.

44. Crafts Tina is making a turning tray in wood shop. She has a circular piece of wood cut and sanded. Now she needs to find the center of the circle so that she can attach it to the base. Explain how she can find the center of the circle.

Mixed Review

45. In $\odot Q$, $\overline{AB}$ is a diameter and $\overline{QC}$ is perpendicular to $\overline{AB}$. If $m\angle CQM = 15$ and $m\angle BQN = 18$, find $m\widehat{MN}$. **(Lesson 9-2)**

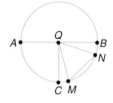

CONSTRUCTION

46. Find the geometric mean between 9 and 21. **(Lesson 8-1)**

47. Draw a segment that is 11 centimeters long. Then by construction separate the segment into three congruent parts. **(Lesson 7-5)**

48. Name the property of equality that justifies the statement "If $7x = 21$, then $x = 3$." **(Lesson 2-4)**

49. C is between A and B. If $BC = 12$ and $BA = 17$, find CA. **(Lesson 1-4)**

Wrap-Up

50. Portfolio Choose the application problem that you find most interesting and place the problem with your solution in your portfolio.

9-4 Inscribed Angles

Objectives

After studying this lesson, you should be able to:

- recognize and find the measures of inscribed angles, and
- use properties of inscribed figures.

Earlier in this chapter, you studied central angles. Recall that these are angles whose vertices are at the center of the circle and whose sides intersect the circle.

Another type of angle connected with circles is an inscribed angle. An **inscribed angle** is an angle whose vertex is on the circle and whose sides contain chords of the circle. We say that $\angle ABC$ intercepts $\overset{\frown}{AC}$. $\overset{\frown}{AC}$ is called the **intercepted arc** of $\angle ABC$. Notice that vertex B must be on the circle.

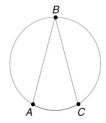

Using a compass, draw a circle. Then with a straightedge, draw a central angle of convenient size. Then draw several inscribed angles that intercept the same arc as the central angle. Use a protractor to measure the central angle and each of the inscribed angles. How is the measure of each of the inscribed angles related to the measure of the central angle?

Repeat this investigation using a different size circle and a different size central angle.

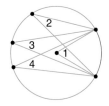

∠1 is the central angle. ∠2, ∠3, and ∠4 are the inscribed angles. All the angles intercept the arc shown in red.

This investigation leads us to the following theorem.

Theorem 9-4

If an angle is inscribed in a circle, then the measure of the angle equals one-half the measure of its intercepted arc.

There are three cases that we must consider when writing a proof of this theorem.

Case 1: The center of the circle lies on one of the rays of the angle.
Case 2: The center of the circle is in the interior of the angle.
Case 3: The center lies in the exterior of the angle.

We will complete the proof for Case 1. You will be asked to complete the proofs for Cases 2 and 3 in Exercises 49 and 50, respectively.

Proof of
Theorem 9-4
Case 1

Given: $\angle PRQ$ inscribed in $\odot T$

Prove: $m\angle PRQ = \frac{1}{2}m\widehat{PQ}$

Paragraph Proof:

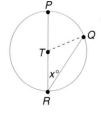

The center, T, lies on $\overline{RP}$. Draw radius $\overline{TQ}$ and let $m\angle PRQ = x$. Since $\angle PTQ$ is a central angle, $m\widehat{PQ} = m\angle PTQ$. Since $\overline{TQ}$ and $\overline{TR}$ are radii, $\triangle TQR$ is isosceles and $m\angle TQR = x$. By the Exterior Angle Theorem, $m\angle PTQ = 2x$. Therefore, $m\widehat{PQ} = 2x$ and $m\angle PRQ = \frac{1}{2}m\widehat{PQ}$.

Example 1

In the figure at the right, $m\widehat{PQ} = 112$, $m\widehat{QS} = 54$, and $m\widehat{ST} = 88$. Find $m\angle 1$, $m\angle 2$, $m\angle 3$, and $m\angle 4$.

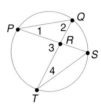

Since there are 360° in a circle, the sum of the measures of arcs $\widehat{PQ}$, $\widehat{QS}$, $\widehat{ST}$, and $\widehat{TP}$ must be 360.

$$m\widehat{PQ} + m\widehat{QS} + m\widehat{ST} + m\widehat{TP} = 360$$
$$112 + 54 + 88 + m\widehat{TP} = 360$$
$$m\widehat{TP} = 106$$

$$m\angle 1 = \frac{1}{2}(m\widehat{QS}) \qquad\qquad m\angle 2 = \frac{1}{2}(m\widehat{TP})$$
$$= \frac{1}{2}(54) \qquad\qquad\qquad = \frac{1}{2}(106)$$
$$= 27 \qquad\qquad\qquad\qquad = 53$$

$$m\angle 3 = m\angle 1 + m\angle 2 \qquad\qquad m\angle 4 = \frac{1}{2}(m\widehat{QS})$$
$$= 27 + 53 \qquad\qquad\qquad = \frac{1}{2}(54)$$
$$= 80 \qquad\qquad\qquad\qquad = 27$$

In Example 1, $m\angle 1 = m\angle 4$, which illustrates the first of the following theorems. The proofs of these theorems are based on Theorem 9-4 and you will be asked to prove them in Exercises 51 and 52.

Theorem 9-5

If two inscribed angles of a circle or congruent circles intercept congruent arcs or the same arc, then the angles are congruent.

Theorem 9-6

If an inscribed angle of a circle intercepts a semicircle, then the angle is a right angle.

Example 2

A carpenter needs to find the center of a circular pattern of a parquet floor. How can he find the center of the circle using only a carpenter's square?

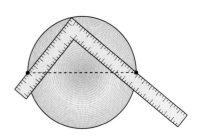

First position the carpenter's square so that the vertex of the right angle of the square is on the circle and the square forms an inscribed angle. Since the inscribed angle is a right angle, the measure of the intercepted arc is 180. So, the points where the square crosses the circle are the ends of a diameter (Theorem 9-6). Mark these points and draw the diameter. Draw another diameter using the same method. The point where the two diameters intersect is the center of the circle.

A polygon is an **inscribed polygon** if each of its vertices lies on a circle. The polygon is said to be inscribed in the circle. In the figure, quadrilateral $PQRS$ is inscribed in $\odot C$. The opposite angles of an inscribed quadrilateral are related in a special way.

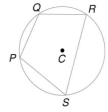

Theorem 9-7

If a quadrilateral is inscribed in a circle, then its opposite angles are supplementary.

You will be asked to prove this theorem in Exercise 53.

CONSTRUCTION

Construct a circle so that a given triangle *ABC* is inscribed in it.

1. Construct perpendicular bisectors for two sides, $\overline{AB}$ and $\overline{AC}$. Call their intersection P.

2. Using P as a center and PA as the measure of the radius, draw $\odot P$. Then $\triangle ABC$ is inscribed in $\odot P$.

$$PA = PB = PC$$

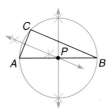

What is the relationship between the perpendicular bisector of a chord and the center of a circle? What kind of segments in a circle do the sides of $\triangle ABC$ represent?

CHECKING FOR UNDERSTANDING

Communicating Mathematics

Read and study the lesson to answer these questions. In the figure, $\overline{CD}$ is a diameter of $\odot A$.

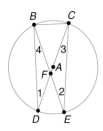

1. If $m\widehat{BC} = 42$, find $m\angle 1$.

2. Explain why $m\angle 1 = m\angle 2$.

3. What type of triangle is $\triangle BCD$? Explain your answer.

4. Ms. O'Connor, a geometry teacher, tells her class that $m\widehat{DE} = 36$. One student claims $m\angle DFE = 36$ and then changes his mind and says that $m\angle DFE = 18$. Explain why both responses are incorrect.

Guided Practice

Determine whether each angle is an inscribed angle. Explain your answer.

5.
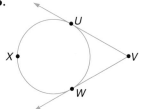

6.

7.
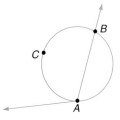

Quadrilateral *UTSK* is inscribed in a circle. Find each measure.

8. $m\widehat{KS}$ 9. $m\angle KTS$ 10. $m\widehat{UK}$

11. $m\angle USK$ 12. $m\angle TSK$ 13. $m\angle TSU$

14. $m\widehat{UT}$ 15. $m\widehat{TS}$ 16. $m\angle TKS$

CONSTRUCTION

17. Draw an acute triangle. Inscribe the triangle in a circle.

EXERCISES

Practice

In $\odot X$, $\overline{AB} \parallel \overline{DC}$, $m\widehat{BC} = 94$, and $m\angle AXB = 104$. Find each measure.

18. $m\widehat{AB}$ 19. $m\angle BAC$ 20. $m\angle BDC$

21. $m\angle BCA$ 22. $m\angle ADB$ 23. $m\angle ADC$

24. $m\angle XAB$ 25. $m\angle ABX$ 26. $m\angle ACD$

27. $m\angle BCD$ 28. $m\angle DEC$ 29. $m\angle AED$

30. $m\angle EAD$ 31. $m\widehat{DC}$ 32. $m\angle BAD$

33. $m\angle DBC$ 34. $m\widehat{AD}$ 35. $m\angle ABD$

**Quadrilateral *GHIJ* is inscribed in the circle,
m∠GHI = 2*x*, *m∠HGJ* = 2*x* − 10, and
m∠IJG = 2*x* + 10. Find each measure.**

36. $m\angle GHI$ **37.** $m\angle IJG$ **38.** $m\angle HGJ$

39. $m\angle JIH$ **40.** $m\widehat{JGH}$ **41.** $m\widehat{GHI}$

42. $m\widehat{GJI}$ **43.** $m\widehat{JIH}$

44. Draw an obtuse triangle. Inscribe the triangle in a circle.

45. Suppose an acute triangle is inscribed in a circle. Can the center of the circle be in the interior of the triangle? Must it be in the interior of the triangle? Explain your answer.

46. A parallelogram is inscribed in a circle. What are the measures of the angles of the parallelogram? Explain your reasoning.

Write a paragraph proof for each.

47. Given: $\overline{MH} \parallel \overline{AT}$
 Prove: $\widehat{AM} \cong \widehat{HT}$
 (Hint: Draw $\overline{MT}$.)

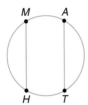

48. Given: inscribed $\angle XAZ$
  $\overline{AY}$ bisects $\angle XAZ$.
 Prove: Y bisects $\widehat{XZ}$.

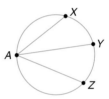

Study the proof for Case 1 of Theorem 9-4. In Case 2 and Case 3, $\overline{PR}$ is not a diameter. To prove Cases 2 and 3, draw diameter $\overline{KR}$ and use Case 1 of Theorem 9-4, the Angle Addition Postulate, and the Arc Addition Postulate.

49. Case 2
 Given: *T* lies inside $\angle PRQ$.
 Prove: $m\angle PRQ = \frac{1}{2}m\widehat{PQ}$

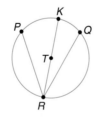

50. Case 3
 Given: *T* lies outside $\angle PRQ$.
 Prove: $m\angle PRQ = \frac{1}{2}m\widehat{PQ}$

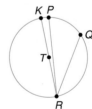

51. Prove that if two inscribed angles of a circle intercept congruent arcs, then the angles are congruent. (Theorem 9-5)

52. Prove that if an angle is inscribed in a semicircle, then the angle is a right angle. (Theorem 9-6)

53. Prove that if a quadrilateral is inscribed in a circle, then the opposite angles of the quadrilateral are supplementary. (Theorem 9-7)

Critical Thinking

54. An equilateral triangle XYZ is inscribed in $\odot O$, point A bisects $\overset{\frown}{XY}$, and point B bisects $\overset{\frown}{XZ}$. What do you know about quadrilateral $ABZY$? Explain.

Application

55. Engineering The Reuleaux Triangle is used in the rotor design of the Wankel engine of some Mazda cars. A Reuleaux Triangle is drawn by starting with an equilateral triangle, ELY. Then $\overset{\frown}{EY}$, $\overset{\frown}{YL}$, and $\overset{\frown}{LE}$ are drawn using L, E, and Y respectively as centers.
  a. Find $m\overset{\frown}{EMY}$.
  b. Suppose the Reuleaux Triangle is inscribed in a circle. Find $m\overset{\frown}{ENY}$.

Mixed Review

56. Suppose a chord of a circle is 50 centimeters long and 60 centimeters from the center of the circle. Find the radius of the circle. **(Lesson 9-3)**

57. Can 6, 9, and 11 be the measures of the sides of a right triangle? Explain your answer. **(Lesson 8-2)**

58. Are two squares always similar? Explain. **(Lesson 7-3)**

59. The median of a trapezoid is 18 inches long. If one base is 29 inches long, what is the length of the other base? **(Lesson 6-6)**

60. Identify the hypothesis and the conclusion of the conditional *If an angle is inscribed in a semicircle, then the angle is a right angle.* **(Lesson 2-2)**

61. Graph the points $A(7, 4)$ and $B(-3, 1)$ on a coordinate plane. Find the measure of $\overline{AB}$. **(Lesson 1-4)**

Wrap-Up

62. Explain the difference between a central angle and an inscribed angle of a circle. If a central angle and an inscribed angle intercept the same arc, how are their measures related?

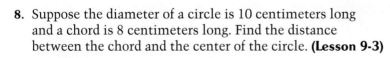

∼∼∼∼∼∼∼ MID-CHAPTER REVIEW ∼∼∼∼∼∼∼

1. Draw a diagram of a circle O with a tangent $\overleftrightarrow{AB}$ and a secant $\overleftrightarrow{BC}$. **(Lesson 9-1)**

In $\odot C$, $m\angle 1 = m\angle 2 = 32$ and $CD = 6$. $\overline{AE}$ and $\overline{BF}$ are diameters. Find each measure. (Lesson 9-2)

2. BF　　　　**3.** $m\overset{\frown}{AFE}$　　　　**4.** $m\overset{\frown}{AD}$

5. $m\overset{\frown}{AF}$　　　　**6.** $m\overset{\frown}{DE}$　　　　**7.** $m\angle 4$

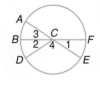

8. Suppose the diameter of a circle is 10 centimeters long and a chord is 8 centimeters long. Find the distance between the chord and the center of the circle. **(Lesson 9-3)**

Equilateral triangle ABC is inscribed in $\odot Q$ and $\odot Q$ has a radius of 12 units. (Lesson 9-4)

9. Find the length of each side.

10. Find the distance from each vertex of the triangle to the center of the circle.

11. Find $m\overset{\frown}{AB}$.

Tangents

Objective

After studying this lesson, you should be able to:

- use properties of tangents to solve problems.

Application

Duane has a tether ball that is attached to a rope. As he walks to the park, he twirls the ball in a circle. When he lets go of the rope, the ball flies off in a path that is tangent to its original circular path.

As you learned in Lesson 9-1, a tangent is a line in the plane of a circle that intersects the circle in exactly one point. Segments and rays that are contained in the tangent and intersect the circle are also said to be tangent to the circle.

In the figure, S is the center of the circle and T is the **point of tangency.** X is in the exterior of the circle, and $\overline{ST}$ is a radius. Thus $SX > ST$. A similar inequality holds for any point in the exterior of the circle.

The shortest segment from a point to a line is a perpendicular segment. Thus, $\overline{ST} \perp \overleftrightarrow{TX}$. This leads to Theorem 9-8, which you will prove in Exercise 37.

Theorem 9-8	**If a line is tangent to a circle, then it is perpendicular to the radius drawn to the point of tangency.**

You can apply Theorem 9-8 to solve practical problems.

Example 1

Aerospace

A spacecraft is 3000 kilometers above Earth's surface. If the radius of Earth is about 6400 kilometers, find the distance between the spacecraft and the horizon.

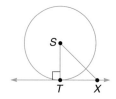

Since $\overline{SH}$ is a tangent segment, $\overline{SH} \perp \overline{EH}$ by Theorem 9-8. Therefore, $\triangle SHE$ is a right triangle. We can use the Pythagorean Theorem to find the distance between the spacecraft and the horizon.

$EH = 6400$, and $SE = 3000 + 6400$ or 9400

$$(EH)^2 + (SH)^2 = (SE)^2 \quad \text{\textit{Pythagorean Theorem}}$$
$$6400^2 + (SH)^2 = 9400^2 \quad \text{\textit{Substitute 6400 for EH and 9400 for SE.}}$$
$$40{,}960{,}000 + (SH)^2 = 88{,}360{,}000$$
$$(SH)^2 = 47{,}400{,}000$$
$$SH \approx 6885$$

The distance from the spacecraft to the horizon is about 6885 kilometers.

The converse of Theorem 9-8 is also true and provides a method for identifying tangents to a circle.

Theorem 9-9	**In a plane, if a line is perpendicular to a radius of a circle at the endpoint on the circle, then the line is a tangent of the circle.**

CONSTRUCTION

Construct a line tangent to a given circle P at a point A on the circle.

1. Draw $\overrightarrow{PA}$.

2. Construct ℓ through A and perpendicular to $\overrightarrow{PA}$. Line ℓ is tangent to $\odot P$ at A.

Since ℓ is perpendicular to radius $\overline{PA}$ at its endpoint, A, on the circle, then line ℓ is tangent to $\odot P$. (Theorem 9-9)

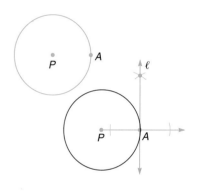

The following construction shows how to construct a tangent from a point outside the circle.

CONSTRUCTION

Construct a line tangent to a given circle C through a point A outside the circle.

1. Draw $\overline{AC}$.

2. Construct the perpendicular bisector of $\overline{AC}$. Call this line ℓ. Call X the intersection of ℓ and $\overline{AC}$.

3. Using X as the center, draw a circle with radius measuring XC. Call D and E the intersection points of the two circles.

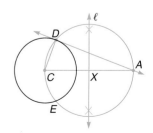

4. Draw $\overleftrightarrow{AD}$. Then $\overleftrightarrow{AD}$ is tangent to $\odot C$.

$\overleftrightarrow{AD}$ is a tangent to $\odot C$ if $\overleftrightarrow{AD} \perp \overline{DC}$. How do you know $\angle CDA$ is a right angle?

In the picture at the left, the ball bearing game demonstrates that it is possible to have two tangents to a circle from the same exterior point.

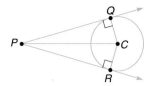

Note that $\overrightarrow{PQ}$ and $\overrightarrow{PR}$ are both tangent to $\odot C$. Also $\overline{PQ}$ and $\overline{PR}$ are **tangent segments.** By drawing $\overline{PC}$, two right triangles are formed. These triangles are congruent by *HL*, which leads us to the following theorem. You will be asked to prove this theorem in Exercise 38.

Theorem 9-10	If two segments from the same exterior point are tangent to a circle, then they are congruent.

A polygon is a **circumscribed polygon** if each side of the polygon is tangent to a circle. The following two statements are equivalent.

The polygons are circumscribed about the circles.

The circles are **inscribed** in the polygons.

CONSTRUCTION

Construct a circle inscribed in a given triangle *ABC*.

1. Construct the angle bisectors of $\angle A$ and $\angle C$. Extend the bisectors to meet at point X.

2. Construct a line from X perpendicular to $\overline{AC}$. Label the intersection of the perpendicular line and $\overline{AC}$, Y.

3. Setting the compass length equal to XY, draw $\odot X$.

$\odot X$ is inscribed in $\triangle ABC$. Point X is called the **incenter.**

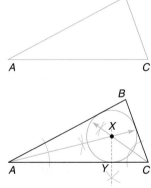

CHECKING FOR UNDERSTANDING

Communicating Mathematics

Read and study the lesson to answer these questions.

1. How many tangents can be drawn to a circle through a point outside the circle? Explain your answer.

2. How many tangents can be drawn to a circle through a point inside the circle? Explain your answer.

3. How many tangents can be drawn to a circle through a point on the circle? Explain your answer.

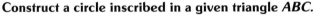

Guided Practice

Identify each polygon as circumscribed, inscribed, or neither.

4.

5.
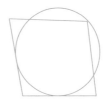

6.

Find the measure of x. Assume that C is the center of the circle.

7.
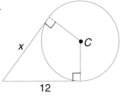

8.
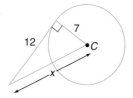

9.

10.
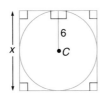

11.
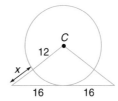

12.
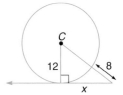

EXERCISES

Practice

In the figure, $\overline{AB}$ and $\overline{CD}$ both are tangent to $\odot P$ and $\odot Q$. Also $AP = 8$, $BQ = 5$, and $m\angle CPE = 45$. Find the measure of each of the following.

13. $\overset{\frown}{CE}$ **14.** $\angle PCG$ **15.** $\angle CGP$

16. $\overline{CG}$ **17.** $\angle QDC$ **18.** $\angle FGD$

19. $\angle FQD$ **20.** $\overset{\frown}{DF}$ **21.** $\overline{DQ}$

22. $\overline{DG}$ **23.** $\overline{DC}$ **24.** $\overline{PG}$

25. $\overline{GQ}$ **26.** $\overline{PQ}$ **27.** $\overline{AB}$

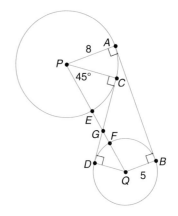

Two circles are tangent if they share one point. In the figure at the right, $\odot O$ and $\odot C$ are tangent at W, and $\overrightarrow{MK}$ is tangent to $\odot O$ at K and to $\odot C$ at L.

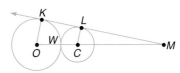

28. What is true about $\overline{OK}$ and $\overline{CL}$? Explain your answer.

29. What is true about $\triangle OKM$ and $\triangle CLM$? Explain your answer.

In the figure at the right, △ABC is circumscribed about the circle.

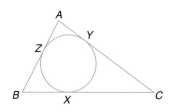

30. Suppose the perimeter of △ABC is 42 units, BX = 6, and CX = 7. Find AB and AC.

31. Suppose the perimeter of △ABC is 50 units, AZ = 10, and CX = 12. Find BC.

32. Suppose the perimeter of △ABC is 48 units, CY = 9.5, and AZ = 6. Find BZ.

In each of the following, the lines are tangent to the circles. Explain why AB = RS.

33.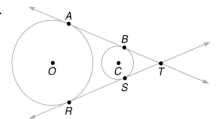

34.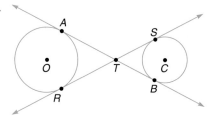

CONSTRUCTION

35. Draw a circle. Label the center P. Locate a point on the circle. Label it A. Construct a tangent to ⊙P at A.

36. Draw a circle. Label the center Q. Draw a point exterior to the circle. Label it B. Construct a tangent to ⊙Q containing B.

Write a paragraph proof for each.

37. If a line is tangent to a circle, then it is perpendicular to the radius drawn to the point of tangency. (Theorem 9-8)

38. If two segments from the same exterior point are tangent to a circle, then they are congruent. (Theorem 9-10)

39. Write an indirect proof.

    Given: $\ell \perp \overline{AB}$
    $\overline{AB}$ is a radius of ⊙A.

    Prove: ℓ is tangent to ⊙A.
    (Theorem 9-9)

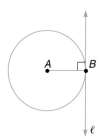

CONNECTION
Algebra

40. The graphs of x = 4 and y = -1 are both tangent to a circle that has its center in the fourth quadrant and a diameter of 14 units.
    a. Write an equation of the circle.
    b. Draw a graph of the circle and the tangents.

Critical Thinking

41. A circle is inscribed in a triangle whose sides are 9 centimeters, 14 centimeters, and 17 centimeters long. If P separates the 14-centimeter side into segments whose ratio is $x{:}y$ with $x < y$, find the values of x and y.

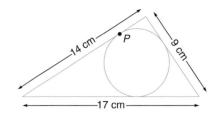

Applications

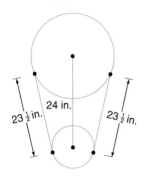

42. Agriculture The distance between the centers of the two wheels on the belt pulley system of a tractor is 24 inches. The length of the belt between the two wheels is $23\frac{1}{2}$ inches. Find the radius of each wheel if the radius of the larger wheel is twice the radius of the smaller wheel.

43. Literature In your own words, explain the meaning of the following poem.

We are, all of us, alone
Though not uncommon
In our singularity.

Touching,
We become tangent to
Circles of common experience,
Co-incident,
Defining in collective tangency
Circles
Reciprocal in their subtle
Redefinition of us.

In tangency
We are never less alone,
But no longer

Only.　　　　by Gene Mattingly

Mixed Review

44. If the measure of inscribed angle ABC is 42, what is the measure of the intercepted arc AC? **(Lesson 9-4)**

45. Find the perimeter of a rhombus whose diagonals are 30 inches and 16 inches long. **(Lesson 8-2)**

46. One right triangle has an acute angle measuring 67°, and a second right triangle has an acute angle measuring 23°. Are the triangles similar? Explain. **(Lesson 7-4)**

47. Can a median of a triangle also be an altitude? If so, what type of triangle is it? **(Lesson 5-1)**

48. Find the slope of a line that passes through the points (2, -9) and (0, 3). **(Lesson 3-5)**

Wrap-Up

49. Write three questions that could be used as a quiz over this lesson. Be sure to include the answers to your questions.

9-6　More Angle Measures

Objective

After studying this lesson, you should be able to:

- find the measures of angles formed by intersecting secants and tangents in relation to intercepted arcs.

Application

Lines intersecting the circular patterns of a stained glass window demonstrate that many different angle relationships exist. The location of the vertex of the angle determines the relationship between the measure of the angle and its intercepted arcs.

In Lesson 9-4, you studied the relationship between the measures of intercepted arcs and inscribed angles. The inscribed angle relationship can be extended to any angle that has its vertex on the circle. This includes angles formed by two secants or a secant and a tangent.

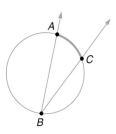

two secants

$$m\angle ABC = \tfrac{1}{2}\,m\widehat{AC}$$

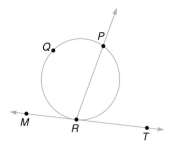

one secant, one tangent

$$m\angle PRT = \tfrac{1}{2}\,m\widehat{PR}$$

$$m\angle MRP = \tfrac{1}{2}\,m\widehat{PQR}$$

The relationship is stated in Theorem 9-11. You will prove this theorem in Exercise 55.

Theorem 9-11

If a secant and a tangent intersect at the point of tangency, then the measure of each angle formed is one-half the measure of its intercepted arc.

However, this lesson is primarily concerned with the measures of angles with vertices not on the circle.

Example 1

If $m\widehat{MH} = 60$ and $m\widehat{AT} = 140$, find $m\angle 1$.

Notice that the vertex of $\angle 1$ is inside the circle. Draw $\overline{HT}$.

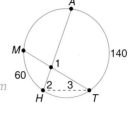

$$m\angle 1 = m\angle 2 + m\angle 3 \qquad \textit{Exterior Angle Theorem}$$
$$= \tfrac{1}{2}m\widehat{AT} + \tfrac{1}{2}m\widehat{MH} \qquad \textit{Theorem 9-4}$$
$$= \tfrac{1}{2}(140) + \tfrac{1}{2}(60) \qquad \textit{m}\widehat{AT} = 140, \textit{m}\widehat{MH} = 60$$
$$= 70 + 30$$
$$= 100$$

$\angle 1$ measures $100°$.

Example 1 illustrates Theorem 9-12. You will be asked to prove this theorem in Exercise 51.

Theorem 9-12	**If two secants intersect in the interior of a circle, then the measure of an angle formed is one-half the sum of the measures of the arcs intercepted by the angle and its vertical angle.**

In the circle at the right, two secants intersect inside the circle. So, $m\angle 1 = \tfrac{1}{2}(m\widehat{MH} + m\widehat{AT})$

Theorem 9-13 states the relationship between the angles and the intercepted arcs when the vertex of the angle is on the exterior of the circle.

Theorem 9-13	**If two secants, a secant and a tangent, or two tangents intersect in the exterior of a circle, then the measure of the angle formed is one-half the positive difference of the measures of the intercepted arcs.**

There are three cases that need to be considered. You will prove these cases in Exercises 52, 53, and 54.

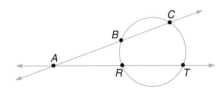

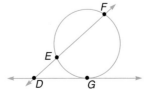

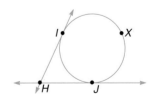

Case 1: two secants
$$m\angle CAT = \tfrac{1}{2}(m\widehat{CT} - m\widehat{BR})$$

Case 2: a secant and a tangent
$$m\angle FDG = \tfrac{1}{2}(m\widehat{FG} - m\widehat{EG})$$

Case 3: two tangents
$$m\angle IHJ = \tfrac{1}{2}(m\widehat{IXJ} - m\widehat{IJ})$$

Example 2

To start a game of pool, 15 balls are placed in a rack as shown below. The rack is tangent to the 9-ball at two points. If a circle represents the ball, find the measure of each arc whose endpoints are the points of tangency.

The rack is an equilateral triangle, so $m\angle A = 60$. The sides of the rack are tangent to the 9-ball at M and N. Let $m\widehat{MN} = x$. Then the measure of the corresponding major arc, $\widehat{MON}$, is $360 - x$.

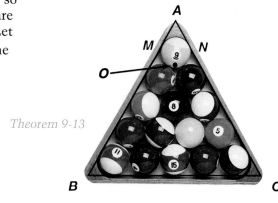

$m\angle A = \frac{1}{2}(m\widehat{MON} - m\widehat{MN})$ *Theorem 9-13*

$m\angle A = \frac{1}{2}[(360 - x) - x]$

$60 = \frac{1}{2}[360 - 2x]$

$60 = 180 - x$

$x = 120$

The measures of the two arcs are 120 and $360 - 120$ or 240.

Example 3

$\overleftrightarrow{TN}$ is tangent to $\odot Q$ at A and $m\widehat{AB} = 124$. Find $m\angle BAN$.

The vertex of $\angle BAN$ is on $\odot Q$. By Theorem 9-11, $m\angle BAN = \frac{1}{2}(m\widehat{BCA})$. So, first find $m\widehat{BCA}$.

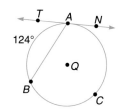

$m\widehat{AB} + m\widehat{BCA} = 360$ *The sum of the measures of a minor arc and its major arc is 360.*

$124 + m\widehat{BCA} = 360$ $m\widehat{AB} = 124$

$m\widehat{BCA} = 236$

$m\angle BAN = \frac{1}{2}(m\widehat{BCA})$

$= \frac{1}{2}(236)$ or 118

The measure of $\angle BAN$ is 118.

CHECKING FOR UNDERSTANDING

Communicating Mathematics

Read and study the lesson to answer these questions.
See margin.
1. Draw examples of the three ways two secants can intersect; in the circle, on the circle, and outside the circle. In each case, explain how to find the measures of the angles in terms of the measures of the arcs.

2. How are inscribed angles and angles formed by a secant and a tangent at the point of tangency alike? How are they different?

3. When you are finding the measure of an angle whose vertex is outside of the circle, how do you know which arc measure to subtract from which arc measure?

Guided Practice

For each circle, measurements of certain arcs are given. Find the measure of each numbered angle. Assume lines that appear to be tangent are tangent.

4.

5.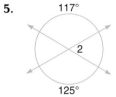

6.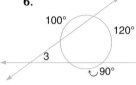

7.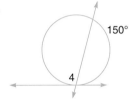

8.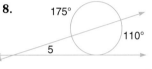

9.

For each figure, write an equation in terms of *x* and the given measures. Then solve for *x*. Assume lines that appear to be tangent are tangent.

10.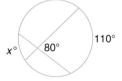

11.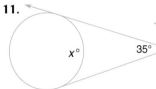

12.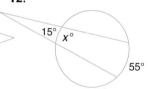

EXERCISES

Practice

In the figure, $m\widehat{BC} = 84$, $m\widehat{CD} = 38$, $m\widehat{DE} = 64$, $m\widehat{EF} = 60$, and $\overleftrightarrow{AB}$ and $\overleftrightarrow{AF}$ are tangents. Find each measure.

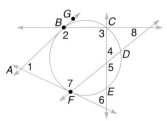

13. $m\widehat{BF}$ **14.** $m\widehat{BDF}$ **15.** $m\angle 1$

16. $m\widehat{BFC}$ **17.** $m\angle 2$ **18.** $m\angle GBC$

19. $m\widehat{BFE}$ **20.** $m\angle 3$ **21.** $m\angle 4$

22. $m\angle 5$ **23.** $m\widehat{FBC}$ **24.** $m\angle 6$

25. $m\widehat{FBD}$ **26.** $m\angle 7$ **27.** $m\angle 8$

In the figure, $m\angle 1 = 2x$, $m\angle 1 = m\angle 2$, $m\overarc{RYT} = 4x + 4$, $m\overarc{YT} = 3x - 20$, $m\angle 4 = 3x + 14$, and $\overrightarrow{ST}$ and $\overrightarrow{SR}$ are tangents. Find each value or measure.

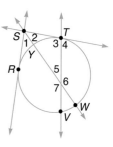

28. x

29. $m\angle 1$

30. $m\overarc{RV}$

31. $m\angle 2$

32. $m\overarc{RYT}$

33. $m\overarc{TRV}$

34. $m\overarc{YT}$

35. $m\overarc{YR}$

36. $m\angle 5$

37. $m\overarc{TW}$

38. $m\overarc{RW}$

39. $m\angle 6$

40. $m\angle 4$

41. $m\overarc{TWV}$

42. $m\overarc{YV}$

43. $m\overarc{VW}$

44. $m\angle 3$

45. $m\angle 7$

46. In the figure below, find $m\angle CED$.

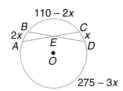

47. In the figure, $\overleftrightarrow{AD}$ passes through the center of $\odot O$. If $m\overarc{SB} = 120$, $m\overarc{AB} = 2(m\overarc{RS})$, and $m\overarc{CD} = 35$, find $m\overarc{UV}$.

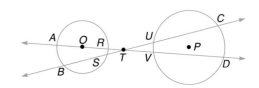

48. In the figure, $\overrightarrow{BT}$ and $\overrightarrow{BP}$ are tangent to the circle. Find the values of x, y, and z.

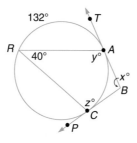

49. In the figure, $\overrightarrow{AC}$ and $\overrightarrow{AS}$ are tangent to the circle. Find the values of x, y, and z.

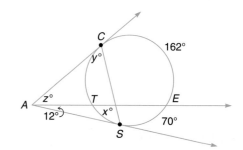

50. In a circle, chords $\overline{AC}$ and $\overline{BD}$ meet at P. If $m\angle CPB = 115$, $m\overarc{AB} = 6x + 16$, and $m\overarc{CD} = 3x - 12$, find x, $m\overarc{AB}$, and $m\overarc{CD}$.

Write a paragraph proof for each.

51. If two secants intersect in the interior of a circle, then the measure of an angle formed is one-half the sum of the measures of the arcs intercepted by the angle and its vertical angle. (Theorem 9-12)

52. Case 1 of Theorem 9-13

   Given: $\overleftrightarrow{AC}$ and $\overleftrightarrow{AT}$ are secants to the circle.

   Prove: $m\angle CAT = \frac{1}{2}(m\widehat{CT} - m\widehat{BR})$

   (Hint: Draw $\overline{CR}$.)

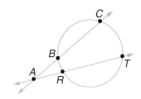

53. Case 2 of Theorem 9-13

   Given: $\overleftrightarrow{DG}$ is a tangent to the circle.
   $\overleftrightarrow{DF}$ is a secant to the circle.

   Prove: $m\angle FDG = \frac{1}{2}(m\widehat{FG} - m\widehat{GE})$

   (Hint: Draw $\overline{FG}$.)

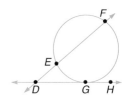

54. Case 3 of Theorem 9-13

   Given: $\overleftrightarrow{HI}$ and $\overleftrightarrow{HJ}$ are tangents to the circle.

   Prove: $m\angle IHJ = \frac{1}{2}(m\widehat{IXJ} - m\widehat{IJ})$

   (Hint: Draw $\overline{IJ}$.)

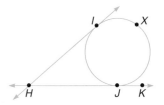

55. Write a paragraph proof of Theorem 9-11. You must prove the theorem for three cases: an acute angle, a right angle, and an obtuse angle.

Critical Thinking

56. In the figure, $\overleftrightarrow{DA}$ and $\overleftrightarrow{DB}$ are tangent to the circle. What is the value of $x + y$? Explain your reasoning.

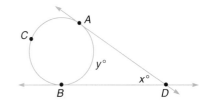

Application

57. Navigation In order to avoid dangerous rocks and shallow waters, a captain must steer the ship around what the sailors call the *danger circle*. Two lighthouses help to locate this danger circle. The measure of an inscribed angle that cuts the arc defined by the lighthouses is published on navigation charts.

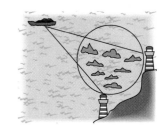

   a. If the measure of the angle formed by the lighthouses with the ship as the vertex is less than the published measure, is the ship safe? Why or why not?

   b. If the measure of the angle formed by the lighthouses with the ship as the vertex is greater than the published measure, is the ship safe? Why or why not?

58. What can you say about a line in the plane of a circle that is perpendicular to a radius of the circle at its endpoint that is on the circle? **(Lesson 9-5)**

59. Solve the proportion $\frac{x}{5} = \frac{7}{2}$. **(Lesson 7-1)**

60. Can 7, 9, and 11 be the measures of the sides of a triangle? **(Lesson 5-6)**

61. The base angles of an isosceles triangle have measures $7x - 3$ and $4x + 12$. What is the measure of the vertex angle? **(Lesson 4-7)**

62. Find the value of x. **(Lesson 4-7)**

63. What guarantees that $\angle B \cong \angle B$? **(Lesson 2-7)**

64. Draw and label a diagram that shows lines ℓ, m, and n that intersect at point X. **(Lesson 1-2)**

Wrap-Up

65. Draw six circles. Use each circle to show a different way an angle can be drawn such that its sides intercept arcs of the circle. Explain how the measure of each angle is related to the measures of intercepted arcs of the circle.

Looking Ahead
Algebra Review ▶

You will need to use the quadratic formula in the next lesson. Use the quadratic equation to solve each equation.

Example: $x^2 - 8x - 4 = 0$

$$x = \frac{-b \pm \sqrt{b^2 - 4ac}}{2a}$$

$$= \frac{-(-8) \pm \sqrt{(-8)^2 - 4(1)(-4)}}{2(1)} \qquad a = 1, b = -8, c = -4$$

$$= \frac{8 \pm \sqrt{64 + 16}}{2}$$

$$= \frac{8 \pm \sqrt{80}}{2}$$

$$= \frac{8 \pm 4\sqrt{5}}{2} \qquad \textit{Simplify the radical.}$$

$$x = 4 + 2\sqrt{5} \text{ or } 4 - 2\sqrt{5}$$

$$x \approx 8.47 \text{ or } -0.47$$

The solution set is $\{4 + 2\sqrt{5}, 4 - 2\sqrt{5}\}$.

66. $x^2 - 7x - 8 = 0$

67. $3x^2 + 14x = 5$

68. $n^2 - 13n - 32 = 0$

69. $-4y^2 + 13 = -16y$

9-7 Special Segments in a Circle

Objective

After studying this lesson, you should be able to:
- use properties of chords, secants, and tangents to solve problems.

Application

An escape wheel from an antique grandfather clock has been broken and only part of the gear remains. In order to replace it, a tool and die maker needs to know the radius of the original wheel. *You will solve this problem in Example 1.*

You will learn about special relationships involving segments of chords, secants, and tangents that will help you solve this and other problems. Similar triangles can be used to prove these relationships.

In the figure, $\overline{AC}$ is separated into $\overline{AE}$ and $\overline{EC}$. $\overline{BD}$ is separated into $\overline{BE}$ and $\overline{ED}$. Study these products.

$$AE \cdot EC = 2 \cdot 6 \qquad BE \cdot ED = 3 \cdot 4$$
$$= 12 \qquad\qquad\quad = 12$$

This is an application of Theorem 9-14.

Theorem 9-14	**If two chords intersect in a circle, then the products of the measures of the segments of the chords are equal.**

Proof of Theorem 9-14

Given: $\overline{AC}$ and $\overline{BD}$ intersect at E.

Prove: $AE \cdot EC = BE \cdot ED$

Proof:

Statements	Reasons
1. Draw $\overline{AD}$ and $\overline{BC}$ forming $\triangle DAE$ and $\triangle CBE$.	1. Through any 2 pts. there is 1 line.
2. $\angle A \cong \angle B$ $\angle D \cong \angle C$	2. If 2 inscribed $\angle$s of a $\odot$ intercept the same arc, then the $\angle$s are $\cong$.
3. $\triangle DAE \sim \triangle CBE$	3. AA similarity
4. $\dfrac{AE}{BE} = \dfrac{ED}{EC}$	4. Definition of similar polygons
5. $AE \cdot EC = BE \cdot ED$	5. Cross products

LESSON 9-7 SPECIAL SEGMENTS IN A CIRCLE 447

Example 1

How can the tool and die maker find the radius of the escape wheel of the clock?

First, measure the length of the chord across the ends of the broken piece. Suppose the chord measures 10 millimeters. Then find the midpoint of the chord and the distance to the arc. Suppose this measure is 4 millimeters.

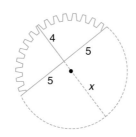

$$4 \cdot x = 5 \cdot 5 \qquad \textit{Theorem 9-14}$$
$$4x = 25$$
$$x = 6\frac{1}{4}$$

The diameter of the circle measures $4 + 6\frac{1}{4}$ or $10\frac{1}{4}$ millimeters. The radius measures $\frac{1}{2}(10\frac{1}{4})$ or $5\frac{1}{8}$ millimeters.

In the figure at the right, both $\overline{RP}$ and $\overline{RT}$ are called **secant segments**. As with all secants, they contain chords of the circle. The parts of these segments that are exterior to the circle are called **external secant segments**. In the figure, $\overline{RQ}$ and $\overline{RS}$ are external secant segments.

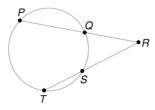

Draw a figure like the one shown above and use a ruler to measure RQ, RP, RS, and RT. Find the products $RQ \cdot RP$ and $RS \cdot RT$. How are they related? Draw several other figures and make a conjecture.

This investigation leads us to Theorem 9-15.

Theorem 9-15

If two secant segments are drawn to a circle from an exterior point, then the product of the measures of one secant segment and its external secant segment is equal to the product of the measures of the other secant segment and its external secant segment.

Plan for
Proof of
Theorem 9-15

To prove Theorem 9-15, start with a figure like the one shown above. Draw $\overline{PS}$ and $\overline{TQ}$. Then show that $\triangle PSR \sim \triangle TQR$. Write the proportion, $\frac{RQ}{RS} = \frac{RT}{RP}$, and find its cross product, $RQ \cdot RP = RS \cdot RT$. *You will be asked to complete this proof in Exercise 33.*

The segments formed by a tangent and a secant also have a special relationship.

The figure shows a tangent segment, $\overline{XY}$, and a secant segment, $\overline{YW}$, drawn to a circle from an exterior point, Y. If you draw $\overline{XW}$ and $\overline{XZ}$, triangles $\triangle YXZ$ and $\triangle YWX$ are formed. $m\angle YXZ = \frac{1}{2}\,m\widehat{XZ}$ and $m\angle XWZ = \frac{1}{2}\,m\widehat{XZ}$, so $\angle YXZ \cong \angle XWZ$. $\angle Y \cong \angle Y$ since congruence of angles is reflexive. Therefore, $\triangle YXZ \sim \triangle YWX$ by AA similarity.

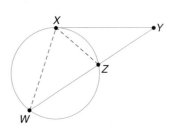

$$\frac{XY}{YW} = \frac{YZ}{XY} \qquad \textit{Definition of similarity}$$

$$XY^2 = YW \cdot YZ \qquad \textit{Cross products}$$

This leads us to Theorem 9-16. You will prove this theorem in Exercise 34.

Theorem 9-16	**If a tangent segment and a secant segment are drawn to a circle from an exterior point, then the square of the measure of the tangent segment is equal to the product of the measures of the secant segment and its external secant segment.**

Example 2

In the figure, $\overline{BC}$ is a tangent segment. Find the value of x.

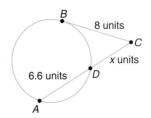

$$AC \cdot DC = (BC)^2 \qquad \textit{Theorem 9-16}$$

$$(x + 6.6)x = 8^2$$

$$x^2 + 6.6x = 64$$

$$x^2 + 6.6\,x - 64 = 0$$

Use the Quadratic Formula.

$$x = \frac{-b \pm \sqrt{b^2 - 4ac}}{2a}$$

$$x = \frac{-6.6 \pm \sqrt{(6.6)^2 - 4(1)(-64)}}{2(1)} \qquad a = 1, b = 6.6, c = -64$$

$$= \frac{-6.6 \pm \sqrt{43.56 + 256}}{2}$$

$$\approx \frac{-6.6 \pm 17.3}{2}$$

$$\approx 5.35 \text{ or } -11.95$$

x is approximately 5.35. *Why is –11.95 not used?*

CHECKING FOR UNDERSTANDING

Communicating Mathematics

Read and study the lesson to answer these questions.

1. Describe the difference between a secant segment and an external secant segment. Use the diagram at the right to give an example of each.

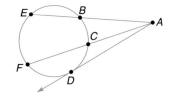

2. Two secants segments are drawn to a circle from an external point. Use the figure to explain the relationship involving the secant segments and exterior secant segments.

3. Define tangent segment. Use the diagram to give an example of a tangent segment.

4. A tangent segment and a secant segment are drawn to a circle from an exterior point. Describe the relationship involving the tangent segment, the secant segment, and the external secant segment.

5. Draw a circle. Draw two intersecting chords in the circle. Label the point of intersection and write an equation involving the segments.

Guided Practice

State the equation you would use to find the value of *x*. Assume segments that appear to be tangent are tangent. Then find the value of *x*.

6.

7.

8.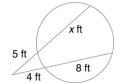

9.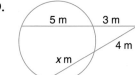

10.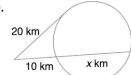

11.

12.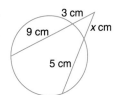

13.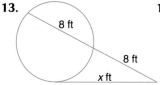

14.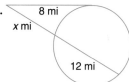

EXERCISES

Practice **Find the value of _x_. Assume segments that appear to be tangent are tangent.**

15.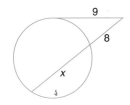

16.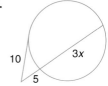

17.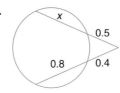

18.

19.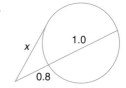

20.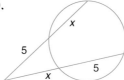

21.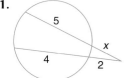

22.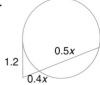

23.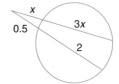

In the figure, $\overline{AB}$ is a tangent segment. Round each answer to the nearest tenth.

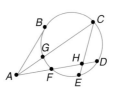

24. If $FH = 6$, $HD = 2$, and $HE = 3$, find CH.

25. If $AD = 16$, $FD = 8$, and $AG = 6$, find GC.

26. If $AC = 21$, $AF = 8$, and $FD = 8$, find AG.

27. If $CH = 20$, $EH = 10.5$, and $DH = 8$, find FH.

28. If $AF = 16$, $FH = 6$, and $DH = 2$, find BA.

Find each value.

29. $\overline{TU}$ is tangent to the circle and $\overline{RT}$ is perpendicular to $\overline{TU}$. Find x.

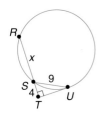

30. $\overline{AB}$ is tangent to the circle. Find x and y.

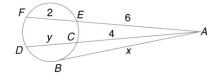

31. Your geometry teacher asks you to draw a circle through points *L, M,* and *N.* The measures are indicated in the figure. Find the measure of the radius of the circle you will draw.

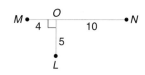

32. A sphere is cut by a plane. The radius of the sphere is 12 units long. The distance between the plane and the farthest point of the dome cut by the plane is 5 units. Find the length of the radius of the circle formed by the intersection of the sphere and the plane.

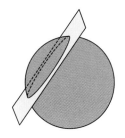

33. Write a two-column proof of Theorem 9-15.

34. Write a paragraph proof of Theorem 9-16.

Critical Thinking

35. The figure shows two concentric circles. Find the values of *x* and *y.*

Applications

36. Carpentry   An arch over a door is 50 centimeters high and 180 centimeters wide. Find the length of the radius of the arch.

37. Space   The spaceshuttle *Discovery* (S) is 150 miles above Earth (⊙E). What is the length of its longest line of sight ($\overline{SA}$) to Earth? The diameter of Earth is about 8000 miles long.

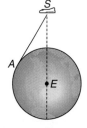

38. Water Management   Water is flowing through a pipe. The width of the surface of the water is 50 centimeters and the maximum depth of the water is 20 centimeters. How long is the diameter of the pipe?

39. Architecture   An architect is designing a new domed stadium. The outside diameter of the stadium measures 600 feet. If the center of the dome is 50 feet higher than the sides of the stadium, how long is the radius of the sphere that forms the dome?

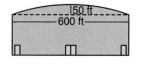

Computer The BASIC program below will find the measure of a segment in a circle. If you know the measures of $\overline{AE}$, $\overline{EB}$, and $\overline{CE}$, the program will find the measure of $\overline{ED}$.

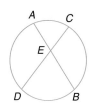

```
10  PRINT "ENTER THE MEASURES OF
    SEGMENTS AE, EB, AND CE."
20  INPUT A, E, C
30  D = (A*E)/C
40  PRINT "THE MEASURE OF SEGMENT ED IS "; D; "."
```

Use the BASIC program to find the measure of $\overline{ED}$ for each situation.

40. $AE = 8$, $EB = 6$, $CE = 3$

41. $AE = 9$, $EB = 2$, $CE = 7$

42. $AE = 10$, $EB = 30$, $CE = 18$

43. $AE = 0.32$, $EB = 0.69$, $CE = 0.22$

44. How could you change the BASIC program to find the measure of $\overline{AE}$ if you know the measures of $\overline{AB}$, $\overline{AC}$, and $\overline{AD}$?

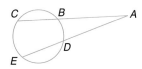

Mixed Review **45.** Find $m\angle ABC$. **(Lesson 9-6)**

46. Find $m\angle ECD$. **(Lesson 9-4)**

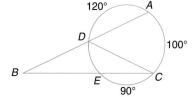

47. A 15-foot ladder leaning against a wall makes a 62° angle with the ground. If the ladder is moved so that the angle is 70°, how far did the top of the ladder move up the wall? **(Lesson 8-4)**

48. Could a trapezoid have a right angle? If so, draw one. **(Lesson 6-6)**

49. Lines ℓ and m are cut by a transversal t, and $\angle 1$ and $\angle 2$ are corresponding angles. If $m\angle 1 = 4x + 3$ and $m\angle 2 = 8x - 7$, what is the value of x if ℓ and m are parallel? **(Lesson 3-4)**

Wrap-Up **50.** Your friend is trying to help you find the value of x in the figure at the right. Your friend works the problem as follows:

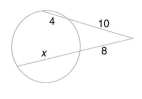

$$4 \cdot 10 = 8x$$
$$40 = 8x$$
$$5 = x$$

Is your friend correct? Write a paragraph explaining to your friend why the work is right or wrong.

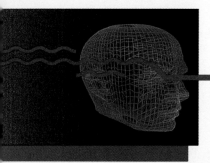

Technology

Circles

BASIC
▶ **Geometric Supposer**
Graphing calculators
LOGO
Spreadsheets

The Geometric Supposer is a powerful tool for investigating geometric relationships. Let's use it to look at some relationships in circles.

Begin by loading The Geometric Supposer: Circles and drawing any circle. Your circle will be labeled ⊙A. Choose *(2) Label* from the main menu, then choose *(4) Random Point* from the Label menu. Plot a random point outside of the circle by choosing *(3)* from this menu. The point will be labeled *B*.

Now draw two segments through *B* that are secants to circle *A*. To do this, choose *(1) Draw* on the main menu, then choose *(2)* for a line through point *B*. Define the length as intersecting the circle by choosing *(3)* from this menu. Repeat the process to draw a second segment. Your screen will show a drawing like the one at the right.

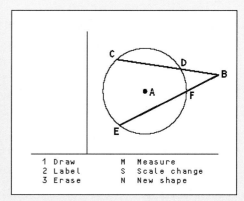

EXERCISES

Use the circle you drew on the Geometric Supposer to answer each question.

1. Find the measure of each segment using the measure option.
   a. $\overline{BD}$
   b. $\overline{BC}$
   c. $\overline{BF}$
   d. $\overline{BE}$

2. Use the measures you found in Exercise 1 to find each product.
   a. $(BD)(BC)$
   b. $(BF)(BE)$

3. Draw a third secant, $\overline{BG}$, through the point *B*. Then, find the measure of each segment and the product of their measures.
   a. $\overline{BH}$
   b. $\overline{BG}$
   c. $(BH)(BG)$

4. The products you found are called the *power of the point B*. Does it appear that the power of the point is the same for every secant drawn from a given point?

5. Will the power of any point outside of the circle be the same as the power of point *B*? Draw other random points and secants to investigate.

9-8

Problem-Solving Strategy: Using Graphs

EXPLORE
PLAN
SOLVE
EXAMINE

Objective

After studying this lesson, you should be able to:

- solve problems by using graphs.

Business men and women, politicians, environmentalists, investors, and a host of other people make many important decisions based on statistics. Since lists of numbers and other raw data are difficult to understand, people often prepare graphs to help them make decisions and to help them convince others of their point of view.

Example

APPLICATION

Sales

Marcus Taylor is a manager for the High Fashion Department Store. He studies the sales reports for jeans from the last three months. His store sold 310 pairs of Prestige Jeans, 482 pairs of Rugged Jeans, 107 pairs of Best Jeans, 256 pairs of Alpha Jeans, and 44 pairs of Fashionable Jeans. Since the store only has a limited amount of space to display the jeans, Mr. Taylor wishes to show his employees how much space to allot for each brand of jeans based on previous sales. Make a graph to help Mr. Taylor instruct his employees.

Mr. Taylor wants to show his employees what part of the available space should be used for each brand of jeans. A circle graph shows how a part is related to the whole. To make a circle graph, Mr. Taylor must first find the total number of jeans sold in the last three months.

$$310 + 482 + 107 + 256 + 44 = 1199$$

Then he must find the percent represented by each brand of jeans and find that percent of 360°.

Prestige	$\frac{310}{1199} \approx 25.9\%$	25.9% of 360° = 93.2°	
Rugged	$\frac{482}{1199} \approx 40.2\%$	40.2% of 360° = 144.7°	
Best	$\frac{107}{1199} \approx 8.9\%$	8.9% of 360° = 32.1°	
Alpha	$\frac{256}{1199} \approx 21.4\%$	21.4% of 360° = 77.0°	
Fashionable	$\frac{44}{1199} \approx 3.7\%$	3.7% of 360° = 13.3°	

Using these figures, Mr. Taylor can draw a circle graph as shown at the right, or use software to create the circle graph.

Fashionable 3.7%
Rugged 40.2%
Prestige 25.9%
Alpha 21.4%
Best 8.9%

CHECKING FOR UNDERSTANDING

Communicating Mathematics

Read and study the lesson to answer these questions.

1. What brand of jeans should take up the most space? Should this brand take up more or less than half the space?

2. If Mr. Taylor decides to discontinue one brand of jeans, what brand would you advise him to discontinue and why?

3. Do you think that Mr. Taylor could sell more Prestige Jeans than Rugged Jeans if he had a sale on Prestige Jeans? Do you think he could sell more Best Jeans than Rugged Jeans if he had a sale on Best Jeans? Explain your answers.

4. How does the circle graph help you to answer Exercises 1-3?

5. What other types of graphs can be used to display information?

Guided Practice

Use graphs to solve each problem.

6. Mindy invests some of her money in stocks. She keeps track of her stocks by graphing the closing price of each stock on the first day of the month. The graph at the right shows the values of one share of stock in Funky Records and one share of stock in Glamorous Fashions.

   a. Which stock was worth more on the first day of March?
   b. Which stock is worth more on the first day of June?
   c. In general, is the value of stock in Funky Records going up or down in value?
   d. In general, is the value of stock in Glamorous Fashions going up or down in value?
   e. What would you advise Mindy to do with her stocks and why?

7. Lisa is concerned about our natural resources. She has made a study about the use of water in the United States. Use the information in the chart at the right to make a circle graph that would help Lisa illustrate the use of water to her classmates.

Water Use in the U.S.	
Agriculture	36%
Public Water	8%
Utilities	33%
Industry	23%

EXERCISES

Practice

Solve. Use any strategy.

8. If you were one billion seconds old, how old would you be in years?

9. An electric utility company is interested in buying some coal to fuel their generator. The table at the right gives the location of the coal resources in the United States.

U.S. Coal Resources	
Rocky Mountains	45%
Appalachia	30%
Midwest	20%
Other	5%

   a. Use the information at the right to make a circle graph.

   b. What other information might the company need to make a decision about buying their coal?

10. When Sherri typed the equation, "$101 - 102 = 1$," she made just one small error. What did she do wrong? Write the correct equation.

11. Find the next number: 4800, 2400, 800, 200, __?__. Describe the pattern.

Refer to the graph at the right for Exercises 12-14.

Highest Paid Players in Football

Salary (in 1990 dollars)

1940 Sammy Baugh
$12,000 ($111,343)

1950 Sid Luckman
$25,000 ($134,751)

1960 Billy Cannon
$46,667 ($204,799)

1970 Joe Namath
$150,000 ($502,191)

1980 Walter Payton
$475,000 ($748,817)

1990 Joe Montana
$4 million

12. Using 1990 dollars, what percent of Joe Montana's salary did Joe Namath make?

13. Using 1990 dollars, what percent of Walter Payton's salary did Sid Luckman make?

14. How many times more is Sammy Baugh's salary in 1990 dollars than it was in 1940?

COOPERATIVE LEARNING PROJECT

Work in groups. Each person in the group should understand the solution and be able to explain it to any person in the class.

Business people often use statistics and graphs to influence people's decisions. So, you must be able to read a graph and make sure that it accurately represents the data it presents. The graphs below show data in a misleading way. Explain how they are misleading and why someone might draw these graphs this way.

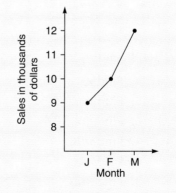

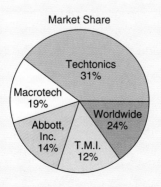

Market Share

Techtonics 31%
Macrotech 19%
Worldwide 24%
Abbott, Inc. 14%
T.M.I. 12%

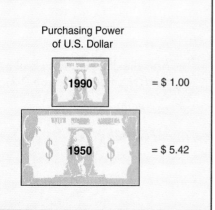

Purchasing Power of U.S. Dollar

$1990 = $ 1.00

1950 = $ 5.42

9 | SUMMARY AND REVIEW

VOCABULARY

Upon completing this chapter, you should be familiar with the following terms:

422 arc of the chord
416 central angle
410 chord
410 circle
436 circumscribed polygon
413 common external tangent
413 common internal tangent
413 common tangent

418 concentric circles
410 diameter
411 exterior of a circle
448 external secant segment
428 inscribed angle
430 inscribed polygon
411 interior of a circle
416 major arc

416 minor arc
434 point of tangency
410 radius
412 secant
448 secant segment
416 semicircle
412 tangent
436 tangent segment

SKILLS AND CONCEPTS

OBJECTIVES AND EXAMPLES	**REVIEW EXERCISES**

Upon completing this chapter, you should be able to:

Use these exercises to review and prepare for the chapter test.

■ name parts of circles. **(Lesson 9-1)**

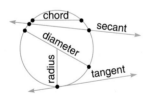

Complete.

1. A radius of a circle is a chord of the circle. Write *yes* or *no*.

2. A diameter of a circle is a chord of the circle. Write *yes* or *no*.

3. Draw two distinct circles that have no common tangents.

■ find the degree measure of arcs and central angles. **(Lesson 9-2)**

$m\angle FOG = 38$

$m\widehat{FG} = 38$

In $\odot P$, $\overline{XY}$ and $\overline{AB}$ are diameters.

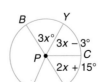

4. Find x.
5. Find $m\widehat{YAX}$.
6. Find $m\angle BPY$.
7. Find $m\widehat{BX}$.
8. Find $m\angle CPA$.
9. Find $m\widehat{BC}$.

OBJECTIVES AND EXAMPLES	REVIEW EXERCISES

■ recognize and use relationships between arcs, chords, and diameters. **(Lesson 9-3)**

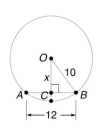

$$CB = \frac{1}{2}AB$$
$$= \frac{1}{2}(12) \text{ or } 6$$

$$(OC)^2 + (CB)^2 = (OB)^2$$
$$x^2 + 6^2 = 10^2$$
$$x^2 + 36 = 100$$
$$x^2 = 64$$
$$x = 8$$

Find each measurement.

10. A chord is 5 centimeters from the center of a circle with radius 13 centimeters. Find the length of the chord.

11. Suppose a 24-centimeter chord of a circle is 32 centimeters from the center of the circle. Find the length of the radius.

■ recognize and find the measure of inscribed angles. **(Lesson 9-4)**

$$m\angle XYZ = \frac{1}{2}m\widehat{XZ}$$
$$= \frac{1}{2}(78)$$
$$= 39$$

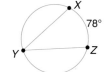

In $\odot P$, $\overline{AB} \parallel \overline{CD}$, $m\widehat{BD} = 72$, and $m\angle CPD = 144$. Find each measure.

12. $m\angle DAB$
13. $m\widehat{CD}$
14. $m\widehat{CA}$
15. $m\angle CDA$
16. $m\widehat{AB}$

■ use properties of tangents. **(Lesson 9-5)**

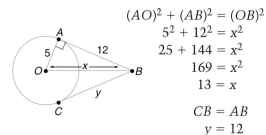

$$(AO)^2 + (AB)^2 = (OB)^2$$
$$5^2 + 12^2 = x^2$$
$$25 + 144 = x^2$$
$$169 = x^2$$
$$13 = x$$

$$CB = AB$$
$$y = 12$$

For each $\odot C$, find the value of x. Assume segments that appear to be tangent are tangent.

17.

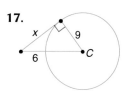

18.

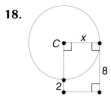

■ find the measure of angles formed by the intersection of secants and tangents in relation to intercepted arcs. **(Lesson 9-6)**

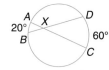

$$m\angle DXC = \frac{1}{2}(m\widehat{DC} + m\widehat{AB})$$
$$= \frac{1}{2}(60 + 20)$$
$$= \frac{1}{2}(80) \text{ or } 40$$

In $\odot P$, $m\widehat{AB} = 29$, $m\angle AEB = 42$, $m\widehat{BG} = 18$, and $\overline{AC}$ is a diameter. Find each measure.

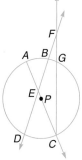

19. $m\angle DEC$
20. $m\widehat{CD}$
21. $m\angle GFD$
22. $m\widehat{AD}$
23. $m\angle AED$
24. $m\widehat{GC}$

■ use properties of chords, secants, and tangents. **(Lesson 9-7)**

Find the value of *x*. Assume segments that appear to be tangent are tangent.

$$(AO)(OB) = (CO)(OD)$$
$$2x = 3 \cdot 4$$
$$2x = 12$$
$$x = 6$$

25.

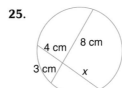

4 cm   8 cm

3 cm   *x*

26.

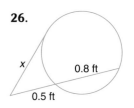

x   0.8 ft

0.5 ft

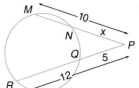

$$(MP)(NP) = (QP)(RP)$$
$$10x = 5 \cdot 12$$
$$10x = 60$$
$$x = 6$$

27.

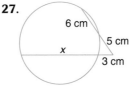

6 cm

x   5 cm

3 cm

28.

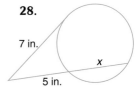

7 in.   *x*

5 in.

Applications and Connections

29. Food   Carlo made a pizza for his friends. He decides to cut the pizza into pie-shaped pieces. If he cuts it into 10 congruent pieces, what is the measure of the central angle of each piece? **(Lesson 9-2)**

30. Crafts   Sara makes wooden paper weights to sell at craft shows. For each paperweight, she starts with a sphere made of wood. She cuts off a flat surface for the base. If the original sphere has a radius of 4 centimeters and the diameter of the flat surface is 6 centimeters, what is the height of the paperweight? **(Lesson 9-3)**

31. Construction   The support of a bridge is in the shape of an arc. The span of the bridge (the length of the chord connecting the endpoints of the arc) is 28 meters. The highest point of the arc is 5 meters above the imaginary chord connecting the endpoints of the arc. How long is the radius of the circle that forms the arc? **(Lesson 9-3)**

32. Ms. Schultz is thinking about opening a music store. She gathers information to help her decide what she should sell. Use the information at the right to make a circle graph showing the portion of sales for each type of recording. **(Lesson 9-8)**

Music Sales	
Cassettes	49%
12-inch singles	2%
LPs	4%
45s	6%
Compact discs	39%

1. Write a definition for the radius of a circle.

2. Graph the circle whose equation is $x^2 + y^2 = 49$. Label the center and the measure of the radius on the graph.

Find the value of *x* for each ⊙C. Assume segments that appear to be tangent are tangent.

3.

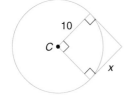

4.

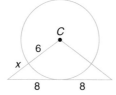

5.

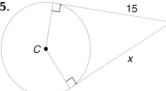

6.

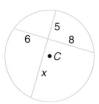

7.

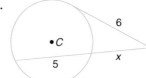

8.

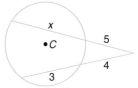

In ⊙P, $\overline{AB} \parallel \overline{CD}$, $m\widehat{BD} = 42$, $m\widehat{BE} = 12$, and $\overline{CF}$ and $\overline{AB}$ are diameters. Find each measure.

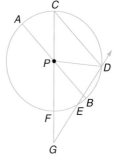

9. $m\angle BPD$

10. $m\widehat{AC}$

11. $m\angle APC$

12. $m\widehat{CD}$

13. $m\angle BPF$

14. $m\widehat{FB}$

15. $m\widehat{AF}$

16. $m\widehat{FE}$

17. $m\angle FCD$

18. $m\angle EDC$

19. Suppose the diameter of a circle is 10 inches long and a chord is 6 inches long. Find the distance between the chord and the center of the circle.

20. Determine whether the sides of an inscribed polygon are chords or tangents of the circle.

Bonus

Each circle has a radius of 2 units. $\triangle ABC$ is equilateral. The sides of the triangle are tangent to the circles. Find AC.

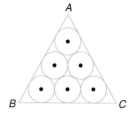

Algebra Review

- Solve compound inequalities and graph the solution sets.

$$2a > a - 3 \quad \text{and} \quad 3a < a + 6$$
$$a > \text{-}3 \qquad\qquad 2a < 6$$
$$\qquad\qquad\qquad a < 3$$

The solution set is $\{a \mid \text{-}3 < a < 3\}$.

- Use factoring and the zero product property to solve equations.

$$n^2 - n = 2$$
$$n^2 - n - 2 = 0$$
$$(n - 2)(n + 1) = 0$$

$n - 2 = 0 \quad \text{or} \quad n + 1 = 0$ *Zero product*
$n = 2 \qquad\qquad n = \text{-}1$ *property*

The solution set is $\{2, \text{-}1\}$.

- Solve systems of equations by graphing.

Graph $y = x$ and $y = 2 - x$ to find the solution to the system of equations.

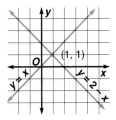

The solution is $(1, 1)$.

- Simplify square roots.

$$\sqrt{450} = \sqrt{2 \cdot 15^2}$$
$$= \sqrt{2} \cdot \sqrt{15^2}$$
$$= \sqrt{2} \cdot 15$$
$$= 15\sqrt{2}$$

$$\frac{\sqrt{18}}{\sqrt{5}} = \frac{\sqrt{18}}{\sqrt{5}} \cdot \frac{\sqrt{5}}{\sqrt{5}}$$
$$= \frac{\sqrt{2 \cdot 3^2 \cdot 5}}{\sqrt{5 \cdot 5}}$$
$$= \frac{\sqrt{2} \cdot \sqrt{3^2} \cdot \sqrt{5}}{\sqrt{5^2}}$$
$$= \frac{3\sqrt{10}}{5}$$

Solve each compound inequality. Then graph its solution set.

1. $x < \text{-}1$ or $x \geq 3$
2. $y + 1 \geq \text{-}3$ and $2y < 0$
3. $4r > 3r + 7$ and $3r + 7 \leq r + 29$
4. $2a + 5 \leq 7$ or $\text{-}2a + 4 \geq \text{-}3a + 1$

Solve each equation. Check the solution.

5. $a^2 = \text{-}17a$
6. $y^2 + 13y + 40 = 0$
7. $2x^2 + 13x = 24$
8. $(x + 6)(x - 1) = 78$
9. $25r^3 + 20r^2 + 4r = 0$

Graph each system of equations. Then find the solution to the system of equations.

10. $x + y = 6$
    $x - y = 2$
11. $y = 4x - 7$
    $x + y = 8$
12. $5x - 3y = 11$
    $2x + 3y = \text{-}25$

Simplify. Use absolute value symbols when necessary to ensure nonnegative results.

13. $\sqrt{108}$
14. $\sqrt{720}$
15. $\dfrac{\sqrt{5}}{\sqrt{55}}$
16. $\sqrt{\dfrac{20}{7}}$
17. $\sqrt{96x^4}$
18. $\sqrt{\dfrac{60}{y^2}}$

■ Multiply rational expressions.

$$\frac{x^2 + 5x - 6}{x^2 - x - 12} \cdot \frac{x + 3}{x + 6}$$

$$= \frac{(x + 6)(x - 1)}{(x - 4)(x + 3)} \cdot \frac{x + 3}{x + 6}$$

$$= \frac{x - 1}{x - 4}$$

Find each product. Assume that no denominator is equal to zero.

19. $\dfrac{5x^2y}{8ab} \cdot \dfrac{12a^2b}{25x}$

20. $\dfrac{r^2 + 3r - 18}{r + 2} \cdot \dfrac{r + 6}{r^2 - r - 6}$

21. $\dfrac{b^2 + 19b + 84}{b - 3} \cdot \dfrac{b^2 - 9}{b^2 + 15b + 36}$

■ Divide rational expressions.

$$\frac{y^2 - 16}{y^2 - 64} \div \frac{y + 4}{y - 8} = \frac{y^2 - 16}{y^2 - 64} \cdot \frac{y - 8}{y + 4}$$

$$= \frac{(y + 4)(y - 4)}{(y + 8)(y - 8)} \cdot \frac{y - 8}{y + 4}$$

$$= \frac{y - 4}{y + 8}$$

Find each quotient. Assume that no denominator is equal to zero.

22. $\dfrac{p^3r}{2q} \div \dfrac{-(p^2)}{4q}$

23. $\dfrac{7a^2b}{x^2 + x - 30} \div \dfrac{3a}{x^2 + 15x + 54}$

24. $\dfrac{n^2 + 4n - 21}{n^2 + 8n + 15} \div \dfrac{n^2 - 9}{n^2 + 12n + 35}$

■ Calculate functional values for a given function.

If $g(x) = 2x - 1$, find $g(-6)$.

$$g(-6) = 2(-6) - 1$$
$$= -12 - 1 \text{ or } -13$$

If $f(x) = x^2 - x + 1$, find each value.

25. $f(2)$

26. $f(-1)$

27. $f\left(\dfrac{1}{2}\right)$

28. $f(a + 2)$

Applications and Connections

29. **Chemistry** How many ounces of a 6% iodine solution need to be added to 12 ounces of a 10% iodine solution to produce a 7% iodine solution?

30. **Geometry** The measure of the area of a rectangle is $4m^2 - 3mp + 3p - 4m$. What are its dimensions?

31. **Sales** Clothes Outlet is having a sale on a certain brand of shirts and ties. Mr. Gill bought 8 shirts and 3 ties for $155. Mr. Ayala bought 5 shirts and 3 ties for $107. What are the sale prices of these shirts and ties?

32. **Construction** A rectangular garden is 24 feet by 32 feet. A sidewalk is built along the inside edges of all four sides. If the garden now has an area of 425 ft^2, how wide is the sidewalk?

33. **Physics** If a ball is thrown upward with an initial velocity of 72 meters per second, its height h, in meters, after t seconds is given by the equation $h = 72t - 4.9t^2$. Make a table of values for this equation to determine how many seconds it takes for the ball to hit the ground, to the nearest second.

CHAPTER 10

Polygons and Area

CHAPTER OBJECTIVES

In this chapter, you will:
- Identify parts of polygons and polyhedrons.
- Find areas of polygons.
- Find areas and circumferences of circles.
- Solve problems involving geometric probability.
- Determine characteristics of networks.

GEOMETRY AROUND THE WORLD
China

China Tower in Hong Kong is probably the only structure on Earth whose design was literally shaped within the architect's hands.

To plan the tower's design, Chinese-born architect I.M. Pei used a square shaft of wood cut diagonally into four equal triangular pieces. He held the four pieces together in one fist and pushed each piece up with the fingers of his other hand. With each thrust of Pei's fingers, each piece rose a bit higher than the preceding piece. The result was an unusual skyscraper design featuring a roof of four triangles rising to different heights.

Pei came to the United States in the 1930s to study at the Massachusetts Institute of Technology and at Harvard. A United States citizen for many years, he has designed buildings all over the world. Among Pei's most famous are the China Tower in Hong Kong, the glass pyramid entrance to the Louvre museum in Paris, and the John F. Kennedy Library in Boston. The entrance to the Louvre museum is pictured on the cover of this book.

GEOMETRY IN ACTION

Pei's design for the Kennedy Library consists of an intersecting square, circle, and triangle. The figure below is a diagram for the library's ground floor. What type of triangle is labeled *A*? What shapes adjoin it on either side?

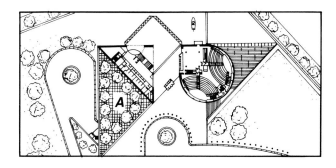

◀ *China Tower in Hong Kong*   Inset: *I.M. Pei*

China Tower's stepped triangular design fits perfectly on the small site I.M. Pei had to work with when he designed the building.

Polygons and Polyhedra

Objectives

After studying this lesson, you should be able to:
- identify and name polygons, and
- identify faces, edges, and vertices of a polyhedron.

Application

The molecular structure represented at the right is benzene, C_6H_6. The benzene molecule, or ring, consists of six carbon atoms arranged in a flat, six-sided shape with a hydrogen atom attached to each carbon atom. The symbol for benzene is ⬡. The figure represents the shape in which the atoms are bonded together.

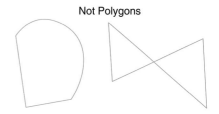

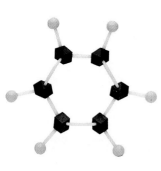

The term **polygon** is derived from the Greek word meaning "many-angled". Look at the figures below on the left. Each polygon is formed by a finite number of coplanar segments (sides) such that:

1. sides that have a common endpoint are noncollinear, and
2. each side intersects exactly two other sides, but only at their endpoints.

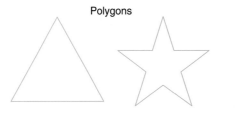

Polygons   Not Polygons

Can you explain why the two nonexamples of polygons shown above at the right fail to satisfy the conditions of the definition of a polygon?

A **convex** polygon is a polygon such that no line containing a side of the polygon contains a point in the interior of the polygon. A polygon that is not convex is called nonconvex or **concave.**

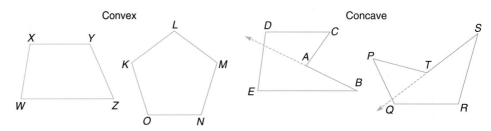

Convex   Concave

Notice that in the nonconvex examples, the extensions of sides $\overline{AB}$ and $\overline{ST}$ fail to satisfy the definition for convex. *Why?*

Polygons may be classified by the number of sides they have. The chart at the right gives some common names for polygons. In general, a polygon with *n* sides is called an ***n*-gon.** Thus, an octagon can also be called an 8-gon. A polygon with 13 sides is called a 13-gon.

Number of Sides	Polygon
3	triangle
4	quadrilateral
5	pentagon
6	hexagon
7	heptagon
8	octagon
9	nonagon
10	decagon
12	dodecagon
n	*n*-gon

FYI···

Polygon *ABCDE* in Example 1 is the flag of Nepal. Nepal is located south of China and is about the size of North Carolina. Many tourists find Nepal inviting for the beautiful views of the Himalayas.

When referring to a polygon, we use its name and list the consecutive vertices in order. Hexagon *ABCDEF* and hexagon *FABCDE* are two possible correct names for the polygon at the right.

Observe that in hexagon *ABCDEF* all the sides are congruent and all the angles are congruent. When this is true, the polygon is called **regular.** A regular polygon is a convex polygon with all sides congruent and all angles congruent.

Example 1

Classify each polygon by the number of sides, as *convex* or *concave,* and as *regular* or *not regular.*

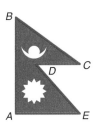

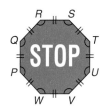

Polygon *ABCDE* has five sides. It is a pentagon.

If $\overline{CD}$ is extended through *D*, it passes through the interior of the pentagon. So, the pentagon is concave.

Since it is concave, pentagon *ABCDE* cannot be regular.

Polygon *PQRSTUVW* has 8 sides, so it is an octagon.

No lines containing sides of the octagon pass through the interior. Therefore, the polygon is convex.

Since all sides and all angles are congruent, octagon *PQRSTUVW* is regular.

Polyhedra are solids formed by regions shaped like polygons that share a common side. An ice cube and a die are examples of polyhedra. The flat surfaces formed by polygons and their interiors are called **faces.** Pairs of faces intersect at line segments called **edges.** Three or more edges intersect at a point called a **vertex.** *Polyhedra is the plural of polyhedron.*

A polyhedron must have at least three edges intersecting at each vertex. Also, the sum of the measures of the angles formed at each vertex must be less than 360.

The table below lists the faces, edges, and vertices for the polyhedron at the right.

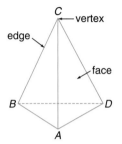

Faces	Edges	Vertices
△ABC, △BCD, △ACD, △ABD	$\overline{AB}, \overline{AC}, \overline{AD},$ $\overline{BD}, \overline{BC}, \overline{CD}$	A, B, C, D

Example 2

Name the faces, edges, and vertices of the polyhedron.

The faces are the regions bounded by the quadrilaterals *AEJF, DEJI, CDIH, BCHG, ABGF,* and pentagons *ABCDE* and *FGHIJ.*

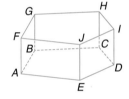

The edges are $\overline{AB}, \overline{BC}, \overline{CD}, \overline{DE}, \overline{EA}, \overline{FG}, \overline{GH},$ $\overline{HI}, \overline{IJ}, \overline{JF}, \overline{FA}, \overline{GB}, \overline{HC}, \overline{ID},$ and $\overline{JE}.$

The vertices are *A, B, C, D, E, F, G, H, I,* and *J.*

A polyhedron is **regular** if all of its faces are shaped like congruent regular polygons. Since all of the faces of a regular polyhedron are regular and congruent, all of the edges of a regular polyhedron are congruent. There are exactly five types of regular polyhedrons. These are called the *Platonic solids,* because Plato described them so fully in his writings.

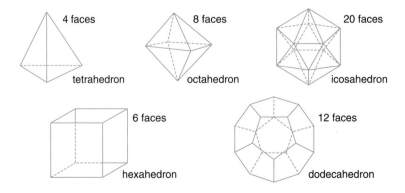

CHECKING FOR UNDERSTANDING

Communicating Mathematics

Read and study the lesson to answer these questions.

1. Decide whether each figure below is a polygon. If the figure is not a polygon, explain why not.

   a. 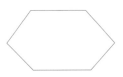 **b.** 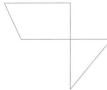 **c.**   **d.**

2. Which of the polygons in Exercise 1 are convex? Explain.

3. Name the polygon at the right. Is more than one name possible? Does the polygon appear to be regular? Explain.

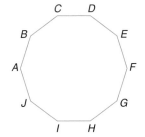

4. In polygon *ABCDE,* all sides are congruent. Is this sufficient for the pentagon to be regular? Draw an example to explain your answer.

5. What is special about the faces of a regular polyhedron?

Guided Practice

Classify each figure as a convex polygon, a concave polygon, or not a polygon.

   6.  **7.**  **8.**  **9.**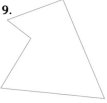

10. **11.**  **12.** **13.**

Name the faces, edges, and vertices of each polyhedron.

14. 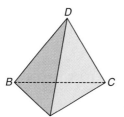 **15.** 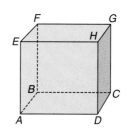 **16.**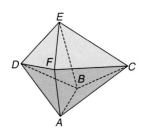

EXERCISES

Practice **Name a polygon with the given number of sides.**

17. 3   **18.** 10   **19.** 20

20. 8   **21.** 6   **22.** x

Classify each figure as a *convex polygon,* a *concave polygon,* or *not a polygon.* If a figure is a polygon, name it according to the number of sides.

23.  **24.**  **25.**

26.  **27.**  **28.**

Use polygon *MNOPQ* to answer each question.

29. Name the vertices of the polygon.

30. Name the angles of the polygon.

31. Name the sides of the polygon.

32. Is the polygon convex or concave?

33. Name the polygon according to the number of sides it has.

34. Is the polygon regular? Explain.

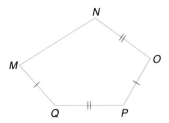

Classify each polygon as either regular or not regular. Explain.

35.   **36.**   **37.**   **38.**

If possible, draw a polygon that fits each description.

39. a regular quadrilateral

40. a quadrilateral with all sides congruent, but not all angles congruent

41. a concave pentagon with all angles congruent

42. a hexagon with all angles congruent

Answer each question for each of the polyhedrons shown below.

a.   **b.**   **c.**   **d.**   **e.**

Tetrahedron   Hexahedron   Octahedron   Dodecahedron   Icosahedron

43. Name the type of polygon that forms the faces.

44. What is the number of polygons that intersect at each vertex?

45. List the number of faces, vertices, and edges for each solid.

46. A Swiss mathematician, Leonhard Euler, stated that the number of faces, *F*, the number of vertices, *V*, and the number of edges, *E*, of a polyhedron had the relationship $F + V = E + 2$. Is this true for the regular polyhedra shown above? Justify your answer.

Critical Thinking   **47.** Describe a counterexample to the statement *Any polygon with all its angles congruent is a regular polygon.*

Application

48. **Mineralogy** Mineral crystals are solids with flat surfaces. The flat surfaces, called faces, are different shapes according to the type of mineral. Each diagram below shows the shape of a single mineral crystal. Name the number of faces and the shapes of the faces. Then identify any faces that appear to be congruent.

a. borax   **b.** quartz   **c.** calcite

 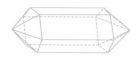 

Mixed Review

49. The national parks in the United States are used for many recreational activities. In 1989, they were used for driving off-road vehicles 65,808 times, for camping 173,597 times, for hunting 46,760 times, for fishing 23,392 times, for boating 18,491 times, and for winter sports 3,119 times. Make a circle graph of this data. **(Lesson 9-8)**

50. The measures of the legs of a right triangle are 4.0 and 5.6. Find the measure of the hypotenuse. Round your answer to the nearest tenth. **(Lesson 8-2)**

51. Determine whether the pair of triangles is congruent. If so, state the postulate or theorem that you used. **(Lesson 5-2)**

52. Find the slope of the line that passes through (5, 0) and (5, -2). **(Lesson 3-5)**

53. Write the conditional *A regular polygon is convex and has all sides congruent* in if-then form. **(Lesson 2-2)**

Wrap-Up

54. Polygons are classified by the number of sides they have and by whether they are concave or convex. Draw and label a concave decagon.

10-2 Angles of Polygons

Objectives

After studying this lesson, you should be able to:
- find the sum of the measures of the interior and exterior angles of a convex polygon,
- find the measure of each interior and exterior angle of a regular polygon, and
- use angle measures of polygons in problem solving.

Application

When an architect designs a window like this one, what should she designate as the measure of each angle of this regular octagon? To answer this question, consider each convex polygon below with all possible diagonals drawn from one vertex. *You will solve this problem in Example 1.*

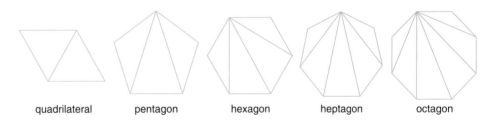

| quadrilateral | pentagon | hexagon | heptagon | octagon |

Notice that in each case, the polygon is separated into triangles. The sum of the measures of the angles of each polygon can be found by adding the measures of the angles of the triangles. Since the sum of the measures of the angles in a triangle is 180, we can easily find this sum. Let's make a chart to find the sum of the angle measures for several convex polygons.

Convex Polygon	Number of Sides	Number of Triangles	Sum of Angle Measures
triangle	3	1	(1 · 180) or 180
quadrilateral	4	2	(2 · 180) or 360
pentagon	5	3	(3 · 180) or 540
hexagon	6	4	(4 · 180) or 720
heptagon	7	5	(5 · 180) or 900
octagon	8	6	(6 · 180) or 1080

Look for a pattern in the angle measures. In each case, the sum of the angle measures is 2 less than the number of sides in the polygon times 180. So in an *n*-gon, the sum of the angle measures will be $(n - 2)180$ or $180(n - 2)$. Our inductive reasoning has led us to a correct conclusion. This conclusion is stated formally in Theorem 10-1.

Theorem 10-1 *Interior Angle Sum* *Theorem*	If a convex polygon has *n* sides and *S* is the sum of the measures of its interior angles, then $S = 180(n - 2)$.

We can use the interior angle sum theorem to find the measure of each angle of the octagonal window.

Example 1

Find the measure of each interior angle of the regular octagonal window.

First use the interior angle sum theorem to find the sum of the angle measures in a convex octagon.

$S = 180(n - 2)$ *Interior angle sum theorem*

$S = 180(8 - 2)$ *An octagon has 8 sides, so substitute 8 for n.*

$S = 180(6)$

$S = 1080$

All the angles in a regular octagon are congruent. So the measure of each angle is $\frac{1080}{8}$ or 135.

Example 2

Two angles of a convex heptagon are congruent. Each of the other five angles has a measure twice that of the other two angles. Find the measure of each angle.

Let $m\angle A = m\angle B = x$. Then the measures of angles *C, D, E, F,* and *G* must be 2*x*. The sum of the measures of all interior angles is $S = x + x + 2x + 2x + 2x + 2x + 2x$. The interior angle sum theorem says that the sum of the angle measures is $180(7 - 2)$ or 900. Write an equation.

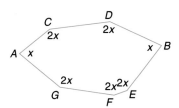

$900 = x + x + 2x + 2x + 2x + 2x + 2x$

$900 = 12x$

$75 = x$

The measures of two angles of the heptagon are 75 each. The measures of the other five angles are 2·75 or 150 each.

The interior angle sum theorem identifies a relationship among the interior angles of a convex polygon. Is there a relationship among the exterior angles of a convex polygon?

1. Draw a convex hexagon.

2. Extend the sides of the hexagon to form one exterior angle at each vertex.

3. Use a protractor to find the measure of each exterior angle.

4. Find the sum of the measures of the exterior angles.

5. Repeat for a triangle, a quadrilateral, and a pentagon. What conjecture can you make?

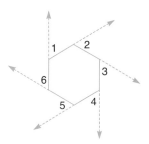

Consider the sum of the measures of the exterior angles for an *n*-gon.

$$\begin{array}{rcccc}
\text{sum of measures} & = & \text{sum of measures} & - & \text{sum of measures} \\
\text{of exterior angles} & & \text{of linear pairs} & & \text{of interior angles} \\
& = & n \cdot 180 & - & 180(n - 2) \\
& = & 180n & - & 180n + 360 \\
& = & 360 & &
\end{array}$$

So, the sum of the exterior angle measures is 360 for *any* convex polygon.

Theorem 10-2
Exterior Angle
Sum Theorem

If a polygon is convex, then the sum of the measures of the exterior angles, one at each vertex, is 360.

Example 3

Use the exterior angle sum theorem to find the measure of an interior angle and an exterior angle of a regular pentagon.

According to the exterior angle sum theorem, the sum of the measures of the five exterior angles is 360.

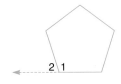

The pentagon is regular, so all of the interior angles are congruent. The interior angles are supplementary to the exterior angles. If two angles are supplementary to congruent angles they are congruent so all of the exterior angles are congruent, and the measure of each exterior angle is $\frac{360}{5}$ or 72. The measure of each interior angle is $180 - 72$ or 108.

CHECKING FOR UNDERSTANDING

Communicating Mathematics

Read and study the lesson to answer these questions.

1. Does the sum of the exterior angles of a convex polygon depend on the number of sides in the polygon? Explain your answer.

2. Your friend is working on a problem and claims that the sum of the measures of the interior angles of a convex polygon is 2070. Explain why this is not possible.

3. The sum of the measures of the interior angles of a regular polygon is 24,840. Write and solve an equation to find the number of sides the polygon has.

Guided Practice

Find the sum of the measures of the interior angles of each convex polygon.

4. pentagon
5. hexagon
6. dodecagon

7. 25-gon
8. 36-gon
9. x-gon

The number of sides of a regular polygon is given. Find the measure of each interior angle of the polygon.

10. 5
11. 7
12. 15
13. x

The measure of an exterior angle of a regular polygon is given. Find the number of sides of the polygon.

14. 45
15. 72
16. 60
17. n

The sum of the measures of the interior angles of a convex polygon is given. Find the number of sides in each polygon.

18. 720
19. 1440
20. 2880
21. 3240

EXERCISES

Practice

Find the sum of the measures of the interior angles of each convex polygon.

22. 17-gon
23. 20-gon
24. 13-gon

25. 15-gon
26. 59-gon
27. $2t$-gon

The measure of an interior angle of a regular polygon is given. Find the number of sides in each polygon.

28. 160 **29.** 120 **30.** 156

31. 165 **32.** 144 **33.** 179

The number of sides of a regular polygon is given. Find the measure of an interior angle and an exterior angle of the polygon.

34. 4 **35.** 8 **36.** 10

37. 20 **38.** 18 **39.** x

40. The sum of the measures of eight interior angles of a convex nonagon is 1190. Find the measure of the ninth angle.

41. The sum of the measures of seven exterior angles of a convex octagon is 339. Find the measure of the eighth angle.

42. The measure of the exterior angles of a convex quadrilateral are x, $2x$, $3x$, and $4x$. Find the value of x and the measure of each exterior angle.

43. The measure of each exterior angle of a regular decagon is $x + 10$. Find the value of x and the measure of each exterior angle.

44. The measure of an exterior angle of a regular polygon is $2x$, and the measure of an interior angle is $4x$. Find the number of sides in the polygon.

45. If you extend the sides of a regular pentagon as shown, you can form a five-pointed star. Find the measure of the angle at each point.

46. Use a regular hexagon to draw a six-pointed star. Are the angles at the points of the star congruent? If so, what is the measure of each angle?

47. The sum of the measures of the interior angles of a convex polygon is between 7300 and 7500.

   a. How many sides does the polygon have?

   b. What is the exact sum of the measures of the interior angles?

48. The sum of the measures of five of the interior angles of an octagon is 890. Exactly two of the three remaining angles are complementary, and exactly two are supplementary. Find the measures of these three angles.

49. Study the following pattern.
$$11 \times 1 = 11$$
$$11 \times 2 = 22$$
$$11 \times 3 = 33$$
$$11 \times 4 = 44$$

   a. If you were to continue the pattern, how would you write 11×10?

   b. What is the actual value of 11×10?

   c. What does this tell you about inductive arguments? Can you prove something with only an inductive argument?

50. Complete each statement.

a. As the number of sides of a convex polygon increases, the sum of the measures of the interior angles __?__.

b. The measure of an interior angle of a regular polygon with n sides is __?__ than the measure of an interior angle of a regular polygon with $n + 1$ sides.

c. As the number of sides of a convex polygon increases, the sum of the exterior angles __?__.

Application

51. Patterns that cover a plane with repeated shapes so that there are no empty spaces are called *tessellations*. A *regular tessellation* uses only one type of regular polygon. You can use squares to make a regular tessellation. This tessellation is often used for floor tiles.

a. Mark a point A on a piece of plain paper. Try to draw several copies of an equilateral triangle, each with point A as a vertex, so that no space is left empty and no two triangles overlap. If you can do it, then a regular triangle can tessellate the plane. Repeat the process with a regular pentagon, a regular hexagon, a regular heptagon, and a regular octagon. Copy the table below. Use your results to complete the table.

Regular Polygon	triangle	square	pentagon	hexagon	heptagon	octagon
Does it tessellate the plane?	?	?	?	?	?	?
Measure, m, of one interior angle	?	?	?	?	?	?
Is m a factor of 360?	?	?	?	?	?	?

b. Make a conjecture about the types of regular polygons that will tessellate the plane.

Mixed Review

52. Classify the figure at the right as a convex polygon, a concave polygon, or neither. If it is a polygon, name it according to the number of sides. **(Lesson 10-1)**

53. A chord of a circle is 10 inches long, and it is 12 inches from the center of the circle. Find the length of the radius of the circle. **(Lesson 9-3)**

54. A tree casts a shadow 75 feet long, and at the same time, a 6-foot fence post casts a shadow 10 feet long. How tall is the tree? **(Lesson 7-6)**

55. Graph $A(7, 6)$, $B(3, 7)$, $C(5, 11)$, and $D(9, 10)$, and then draw quadrilateral $ABCD$. Determine if $ABCD$ is a parallelogram. Justify your answer. **(Lesson 6-3)**

56. The vertex angle of isosceles triangle ABC measures 88°. Find the measures of the base angles. **(Lesson 4-7)**

57. The supplement of an angle measures 58° more than the angle. Find the measure of the angle. **(Lesson 1-8)**

Wrap-Up

58. Draw a convex hexagon with one exterior angle at each vertex. Cut out the exterior angles and arrange them so that they all have a common vertex. What is the sum of the measures of the exterior angles? Do the results agree with the exterior angle sum theorem?

HISTORY CONNECTION

Poets often write that no two snowflakes are alike, but did you know that all snowflakes are shaped like hexagons? In 1591, Thomas Hariot, for the first time in European history, recognized that snowflakes are hexagonal. He jotted his findings down in his private manuscripts, but never published them. In 1611, Johannes Kepler made the first European publication about snowflakes. However, the Chinese knew of the hexagonal structure of snowflakes in or before the second century B.C. In his book *Moral Discourses Illustrating the Han Text of the 'Book of Songs,'* Han Ying wrote "Flowers of plants and trees are generally five-pointed, but those of snow, which are called *ying*, are always six-pointed."

Snowflakes have a hexagonal structure because of the arrangement of the hydrogen and oxygen atoms in a water molecule. The crystalline structure of a snowflake is an intricate arrangement of tiny crystal prisms, each one reflecting a rainbow of bright colors to give snow its shimmering beauty.

10-3 Problem-Solving Strategy: Guess and Check

Objective

After studying this lesson, you should be able to:

- solve problems by using guess and check.

The problem-solving strategy guess and check, or trial and error, is a powerful strategy. To use this strategy, guess the answer to a problem and then use the conditions of the problem to check to see if the answer is correct. If the first guess is not correct, use the information gathered from that guess to continue guessing until you find the correct answer.

Example 1

The license plate on Anna Silver's car has a five-digit number. The plate was installed upside down, but it still shows a five-digit number. The number that shows now exceeds the number that is supposed to appear on the plate by 63,783. What is the original license number?

Since the license plate still shows a five-digit number when it is upside down, the number must consist of digits that are readable when they are upside down. These are 0, 1, 6, 8, and 9.

The units digit in the difference is 3, so they must have subtracted 6 from 9 or 8 from 1. Let's guess that they subtracted 6 from 9. So the last digit in the new number would be 9 and the last digit in the original number is 6. That makes the first digits 6 in the new number and 9 in the original number, since the digits are upside down.

$$
\begin{array}{r}
6\ \square\ \square\ \square\ 9 \\
-\ 9\ \square\ \square\ \square\ 6 \\
\hline
6\ 3\ 7\ 8\ 3
\end{array}
$$
← *new number*
← *original number*

But, if the first digits are 9 and 6, their difference would be -3, not 6. The last digits must be 1 and 8.

$$
\begin{array}{r}
8\ \square\ \square\ \square\ 1 \\
-\ 1\ \square\ \square\ \square\ 8 \\
\hline
6\ 3\ 7\ 8\ 3
\end{array}
$$

The difference in the tens place is 8. But one had to be regrouped to subtract the 8 from 1 in the ones digits. The difference in the tens place must be 9. This is possible only if the digits are 9 and 0. *Remember to write the digits in the thousands places as the tens digits would appear upside down.*

$$\begin{array}{r} 8\ 0\ \square\ 9\ 1 \\ -\ 1\ 6\ \blacksquare\ 0\ 8 \\ \hline 6\ 3\ 7\ 8\ 3 \end{array}$$

The digits in the center have a difference of 7. Notice that one must have been regrouped in the thousands place for the subtraction to be correct. The only possible digits are 6 and 9.

$$\begin{array}{r} 8\ 0\ 6\ 9\ 1 \\ -\ 1\ 6\ 9\ 0\ 8 \\ \hline 6\ 3\ 7\ 8\ 3 \end{array}$$

The license number is 16908.

Be sure to use the information that you can gather from your guesses to help you make better guesses the next time.

Example 2

Place operation symbols in the equation below to make it correct.

$$9\ \underline{\ ?\ }\ 6\ \underline{\ ?\ }\ 2\ \underline{\ ?\ }\ 1\ \underline{\ ?\ }\ 4 = 6$$

Guess at a combination of symbols and evaluate the sentence.

$$9 + 6 \div 2 - 1 + 4 = 15$$

The result was 15, which is 9 greater than 6. So we must change the symbols to make the result lower.

$$9 - 6 \div 2 - 1 - 4 = 1 \qquad \textit{Now the result is too low.}$$

$$9 - 6 - 2 - 1 + 4 = 4 \qquad \textit{Getting closer.}$$

$$9 - 6 - 2 + 1 + 4 = 6 \qquad \textit{Correct!}$$

CHECKING FOR UNDERSTANDING

Communicating Mathematics

Read and study the lesson to answer these questions.

1. Describe a situation in which you might use the strategy guess and check.

2. How can you use the results of incorrect guesses to speed up the guessing process?

Solve by using guess and check.

3. What is the least three-digit number that is divisible by the first three prime numbers and by the first three composite numbers?

4. Place two addition symbols and two subtraction symbols in the equation below to make it a true equation.

3 5 6 4 7 5 2 6 1 7 = 712

5. Copy the figure at the right and fill in circles with prime numbers so that the sum of all six numbers is 20, and the sum of the numbers on every small triangle is the same.

EXERCISES

Strategies

Look for a pattern.
Solve a simpler problem.
Act it out.
Guess and check.
Draw a diagram.
Make a chart.
Work backward.

Solve. Use any strategy.

6. Which two consecutive whole numbers have squares that differ by 75?

7. Fill in the blank in the following pattern. 25, 30, 20, ___?___, 15, 40, 10

8. In the United States, dates are usually written month / day, so 4/6 would represent April sixth. However, in Canada, dates are written day / month, so 4/6 would represent the fourth of June. How many dates can be written so that they mean one date to an American and another to a Canadian?

9. Find the pairs of consecutive whole numbers less than 50 that have squares that differ by a perfect square.

10. Substitute a different digit for each letter in the equation below to make it true.

$$\begin{array}{r} H A L F \\ + \ \underline{H A L F} \\ W H O L E \end{array}$$

COOPERATIVE LEARNING PROJECT

Work in groups. Each person in the group must understand the solution and be able to explain it to any person in class.

A set of dominoes has 28 rectangular pieces with two numbers on each one. Each number 0 through 6, is paired with every other number, including itself, on exactly one domino. The grid at the right was made by arranging the dominoes and then recording the numbers. Four dominoes are shown. Copy the grid and draw in the remaining dominoes.

(Hint: If you come to a point where you can't deduce which numbers are on a domino, guess and check to see if it leads you to a contradiction.)

1	0	2	0	0	5	4	1
1	1	5	3	6	2	4	2
3	3	1	0	3	5	3	4
0	6	6	4	6	5	1	1
0	4	0	2	5	4	2	6
1	2	3	2	6	4	5	2
3	5	5	0	3	4	6	6

10-4 Area of Parallelograms

Objective

After studying this lesson, you should be able to:
- find areas of parallelograms.

Application

In 1991, Worthington Kilbourne High School was built in Worthington, Ohio. To use the empty lot behind the school for football practice fields, they had to have sod installed. How much sod was needed? If the sod cost 85¢ a square yard, how much did the sod for the fields cost? *You will solve this problem in Example 1.*

To find the amount of sod that the school needed, we must find the area of the lot. The **area** of a figure is the number of square units contained in the interior of the figure. The area of a rectangle, A square units, can be found using the formula $A = \ell w$, where ℓ units is the length of the rectangle and w units is the width.

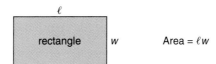

ℓ

rectangle w Area = ℓw

As we learned in Chapter 6, a square is a special rectangle in which the sides are all congruent. So, we can use the formula for the area of a rectangle to find the area of a square. If the length of a side of the square is s units, the area is $A = s \cdot s$ or s^2.

To find the area of a figure that is made up of several figures, you must find the area of each figure. Then add these to find the total area. This is stated in the postulate below.

Postulate 10-1

The area of a region is the sum of the areas of all of its nonoverlapping parts.

We can use this property with the formulas for the area of a rectangle and a square to find the amount of sod that was needed for the practice football fields at Worthington.

Example 1

Refer to the opening application.

a. How much sod did the school need to order for the lot?

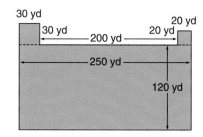

The field can be separated into a rectangle and two squares. Since these parts of the lot are nonoverlapping, we can find the area of the lot by finding the sum of the areas of the parts.

The rectangle is 250 yards long and 120 yards wide. One square is 30 yards long on each side, and the other is 20 yards long on each side. Use the formulas for the area of a rectangle and a square to find the area of each part.

$$\text{Area of rectangle} = \ell w$$
$$= 250 \cdot 120 \qquad \textit{Substitute 250 for } \ell \textit{ and 120 for w.}$$
$$= 30,000 \text{ sq yd}$$

$$\text{Area of square 1} = s^2 \qquad\qquad \text{Area of square 2} = s^2$$
$$= 30^2 \quad \textit{s = 30} \qquad\qquad\qquad = 20^2 \quad \textit{s = 20}$$
$$= 900 \text{ sq yd} \qquad\qquad\qquad\quad = 400 \text{ sq yd}$$

The area of the lot was $30,000 + 900 + 400$ or 31,300 square yards. The school needed to order 31,300 square yards of sod.

b. If the sod cost 85¢ a square yard, how much did the sod cost?

The sod cost $0.85(31,300)$ or \$26,605.

The formula for the area of a parallelogram is closely related to the formula for the area of a rectangle. Before we derive the formula, we must discuss some parts of a parallelogram.

Any side of a parallelogram can be called a **base.** For each base, there is a corresponding **altitude.** In $\square ABCD$, $\overline{AB}$ is considered to be the base. A corresponding altitude is any perpendicular segment between the parallel lines $\overleftrightarrow{AB}$ and $\overleftrightarrow{DC}$. So in $\square ABCD$, $\overline{CE}$ and $\overline{DB}$ are altitudes. The length of the altitude is called the **height.**

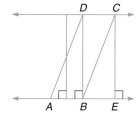

Copy □*PQRS.*

Use a compass and straightedge to construct an altitude from *S* to $\overline{QR}$. Label the point of intersection *T*. Let *h* = *ST* and *b* = *PS*.

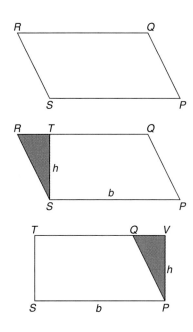

Cut out the parallelogram. Then cut Δ*STR* from the parallelogram and place it so that $\overline{RS}$ lies on $\overline{PQ}$.

The new figure you have formed is a rectangle. It has the same area as □*PQRS*. Since *VP* = *TS* = *h*, the area of the rectangle is *bh* square units. So, the area of □*PQRS* is also *bh* square units.

The area of a parallelogram is the same as the area of a rectangle that has the same base and height.

Area of a Parallelogram	**If a parallelogram has an area of *A* square units, a base of *b* units, and a height of *h* units, then *A* = *bh*.**

Example 2

Find the area of □*MNOP.* **Round your answer to the nearest hundredth.**

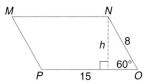

Recall that in a 30°-60°-90° triangle, the length of the hypotenuse is twice the length of the shorter leg and the length of the longer leg is $\sqrt{3}$ times the length of the shorter leg. Therefore, the length of the shorter leg is $\frac{8}{2}$ or 4 units and *h* = $4\sqrt{3}$.

$$A = bh \qquad \textit{Formula for the area of a parallelogram}$$
$$A = 15(4\sqrt{3})$$
$$A = 60\sqrt{3}$$

The area of □*MNOP* is $60\sqrt{3}$ or about 103.92 square units.

CHECKING FOR UNDERSTANDING

Communicating Mathematics

Read and study the lesson to answer these questions.

1. Why are the formulas for the area of a square and the area of a rectangle related?

2. Region Z is made up of nonoverlapping regions X and Y. If the area of region X is a square units and the area of region Y is b square units, what is the area of region Z?

3. Is the area of the parallelogram at the right 48 square units? Explain.

Guided Practice

Find the area of each figure.

4.

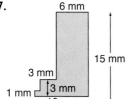

5 m

5.

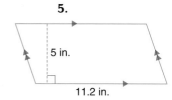

5 in.

11.2 in.

6.

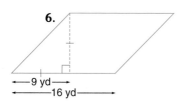

9 yd

16 yd

Find the area of each shaded region. Assume that angles that appear to be right are right angles.

7.

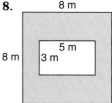

8.

8 m

8 m

5 m

3 m

9.

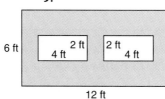

6 ft

2 ft

4 ft

2 ft

4 ft

12 ft

EXERCISES

Practice

Find the missing measure of each quadrilateral.

10. A rectangle is 6 feet long and 2 feet wide. Find its area.

11. A parallelogram has an area of 36 m² and a base of 9 m. Find its height.

12. The area of a rectangle is 20 square yards. If it is 4 yards wide, find the length.

13. The area of parallelogram $ABCD$ is 3810 mm². If the base is 120 mm long, find the height.

The coordinates of the vertices of a quadrilateral are given. Graph the points and draw the quadrilateral and an altitude. Then identify the quadrilateral as a square, a rectangle, or a parallelogram and find its area.

14. (3, 4), (2, 1), (8, 4), (9, 7)

15. (3, 7), (-3, 3), (1, -3), (7, 1)

16. (6, 2), (1, 7), (-5, 5), (0, 0)

17. (0, -5), (3, -4), (1, 2), (-2, 1)

18. The areas of a rectangle and a parallelogram are equal. The rectangle has a length of 8 meters and a width of 6 meters. If the parallelogram has a base of 12 meters, find the height.

19. The area of a parallelogram is $2x^2 + 9x + 4$ cm^2. If the length of the base is $2x + 1$ cm, find the length of the altitude.

20. A rectangle is 4 cm longer than it is wide. The area of the rectangle is 117 cm^2. Find the length and the width.

21. If the length of each side of a square is doubled, the area of the resulting square is increased by 363 in^2. Find the length of the original square.

22. When the length of each side of a square is increased by 5 inches, the area of the resulting square is 2.25 times the area of the original square. What is the area of the original square?

23. The vertices of square *MNOP* are the midpoints of square *ABCD*. Explain why the area of *ABCD* is twice the area of *MNOP*.

Critical Thinking

24. The figure at the right consists of six congruent squares and has an area of 486 square centimeters. Find the perimeter of the figure.

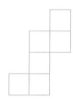

Applications

25. **Engineering**   A metal part is under stress when force is applied to stretch or compress it. When this force is applied perpendicular to the cross sectional area, the stress, S, in pounds per square inch that the part is under is found by the formula $S = \frac{F}{A}$, where F is the force in pounds and A is the cross sectional area. Find the stress on a rod whose cross section is a square with length 1.5 inches if a squeezing force of 3550 pounds is applied.

26. **Interior Design** Dale and Kathy Husted are having new carpet installed in their living room and dining room. A scale drawing of the rooms is shown at the right.
    a. Find the total area to be carpeted.
    b. The carpet store will not sell a fraction of a square yard. Also, extra yardage must be allowed for going around corners. If the Husteds add 4 square yards of carpet for waste, how many square yards of carpet will they buy?
    c. If the carpet Mr. and Mrs. Husted have chosen is $16.99 a square yard, find the cost of the carpeting.

Mixed Review

27. Insert one set of parentheses in the equation $4 \cdot 5 - 2 + 7 = 19$ so that the equation is true. **(Lesson 10-3)**

28. The measure of an interior angle of a regular polygon is 160. How many sides does the polygon have? **(Lesson 10-2)**

29. Write a two-column proof. **(Lesson 6-6)**
    Given: trapezoid $ABCD$
    $\overline{AB} \parallel \overline{DC}$
    Prove: $\angle A$ and $\angle D$ are supplementary.

Wrap-Up

30. Write a word problem that involves finding the area of a figure containing squares, rectangles, or parallelograms. Be sure to provide the answer to your problem.

～～～ MID-CHAPTER REVIEW ～～～

Classify each figure as a *convex polygon*, a *concave polygon*, or *not a polygon*. If a figure is a polygon, name it according to the number of sides. (Lesson 10-1)

1.  2.   3.   4.

The number of sides of a regular polygon is given. Find the measure of an interior angle and an exterior angle of the polygon. (Lesson 10-2)

5. 6 6. 16 7. 30

8. Use the problem-solving strategy guess and check, to find the least prime number greater than 720. **(Lesson 10-3)**

9. **Horticulture** Mr. Jackson is going to apply fertilizer to his lawn. If a 20-pound bag of fertilizer covers 500 square yards, will he need more than one bag? **(Lesson 10-4)**

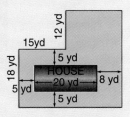

10-5 Area of Triangles, Rhombi, and Trapezoids

Objective

After studying this lesson, you should be able to:

- find the areas of triangles, rhombi, and trapezoids.

Application

When developers lay out the streets and lots in a new development, they often make lots that are not rectangular or square. Since the area of a lot must be listed in the legal description of the property, the developer must use geometry to find the area.

An important property of area is described in Postulate 10-2.

Postulate 10-2 | **Congruent figures have equal areas.**

In the previous lesson, we found the area of a parallelogram by multiplying the measure of the base by the height. The formula for the area of a triangle is derived from this formula.

Suppose we are given a parallelogram $ABCD$ with diagonal $\overline{AC}$. The opposite sides of a parallelogram are congruent, so $\overline{AB} \cong \overline{CD}$ and $\overline{BC} \cong \overline{DA}$. Also, $\overline{AC} \cong \overline{AC}$ because congruence of segments is reflexive. Therefore, $\triangle ABC \cong \triangle CDA$ by SSS. Postulate 10-2 says that congruent figures have equal areas, so the area of $\triangle ABC$ is equal to the area of $\triangle CDA$. Since the area of parallelogram $ABCD$ is the sum of the areas of its nonoverlapping parts, we can write the equation below.

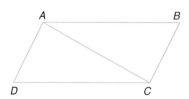

area of $\triangle ABC$ + area of $\triangle CDA$ = area of $\square ABCD$

$\qquad$ 2(area of $\triangle ABC$) = area of $\square ABCD$ $\qquad$ *area of $\triangle ABC$*

$\qquad$ 2(area of $\triangle ABC$) = bh $\qquad\qquad\qquad$ *= area of $\triangle CDA$*

$\qquad\qquad$ area of $\triangle ABC$ = $\frac{1}{2}bh$

This leads us to the formula for the area of a triangle.

Area of a Triangle | **If a triangle has an area of A square units, a base of b units, and a corresponding height of h units, then $A = \frac{1}{2}bh$.**

Example 1

Find each area.

a.

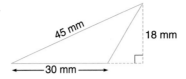

$A = \frac{1}{2}bh$ *Formula for the area of a triangle*

$= \frac{1}{2}(30)(18)$

$= 270 \text{ mm}^2$

b.

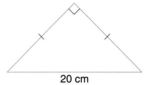

20 cm

An isosceles right triangle has base angles of 45° each. The length of the hypotenuse of a 45°-45°-90° triangle is $\sqrt{2}$ times the length of a leg. So the length of a leg is $\frac{20}{\sqrt{2}}$ cm.

$A = \frac{1}{2}bh$

$= \frac{1}{2}\left(\frac{20}{\sqrt{2}}\right)\left(\frac{20}{\sqrt{2}}\right)$ *The legs of a right triangle are the base and the height.*

$= \frac{1}{2}\left(\frac{400}{2}\right)$

$= 100 \text{ cm}^2$

The formula for the area of a parallelogram can be used to derive the formula for the area of a trapezoid.

INVESTIGATION

Make two copies of the trapezoid below and cut them out.

Position the trapezoids to form a parallelogram. If the lengths of the bases of the original trapezoid are b_1 units and b_2 units and its height is h units, what is the area of the parallelogram?

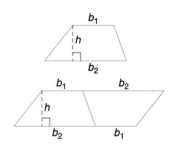

Since the two trapezoids are congruent, the area of each trapezoid is half the area of the parallelogram. What is the area of each trapezoid?

Area of a Trapezoid

If a trapezoid has an area of A square units, bases of b_1 units and b_2 units, and an height of h units, then $A = \frac{1}{2}h(b_1 + b_2)$.

Building lots are often in the shape of a trapezoid. So, the formula for the area of a trapezoid can be very useful to developers.

Example 2

Charlotte Burrows is an engineer laying out a new housing development. A scale drawing of lot 495 is shown at the right. Find the area of lot 495.

Since the bases are parallel and the 106-foot side is perpendicular to both bases, it is an altitude. So, $h = 106$, $b_1 = 94$, and $b_2 = 100$.

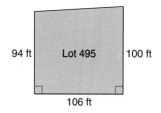

$$A = \frac{1}{2}h(b_1 + b_2)$$
$$= \frac{1}{2}(106)(94 + 100)$$
$$= 53(194)$$
$$= 10{,}282$$

The area of lot 495 is 10,282 square feet.

The formula for the area of a triangle can be used to derive the formula for the area of a rhombus. You will be asked to do this in Exercise 25.

Area of a Rhombus	If a rhombus has an area of A square units, and diagonals of d_1 and d_2 units, then $A = \frac{1}{2}d_1d_2$.

Example 3

A rhombus with a 48-inch diagonal has an area of 768 square inches. Find the length of a side of the rhombus.

$A = \frac{1}{2}d_1d_2$ *Formula for the area of a rhombus*

$768 = \frac{1}{2}(48)d_2$

$768 = 24d_2$

$32 = d_2$

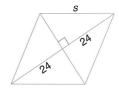

The diagonals are 48 and 32 inches long.

Since the diagonals are perpendicular and bisect each other, we can use the Pythagorean Theorem to find the length of a side.

$s^2 = 24^2 + 16^2$

$s^2 = 576 + 256$

$s^2 = 832$

$s \approx 28.8$

Each side is about 28.8 inches long.

Summary of Area Formulas	
Area of a square	$A = s^2$
Area of a rectangle	$A = \ell w$
Area of a parallelogram	$A = bh$
Area of a triangle	$A = \frac{1}{2}bh$
Area of a trapezoid	$A = \frac{1}{2}h(b_1 + b_2)$
Area of a rhombus	$A = \frac{1}{2}d_1 d_2$

CHECKING FOR UNDERSTANDING

Communicating Mathematics

Read and study the lesson to answer these questions.

1. The area of quadrilateral $FGHI$ is 76 square units. If quadrilateral $OPQR$ is congruent to quadrilateral $FGHI$, what is its area? Why?

2. An engineering book gives the formula for the area of a trapezoid as the product of the height and the length of the median. Is this formula valid? Justify your answer.

3. If the side of the trapezoid in Example 2 was not perpendicular to the bases, could Ms. Burrows find the area in the same way? Explain.

Guided Practice

Find the area of each figure.

4.

13 cm

8 cm

5.

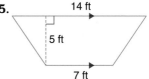

14 ft

5 ft

7 ft

6.

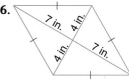

7 in. 4 in.

4 in. 7 in.

7.

6 ft

5 ft

8.

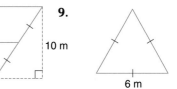

14 m

10 m

9.

6 m

10. A rhombus has a 120° angle and its longer diagonal is 10 inches long. Find the area of the rhombus.

11. The perimeter of a trapezoid is 29 inches. Its nonparallel sides are 4 inches and 5 inches long. If the height of the trapezoid is 3 inches, find its area.

EXERCISES

Practice   **Find each missing measure.**

12. The area of a triangle is 88 square units. If the height is 16 units, what is the length of the base?

13. The diagonals of a rhombus are 19 and 12 centimeters long. Find the area of the rhombus.

14. The area of a trapezoid is 96 square units. If its altitude is 6 units long, find the length of its median.

15. The altitude of a trapezoid is 11 m long. The bases are 16 m and 11 m long. Find the area.

16. A trapezoid has an area of 997.5 cm². If the altitude measures 21 cm and one base measures 40 cm, find the length of the other base.

17. A rhombus has a perimeter of 52 units and a diagonal 24 units long. Find the area of the rhombus.

18. The bases of a trapezoid are 30 miles and 20 miles. If the area is 850 square miles, find the length of the altitude.

19. One side of a triangle is 7 inches long and its corresponding altitude is 4 inches long. Find the length of the corresponding altitude of another side of the triangle whose length is 8 inches.

20. The measures of the consecutive sides of an isosceles trapezoid are in the ratio of 10:5:2:5. The perimeter of the trapezoid is 88 inches. If its height is 12 inches, find the area of the trapezoid.

21. The area of an isosceles trapezoid is 36 cm². The perimeter is 28 cm. If a leg is 5 cm long, find the height of the trapezoid.

22. Describe what the figure below represents.

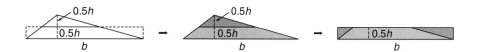

23. A triangle and a parallelogram of the same height have equal areas. How do their bases compare?

24. Trapezoid *TRAP* has diagonal $\overline{RP}$. Use the formula for the area of a triangle to derive the formula for the area of a trapezoid.

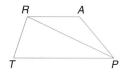

25. Use the formula for the area of a triangle to derive the formula for the area of a rhombus.

26. $\overline{QR}$ is the median of trapezoid $MNOP$. Are the areas of trapezoids $MQRP$ and $QNOR$ equal? Explain.

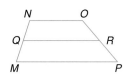

Critical Thinking **27.** The ancient Babylonians found the area of a quadrilateral whose sides had lengths of a, b, c, and d units using the formula $A = \dfrac{(a + c)(b + d)}{4}$. Does the formula work for a rectangle? Does it work for a rhombus?

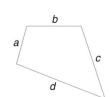

Application **28.** **Gardening** Hector is applying fertilizer to his lawn. According to the instructions on the package, he is to mix one scoop of fertilizer to one gallon of water for every fifty square feet of grass. Use the diagram at the right to determine how many gallons of fertilizer Hector should mix.

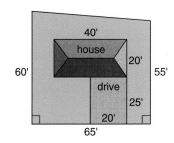

Computer **29.** The BASIC program at the right finds the area of a trapezoid given the height and the measures of the bases. Use the program to find the area of each trapezoid.

a. $h = 5$, $b_1 = 8$, $b_2 = 6$

b. $h = 3.5$, $b_1 = 7.1$, $b_2 = 8.4$

```
10 INPUT "ENTER THE HEIGHT OF THE
   TRAPEZOID."; H
20 INPUT "ENTER THE LENGTH OF ONE
   BASE OF THE TRAPEZOID."; B1
30 INPUT "ENTER THE LENGTH OF THE
   OTHER BASE OF THE TRAPEZOID.";
   B2
40 A = 0.5*H*(B1 + B2)
50 PRINT "THE AREA OF THE
   TRAPEZOID IS "; A; "SQUARE
   UNITS."
60 END
```

Mixed Review **30.** The sides of a parallelogram are 32 and 18 inches long. One angle of the parallelogram measures 45°. Find the area of the parallelogram. **(Lesson 10-4)**

31. Find the geometric mean between 7 and 14. **(Lesson 8-1)**

32. Find the next number in the pattern. **(Lesson 6-2)**

$$6, -2, \frac{2}{3}, -\frac{2}{9}, \underline{\ \ ?\ \ }$$

33. Can segments of measure 4, 9, and 21 form a triangle? Explain. **(Lesson 5-6)**

Wrap-Up **34.** **Journal Entry** Write a few sentences in your journal to describe what you think is the most useful formula you learned in this lesson. Explain.

10-6 Area of Regular Polygons

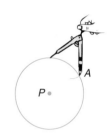

Objective After studying this lesson, you should be able to:

■ find areas of regular polygons.

Application

Keith Heyen built a deck in the shape of a regular hexagon with sides 12 feet long on the back of his house. He now needs to apply water sealant to the surface of the deck to protect the wood from the weather. Mr. Heyen needs to determine the total area of the deck in order to know how much sealant to buy.

Fortunately for Mr. Heyen, his daughter Lauren is taking geometry. Together they explored the area of a regular hexagon. To begin, Lauren remembered that all regular polygons can be inscribed in a circle. She showed her dad a construction she learned in class.

CONSTRUCTION

Construct a regular hexagon.

1. Use your compass to draw ⊙*P*. *P* will also be the **center** of the hexagon. The radius of ⊙*P* is the **radius** of the hexagon and is congruent to a side of the hexagon.

2. Using the same compass setting, place the compass point on the circle and draw an arc, labeling the point of intersection with the circle point *A*.

3. Place the compass point on *A* and draw another arc. Label the point of intersection with the circle *B*.

4. Continue this process, labeling points *C*, *D*, *E*, and *F* on the circle. Point *F* should be where the point of the compass was placed to locate point *A*.

5. Use a straightedge to connect *A*, *B*, *C*, *D*, *E*, and *F* consecutively. *ABCDEF* is a regular hexagon inscribed in ⊙*P*.

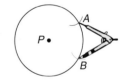

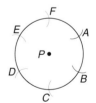

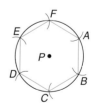

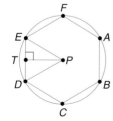

You can draw $\overline{PE}$ and $\overline{PD}$. You can also draw $\overline{PT}$ so that $\overline{PT}$ is perpendicular to $\overline{ED}$. A segment like $\overline{PT}$ that is drawn from the center of a regular polygon perpendicular to a side of the polygon is called an **apothem.**

$\triangle PED$ is an isosceles triangle, since sides $\overline{PE}$ and $\overline{PD}$ are radii of $\odot P$. If Lauren were to draw in all of the radii of hexagon $ABCDEF$, they would separate the hexagon into six triangles all congruent to $\triangle PED$.

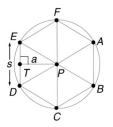

Now, since the area of a region is the sum of the areas of its nonoverlapping parts, Lauren and her dad can find the area of the hexagon by adding the areas of the triangles. Since $\overline{PT}$ is perpendicular to $\overline{ED}$, it is an altitude of $\triangle PED$ as well as an apothem of hexagon $ABCDEF$. Let a represent the measure of $\overline{PT}$ and s represent the length of a side of the hexagon.

$$\text{Area of } \triangle PED = \tfrac{1}{2}\,bh \qquad \textit{Formula for the area of a triangle}$$
$$= \tfrac{1}{2}\,sa$$

The area of one of the six triangles is $\frac{1}{2}sa$ units^2. So the area of the hexagon is $6\left(\frac{1}{2}\right)sa$ units^2. Notice that the perimeter of hexagon $ABCDEF$ is $6s$ units. Therefore, if the perimeter of the hexagon is P units, the area will be $\frac{1}{2}Pa$ units^2. This area formula can be used for any regular polygon.

Area of a Regular Polygon	**If a regular polygon has an area of A square units, a perimeter of P units, and an apothem of a units, then $A = \frac{1}{2}Pa$.**

Example 1

Refer to the opening application.

a. Find the area of Mr. Heyen's deck.

Look at the diagram of $\triangle PED$ with apothem $\overline{PT}$ at the top of the page. $\overline{PT}$ separates $\triangle PED$ into two 30°-60°-90° triangles. The length of the longer leg in a 30°-60°-90° triangle is $\sqrt{3}$ times the length of the shorter leg. The shorter leg of the triangle is $\frac{1}{2}(12)$ or 6 feet long, so the longer leg is $6\sqrt{3}$ feet long. Therefore, $a = 6\sqrt{3}$.

$$\text{Area of hexagonal deck} = \tfrac{1}{2}\,Pa \qquad \textit{Formula for the area of a regular polygon}$$
$$= \tfrac{1}{2}(6 \cdot 12)(6\sqrt{3}) \qquad \textit{The perimeter is six times the length of one side.}$$
$$= 36(6\sqrt{3})$$
$$= 216\sqrt{3}$$

The deck has an area of $216\sqrt{3}$ or about 374 square feet.

b. If one gallon of water sealant covers 200 square feet, how many gallons will Mr. Heyen need? If a one-gallon can of water sealant costs $11.99, how much will the water sealant for the deck cost?

Mr. Heyen will need $\frac{374}{200}$ or 1.87 gallons of water sealant. He should buy 2 one-gallon cans at a total cost of $23.98.

The angle formed by two radii drawn to consecutive vertices of a regular polygon is called a **central angle** of the polygon. All of the central angles of a regular polygon are congruent.

Example 2

Find the area of a regular pentagon whose perimeter is 45 inches.

Since the perimeter is 45 inches and the pentagon is regular, the length of each side is $\frac{45}{5}$ or 9 inches.

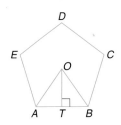

To use the formula for the area of a regular polygon, we must find the length of the apothem $\overline{OT}$. The central angles of $ABCDE$ are all congruent. Therefore, the measure of each one is $\frac{360}{5}$ or 72. Since $\overline{OT}$ is an apothem of pentagon $ABCDE$, it is perpendicular to $\overline{AB}$. It forms right triangle BOT with $m\angle BOT = 36$ and BT is 4.5 inches long.

Use the trigonometric ratios that we learned in Chapter 8.

$$\tan O = \frac{BT}{OT} \qquad tan = \frac{opposite}{adjacent}$$

$$\tan 36° = \frac{4.5}{OT}$$

$$\tan 36° \,(OT) = 4.5$$

$$OT = \frac{4.5}{\tan 36°} \qquad \textit{Use your calculator}$$

$$OT \approx 6.19$$

Now use the formula for the area of a regular polygon.

$$A = \tfrac{1}{2}Pa \qquad \textit{Area of a regular polygon}$$

$$A \approx \tfrac{1}{2}(45)(6.19) \qquad \textit{P = 45 and a ≈ 6.19.}$$

$$A \approx 139.3$$

The area of a regular pentagon with a perimeter of 45 inches is about 139 square inches.

CHECKING FOR UNDERSTANDING

Communicating Mathematics

Read and study the lesson to answer these questions.

1. Describe the difference between a radius and an apothem in a regular polygon.

2. What is the measure of a central angle of a regular polygon with 10 sides? a regular polygon with 24 sides? a regular polygon with n sides?

3. How is the formula for the area of a regular polygon related to the formula for the area of a triangle?

4. Describe how to find the measure of the apothem of a polygon if you know its radius.

Guided Practice

Use the figure below to answer each question. Triangle *ABC* is equilateral and inscribed in circle *O*, and $\overline{AB}$ is 10 cm long.

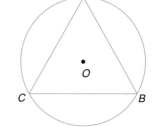

5. Find the perimeter of $\triangle ABC$.

6. Copy the figure. Draw radius $\overline{OA}$ and construct apothem $\overline{OT}$ to side $\overline{AC}$.

7. What type of right triangle is $\triangle OTA$? Explain.

8. What is the length of $\overline{AT}$? Why?

9. Draw right triangle *OTA* separately and label the measure of the angles. Use the properties of right triangles to find the lengths of $\overline{OT}$ and $\overline{OA}$.

10. Find the area of $\triangle ABC$ using the formula for regular polygons.

Find the perimeter, the measure of a central angle, the length of an apothem, and the area for each regular polygon.

11.

6 cm

12.

9 in.

13.

2 m

EXERCISES

Find the area of each regular polygon. Round your answers to the nearest tenth.

14. an equilateral triangle with an apothem 5.8 centimeters long and a side 20 centimeters long

15. a square with a side 16 inches long and an apothem 8 inches long

16. a hexagon with a side 19.1 millimeters long and an apothem 16.5 millimeters long

17. a pentagon with an apothem 8.9 miles long and a side 13.0 miles long

18. a hexagon with an apothem 8.7 meters long and a side 10 meters long

19. an octagon with an apothem 7.5 feet long and a side 6.2 feet long

Find the area of each orange region. Assume that polygons that appear to be regular are regular.

20.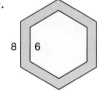

21.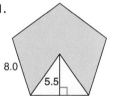

22.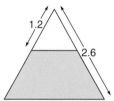

23.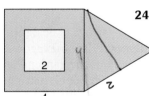

24.

25.

Find the perimeter, the length of the apothem, and the area of each regular polygon.

26.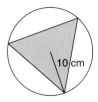

27.

28.

29. *ABCDEF* is a regular hexagon. Find the length of each side.

30. Find the ratio of the area of square *ABCD* to the area of square *BFCE*.

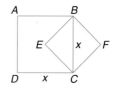

31. Find the perimeter and area of a regular decagon with a radius of 4 feet.

32. A circle inscribes a regular hexagon and circumscribes another. If the radius of the circle is 10 units long, find the ratio of the area of the smaller hexagon to the area of the larger hexagon.

33. Draw a regular octagon with sides 1 inch long.
 a. Draw segments to join the midpoints of four nonadjacent sides of the octagon. What figure is produced?
 b. Find the area of the figure formed by joining the midpoints of the nonadjacent sides of the octagon. Round your answer to the nearest hundredth.
 c. Find the area of the region that is inside of the octagon but outside of the figure.

Critical Thinking

34. Use the formula for the area of a regular polygon to show that the area of an equilateral triangle with side length *s* units is $\frac{\sqrt{3}}{4} s^2$ units^2.

Applications

35. Gardening Dave and Kelly Rea want to install a fence to keep animals out of their rose garden. Each side of the triangular garden is 18 feet long.
 a. Find the amount of fencing material that they will need to buy.
 b. What is the area of their garden to the nearest tenth of a square foot?

36. Construction Carol Nystrom is designing the garden area of the new Inniswood Park. There will be a hexagonal bench with a hexagonal opening around each tree in the garden. The perimeter of each bench is 36 feet and the opening in each bench has a perimeter of 12 feet. How many cans of stain should Ms. Nystrom order for the seats of the benches if there will be 14 benches in the garden and one can of stain covers 175 square feet?

Mixed Review

37. Find the area of a rhombus whose diagonals are 17 and 24 feet long. **(Lesson 10-5)**

38. Can a triangle have sides with lengths of 18, 32, and 67 inches? If not, why not? **(Lesson 5-6)**

39. Write the conditional *Congruent figures have equal areas* in if-then form. Then identify the hypothesis and the conclusion. **(Lesson 2-2)**

Wrap-Up

40. Name three terms from this lesson connected to regular polygons. Which term was important in finding the area of a regular polygon?

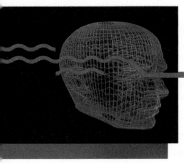

Technology

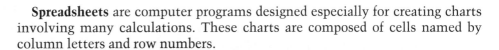

Area

BASIC
Geometric Supposer
Graphing calculators
LOGO
▶ **Spreadsheets**

Spreadsheets are computer programs designed especially for creating charts involving many calculations. These charts are composed of cells named by column letters and row numbers.

The spreadsheet below is set up to find the area of a regular polygon. Cells A1, B1, C1, D1, and E1 hold the labels N, S, P, A, and AREA for the number of sides, the length of a side, the perimeter, the length of the apothem, and the area of the polygon. Cells A3 to A7 hold the values for the number of sides of each polygon, cells B3 to B7 hold the values for the length of a side of each polygon, cells C3 to C7 hold the values for the perimeter of each polygon, cells D3 to D7 hold the length of the apothem of each polygon, and cells E3 to E7 hold the area of the polygon.

```
=====A=======B=========C=========D=======E=====
1:   N        S         P         A        AREA
2: -------------------------------------------------
3:   4        6         24        4.2      50.4
4:   6        6         36        5.2      93.6
5:   3        12        36        3.5      63
6:   7        14        98        14.5     710.5
7:   10       8         80        12.3     492
```

Each cell in column C holds a formula to find the perimeter of the regular polygon described in that row. Since cells A3 and B3 hold the values for the number of sides in the polygon and the measure of each side, the computer will multiply these together to find the perimeter. So, cell C3 holds the formula A3 ⋆ B3. The cells in column E hold a formula to compute the area of the regular polygon.

EXERCISES

1. What formula does cell E3 of the spreadsheet hold for finding the area of a regular polygon?

2. Why does the formula contained in cell C3 of the spreadsheet work for finding the perimeter of the polygon?

3. Set up a spreadsheet program to find the area of a trapezoid.

10-7 Area and Circumference of a Circle

Objectives

After studying this lesson, you should be able to:
- find the circumference of a circle, and
- find the area of a circle.

Application

Each time Jessica's bicycle wheel makes one revolution, the distance it travels is the same as the **circumference** of the wheel. The circumference of a circle is the distance around the circle. If the wheel on Jessica's bike has a radius of 12 inches, how many revolutions will the wheel make while Jessica travels down the block 500 feet? *You will answer this question in Example 1.*

Start   Turning   All the way around

The circumference of a circle can be approximated by considering the perimeter of regular polygons. Notice that as the number of sides of the inscribed polygon increases, the perimeter of the polygon becomes closer to the circumference of the circle.

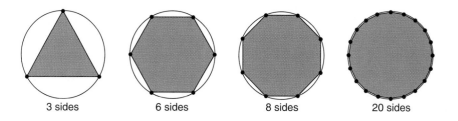

3 sides   6 sides   8 sides   20 sides

The chart below shows the approximate perimeter and area of each regular polygon inscribed in a circle of radius r.

Number of Sides	3	4	5	6	8	10	20	50	100
Measure of a Side	$1.73r$	$1.41r$	$1.18r$	r	$0.77r$	$0.62r$	$0.31r$	$0.126r$	$0.0628r$
Perimeter	$5.20r$	$5.64r$	$5.90r$	$6.00r$	$6.16r$	$6.20r$	$6.20r$	$6.30r$	$6.28r$
Measure of Apothem	$0.5r$	$0.71r$	$0.81r$	$0.87r$	$0.92r$	$0.95r$	$0.99r$	$0.998r$	$0.9995r$
Area	$1.30r^2$	$2.00r^2$	$2.39r^2$	$2.61r^2$	$2.83r^2$	$2.95r^2$	$3.07r^2$	$3.14r^2$	$3.14r^2$

Look at the rows for perimeter and area. Notice that as the number of sides of the inscribed polygon increases, the perimeter and the area both approach a different limiting number. The perimeter approaches the circumference of the circle, and the area approaches the area of the circle. Notice that 6.28 = 2 · 3.14 and 3.14 is an approximation of the irrational number called π (pi). So, $6.28r \approx 2\pi r$ and $3.14r^2 \approx \pi r^2$.

Circumference of a Circle	**If a circle has a circumference of C units and a radius of r units, then $C = 2\pi r$.**
Area of a Circle	**If a circle has an area of A square units and a radius of r units, then $A = \pi r^2$.**

Example 1

Bicycling

The wheel on Jessica's bike has a radius of 12 inches. How many revolutions will the wheel make while Jessica travels 500 feet?

For each revolution that the wheel makes, it travels a distance equal to the circumference of the wheel. So, we must first find the circumference of the wheel.

$$C = 2\pi r \qquad \textit{Formula for the circumference of a circle}$$
$$C = 2\pi(12)$$

Use a calculator to find the circumference. *If your calculator doesn't have a π key, use 3.14 to approximate π.*

Enter: 2 ⊠ π ⊠ 12 = 75.398224

The circumference is about 75 inches or 6.25 feet. The wheel will travel about 6.25 feet each time it makes a revolution. Now, find the number of revolutions that the wheel will make in 500 feet.

$$\text{number of revolutions} = \text{total distance} \div \text{distance per revolution}$$
$$\approx 500 \div 6.25$$
$$\approx 80$$

The wheel will make about 80 revolutions as it travels 500 feet.

Example 2

Find the area of the shaded region.

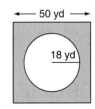

← 50 yd →

18 yd

$$\text{area of shaded region} = \text{area of square} - \text{area of circle}$$
$$= s^2 \qquad - \qquad \pi r^2$$
$$= (50)^2 - \pi(18)^2$$
$$= 2500 - 324\pi$$
$$\approx 1482.12$$

The area of the shaded region is about 1482 square yards.

You can use what you learned about the general equation for a circle to find the area of the circle.

Example 3

Find the area of a circle with equation $(x - 2)^2 + (y + 5)^2 = 81$.

The general form of the equation for a circle with center (h, k) is $(x - h)^2 + (y - k)^2 = r^2$, where r is the radius of the circle. Therefore, the radius of the given circle is $\sqrt{81}$ or 9. Now use the formula for the area of a circle.

$$\begin{aligned} A &= \pi r^2 \quad \text{\textit{Formula for area of a circle}} \\ &= \pi(9)^2 \\ &= 81\pi \end{aligned}$$

The area of the circle is 81π or about 254.5 square units.

CHECKING FOR UNDERSTANDING

Communicating Mathematics

Read and study the lesson to answer these questions.

1. π is the first letter in a Greek word that means "measure around". Why do you think π is used in the formula for the circumference of a circle?

2. The circumference of a circle is sometimes expressed as $C = \pi d$, where d is the measure of the diameter of the circle. Why is this formula equivalent to $C = 2\pi r$?

3. In stating the exact values of the circumference and area for a circle with diameter 24 feet, Alan tells Mary that the circumference is 24π feet and the area is 576π square feet. Explain why Alan is only partially correct.

Guided Practice

Copy and complete the chart from the information given about different circles. Give exact answers using π when necessary.

	r	d	C	A
4.	12 cm	?	?	?
5.	?	4.8 km	?	?
6.	?	?	20π in.	?
7.	?	?	?	81π ft^2

8. The area of a circular pool is approximately 7,850 ft^2. The owner has decided to place a fence around the pool. If the fence is five feet from the edge of the pool, how many feet of fencing will be needed?

EXERCISES

Practice

Find the circumference of a circle with a radius of the given length. Round your answers to the nearest tenth.

9. 10 m

10. 4 in.

11. 7 yd

12. 3.6 km

13. 1.1 mm

14. $\frac{1}{4}$ mi

Find the area of a circle with a radius of the given length. Round your answers to the nearest tenth.

15. 18 in.

16. 5.8 m

17. 9.7 km

18. 0.4 in.

19. $3\frac{1}{3}$ yd

20. $4\sqrt{5}$ cm

State the measure of the radius of a circle with each equation. Then find the area of each circle. Give exact answers using π.

21. $x^2 + y^2 = 121$

22. $(x - 3)^2 + y^2 = 49$

23. $(x - 6)^2 + y^2 = 11$

24. $(x - 8)^2 + (y - 6)^2 = 15$

25. $x^2 + (y + 1)^2 = 625$

26. $(x + 1)^2 + (y + 2)^2 = 68$

27. Find the circumference and the area of a circle inscribed in a square whose sides are 6 meters long.

28. The diagonal of a square is 8 feet long. Find the circumference and the area of a circle inscribed in the square.

Find the area of each shaded region. Assume that all polygons are regular. Express each answer as an exact number involving π and a decimal rounded to the nearest hundredth.

29.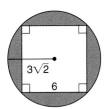

30.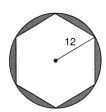

31.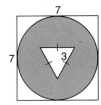

32.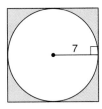

33.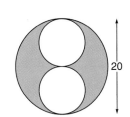

34.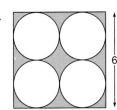

35.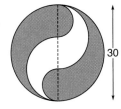

36.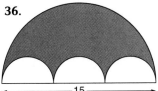

37.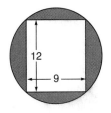

38. A circle is inscribed in a rhombus whose diagonals are 24 and 32 feet long.
   a. Find the area of the circle.
   b. What is the area of the region that is inside the rhombus but outside the circle?

Critical Thinking

39. Find the area of the shaded region if $r = 2, 3, 4, 5, 6,$ and 8. What is the relationship between the area of the shaded region and the area of the large circle?

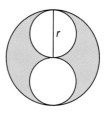

40. The ratio of the circumferences for two circles is 3:5. Is the ratio for the areas the same? Explain.

Applications

41. **City Planning**   A new system of sirens is being considered for the city of Grandview. The system will be used to alert the city's residents in the event of a tornado or other severe weather. The sound emitted from each siren will travel up to 1.5 miles. Find the area of the city that will benefit from each siren.

42. **Sports**   A diagram of the new practice track and football field at Edison High School is shown at the right. Find the amount of sod it would take to cover the area inside the track.

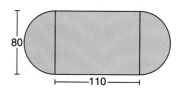

43. **Home Economics**   A recipe for a 14-inch pizza calls for a 15-ounce can of pizza sauce. How much sauce would be needed for a 10-inch pizza? (A 14-inch pizza has a diameter of 14 inches.)

44. **Sewing**   Carolyn is making a circular tablecloth for a table whose diameter is 48 inches. The top of the table is 28 inches from the floor. If the tablecloth is to reach to the floor, how many yards of lace should Carolyn buy to place around the edge of the tablecloth?

Mixed Review

45. Find the area of a regular hexagon with sides 10 inches long. Round your answer to the nearest hundredth. **(Lesson 10-6)**

46. A chord that is 16 centimeters long is 6 centimeters from the center of a circle. Find the length of the radius of the circle. **(Lesson 9-3)**

47. **Gardening**   A flower bed is in the shape of an obtuse triangle. One angle measures 103°, and the side opposite is 14 feet long. The shortest side is 7.5 feet long. Find the measures of the remaining side and angles. **(Lesson 8-6)**

48. The measures of the angles in a triangle are $x + 16$, $8x + 7$, and $11x - 3$. Is the triangle acute, obtuse, right, or equiangular? **(Lesson 4-2)**

49. Two supplementary angles have measures of $6y + 14$ and $22y - 2$. Find the value of y. **(Lesson 1-8)**

Wrap-Up

50. **Portfolio**   Choose the application problem that you think is most interesting. Write a complete solution and explanation of this problem and place it in your portfolio.

10-8 Geometric Probability

Objective   After studying this lesson, you should be able to:
- use area to solve problems involving geometric probability.

Application   WGEO radio station is having a "Call In To Win" Contest. The song of the day is announced at 7:00 A.M. each day. Then, sometime during each hour of the day, the song of the day is played. The first person to call the station when the song of the day begins to play wins $100. If you turn on your radio at 12:20 P.M., what is the probability that you have not missed the start of the song of the day played sometime between 12:00 P.M. and 1:00 P.M.?

This problem can be solved using **geometric probability.** Geometric probability involves using the principles of length and area to find the probability of an event. One of the principles of geometric probability is stated in Postulate 10-3.

Postulate 10-3 *Length Probability* *Postulate*	**If a point on $\overline{AB}$ is chosen at random and C is between A and B, then the probability that the point is on $\overline{AC}$ is $\dfrac{\text{length of } \overline{AC}}{\text{length of } \overline{AB}}$.**

To find the probability that you have not missed the start of the song of the day, draw a line segment to represent the time from 12:00 to 1:00. Point C represents the time that you started listening.

The length of $\overline{AB}$ represents the entire hour. The length of $\overline{AC}$ represents the time from 12:00 to 12:20, and the length of $\overline{CB}$ represents the time from 12:20 to 1:00.

You have not missed the song of the day if the station plays the song after 12:20. So, the probability of having not missed the song is $\dfrac{\text{length of } \overline{CB}}{\text{length of } \overline{AB}}$.

$$P(\text{not missed the song}) = \frac{\text{length of } \overline{CB}}{\text{length of } \overline{AB}}$$
$$= \frac{40 \text{ units}}{60 \text{ units}}$$
$$= \frac{2}{3}$$

The probability that you have not missed the song of the day is $\frac{2}{3}$.

Postulate 10-4 states a second principle of geometric probability.

Postulate 10-4 Area Probability Postulate	If a point in region A is chosen at random, then the probability that the point is in region B, which is in the interior of region A, is $\dfrac{\text{area of region B}}{\text{area of region A}}$.

Example 1 uses the area probability postulate.

Example 1

APPLICATION

Entertainment

The children at Joshua's birthday party are playing a game for prizes. Each child tosses a beanbag at the target on the floor. Depending on where one marked corner of the beanbag lands, prizes are given. Red gets a jumbo squirtgun; blue gets a yo-yo; green gets a candy bar. If the corner of a beanbag lands on the target, find the probability of a child winning each prize. Round your answers to the nearest hundredth.

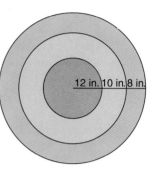

12 in. 10 in. 8 in.

Use geometric probability to find the probability of the corner of the bag landing in each region.

Red area: probability $= \dfrac{\text{area of red circle}}{\text{area of target}}$ *Area probability postulate*

$= \dfrac{\pi(12)^2}{\pi(30)^2}$ *Formula for area of a circle*

$= \dfrac{144\pi}{900\pi}$

$= \dfrac{4}{25}$ or 0.16

Blue area: probability $= \dfrac{\text{area of blue ring}}{\text{area of target}}$

$= \dfrac{\pi(22)^2 - \pi(12)^2}{\pi(30)^2}$ *Area of blue ring = area of blue*

$= \dfrac{484\pi - 144\pi}{900\pi}$ *and red circle −*

area of red circle.

$= \dfrac{340\pi}{900\pi}$

$= \dfrac{17}{45}$ or about 0.38

Green area: probability $= \dfrac{\text{area of green ring}}{\text{area of target}}$

$= \dfrac{\pi(30)^2 - \pi(22)^2}{\pi(30)^2}$ *Area of green ring = area of target −*

area of blue

$= \dfrac{900\pi - 484\pi}{900\pi}$ *and red circle.*

$= \dfrac{416\pi}{900\pi}$

$= \dfrac{104}{225}$ or about 0.46

The probability of winning a jumbo squirtgun is 0.16, the probability of winning a yo-yo is about 0.38, and the probability of winning a candy bar is about 0.46.

Sometimes when you are finding a geometric probability, you will need to find the area of a **sector of a circle.** A sector of a circle is a region of a circle bounded by a central angle and its intercepted arc.

Area of a Sector of a Circle	If a sector of a circle has an area of A square units, a central angle measuring $N°$, and a radius of r units, then $A = \frac{N}{360}\pi r^2$.

Example 2

If Mitzi gets a 3 on her next spin, she will win the game. What is the probability that Mitzi will spin a 3?

Find the area of the sector of the circle and the area of the circle.

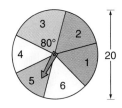

Area of sector	*Area of circle*
$A = \dfrac{N}{360}\pi r^2$	$A = \pi r^2$
$= \dfrac{80}{360}(\pi)(10)^2$	$= \pi(10)^2$
$= \dfrac{200}{9}\pi$ or $22\frac{2}{9}\pi$	$= 100\pi$

Now find the geometric probability.

$$\text{probability} = \frac{\text{area of sector}}{\text{area of circle}}$$
$$= \frac{22\frac{2}{9}\pi}{100\pi}$$
$$= \frac{2}{9}$$

The probability Mitzi will spin a 3 is $\frac{2}{9}$.

CHECKING FOR UNDERSTANDING

Communicating Mathematics

Read and study the lesson to answer these questions.

1. A point is chosen at random from the interior of region X. Explain how to find the probability that the point is in region Y if region Y is in region X.

2. Suppose that the radius of the red circle of the target in Example 1 is increased to 14 inches and the widths of the blue and green rings remain the same. Find the probability of winning each prize.

3. Describe a situation where geometric probability may be used in everyday life.

Find the probability that a point chosen at random on $\overline{AG}$ is on each segment.

A		B	C				D		E	F	G
0	1	2	3	4	5	6	7	8	9	10	

4. $\overline{AC}$ **5.** $\overline{AE}$ **6.** $\overline{CF}$

7. $\overline{DE}$ **8.** $\overline{AG}$ **9.** $\overline{GC}$

Find the probability that a point chosen at random in each figure is in the shaded region. Assume polygons that appear to be regular are regular. Round your answer to the nearest hundredth.

10.

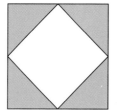

11.

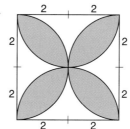

12.

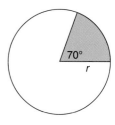

EXERCISES

13. A point is chosen at random on $\overline{XY}$. If Z is the midpoint of $\overline{XY}$, W is the midpoint of $\overline{XZ}$, and V is the midpoint of $\overline{WY}$, what is the probability that the point is on $\overline{XV}$?

14. Points A, B, C, D, and E are collinear. $\overline{CA} \cong \overline{AE}$, $\overline{AD} \cong \overline{DE}$, and $\overline{CA} \cong \overline{EB}$.
   a. Draw a diagram of the figure.
   b. If another point M is on the line above, what is the probability that M is between C and D?
   c. If another point N is on the line above, what is the probability that N is between A and B?

15. To win at a carnival game, you must throw a dart at a board that is 6 feet by 3 feet and hit one of the 25 playing cards on the board. The playing cards are each $2\frac{1}{2}$ by $3\frac{1}{2}$ inches.
   a. Draw a diagram of the dartboard.
   b. What is the probability that a randomly thrown dart that hits the board hits a playing card? Round your answer to the nearest hundredth.
   c. Does the arrangement of the cards on the board affect the probability? Explain.

16. Find the probability for each outcome on the game spinner. Round your answers to the nearest hundredth.

a. lose a turn
b. choose a card
c. advance 2
d. collect $500

17. You tell a friend that you will arrive at the mall sometime between 12:00 and 12:30. Since she is not sure she will be able to come, your friend says not to wait for her for longer than 10 minutes. If your friend comes at 12:25, what is the probability that she missed you?

18. The method of using probability to approximate the area of a region is called the *Monte Carlo method*. To use the Monte Carlo method, choose points randomly in a figure. The percentage of the points that fall in a region is the same as the percentage of the total area that is in the region. For example, suppose you throw 100 darts at a square with sides 1 foot long and 22 of those darts land in the red region. The area of the red region is approximately 22% of the total area or 0.22 square feet.

a. Peg cut out a map of the 48 contiguous United States and placed it on a dartboard. She randomly threw 300 darts at the board and 252 of them landed on the map. Of the 252, 25 darts landed in Texas. If the area of the contiguous United States is approximately 2,962,000 square miles, what is the approximate area of Texas?

b. Thirteen of the 252 darts landed in Montana. What is the approximate area of Montana?

c. If Peg threw 100 more darts, do you think her approximations would get better? Explain.

Critical Thinking

19. The bull's eye on Jim's dartboard has one-sixteenth of the area of the dartboard. If Jim hits the dartboard on two out of every three throws, what is the probability that Jim will hit the bull's eye in one throw? *Hint: probability of hitting bull's eye in one throw = $\dfrac{\text{probability of hitting bull's eye}}{\text{probability of hitting dartboard}}$*

Applications

20. **Gardening** A sprinkler with a range of 5 meters waters a portion of the lawn. If the head of the sprinkler can rotate 150°, find the area of the lawn that is watered. Round to the nearest tenth of a square meter.

21. **Entertainment** You can win a large stuffed animal at a game on the State Fair Midway if you can toss a quarter onto a grid board so that it doesn't touch a line. If the sides of the squares of the grid are 32 millimeters long and the radius of a quarter is 12 millimeters, what is the probability that you will win? *Hint: Look at the area in which the center of the coin could land so that the edges will not touch a line.*

Mixed Review

22. Find the circumference and the area of a circle with a diameter of 1.6 in. Round your answers to the nearest tenth. **(Lesson 10-7)**

23. Find the geometric mean between 26 and 44. **(Lesson 8-1)**

Wrap-Up

24. Write an application problem that involves geometric probability. Be sure to give the answer to your problem.

Objectives

After studying this lesson, you should be able to:
- recognize nodes and edges as used in graph theory,
- determine if a network is traceable, and
- determine if a network is complete.

Application

Corporate Air provides daily air service between Richmond and Washington, D.C., Richmond and New York City, New York City and Washington, D.C., Washington, D.C. and Philadelphia, and Philadelphia and Pittsburgh. Their daily routes can be represented using a **network** like the one at the right. Such a diagram illustrates a branch of mathematics called **graph theory.** In a network, the points are called **nodes** and the paths connecting the nodes are called **edges.** Edges may be straight or curved.

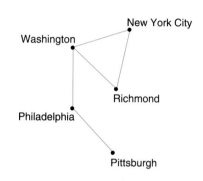

Straight edges can be used to form *open* or *closed* graphs. If an edge of a closed graph intersects exactly two other edges only at their endpoints, then the graph forms a polygon.

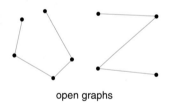

open graphs

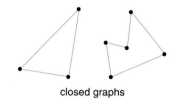

closed graphs

The study of networks began with a famous problem about the bridges in the city of Königsberg, which is now Kaliningrad. The city had seven bridges connecting both sides of the Pregel River to two islands in the river. The problem was to find a path that would take you over all seven bridges without crossing the same bridge twice.

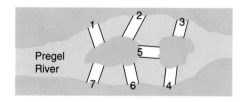

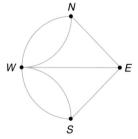

W and E represent the islands, N and S are the sides of the river.

The problem caught the attention of eighteenth century mathematician Leonhard Euler (pronounced OY-lur). He studied the problem by modeling it as a network. The edges of the network on the left represent the bridges and the nodes represent the sides of the river and the two islands. Can you trace this network with your finger without lifting your finger or retracing an edge? If you can, then the network is **traceable.**

The degree of a node is the number of edges that are connected to that node. The traceability of a network is related to the degrees of the nodes in the network.

Example 1

Find the degree of each node in the network of the Königsberg bridges.

Node	Edges at the node	Degree of node
N	B1, B2, B3	3
E	B3, B5, B4	3
S	B4, B6, B7	3
W	B1, B2, B5, B6, B7	5

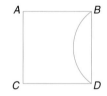

You probably concluded that the network that represents the Königsberg bridges is not traceable. Two tests for traceability are described below.

A network is traceable if and only if one of the following is true:
  1. All of the nodes in the network have even degrees.
  2. Exactly two nodes in the network have odd degrees.

Look back at Example 1. All of the nodes in the network for the bridge problem have an odd degree. Neither of the conditions for traceability is met. Therefore, there is no path that you could walk to travel over all seven bridges without recrossing at least one of the bridges.

Example 2

Determine if each network is traceable.

a.

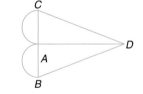

Node	Degree
A	5
B	3
C	3
D	3

No; all of the nodes have odd degrees.

b.

Node	Degree
A	2
B	3
C	2
D	3

Yes; exactly two nodes, B and D, have odd degrees.

Try to trace the network in Example 2b with your finger. Does it matter where you start to trace? If you start at node A it doesn't seem to work. Euler noticed that when a path goes through a node, it uses two edges. When a traceable network has an odd node, it must be a starting or finishing point of the traceability path.

Not all the pairs of nodes are connected by an edge in some networks. A network like this is called **incomplete**. A **complete network** has at least one path between each pair of nodes.

Example 3

Determine if each network is complete. If a network is not complete, name the edges that must be added to make the network complete.

a.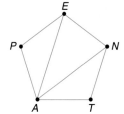

b.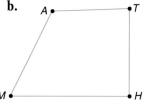

c.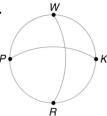

No; edges must be added between nodes *P* and *N*, *P* and *T*, and *E* and *T*.

No; edges must be added between nodes *A* and *H* and nodes *M* and *T*.

Yes; all the pairs of nodes are connected by an edge.

CHECKING FOR UNDERSTANDING

Communicating Mathematics

Read and study the lesson to answer these questions.

1. Draw an example of a network. Is your network traceable? Is it complete?

2. List the two tests for the traceability of a network.

3. Is the network at the right complete? Explain.

4. Mark and Judy are discussing where they should start to trace the network at the right. Mark says node *M* and Judy says node *L*. Who is right and why?

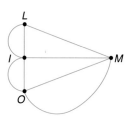

Guided Practice

Find the degree of each node in each network.

5.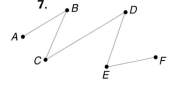

6.

7.

Determine if each network is traceable.

8.

9.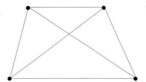

10.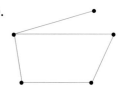

Determine if each network is complete. If a network is not complete, name the edges that must be added to make the network complete.

11.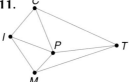

12.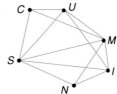

13.

EXERCISES

Practice   **Find the degree of each node in the network.**

14.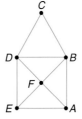

15.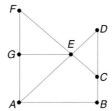

16.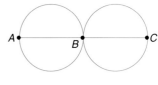

Name the additional edges that must be drawn for each network to be complete.

17.   P •

    E •

        • N

18.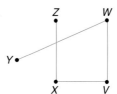

19.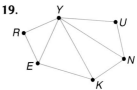

20. Each of the networks below are complete.

  a. List the number of nodes and edges in each network.
  b. Do you notice a pattern in the number of edges in the complete networks? Describe the pattern.
  c. How many edges would a complete network with 10 nodes have?

Determine whether each network is traceable. If a network is traceable, copy it on your paper and use arrows to show the direction of the paths taken to trace the network.

21.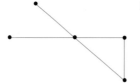

22.

23.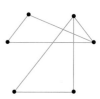

24. In 1875, the town of Königsberg built an eighth bridge. The additional bridge is shown in red in the network at the right. Did this bridge allow the people to walk over all eight bridges without crossing the same bridge twice? Explain.

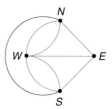

25. Add one or more edges to the network at the right to make the network traceable.

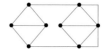

26. Construct a traceable network with 11 nodes and 18 edges.

27. If you can use your pencil to draw a path through a network so that the path starts and ends at two different nodes and no edge is passed more than once, the path is called an *Euler path*. If the path can be drawn so that the path starts and ends at the same node and no edge is passed more than once, the path is called an *Euler circuit*.

   a. Does the network at the right have an Euler path?

   b. Does the network contain an Euler circuit?

   c. Do you think a network could contain both an Euler path and an Euler circuit? Explain.

Critical Thinking 28. Find the degree of each node in the network at the right. Then find the sum of the degrees of all the nodes. Compare this sum to the number of edges in the network. Is there a relationship between the number of edges and the sum of the degrees of the nodes? Draw several networks to verify your answer.

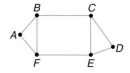

Applications

29. Travel The Budgetminded Bus Company offers service between Chicago and Bloomington, Chicago and Springfield, Springfield and Fort Wayne, Columbus and Fort Wayne, and Columbus and Chicago.
 a. Draw a network to represent the bus routes.
 b. Could you take a bus from Columbus to Springfield by passing through one other city?

30. Business The network to the right represents streets on Tom's newspaper route. To maximize efficiency, he would like to ride his bike down each street exactly once. Is this possible? Why or why not?

31. Construction The new campus for the Rosepoint Medical Center will have seven buildings built around the edge of a circular courtyard. There will be a sidewalk between each pair of buildings.
 a. What type of network does this represent?
 b. How many sidewalks will there be?

Mixed Review

32. Find the probability that a randomly thrown dart that lands on the board will land in the red region. Round your answer to the nearest hundredth. **(Lesson 10-8)**

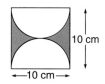

10 cm

←—10 cm—→

33. Write the given statement and draw the figure for a proof of the statement *If two lines intersect, then at least one plane contains both lines.* Then write the assumption you would make to write an indirect proof. **(Lesson 5-4)**

34. What property justifies the statement $m\angle K = m\angle K$? **(Lesson 2-4)**

35. Draw and label a figure showing planes *M*, *N*, and *L* that do not intersect. **(Lesson 1-2)**

Wrap-Up

36. Journal Entry Write a sentence or two in your journal to describe what you have learned about networks.

∼∼∼ DEVELOPING REASONING SKILLS ∼∼∼

The game of Sprouts begins with any number of dots or nodes. Each player takes a turn by connecting one of the nodes with another node or with itself and then placing another node on the edge connecting the nodes. No edge can intersect itself or another edge, and a maximum of three edges can meet at any node. The last person to be able to move wins.
A three-dot game of Sprouts is shown below.

| Begin | First Move | Second Move | Third Move | Fourth Move | Fifth Move | Sixth Move | Last Move |

10 SUMMARY AND REVIEW

VOCABULARY

Upon completing this chapter, you should be
familiar with the following terms:

484 altitude	**513** degree of a node	**512** nodes
495 apothem	**468** edge	**466** polygon
483 area	**468** face	**468** polyhedron
484 base	**507** geometric probability	**467** regular
497 central angle	**484** height	**509** sector of a circle
502 circumference	**514** incomplete network	**501** spreadsheet
514 complete network	**512** network	**512** traceable
466 concave	**467** *n*-gon	**468** vertex
466 convex		

SKILLS AND CONCEPTS

OBJECTIVES AND EXAMPLES

Upon completing this chapter, you should
be able to:

- classify and identify parts of
  polygons and polyhedrons
  (Lesson 10-1)

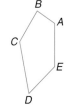

The vertices of the polygon
are *A, B, C, D,* and *E.* The
edges are $\overline{AB}, \overline{BC}, \overline{CD},$
$\overline{DE},$ and $\overline{AE}.$ It is convex and is not regular.

- find measure of angles in polygons.
  (Lesson 10-2)

*Interior Angle Sum Theorem (Theorem
10-1)* If a convex polygon has *n* sides and *S*
is the sum of the measures of its interior
angles, then $S = 180(n - 2)$.

*Exterior Angle Sum Theorem (Theorem
10-2)* If a polygon is convex, then the sum
of the measures of the exterior angles, one
at each vertex, is 360.

REVIEW EXERCISES

Use these exercises to review and prepare
for the chapter test.

**Refer to the polyhedron
at the right.**

1. Name the vertices and
   edges.
2. Name the faces.
3. Is the polyhedron
   regular? Explain.

4. Find the measure of an interior angle of
   a regular decagon.

5. The sum of the measures of the interior
   angles of a convex polygon is 1980.
   How many sides does the polygon
   have?

6. A regular polygon has 20 sides. Find the
   measure of an interior angle and an
   exterior angle of the polygon.

OBJECTIVES AND EXAMPLES	REVIEW EXERCISES

■ find areas of parallelograms. **(Lesson 10-4)**

Find the area of the parallelogram.

$A = bh$ *Formula for area of a parallelogram*

$A = (65)(36)$ *b = 65, h = 36.*

$A = 2340$ square centimeters

Find the missing measure of each quadrilateral.

7. A rectangle has a base of 7 feet and an altitude of 2.9 feet. Find the area.

8. The area of parallelogram $ABCD$ is 134.19 in². If the base is 18.9 inches long, find the height.

9. If the length of each side of a square is tripled, the area of the resulting square is increased by 648 m². Find the length of the original square.

■ find areas of triangles, rhombi, and trapezoids. **(Lesson 10-5)**

Area Formulas	
Triangle	$A = \frac{1}{2}bh$
Trapezoid	$A = \frac{1}{2}h(b_1 + b_2)$
Rhombus	$A = \frac{1}{2}d_1d_2$

Find each missing measure.

10. The area of a triangle is 48 mm². If the height is 6 mm, find the length of the base.

11. A rhombus has diagonals 8.6 cm and 6.3 cm long. What is its area?

12. The area of an isosceles trapezoid is 7.2 ft². The perimeter is 6.8 ft. If a leg is 1.9 ft long, find the height.

■ find areas of regular polygons. **(Lesson 10-6)**

Find the area of a regular hexagon with an apothem 12 cm long.

Since $\triangle ABC$ is a 30°-60°-90° triangle, $\frac{1}{2}s = \frac{12}{\sqrt{3}}$ or $4\sqrt{3}$. So, $s = 8\sqrt{3}$ and $P = 48\sqrt{3}$.

$A = \frac{1}{2}Pa$

$A = \frac{1}{2}(48\sqrt{3})(12)$

$A = 288\sqrt{3}$ or about 498.8 cm²

Find the area of each regular polygon. Round your answers to the nearest tenth.

13. an equilateral triangle with an apothem 8.9 inches long

14. a pentagon with an apothem 0.4 feet long

15. a hexagon with sides 64 millimeters long

16. a square with an apothem n centimeters long

■ find the circumference and area of a circle. **(Lesson 10-7)**

If a circle has a radius of r units,

$$\text{Area} = \pi r^2$$
$$\text{Circumference} = 2\pi r$$

Find the circumference and area of a circle with a radius of the given length. Round your answers to the nearest tenth.

17. 7 mm

18. 19 in.

19. 0.9 ft

OBJECTIVES AND EXAMPLES	REVIEW EXERCISES

■ find geometric probabilities. **(Lesson 10-8)**

Length Probability Postulate (Postulate 10-3)
If a point on $\overline{AB}$ is chosen at random and C is between A and B, then the probability that the point is on $\overline{AC}$ is $\dfrac{\text{length of } \overline{AC}}{\text{length of } \overline{AB}}$.

Area Probability Postulate (Postulate 10-4)
If a point in region A is chosen at random, then the probability that the point is in region B, which is in region A, is
$\dfrac{\text{area of region } B}{\text{area of region } A}$.

20. During the morning rush hour, a bus arrives at the bus stop at Indianola and Morse Roads every seven minutes. A bus will wait for thirty seconds before departing. If you arrive at a random time, what is the probability that there will be a bus waiting?

■ determine if a network is traceable or complete. **(Lesson 10-9)**

The network at the right is traceable, since exactly two nodes have odd degrees. However, it is not complete since there is no edge between M and A.

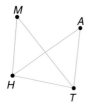

Use the network at the right to answer each question.

21. Name the nodes and edges in the network.
22. Is the network traceable?
23. Is the network complete? If not, what edges need to be added for the network to be complete?

APPLICATIONS AND CONNECTIONS

24. **Number Theory** Using the digits 1, 2, 3, 4, 5, and 6 only once, find two whole numbers whose product is as great as possible. **(Lesson 10-3)**

25. **Interior Design** John and Cynthia are buying paint for the walls of their kitchen. Three of the walls are 8 feet long and one is ten feet long. The ceiling is 8 feet high. If a gallon of paint covers 400 square feet, will one gallon be enough to paint the kitchen? **(Lesson 10-4)**

26. **Manufacturing** A wooden planter has a square base and sides that are shaped like trapezoids. An edge of the base is 8 inches long and the top edge of each side is 10 inches long. If the height of a side is 12 inches, how much wood does it take to make a planter? **(Lesson 10-5)**

Use the polygon at the right for 1-7.

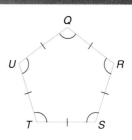

1. Name the vertices of the polygon.

2. Name the sides of the polygon.

3. Classify the polygon by the number of sides.

4. Classify the polygon as convex or concave.

5. Classify the polygon as regular or not regular. Explain.

6. Find the sum of the measures of the interior angles of the polygon.

7. Find the measure of one exterior angle of the polygon.

8. Use the strategy of guess and check to find the least prime number greater than 720.

9. Find the area of a parallelogram with a base of 4.95 feet and a height of 0.51 feet.

10. A square has a perimeter of 258 inches. Find the length of one side and the area of the square.

11. A triangle has a base of 16 feet and a height of 30.6 feet. Find its area.

12. The area of triangle ABC is $3x^2 + 6x$. If the height is $3x + 6$ units, find the measure of the base.

13. The median of a trapezoid is 13 feet 6 inches long. If the height is 10 feet, what is the area of the trapezoid?

14. A regular hexagon has sides 10 centimeters long. What is the area?

Circle O has a diameter of 4 inches. Find the missing measures. Round your answers to the nearest tenth.

15. What is the area of circle O?

16. What is the circumference of circle O?

17. A sector of circle O has a central angle measuring 30°. Find the area of the sector.

A circular dartboard has a diameter of 18 inches. The bull's eye has a diameter of 3 inches and the blue ring around the bull's eye is 4 inches wide.

18. What is the probability that a randomly thrown dart that hits the dartboard will land in the bull's eye?

19. What is the probability that a randomly thrown dart that hits the dartboard will land in the blue ring?

20. Draw a complete network with 6 nodes. How many edges does it have?

Bonus Ed dropped 100 ball bearings from a ladder and 92 of them landed on the figure shown at the right. How many of the ball bearings do you think fell in the green region?

College Entrance Exam Preview

The questions on these pages involve comparing two quantities, one in Column A and one in Column B. In certain questions, information related to one or both quantities is centered above them.

Directions:
Write A if the quantity in Column A is greater. Write B if the quantity in Column B is greater. Write C if the quantities are equal. Write D if there is not enough information to determine the relationship.

Column A	Column B
1.	$x = 4$
$x^2 + 3$	$5x - 1$
2.	$x > y$
x^2	y^2
3.	$A \# B = 5A - 2B$
$\left(\frac{3}{4}\right) \# 6$	$5 \# \left(\frac{3}{4}\right)$
4. a given chord in a given circle	the radius of the same circle
5.	
the area of the shaded region	the area of the unshaded region
6. $\lvert -a \rvert$	a
7.	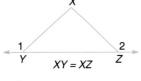
$m\angle 1$	$m\angle 2$

Column A	Column B
8. circumference of circle with radius $2x$	perimeter of square with side πx
9.	$s > 0, t < 0$
st	$-\frac{s}{t}$
10.	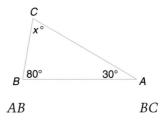
AB	BC
11. the area of a circle with diameter 10	the area of a right triangle with hypotenuse 10
12.	$0, -1, 2, -3, 4, -5, \ldots$
the product of the first and twentieth numbers of the sequence	the product of the fourth and nineteenth numbers of the sequence
13.	$0, -1, 2, -3, 4, -5, \ldots$
the sum of the first twenty numbers of the sequence	0
14.	$\overline{AC} \cong \overline{DF}, \angle A \cong \angle F, \angle D \cong \angle C$
	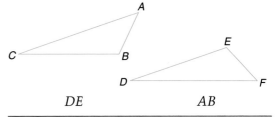
DE	AB
15.	$x > y$
$x - y$	$y - x$

Solve. You may use a calculator.

16. The figure below is formed from a semicircle and a rectangle. The diameter of the semicircle is the length of the rectangle. Write a formula for the area of the figure if the length of the rectangle is $3x$ and the width is x.

17. An airplane can travel 1716 feet in 3 seconds. Find the speed of the plane in miles per hour.

18. In the figure below, P and Q lie on circle O. $\overline{OP}$ is 8 units long, and $\angle POQ$ measures 80°. Find the length of minor arc PQ.

19. Find the solution set for $\dfrac{7}{x-3} - \dfrac{1}{2} > \dfrac{3}{x-4}$.

20. A survey of 100 adults showed that 65% of the women were registered to vote and 55% of the men were registered to vote. If 40 men were surveyed, how many people surveyed were not registered to vote?

CHAPTER 11

Surface Area and Volume

CHAPTER OBJECTIVES

In this chapter, you will:
- Solve problems by making models.
- Draw three-dimensional figures.
- Find lateral areas, surface areas, and volumes of solids.

GEOMETRY AROUND THE WORLD
Egypt

Most people know that the three Great Pyramids near Cairo, Egypt, house the bodies of the pharaohs who ordered them to be built. But few realize that the structures also show some of what the ancient Egyptians knew about geometry.

Originally coated with white casing stones, which have long since worn away, the Great Pyramids are named for the Pharaohs Chephren, Mycerine, and Cheops. The pyramid Cheops, especially, has great geometric significance.

Dividing the perimeter of Cheops' base by twice its height results in the value 3.14—remarkably close to the value of π (pi). Also significant is the fact that the measure of the area of each lateral face of the pyramid is approximately equal to the measure of its height. Furthermore, the slant height of each lateral face, when divided by one half of its base length, comes exceptionally close to 1.618, or ϕ (phi). This value has for centuries been associated with the "golden rectangle"—one whose ratio of length to width equals ϕ.

GEOMETRY IN ACTION

The figure below shows the dimensions of the Cheops Pyramid. Find the perimeter of the base. Then divide this value by twice the height of the pyramid to verify that it is close to the value of π.

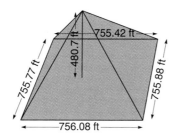

King Cheops' Great Pyramid was the single largest structure of its day. Workers spent 20 years fashioning it from more than 2 million stone blocks. It is the tomb of Cheops, who died in 2567 B.C.

◄ *The Great Pyramids* Inset: *Archaeologists working in Egypt*

11-1 Problem-Solving Strategy: Make a Model

Objective

After studying this lesson, you should be able to:

• solve problems by making a model.

Models are often useful in solving problems. A model can be, among other things, a simple sketch, a precise scale drawing, or an actual three-dimensional object.

Example

Selvi has 24 feet of fencing to make a rectangular outdoor pen for her dog Snickers. She wants Snickers to have the largest amount of space possible. What dimensions should she use to make the pen?

To solve this problem, Selvi draws sketches of possible dimensions for the pen. She then calculates the area of each possible pen.

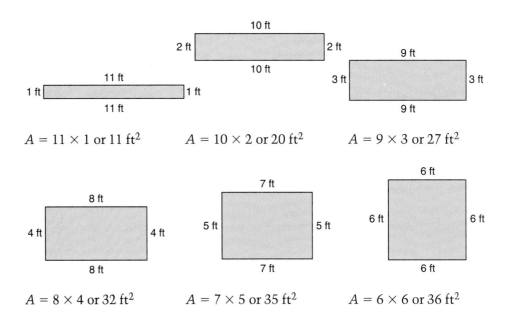

$A = 11 \times 1$ or 11 ft^2 $A = 10 \times 2$ or 20 ft^2 $A = 9 \times 3$ or 27 ft^2

$A = 8 \times 4$ or 32 ft^2 $A = 7 \times 5$ or 35 ft^2 $A = 6 \times 6$ or 36 ft^2

A square pen 6 feet on each side would have the largest area, 36 ft^2. *Check your answer by graphing the area function, $y = x(12 - x)$, on your graphing calculator. Is 36 the maximum value of y?*

In this chapter, models of three-dimensional figures such as cubes, rectangular prisms, cylinders, cones, and spheres may help you to understand concepts and to solve problems.

CHECKING FOR UNDERSTANDING

Communicating Mathematics

Read and study the lesson to answer these questions.

1. In the example, Selvi only used whole numbers for the measures of the sides. Do you think Selvi could get a larger area if she used fractional measurements? Why or why not?

2. Ron has 16 square tiles. Each side of each tile is 1 inch long. Ron wants to arrange the tiles so that the perimeter is 22 inches. Describe a way to solve this problem using models.

3. An interior decorator is planning a furniture arrangement for a living room. He makes a model for possible arrangements by drawing a scale drawing of the room and cutting out pieces of paper to represent the furniture using the same scale.

   a. What are the advantages of using such a model?

   b. What are the disadvantages of using models to plan furniture arrangements?

Guided Practice

Make a model to solve each problem.

4. A circular fountain has a diameter of 9 feet. A 3-foot wide flower garden surrounds the fountain. A sidewalk circles the garden. If the sidewalk is 2 feet wide, find the total area of the fountain, garden, and sidewalk.

5. There are 32 volleyball teams playing in a single-elimination tournament. In other words, any team that loses a game is out of the tournament. How many games will be played in the tournament?

6. A train is 1 mile long. It travels through a tunnel 1 mile long at the speed of 60 miles per hour. How much time elapses from the time the engine enters the tunnel until the entire train is out of the tunnel?

EXERCISES

Practice

Solve. Use any strategy.

Strategies
Look for a pattern.
Solve a simpler problem.
Act it out.
Guess and check.
Draw a diagram.
Make a chart.
Work backward.

7. Insert two addition signs and two subtraction signs on the left side of the equals sign to make the statement true.

$$1\ 2\ 3\ 4\ 5\ 6\ 7\ 8\ 9 = 116$$

8. The escalator between the first and second floors of Cole's Department Store travels at a rate of x steps per second. Kelly walked up the escalator at a rate of one step per second and reached the top in 10 seconds. Dominic walked up the escalator at a rate of three steps per second and reached the top in 6 seconds. If the escalator is turned off, how many steps would you have to climb to reach the second floor?

9. Lisa is studying a certain type of bacteria in the laboratory. She knows this bacteria doubles its population every 8 hours. At the end of the second full day, there are 1600 bacteria. How many bacteria did Lisa have at the beginning of the first day?

10. Consider the perfect squares between 0 and 5000. What percent of these numbers are odd numbers?

11. Consider the following game. The red markers must always move to the right. The blue markers must always move to the left. A marker can move into an empty space next to it or jump one marker of the other color if it lands on an empty space. How many moves will it take to switch the positions of the red and blue markers?

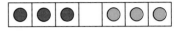

12. Consider a variation of the game in Exercise 11 where there are 4 red markers, 4 blue markers, and 9 spaces. How many moves will it take to switch the positions of the markers?

13. During a certain year, the community of Oxford spent 65% of its income on services, 25% on public safety, 4% on recreation, and 6% on administration. Construct a circle graph to show how Oxford spent its income.

14. There are four children in the Monroe family: Andy, Sally, Lisa, and John. Andy is a middle child. Lisa is one of the two older children. John is the "baby" of the family. Sally always looks up to her older sister and her older brother. List the siblings from the youngest to the oldest.

15. Two pennies, a nickel, a dime, and a quarter are in a change purse. Three coins are removed at random. What are the possible values of the coins?

COOPERATIVE LEARNING PROJECT

Work in groups. Each person in the group must understand the solution and be able to explain it to any person in class.

The diagram at the right shows how a 9×4 rectangle can be cut into two pieces that can be arranged to form a 6×6 square. Draw a 16×9 rectangle and a 25×16 rectangle and cut each into two pieces to form a 12×12 square and a 20×20 square respectively.

11-2 Exploring Surface Area

Objectives

After studying this lesson, you should be able to:
- create, draw, and fold three-dimensional figures, and
- make two-dimensional nets for three-dimensional solids.

Application

Aircraft manufacturers take surface area very seriously. The **surface area** of a three-dimensional object or **solid** is the sum of the areas of its faces. It is used to determine the amount of material it takes to cover the outer surface of an airplane and to determine the design of the aircraft.

The surface area of the wing on an aircraft is used to determine a design factor known as wing-loading. Wing-loading is the total weight of the aircraft and its load on take-off divided by the total surface area of its wings. If the wing-loading factor is exceeded, the pilot must either reduce the fuel load or remove passengers or cargo.

Surface area is important in other areas of everyday life. For example, we need to consider it when we resurface a highway, wallpaper a room, wrap a birthday gift, fertilize a field, or paint a house.

In this lesson, we will consider surface areas that are related to geometric solids with flat surfaces or faces. Solids with flat surfaces are called **polyhedrons** or **polyhedra.** The faces or flat surfaces are polygons, and the lines where the faces intersect are called edges.

The polyhedron at the right is called a **pyramid.** Its faces are $\triangle ABC$, $\triangle BCD$, $\triangle ACD$, and $\triangle ABD$. Points A, B, C, and D are vertices of the polyhedron. Its edges are $\overline{AB}$, $\overline{BD}$, $\overline{DA}$, $\overline{CA}$, $\overline{CB}$, and $\overline{CD}$.

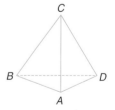

Some other polyhedrons are shown below.

Rectangular Solid

Square Pyramid

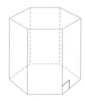

Hexagonal Solid

A sketch of a rectangular solid is more complex than a sketch of a plane figure because it is three-dimensional. The use of isometric dot paper makes it easier to draw some three-dimensional figures.

Using isometric dot paper, sketch a rectangular solid 4 units high, 6 units long, and 2 units wide.

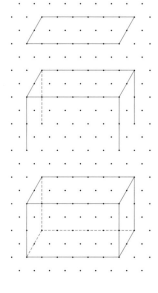

Step 1:
Draw the top of the solid 6 × 2 units.

Step 2:
Draw a segment 4 units down from each vertex. Hidden edges are shown by dashed lines.

Step 3:
Connect the corresponding vertices.

When a polyhedron such as the one at the right is unfolded, the result is a two-dimensional figure known as a **net.** Nets are very useful in helping us see the polygonal shapes whose areas must be computed in finding the total surface area of the polyhedron.

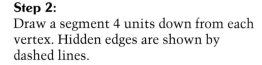

Example 1

Use rectangular dot paper to draw a net for the rectangular solid in the Investigation.

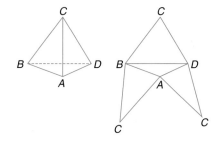

Often there are several different nets that could be used to produce the same solid. Could you design a different one for Example 1?

CHECKING FOR UNDERSTANDING

Communicating Mathematics

Read and study the lesson to answer these questions.

1. What is the difference between isometric and rectangular dot paper? Why is each type of paper important?

2. Describe a net and explain why nets are important.

3. How is wing-loading in the aircraft industry related to the geometric concept of surface area?

4. Why won't the figure at the right produce a cube when folded?

Guided Practice

Use isometric dot paper to draw each polyhedron.

5. a rectangular solid 3 units high, 6 units long, and 4 units wide

6. a cube 5 units on each edge

Given each polyhedron, copy its net and label the remaining vertices.

7.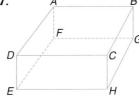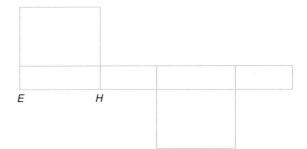

8.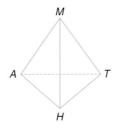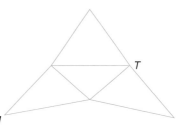

Identify the number and type of polygons that are faces in each polyhedron.

9.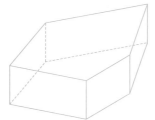

10.

Match each net to one of the solids at its right.

11.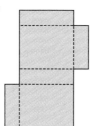

a.

b.

c.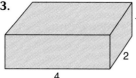

12.

a.

b.

c.

Use rectangular dot paper to draw a net for each solid.

13.

14.

EXERCISES

Practice   Determine whether figure is a net for a rectangular solid.
Write *yes* or *no.*

15.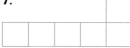

16.

17.

18.

19.

20.

Match each net to one of the solids at its right.

21.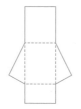

a.

b.

c.

22.

a.

b.

c.

Identify the number and types of polygons that are faces in each polyhedron.

23.

24.

25.

Nets are not drawn to scale.

Given each polyhedron, copy its net and label the remaining vertices.

26.
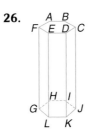

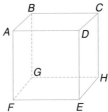

27.

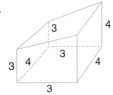

Use rectangular dot paper to draw a net for each solid.

28.
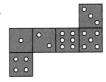

29.
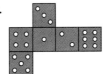

30.

A standard die as used in most board games is shown at the right. The opposite faces are numbered so that their sum is 7. Determine whether each numbered net can be folded to result in a standard die.

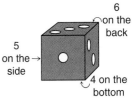

31.
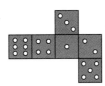

32.
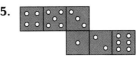

33.
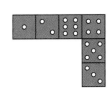

34.
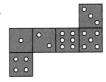

35.
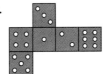

36.

Critical Thinking

37. Design three different nets that could produce a rectangular solid 4 units long, 2 units wide, and 1 unit high. Then lay out a cutting pattern on an 18 × 30 rectangular grid using the net that will allow you to produce the largest number of complete rectangular solids and at the same time minimize the waste in the grid.

Applications

38. Manufacturing When cardboard boxes are manufactured, the flat cutout resembles a net. However there are a few differences. Examine the pattern for a cardboard box. Sketch a net for the same geometric solid. Explain the differences between the cardboard box pattern and your net. Give a reason for these differences.

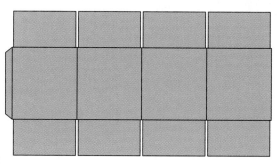

39. Aircraft Compute the wing-loading factor for the following aircraft based on their maximum takeoff weight and total surface area of their wings.

Aircraft	Maximum Takeoff Weight	Surface Area of Wings	Wing-loading Factor in lb/ft^2
Wright brothers' plane	750 lb	532 ft^2	_?_
Four-passenger plane	1150 lb	117 ft^2	_?_
Supersonic transport	750,000 lb	7700 ft^2	_?_

Mixed Review

40. A rectangular painting is 30 inches by 20 inches. The painting is surrounded by a 3-inch matting and placed in a frame that adds another 5 inches on each side. What is the total area of the painting, matting, and frame? **(Lesson 11-1)**

41. Brianna's little brother's tricycle tire has a diameter of 12 inches. About how far does the tricycle travel in one turn of the wheel? **(Lesson 10-7)**

42. Find the area of a triangle with a base 7 feet long and a height of 2 feet. **(Lesson 10-5)**

43. Complete each statement. Refer to the figure at the right. **(Lesson 5-7)**

   a. If $AB = BC$, and $m\angle 1 < m\angle 2$, then $AD \underline{\ ?\ } CD$.

   b. If $AD = DC$ and $AB \underline{\ ?\ } BC$, then $m\angle 3 > m\angle 4$.

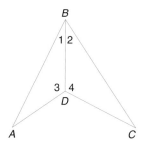

Wrap-Up

44. Describe the connection between finding the total surface area of a geometric solid and what you studied in this lesson.

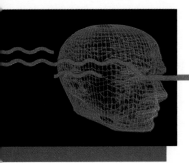

Technology

BASIC
Geometric Supposer
Graphing calculators
▶ **LOGO**
Spreadsheets

Drawing Three-Dimensional Figures

LOGO can be used to draw three-dimensional figures. Remember that the drawing of many solid figures involves the use of parallel lines. By using special pairs of angles and what you know about supplementary angles, you can draw the figure below using LOGO commands.

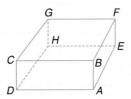

FD 40 LT 90 FD 80 LT 90 FD 40 LT 90 FD 80 LT 90   *Draws ABCD.*
RT 30 FD 70 BK 70   *Rotates the turtle and draws $\overline{AE}$.*

PU LT 30 FD 40 PD   *Rotates the turtle and moves it to B without drawing.*
RT 30 FD 70 BK 70   *Rotates the turtle and draws $\overline{BF}$.*

PU LT 120 FD 80 RT 90 PD   *Rotates turtle ; moves it to C without drawing.*
RT 30 FD 70 BK 70   *Rotates the turtle and draws $\overline{CG}$.*

LT 30 BK 40 RT 30 FD 70 LT 30   *Moves the turtle to D and draws $\overline{DH}$.*

FD 40 RT 90 FD 80 RT 90 FD 40 RT 90 FD 80   *Draws HGFE.*

Study the commands to draw the prism. Do you see any patterns in the groups of steps?

EXERCISES

1. Draw another rectangular solid using different lengths for the sides and different angle measures to position the parallel lines.

2. Experiment with the LOGO commands to draw different parallelograms. Devise a plan for drawing a solid that has parallelograms for its bases.

3. Use LOGO commands to draw a triangle. Then draw a solid that has triangles for its bases.

11-3 Surface Area of Prisms and Cylinders

Objectives

After studying this lesson, you should be able to:
- identify the parts of prisms and cylinders, and
- find the lateral areas and surface areas of right prisms and right cylinders.

Application

A certain inn provides shower caps for its guests. Each shower cap is packaged in a box such as the one shown at the right. The box is a special kind of polyhedron called a **prism.** More specifically, it is a rhombohedron.

Other examples of prisms are cubes and rectangular solids. Prisms have the following characteristics.

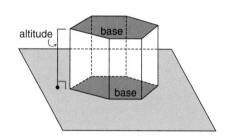

- Two faces, called **bases,** are formed by congruent polygons that lie in parallel planes.

- The faces that are not bases, called **lateral faces,** are formed by parallelograms.

- The intersections of two adjacent lateral faces are called **lateral edges** and are parallel segments.

An **altitude** of a prism is a segment perpendicular to the planes containing the two bases, with an endpoint in each plane. The length of an altitude of a prism is called the **height** of the prism. If the lateral edges of a prism are also altitudes, then the prism is a **right prism.** Otherwise, the prism is an **oblique prism.**

A prism can be classified by the shape of its bases.

Lateral faces of right prisms are rectangles.

Right triangular prism

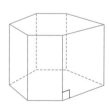

Right hexagonal prism

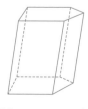

Oblique pentagonal prism

Lateral edges of oblique prisms are not altitudes.

One way to find the **surface area** (T) of a prism is to add the areas of each of the faces. Another way is to develop and use a formula. We must first find a formula for the area of the lateral faces.

The **lateral area** (L) of a prism is the area of all the lateral faces. As a result of the Distributive Property, the lateral area of a right prism can be found by multiplying the height by the perimeter (P) of the base as shown below. Note that a, b, c, d, e, and f are the measures of the sides of the base.

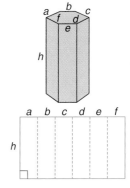

$$L = ah + bh + ch + dh + eh + fh$$
$$= (a + b + c + d + e + f)h$$
$$= Ph$$

The formula for the lateral area of any right prism is generalized as follows.

Lateral Area of a Right Prism	**If a right prism has a lateral area of L square units, a height of h units, and each base has a perimeter of P units, then $L = Ph$.**

Since the bases are congruent, they have the same area (B). The surface area (T) of a right prism is found by adding the lateral area to the area of both bases ($2B$).

Surface Area of a Right Prism	**If the total surface area of a right prism is T square units, its height is h units, and each base has an area of B square units and a perimeter of P units, then $T = Ph + 2B$.**

Example 1

Find the lateral area and surface area of a right triangular prism with a height of 15 inches and a right triangular base with legs of 4 and 3 inches.

First, use the Pythagorean Theorem to find c, the measure of the hypotenuse. Then use the value of c to find the measure of the perimeter.

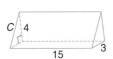

$c^2 = 3^2 + 4^2$ $P = 4 + 3 + c$ $B = \frac{1}{2} bh$
$c^2 = 25$ $\quad = 4 + 3 + 5$ $\quad = \frac{1}{2}(4)(3)$
$\;\; c = 5$ $\quad = 12$ $\quad = 6$

$L = Ph$ $T = L + 2B$ Thus, the lateral area is
$\quad = (12)(15)$ $\quad = 180 + 2(6)$ 180 in^2, and the surface
$\quad = 180$ $\quad = 192$ area is 192 in^2.

Another type of solid is a **cylinder.** The **bases** of a cylinder are formed by two congruent circles that lie in parallel planes. The segment whose endpoints are centers of these circles is called the **axis** of the cylinder.

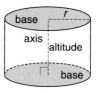
Right cylinder

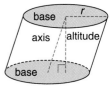
Oblique cylinder

An altitude of a **cylinder** is a segment perpendicular to the planes containing the bases with an endpoint in each plane. If the axis of a cylinder is also an altitude of the cylinder, then the cylinder is called a **right cylinder.** Otherwise, the cylinder is an **oblique cylinder.**

A net for a right cylinder would resemble the figure at the right.

The lateral area of a cylinder is the area of the curved surface. The curved surface is a rectangle whose width is the height of the cylinder, h units, and whose length is the circumference of one of its bases, $2\pi r$ units.

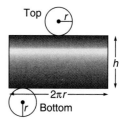

Lateral Area of a Right Cylinder	If a right cylinder has a lateral area of L square units, a height of h units, and the bases have radii of r units, then $L = 2\pi rh$.

The surface area of a cylinder is the sum of the lateral area and the areas of the bases.

Surface Area of a Right Cylinder	If a right cylinder has a total surface area of T square units, a height of h units, and the bases have radii of r units, then $T = 2\pi rh + 2\pi r^2$.

Example 2

Find the number of square feet of cardboard needed to make the sides of 30,000 frozen concentrate orange juice cans if each can is 6 inches tall and 3 inches in diameter.

If the diameter of the base of a can is 3 inches long, the radius is 1.5 inches long.

$L = 2\pi rh$

$\quad = 2\pi(1.5)(6)$ *Substitute 1.5 for r and 6 for h.*

$\quad = 18\pi$

$\quad \approx 56.5$

The lateral area of each can is about 56.5 square inches. So, the lateral area of 30,000 cans is about 30,000 × 56.5 or about 1,695,000 square inches. Since there are 144 square inches in a square foot, about 1,695,000 ÷ 144 or 11,771 square feet of cardboard are needed.

CHECKING FOR UNDERSTANDING

Communicating Mathematics

Read and study the lesson to answer these questions.

1. Refer to the shower cap box on page 536.
   a. What are the shapes of the polygonal faces that make up the shower cap box? Be specific.
   b. Why do you think it is called a rhombohedron?
   c. Draw a net of the shower cap box on isometric dot paper.

2. Describe the difference between lateral area and surface area.

3. If a printer is making labels for canned soup, what dimensions are important for him or her to know? The area of the label is closely related to which concept you studied in this lesson?

4. Describe in your own words the difference between a right prism and an oblique prism.

5. Where are the endpoints of the axis of a cylinder located in relation to its bases? Illustrate your answer with a diagram.

Guided Practice

Find the surface area of each right rectangular prism. Round your answers to the nearest tenth.

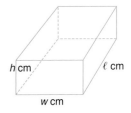

6. $\ell = 20$, $w = 5$, and $h = 3$

7. $\ell = 14.5$, $w = 7.5$, and $h = 8$

Find the surface area of each right triangular prism. The triangular bases are equilateral. Round your answers to the nearest tenth.

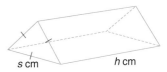

8. $s = 5$ and $h = 12$

9. $s = 6.2$ and $h = 20.4$

Find the surface area of each cylinder. Round your answers to the nearest tenth.

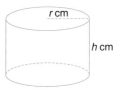

10. $r = 4$ and $h = 8$

11. $r = 6.8$ and $h = 17$

12. Find the radius of a cylinder that has a height of 35 inches and a lateral area of 630 square inches.

EXERCISES

Use the prism at the right to answer each of the following.

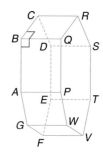

13. Is the prism a right prism or an oblique prism?

14. What is the polygonal shape of its bases and lateral faces?

15. If the bases are regular polygons with sides 6 units long, find the perimeter of the base.

16. If the perimeter of the base is 56 units and the length of a lateral edge is 7 units, find the lateral area of the prism.

Use the right cylinder at the right to answer each of the following. Express all answers in terms of π.

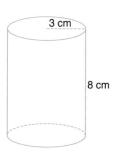

17. Find the circumference of a base.

18. Find the lateral area.

19. Find the area of the base.

20. Find the surface area.

Find the lateral area and surface area of each right prism. Round your answers to the nearest tenth. Assume that polygons that appear regular are regular.

21.

22.

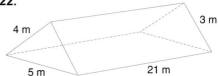

23.

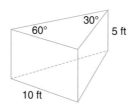

24.

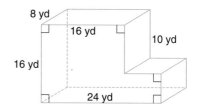

25. Find the surface area of a cube whose lateral edge is 8 units long.

26. If the lateral area of a right rectangular prism is 144 square centimeters, its length is three times its width, and its height is twice its width, find its surface area.

27. Find the surface area of a cylindrical water tank that is 8 meters tall and has a diameter of 8 meters.

Find the surface area of each oblique rectangular prism. Round your answers to the nearest tenth.

28.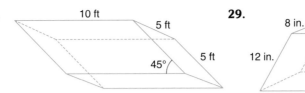

29.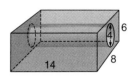

30. Find the total surface area of the solid at the right, including the surface area inside the cylindrical hole.

Critical Thinking

31. Suppose you are designing a solid figure with two congruent parallel bases. The distance around each base is fixed, and the height is always the same. If you wanted a maximum surface area, what shape would you make the base? Justify your answer.

Applications

32. **Manufacturing** A rectangular cake pan is 9 inches by 13 inches and 2 inches deep. It needs to be coated with a non-stick coating. What is the area of the surface to be coated?

33. **Agriculture** Over the years the acid associated with filling a silo with silage eats away at the cement sides, resulting in spoilage. The acid also weakens the cement walls and may seriously damage its structural integrity. In order to properly maintain a silo, a farmer must have the inside of the silo resurfaced. The cost of the process is directly related to the lateral area of the inside of the silo. Find the lateral area of a silo 40 feet tall with an interior diameter of 16 feet.

34. **Manufacturing** Gum manufacturers always cover sticks of gum with paper.

    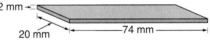

    a. Find the surface area of the stick of gum at the right.
    b. If the foil wrapper for the gum is designed to overlap 2 millimeters and have foldover ends of 6.5 millimeters on each of the shorter ends, what are the dimensions of the foil wrapper and its area?

Mixed Review

35. Draw a net for the prism at the right.
    (Lesson 11-2)

If $\overline{AE} \parallel \overline{BD}$ and $\overline{AE} \perp \overline{ED}$, find each measure.

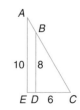

36. *EC* **(Lesson 7-3)**

37. *BC* **(Lesson 8-2)**

38. *AB* **(Lesson 7-5)**

39. area of *ABDE* **(Lesson 10-5)**

Wrap-Up

40. What are the essential ideas in this lesson that you should know in order to be successful on the next quiz or test?

11-4 Surface Area of Pyramids and Cones

Objectives

After studying this lesson, you should be able to:
- find the lateral area of a regular pyramid, and
- find the lateral area and surface area of a right circular cone.

Application

In large metropolitan areas that experience severe winter weather, state and local highway departments frequently spread salt to help keep ice from forming on the roads. In some states, buildings like this one are strategically located to provide highway crews with a ready supply of dry salt. This building is designed in this way, because the salt supply stored inside is piled up in the shape of a cone. The shape of the building closely resembles that of a pyramid.

A pyramid has the following characteristics.

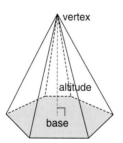

- All the faces, except one, intersect at a point called the **vertex.**

- The face that does not intersect the other faces at the vertex is called the **base** and is a polygon.

- The faces intersecting at the vertex are called **lateral faces** and form triangles. The edges of the lateral faces that have the vertex as an endpoint are called **lateral edges.**

- The segment from the vertex perpendicular to the base is called the **altitude.**

A pyramid is a **regular pyramid** if its base is a regular polygon and the segment whose endpoints are the center of the base and the vertex is perpendicular to the base. This segment is called the **altitude.**

All of the lateral faces are congruent isosceles triangles. The height of each lateral face is called the **slant height, ℓ,** of the pyramid.

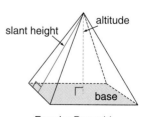

Regular Pyramid

The figure at the right is a regular hexagonal pyramid. Its lateral area (L) can be found by adding the areas of all its congruent triangular faces as shown in its net below.

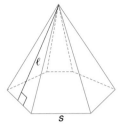

$$L = \tfrac{1}{2} s\ell + \tfrac{1}{2} s\ell + \tfrac{1}{2} s\ell + \tfrac{1}{2} s\ell + \tfrac{1}{2} s\ell + \tfrac{1}{2} s\ell$$

$$= \tfrac{1}{2}(s + s + s + s + s + s)\ell$$

$$= \tfrac{1}{2} P\ell \qquad P = s + s + s + s + s + s$$

This suggests the following formula.

Lateral Area of a Regular Pyramid	If a regular pyramid has a lateral area of L square units, a slant height of ℓ units, and its base has a perimeter of P units, then $L = \tfrac{1}{2} P\ell$.

The **surface area** of a regular pyramid is the sum of its lateral area and the area of its base.

Example 1

A regular square pyramid has a slant height of 15 units and a lateral edge of 17 units. Find its lateral area.

Use the Pythagorean Theorem to find the measure of half of each edge of the base (s).

$$\tfrac{1}{2} s = \sqrt{17^2 - 15^2}$$

$$\tfrac{1}{2} s = \sqrt{64}$$

$$\tfrac{1}{2} s = 8$$

$$s = 16$$

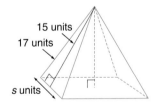

Thus, the length of each edge of its base is 16 units, and the perimeter of its base is 4×16, or 64 units.

$$L = \tfrac{1}{2} P\ell$$

$$= \tfrac{1}{2}(64)(15)$$

$$= 480$$

The lateral area of the pyramid is 480 square units.

The figure at the right is a **right circular cone.** It has a circular **base** and a **vertex** at T. Its **axis**, $\overline{TC}$, is the segment whose endpoints are the vertex and the center of the base. The **altitude** of a cone is the segment that has the vertex as one endpoint and is perpendicular to the base.

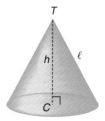

If the axis of a cone is also an altitude, then the cone is called a **right cone.** Otherwise it is called an **oblique cone.** The measure of any segment joining the vertex of a right cone to the edge of the circular base is called its **slant height,** ℓ. The measure of the altitude is the **height,** h, of the cone.

Finding the lateral area and surface area of a right cone is similar to finding those same measures for a regular pyramid. If a cone is cut along its slant height and unfolded, the resulting surface is a sector of a circle whose radius is the slant height ℓ. The area of the sector is proportional to the area of the circle. Notice that the arc length of this sector is equal to the circumference of the base of the original cone, $2\pi r$.

$$\frac{\text{area of sector}}{\text{area of circle}} = \frac{\text{measure of arc}}{\text{circumference of circle}}$$

$$\frac{\text{area of sector}}{\pi \ell^2} = \frac{2\pi r}{2\pi \ell}$$

$$\text{area of sector} = \pi r \ell$$

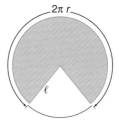

Lateral Area and Surface Area of a Right Cone	If a right circular cone has a lateral area of L square units, a surface area of T square units, a slant height of ℓ units, and the radius of the base is r units, then $L = \pi r \ell$ and $T = \pi r \ell + \pi r^2$.

Example 2

APPLICATION

Manufacturing

The National Paper Products Company makes cone-shaped cups for snow cones. If the diameter of the top of a cup is 8 centimeters long and its slant height is 11 centimeters, what is the lateral area of a cup?

If the diameter is 8 centimeters, the radius is $\frac{1}{2} \times 8$, or 4 centimeters.

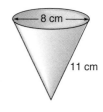

$$L = \pi r \ell$$
$$= \pi(4)(11) \quad \textit{Substitute 4 for r and 11 for } \ell.$$
$$= 44\pi$$
$$\approx 138.2$$

The lateral area is about 138.2 square centimeters.

CHECKING FOR UNDERSTANDING

Communicating Mathematics

Read and study the lesson to answer these questions.

1. Describe the differences between the lateral edges of a pyramid and those of a prism.

2. As the number of sides of the base of a regular pyramid increases, the base begins to resemble a familiar shape. What shape is it? As this transformation takes place, what happens to the shape of the pyramid?

3. For a regular pyramid, which is longer, one of its lateral edges or its slant height? Sketch a pyramid to illustrate your answer.

4. In what case would the axis of a cone not be the altitude of the cone?

Guided Practice

Determine whether the condition given is characteristic of all pyramids or all prisms, both, or neither.

5. It has only one base.
6. It can have as few as five faces.
7. Its lateral faces are parallelograms.
8. It has the same number of lateral faces as vertices.

Find the lateral area of each regular pyramid or right cone. Round your answers to the nearest tenth.

9.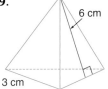
6 cm
3 cm

10.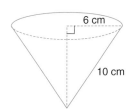
6 cm
10 cm

11.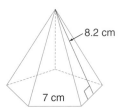
8.2 cm
7 cm

Find the surface area of each solid. Round your answers to the nearest tenth.

12.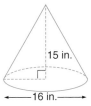
15 in.
16 in.

13.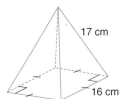
17 cm
16 cm

14.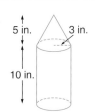
5 in.
3 in.
10 in.

EXERCISES

Practice

Determine whether the condition given is characteristic of all pyramids or all prisms, both, or neither.

15. It has two bases.
16. Its lateral faces are triangles.
17. It always has an even number of faces.
18. It can have as few as four faces.
19. It always has an even number of edges.

Use the regular pyramid at the right to answer each of the following.

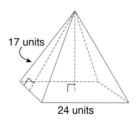

17 units

24 units

20. Find the slant height of each lateral face.

21. Find the area of each lateral face.

22. Find the pyramid's lateral area.

23. Find the pyramid's surface area.

Find the lateral area of each regular pyramid or right cone.

24.

$5\frac{3}{8}$ in.

$3\frac{1}{2}$ in.

25.

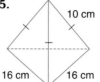

10 cm

16 cm   16 cm

26.

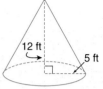

12 ft

5 ft

Find the surface area of each solid.

27.

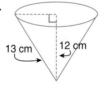

13 cm   12 cm

28.

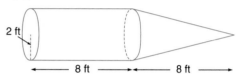

2 ft

8 ft   8 ft

29. A regular pyramid has a slant height of 16 feet. The area of its square base is 100 square feet. Find its surface area.

30. Suppose that the vertices *T, O,* and *P* on a cube form the base of a pyramid with vertex *V.* If *TV* = 16, find the lateral area and the surface area of the pyramid.

O

T   *V*

P

31. The base of a rectangular pyramid is 30 inches by 12 inches. The altitude is 8 inches. All lateral edges are congruent. Find its surface area.

32. A frustum of a pyramid is the part of the pyramid that remains after the top portion of the pyramid has been cut off by a plane parallel to the base. The figure at the right is a frustum of a regular pyramid. Find its lateral area.

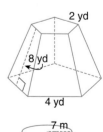

2 yd

8 yd

4 yd

33. Find the surface area of the frustum of a cone shown at the right.

7 m

13 m

12 m

Critical Thinking

34. As the vertex of a right cone moves down the axis toward the center of the base, describe what happens to the lateral area of the cone. Be as specific as possible and demonstrate the validity of your answer with a series of diagrams.

Applications

35. **Buildings** The Transamerica Tower in San Francisco is a pyramid with a square base that is 149 feet on each side and a height of 853 feet. Find its lateral area.

36. **Mining** Open pit mines approximate inverted right cones. If the base of the cone of an open pit mine is to be 420 feet across and the depth of the pit is to be 250 feet, what will the lateral area of the exposed surface be?

37. **Highway Management** A building used to store salt is in the shape of a pyramid. It has a height of 32 feet and a slant height of 56 feet. If the salt is piled up in the shape of a cone, what is the longest possible radius for the base of the pile?

Mixed Review

38. Is the statement *The axis is an altitude in an oblique cylinder* true or false? **(Lesson 11-3)**

39. A circular area rug has a diameter of 4 yards. What is the area of the rug? **(Lesson 10-7)**

40. How are the measures of a central angle and an inscribed angle that intercept the same arc related? **(Lesson 9-4)**

41. How are the two diagonals of a rhombus related? **(Lesson 6-1)**

Wrap-Up

42. Write a three-question quiz for this lesson. Be sure to include complete solutions to your quiz questions. Trade quizzes with another student and take each other's quizzes. Then grade each other's work.

~~~~~~~~~~~ **MID-CHAPTER REVIEW** ~~~~~~~~~~~

Use the regular pyramid at the right for Exercises 1-4. (Lesson 11-1)

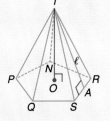

1. Name its vertex.
2. Name its altitude.
3. How many faces does it have?
4. Draw a net for the pyramid.

Find the total surface area of each solid described below in terms of π.

5. a cylinder with height of 3 inches and a diameter of 12 inches **(Lesson 11-3)**

6. a cone with slant height of 14 centimeters and radius of 8 centimeters **(Lesson 11-4)**

7. Find the lateral area of a regular square pyramid with a base edge of 24 units and slant height of 15 units. **(Lesson 11-4)**

11-5 | Volume of Prisms and Cylinders

Objective

After studying this lesson, you should be able to:
- find the volume of a right prism and a right cylinder.

Application

As you drive by a new housing development and see workers pouring cement sidewalks, garage floors, and driveways, you probably have not given much thought to all the geometry that is involved in their work. Their ability to correctly compute the volume of cement needed for each job is absolutely critical. Being able to determine volume exactly often means the difference between making or losing money on the job. For a small contractor, making mistakes in computing volume may mean the end of the business.

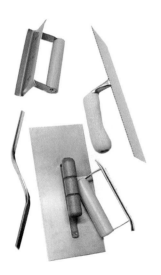

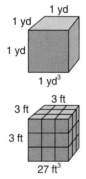

FYI···

Artificial cement was invented in 1824 in London, England, by Joseph Aspdin. It was called Portland cement.

Cement is ordered in cubic yards, usually simply called yards, from ready-mix companies. A cubic unit, such as a cubic yard, is a unit of **volume**. A cubic yard is equivalent to the volume of a cube 1 yard, or 3 feet, on each edge. Thus, a cube 3 feet by 3 feet by 3 feet has a volume of 27 cubic feet and is equivalent to 1 cubic yard. If two solid regions are the same size and the same shape, they have equal volumes.

Observe that in the second cube shown above, each layer of small cubes has 3 by 3 or 9 cubes. Notice that the base has 9 cubes and there are three layers each with the same number of cubes. Therefore, there are 3×9 or 27 cubes in the solid.

Volume of a Right Prism	If a right prism has a volume of V cubic units, a base of B square units, and a height of h units, then $V = Bh$.

Example 1

APPLICATION

Construction

Find the cubic yards of cement that are required for a 60-foot long driveway that is 8 inches thick and 20 feet wide.

Since this driveway can be thought of as a right prism, we can use the formula $V = Bh$.

The area of the base is 20 feet times 60 feet or 1200 square feet. The height in feet is $\frac{8}{12}$ or $\frac{2}{3}$. Thus, the volume is $1200 \times \frac{2}{3}$ or 800 cubic feet. Since a cubic yard equals 27 cubic feet, they will need $\frac{800}{27}$ or about 29.6 cubic yards of cement. The cement contractor will order 30 cubic yards.

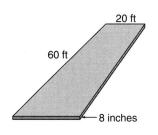

The volume of a right cylinder can be found using the same process we used to find the volume of a right prism.

Consider the stack of coins shown at the right. Each coin represents a circular layer. Taken together, the stack of coins represents the volume of a right cylinder. In a similar way, the volume of a right cylinder is the area of each layer times the height of the stack. Since the base of a cylinder is a circle, the area of the base is πr^2.

Volume = area of base $\times$ height
$$V = \pi r^2 \times h$$

Volume of a Right Cylinder	If a right cylinder has a volume of V cubic units, a height of h units, and a radius of r units, then $V = \pi r^2 h$.

Example 2

Manufacturing

A certain metal pipe is made by boring a cylindrical hole in a metal cylinder. The outer diameter is 2.5 centimeters and the inside diameter is 1.8 centimeters. If the metal pipe is 30 centimeters long, find the volume of metal contained in the pipe to the nearest cubic centimeter.

The volume of the metal pipe equals the volume of the cylinder minus the volume of the cylindrical hole.

Let r_1 represent the radius of the cylinder.
$$r_1 = (0.5)(2.5) \text{ or } 1.25 \text{ cm}$$

Let r_2 represent the radius of the hole.
$$r_2 = (0.5)(1.8) \text{ or } 0.9 \text{ cm}$$

$$V = \pi(r_1)^2 h - \pi(r_2)^2 h$$
$$= \pi(1.25)^2(30) - \pi(0.9)^2(30)$$
$$\approx 70.9$$

The volume of the metal in the pipe is about 70.9 cm^3.

CHECKING FOR UNDERSTANDING

Communicating Mathematics

Read and study the lesson to answer these questions.

1. Besides cement, what other items are purchased by volume? State the units of measure used in their purchase.

2. How many cubic feet are in one cubic yard?

3. Recall how coins were used to develop the volume of the right cylinder. What objects might be used in a similar way to develop the formula for the volume of a right prism?

4. Describe the relationship between the volumes of two geometric solids that are the same size and shape.

Guided Practice

Find the volume of each solid. Round your answers to the nearest tenth.

5.
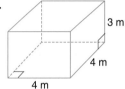

6.
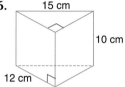

7.
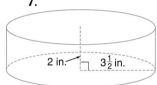

8. Find the volume of a right prism whose base has an area of 12 square meters and a height of 3.5 meters.

9. Find the volume of a right cylinder whose radius is 2 meters long and has a height of 8 meters.

10. Find the volume of a right hexagonal prism that has a height of 20 centimeters and whose base is a regular hexagon with sides of 8 centimeters. Round the answer to the nearest cubic unit.

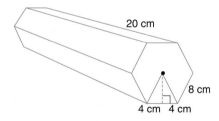

11. Find the volume of the partial right cylinder shown at the right. Round your answer to the nearest tenth.

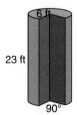

12. A right prism is formed by folding this net. Find its volume.

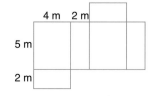

EXERCISES

Practice

Find each of the following.

13. The area of the shaded base of the right prism is 68 square centimeters. Find the volume of the prism.

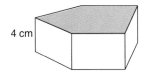

14. Find the volume of a cube that has an edge of 9 inches.

15. Find the length of a lateral edge of a right prism with a volume of 962 cubic centimeters and a base whose area is 52 square centimeters.

16. Find the volume of a right prism that has a base with an area of 17.5 square centimeters and a height of 14 centimeters.

17. Find the volume of a right prism that has a base with an area of 16 square feet and a height of 4.2 feet.

18. Find the volume of a right cylinder whose radius is 3.2 centimeters and height is 10.5 centimeters.

19. Find the volume of a right cylinder whose diameter is 3 feet and height is 4 feet.

Find the volume of each solid. Round your answers to the nearest tenth.

20.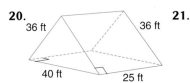

21.

22.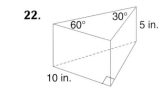

23.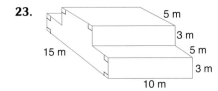

24.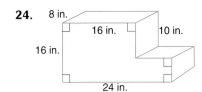

25. Find the volume of the regular hexagonal right prism at the right. Round your answer to the nearest tenth.

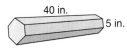

26. Find the volume of a trapezoidal prism that has a height of 20 centimeters. The trapezoidal base has a height of 3 centimeters, and the two bases of the trapezoid measure 5 and 9 centimeters.

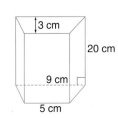

27. A hole with a diameter of 4 millimeters is drilled through a block of copper as shown at the right.

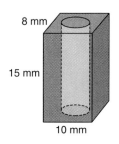

8 mm

15 mm

10 mm

   a. Find the volume of the resulting solid.

   b. The *density* of a substance is its mass per unit of volume. At room temperature, the density of copper is 8.9 grams per cubic centimeter. What is the mass of this block?

28. Find the volume of a cube for which a diagonal of one of its faces measures 12 centimeters.

The nets below form a right prism and a right cylinder. Find the volume of the solid formed by each net.

29.

12 cm

3 cm

10 cm

30.

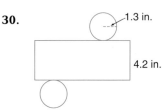

1.3 in.

4.2 in.

31. A cylindrical glass pitcher full of water is poured into a rectangular cake pan with dimensions 9 inches by 12 inches by 2 inches. If the pitcher has a diameter of 6 inches and a height of 14 inches, will the pan hold all the water?

32. Find the volume of a cube whose surface area is 54 square inches.

33. Find the length of the edge of a cube whose surface area and volume have exactly the same measure.

34. Describe the effect upon the volume of a rectangular solid when one dimension is doubled and the other two remain the same. What happens when two dimensions are doubled?

Critical Thinking

35. Develop an argument that would verify the conjecture *The volume of every oblique prism or oblique cylinder is equal to the area of its base times its height.*

Applications

36. Food A wedge of cheese is cut from a fresh wheel of cheese (cylindrical block) which is 6 inches thick. The vertex of the wedge is at the center of the wheel whose radius is 10 inches. If the central angle of the wedge measures 50°, find the volume of this wedge of cheese.

37. Business If the trunk of an oak tree 30 feet long and 4 feet in diameter is split up for firewood, how many cords will it produce? A cord of firewood makes a stack 4 × 4 × 8 feet.

38. Agriculture A farmer stores water in a rectangular tank with dimensions as shown at the right. The tank was filled with water, but developed a leak. If it loses 0.1 cubic feet of water each second, what percent of the water remains in the tank after 1 hour?

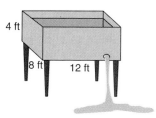

39. Construction A drawing of the cement driveway for a new home is shown at the right.

  a. Boards are laid out in the shape of the driveway to contain the cement when it is poured. What is the total length of lumber required?

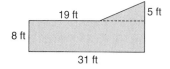

  b. The driveway is to be 6 inches thick. What is the volume of cement that will be needed?

  c. If cement costs $2.65 per cubic foot, how much will the cement cost?

Mixed Review

40. The diameter of the base of a right circular cone measures 6.4 millimeters, and the slant height measures 5.2 millimeters. Find the lateral area. **(Lesson 11-4)**

41. The diameter of a circle measures 10 centimeters, and the length of a chord is 8 centimeters. Find the distance from the chord to the center of the circle. **(Lesson 9-3)**

Find the value of x.

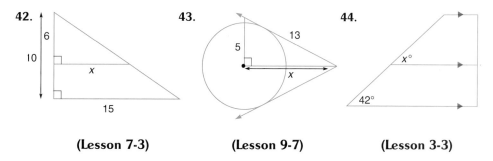

       (Lesson 7-3)   **(Lesson 9-7)**   **(Lesson 3-3)**

Wrap-Up

45. Describe the difference between the concepts of area and volume. Give the formulas for lateral area, surface area, and volume of right prisms and cylinders.

Volume of Pyramids and Cones

Objective

After studying this lesson, you should be able to:
- find the volume of a pyramid and a circular cone.

Application

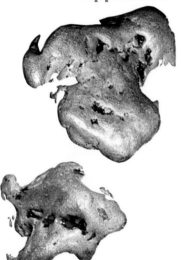

A mining company is trying to decide whether to develop an open pit gold mining operation in the Elk mountain range located in Colorado. Mining companies will usually use open pit mines whenever the ore is relatively close to the surface because they are cheaper to operate than underground mines. An open pit gold mine usually has a shape that is close to an inverted right cone.

In order for the company to decide whether to commit to an investment of several million dollars, it must calculate its return on investment. To do that, it must be able to estimate the amount of gold it's likely to mine. This can be determined from assay reports on core samples taken from the potential mining site. Assay reports will reveal the number of ounces of gold per ton of the material (ore and waste) mined.

FYI ···

The Lost Dutchman's gold mine, somewhere in the mountains of Arizona, is said to have been so rich that, if you were to tap on the walls with a hammer, nuggets of gold would come tumbling down.

A report indicates that each ton of material will have a volume of approximately 10 cubic feet. Now all the company needs to do is compute the volume of the material they expect to remove from this cone-shaped mine. Then they can estimate the possible amount of gold that can be extracted.

A formula to compute the volume of a cone is needed. In the figures at the right, the cone and cylinder have the same base and height, and the pyramid and prism have the same base and height. You can see that the volume of the cone is less than the volume of the cylinder and that the volume of the pyramid is less than the volume of the prism. As a matter of fact, the ratio of the volumes in each case is 1:3.

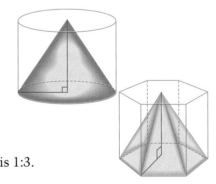

Volume of a Right Circular Cone	If a right circular cone has a volume of V cubic units, a height of h units, and the area of the base is B square units, then $V = \frac{1}{3}Bh$.
Volume of a Right Pyramid	If a right pyramid has a volume of V cubic units, a height of h units, and the area of the base is B square units, then $V = \frac{1}{3}Bh$.

Example 1

Find the volume of the right circular cone with a radius of 2.7 centimeters and a slant height of 9.5 centimeters. Round your answer to the nearest tenth.

Use the Pythagorean Theorem to find the height.

$$h^2 + r^2 = \ell^2$$
$$h^2 + 2.7^2 = 9.5^2 \qquad \textit{Substitute 2.7 for r and 9.5 for } \ell.$$
$$h^2 = 82.96$$
$$h \approx 9.1 \text{ cm}$$

Now find the volume.

$$V = \frac{1}{3} Bh \qquad B = \pi r^2$$
$$\approx \frac{1}{3} \pi (2.7)^2 (9.1)$$
$$\approx 69.5 \text{ cm}^3 \qquad \textit{Use your calculator.}$$

Example 2

Find the volume of a regular hexagonal pyramid if each edge of the base is 8 centimeters long and the height is 12 centimeters.

$$V = \frac{1}{3} Bh$$

The area of a regular polygon is $A = \frac{1}{2} Pa$.

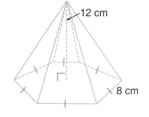

$$V = \frac{1}{3} \left(\frac{1}{2} \cdot 48 \cdot 4\sqrt{3} \right) 12 \qquad \textit{P is the perimeter of}$$
$$= 384\sqrt{3} \qquad\qquad\qquad \textit{the regular hexagon.}$$
$$\approx 665.1 \qquad\qquad\qquad \textit{a is the apothem.}$$

The volume of this pyramid is about 665.1 cubic centimeters.

So far, only the formulas for the volume of right prisms, cylinders, pyramids, and cones have been presented in this chapter. However, an exercise in the last lesson asked you to develop an argument in support of the conjecture that the volumes of some oblique solids are equal to the volumes of corresponding right solids. You may have wondered if the same formulas can be applied to all oblique solids.

The photograph at the right shows two matching decks of cards. One represents a right prism, and the other represents an oblique prism. Since the decks have the same number of cards with all cards the same size and shape, the two prisms represented by the decks have the same volume. This observation was first made by Cavalieri, an Italian mathematician of the seventeenth century. It is known as **Cavalieri's Principle.**

Cavalieri's Principle	If two solids have the same cross-sectional area at every level and the same height, then they have the same volume.

As a result of Cavalieri's Principle, we know that if a prism has a base with an area of B square units and a height of h units, then its volume is Bh cubic units, whether it is right or oblique. Similarly, the volume formulas for cylinders, cones, prisms, and pyramids hold whether they are right or oblique.

Example 3

Using Cavalieri's Principle, find the volume of the cone at the right.

$$V = \frac{1}{3} Bh$$

$$= \frac{1}{3} \pi (6)^2 (13)$$

$$= 156\pi$$

$$\approx 490.1$$

The volume is about 490.1 cubic meters.

13 m

12 m

CHECKING FOR UNDERSTANDING

Communicating Mathematics

Read and study the lesson to answer these questions.

1. How would you find the volume of an ice cream cone?

2. How is the volume of a cone related to that of a cylinder with the same altitude and a base congruent to that of the cone?

3. Devise and describe an experiment that would help convince a friend that the volume of a pyramid is one third the volume of a prism that has the same base and height as the pyramid.

Guided Practice

Find the volume of each solid. Round your answers to the nearest tenth.

4.
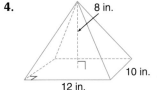
8 in.
10 in.
12 in.

5.
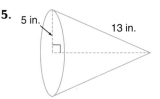
5 in.
13 in.

6.
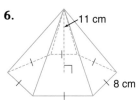
11 cm
8 cm

7. Using the figure at the right, find the volume of the resulting solid if the smaller cone is removed from the larger cone.

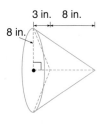
3 in.   8 in.
8 in.

8. If a pyramid has a height of 30 meters and its base is an equilateral triangle with sides of 8 meters, find its volume.

9. Find the volume of a right cone with a slant height of 18 and an angle of 60° at the vertex of the cone.

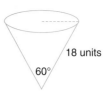

18 units

60°

EXERCISES

Practice

Find the volume of each pyramid. Round your answers to the nearest tenth.

10. The base has an area of 27.9 square millimeters, and the height is 18.5 millimeters.

11. The base has an area of 15 square feet, and the height is 7 feet.

Find the volume of each cone. Round your answers to the nearest tenth.

12. The base has a radius of 5 feet, and the height is 16 feet.

13. The base has a diameter of 8.4 meters, and the height is 10.3 meters.

14. The volume of a pyramid is 729 cubic units. If the area of the base is 243 square units, find the height of the pyramid.

15. Find the volume of the oblique cone whose height is 21 units and whose base has a radius of 8 units.

16. The base of a rectangular pyramid is 30 units by 12 units. If the altitude is 8 units, find its volume.

Find the volume of each solid. Round your answers to the nearest tenth.

17.

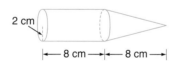

2 cm

|← 8 cm →|← 8 cm →|

18.

12 in.

13 in.

19.

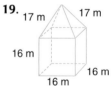

17 m   17 m

16 m

16 m

16 m

20. A regular square pyramid has a slant height of 15 meters and a lateral edge of 17 meters. What is the volume of this pyramid?

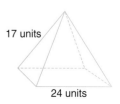

15 m

17 m

21. Find the volume of the regular pyramid at the right if its lateral edges are each 17 units long and its base is 24 units on a side.

17 units

24 units

22. The eight faces of the regular octahedron at the right are congruent equilateral triangles. If each edge is 12 centimeters, find its volume.

23. Refer to the figure at the right.

  a. Find the volume of the shaded pyramid cut from the rectangular solid shown.

  b. What is the ratio of the volume of the pyramid to the volume of the rectangular solid from which it is cut?

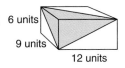

24. A pyramid with a rectangular base has a volume of 80 cubic inches and a height of 5 inches. How many different rectangles are possible for the base if its edge measures are whole numbers?

25. Two right circular cones have the same axis, the same vertex, and the same height. The smaller cone lies within the larger one. Find the volume of the space between the two cones if the diameter of one cone is 6 inches, the diameter of the other is 9 inches, and the height of both is 5 inches.

26. Find the volume of the regular square pyramid shown at the right. Give your answer to the nearest cubic unit.

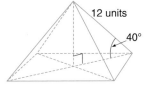

27. Find the volume of the pyramid shown at the right.

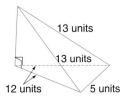

28. A pyramid with four congruent equilateral triangles as its only faces is known as a regular tetrahedron. If the length of one of its edges is 12 units, find the volume of the tetrahedron.

29. Find the volume of the frustum of a cone shown below.

30. Find the volume of the frustum of a pyramid shown below.

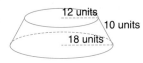

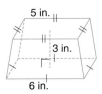

31. A cone with a slant height equal to the length of the diameter of the base is inscribed in a sphere with a radius of 8 inches. What is the volume of the cone? *Hint: The definition of sphere appears on page 560.*

Application

32. Mining A certain open pit mine being dug in the Elk mountain range is to be 420 feet across with a depth of 250 feet. Suppose the assay report projects that the company will be able to retrieve an average of 0.115 ounces of gold per ton of ore and waste mined.

 a. What would be the volume of ore and waste material removed during its mining operation?

 b. How many ounces of gold should the company expect to retrieve from this mine? (Assume one ton is equivalent to 10 cubic feet.)

 c. If the price of gold is $350 per ounce, what is the estimated value of the gold that will be extracted from this mine?

33. History One of the Great Pyramids in Egypt has a square base that is 750 feet on a side. If its original height is estimated to have been 481 feet tall, what was its original volume?

34. Landscaping A certain landscaping company has piled a quantity of dry, loose soil against a building. The highest point of the soil is 5 feet above the ground. The base of the pile is semicircular in shape and 12 feet in width along the building. About how many cubic feet of soil are in the pile?

Computer

The BASIC computer program below will find the volume of a right circular cone rounded to the nearest cubic unit.

```
10 INPUT "ENTER THE RADIUS OF THE BASE OF THE CONE."; R
20 INPUT "ENTER THE HEIGHT OF THE CONE."; H
30 V = INT ((3.14159 * R^2 * H)/3 + 0.5)
40 PRINT "THE VOLUME OF THE CONE IS ABOUT "; V; " CUBIC
   UNITS."
50 END
```

35. What does line 30 of the program represent?

36. How could you change the program to find the volume of a cylinder?

37. Use the computer program to check your answers for Exercises 12 and 13.

Mixed Review

38. Find the volume of a rectangular prism that is 3 feet by 5 feet by 8 feet. **(Lesson 11-5)**

39. Find the probability that a point chosen at random in the figure at the right is in the shaded region. Round your answer to the nearest hundredth. **(Lesson 10-8)**

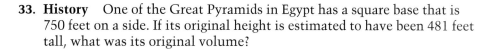

40. What must be true of the two acute angles of a right triangle? **(Lesson 4-1)**

41. Find the distance between the graphs of $A(0, 4)$ and $B(-8, 1)$. **(Lesson 1-4)**

Wrap-Up

42. Compare the volumes of cones and cylinders and the volumes of pyramids and prisms.

11-7 Surface Area and Volume of Spheres

Objectives

After studying this lesson, you should be able to:
- recognize and define basic properties of spheres,
- find the surface area of a sphere, and
- find the volume of a sphere.

Application

Have you ever noticed how small a basketball looks in the hands of someone like Michael Jordan? Great players like Jordan make the ball look like a toy, to do with as they like!

How big is a basketball anyway? That question could be answered by giving its surface area, its volume, its weight, its diameter, or all four.

In this lesson, we will use a basketball as a model of a sphere, and our attention will be focused on finding the surface area and volume of spheres.

For a moment, let's consider infinitely many congruent circles in space, all with the same point for their center. Considered together, all these circles form a **sphere**. In space, a sphere is the set of all points that are a given distance from a given point called its center.

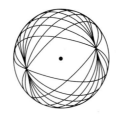 

Several special segments and lines related to spheres are defined and illustrated below.

FYI···

The final score of the first game of basketball was 1 - 0 after 30 minutes of play. The game was stopped so the players could climb up and retrieve the soccer ball from the peach basket.

- A **radius** of a sphere is a segment whose endpoints are the center of the sphere and a point on the sphere. In the figure, $\overline{CR}$, $\overline{CP}$, and $\overline{CQ}$ are radii.

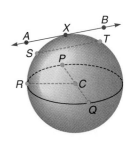

- A **chord** of a sphere is a segment whose endpoints are points on the sphere. In the figure, $\overline{TS}$ and $\overline{PQ}$ are chords.

- A **diameter** of a sphere is a chord that contains the sphere's center. In the figure, $\overline{PQ}$ is a diameter.

- A **tangent** to a sphere is a line that intersects the sphere in exactly one point. In the figure, $\overleftrightarrow{AB}$ is tangent to the sphere at X.

560   CHAPTER 11   SURFACE AREA AND VOLUME

A point

A circle

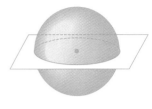

A great circle

FYI ···

Great circles are used by airlines for their transoceanic flights. These flights travel along a great circle path to reduce time of travel and cost.

A plane can intersect a sphere in a point or in a circle. When a plane intersects a sphere so that it contains the center of the sphere, the intersection is called a **great circle.** A great circle has the same center as the sphere, and its radii are also radii of the sphere. On the surface of a sphere, the shortest distance between any two points is the length of the arc of a great circle passing through those two points. Each great circle separates a sphere into two congruent halves called **hemispheres.**

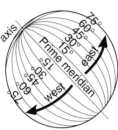

Lines of longitude

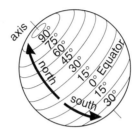

Lines of latitude

The surface of a globe is a good model of a sphere. Each north-south meridian goes halfway around the globe and meets another line at both poles to make a great circle. Of the parallel latitudes, only the equator is a great circle.

In order to determine the amount of leather needed to cover an NBA (National Basketball Association) basketball, we need to find its surface area. A formula for the surface area of a sphere follows.

Surface Area of a Sphere	If a sphere has a surface area of T square units and a radius of r units, then $T = 4\pi r^2$.

Example 1

APPLICATION

Sports

The diameter of an NBA basketball is about 9.5 inches. Find its surface area.

The radius of the basketball is about 4.75 inches.

$$T = 4\pi r^2$$
$$= 4\pi(4.75)^2 \qquad \textit{Use a calculator.}$$
$$\approx 283.5 \text{ in}^2$$

The surface area of an NBA basketball is about 283.5 square inches.

The development of a formula for the volume of a sphere can be related to the volume of a right pyramid and the surface area of a sphere.

Imagine separating the space inside a sphere into infinitely many small pyramids, all with their vertices located at the center of the sphere. Collectively, all their bases equal the surface of the sphere shown below. Observe that the height of these very small pyramids is equal to the radius, r, of the sphere.

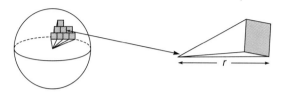

The volume of each pyramid is $\frac{1}{3} Bh$, where B is the area of its base and h is its height. The volume of the sphere is equal to the sum of the volumes of all the infinitely many small pyramids. The volume, V, of the sphere can then be represented as follows.

$$
\begin{aligned}
V &= \tfrac{1}{3} B_1 h_1 + \tfrac{1}{3} B_2 h_2 + \tfrac{1}{3} B_3 h_3 + \ldots + \tfrac{1}{3} B_n h_n \\
&= \tfrac{1}{3} B_1 r + \tfrac{1}{3} B_2 r + \tfrac{1}{3} B_3 r + \ldots + \tfrac{1}{3} B_n r \\
&= \tfrac{1}{3} r (B_1 + B_2 + B_3 + \ldots + B_n) \\
&= \tfrac{1}{3} r (4\pi r^2) \qquad \textit{$B_1 + B_2 + B_3 + \ldots + B_n$ is the surface area of the sphere.} \\
&= \tfrac{4}{3} \pi r^3
\end{aligned}
$$

Volume of a Sphere	If a sphere has a volume of V cubic units and a radius of r units, then $V = \frac{4}{3} \pi r^3$.

Example 2

Find the volume of the empty space in a tennis ball can containing three tennis balls. The inside diameter of the can and the diameter of each ball is about 6.5 cm.

V(empty space) $= V$(cylinder) $- V$(tennis balls)

$$V = \pi r^2 h - 3\left(\tfrac{4}{3} \pi r^3\right)$$

$$= \pi (3.25)^2 (19.5) - 3\left(\tfrac{4}{3}\right) \pi (3.25)^3$$

$$\approx 215.7$$

There are about 215.7 cubic centimeters of empty space in the can.

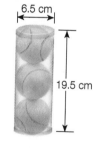

6.5 cm

19.5 cm

CHECKING FOR UNDERSTANDING

Communicating Mathematics

Read and study the lesson to answer these questions.

1. Transoceanic flights by major airlines always fly a path that is related to a sphere. Describe its relation to a sphere.

2. Describe all the different ways a plane and a sphere can intersect.

3. What is a great circle?

4. Describe how the formula for the volume of a sphere was developed in this lesson.

Guided Practice

Determine whether each statement is *true* or *false.*

5. A diameter of a sphere is a chord of the sphere.

6. A radius of a sphere is a chord of the sphere.

7. All great circles of the same sphere are congruent.

In the figure, *P* is the center of the sphere, and plane $\mathcal{B}$ intersects the sphere in $\odot R$.

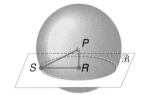

8. Suppose $PS = 25$ and $PR = 7$, find RS.

9. Suppose $PS = 13$ and $RS = 12$, find PR.

Find the surface area and volume of each sphere described below.

10. The radius is 10 centimeters long.

11. The area of one of its great circles is 50.24 square centimeters.

12. Find the volume of a sphere whose surface area is 256π square centimeters.

13. Find the volume of air that is contained in the NBA basketball described in Example 1. Ignore the thickness of the ball.

EXERCISES

Practice

Describe each object as a model of a *circle*, a *sphere*, or *neither*.

14. ball bearing
15. basketball hoop
16. orbit of an electron
17. telephone dial
18. ping pong ball
19. record
20. Jupiter
21. chicken egg
22. an orange
23. football

Determine whether each statement is *true* or *false*.

24. All radii of a sphere are congruent.

25. The chord of greatest length in a sphere will always pass through the sphere's center.

26. A plane and a sphere may intersect in exactly two points.

27. All diameters of a sphere are congruent.

28. A diameter of a great circle is a diameter of the sphere.

29. Two spheres may intersect in exactly one point.

30. Two different great circles of a sphere intersect in exactly one point.

31. The intersection of two spheres may be a circle.

32. The intersection of two spheres with congruent radii may be a great circle.

In the figure, *P* is the center of the sphere, and plane $\mathcal{B}$ intersects the sphere in ⊙*R*.

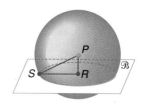

33. Suppose *PS* = 15 and *PR* = 9, find *RS*.

34. Suppose *PS* = 26 and *RS* = 24, find *PR*.

Find the surface area and volume of each sphere described below.

35. The diameter is 4000 feet long.

36. The circumference of one of its great circles is 18.84 meters.

37. If the volume of a sphere is $\frac{32}{3}\pi$ cubic inches, what is the radius?

38. What is the ratio of the radii of two spheres if the surface area of one is 4 times the surface area of the other?

39. Find the ratio of the volumes of a hemisphere and a cone having congruent bases and equal heights.

For Exercises 40 and 41, the volume of the cube is 1728 cubic centimeters.

40. Find the volume of the sphere that can be inscribed inside the cube.

41. Find the volume of the sphere that can be circumscribed about the cube.

Critical Thinking

42. The edge of a cube, the diameter and height of a cylinder, and the diameter of a sphere all have the same measure. Which has the least surface area? Which has the least volume?

Applications

43. Travel Name some cities, countries, or other significant geographic features that a transoceanic great circle route from Chicago, Illinois, to Frankfurt, Germany, would fly over.

44. Sports Find the surface area and volume of a soccer ball that has a circumference of 27 inches.

45. Sports A hemispherical plastic dome is used to cover several indoor tennis courts. If the diameter of the dome measures 400 feet, find the volume enclosed by the dome in cubic yards.

46. Food An ice cream cone is 10 centimeters deep and has a diameter of 4 centimeters. A scoop of ice cream with a diameter of 4 centimeters rests on the top of the cone.

   a. If all the ice cream melts into the cone, will the cone overflow?

   b. If the cone does not overflow, what percentage of the cone will be filled?

47. Architecture If the hemispherical dome of the Iowa State Capitol is 80 feet in diameter, find the surface area of the dome.

48. **Science** Consider Earth to be a sphere with a radius of 4000 miles.
    a. Find the approximate surface area of Earth.
    b. The area of the land on Earth is about 57,900,000 square miles. What percentage of Earth's surface is land?

Mixed Review

49. Find the volume of a right circular cone if the radius of the base is 5 feet and the height of the cone is 16 feet. **(Lesson 11-6)**

50. Find the area of a trapezoid whose median is 8.5 feet long and whose altitude is 7.1 feet long. **(Lesson 10-5)**

51. Can 2.7, 3.0, and 5.3 be the measures of the sides of a right triangle? Explain. **(Lesson 8-2)**

52. Two isosceles triangles have congruent legs. Must the triangles be congruent? **(Lesson 4-4)**

53. In the figure, find the value of x. **(Lesson 3-2)**

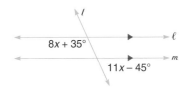

Wrap-Up

54. As the radius of a sphere gets longer, which of the sphere's measurements increases more, the surface area or the volume? Defend your answer.

~~~~~ HISTORY CONNECTION ~~~~~

Why were the European sailors of the fourteenth and fifteenth centuries afraid of plunging off the edge of a flat Earth? At least sixteen centuries earlier, Greek mathematicians not only knew that the Earth was round, but had calculated its circumference!

Eratosthenes of Cyrene (275-194 B.C.) was director of the Alexandrian Library. He knew that on the first day of summer, the Sun was directly over the Egyptian city of Syene, near present-day Aswan. He also knew that Syene was 5000 stades directly south of Alexandria. (An Egyptian stade is roughly 0.1575 kilometers.)

Eratosthenes assumed that the Sun was far enough away that its rays of light arrived at Earth in parallel lines. He measured the angle, α, formed by the top of a pole in Alexandria and the pole's shadow as $7\frac{1}{5}°$. He concluded that $\angle AOS$ was equal to α, where O represented the center of a spherical Earth. He used the following proportion to calculate the circumference of Earth.

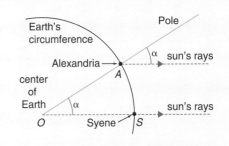

$$\frac{\text{distance from Syene to Alexandria}}{\text{Earth's circumference}} = \frac{\text{measure of angle } \alpha}{360°}$$

$$\frac{5000 \text{ stades}}{\text{Earth's circumference}} = \frac{1}{50}$$

$$\text{Earth's circumference} = 250,000 \text{ stades}$$

Eratosthenes' measurement is 39,375 kilometers. This is remarkably close to the current accepted value of 40,075 kilometers.

11 SUMMARY AND REVIEW

VOCABULARY

Upon completing this chapter, you should be
familiar with the following terms:

536 altitude	**537** lateral area	**544** right circular cone
544 axis	**536** lateral edge	**538** right cylinder
536 base	**536** lateral face	**536** right prism
555 Cavalieri's Principle	**530** net	**542** slant height
560 chord	**544** oblique cone	**529** solid
544 cone	**538** oblique cylinder	**560** sphere
538 cylinder	**536** oblique prism	**529** surface area
560 diameter	**536** prism	**560** tangent
561 great circle	**542** pyramid	**542** vertex
536 height	**560** radius	**548** volume
561 hemisphere	**542** regular pyramid	

SKILLS AND CONCEPTS

OBJECTIVES AND EXAMPLES

Upon completing this chapter, you should
be able to:

- draw two-dimensional nets for three-dimensional solids. **(Lesson 11-2)**

A polyhedron and one of its two-dimensional nets are shown below.

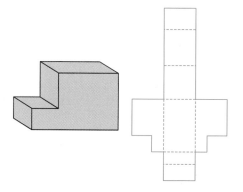

REVIEW EXERCISES

Use these exercises to review and prepare
for the chapter test.

Match each polyhedron with its net.

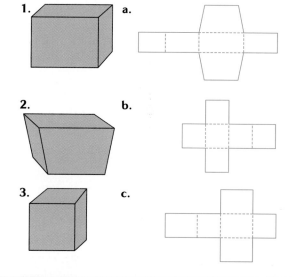

OBJECTIVES AND EXAMPLES	REVIEW EXERCISES

■ find the lateral areas and surface areas of right prisms and right cylinders. **(Lesson 11-3)**

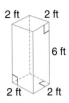 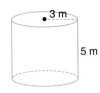

$L = Ph$
$\quad = (8)(6) = 48 \text{ ft}^2$
$T = Ph + 2B$
$\quad = 48 + 2(4)$
$\quad = 56 \text{ ft}^2$

$L = 2\pi rh$
$\quad = 2\pi(3)(5) \approx 94.2 \text{ m}^2$
$T = 2\pi rh + 2\pi r^2$
$\quad \approx 94.2 + 2\pi 3^2$
$\quad \approx 150.7 \text{ m}^2$

Find the lateral area and the surface area of each right prism or right cylinder.

4.

5.

6.

7.
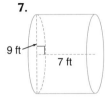

■ find the lateral areas and surface areas of regular pyramids and right circular cones. **(Lesson 11-4)**

$L = \frac{1}{2} P\ell$
$\quad = \frac{1}{2}(8)(5) \text{ or } 20 \text{ in}^2$
$T = L + B$
$\quad = 20 + 4$
$\quad = 24 \text{ in}^2$

$L = \pi r\ell$
$\quad = \pi(3)(5) \text{ or } 47.1 \text{ ft}^2$
$T = \pi r\ell + \pi r^2$
$\quad \approx 47.1 + \pi 3^2$
$\quad \approx 75.4 \text{ ft}^2$

Find the lateral area and the surface area of each regular pyramid or right circular cone.

8.

9.

10.

11.

■ find the volume of a right prism and a right cylinder. **(Lesson 11-5)**

 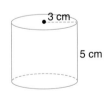

$V = Bh$
$\quad = (4)(5)$
$\quad = 20 \text{ m}^3$

$V = \pi r^2 h$
$\quad = \pi 3^2(5)$
$\quad \approx 141.4 \text{ cm}^3$

Find each of the following.

12. Find the volume of a regular hexagonal prism if its radius is 10 centimeters and its height is 20 centimeters.

13. Find the volume of a right cylinder if its radius is 10 centimeters and its height is 20 centimeters.

14. Find the volume of a right cylinder if its diameter is 10 feet and its height is 13 feet.

OBJECTIVES AND EXAMPLES	REVIEW EXERCISES

▪ find the volume of a circular cone and a pyramid. **(Lesson 11-6)**

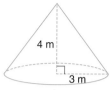

$$V = \frac{1}{3} Bh$$
$$= \frac{1}{3} \pi (3^2)(4)$$
$$= 12\pi$$
$$\approx 37.7 \text{ m}^3$$

$$V = \frac{1}{3} Bh$$
$$= \frac{1}{3}(4)(5)$$
$$\approx 6.7 \text{ in}^3$$

Find each of the following.

15. Find the volume of a triangular pyramid if its base is an equilateral triangle with sides 9 centimeters long and its height is 15 centimeters.

16. Find the volume of a right circular cone if its height is 22 centimeters and its radius is 11 centimeters.

17. Find the volume of a right circular cone if the circumference of its base is 62.8 millimeters and its height is 15 millimeters.

▪ find the surface area and volume of a sphere. **(Lesson 11-7)**

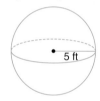

$$T = 4\pi r^2$$
$$= 4\pi 5^2 \text{ or } \approx 314.2 \text{ ft}^2$$

$$V = \frac{4}{3} \pi r^3$$
$$= \frac{4}{3} \pi 5^3 \text{ or } \approx 523.6 \text{ ft}^3$$

Answer each of the following.

18. All great circles of a sphere are congruent. Write *yes* or *no*.

19. Find the surface area of the moon if its diameter is approximately 2160 miles.

20. Find the volume of a sphere with a radius of 20 feet.

APPLICATIONS AND CONNECTIONS

21. Make a model to solve this problem. Brett has a number of cubic blocks. He makes a square prism one layer high by putting them all together on a flat surface. Then he takes the same blocks and makes a cube. What is the least number of blocks greater than one that Brett can have? **(Lesson 11-1)**

22. **Science** Suppose two different crystals occur in the shape of right prisms. One crystal has a rectangular base of 2 units by 3 units and a height of 2.5 units. The other crystal has a regular hexagonal base with sides of 2.1 units and a height of 2.3 units. Which crystal has the greater surface area? **(Lesson 11-3)**

23. A waterbed is 6.5 feet long, 5.5 feet wide, and 1 foot thick. If water weighs about 60 pounds per cubic foot, what is the total weight of the water in the waterbed? **(Lesson 11-5)**

24. **Manufacturing** A hollow ball is formed by molding rubber in a form that gives it an inside diameter of 8 centimeters and an outside diameter of 85 millimeters. What volume of rubber is used to produce this ball? **(Lesson 11-7)**

1. Draw a rectangular prism with a square base that is 5 units on each side and a height that is 10 units.

2. Sketch a net of a right circular cylinder.

Find the surface area of each solid figure. Round answers to the nearest tenth. Assume that the bases of each pyramid or prism are regular.

3.
10 ft
5 ft

4.
10 ft
5 ft

5.
40 cm
23 cm

6. a right circular cone with a radius of 27 millimeters and a height of 30 millimeters

7. a sphere with diameter of 6 inches

Find the volume of each solid figure. Round answers to the nearest tenth. Assume that the bases of each pyramid or prism are regular.

8.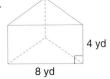
4 yd
8 yd

9.
6 in.
4 in.

10.
50 cm
39 cm

11. a right cylinder with diameter of 39 centimeters and a height of 50 centimeters

12. a sphere with radius of 18 millimeters

Answer each of the following.

13. A rectangular swimming pool is 4 meters wide and 10 meters long. A concrete walkway is poured around the pool. The walkway is 1 meter wide and 0.1 meter deep. What is the volume of the concrete?

14. The base of a right prism is a right triangle with legs 6 inches and 8 inches. If the height of the prism is 16 inches, find the lateral area of the prism.

15. Find the volume of the figure at the right. Round your answer to the nearest tenth.

13 cm
5 cm
15 cm
5 cm

Bonus

The base of a right prism is a rectangle. The length of the rectangular base is twice its width. The height of the prism is twice the longest side of its base. If the volume of the prism is 216 cubic feet, find its surface area.

Algebra Review

OBJECTIVES AND EXAMPLES	REVIEW EXERCISES

OBJECTIVES AND EXAMPLES

■ Solve open sentences involving absolute value and graph the solution sets.

$$|2x - 1| > 1$$

$$2x - 1 > 1 \quad \text{or} \quad 2x - 1 < -1$$
$$2x > 2 \qquad\qquad 2x < 0$$
$$x > 1 \qquad\qquad x < 0$$

The solution set is $\{x \,|\, x > 1 \text{ or } x < 0\}$.

REVIEW EXERCISES

Solve each open sentence. Then graph its solution set.

1. $|y - 1| \le 5$
2. $|2 - n| = 5$
3. $\left|2p - \frac{1}{2}\right| > \frac{9}{2}$
4. $|7a - 10| < 0$

■ Add and subtract rational expressions.

$$\frac{m-1}{m+1} + \frac{4}{2m+5}$$
$$= \frac{2m+5}{2m+5} \cdot \frac{m-1}{m+1} + \frac{m+1}{m+1} \cdot \frac{4}{2m+5}$$
$$= \frac{2m^2+3m-5}{2m^2+7m+5} + \frac{4m+4}{2m^2+7m+5}$$
$$= \frac{2m^2+7m-1}{2m^2+7m+5}$$

Find each sum or difference.

5. $\dfrac{x}{x^2+3x+2} + \dfrac{1}{x^2+3x+2}$
6. $\dfrac{2x}{4x^2-9} - \dfrac{3}{9-4x^2}$
7. $\dfrac{x}{x+3} - \dfrac{5}{x-2}$
8. $\dfrac{2x+3}{x^2-4} + \dfrac{6}{x+2}$

■ Solve systems of equations by the substitution method.

Use substitution to solve the system of equations $y = x - 1$ and $4x - y = 19$.

$$4x - y = 19 \qquad y = x - 1$$
$$4x - (x - 1) = 19 \qquad y = 6 - 1$$
$$3x + 1 = 19 \qquad y = 5$$
$$3x = 18$$
$$x = 6$$

The solution is $(6, 5)$.

Use substitution to solve each system of equations.

9. $x = 2y$
   $x + y = 6$
10. $2m + n = 1$
    $m - n = 8$
11. $3a - 2b = -4$
    $3a + b = 2$
12. $3x - y = 1$
    $2x + 4y = 3$

■ Find the probability of a simple event.

A soccer team consists of 8 seniors, 7 juniors, 3 sophomores, and 2 freshmen. The probability that a player chosen at random is a senior is $\frac{8}{20}$ or 0.4.

13. For the soccer team listed at the left, find the probability that a player chosen at random is not a junior or a freshman.

14. What is the probability that a number chosen at random from the first 100 positive integers is prime?

- Graph inequalities in the coordinate plane.

  Graph $2x + 7y < 9$.

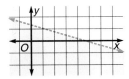

Graph each inequality.

15. $x + 2y > 5$

16. $4x - y \leq 8$

17. $3x - 2y < 6$

18. $\frac{1}{2}y \geq x + 4$

- Simplify radical expressions involving addition and subtraction.

$$\sqrt{48} + \sqrt{54} - 2\sqrt{3}$$
$$= \sqrt{4^2 \cdot \sqrt{3}} + \sqrt{3^2 \cdot \sqrt{6}} - 2\sqrt{3}$$
$$= 4\sqrt{3} + 3\sqrt{6} - 2\sqrt{3}$$
$$= 2\sqrt{3} + 3\sqrt{6}$$

Simplify.

19. $2\sqrt{13} + 8\sqrt{15} - 3\sqrt{15} + 3\sqrt{13}$

20. $4\sqrt{27} + 6\sqrt{48}$

21. $5\sqrt{18} - 3\sqrt{112} - 3\sqrt{98}$

22. $\sqrt{8} + \sqrt{\frac{1}{8}}$

- Solve quadratic equations by using the quadratic formula.

  Solve $2x^2 + 7x - 16 = 0$.

$$x = \frac{-b \pm \sqrt{b^2 - 4ac}}{2a} \quad a = 2, b = 7, c = -16$$

$$= \frac{-7 \pm \sqrt{7^2 - 4(2)(-16)}}{2(2)} \quad \text{or} \quad \frac{-7 \pm \sqrt{177}}{4}$$

The roots are $\dfrac{-7 + \sqrt{177}}{4}$ and $\dfrac{-7 - \sqrt{177}}{4}$.

Solve each quadratic equation by using the quadratic formula.

23. $x^2 - 8x = 20$

24. $5b^2 + 3 = -9b$

25. $9k^2 = 12k + 1$

26. $2m^2 - \dfrac{17m}{6} + 1 = 0$

Applications and Connections

27. **Finance**   Last year, Jodi invested $10,000, part at 8% annual interest and the rest at 6% annual interest. If she received $760 in interest at the end of the year, how much did she invest at each rate?

28. **Physics**   The height h, in feet, of a rocket t seconds after blast-off is given by the formula $h = 1440t - 16t^2$. After how many seconds will this rocket reach a height of 25,000 feet? 35,000 feet?

29. **Sales**   When you use Ray's Taxi Service, a two-mile trip costs $6.30, a five-mile trip costs $11.25, and a ten-mile trip costs $19.50. Write an equation to describe this relationship and use it to find the cost of a one-mile trip.

30. **Geometry**   Find the measure of the area of the shaded region below.

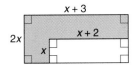

More Coordinate Geometry

CHAPTER OBJECTIVES

In this chapter, you will:
- Write and graph linear equations.
- Prove theorems using coordinate proofs.
- Perform operations with vectors.
- Locate points in space.

GEOMETRY AROUND THE WORLD
Mexico

If you like unusual architecture, perhaps some day you can take a trip to Mexico to feast your eyes on the graceful concrete churches, office buildings, restaurants, and warehouses designed by Felix Candela. Candela, an architect and engineer born in 1910, based his designs on the geometry of the *hyperbolic paraboloid.*

This design, featuring curves similar to those in a horse saddle, combines two geometric shapes. A *hyperbola* is a plane curve having two branches formed by a plane intersecting both halves of a pair of right circular cones. A *paraboloid* is a surface having cross-sections that are *parabolas*—a plane curve formed by a plane intersecting a right circular cone.

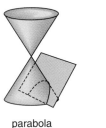
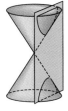

parabola   hyperbola

A hyperbolic paraboloid building is pleasing to the eye. Its open-air design, thin walls, and light roof are well suited to the hot Mexican climate. And because it requires less construction material than more traditional designs, it's less expensive to build.

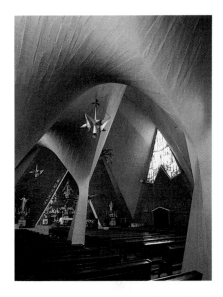

The arches inside this Mexican church designed by Felix Candela make the building beautiful and structurally sound.

GEOMETRY IN ACTION

With your finger, trace the lines that form the parabolas in the figure at the right.

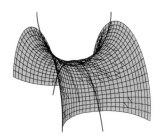

◀ *Mexican Church*

573

12-1 **Graphing Linear Equations**

Objectives

After studying this lesson, you should be able to:

- graph linear equations using the intercepts method, and
- graph linear equations using the slope-intercept method.

Application

The maximum road grade recommended by the Federal Highway Commission is 12%. This means that the slope of a road should be no more than $\frac{12}{100}$ or 0.12. At the maximum grade, a road would change 12 feet vertically for every 100 feet horizontally.

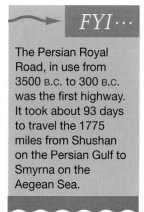

FYI ···

The Persian Royal Road, in use from 3500 B.C. to 300 B.C. was the first highway. It took about 93 days to travel the 1775 miles from Shushan on the Persian Gulf to Smyrna on the Aegean Sea.

A line that represents this road grade could be drawn on a coordinate grid using the points with coordinates (0, 0) and (100, 12).

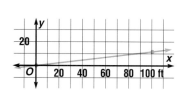

Every line drawn on a coordinate plane has a corresponding **linear equation** that describes the line algebraically. A linear equation can be written in the form $Ax + By = C$, where A, B, and C are any real numbers, and A and B are not both zero. A linear equation written in the form $Ax + By = C$ is said to be in **standard form.**

The equations $x + 3y = 12$, $4x = 11y + 8$, $7x - y = \frac{1}{2}$, and $y = 4$ are all linear equations. Each can be written in the form $Ax + By = C$. The equations $2x + 3y^2 = 8$ and $\frac{1}{y} + x = 4$ are not linear equations. *Why not?*

The graph of a linear equation is the set of all points with coordinates (x, y) that satisfy the equation $Ax + By = C$. One technique for graphing a linear equation is called the **intercepts method.** This method uses two special values, the *x-intercept* and the *y-intercept*. The x-intercept is the value of x when y equals 0. The y-intercept is the value of y when x equals 0. These values can be used as long as the line is not parallel to either axis.

Example 1

APPLICATION

Manufacturing

The Family Treasures Furniture Company makes chairs and tables. The equation $15x + 30y = 30,000$ represents the cost of making x chairs and y tables. Graph the equation by finding the intercepts.

Let $x = 0$ to find the y-intercept.
$$15(0) + 30y = 30,000$$
$$30y = 30,000$$
$$y = 1000$$

Let $y = 0$ to find the x-intercept.
$$15x + 30(0) = 30,000$$
$$15x = 30,000$$
$$x = 2000$$

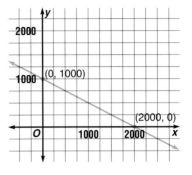

The x-intercept is 2000, and the y-intercept is 1000. To graph $15x + 30y = 30,000$, plot (0, 1000) and (2000, 0) and draw a line through the points.

Another method for graphing is more convenient to use if the equation is rewritten so one side has y by itself. Use a graphing calculator, graphing software, or graph paper to graph the family of equations that appear below.

$$y = 2x + 8$$
$$y = 2x + 0$$
$$y = 2x - 4$$

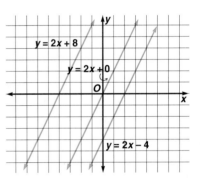

Recall from algebra that the coefficient is the numerical part of a term and a constant term contains no variables.

Notice that the graphs are parallel lines and that the slope of each line is 2. Compare the slope of each line to its equation. Did you notice that the slope of each line is equal to the coefficient of x in its equation? Also the y-intercept of each line is equal to the constant term of the equation. A linear equation written in the form $y = mx + b$ is called the **slope-intercept form** of the equation.

Theorem 12-1 *Slope-Intercept Form*	If the equation of a line is written in the form $y = mx + b$, m is the slope of the line and b is the y-intercept.

Proof of Theorem 12-1

The slope, m, of a line passing through points with coordinates (x_1, y_1) and (x_2, y_2) is given by
$$m = \frac{y_2 - y_1}{x_2 - x_1}.$$

Notice that if $x = 0$, $y = b$. Therefore, b is the y-intercept. If $x = 1$, $y = m + b$. Since the ordered pairs for two points on the line have coordinates $(0, b)$ and $(1, m + b)$, the slope is $\frac{(m + b) - b}{1 - 0}$ or m.

Horizontal and vertical lines are special cases. The graph of an equation of the form $x = a$ is a vertical line and has an undefined slope. The graph of an equation of the form $y = b$ is a horizontal line and has a slope of 0.

You can graph a line if you know its slope and *y*-intercept.

Example 2

Graph $y = \frac{1}{4}x + 5$ using the slope and the *y*-intercept.

Since the equation is in slope-intercept form, the slope is $\frac{1}{4}$ and the *y*-intercept is 5.

Plot the point with coordinates (0, 5) and from this point move up 1 unit and to the right 4 units. This point, that has coordinates (4, 6), must also lie on the line. Draw the line containing the two points.

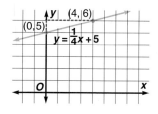

The slope-intercept form of a linear equation can also be used to determine if two lines are parallel or perpendicular without graphing the equations. Remember that if two lines are parallel, they have the same slope. If two lines are perpendicular, the product of their slopes is -1.

Example 3

CONNECTION

Algebra

Determine if the graphs of $2x + y = 3$ and $4x + 2y = 5$ are parallel, perpendicular, or neither.

Write each equation in slope-intercept form.

$$2x + y = 3 \qquad\qquad 4x + 2y = 5$$
$$y = -2x + 3 \qquad\qquad 2y = -4x + 5$$
$$\qquad\qquad\qquad\qquad y = -2x + \frac{5}{2}$$

Since for each equation $m = -2$ and the *y*-intercepts are not equal, the graphs are parallel lines.

CHECKING FOR UNDERSTANDING

Communicating Mathematics

Read and study the lesson to answer these questions.

1. A vertical line contains the point with coordinates (4, -7). What is the equation of the line? Explain your answer.

2. *True* or *false*: Horizontal lines are perpendicular to the *x*-axis. Explain your reasoning.

3. The graph at the right displays a "family" of parallel lines. What must be true about the equations of these lines?

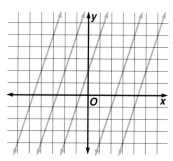

4. Describe two methods of graphing a linear equation. Which method is more convenient for graphing $2x - 3y = 10$? Explain.

5. One way to graph the equation $y = \frac{1}{2}x + 3$ is to start at the point with coordinates $(0, 3)$ and move up 1 and to the right 2 to determine the coordinates of a second point on the line. What other ways can you use to locate other points on the line?

Guided Practice

Determine whether each equation is a linear equation. Explain your reasoning.

6. $y = 2x - 1$ **7.** $3x + 4y^2 = 9$ **8.** $y = x^3$

9. $x = 8$ **10.** $x^2 + y^2 = 10$ **11.** $4x = 9 - y$

Find the *x*- and *y*-intercepts of the graph of each equation.

12. $y = x$ **13.** $4x - y = 4$ **14.** $x + 2y = 6$

15. $x = 4$ **16.** $y + 3 = x$ **17.** $3x - 6y = 6$

Graph each equation. Determine if this line and the line already drawn on the coordinate axes are parallel, perpendicular, or neither. Verify by finding the slope of each line.

18. $3x + y = 6$ **19.** $x - y = 4$ **20.** $y = \frac{1}{2}x + 3$

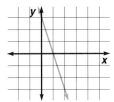   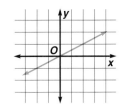 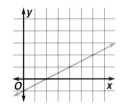

21. $y = 3$ **22.** $3x = 4y + 6$ **23.** $y = 8 - x$

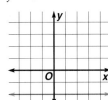   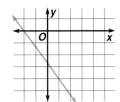 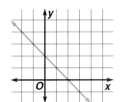

EXERCISES

Practice

Graph each equation. Explain the method you used to draw the graph.

24. $y = 4x - 2$ **25.** $y = 3x$ **26.** $5x + 2y = 0$

27. $y = 2x - 10$ **28.** $x + 4y = 16$ **29.** $3x - 4y = 6$

Graph each pair of linear equations on the same coordinate plane. Determine if the lines are parallel, perpendicular, or neither by finding the slope of each line.

30. $y = 6 - x$
$x + y = 2$

31. $5x - y = 0$
$y = 5x - 3$

32. $2y + x = 4$
$y = 2x - 5$

If they exist, find the slope and *y*-intercept of the graph of each equation.

33. $y = x$

34. $3x + y = 5$

35. $3x + 4y = 8$

36. $y = 6$

37. $x = 9$

38. $2x + y = 6$

Draw each line described on a separate coordinate plane. Then write the equation of each line in slope-intercept form.

39. $m = -\frac{1}{3}$, *y*-intercept = 2

40. $m = 4$, passes through $(0, 5)$

41. $m = 4$, *y*-intercept = -3

42. parallel to *y*-axis through $(2, 0)$

43. $m = 2$, passes through $(0, 3)$

44. parallel to *x*-axis through $(0, -6)$

45. $m = \frac{1}{3}$, passes through $(0, 4)$

46. $m = 4$, passes through $(0, 7)$

47. perpendicular to the *x*-axis, passes through $(2, 0)$

48. Draw several lines having a slope of -2. Describe the similarities and the differences between the equations of these lines.

49. Draw several lines passing through the point $(0, 3)$. Describe the similarities and the differences between the equations of these lines.

50. What is the general equation for the lines parallel to the *y*-axis?

Critical Thinking

51. The graphs of $3x + 2y = 10$ and $3x + 2y = 4$ are parallel lines. Find the equation of the line that is parallel to both lines and lies midway between them.

Applications

52. Road Grade A certain road has an 8% grade. The Hendersons' mailbox and the Pauls' mailbox are located along this road and are 650 feet apart. What is the vertical change in distance between these mailboxes?

53. Business Just Like Grandma's Bake Shop spends $1400 a month for rent and utilities. For each day of operation, they spend $200 for employees' wages, benefits, and baking supplies. If x represents the number of days of operation in a month, $y = 200x + 1400$ represents the cost of operation for the month.

    a. What is the slope of the line that represents the bake shop's cost of operation for the month?

    b. What is the y-intercept?

    c. Graph the equation.

Mixed Review

54. Find the volume of a sphere with a radius of 5 inches. **(Lesson 11-7)**

55. Find the surface area of a sphere with radius of 3 feet. **(Lesson 11-7)**

56. Find the slope of the line that passes through the points with coordinates (3, 0) and (8, -2). **(Lesson 3-4)**

57. Draw a figure to illustrate two lines that are perpendicular to a third line, but are not parallel to each other. **(Lesson 3-1)**

58. Algebra Find the measures of $\angle 1$ and $\angle 2$, if $m\angle 1 = 2x + 15$ and $m\angle 2 = 8x - 5$. **(Lesson 2-7)**

59. Determine if a valid conclusion can be made from the statements *If it snows Saturday, then we will go skiing,* and *If we go skiing, then we will need to rent skis.* State the law of logic that you used. **(Lesson 2-3)**

Wrap-Up

60. Write an example of a linear equation in standard form and another linear equation in slope-intercept form. Describe the graph of each equation.

~~~ HISTORY CONNECTION ~~~

Did you know that sound travels at a different speed in water than it does in air? Berthel Carmichael, the first African-American woman to go to sea on a military sealift oceanographic research ship, is a research mathematician who studied the properties of sound waves under water.

Ms. Carmichael is a native of Richmond, Virginia, and a graduate of Virginia Union University. She taught in the public school system before joining the Naval Research Laboratory in Washington, D.C. While working in the Acoustics Division Propagation Branch, Ms. Carmichael spent time aboard the U.S.S. Hayes and the U.S.S. Mizar in the Arctic Ocean.

12-2 Writing Equations of Lines

Objective

After studying this lesson, you should be able to:

- write an equation of a line given information about its graph.

Application

There are about 90,100,000 televisions in the United States. 98% of American households own at least one T.V.

In 1980, there were 15 million cable television subscribers. If the number of subscribers increases by 5 million each year, the equation $y = 5x + 15$ can be used to find y, the number of cable subscribers (in millions) for any number of years, x, after 1980.

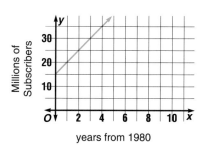

years from 1980

The y-intercept, 15, represents the number of subscribers in 1980. The slope, 5, represents the yearly increase in the number of subscribers.

In this problem, the slope and y-intercept were used to write the equation in slope-intercept form. This equation can be useful for making predictions regarding the number of cable subscribers in the years after 1980. For example, to find the number of subscribers in the year 2000, we would substitute 20 for x and solve for y. *Why 20?*

$$y = 5x + 15$$
$$y = 5(20) + 15 \quad \textit{Substitute 20 for x.}$$
$$y = 115$$

If the yearly increase in subscribers remains constant, there will be 115 million cable subscribers in the year 2000. *Do you think this is a reasonable answer? Explain.*

In general you can write an equation of a line if you are given:

- **Case 1:** the slope and the coordinates of a point on the line, or
- **Case 2:** the coordinate of two points on the line.

Consider Case 1. To write an equation of a line given the slope and the coordinates of a point on the line, substitute the slope in the equation $y = mx + b$. Then substitute the coordinates of the point for x and y and solve for b. Finally, write an equation of the line by substituting the values for m and b into $y = mx + b$.

Example 1

Write an equation of the line whose slope is 3 and *x*-intercept is -10.

Since we know that the slope is 3, we can substitute 3 for m in $y = mx + b$.

$$y = 3x + b$$

Now, since the *x*-intercept is -10, the point (-10, 0) is on the line. Substitute the coordinates of this point into the equation to find the value of b.

$$y = 3x + b$$
$$0 = 3(-10) + b \qquad \textit{y = 0 and x = -10}$$
$$0 = -30 + b$$
$$30 = b \qquad\qquad \textit{Solve for b.}$$

The slope-intercept form of the equation of the line is $y = 3x + 30$.

The following example illustrates Case 2. You can find an equation of a line given the coordinates of two points on the line using the **point-slope form** of a linear equation. The point-slope form is $y - y_1 = m(x - x_1)$, where (x_1, y_1) are the coordinates of a point on the line and m is the slope of the line.

Example 2

With a certain long-distance carrier, the price of a 4-minute long-distance telephone call is $1.90. A 13-minute call with the same carrier costs $5.05.
a. Write a linear equation that describes the cost of telephone calls.

The line passes through the points (4, 1.9) and (13, 5.05). Find the slope of the line.

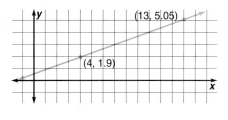

$$m = \frac{y_2 - y_1}{x_2 - x_1}$$
$$= \frac{5.05 - 1.9}{13 - 4}$$
$$= \frac{3.15}{9} \text{ or } 0.35$$

Now use the point-slope form to write the linear equation.

$$y - y_1 = m(x - x_1) \qquad \textit{Use point-slope form.}$$
$$y - 1.9 = 0.35(x - 4) \qquad \textit{Use the coordinates of either point for } (x_1, y_1).$$
$$y - 1.9 = 0.35x - 1.4 \qquad \textit{We chose (4, 1.9).}$$
$$y = 0.35x + 0.5$$

The slope-intercept form of the equation of the line is $y = 0.35x + 0.5$.
What do 0.35 and 0.5 represent?

b. How much would a 30-minute telephone call cost?

Use the equation to find the cost of a 30-minute telephone call.

$$y = 0.35x + 0.5$$
$$y = 0.35(30) + 0.5 \qquad \textit{Substitute 30 for x.}$$
$$y = 10.5 + 0.5$$
$$y = 11$$

A 30-minute call costs $11.

Example 3

Write an equation of the line that is the perpendicular bisector of $\overline{AB}$ with endpoints at $A(8, 1)$ and $B(-10, -1)$.

Find the slope of $\overline{AB}$ by using the coordinates of A and B.

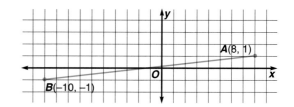

$$m = \frac{y_2 - y_1}{x_2 - x_1}$$
$$= \frac{1 - (-1)}{8 - (-10)}$$
$$= \frac{2}{18} \text{ or } \frac{1}{9}$$

Since the slope of $\overline{AB}$ is $\frac{1}{9}$, the slope of the perpendicular bisector is -9. *Why?*

To find a point on the perpendicular bisector, recall that the perpendicular bisector must pass through the midpoint of the segment. Use the midpoint formula to find the midpoint of $\overline{AB}$.

$$M = \left(\frac{x_1 + x_2}{2}, \frac{y_1 + y_2}{2} \right)$$
$$= \left(\frac{8 + (-10)}{2}, \frac{1 + (-1)}{2} \right) \quad \textit{Substitute (8, 1) for } (x_1, y_1) \textit{ and (-10, -1) for } (x_2, y_2).$$
$$= (-1, 0)$$

Now write an equation.

$$y - y_1 = m(x - x_1) \quad \textit{Use the point-slope form of the equation.}$$
$$y - 0 = -9(x - (-1)) \quad \textit{The slope is -9 and the point with coordinates (-1, 0)}$$
$$y = -9x - 9 \quad \textit{is on the line.}$$

An equation of the perpendicular bisector of $\overline{AB}$ is $y = -9x - 9$.

CHECKING FOR UNDERSTANDING

Communicating Mathematics

Read and study the lesson to answer these questions.

1. Determine if each of the following is sufficient information to write an equation of exactly one line. Explain your answer.

   a. a point on the line   **b.** the slope of the line

   c. two points on the line   **d.** the slope and a point on the line

2. There are currently 5.3 million cellular telephone users. Each year the number of cellular phone users increases by 1.2 million.

   a. Write a linear equation that represents y, the total number of cellular telephone users x years from now if the rate of increase stays the same.

   b. Approximately how many cellular telephone users will there be in 10 years if the rate of increase stays the same?

3. Find the slope of $\overleftrightarrow{AB}$.

4. Write an equation for $\overleftrightarrow{CD}$.

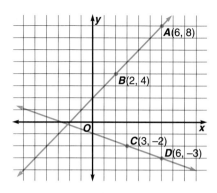

Guided Practice

Write the equation in slope-intercept form of the line that satisfies the given conditions.

5. $m = 4$, y-intercept $= 2$

6. $m = -\frac{1}{2}$, y-intercept $= 1$

7. $m = 5$, passes through $(-1, 3)$

8. $m = -3$, x-intercept $= 6$

9. passes through $(-7, 4)$ and $(-5, -6)$

10. passes through $(6, -1)$ and $(-3, -7)$

11. parallel to $y = 3x + 4$, passes through $(3, 7)$

12. perpendicular to $y = -\frac{1}{3}x - 2$, passes through $(1, 2)$

EXERCISES

Practice

State the slope and y-intercept for each line. Then write the equation of the line in slope-intercept form.

13. a

14. b

15. c

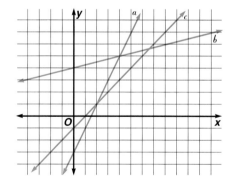

Write the equation in slope-intercept form of the line having the given slope and passing through the given point.

16. 3, $(1, -4)$

17. $\frac{1}{6}$, $(12, -3)$

18. $-\frac{2}{3}$, $(-3, 2)$

19. -4, $(-3, -2)$

20. -1, $(5, 0)$

21. 0, $(6, 7)$

Write the equation in slope-intercept form of the line satisfying the given conditions.

22. $m = 0$, y-intercept = 7

23. $m = -2$, y-intercept = 1

24. $m = \frac{3}{4}$, y-intercept = 8

25. $m = -\frac{1}{2}$, passes through (5, 3)

26. $m = 6$, passes through (-3, 1)

27. parallel to $y = -4x + 1$, passes through (-3, 1)

28. parallel to the y-axis, passes through (3, 9)

29. perpendicular to the y-axis, passes through (-8, 2)

30. passes through points (0, 3) and (4, -3)

31. passes through (9, -4) and (-2, 7)

32. the perpendicular bisector of the segment whose endpoints have coordinates (2, 5) and (-2, -1)

33. Write the equation in slope-intercept form of the line that has a slope of -5 and bisects a segment whose endpoints have coordinates (-4, 10) and (5, -7).

34. Write the equation in slope-intercept form of the line that is the perpendicular bisector of a segment whose endpoints have coordinates (-3, -7) and (5, 1).

35. A line is parallel to the line whose equation is $3x - 4y = 11$ and passes through (-4, 5). Write the equation of the line in slope-intercept form.

36. Write the equation in slope-intercept form of the line tangent to the circle with equation $x^2 + (y - 3)^2 = 25$ at the point with coordinates (4, 0).

Critical Thinking

37. The x-intercept of a line is s and the y-intercept is t. Write the equation of the line in slope-intercept form in terms of s and t.

Applications

38. **Business** Handy Helper Hardware sells about 360 gallons of interior house paint in a week.

    a. How many gallons of paint will they sell in x weeks?

    b. The store has 2880 gallons of paint on hand. Write an equation that describes how many gallons they will have in stock in x weeks if no new stock is added.

    c. Draw a graph that represents the number of gallons of paint the store will have at any given time.

    d. If it takes three weeks to receive a shipment of paint from the warehouse after it is ordered, when should the manager order more paint so that the store will not run out?

39. Sports It costs about $900 to equip an Olympic skier with skis, poles, and boots. The costs are expected to increase $12 per year. Write the linear equation in slope-intercept form that represents the approximate cost of equipping a skier *x* years from now. Assume that the rate of increase remains constant.

Computer

40. The BASIC program given below will write an equation for the line passing through two points. Use the program to check your answers to Exercises 22-36.

```
10   PRINT "ENTER THE COORDINATES OF THE FIRST POINT."
20   INPUT X1, Y1
30   PRINT "ENTER THE COORDINATES OF THE SECOND POINT."
40   INPUT X2, Y2
50   IF (X2 − X1) = 0 THEN PRINT "AN EQUATION OF THE LINE
     IS  X = "; X1; ".": GOTO 130
60   M = (Y2 − Y1)/(X2 − X1)
70   IF M = 0 THEN PRINT "AN EQUATION OF THE LINE IS Y =
     "; Y1; ".": GOTO 130
80   B = Y1 − M*X1
90   A$ = "+"
100  IF B < 0 THEN B = ABS(B): A$ = "−"
120  PRINT "AN EQUATION OF THE LINE IS Y = "; M; "X"; A$;
     " "; B; "."
130  END
```

Mixed Review

41. Graph the line $6x - 4y = 3$ and state its slope and *y*-intercept. **(Lesson 12-1)**

42. Find the missing measures in $\triangle ABC$. **(Lesson 8-6)**

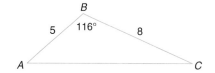

43. Prove that if one pair of alternate interior angles formed by two lines cut by a transversal is congruent, then the other pair of alternate interior angles is congruent also. **(Lesson 3-1)**

44. Given $A(5, -1)$ and $B(-2, -6)$, find the measure of $\overline{AB}$. **(Lesson 1-4)**

45. Given $M(2, 3)$ and $N(-5, -1)$, find the coordinates of the midpoint of $\overline{MN}$. **(Lesson 1-5)**

Wrap-Up

46. Write a few sentences to explain how to write an equation of a line given the coordinates of two points on the line.

12-3 Connections to Algebra and Statistics

Objective

After studying this lesson, you should be able to:
- relate equations of lines and statistics to geometric concepts.

Application

Can you predict the temperature by how much noise crickets make? Crickets make their sounds by moving one wing over the other. Scientists have noticed that crickets move their wings faster in warm weather than in cold weather. The table gives the number of chirps per minute at different temperatures.

Chirps per minute	20	16	20	18	18	16	15	17	15	16	15	17	16	17	14
Temperature (°F)	88	72	93	84	82	75	70	82	69	83	73	83	81	84	76

A **scatter plot**, like the one at the right, shows the relationship between the variables by plotting a set of data as points. In the graph at the right, we can fit a line as a way to summarize the data, and find an equation that will express the approximate temperature in terms of the number of chirps. Notice that the points do not lie in a straight line, but they do suggest a linear pattern. Using the points with coordinates (20, 88) and (15, 73), you can determine an equation for this line using techniques from algebra.

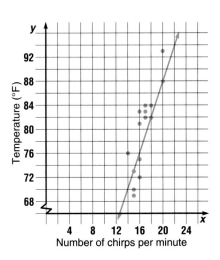

You can use a graphing calculator to find the equation of the line that summarizes this data best. This line is called a least squares line of regression.

First, find the slope of the line.

$$m = \frac{y_2 - y_1}{x_2 - x_1}$$

$$= \frac{88 - 73}{20 - 15} \qquad (x_1, y_1) = (15, 73), (x_2, y_2) = (20, 88)$$

$$= \frac{15}{5} \text{ or } 3$$

Substitute the slope and the coordinates of one point into the point-slope form to find an equation.

$$y - y_1 = m(x - x_1)$$
$$y - 88 = 3(x - 20) \qquad \textit{The slope is 3 and one point has coordinates (20, 88).}$$
$$y - 88 = 3x - 60$$
$$y = 3x + 28$$

Therefore, an equation that relates the temperature and number of chirps per minute is $y = 3x + 28$.

You can use the equation to predict the temperature if a cricket chirps a certain number of times or predict the number of chirps that would occur given the temperature. Find the approximate temperature if a cricket chirps 19 times per minute by substituting 19 for *x*.

$y = 3x + 28$

$y = 3 \cdot 19 + 28$

$y = 57 + 28 \text{ or } 85$

The temperature will be approximately 85° F.

Whenever you work with coordinates, it is a good idea to begin by graphing the information you are given.

Example 1

Determine whether A(4, 3), B(-2, -9), and C(7, 9) are collinear.

One way to approach the problem is to find the equation of $\overleftrightarrow{AB}$ and see if the coordinates of *C* satisfy the equation. First find the slope of $\overleftrightarrow{AB}$.

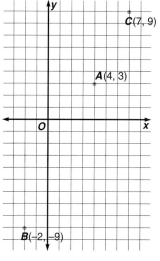

$m = \dfrac{y_2 - y_1}{x_2 - x_1}$

$\quad = \dfrac{3 - (-9)}{4 - (-2)}$ *A(4, 3) = (x₂, y₂),*
B(-2, -9) = (x₁, y₁)

$\quad = \dfrac{12}{6} \text{ or } 2$

$y - y_1 = m(x - x_1)$ *Use point-slope form.*

$y - 3 = 2(x - 4)$ *The slope is 2 and (4, 3) are the*
coordinates of a point on the line.

$y - 3 = 2x - 8$

$\quad\quad y = 2x - 5$

The equation of $\overleftrightarrow{AB}$ is $y = 2x - 5$. Since $9 = 2(7) - 5$, $C(7, 9)$ satisfies the equation, the points are collinear.

Example 2

Given $\triangle ABC$ with A(-6, -8), B(6, 4), and C(-6, 10), write the equation of the line containing the altitude from A.

The altitude from *A* is perpendicular to $\overline{BC}$. So the slopes of the altitude and of $\overline{BC}$ are opposite reciprocals. To find the equation of the line containing the altitude from *A*, first find the slope of $\overline{BC}$.

$$m = \frac{y_2 - y_1}{x_2 - x_1}$$
$$= \frac{10 - 4}{-6 - 6}$$
$$= \frac{6}{-12} \text{ or } -\frac{1}{2}$$

The slope of $\overline{BC}$ is $-\frac{1}{2}$. So the slope of the line containing the altitude drawn to $\overline{BC}$ is 2. Find the equation of the line that has a slope of 2 and passes through the point with coordinates (-6, -8).

$y - y_1 = m(x - x_1)$ *Use point-slope form.*
$y - (-8) = 2(x - (-6))$ *The slope is 2 and the point with coordinates (-6, -8) is*
$y + 8 = 2x + 12$ *on the line.*
$y = 2x + 4$

The equation of the line containing the altitude is $y = 2x + 4$.

CHECKING FOR UNDERSTANDING

Communicating Mathematics

Read and study the lesson to answer these questions.

1. On a coordinate plane, draw examples of two lines that are parallel and two lines that are perpendicular. How are their slopes related?

2. Describe two ways to find the y-intercept of the graph of $2x + 5y = 10$.

3. What do you call the graphs of paired data points?

Guided Practice

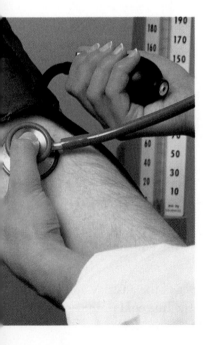

The vertices of $\triangle RST$ are $R(-4, -3)$, $S(2, 3)$, and $T(-4, 5)$.

4. Find the equations of the lines containing the medians to each of the sides of $\triangle RST$.

5. Find the equations of the lines containing perpendicular bisectors of the sides of $\triangle RST$.

6. The table below shows the age and systolic blood pressure for a group of people who recently donated blood.

Age	35	24	48	50	34	55	30	26	41	37
Blood Pressure	128	108	140	135	119	146	132	104	132	121

a. Draw a scatter plot to show how age, x, and systolic blood pressure, y, are related.

b. Write an equation that relates a person's age to their approximate systolic blood pressure.

c. Find the approximate blood pressure of a person 54 years old.

d. List a few factors that may affect a person's actual systolic blood pressure.

EXERCISES

Practice

Determine whether the three points listed are collinear.

7. $A(9, 0)$, $B(4, 2)$, $C(2, -1)$ **8.** $X(6, 9)$, $Y(3, -1)$, $Z(4, 0)$

9. $L(0, 4)$, $M(2, 3)$, $N(-4, 6)$ **10.** $D(9, -3)$, $E(4, 8)$, $F(0, 0)$

11. The table below lists the Federal Income Tax due from a single person with the given taxable income for 1991.

Taxable Income	11,905	7,412	22,898	19,054	10,995	3,268	18,753
Tax due	1,789	1,114	3,760	2,861	1,646	491	2,816

     a. Draw a scatter plot to show how taxable income, x, and Federal Income Tax due, y, are related.

     b. Write an equation that relates a single person's taxable income and their approximate Federal Income Tax due.

     c. Angela's taxable income for 1991 was $12,982. Approximately how much did she owe in Federal Income Tax?

12. Find the equation of the line containing the perpendicular bisector of the segment whose endpoints have coordinates $(2, 5)$ and $(-2, -1)$.

13. Find the equations of the lines containing the sides of an isosceles triangle if the vertex is at the y-intercept of the line whose equation is $y = -2x + 6$ and a vertex of a base angle is at $(4, -2)$.

The vertices of $\triangle ABC$ are $A(0, 14)$, $B(2, -4)$, and $C(6, 2)$.

14. Write the equation of the line containing the altitude to $\overline{BC}$.

15. Write the equation of the line containing the perpendicular bisector of $\overline{AB}$.

16. Find the equation of the line containing $\overline{BE}$ if $\overline{BE}$ is a median of $\triangle ABC$.

The vertices of $\triangle RST$ are $R(-6, -8)$, $S(6, 4)$, and $T(-6, 10)$.

17. Write the equations of the lines containing the sides of $\triangle RST$.

18. Write the equations of the lines containing the medians of $\triangle RST$.

19. Write the equations of the lines containing the altitudes of $\triangle RST$.

20. The sum of the measures of the angles of a triangle is 180°. This can be represented by the ordered pair (3, 180). The relationship between the number of sides and the sum of the measures of the angles of a rectangle is represented by (4, 360). This relationship for pentagons is represented by (5, 540).

  a. Plot the points and prove they are collinear.

  b. Write the equation of the line determined by the points.

  c. Use the equation of the line to predict the sum of the measures of the angles in a 10-sided polygon.

The equation of a circle is $(x + 1)^2 + (y - 2)^2 = 25$.

21. Show that the point with coordinates (3, 5) is on the circle.

22. Find the equation of the tangent to the circle at (3, 5).

Critical Thinking

23. The equations of two parallel lines are $y = 2x + 10$ and $y = 2x + 3$. Jason says that the two lines are 7 units apart. Do you agree? Why or why not?

Applications

24. Investments Terry can invest his money in an investment that pays 7% interest or in a higher risk investment that pays 10% interest. He has decided to place some of his money in each investment so that he can earn fairly high interest and still have a relatively safe investment.

  a. If he would like to make $210 in interest after one year, write an equation that would describe his investment.

  b. Graph the equation.

  c. Find the intercepts and explain what they mean.

  d. If Terry would like to make at least $210 in interest in one year, how would this change the graph?

25. Health The ages and optimum exercise heart rates for several people are listed in the table below.

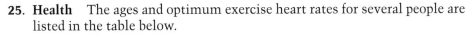

Age	31	42	18	24	55	61	44	15
Heart Rate	117	107	127	122	104	99	109	127

  a. Draw a scatter plot to show how age, x, and optimum exercise heart rate, y, are related.

  b. Write an equation that relates a person's age to their optimum exercise heart rate.

  c. Miguel is 19 years old. What is his optimum exercise heart rate?

26. Belinda Jackson bought a microwave oven for $60 more than half its original price. She paid $274 for the oven. What was the original price of the oven? **(Lesson 12-2)**

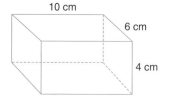

27. Find the total surface area of the solid shown at the right. **(Lesson 11-3)**

28. What can you conclude about *RS* and *QR* from the figure below? State whether the SAS or the SSS Inequality can be used to justify your answer. **(Lesson 5-7)**

Wrap-Up

29. Write a three-question quiz that covers the material in this lesson. Exchange quizzes with a partner and take each other's quiz.

HISTORY CONNECTION

Emilie du Chatelet (1706-1749) was a well-published mathematician, scientist, classicist, and translator because when she was six or seven years old, her father thought that she would be a homely woman and would never marry. So, he provided the best tutors for her studies and riding and fencing instructors as well. By the age of fifteen, Emilie had grown to be a beautiful woman. And because of her exceptional intelligence and education, she had both beauty and brains.

Emilie du Chatelet's greatest work was a translation and analysis of Isaac Newton's *Principia.* This translation brought modern science and mathematics to Europe. To this day, her work is the only French translation of this important scientific document.

12-4 Problem-Solving Strategy: Write an Equation

Objective

After studying this lesson, you should be able to:

- solve problems by using equations.

Application

Neva is lining a soccer field. The length should be 75 yards shorter than 3 times its width. The perimeter of the field is 370 yards. What are the length and the width of the field? An equation can help you solve Neva's problem.

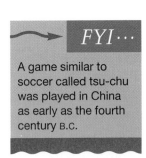

FYI⋯

A game similar to soccer called tsu-chu was played in China as early as the fourth century B.C.

To solve problems using equations, you should follow five steps.

1. Define a variable.
2. Write an equation.
3. Solve the equation.
4. Check the solution.
5. Answer the problem.

Example 1

APPLICATION

Sports

Find the length and width of the soccer field that Neva is lining.

1. Define a variable.

   Let x represent the width in yards.
   Then the length will be $3x - 75$ yards.

2. Write an equation.

   The perimeter of a rectangle equals 2 times the length plus 2 times the width.

   $$370 = 2(3x - 75) + 2x$$

3. Solve the equation.

   $$370 = 2(3x - 75) + 2x$$
   $$370 = 6x - 150 + 2x$$
   $$520 = 8x$$
   $$65 = x$$

   If $x = 65$, $3x - 75 = 120$.

4. Check your solution.

If the length is 120 yards and the width is 65 yards, will the perimeter equal 370?

$$2(120) + 2(65) = 370$$
$$240 + 130 = 370$$
$$370 = 370 \;\vee$$

5. Answer the problem.

The length of the soccer field is 120 yards and the width is 65 yards.

CHECKING FOR UNDERSTANDING

Communicating Mathematics

Read and study the lesson to answer these questions.

1. Why should you define the variable before you write an equation?

2. Explain the steps used to solve the equation in Example 1.

3. Define a variable and write an equation for the following problem.

Felipe's bowling handicap is 7 less than half his average. His handicap is 53. What is Felipe's bowling average?

Guided Practice

Solve. Use an equation.

4. To estimate when to harvest her early pea crop, Dorothy counts heat units. As of June 1 she has counted 835 heat units. There are usually 30 heat units per day in June. Early peas require 1165 heat units to mature. On what day can Dorothy plan to harvest her crop?

5. Ron is on his way to San Diego, 300 miles away. He drives 45 miles per hour for 3 hours. He drives 55 miles per hour for the rest of the trip. How long does Ron drive at 55 miles per hour?

6. The length of a rectangular garden is 40 meters less than 2 times its width. Its perimeter is 220 meters. Find its length and width.

7. Jenny sold tickets for the annual spring concert at Middletown High School. The concert tickets cost $3.50 for adults and $2.50 for students. If Jenny sold 4 more student tickets than adult tickets and she has a total of $58.00 in sales, how many of each type of ticket did she sell?

EXERCISES

Practice

Strategies

Look for a pattern.
Solve a simpler problem.
Act it out.
Guess and check.
Draw a diagram.
Make a chart.
Work backward.

Solve. Use any strategy.

8. Turn this triangle upside down by moving just 3 coins.

9. Suppose a student who is 5 feet tall could walk around the equator. How much farther would his head travel than his feet?

10. Assign each of the numbers from 1 to 8 to one of the vertices of a cube so that the sum of the numbers assigned to the vertices of each face is 18.

11. San Francisco and Los Angeles are 470 miles apart by train. An express train leaves Los Angeles at the same time a passenger train leaves San Francisco. The express train travels 10 miles per hour faster than the passenger train. The 2 trains pass each other in 2.5 hours. How fast is each train traveling?

12. A certain bacteria doubles its population every 8 hours. After 3 days there are 12,800 bacteria. How many bacteria were there at the beginning of the first day?

13. The sum of the digits of a two-digit number is 10 and three times its tens digit is twice its ones digit. What is the number?

14. Six equilateral triangles are placed together to form a hexagon. The length of a side of one triangle is 2 inches. What is the diameter of the smallest circle that includes all six vertices of the hexagon?

COOPERATIVE LEARNING PROJECT

Work in groups. Each person in the group must understand the solution and be able to explain it to any person in class.

A helium tank fills a balloon at a steady rate. It takes 3 seconds for the radius of a balloon to become 5 centimeters long. How many seconds would it take for the radius of an empty balloon to become 10 centimeters long? (Assume that a balloon is approximately spherical.)

12-5 Coordinate Proof

Objective

After studying this lesson, you should be able to:

- prove theorems using coordinate proofs.

Application

The stage of the Palace Theater is formed by a semicircle and a rectangle. When the theater was renovated, the architect made a scale drawing of the stage and assigned coordinates to the drawing. Then the architect used algebra to find the best locations for the microphones and lights.

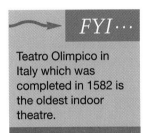

FYI···

Teatro Olimpico in Italy which was completed in 1582 is the oldest indoor theatre.

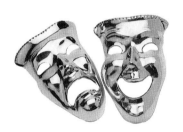

The diameter of the semicircle that forms part of the stage of the Palace Theater is 30 feet long. The width of the rectangle is 10 feet. Using the placement of the stage as shown, you can name the coordinates of the front center point of the stage.

Start by naming the vertices of the rectangle as shown. The x-coordinate of the front center is 15, and the y-coordinate is $10 + 15$ or 25. So the coordinates of the front center point are (15, 25).

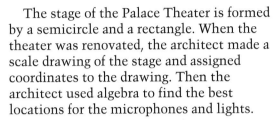

In geometry, we can assign coordinates to figures and use the coordinates to prove theorems. An important part of planning a **coordinate proof** is the placement of the figure on the coordinate plane.

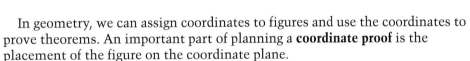

Guidelines for Placing Figures on a Coordinate Plane	1. **Use the origin as a vertex or center.** 2. **Place at least one side of a polygon on an axis.** 3. **Keep the figure within the first quadrant if possible.** 4. **Use coordinates that make computations simple.**

Example 1

Position and label a right triangle with legs of *a* units and *b* units on the coordinate plane.

Use the origin as the vertex of the right angle. Place the legs of the triangle on the positive x- and y-axes. Label the vertices P, Q, and R. Since Q is on the y-axis, its x-coordinate is 0. Its y-coordinate is b, because the leg is b units long. Since R is on the x-axis, its y-coordinate is 0 and its x-coordinate is a because it is a units long. P is at the origin, so both coordinates are 0.

Some examples of figures placed on the coordinate plane are given below.

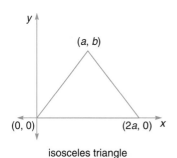

isosceles triangle

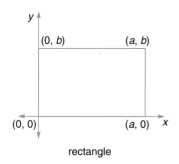

rectangle

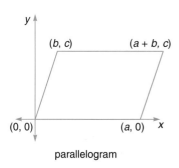

parallelogram

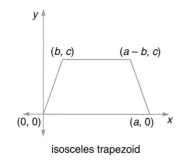

isosceles trapezoid

Example 2

Use a coordinate proof to prove that the midpoint of the hypotenuse of a right triangle is equidistant from the vertices.

Given: $\triangle QPR$ is a right triangle.

M is the midpoint of $\overline{QR}$.

Prove: M is equidistant from Q, P, and R.

Proof:

Place right $\triangle QPR$ on the coordinate plane and label coordinates as shown. (Using coordinates that are multiples of 2 for Q and R will make the computation easier.) By the Midpoint Formula, the coordinates of M are $\left(\frac{2a}{2}, \frac{2b}{2}\right) = (a,b)$.

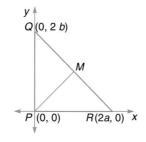

Use the Distance Formula to find MR and PM.

$$MR = \sqrt{(a - 2a)^2 + (b - 0)^2} \qquad PM = \sqrt{(a - 0)^2 + (b - 0)^2}$$
$$= \sqrt{(-a)^2 + (b)^2} \qquad\qquad = \sqrt{(a)^2 + (b)^2}$$
$$= \sqrt{a^2 + b^2} \qquad\qquad\quad = \sqrt{a^2 + b^2}$$

Thus, $MR = PM$. Also, by the definition of midpoint, $QM = MR$. By the transitive property, $QM = MR = PM$, and M is equidistant from Q, P, and R.

CHECKING FOR UNDERSTANDING

Communicating Mathematics

Read and study the lesson to answer these questions.

1. When planning a coordinate proof, why is it helpful to place at least one of the sides of a polygon on an axis?

2. Show how to place an isosceles triangle on a coordinate plane in two ways.

3. Two ways of positioning a rectangle on coordinate planes are shown below. Which drawing would be a better way to start a coordinate proof? Explain your answer.

a.

b.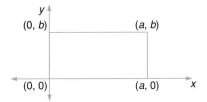

Guided Practice

Name the missing coordinates in terms of the given variables.

4. $\triangle BAT$ is isosceles.

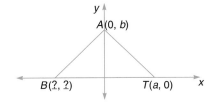

5. $ABCE$ is a parallelogram.

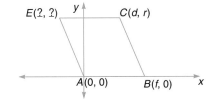

6. Use $\triangle LPT$ with coordinates as indicated in the figure to answer each of the following.

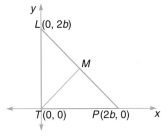

   a. What kind of triangle is $\triangle LPT$? Explain your answer.
   b. If $\overline{TM}$ is a median, find the coordinates of M.
   c. Find the slope of $\overline{TM}$.
   d. Find the slope of $\overline{LP}$.
   e. What conclusion can you make about $\overline{TM}$ and $\overline{LP}$?

Use the quadrilateral at the right to prove each of the following.

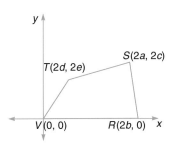

7. The segments joining the midpoints of the sides of a quadrilateral form a parallelogram.

8. The segments joining the midpoints of opposite sides of a quadrilateral bisect each other.

Position and label each figure on the coordinate plane. Then write a coordinate proof for each of the following.

9. The diagonals of an isosceles trapezoid are congruent.

10. The medians to the legs of an isosceles triangle are congruent.

EXERCISES

Practice   **Name the missing coordinates in terms of the given variables.**

11. $\triangle DAY$ is isosceles and right.

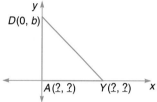

12. *TUES* is a square.

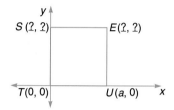

13. $\triangle RUN$ is isosceles and right.

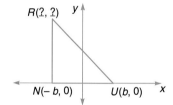

14. *RIDE* is an isosceles trapezoid.

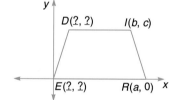

Prove using a coordinate proof.

15. $\triangle ABC$ is isosceles.

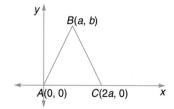

16. *HIJK* is a parallelogram.

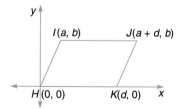

17. $\triangle DEF$ is equilateral.

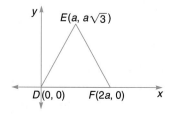

18. $\triangle PQR$ is a right triangle.

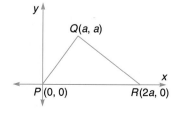

19. The diagonals of a square are perpendicular.

20. The three segments joining the midpoints of the sides of an isosceles triangle form another isosceles triangle.

21. The diagonals of a parallelogram bisect each other.

22. The segments joining the midpoints of the sides of an isosceles trapezoid form a rhombus.

23. If a line segment joins the midpoints of two sides of a triangle, then it is parallel to the third side.

24. If a line segment joins the midpoints of two sides of a triangle, then its length is equal to one-half the length of the third side.

25. The line segments joining the midpoints of the sides of a rectangle form a rhombus.

26. If the diagonals of a parallelogram are congruent, then it is a rectangle.

27. If the diagonals of a parallelogram are perpendicular, then the parallelogram is a rhombus.

Critical Thinking

28. Position a regular hexagon on a coordinate plane and label the vertices using as few variables as possible.

Applications

29. **Theater** A stage is in the shape of a semicircle with a 40-foot diameter. If a scale drawing of the stage is assigned coordinates as shown at the right, find the equation of the line that bisects the stage into two congruent parts.

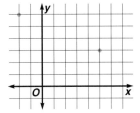

30. **Interior Decorating** When planning the design of a room, a decorator makes a scale drawing of the room on the coordinate plane. The vertices are given as points $A(4, 8)$, $B(0, 8)$, $C(4, 0)$, and $D(0, 0)$. What are the coordinates of the center of the room?

31. **Air Traffic Control** An airplane is 5 kilometers east and 3 kilometers north of the airport, while a second airplane is 2 kilometers west and 6 kilometers north. Use your knowledge of coordinate geometry to find the distance between the airplanes.

Mixed Review

Determine whether each statement is *true* or *false*.

32. To find an equation for the line that passes through $(-1, 4)$ and $(5, 0)$, first find the y-intercept and then find the slope. **(Lesson 12-2)**

33. An obtuse triangle cannot be isosceles. **(Lesson 4-1)**

34. The altitude of a triangle can sometimes be located outside the triangle. **(Lesson 5-1)**

35. AAS is a test for congruent triangles. **(Lesson 4-5)**

36. The measures of the three sides of a triangle can be 5, 4, and 6. **(Lesson 5-6)**

37. The opposite angles of a parallelogram are supplementary. **(Lesson 6-1)**

Determine whether each statement is *true* or *false*.

38. The diagonals of an isosceles trapezoid bisect each other. **(Lesson 6-6)**

39. An equilateral polygon must also be an equiangular polygon. **(Lesson 10-1)**

Wrap-Up 40. **Journal Entry** Show how you would position each of the following on a coordinate plane to start a proof. Label the vertices with as few variables as possible.

    a. an isosceles triangle

    b. a parallelogram

    c. a right triangle

~~~~~~~~~~~~~~~~~ **MID-CHAPTER REVIEW** ~~~~~~~~~~~~~~~~~

Find the *x*-intercept and *y*-intercept of the graph of each equation if they exist. (Lesson 12-1)

1. $4x - 3y = 12$ 2. $y = 15$

Write the equation in slope-intercept form of the line that satisfies the given conditions. (Lesson 12-2)

3. passes through $(-1, 4)$ and $(2, 2)$

4. parallel to $y = 3x - 4$, passes through $(0, 0)$

5. $m = -\frac{1}{2}$, *x*-intercept $= 6$

Determine whether the three points listed are collinear. (Lesson 12-3)

6. $A(8, 0)$, $B(9, -2)$, $C(0, 16)$ 7. $X(-2, 2)$, $Y(8, -1)$, $Z(4, 3)$

8. Ms. Kelly is driving the 88 miles from her office to an appointment at her client's office. If she averages 40 miles per hour on the trip and is fifteen minutes early for her 1:00 appointment, what time did she leave her office? **(Lesson 12-4)**

Name the missing coordinates in terms of the given variables. (Lesson 12-5)

9. *ABCD* is a square.   10. *EFGH* is a parallelogram.

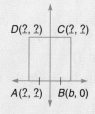   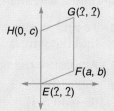

11. Use a coordinate proof to show that the medians drawn to the congruent sides of an isosceles triangle are congruent.

12-6 Vectors

Objectives

After studying this lesson, you should be able to:
- find the magnitude and direction of a vector,
- determine if two vectors are equal, and
- perform operations with vectors.

Application

The speed and direction of a rocket can be represented by a directed segment called a **vector**. A vector is any quantity that has both **magnitude** (length) and **direction.** In this case, the length of the segment represents the speed of the rocket. The direction of the rocket is indicated by the direction of the segment.

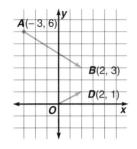

In symbols, a vector is written as $\vec{v}$ or $\overrightarrow{AB}$. A vector can be represented by an ordered pair (change in x, change in y). In the diagram above, vector OD $(\overrightarrow{OD})$ is in standard position: that is, its initial point is at the origin. $\overrightarrow{OD}$ can be represented by the ordered pair (2, 1). To represent $\overrightarrow{AB}$ as an ordered pair, find the change in x and the corresponding change in y and write as an ordered pair.

$$\overrightarrow{AB} = (x_2 - x_1, y_2 - y_1)$$
$$= (2 - (-3), 3 - 6) \qquad (x_1, y_1) = A(-3, 6), (x_2, y_2) = B(2, 3)$$
$$= (5, -3)$$

Because the magnitude and direction are not changed by moving, or translating, a vector, (5, -3) represents the same vector as $\overrightarrow{AB}$.

You can use the distance formula to find the magnitude, or length, of a vector. The symbol for the magnitude of $\overrightarrow{AB}$ is $|\overrightarrow{AB}|$. The direction of a vector is the measure of the angle that the vector forms with the positive x-axis or another horizontal line. You can use the trigonometric ratios to find the direction of a vector.

Example 1

Given $A(1, 2)$ and $B(3, 5)$, find the magnitude and direction of $\overrightarrow{AB}$.

magnitude
$$|\overrightarrow{AB}| = \sqrt{(3-1)^2 + (5-2)^2}$$
$$= \sqrt{13} \text{ or about 3.6 units}$$

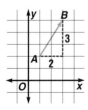

direction
$$\tan A = \frac{3}{2} \qquad \tan A = \frac{opposite}{adjacent}$$
$$m\angle A \approx 56.3$$

$\overrightarrow{AB}$ has a magnitude of about 3.6 units and a direction of about 56.3°.

Two vectors are equal if they have the same magnitude and direction. They are parallel if they have the same direction or slope.

$\overrightarrow{BA} \parallel \overrightarrow{CD}$ Both have a slope of 2, but have different lengths.

$\overrightarrow{BA} = \overrightarrow{EF}$ Both have a length of $2\sqrt{5}$ units and a slope of 2.

$\overrightarrow{BA} \neq \overrightarrow{CG}$ Both have a length of $2\sqrt{5}$ units, but have different directions.

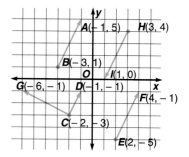

$\overrightarrow{BA}$ is *not* considered to be parallel to $\overrightarrow{HI}$ because they have opposite directions.

A vector can be multiplied by a constant that will change the magnitude of the vector but not affect the direction. If $\vec{v} = (2, 6)$, then $2\vec{v} = (2 \times 2, 2 \times 6)$ or $(4, 12)$. Now compare the magnitudes of $\vec{v}$ and $2\vec{v}$.

$$|\vec{v}| = \sqrt{2^2 + 6^2} \qquad |2\vec{v}| = \sqrt{4^2 + 12^2}$$
$$= \sqrt{4 + 36} \qquad = \sqrt{16 + 144}$$
$$= \sqrt{40} \qquad = \sqrt{160}$$
$$= 2\sqrt{10} \qquad = 4\sqrt{10}$$

Notice that multiplying the vector by 2 doubled its magnitude. Multiplying a vector by a constant is called **scalar multiplication.**

It is also possible to add vectors. Suppose a plane flew from Chicago to Nashville and then from Nashville to Washington, D.C. This has the same result as flying directly from Chicago to Washington, D.C. In terms of vectors, $\overrightarrow{CN} + \overrightarrow{NW} = \overrightarrow{CW}$.

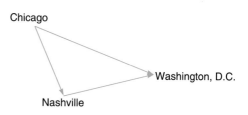

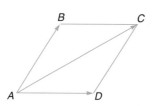

To add vectors, you can use the **parallelogram law.** The **resultant** or sum of two vectors is the diagonal of the parallelogram made by using the given vectors as sides. To add $\overrightarrow{AB}$ and $\overrightarrow{AD}$, draw parallelogram $ADCB$ using the magnitude and direction of $\overrightarrow{AB}$ for side DC and that of $\overrightarrow{AD}$ for side BC. The sum of the vectors is the diagonal of the parallelogram.

Vectors can also be added by adding their coordinates.

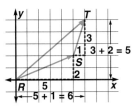

$$(a, b) + (c, d) = (a + c, b + d)$$
$$\overrightarrow{RS} + \overrightarrow{ST} = \overrightarrow{RT}$$
$$(5, 2) + (1, 3) = (5 + 1, 2 + 3)$$
$$= (6, 5)$$

Example 2

If $\vec{v} = (5, 1)$ and $\vec{w} = (1, 3)$, find the coordinates of $2\,\vec{v} + \vec{w}$.

$$2\,\vec{v} + \vec{w} = (2 \times 5, 2 \times 1) + (1, 3)$$
$$= (10, 2) + (1, 3)$$
$$= (10 + 1, 2 + 3)$$
$$= (11, 5)$$

Example 3

APPLICATION

Physics

A rocket is traveling straight up at the speed of 600 miles per hour. A 50-mile per hour wind is blowing at a right angle on the path of the rocket. How does the wind affect the speed and direction of the rocket?

Use coordinates to make a model. Let each unit represent 50 miles. $\overrightarrow{OR}$ is the vector that represents the speed and direction of the rocket. $\overrightarrow{OW}$ is the vector that represents the speed and direction of the wind.

$$\overrightarrow{OR} = (0, 600)$$
$$\overrightarrow{OW} = (50, 0)$$

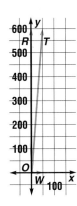

$\overrightarrow{OT}$ is the resultant of $\overrightarrow{OR} + \overrightarrow{OW}$.

$$\overrightarrow{OT} = \overrightarrow{OR} + \overrightarrow{OW}$$
$$= (0 + 50, 600 + 0) \text{ or } (50, 600)$$

Find the magnitude of $\overrightarrow{OT}$.

$$|\overrightarrow{OT}| = \sqrt{50^2 + 600^2}$$
$$= \sqrt{2500 + 360{,}000}$$
$$= \sqrt{362{,}500} \approx 602.1$$

Find the direction of $\overrightarrow{OT}$.

$$\tan x = \frac{50}{600} \qquad \textit{tan x} = \frac{OW}{OR}$$
$$\tan x \approx 0.0833$$
$$x \approx 4.8$$

The wind blows the rocket off course by about 4.8° and increases its speed to about 602.1 miles per hour.

CHECKING FOR UNDERSTANDING

Communicating Mathematics

Read and study the lesson to answer these questions.

1. What is the difference between a vector and a line segment?

2. Is $\overrightarrow{AB}$ the same as $\overrightarrow{BA}$? Explain your answer.

3. What is the difference between two parallel vectors and two equal vectors?

4. Describe the effect of scalar multiplication on a vector.

5. Draw two different diagrams showing addition of vectors.

6. Draw two vectors with the same magnitude but different directions.

7. Draw two vectors with the same direction but different magnitudes.

Guided Practice

Sketch each vector. Then find the magnitude and direction to the nearest degree.

8. $\overrightarrow{SL} = (5, 1)$

9. $\overrightarrow{AB} = (3, 8)$

10. $\overrightarrow{RT}$ if $R(-2, -5)$ and $T(1, 7)$

Given $A(0, 0)$, $B(2, 4)$, $C(7, 4)$, and $D(5, 0)$, draw vectors $\overrightarrow{AB}$, $\overrightarrow{BC}$, $\overrightarrow{DC}$, $\overrightarrow{DA}$, $\overrightarrow{AC}$, and $\overrightarrow{BD}$. Use your diagram to answer each question. Explain your answers.

11. Which vectors are parallel?

12. Which vectors are equal?

13. Name a pair of vectors that are not equal.

14. Name a pair of vectors that are the same magnitude, but different directions.

Copy each pair of vectors and draw a resultant vector.

15.

16.

17.

Given $\overrightarrow{v} = (3, 2)$, $\overrightarrow{s} = (2, 7)$, $\overrightarrow{t} = (-5, 6)$, and $\overrightarrow{w} = (2, -4)$, find the coordinates of each of the following.

18. $\overrightarrow{w} + \overrightarrow{s}$

19. $2\overrightarrow{v} + \overrightarrow{t}$

20. $3\overrightarrow{v} + \overrightarrow{s}$

21. $\overrightarrow{w} + \overrightarrow{s} + \overrightarrow{t}$

EXERCISES

Given a path from *C* north 6 units to *A*, then west 6 units to *B*, answer each question.

22. What is the total length of the path?

23. What is the magnitude of $\overrightarrow{CB}$?

24. Describe the direction of *C* to *B*.

Sketch each vector. Then find the magnitude of $\overrightarrow{AB}$ to the nearest tenth and the direction to the nearest degree.

25. *A*(4, 2), *B*(8, 6)

26. *A*(-2, 4), *B*(5, 10)

Given *A*(-6, -4), *B*(4, 2), *C*(6, 8), and *D*(-4, 2), draw the vectors, $\overrightarrow{AB}$, $\overrightarrow{BC}$, $\overrightarrow{CD}$, and $\overrightarrow{AD}$. Use your diagram to answer each question. Explain your answers.

27. Which vectors are parallel?

28. Which vectors are equal?

Given the quadrilateral *RPTQ*, complete each statement.

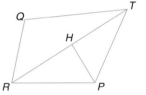

29. $\overrightarrow{RP} + \overrightarrow{PT} = \underline{\ ?\ }$

30. $\overrightarrow{TQ} + \overrightarrow{QR} = \underline{\ ?\ }$

31. $\overrightarrow{PH} + \overrightarrow{HR} = \underline{\ ?\ }$

32. $\overrightarrow{RT} + \overrightarrow{TQ} = \underline{\ ?\ }$

Given $\overrightarrow{v} = (2, 5)$ and $\overrightarrow{u} = (7, 1)$, find the coordinates of each of the following.

33. $\overrightarrow{v} + \overrightarrow{u}$

34. $\overrightarrow{v} + 2\overrightarrow{u}$

35. $2\overrightarrow{u} + 3\overrightarrow{v}$

36. a. Supply the reasons for the following proof.

Given: $\overrightarrow{AB} = \overrightarrow{BC}$ and $\overrightarrow{CD} = \overrightarrow{DE}$

Prove: $\overrightarrow{BD} = \frac{1}{2}\overrightarrow{AE}$

Proof:

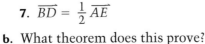

Statements	Reasons
1. $\overrightarrow{BC} + \overrightarrow{CD} = \overrightarrow{BD}$	**1.** $\underline{\ ?\ }$
2. $\overrightarrow{AC} + \overrightarrow{CE} = \overrightarrow{AE}$	**2.** $\underline{\ ?\ }$
3. $\overrightarrow{AB} = \overrightarrow{BC}, \overrightarrow{CD} = \overrightarrow{DE}$	**3.** $\underline{\ ?\ }$
4. $2\overrightarrow{BC} + 2\overrightarrow{CD} = \overrightarrow{AE}$	**4.** $\underline{\ ?\ }$
5. $2(\overrightarrow{BC} + \overrightarrow{CD}) = \overrightarrow{AE}$	**5.** $\underline{\ ?\ }$
6. $2\overrightarrow{BD} = \overrightarrow{AE}$	**6.** $\underline{\ ?\ }$
7. $\overrightarrow{BD} = \frac{1}{2}\overrightarrow{AE}$	**7.** $\underline{\ ?\ }$

b. What theorem does this prove?

37. Given $A(2, 5)$, $B(x, -4)$, $C(6, 1)$, and $D(4, 3)$, find x if $\overrightarrow{AB} \parallel \overrightarrow{CD}$.

38. How do the magnitude and direction of $\overrightarrow{BO}$ compare with $\overrightarrow{OB}$?

Given $A(0, 0)$, $B(2, 5)$, $R(0, -3)$, $S(-8, -6)$, and $L(8, 3)$, draw $\overrightarrow{AB}$, $2\overrightarrow{AB}$, $\overrightarrow{AB} + \overrightarrow{BL}$, $\overrightarrow{SR}$, and $\overrightarrow{BL}$. Use your diagram to answer each question.

39. Which vectors are parallel?

40. Which vectors are equal?

41. If two vectors are represented by a single ordered pair, it is possible to determine whether they are perpendicular by using the dot product test. If $\overrightarrow{a} = (x_1, y_1)$ and $\overrightarrow{b} = (x_2, y_2)$, their dot product $\overrightarrow{a} \cdot \overrightarrow{b}$ is $x_1 \cdot x_2 + y_1 \cdot y_2$. If the dot product is 0, the vectors are perpendicular. Determine which of the following vectors are perpendicular.
$\overrightarrow{v} = (7, -2)$ $\overrightarrow{w} = (4, 14)$ $\overrightarrow{u} = (-2, 7)$ $\overrightarrow{t} = (2, 7)$

Critical Thinking

42. In this lesson, you studied vector addition. Find an algebraic way to perform vector subtraction and show how this would look geometrically.

Applications

43. Navigation An airplane flies due west at 240 kilometers per hour. At the same time, the wind is blowing due south at 70 kilometers per hour. How does the wind affect the speed and direction of the plane?

44. Recreation A hiker leaves camp and walks 15 kilometers due north. The hiker then walks 15 kilometers due east. What is the hiker's direction and distance from the starting point?

45. Sports Two soccer players kick the ball at the same time. One player's foot exerts a force of 70 newtons west. The other's foot exerts a force of 50 newtons north. What is the magnitude and direction of the resultant force on the ball?

Mixed Review

46. $\triangle ABC$ is a right isosceles triangle. M is the midpoint of $\overline{AB}$. Use a coordinate proof to show that $\overline{CM}$ is perpendicular to $\overline{AB}$. **(Lesson 12-3)**

47. Find the volume of a cylinder with a radius of 5 m and a height of 2 m. **(Lesson 11-5)**

48. A car tire has a radius of 8 inches. How far does the car travel in one revolution of the tire? **(Lesson 10-7)**

49. The measures of the sides of a triangle are 6, 9, and 11. If the shortest side of a similar triangle measures 12, what are the measures of its other two sides? **(Lesson 7-4)**

Wrap-Up

50. Write a few sentences about vectors. Include a definition of a vector, the meaning of equal vectors, and a description of how to add vectors.

12-7   Coordinates in Space

Objectives

After studying this lesson, you should be able to:
- locate a point in space,
- use the distance and midpoint formulas for points in space, and
- determine the center and radius of a sphere.

Application

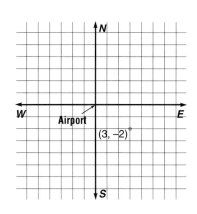

Wanda Burgess is an air traffic controller. She needs to keep accurate information about the location of the aircraft around the airport. Suppose an aircraft is 3 miles east and 2 miles south of the airport. If you think of the airport as being at the origin of a coordinate plane, the coordinates of the aircraft could be (3, -2). However, the aircraft is actually above this point. If the aircraft is 1 mile above ground, the coordinates of the aircraft would be (3, -2, 1).

FYI ···

Egyptian engineers were experimenting with airplanes 2300 years ago. An airplane model similar to the American Hercules transport aircraft was found in Sakkara in 1898.

In a coordinate plane, the ordered pair for each point has two numbers, or coordinates, to describe its location because a plane has two dimensions. In space, each point requires three numbers, or coordinates, to describe its location because space has three dimensions. In space, the x-, y-, and z-axes are perpendicular to each other.

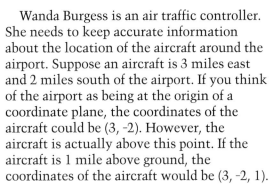

A point in space is represented by an **ordered triple** of real numbers (x, y, z). In the figure at the left, the ordered triple $(-2, -3, 1)$ locates point P. Notice that a parallelogram is used to help convey the idea of the third dimension.

Just as the Pythagorean Theorem can be used to find the formula for the distance between two points in a plane, it can also be used to find the formula for the distance between two points in space.

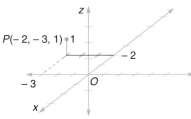

The distance between points A and B on $\triangle ABD$ can be found as follows.

$(AD)^2 = (AC)^2 + (CD)^2$ and
$(AB)^2 = (AD)^2 + (BD)^2$ *Pythagorean Theorem*
$(AB)^2 = (AC)^2 + (CD)^2 + (BD)^2$ *Substitute $(AC)^2 + (CD)^2$ for $(AD)^2$.*
$AB = \sqrt{(AC)^2 + (CD)^2 + (BD)^2}$ *Take the square root of each side.*
$AC = |x_2 - x_1|$, $CD = |y_2 - y_1|$, and $BD = |z_2 - z_1|$.

Therefore, $AB = \sqrt{(x_2 - x_1)^2 + (y_2 - y_1)^2 + (z_2 - z_1)^2}$ by substitution.

	Given two points $A(x_1, y_1, z_1)$ and $B(x_2, y_2, z_2)$ in space, the
Theorem 12-2	distance between A and B is given by the following equation.
	$$AB = \sqrt{(x_2 - x_1)^2 + (y_2 - y_1)^2 + (z_2 - z_1)^2}$$

This formula is an extension of the distance formula in the two-dimensional coordinate system.

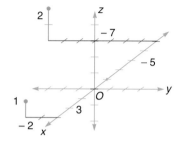

Example 1

Aviation

With reference to an airport, an airplane is 3 miles east, 2 miles south, and 1 mile up. Another airplane is 5 miles west, 7 miles south, and 2 miles up. Find the distance between the two airplanes.

Relative to the airport, the coordinates of the first airplane are (3, -2, 1), and the coordinates of the other airplane are (-5, -7, 2). Use Theorem 12-2 to find the distance between the two airplanes.

$$\sqrt{(3 - (-5))^2 + (-2 - (-7))^2 + (1 - 2)^2} = \sqrt{(3 + 5)^2 + (-2 + 7)^2 + (1 - 2)^2}$$
$$= \sqrt{(8)^2 + (5)^2 + (-1)^2}$$
$$= \sqrt{64 + 25 + 1}$$
$$= \sqrt{90} \approx 9.5$$

The airplanes are about 9.5 miles apart.

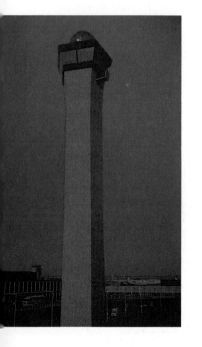

Suppose M is the midpoint of $\overline{PQ}$, a segment in space. The midpoint has the following coordinates.

$$\left(\frac{x_1 + x_2}{2}, \frac{y_1 + y_2}{2}, \frac{z_1 + z_2}{2}\right)$$

This formula is an extension of the midpoint formula in the two-dimensional coordinate system.

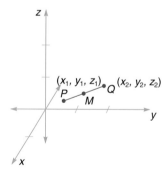

Example 2

Find the coordinates of the midpoint of a segment in space whose endpoints P and Q have coordinates (3, -7, 0) and (5, 1, 7).

Let (3, -7, 0) be (x_1, y_1, z_1) and (5, 1, 7) be (x_2, y_2, z_2).

Then, $\left(\frac{3 + 5}{2}, \frac{-7 + 1}{2}, \frac{0 + 7}{2}\right) = \left(\frac{8}{2}, -\frac{6}{2}, \frac{7}{2}\right)$
$$= (4, -3, 3.5)$$

The coordinates of the midpoint of $\overline{PQ}$ are (4, -3, 3.5).

Example 3

Find the volume of the rectangular solid with vertices $A(0, 0, 0)$, $B(0, 6, 0)$, $C(4, 6, 0)$, $D(4, 0, 0)$, $E(4, 0, 7)$, and $F(4, 6, 7)$.

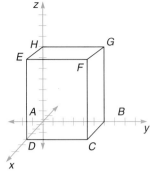

First, find the measures of the length, ℓ, width, w, and height, h.

$w = EF$
$= \sqrt{(4 - 4)^2 + (0 - 6)^2 + (7 - 7)^2}$
$= \sqrt{36}$ or 6

$h = FC$
$= \sqrt{(4 - 4)^2 + (6 - 6)^2 + (0 - 7)^2}$
$= \sqrt{49}$ or 7

$\ell = CB$
$= \sqrt{(4 - 0)^2 + (6 - 6)^2 + (0 - 0)^2}$
$= \sqrt{16}$ or 4

$$V = \ell \cdot w \cdot h$$
$$= 4 \cdot 6 \cdot 7$$
$$= 168$$

The volume is 168 cubic units.

The formula for the equation of a sphere is an extension of the formula for the equation of a circle.

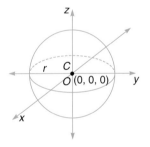

The equation of a sphere whose center is at $(0, 0, 0)$ and whose radius is r units long is as follows.

$$x^2 + y^2 + z^2 = r^2$$

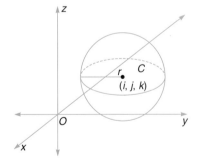

The equation of a sphere whose center is at (i, j, k) and whose radius is r units long is as follows.

$$(x - i)^2 + (y - j)^2 + (z - k)^2 = r^2$$

Example 4

Write the equation of a sphere whose center is at $(3, -2, 4)$ and that has a radius of 6 units.

Let $i = 3$, $j = -2$, $k = 4$, and $r = 6$.

$(x - i)^2 + (y - j)^2 + (z - k)^2 = r^2$
$(x - 3)^2 + (y - (-2))^2 + (z - 4)^2 = 6^2$ *Substitute 3 for i, -2 for j, 4 for k,*
$(x - 3)^2 + (y + 2)^2 + (z - 4)^2 = 36$ *and 6 for r.*

The equation of the sphere is $(x - 3)^2 + (y + 2)^2 + (z - 4)^2 = 36$.

CHECKING FOR UNDERSTANDING

**Communicating
Mathematics**

Read and study the lesson to answer these questions.

1. Describe the position of the three axes in space.

2. What is the difference between the distance formula for two points in a plane and the distance formula for two points in space?

3. Describe how to find the midpoint of a line segment in space.

4. What is the difference between a circle and a sphere? How does the equation of a circle differ from the equation of a sphere?

Guided Practice

Plot each point in a three-dimensional coordinate system.

5. $(1, 5, 7)$ 6. $(-2, 3, 5)$

Determine the distance between each pair of points.

7. $A(0, 0, 0)$ and $B(0, 5, 0)$ 8. $P(0, 2, 3)$ and $Q(1, 0, -3)$

Determine the coordinates of the midpoint of the line segment whose endpoints are given.

9. $A(0, -4, 2)$ and $B(3, 0, 2)$ 10. $S(-6, 3, -1)$ and $T(6, 3, 1)$

Determine the coordinates of the center and the measure of the radius for each sphere whose equation is given.

11. $x^2 + (y - 3)^2 + (z - 4)^2 = 81$

12. $(x + 2)^2 + (y + 4)^2 + (z - 4)^2 = 25$

Write the equation of the sphere given the coordinates of the center and the measure of the radius.

13. $(4, 1, -2)$, 6   14. $(0, 0, 4)$, 16

15. Find the perimeter of the triangle whose vertices are $P(0, 0, 0)$, $Q(3, 4, \sqrt{11})$ and $R(0, 5, 0)$.

16. Find the radius of a sphere whose diameter has endpoints $A(4, 7, -3)$ and $B(0, -2, 9)$.

EXERCISES

Practice

Plot each point in a three-dimensional coordinate system.

17. $(0, 0, 5)$ 18. $(2, -1, 4)$ 19. $(3, 1, -4)$

Determine the distance between each pair of points.

20. $A(2, 4, 5)$ and $B(2, 4, 7)$

21. $A(0, -2, 5)$ and $B(-3, 4, -2)$

22. $A(9, 1, 0)$ and $B(5, -7, 4)$

23. $A(8, 10, -3)$ and $B(1, 12, 6)$

Determine the coordinates of the midpoint of each line segment whose endpoints are given.

24. $A(1, 3, -2)$, $B(7, -3, 2)$

25. $C(-5, 4, -2)$, $D(5, -4, 2)$

26. $E(5, -6, 3)$, $F(11, -2, 7)$

27. $G(22, 5, -1)$, $H(0, -3, 6)$

Determine the coordinates of the center and the measure of the radius for each sphere whose equation is given.

28. $(x - 6)^2 + (y + 5)^2 + (z - 1)^2 = 81$

29. $(x + 2)^2 + (y + 3)^2 + (z - 2)^2 = 100$

30. $x^2 + (y - 2)^2 + (z - 4)^2 = 4$

31. $(x + 8)^2 + y^2 + (z + 4)^2 = 18$

Write the equation of the sphere given the coordinates of the center and the measure of the radius.

32. $(-1, 2, 4)$, 3

33. $(6, -1, 3)$, 12

34. $(0, 3, -2)$, 11

35. $(-2, 4, 1)$, $\sqrt{13}$

36. Find the perimeter of a triangle whose vertices are $A(6, 4, 1)$, $B(4, 6, 0)$ and $C(3, -2, 3)$.

37. The diameter of a sphere has endpoints $A(-3, 5, 7)$ and $B(5, -1, 5)$.

    a. Determine the coordinates of the center of the sphere.

    b. Determine the radius of the sphere.

    c. Write an equation of the sphere.

    d. Sketch a graph of the sphere.

    e. Find the surface area of the sphere.

    f. Find the volume of the sphere.

38. Find the surface area and volume of the figure at the right.

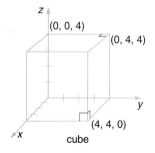

cube

39. Find t if the distance between $A(3t, 5, -t)$ and $B(5, -3, t)$ is 9 units.

40. Find the measures of the medians of the triangle whose vertices are $R(9, 4, 11)$, $S(-3, 8, 1)$ and $T(7, 2, -3)$.

One sphere has the equation $x^2 + y^2 + z^2 = 36$, and another has the equation $x^2 + y^2 + z^2 = 100$.

41. What is the ratio of the radii of the two spheres?

42. What is the ratio of the areas of the two spheres?

43. What is the ratio of the volumes of the two spheres?

Critical Thinking

44. Find the other endpoint of a diameter of the sphere if one endpoint is at (4, -6, 10) and the center of the sphere is at (0, 1, -4).

Applications

45. Air Traffic Control   One airplane is 7 miles east and 9 miles south of the airport and 2 miles above the ground. Another airplane is 4 miles west and 4 miles south from the airport and 1 mile above the ground. Find the distance between the airplanes.

46. Recreation   Two children are playing a three-dimensional tick-tack-toe game. Three Xs or three Os in any row wins the game. The positions of the Xs are (1, 1, 1) and (1, 3, 3) and the position of the O is (2, 2, 2). Where should the next O be placed? Explain your answer.

Mixed Review

47. Given $\vec{t} = (1, -5)$ and $\vec{u} = (-4, 2)$, find the ordered pair for $2\vec{t} + \vec{u}$. **(Lesson 12-6)**

Find the area of each region to the nearest tenth. (Lessons 10-5, 10-6, and 10-7)

48.

10 cm

49.

5.5 cm

50.
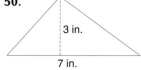
3 in.

7 in.

51. Write the assumption you would make to start an indirect proof of the statement *If two lines intersect, then no more than one plane contains them.* **(Lesson 5-4)**

52. Find the value of x. **(Lesson 3-3)**

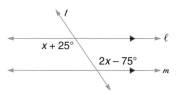

$x + 25°$

$2x - 75°$

Wrap-Up

53. Journal Entry   Explain in your own words the relationship between the formulas involving distance and midpoints in two-dimensions and in three-dimensions.

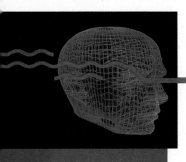

Technology

Perspective Drawing

BASIC
Geometric Supposer
▶ **Graphing Calculators**
LOGO
Spreadsheets

Perspective drawing, drawing three-dimensional objects on a two-dimensional surface, was highly refined in the Renaissance by artists like Leonardo da Vinci and Albrecht Dürer. Today, computers can make perspective drawings. This technique is used extensively for drawing graphics for television and movies.

Computers make perspective drawings by finding the three-dimensional coordinates of the object being drawn. Then they use algebra to transform these into two-dimensional coordinates. The graph of these two-dimensional coordinates is called a *projection*.

The formulas below will draw one type of projection in which the y-axis is drawn horizontally, the z-axis is drawn vertically, and the x-axis is at an angle of $a°$ with the y-axis. In this system if the three-dimensional coordinates of a point are (x, y, z), then the projection coordinates (X, Y) are

$$X = x(\text{-cos } a°) + y$$
$$Y = x(\text{-sin } a°) + z.$$

This projection will give a good perspective drawing, but some lengths may be slightly distorted.

EXERCISES

1. The cube at the right has vertices $A(5, 0, 5)$, $B(5, 5, 5)$, $C(5, 5, 0)$, $D(5, 0, 0)$, $E(0, 0, 5)$, $F(0, 5, 5)$, and $G(0, 5, 0)$ and $H(0, 0, 0)$.

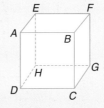

   a. Use the formulas above to find the projection coordinates of each vertex using $a = 45$. Round each coordinate to the nearest whole number.

   b. Use the projection coordinates you found for the vertices of the cube to graph the perspective drawing on your graphing calculator. Sketch the display.

2. $A(10, 2, 0)$, $B(10, 10, 0)$, $C(2, 10, 0)$, and $D(3, 3, 4)$ are the vertices of a pyramid.

   a. Find the projection coordinates of each vertex using $a = 30$. Round each coordinate to the nearest whole number.

   b. Make a perspective of the pyramid on your graphing calculator by graphing the segments $\overline{AB}$, $\overline{BC}$, $\overline{CD}$, $\overline{DA}$, and $\overline{DB}$. Sketch the display.

SUMMARY AND REVIEW

VOCABULARY

Upon completing this chapter, you should be familiar with the following terms:

coordinate proof **595**
direction of a vector **601**
intercepts method **574**
linear equation **574**
magnitude of a vector **601**
ordered triple **607**
parallelogram law **602**

581 point-slope form
602 resultant
602 scalar multiplication
586 scatter plot
575 slope-intercept form
574 standard form
601 vector

SKILLS AND CONCEPTS

OBJECTIVES AND EXAMPLES	REVIEW EXERCISES

Upon completing this chapter, you should be able to:

Use these exercises to review and prepare for the chapter test.

■ graph linear equations. **(Lesson 12-1)**

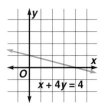

$x + 4y = 4$

Intercepts Method
x-intercept = 4
y-intercept = 1

Slope-Intercept Form
$y = -\frac{1}{4}x + 1$
$m = -\frac{1}{4}$
y-intercept = 1

Graph each equation.

1. $2x - y = 5$ **2.** $y = 3$

Graph each pair of linear equations on the same coordinate plane. Determine if the lines are parallel, perpendicular, or neither by finding the slope of each line.

3. $2x - y = 5$ **4.** $y = 5x - 3$
    $x + 2y = 4$ $5x - y = 0$

■ write an equation of a line given information about its graph. **(Lesson 12-2)**

The slope of the line that passes through $(4, -3)$ and $(2, 1)$ is $\frac{1 - (-3)}{2 - 4}$ or $\frac{4}{-2}$ or -2. To find an equation of the line, use the point-slope form.

$y - y_1 = m(x - x_1)$ *Point-Slope Form*
$y - 1 = -2(x - 2)$ *$m = -2$, $(x_1, y_1) = (2, 1)$*
$y - 1 = -2x + 4$
$y = -2x + 5$

Write an equation of the line satisfying the given conditions.

5. parallel to the graph of $y = x - 5$, passes through $(0, 8)$

6. perpendicular to the graph of $y = 3x - 1$, passes through $(6, 0)$

7. parallel to x-axis, passes through $(5, 2)$

8. perpendicular to the x-axis, passes through $(5, 2)$

OBJECTIVES AND EXAMPLES

■ relate equations of lines to geometric concepts. **(Lesson 12-3)**

To find the equation of the perpendicular bisector of a segment whose endpoints have coordinates (5, -3) and (-1, 1), first find the coordinates of the midpoint.

$$\left(\frac{5-1}{2}, \frac{-3+1}{2}\right) = \left(\frac{4}{2}, \frac{-2}{2}\right) \text{ or } (2, -1)$$

Then find the slope of the segment.

$$\frac{-3-1}{5-(-1)} = \frac{-4}{6} \text{ or } -\frac{2}{3}$$

The perpendicular bisector will pass through (2, -1) and have a slope of $\frac{3}{2}$.

$$y - y_1 = m(x - x_1)$$
$$y - (-1) = \frac{3}{2}(x - 2)$$
$$y = \frac{3}{2}x - 4$$

The equation is $y = \frac{3}{2}x - 4$.

■ prove theorems using coordinate proofs. **(Lesson 12-5)**

When planning a coordinate proof, use the following guidelines to position the figure on a coordinate plane.
■ Use the origin as a vertex or center.
■ Position at least one side of a polygon on a coordinate axis.
■ Keep the figure in the first quadrant if possible.
■ Use coordinates which make computations simple.

■ find the magnitude and direction angle of a vector. **(Lesson 12-6)**

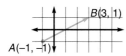

$$\tan A = \frac{2}{4} \text{ or } 0.5$$
$$m\angle A \approx 26.6$$

$$|\overrightarrow{AB}| = \sqrt{(3-(-1))^2 + (1-(-1))^2}$$
$$= \sqrt{4^2 + 2^2}$$
$$= \sqrt{20} \approx 4.5$$

$\overrightarrow{AB}$ has a magnitude of about 4.5 units and a direction of about 26.6°.

REVIEW EXERCISES

The vertices of $\triangle XYZ$ are X(2, -1), Y(6, 1), and Z(0, -3).

9. Find the equations of the lines containing the sides of $\triangle XYZ$.

10. Find the equations of the lines containing the medians of $\triangle XYZ$.

11. Find the equations of the lines containing the altitudes of $\triangle XYZ$.

12. Find the equations of the lines containing the perpendicular bisectors of the sides of $\triangle XYZ$.

Prove using a coordinate proof.

13. The segment through the midpoints of the nonparallel sides of a trapezoid is parallel to the bases.

14. The length of the segment through the midpoints of the nonparallel sides of a trapezoid is one-half the sum of the lengths of the bases.

Find the magnitude and direction angle of each vector.

15. $\overrightarrow{v} = (7, 1)$

16. $\overrightarrow{AB}$ with A(1, 0) and B(7, 5)

Find the resultant of each pair of vectors.

17. $\overrightarrow{a} = (2, 4), \overrightarrow{b} = (5, -3)$

18. $\overrightarrow{r} = (0, 8), \overrightarrow{s} = (4, 0)$

■ use the distance and midpoint formulas for points in space. **(Lesson 12-7)**

The distance between (2, 2, 2) and (-6, 0, 5) is $\sqrt{(-6 - 2)^2 + (0 - 2)^2 + (5 - 2)^2}$ or $\sqrt{77}$ units.

The coordinates of the midpoint of a line segment whose endpoints are (2, 2, 2) and (-6, 0, 5) are

$$\left(\frac{2 - 6}{2},\ \frac{2 + 0}{2},\ \frac{2 + 5}{2}\right) \text{ or } (-2, 1, 3.5).$$

Determine the distance between each pair of points and the coordinates of the midpoint of the segment whose endpoints are given.

19. (3, -3, 1), (7, -3, 5)

20. (2, 4, 6), (0, 2, 4)

Write the equation of the sphere given the coordinates of the center and measure of the radius.

21. (0, 0, 0), 5   **22.** (-1, 2, -3), 4

APPLICATIONS AND CONNECTIONS

23. Manufacturing A factory makes dresses and suits. If x represents the number of dresses made in one week and y represents the number of suits made that same week, then $10x + 20y$ represents the number of worker-hours needed to make the items. If $10x + 20y = 500$, draw a graph that represents the production for that week. What do the intercepts represent? **(Lesson 12-1)**

25. Theater An architect is planning the lighting for a stage that is in the shape of an isosceles trapezoid. The front of the stage is 30 feet long and the back of the stage is 40 feet long. The stage is 20 feet deep. If a scale drawing of the stage is assigned coordinates as shown below, what are the coordinates of the front center of the stage? **(Lesson 12-5)**

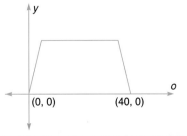

24. Meteorology Sound travels in air at about 0.21 miles per second. Kara counted 6 seconds between when she saw some lightning strike to when she heard the thunder. How far away did the lightning strike? **(Lesson 12-4)**

26. Air Traffic Control One airplane is 8 miles west and 10 miles south of the airport and 2 miles above the ground. Another airplane is 4 miles east and 8 miles south of the airport and 1 mile above the ground. Find the distance between the airplanes. **(Lesson 12-6)**

Graph each equation. State the slope and *y*-intercept.

1. $x + 2y = 6$ 2. $x = 3$ 3. $y = -3x$

Write the equation in slope-intercept form of the line satisfying the given conditions.

4. $m = -4$, passes through $(3, -2)$ 5. passes through $(-4, 11)$ and $(-6, 3)$
6. parallel to the graph of $y = 2x - 5$, passes through $(-1, -4)$
7. parallel to the *y*-axis, passes through $(-4, -2)$

Copy each pair of vectors and draw a resultant vector.

8.   9.

10. What is the resultant of $\vec{a} + \vec{b}$ if $\vec{a} = (-3, 5)$, and $\vec{b} = (0, 7)$?

Determine the coordinates of the center and the measure of the radius for each sphere whose equation is given.

11. $(x - 4)^2 + (y - 5)^2 + (z + 2)^2 = 81$ 12. $x^2 + y^2 + z^2 = 7$

13. Given $\vec{v} = (-5, -3)$, find the magnitude of $\vec{v}$.
14. Given $A(3, 7)$ and $B(-2, 5)$, find the magnitude of $\overrightarrow{AB}$.
15. Determine the distance between points at $(2, 4, 5)$ and $(2, 4, 7)$.
16. Determine the midpoint of the segment whose endpoints are $X(0, -4, 2)$ and $Y(3, 0, 2)$.
17. Write the equation of the sphere whose diameter has endpoints at $P(-3, 5, 7)$ and $Q(5, -1, 5)$.
18. Write an equation of the perpendicular bisector whose endpoints are at $(5, 2)$ and $(1, -4)$.

19. Name the missing coordinates for the parallelogram in terms of the given variables.

20. Use a coordinate proof to show that $\overleftrightarrow{NK}$ is a perpendicular bisector of a side of $\triangle LOM$.

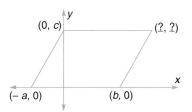

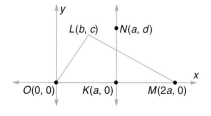

Bonus

A secant segment $\overline{MN}$ intersects sphere S at R and M. Given the points $N(7, 5, -1)$, $R(3, 2, -1)$, and $M(-5, -4, -1)$, find the length of a tangent segment $\overline{NT}$.

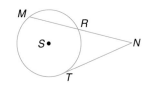

College Entrance Exam Preview

Directions: Choose the one best answer. Write A, B, C, or D. You may use a calculator.

1. A box with a closed top has a square base with sides x cm long and a height of h cm. Write an expression for the surface area of the box.

   (A) $8x + 4x$ (B) hx^2

   (C) $4h + x$ (D) $2x^2 + 4xh$

2. The graph of the equation $y = 4x + 8$ is

   (A) a horizontal line.

   (B) a vertical line.

   (C) a line that rises to the right.

   (D) a line that falls to the right.

3. Suppose the lengths of each of the four sides of a rectangle are tripled. By what factor will the area of the rectangle increase?

   (A) 3   (B) 6

   (C) 8   (D) 9

4. Solve $\dfrac{s}{3s + 6} - \dfrac{s}{5s + 10} = \dfrac{2}{5}$ for s.

   (A) -3   (B) -2

   (C) -3 or -2   (D) 3 or 2

5. The largest possible circle is cut from a square piece of cardboard. If one side of the square measures 5 inches, what is the approximate area of the scrap cardboard?

   (A) 47.1 in^2 (B) 6.2 in^2

   (C) 5.4 in^2 (D) 53.5 in^2

6. What is the average of the expressions $8a + 7$, $2a + 4$, $a - 3$, and $5a$?

   (A) $16a + 8$ (B) $4a + 2$

   (C) $8a + 4$ (D) $2a - 4$

7. What are the coordinates of point B if the coordinates of point A are (5, 8) and $x = -1$ is the equation of the perpendicular bisector of $\overline{AB}$?

   (A) (5, -10)   (B) (-7, 8)

   (C) (-5, -8)   (D) (4, 8)

8. Which of the following represents the solution set of the inequality $-3 - x < 2x < 3 + x$?

   (A) $\{x \mid -1 < x < 3\}$

   (B) $\{x \mid 1 < x < 3\}$

   (C) $\{x \mid -3 < x < 1\}$

   (D) $\{x \mid -3 < x < -1\}$

9. A pulley having a 4-inch diameter is belted to a pulley having a 6-inch diameter as shown below. If the smaller pulley is running at 180 rpm, how fast is the larger pulley running?

   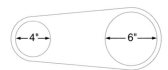

   (A) 270 rpm   (B) 160 rpm

   (C) 240 rpm   (D) 120 rpm

10. If x is between 0 and 1, which of the following increases as x increases?

    I. $x + 1$
    II. $1 - x^2$
    III. $\dfrac{1}{x}$

    (A) I only   (B) III only

    (C) I and II only   (D) I, II, and III

11. If -3 is one solution of the equation $x^2 + kx - 51 = 0$, what is the value of k?

    (A) 14   (B) 17

    (C) -20   (D) -14

Solve. You may use a calculator.

12. One angle of a triangle measures 68°. The measures of the other two angles are in a ratio of 1 to 3. Find the degree measures of the other two angles.

13. The average of six numbers is 10, the average of ten other numbers is 6. Find the average of all sixteen numbers.

14. A swimming pool measures 80 feet by 45 feet. The water level needs to be raised 3 inches. If it takes about 7.5 gallons of water to fill one cubic foot of space, about how much water must be added to the pool?

15. The charge for shipping a package is 75 cents for the first 4 ounces and 9 cents for each additional ounce. Find the weight of a package that costs $1.83 to ship.

16. In the figure below, $ABCD$ is a square with sides 2 inches long. M and N are the midpoints of sides $\overline{AB}$ and $\overline{BC}$ respectively. What is the ratio of the area of triangle MND to the area of square $ABCD$?

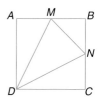

17. Simplify $\dfrac{8x^5y^{-2}z}{16x^{-2}yz^2}$.

TEST-TAKING TIP

Time Management

If you finish a college entrance exam before the allotted time is up, use the remaining time wisely.

Check to be sure that you have marked all of your answers in the correct place on the answer sheet.

Be certain that you have answered all of the questions. If guessing is not penalized, answer the questions you can't complete by eliminating as many answers as you can and then guessing.

Make sure that you have marked only one answer for each question. If two answers are marked, the question will be marked wrong even if one of the answers was the correct choice.

18. A cylindrical can has a diameter of 12 inches and a height of 8 inches. If one gallon of liquid occupies 231 cubic inches, what is the approximate capacity of the can in gallons?

19. Carrie can paint a room alone in 50 minutes. If she works with Ty, they can paint it in 30 minutes. How long would it take Ty to paint the room alone?

20. When the Downtown Metro travels at 45 miles per hour, it arrives at the first stop on time. When it travels at 50 miles per hour, the Metro arrives 2 minutes early. How far is it from the station to the first metro stop?

CHAPTER 13

Loci and Transformations

CHAPTER OBJECTIVES

In this chapter, you will:
- Draw, locate, or describe a locus in a plane or in space.
- Draw reflection images, lines of symmetry, and points of symmetry.
- Draw translation, rotation, or dilation images.

GEOMETRY AROUND THE WORLD
United States

Do you have access to a computer at school or in your home? If so, you have probably used it to write reports, play games, or enter data. But did you know that computers have long been used to create the stunning graphics that introduce some of your favorite music videos and TV programs?

For years, computer graphics have also played an important role in films. In 1982, computer graphics created the exploding planets and other special effects in "Star Trek: The Wrath of Kahn," directed by Nicholas Meyer. Since then, computer graphics have been used in many other movies produced in the United States and abroad.

In the past, creating computer-generated images was a tedious task. First, a programmer entered the mathematical equations for basic two- and three-dimensional forms into the computer's memory. The computer stored the equations, which were then manipulated to create a variety of moving shapes.

Now, sophisticated computer software automatically performs these functions, and more—including showing objects from different angles, or points of view via **geometric transformations**, operations that move and position objects in either two- or three-dimensional space.

GEOMETRY IN ACTION

One type of geometric transformation is called a **rotation**, which specifies a pivoting or angular displacement about an axis. Use your protractor to measure how many degrees the object in the figure at the right has been rotated on its axis.

◀ *Graphic images from "Star Trek: The Wrath of Kahn"*

Computer-generated objects in space are manipulated by applying geometric transformations.

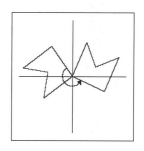

13-1 Locus

Objective

After studying this lesson, you should be able to:

- locate, draw, and describe a locus in a plane or in space.

Application

FYI...

The first artificial satellite was called *Sputnik I*. It was launched by the Soviet Union on October 4, 1957.

The path of a communications satellite orbiting Earth is controlled by a very strict set of conditions that define its position above Earth at all times. This path can be thought of as a **locus** of points. The word *locus* comes from a Latin word meaning "location" or "place." The plural of locus is loci. Loci is pronounced *low-sigh.*

In geometry, a figure is a locus if it is the set of all the points that satisfy a given condition or set of given conditions. A locus may also be defined as the path of a moving point satisfying a set of given conditions. A locus may be one or more points, lines, planes, surfaces, or any combination of these. A locus in a plane may be different than a locus with the same description that is in space.

Circle P is the locus of points in a plane that are 4 centimeters from P. Sphere P is the locus of points in space that are 4 centimeters from P.

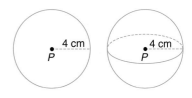

In order to describe a certain locus, you should follow these steps.

- Read the problem carefully.

  Find the locus of all points in a plane that are 15 millimeters from a given line ℓ.

- Draw the given figure.

  The given figure is line ℓ.

- Locate points that satisfy the given conditions.

  Draw points that are 15 mm from line ℓ. Locate enough points to suggest the shape of the locus.

- As soon as the shape of a geometric figure begins to appear, draw a smooth geometric figure that contains the points.

  The points suggest parallel lines.

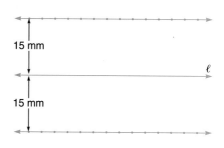

- Describe the locus in words.

  The locus of all points in a plane that are 15 millimeters from a given line ℓ is a pair of parallel lines, one on each side of ℓ, and each 15 millimeters from ℓ.

The steps below summarize the procedure for determining a locus.

Procedure for Determining Locus

1. **Read the problem carefully.**
2. **Draw the given figure.**
3. **Locate the points that satisfy the given conditions.**
4. **Draw a smooth geometric figure.**
5. **Describe the locus.**

Example 1

Determine the locus of all points in space that are 6 millimeters from a given line ℓ.

Draw the given figure.

The given figure is line ℓ.

Locate points that satisfy the given conditions.

Draw enough points to determine the shape. Be sure to consider all possibilities.

Draw a smooth curve or line.

Describe the locus.

The locus of points in space that are 6 millimeters from given line ℓ is a cylindrical surface with line ℓ as the axis and a radius of 6 millimeters.

Example 2

APPLICATION

Agriculture

If a horse is tied with a 15-foot rope to the corner of a shed that is 12 feet by 20 feet, find the locus of the boundary of its grazing area.

The locus is a path composed of $\frac{3}{4}$ of a circle with radius of 15 feet and center at the corner of the building, $\frac{1}{4}$ of a circle with radius of 3 feet and center at the adjacent corner of the building on the 12-foot side, 15 feet of line segments along one side, 12 feet along the end, and 3 feet along the other side of the shed.

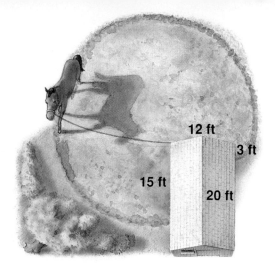

12 ft

3 ft

15 ft

20 ft

CHECKING FOR UNDERSTANDING

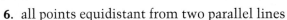

Communicating Mathematics

Read and study the lesson to answer these questions.

1. In your own words, describe the meaning of a locus of points.

2. Describe the locus of points in a plane 3 inches from a given point C.

3. How will the locus in Exercise 2 differ if the words *in a plane* are replaced by *in space*?

4. Describe the locus of points in space 5 centimeters from a given line AB.

5. How will the locus in Exercise 4 differ if the words *in space* are replaced by *in a plane*?

Guided Practice

For each exercise, draw a figure showing the locus of points in a plane. Then describe the locus.

6. all points equidistant from two parallel lines

7. all points equidistant from the endpoints of a given line segment

8. all points on or in the interior of an acute angle and equidistant from the rays that form the angle

9. all points that are the third vertices of triangles having a given base and a given altitude

For each exercise, draw a figure showing the locus of points in space. Then describe the locus.

10. all points equidistant from two parallel lines

11. all points equidistant from the endpoints of a given line segment

12. all points equidistant from three noncollinear points

13. all points that are the centers of spheres with radii r units long and tangent to a given plane

Describe the locus of points in the classroom that meet the following conditions.

14. all points equidistant from the floor and the ceiling

15. all points equidistant from all four corners of the floor

EXERCISES

Practice

Answer each of the following.

16. Describe the locus of points in a plane that are 4 meters from a given line *m*.

17. Describe the locus of points on a football field that are equidistant from the two goal lines.

18. Describe the locus of points in space that are 1 meter from a given point *C*.

19. Describe the locus of points in space that are 4 inches from a given line ℓ.

20. Describe the locus of points on a road that are equidistant from the edges of the road.

For each exercise, draw a figure showing the locus of points in a plane. Then describe the locus.

21. all midpoints of the radii of a circle

22. all points that are the midpoints of parallel chords of a given circle

23. all points that are 3 inches from a circle with radius of 6 inches

24. all points that are less than 6 centimeters from a given point

25. all points of the path of the center of a gear as it rotates around the circumference of a larger gear

26. all points that are equidistant from the points of intersection of two given circles

27. all midpoints of all chords of a given measure that is less than the measure of the diameter in a given circle

28. all points that are equidistant from two intersecting lines

29. all centers of all circles passing through two distinct points

30. all centers of all circles tangent to two intersecting lines

31. all midpoints of all chords formed by secants drawn to a circle from a point outside the circle

32. all midpoints of all chords that have one given endpoint on a circle

For each exercise, draw a figure showing the locus of points in space. Then describe the locus.

33. all points that are a given distance greater than 0 units from a plane
34. all points that are equidistant from two intersecting planes
35. all centers of spheres that are tangent to a plane at a given point in the plane
36. all points equidistant from the vertices of a rectangular prism
37. all points equidistant from all points on a circle

Use square *ABCD* in plane $\mathcal{R}$ to describe each locus of points.

38. all points in $\mathcal{R}$ equidistant from the midpoints of the sides of the square

39. all points in space equidistant from the midpoints of the sides of the square

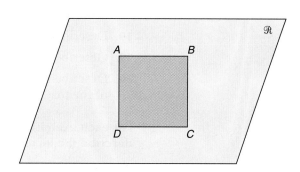

40. all points in $\mathcal{R}$ equidistant from the diagonals of the square

41. all points in space equidistant from the diagonals of the square

42. Describe the locus of points in $\mathcal{R}$ equidistant from $\overline{AB}$ and $\overline{DC}$.

Complete each of the following.

43. Describe the locus of points in a plane formed by the midpoints of all hypotenuses of a given length given the line containing one of the legs and the vertex of one acute angle of a right triangle on that line.

44. Describe the locus of points a ship will travel when it leaves New York and sails a straight course to London, England.

45. Describe the locus of points made by a point on the outermost part of the wheel of a bicycle as it travels down the road.

46. A weight is attached to the end *A* of a string represented by $\overline{OA}$ that is 100 centimeters long and hangs vertically from *O*. The weight is pulled to one side and then allowed to swing back and forth. If an obstacle is placed in the path of the string at *B* which is 50 centimeters below *O*, describe the locus of points traveled by point *A*.

47. What is the locus of points in a plane such that the sum of the distances from two given points *A* and *B* is always 6? *(Hint: Place thumbtacks through points A and B on your paper. Then tie the ends of a piece of string 6 units long to each tack. Draw the locus with your pencil inside of the loop of the string.)*

Applications

48. Sports A 3-point shot can be made in high school basketball if the shot is made from a distance greater than 19 feet 9 inches from the basket. Describe the locus of points on a basketball court that are more than 19 feet 9 inches from the basket.

49. Geography If Earth is assumed to be a sphere, what is the locus of points on Earth's surface equidistant from Chicago and London?

50. Communications A radio station has a broadcast range of 80 miles. What is the locus of points that can receive that station?

51. Teaching A physical education teacher places two students 60 feet apart. The students are supposed to race to a pylon. Describe all the possible placements of the pylon that will make the race fair.

Mixed Review

Complete.

52. In the linear equation $y = mx + b$, m represents the __?__ of the line. **(Lesson 12-1)**

53. The surface area of a sphere whose radius measures 10 centimeters is __?__. **(Lesson 11-7)**

54. The measure of an inscribed angle which intercepts a semicircle is __?__. **(Lesson 9-4)**

55. The longest chords of a circle are __?__. **(Lesson 9-1)**

56. If the measures of the sides of a triangle satisfy the Pythagorean Theorem, then the triangle is a __?__. **(Lesson 8-3)**

57. If two lines do not intersect, then they are parallel or __?__. **(Lesson 3-1)**

Wrap-Up

58. Write three locus problems for which the solutions are a plane, a line, a sphere.

∼∼∼ TECHNOLOGY CONNECTION ∼∼∼

Can you imagine traveling at two or three thousand miles per hour? Christine Darden, an aerospace engineer in the Advanced Vehicles Division of NASA, is working on designing *supersonic aircraft* that will do just that! She expects that by the year 2000 we will be able to fly from the United States to Europe in about two hours and to Australia in about 4 hours.

Before we can begin to travel at these speeds, however, the United States Congress insists that the sonic boom that occurs on the ground beneath a plane traveling faster than the speed of sound must be reduced or eliminated. Ms. Darden is currently researching changes in the design of the nose and wings of airplanes that may reduce or eliminate the sonic boom.

13-2 Locus and Systems of Equations

Objective

After studying this lesson, you should be able to:
- find the locus of points that solve a system of equations by graphing, substitution, or by elimination.

Application

Parts of straight lines are seen everywhere. Fields are plowed in straight rows. Straight lines are used in designing buildings both for support and appearance. In mathematics, graphs of lines are representations of relationships between two variables. These graphs also represent a locus of points that satisfy a certain set of conditions that are often expressed as equations.

Example 1

APPLICATION

Social Studies

In 1990, the population of Wichita, Kansas, was approximately 304,000 and was growing at the rate of 2400 people per year. If the growth continued at the same rate, draw a graph of the locus of points that represent the population of Wichita in x number of years.

Let y equal the population of Wichita and x equal the number of years of growth. The locus of points representing the population can be described by an equation of the line that has a slope of 2400 and contains the point (0, 304000). Remember that the equation of a line can be written in the form $y = mx + b$, where m is the slope and b is the y-intercept.

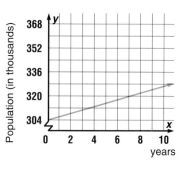

$$y = mx + b$$
$$y = 2400x + 304,000$$

Substitute 2400 for m and 304,000 for b.

The locus of points representing the population of Wichita is the line determined by the equation $y = 2400x + 304,000$.

Example 2

In 1990, St. Louis, Missouri, had a population of about 368,000, which was decreasing at the rate of 5600 people per year.

a. If the rate of decrease continued, write an equation representing the population of St. Louis.

Write an equation that represents the population of St. Louis using the point (0, 368,000) and the slope -5600.

$$y = mx + b$$

$$y = -5600x + 368,000$$

b. Graph this equation on the same coordinate plane as the equation for the population of Wichita found in Example 1.

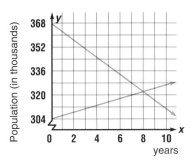

c. Describe the locus of points that represent the time at which the populations of Wichita and St. Louis will be equal.

Study the table of values at the right. The graphs intersect at $P(8, 323200)$. Therefore, P is the locus of points that satisfy both equations $y = 2400x + 304,000$ and $y = -5600x + 368,000$.

In eight years, the populations of both cities will be approximately 323,200.

x	Wichita $2400x + 304,000$	St. Louis $-5600x + 368,000$
1	306,400	362,400
2	308,800	356,800
3	311,200	351,200
4	313,600	345,600
5	316,000	340,000
6	318,400	334,400
7	320,800	328,800
8	323,200	323,200

Notice that the graphs in Example 2 intersect at the point with coordinates (8, 323200). Since this point lies on both graphs, its coordinates satisfy both $y = 2400x + 304,000$ and $y = -5600x + 368,000$. The equations $y = 2400x + 304,000$ and $y = -5600x + 368,000$ together are called a **system of equations.** The solution of this system is (8, 323200). You can check this algebraically by substituting 8 for x and 323,200 for y in both equations.

You may remember from algebra that a system of equations can be solved by algebraic methods as well as by graphing. Two algebraic methods are the substitution method and the elimination method.

Example 3

Use substitution to find the locus of points that satisfy both equations $y = x - 3$ and $3x + 5y = 9$.

By the first equation, y is equal to $x - 3$. Therefore, $x - 3$ can be substituted for y in the second equation.

$$3x + 5y = 9$$
$$3x + 5(x - 3) = 9 \qquad \text{\textit{Substitute } x - 3 \text{ \textit{for} } y.}$$

The resulting equation has only one variable, x. Solve that equation.

$$3x + 5(x - 3) = 9$$
$$3x + 5x - 15 = 9$$
$$8x - 15 = 9$$
$$8x = 24$$
$$x = 3 \qquad \text{\textit{The x-coordinate is 3.}}$$

Find y by substituting 3 for x in the first equation.

$$y = x - 3$$
$$y = 3 - 3 \qquad \text{\textit{Substitute 3 for x.}}$$
$$y = 0 \qquad \text{\textit{The y-coordinate is 0.}}$$

The point with coordinates (3, 0) is the locus of points that satisfy both equations $y = x - 3$ and $3x + 5y = 9$.

Example 4

Use elimination to find the locus of points that satisfy both equations $3x + 4y = 17$ and $2x + 3y = 11$.

Sometimes adding or subtracting two equations will eliminate a variable. In this case, adding or subtracting the two equations will not eliminate a variable. If both sides of the first equation are multiplied by 2 and both sides of the second equation are multiplied by -3, the system can be solved by adding the equations.

$3x + 4y = 17$ *Multiply by 2.* $6x + 8y = 34$

$2x + 3y = 11$ *Multiply by -3.* $\underline{-6x - 9y = -33}$

$$-y = 1 \qquad \text{\textit{Add to eliminate x.}}$$
$$y = -1 \qquad \text{\textit{The y-coordinate is -1.}}$$

Finally, substitute -1 for *y* in the first equation. Then solve for *x*.
-1 could also be substituted for y in the second equation.

$$3x + 4y = 17$$
$$3x + 4(-1) = 17$$
$$3x - 4 = 17$$
$$3x = 21$$
$$x = 7 \qquad \textit{The x-coordinate is 7.}$$

The point with coordinates (7, -1) is the locus of points that satisfy both equations $3x + 4y = 17$ and $2x + 3y = 11$.

CHECKING FOR UNDERSTANDING

Communicating Mathematics

Read and study the lesson to answer these questions.

1. Describe the possibilities for the locus of points that belong to the intersection of two lines.

2. Explain how you know that the graph of (-2, 3) is the locus of points that satisfies both of the equations $3x + 4y = 6$ and $2x + 3y = 5$.

3. Determine which method – graphing, substitution, or elimination – you would use to solve each system of equations. Explain your choice.
   a. $y = x + 5$
   $y = 3x - 5$
   b. $3x - 5y = 11$
   $4x + 7y = -2$
   c. $x = 2y - 3$
   $3x - 5y = 11$

Guided Practice

Graph each pair of equations to find the locus of points that satisfy both equations.

4. $x + y = 6$
   $x - y = 2$
5. $y = x - 1$
   $x + y = 11$
6. $y = 4x$
   $x + y = 5$

Use an algebraic method to find the locus of points that satisfy both equations.

7. $y = 3x$
   $x + 2y = -21$
8. $x - y = 5$
   $x + y = 25$
9. $12 - 3y = -4x$
   $40 + 4x = 10y$

EXERCISES

Practice

State the locus of points that satisfy the intersection of each pair of lines.

10. *c* and *d*

11. *a* and *c*

12. *a* and *d*

13. *b* and *d*

14. *b* and *c*

15. *d* and the *y*-axis

16. *c* and the *x*-axis

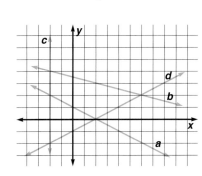

Determine which ordered pairs satisfy each equation or locus of points.

17. $x + 3y = 6$ a. $(0, 2)$ b. $(-1, 4)$ c. $(6, 0)$ d. $(-3, 3)$

18. $2x - 5y = -1$ a. $(0, 5)$ b. $(2, 1)$ c. $(-0.5, 0)$ d. $(-2, -1)$

19. $3x = 15$ a. $(5, 1)$ b. $(5, 0)$ c. $(0, 5)$ d. $(5, 8)$

Graph each pair of equations to find the locus of points that satisfy both equations.

20. $3x - 2y = 10$
    $x + y = 0$

21. $x + 2y = 7$
    $y = 2x + 1$

22. $y = x + 3$
    $3y + x = 5$

Use an algebraic method to find the locus of points that satisfy both equations.

23. $x - y = 6$
    $x + y = 5$

24. $x + 2y = 5$
    $2x + y = 7$

25. $3x + 4y = -7$
    $2x + y = -3$

26. $y = x - 1$
    $4x - y = 19$

27. $x = y + 10$
    $2y = x - 6$

28. $9x + y = 20$
    $3x + 3y = 12$

29. $x - 2y = 5$
    $3x - 5y = 8$

30. $3x + 7y = 16$
    $x - 6y = 11$

31. $6x + 5 = y$
    $x - y = 0$

32. $9x + 7y = 4$
    $6x - 3y = 18$

33. $2x - 5y = -6$
    $6x - 6y = 18$

34. $6x + 7y = -9$
    $-9x + 11y = 78$

35. The graphs of $y = 2$, $x - y = 0$, and $3y = -2x + 30$ intersect to form a triangle.
    a. Find the coordinates of the vertices of the triangle.
    b. Find the area of the triangle.

36. In $\triangle IBM$, $I(-8, 6)$, $B(5, 6)$, and $M(-1, -6)$ are coordinates of the vertices. Find the coordinates of the intersection of the altitude from I to $\overline{MB}$ and $\overline{MB}$.

37. Find the locus of points that satisfy both $x - y = 0$ and $(x + 1)^2 + (y - 4)^2 = 25$.

38. Find the locus of points that satisfy both $4y - 3x = 26$ and $(x - 1)^2 + (y - 1)^2 = 25$.

39. Find the distance between the parallel lines whose equations are $y = 3x + 1$ and $y = 3x - 8$.

Critical Thinking   40. A line is determined by the equation $y = 2x - 3$. Find equations for each of the following.
    a. a line such that the locus of points that satisfy both equations is the point $(1, -1)$
    b. a line such that there are no points that satisfy both equations
    c. a line such that the locus of points that satisfy both equations is an infinite number of points

Applications

41. Social Studies The population of Manchester, United Kingdom, is about 4,050,000 and is decreasing at a rate of 22,300 people per year. The population of Guadalajara, Mexico, is about 3,262,000 and is increasing at a rate of 118,900 people per year. Find the locus of points that describes the number of years before the populations of these two cities will be equal.

42. Consumerism The rate for renting a car from Rockwell Cars is $20 plus a fee of $0.25 per mile. Speedy Car rents cars for $25 plus a fee of $0.20 per mile. Find the locus of points that describes the number of miles traveled for which it is less expensive to rent from Speedy Car.

43. Communications On a coordinate grid of a region of the state, radio station WRAY is located at (-18, 0) and station WARC at (20, 10). If WRAY has a range of 22 units and WARC has a range of 20 units, will they ever reach the same audience? Explain your reasoning.

Computer

44. The BASIC computer program given below will determine whether an ordered pair satisfies a linear equation or locus of points. Use the program to check your answers for Exercises 17-19.

```
10 PRINT "WRITE YOUR EQUATION IN THE FORM Y = MX + B.
   ENTER THE VALUES FOR M AND B."
20 INPUT M, B
30 PRINT "ENTER THE ORDERED PAIR YOU WISH TO TEST."
40 INPUT X, Y
50 IF Y = M * X + B THEN PRINT "THE ORDERED PAIR
   SATISFIES THE EQUATION.": GOTO 70
60 PRINT " THE ORDERED PAIR DOES NOT SATISFY THE
   EQUATION."
70 END
```

Mixed Review

45. Describe the locus of all points in a plane that are 3 inches from a circle with radius measuring 5 inches. **(Lesson 13-1)**

46. Using the information in the figure at the right, find the values of *x*, *y*, and *z*. **(Lesson 8-1)**

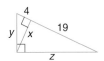

47. Give a counterexample to *All rhombi are squares.* **(Lesson 2-2)**

48. Write the converse of the statement *If three points are coplanar, then they lie in the same plane.* **(Lesson 2-2)**

49. If the supplement of an angle measures 4 times that of the angle itself, find the measure of the angle. **(Lesson 1-8)**

Wrap-Up

50. Describe three different methods you could use to find the locus of points that satisfy both equations $x = y + 10$ and $2y = x - 6$.

13-3 Intersection of Loci

Objective

After studying this lesson, you should be able to:
- solve locus problems that satisfy more than one condition.

Application

FYI…

Chicago's O'Hare International Airport and Atlanta's Hartsfield Airport are the busiest airports in the world with a takeoff or landing about every 40 seconds.

Air traffic controllers track the position of airplanes using radar. The flight path of an airplane can be thought of as a locus of points. It is important that the controller realize that the loci of several airplanes might intersect.

Often loci satisfy several conditions. Such loci can usually be determined by finding the intersection of the loci that meet each separate condition.

Example 1

A city manager wishes to locate a fire station equidistant from two schools and equidistant from two parallel streets. Determine the locus of points that could be locations for the fire station.

Divide the problem into two separate locus problems.

The locus of all points in a plane equidistant from the two schools is the perpendicular bisector of the segment joining the two schools.

The locus of all points in a plane that are equidistant from the two parallel streets is a line parallel to the given streets and midway between them.

Consider the intersection of the loci of the separate problems.

If A and B are the given schools, and ℓ and m are the given streets, then C is the only point that is equidistant from A and B and equidistant from ℓ and m.

The fire station should be located at point C.

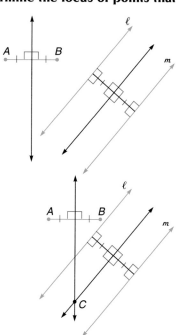

Some construction problems involve finding a point that satisfies several conditions.

CONSTRUCTION

Given *AB*, *AC*, and *CD*, such that *AB* > *AC* > *CD*, construct △*ABC* so that $\overline{CD}$ is an altitude of the triangle.

1. Note that vertex *C* of the triangle will lie *AC* units from *A*. The locus of all points *AC* units from *A* is a circle with center *A* and a radius of *AC* units. Vertex *C* will be somewhere on ⊙*A*.

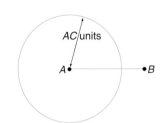

2. Endpoint *D* of altitude $\overline{CD}$ will lie on side $\overline{AB}$ or on the line containing $\overline{AB}$. Since $\overline{CD}$ is an altitude, then $\overline{CD} \perp \overleftrightarrow{AB}$. This means endpoint *C* will lie on a line parallel to $\overleftrightarrow{AB}$, *CD* units from $\overleftrightarrow{AB}$. The locus of all points meeting these conditions are two lines on either side of $\overleftrightarrow{AB}$, parallel to $\overleftrightarrow{AB}$ at a distance of *CD* units.

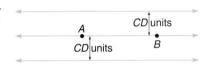

3. Point *C* must satisfy the conditions for being a vertex, Step 1, and the conditions for being an endpoint of an altitude, Step 2. Only four points satisfy these conditions.

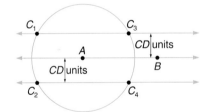

There are four possible ways to draw △*ABC*.

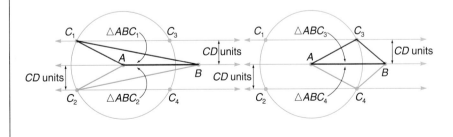

Algebraic equations can be used to represent a locus of points that must satisfy one or more conditions.

Example 2

Find the locus of points that are 10 units from the origin in the coordinate plane and that satisfy the equation $y = -x + 10$.

The locus of points 10 units from the origin in the coordinate plane is a circle with its center at $(0,0)$ and radius of 10 units.

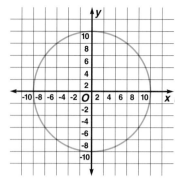

$(x - h)^2 + (y - k)^2 = r^2$ *Equation for a circle*
$(x - 0)^2 + (y - 0)^2 = 10^2$ *$(h, k) = (0, 0), r = 10$*
$$x^2 + y^2 = 100$$

The locus of points that satisfy the equation $y = -x + 10$ is the set of points on the line with slope -1 and y-intercept 10.

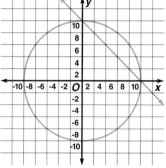

The locus of points that satisfy both conditions will be the points of intersection of the line and the circle. To find this locus, solve the system of equations $x^2 + y^2 = 100$ and $y = -x + 10$.

$$x^2 + y^2 = 100$$
$$x^2 + (-x + 10)^2 = 100$$ *Substitute -x + 10*
$$x^2 + x^2 - 20x + 100 = 100$$ *for y.*
$$2x^2 - 20x = 0$$
$$2x(x - 10) = 0$$
$$x = 0 \text{ or } x = 10$$

When $x = 0$, $y = 10$ and when $x = 10$, $y = 0$.

The locus of points that satisfy both conditions are $(0, 10)$ and $(10, 0)$.
Check your answer by graphing both equations.

CHECKING FOR UNDERSTANDING

Communicating Mathematics

Read and study the lesson to answer these questions.

1. Explain how to determine the locus of points that are equidistant from two given points and equidistant from two parallel lines.

2. Describe the difference between the locus of points in a plane determined by the equation $(x - 4)^2 + (y - 1)^2 = 25$ and the locus of points determined by the equation $(x - 4)^2 + (y - 1)^2 = 0$.

3. Describe the difference between the locus of points determined by the equation $(x - 4)^2 + (y - 1)^2 = 25$ and the locus of points determined by the equation $(x - 4)^2 + (y - 1)^2 + (z + 2)^2 = 25$.

4. Describe the possible ways two parallel lines and a circle can intersect.

Guided Practice

Describe the geometric figure whose locus in space satisfies each condition. Be specific.

5. $(x - 2)^2 + (y + 6)^2 + (z - 5)^2 = 49$ **6.** $x = 2$

7. $(x - 5)^2 + (y + 1)^2 = 9$ **8.** $y + x = 0$

Describe all possible ways the given figures can intersect.

9. a sphere and a plane   **10.** two circles

11. two concentric circles and two parallel lines

Draw a diagram to find the locus of points that satisfy the conditions. Then describe the locus.

12. all points in a coordinate plane that are 5 units from the graph of $x = 6$ and equidistant from the graphs of $y = 1$ and $y = 7$

13. all the points in a plane that are 1.5 centimeters from a given line and 3 centimeters from a given point on the line

14. all the points in space that are 2 inches from a given line and 3 inches from a given point on the line

15. all points in a plane that are equidistant from the rays of an angle and equidistant from two points on one of the sides of the angle

CONSTRUCTION

Complete the following construction.

16. Draw two parallel lines, ℓ and m. Choose a point between ℓ and m and label it B. Construct a circle tangent to the 2 lines and containing B.

EXERCISES

Practice

Describe the geometric figure whose locus in space satisfies each condition. Be specific.

17. $(x - 2)^2 + (y + 4)^2 = 36$ **18.** $y = 2x - 10$

19. $(x - 3)^2 + (y - 4)^2 + (z - 5)^2 = 16$ **20.** $y = 6$

Complete each of the following.

21. Write the equation for the locus of all points in the coordinate plane 4 units from the graph of (-1, -6).

22. Write the equation of the locus of all points in space 6 units from the graph of (-2, 5, 1).

Describe all possible ways the given figures can intersect.

23. two concentric circles and a line   **24.** two spheres

25. a sphere and two parallel lines   **26.** a circle and a sphere

27. a sphere and two parallel planes   **28.** a circle and a plane

Draw a diagram to find the locus of points that satisfy the conditions. Then describe the locus.

29. all the points in a plane that are 2 centimeters from a given line and 5 centimeters from a given point on the line

30. all the points in space that are 2 inches from a given line and 5 inches from a given point on the line

31. all the points in a plane on or inside of a given angle that are equidistant from the sides of the angle and 4 inches from the vertex of the angle

32. all the points in a plane equidistant from two given parallel lines and a given distance from an intersecting line

33. all points in a coordinate plane that are 3 units from the graph of $y = 4$ and equidistant from the graph of $x = 6$ and $x = -2$

34. all the points in space that are 4 centimeters from point A and 2 centimeters from point B if A and B are 5 centimeters apart

35. all points in space that are equidistant from two intersecting lines and 3 units from the point of intersection

CONSTRUCTION

Complete each construction.

36. Draw a line m and two points, D and C, on one side of the line. Then construct the locus of points in the plane that are equidistant from C and D and on line m.

37. Draw two segments $\overline{PQ}$ and $\overline{RS}$. Choose a point between P and Q and label it A. Then construct a circle tangent to $\overline{PQ}$ at A and whose center lies on $\overline{RS}$.

38. Draw a line ℓ and a point P on the line. Draw point Q not on line ℓ. Then construct the locus of the centers of all circles in the plane that contain P and Q.

Find the locus of points in the coordinate plane that satisfy both equations.

39. $x + y = 4$ and $x^2 + (y + 1)^2 = 4$

40. $(x - 1)^2 + y^2 = 1$ and $(x - 2)^2 + (y - 2)^2 = 4$

In the figure below, $\ell \parallel m$, and the distance between ℓ and m is less than $\frac{1}{2}(AB)$.

41. How many points are there in the locus of all points P on line m such that $\triangle PAB$ is an isosceles triangle?

42. How many points are there in the locus of all points P on line m such that $\triangle PAB$ is a right triangle?

43. How many points are there in the locus of all points P on line m such that $\triangle PAB$ is an equilateral triangle?

Critical Thinking

44. Draw and describe the locus of all points in space equidistant from two parallel planes $\mathcal{A}$ and $\mathcal{B}$ and 5 units from a fixed point R.

Applications

45. City Management Hales Corners is a triangular shaped piece of land approximately 3 miles by 3 miles by 3 miles. The community wants to place a fire station equidistant from all three corners of Hales Corners. Draw a diagram to find the locus of points that satisfy the requirements.

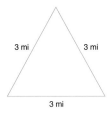

3 mi   3 mi

3 mi

46. Communications Two cities, Attica and Rochester, are 70 miles apart. A television station is building a transmitter and, because of the terrain, would like to locate the transmitter 50 miles from Attica and 60 miles from Rochester. Draw a diagram to find the locus of points that could be used as a site for the transmitter.

47. Water Management A house faces east toward a north-south street. One hundred feet from the southeast corner of the house, under the center of the street, is the city water connection. Draw diagrams of the various possibilities for representing the location of the water connection.

48. Food What is the locus of the intersection of the blade of a delicatessen's slicing machine and a cylinder of lunch meat?

Mixed Review

Classify each statement as *always*, *sometimes*, or *never true*.

49. A radius of a circle is perpendicular to a line tangent to the circle. **(Lesson 9-5)**

50. An angle inscribed in a semicircle is obtuse. **(Lesson 9-4)**

51. Chords that are equidistant from the center of a circle are congruent. **(Lesson 9-3)**

52. A chord of a circle is a radius of the circle. **(Lesson 9-1)**

53. The cosine of an acute angle is less than 1. **(Lesson 8-4)**

54. If the measures of the sides of a triangle are a, b, and c, then $a^2 + b^2 = c^2$. **(Lesson 8-2)**

55. The diagonals of a rhombus are congruent. **(Lesson 6-5)**

56. The diagonals of a rectangle are congruent. **(Lesson 6-4)**

Wrap-Up

57. Describe the steps you should use to find the locus of points that satisfy two conditions. Give an example.

Mappings

Objectives

After studying this lesson, you should be able to:

- name the image and preimage of a mapping, and
- recognize an isometry or congruence transformation.

Application

A Dutch artist Mauritz Escher (1898-1972) created art by using patterns. Study the example of his work at the right. What patterns do you see?

Escher moved figures according to certain rules. Some of the rules are illustrated below.

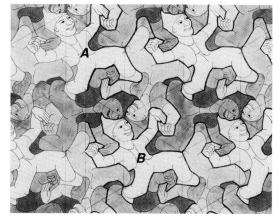

©1938 M.C. Escher/Cordon Art-Baarn, Holland

A figure can be reflected.

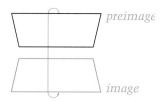

A figure can be rotated.

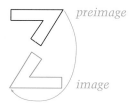

A figure can be slid.

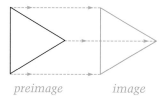

A figure can be enlarged or reduced.

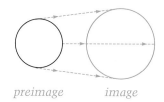

Notice that each point of one figure is paired with exactly one point of the corresponding figure. These are called mappings or **transformations.** A transformation maps a **preimage** onto an **image.**

Definition of Transformation

In a plane, a mapping is a transformation if each preimage point has exactly one image point, and each image point has exactly one preimage point.

A transformation is a one-to-one mapping. That is, each point is mapped to a unique point.

Example 1

Describe the transformation that moves person *A* to person *B* in the Escher art on page 640.

Person *A* can be moved to person *B* by sliding person *A* diagonally down and to the right.

The symbol → is used to indicate a mapping. For example, $\triangle ABC \rightarrow \triangle PQR$ means $\triangle ABC$ is mapped onto $\triangle PQR$. $\triangle ABC$ is the preimage and $\triangle PQR$ is the image. The order of the letters indicates the correspondence of the preimage to the image. The first vertices, *A* and *P*, are corresponding vertices. Similarly, *B* and *Q* are corresponding vertices and *C* and *R* are corresponding vertices.

In the figure, $\overline{AC} \rightarrow \overline{DF}$, and *A* and *D* are corresponding points. *C* and *F* are corresponding points. *B* and *E* are corresponding points. Every point on $\overline{AC}$ corresponds to a point on $\overline{DF}$, and every point on $\overline{DF}$ corresponds to a point on $\overline{AC}$. Thus, the mapping is a transformation.

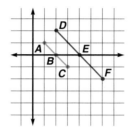

When a geometric figure and its transformation image are congruent, the mapping is called an **isometry** or a **congruence transformation**. When a figure and its transformation image are similar, the mapping is called a **similarity transformation**.

Example 2

Suppose $\triangle ABC \rightarrow \triangle PQR$. Show that this mapping is an isometry.

Each side of $\triangle ABC$ is mapped to the corresponding side of $\triangle PQR$. If these corresponding sides are congruent, then the mapping is an isometry. Since their lengths are the same, $\overline{AB} \cong \overline{PQ}$, $\overline{BC} \cong \overline{QR}$, and $\overline{AC} \cong \overline{PR}$. Therefore, $\triangle ABC \cong \triangle PQR$ by SSS. $\triangle ABC$ is the preimage of the isometry, and $\triangle PQR$ is the image.

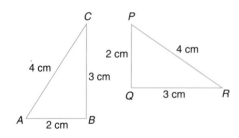

CHECKING FOR UNDERSTANDING

Communicating Mathematics

Read and study the lesson to answer these questions.

1. What is a transformation?
2. Describe how a preimage is related to an image.
3. Name four ways of creating a transformation. Draw an example of each.
4. Explain the difference between an isometry and a similarity transformation.

Guided Practice

Answer the following, if quadrilateral *RSTU* → quadrilateral *ABCD*.

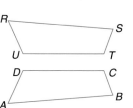

5. Name the image of $\overline{UT}$.

6. Name the preimage of *D*.

7. Name the preimage of ∠*B*.

8. Name the image of ∠*T*.

9. Name the preimage of $\overline{AB}$.

For Exercises 10–21, △*ABC* → △*EBD*. Name the image of each of the following.

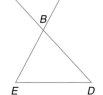

10. *A*   **11.** *B*   **12.** *C*

13. ∠*CAB*   **14.** ∠*BCA*   **15.** $\overline{AC}$

Name the preimage of each of the following.

16. *E*   **17.** *B*   **18.** *D*

19. $\overline{BE}$ **20.** $\overline{DE}$ **21.** ∠*DBE*

EXERCISES

Practice

Answer each of the following, if pentagon *ABCDE* → pentagon *PQRST*.

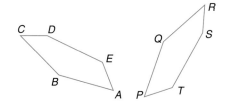

22. Name the image of $\overline{CD}$.

23. Name the image of ∠*E*.

24. Name the preimage of $\overline{PT}$.

25. Name the preimage of ∠*Q*.

26. Name the image of $\overline{BC}$.

Each figure below is the preimage or the image for an isometry. Write the image of each given preimage.

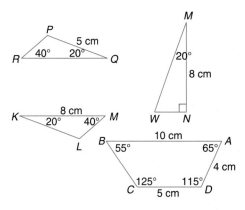

27. △*MWN*   **28.** △*PQR*   **29.** △*SRQ*

30. △*LMK*   **31.** △*MKL*   **32.** △*ZYX*

33. quadrilateral *RSTU*   **34.** quadrilateral *BADC*

In the figure at the right, $\triangle XYW \cong \triangle ZYW$.

35. Name a segment that is its own preimage.

36. Name two points that are their own preimages.

37. Name a mapping that describes the congruence.

38. The L-shaped tile at the right is gray on top and blue underneath. Suppose two of these L-shaped tiles are used to form each of the following figures. Name the colors for Section *A* and Section *B*.

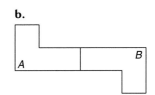
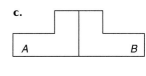

Critical Thinking

39. A transformation maps each point (x, y) to $(x + 3, y - 1)$. Is this transformation an isometry? *(Hint: Choose three points that form a triangle and find their images. Do the images form a congruent triangle?)*

Applications

40. **Art**   Study the Escher drawing at the right. Describe the transformation that moves angel *A* to angel *B*.

41. **Art**   Describe the transformation that moves angel *D* to angel *C*.

42. **Recreation**   Sam is walking down the beach in his bare feet. He looks back at his footprints in the sand. Describe what he sees in terms of transformations.

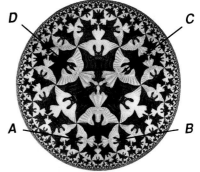

©1941 M.C. Escher/Cordon Art-Baarn, Holland

43. **Cartoons**   The movement you see when you watch a cartoon is created by a series of pictures, each one slightly different than the last one. Suppose you are making a cartoon of a bouncing ball, how would you use transformations to create the illusion of a moving ball?

Mixed Review

44. Describe the locus of all points in space that are a given distance from a line segment and equidistant from the endpoints of the line segment. **(Lesson 13-3)**

45. A trapezoid has a median 17.5 cm long. If the height is 18 cm, find the area of the trapezoid. **(Lesson 10-5)**

46. Solve the proportion $\frac{x + 1}{8} = \frac{3}{4}$. **(Lesson 9-1)**

47. Find the geometric mean between 18 and 31. **(Lesson 8-1)**

48. Write the converse of the conditional *If a mapping is one-to-one, then it is a transformation.* **(Lesson 2-2)**

Wrap-Up

49. Draw a simple figure. Draw four images of the figure showing different transformations.

Reflections

Objectives

After studying this lesson, you should be able to:

- name a reflection image with respect to a line,
- recognize line symmetry and point symmetry, and
- draw reflection images, lines of symmetry, and points of symmetry.

Application

All of us are familiar with scenes like the one shown at the left. When the water is perfectly calm, you can see a mirror-like reflection of the things on shore in the water.

A **reflection** is a type of transformation. For example, in the figure below, A is the reflection image of X with respect to ℓ. The line of reflection, ℓ, is a perpendicular bisector of the segment drawn from X to A. Since P is on the line of reflection, its image is P itself.

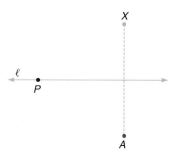

It is also possible to have a reflection image with respect to a point.

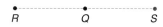

S is the reflection of R with respect to Q. The point of reflection, Q, is the midpoint of the segment drawn from R to S.

The reflection images of collinear points are collinear. The images of collinear points A, B, and C are collinear points P, Q, and R, respectively. Therefore, it is said that reflections preserve collinearity.

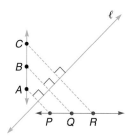

The reflection image of Y is between the image of X and Z if and only if Y is between X and Z. Thus, reflections preserve betweenness of points.

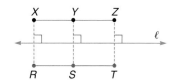

Reflections also preserve angle and distance measure. In the figure below, $\triangle XYZ$ is the reflection image of $\triangle ABC$ and $\triangle ABC$ is the reflection image of $\triangle XYZ$. By measuring the corresponding parts of $\triangle ABC$ and $\triangle XYZ$, it appears that $\triangle ABC$ is congruent to $\triangle XYZ$.

Points A, B, and C can be read in a clockwise order. $\triangle ABC$ is said to have a clockwise orientation.

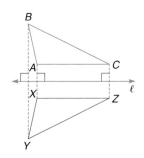

Corresponding points X, Y, and Z are then in a counterclockwise orientation. $\triangle XYZ$ is said to have a counterclockwise orientation.

Suppose $\triangle CBA$ has a counterclockwise orientation, then what is the orientation of the reflection image?

Because only the orientation of a geometric figure is changed, a reflection is an isometry.

CONSTRUCTION

Construct the reflection image of $\triangle ABC$ with respect to line ℓ.

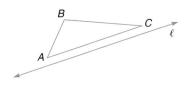

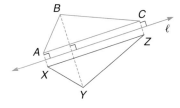

Construct perpendiculars from A, B, and C through line ℓ. Locate X, Y, and Z so that line ℓ is the perpendicular bisector of $\overline{AX}$, $\overline{BY}$, and $\overline{CZ}$. X, Y, and Z are the corresponding vertices of A, B, and C. The reflection image of $\triangle ABC$ is found by connecting the vertices X, Y, and Z.

$$\triangle ABC \rightarrow \triangle XYZ$$

Example 1

Tess is playing miniature golf. If the tee is at point _B_, how can she make a hole-in-one for the situation shown at the right?

Tess must visualize the reflection image _H′_ of hole _H_ with respect to line _p_. She then aims for _H′_. The ball hits the side at _X_ and rebounds along the preimage of $\overline{XH'}$ making a hole-in-one.

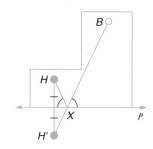

A line can be drawn through many plane figures so that the figure on one side is a reflection image of the figure on the opposite side. In such a case, the line of reflection is called a **line of symmetry.**

Example 2

Test the given lines to see if they are lines of symmetry for rectangle _ABCD_.

By measuring, it appears that line ℓ is the perpendicular bisector of both $\overline{AB}$ and $\overline{CD}$. Any point to the left of line ℓ has its reflection image to the right of line ℓ. Likewise, by measuring, it can be seen that line _m_ is the perpendicular bisector of $\overline{AD}$ and $\overline{BC}$. Also any point above line _m_ has its reflection image below line _m_. Therefore, lines ℓ and _m_ are lines of symmetry for rectangle _ABCD_.

You could also test these lines by copying the figures, cutting them out, and folding them along the lines. If the vertices correspond, then the lines are lines of symmetry.

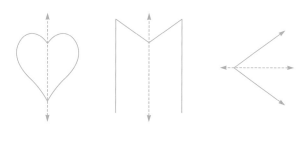

For many figures, a point can be found that is a point of reflection for all points on the figure. This point of reflection is called a **point of symmetry.**

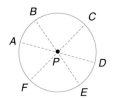
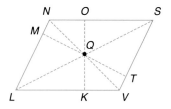
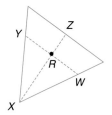

P and Q are points of symmetry.

R is not a point of symmetry.

A point of symmetry must be a midpoint for all segments with endpoints on the figure. In the two figures above at the left, P and Q are midpoints of the segments drawn. In the figure above at the right, R is not a point of symmetry because R is not the midpoint of $\overline{XZ}$.

CHECKING FOR UNDERSTANDING

Communicating Mathematics

Read and study the lesson to answer these questions.

1. Describe the relationship between a line of reflection and the line segment joining an image point and its preimage point.

2. What does it mean for a point to have a reflection image with respect to a point?

3. Why is a reflection an isometry?

4. Name some objects that have lines of symmetry.

5. Name some objects that have points of symmetry.

Guided Practice

For the figure at the right, name the reflection image with respect to line ℓ.

6. A 7. B 8. $\overline{AB}$

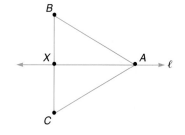

Copy each figure below. Then draw the reflection image of each figure with respect to line m.

9.

10.

Copy the following letters. Draw all possible lines of symmetry. If none exist, write *none*.

11.

12.

13.

14.

Copy each figure below. Then draw the reflection image if *R* is the point of reflection.

15.

16.

17.

EXERCISES

Practice

For each of the figures at the right, name the reflection image with respect to line ℓ.

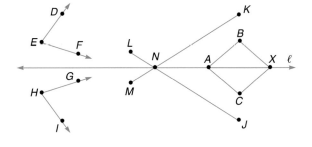

18. *M*   19. *K*

20. $\overline{KM}$ 21. $\angle DEF$

22. $\overline{EF}$ 23. $\triangle BXA$

24. *N*   25. $\overline{DE}$

For each of the following figures, determine whether ℓ is a line of symmetry. Write *yes* or *no*. Then explain your answer.

26.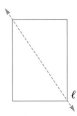

27.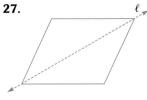

28.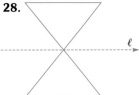

29.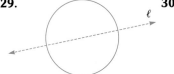

30.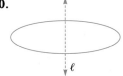

31.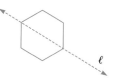

For each of the following figures, indicate if the figure has line symmetry, point symmetry, or both.

32.    **33.**   **34.**

Copy each figure below. Then draw the reflection image of each figure with respect to line _m_.

35. 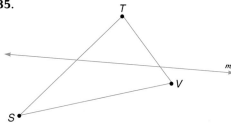   **36.**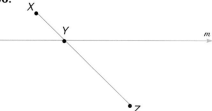

37. 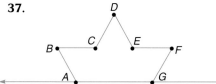   **38.**

Copy the following letters. Draw all possible lines of symmetry. If none exist, write _none_.

39.   **40.**   **41.**   **42.**

Copy each figure below. Then draw the reflection image if _R_ is the point of reflection.

43.   **44.**   **45.**

Copy each figure below. Indicate any points of symmetry. If none exist, write none.

46.

47.

48.

49.

50.

51.

52.

53.

54.

Critical Thinking

55. Copy the figure at the right. Draw the reflection image with respect to line *m* of the reflection image of pentagon *ABCDE* with respect to line *ℓ*. Label the vertices *A′*, *B′*, *C′*, *D′*, and *E′* to correspond to *A*, *B*, *C*, *D*, and *E* respectively.

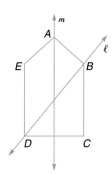

Applications

56. **Recreation**   Amos is playing pool. He needs to hit the 3 ball with the cue ball (all white) without hitting the 8 ball. How can Amos plan his shot?

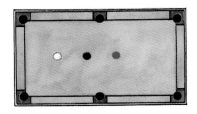

57. **Grooming**   Alison uses two mirrors to check the appearance of the hair on the back of her head. Using the measurements as indicated in the figure, how far away does the image of the back of her head appear to Alison?

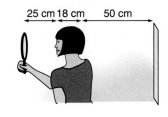

Mixed Review

Answer each of the following, if quadrilateral *ABCD* → quadrilateral *RSTU*. (Lesson 13-4)

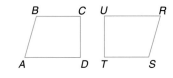

58. Name the image of ∠*B*.

59. Name the preimage of $\overline{TU}$.

Determine whether each pair of triangles is congruent. If so, state the postulate or theorem used. If there is not enough information, write *not enough information*. (Lessons 4-5, 4-6, and 5-3)

60. 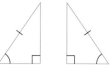   61. 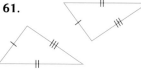   62.

63. 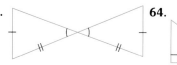   64.   65.

Wrap-Up 66. Find a picture in a magazine that has a line of symmetry and a picture that has a point of symmetry. Show the line of symmetry and the point of symmetry.

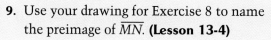

MID-CHAPTER REVIEW

1. What are the five steps in the procedure for determining a locus of points? **(Lesson 13-1)**

2. Describe the locus of points in a plane that are equidistant from two given points. **(Lesson 13-1)**

3. Describe the locus of points in space that are equidistant from two given points. **(Lesson 13-1)**

4. Select the coordinates of the locus of points that satisfy both equations. **(Lesson 13-2)**
   $y = x + 4$
   $y = 2x - 5$ **a.** $(9, 13)$ **b.** $(13, 18)$ **c.** $(13, 9)$ **d.** $(9, 5)$

5. Select the coordinates of the locus of points that satisfy both equations. **(Lesson 13-2)**
   $x = y - 6$
   $x = 30 - y$ **a.** $(18, 12)$ **b.** $(12, 18)$ **c.** $(30, 36)$ **d.** $(6, 12)$

6. Describe the locus of points in space that are 3 centimeters from a circle of radius 7 centimeters. **(Lesson 13-3)**

7. Describe the possible locus of points in space that are 5 units from a given point and 3 units from another given point. **(Lesson 13-3)**

8. Copy the figure at the right. Then draw the reflection image with respect to line ℓ. Label the reflection image so that $\triangle ABC \rightarrow \triangle LMN$. **(Lesson 13-4)**

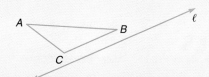

9. Use your drawing for Exercise 8 to name the preimage of $\overline{MN}$. **(Lesson 13-4)**

10. Use your drawing for Exercise 8 to name the preimage of $\angle M$. **(Lesson 13-4)**

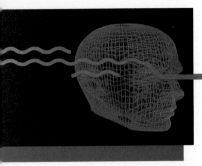

Technology

Reflections

BASIC
Geometric Supposer
Graphing calculators
▶ **LOGO**
Spreadsheets

You can draw figures and their reflections of figures using LOGO. The procedure XYAXES given below will draw the coordinate axes.

```
TO XYAXES
  SETY 119        Draws from HOME to the top of the screen.
  HOME
  SETY -119       Draws from HOME to the bottom of the screen.
  HOME
  SETX 139        Draws from HOME to the right edge of the screen.
  HOME
  SETX -139       Draws from HOME to the left edge of the screen.
  HOME
  FULLSCREEN HT   Hides the turtle.
  END
```

The procedure TRIANGLE will draw the triangle ABC. The triangle $A'B'C'$ is the reflection of $\triangle ABC$ with respect to the y-axis. $\triangle A'B'C'$ can be drawn with the procedure REFLECTION.

```
TO TRIANGLE
  SETXY 0 40
  SETXY 50 0
  SETXY 0 0
  END

TO REFLECTION
  XYAXES
  TRIANGLE
  SETXY 0 40
  SETXY -50 0
  SETXY 0 0
  END
```

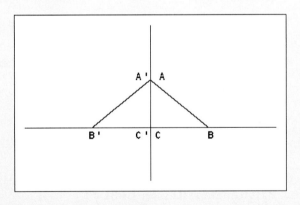

Enter all three procedures. Then type REFLECTION to draw the triangle and its reflection.

EXERCISES

1. Use LOGO to draw $\triangle FGH$ with vertices $F(0, 0)$, $G(0, -60)$, and $H(50, -60)$ and its reflection $\triangle F'G'H'$ with respect to the y-axis.

2. Use LOGO to draw quadrilateral $MNPQ$ with vertices $M(-20, -20)$, $N(-20, -65)$, $P(-55, -65)$, and $Q(-55, -20)$ and its reflection $M'N'P'Q'$ with respect to the x-axis.

13-6 Translations

Objective

After studying this lesson, you should be able to:

■ name and draw translation images of figures with respect to parallel lines.

FYI···

The first auto race was a 201-mile race from Green Bay to Madison, Wisconsin. It was won by an Oshkosh steamer.

Application

A race car speeds along a track to the finish line. The result of a movement in one direction is a transformation called a **translation.** A transformation moves all points the same distance in the same direction.

To find a translation image, perform two reflections in a row with respect to two parallel lines. For example, the translation image of the blue figure with respect to the parallel lines, *n* and *l*, is the red figure. First the blue figure is reflected onto the green figure with respect to line *n*. Then the green figure is reflected onto the red figure with respect to line *l*. Two successive reflections, such as this one, are called a **composite of reflections.**

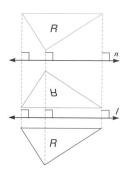

Example 1

Draw the translation image of △*ABC* with respect to the parallel lines *n* and *l*.

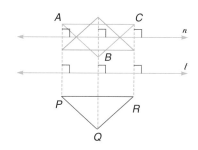

First draw the reflection image of △*ABC* with respect to line *n*. Then draw the reflection image of that figure with respect to line *l*. Thus, △*PQR* is the translation image of △*ABC*.

Since translations are composites of two reflections, all translations are isometries, and all properties preserved by reflections are preserved by translations. These properties include collinearity, angle and distance measure, and betweenness of points.

Use a translation to draw a prism with triangles as bases.

1. Cut a triangle out of a piece of thin cardboard. Label its angles 1, 2, and 3.

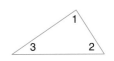

2. Place the triangle on a piece of notebook paper aligning the common side of ∠2 and ∠3 with a horizontal rule. Trace the triangle on the paper. Label the vertices so that the vertex of ∠1 is X, the vertex of ∠2 is Y, and the third vertex is Z.

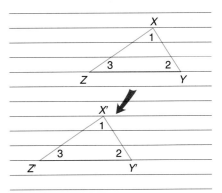

3. Now slide the cutout triangle to another place on the paper, making sure that the common side of ∠2 and ∠3 is still aligned with a horizontal rule. Trace the cutout triangle again. Label the vertices of this triangle X', Y', and Z' so that they correspond to the vertices of the first triangle you drew.

$$\Delta XYZ \rightarrow \Delta X'Y'Z'$$

4. Use a straightedge to draw $\overline{XX'}$, $\overline{YY'}$, and $\overline{ZZ'}$.

5. Now look at the figure. Remember that when drawing 3-dimensional figures, the edges that are not visible when looking at the solid from a certain viewpoint should be represented by dashed segments.

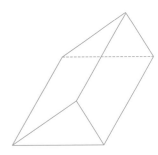

CHECKING FOR UNDERSTANDING

Communicating Mathematics

Read and study the lesson to answer these questions.

1. Emma says that the designs of wallpaper are usually examples of translations. Do you agree with her? Why or why not?

2. Explain why a translation is really the composite of two reflections over parallel lines.

3. List the properties that are preserved by translations.

Guided Practice

For each of the following, lines ℓ and m are parallel. Determine whether each red figure is a translation image of the blue figure. Write *yes* or *no*. Then explain your answer.

4.

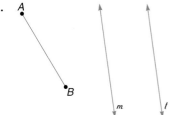

5.

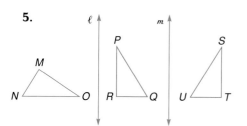

6. Name the reflection of $\overline{AB}$ with respect to line ℓ. If there is none, write *none*.

7. Name the reflection of $\triangle MNO$ with respect to line ℓ. If there is none, write *none*.

8. Name the reflection of $\overline{AB}$ with respect to line m. If there is none, write *none*.

9. Name the reflection of $\triangle MNO$ with respect to line m. If there is none, write *none*.

Copy each figure. Then find the translation image of each geometric figure with respect to the parallel lines m and ℓ.

10.

11.

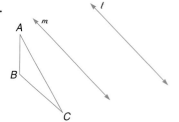

EXERCISES

Practice

In the figure below, lines m and n are parallel. For each of the following, name the reflection image with respect to the given line.

12. A, m 13. F, m 14. E, m

15. B, n 16. H, n 17. G, n

18. A, n 19. F, n 20. E, n

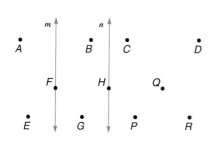

For each of the following, name the translation image with respect to line m, then line n.

21. A 22. F 23. E

LESSON 13-6   TRANSLATIONS   655

For each of the following, lines ℓ and m are parallel. Determine whether each red figure is a translation image of the blue figure. Write *yes* or *no*. Then explain your answer.

24.

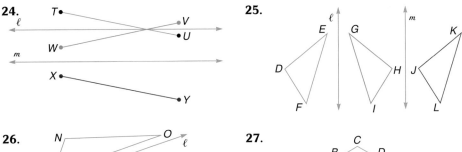

25.

26.

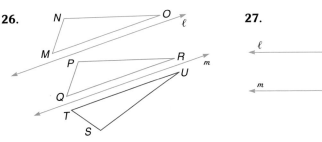

27.

28. Name the reflection of $\overline{TU}$ with respect to line ℓ. If none is drawn, write *none*.

29. Name the reflection of $\triangle DEF$ with respect to line ℓ. If none is drawn, write *none*.

30. Name the reflection of $\triangle MNO$ with respect to line ℓ. If none is drawn, write *none*.

31. Name the reflection of pentagon $ABCDE$ with respect to line ℓ. If none is drawn, write *none*.

32. Name the reflection of $\overline{WV}$ with respect to line m. If none is drawn, write *none*.

33. Name the reflection of $\triangle HGI$ with respect to line m. If none is drawn, write *none*.

34. Name the reflection of $\triangle PQR$ with respect to line m. If none is drawn, write *none*.

35. Name the reflection of pentagon $AHGFE$ with respect to line m. If none is drawn, write *none*.

Copy each figure. Then find the translation image of each geometric figure with respect to the parallel lines n and ℓ.

36.

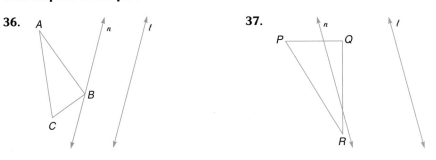

37.

38.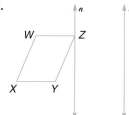

39.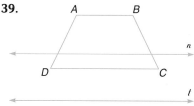

Use the figures below to name each triangle. Assume $n \parallel l$.

  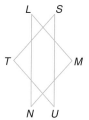

40. reflection image of $\triangle ABC$ with respect to n

41. reflection image of $\triangle XYZ$ with respect to n

42. reflection image of $\triangle PQR$ with respect to l

43. translation image of $\triangle ABC$ with respect to n and l

44. translation image of $\triangle PQR$ with respect to n and l

45. Plan a proof to show that the translation image of $\triangle ABC$ with respect to parallel lines ℓ and m preserves collinearity.

46. Plan a proof to show that the translation image of $\triangle ABC$ with respect to parallel lines ℓ and m preserves betweenness of points.

47. Plan a proof to show that the translation image of $\triangle ABC$ with respect to parallel lines ℓ and m preserves angle and distance measure.

Critical Thinking

48. Draw $\triangle ABC$ with coordinates $A(3, 3)$, $B(7, 6)$, and $C(9, 2)$ and $\triangle RST$ with coordinates $R(3, -7)$, $S(7, -4)$, and $T(9, -8)$. If $\triangle RST$ is the translation image of $\triangle ABC$ with respect to two parallel lines, give the equations of two possible lines.

Applications

49. Art Use a translation to draw a prism with pentagons as bases.

50. Art Use a translation to draw a prism with hexagons as bases.

51. Environment A cloud of smoke blows 40 miles north and then 30 miles east. Make a sketch to show the translation of the smoke particles. Indicate the shortest path that would take the particles to the same position.

52. Environment A tidal flow of water moves 25 kilometers south and then 15 kilometers southeast. Make a sketch to show the translation of the water particles. Indicate the shortest route that would take the water particles to the same position.

Mixed Review

53. Copy the letters at the right. Draw all possible lines of symmetry. If none exist, write *none*. **(Lesson 13-5)**

Find the area of each figure. (Lessons 10-4, 10-5, and 10-6)

54.

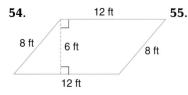
12 ft
8 ft
6 ft
12 ft

55.

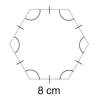
8 ft
8 cm

56.

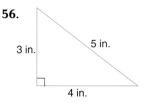
5 in.
3 in.
4 in.

57. The vertices of $\triangle PQR$ have coordinates $P(6, 4)$, $Q(8, -2)$, and $R(0, -1)$. Which angle of the triangle has the greatest measure? **(Lesson 5-5)**

58. An isosceles triangle has a vertex angle measuring 67°. What are the measures of the base angles? **(Lesson 4-7)**

Wrap-Up

59. Draw a simple figure and two parallel lines. Draw the translation image of your drawing with respect to the two lines.

～～～ DEVELOPING REASONING SKILLS ～～～

One of the curiosities of geometry is the *Möbius Strip*. To construct one, cut a long strip of paper as shown at the right. Your strip of paper has two surfaces. If it is laying flat on your desk, one is facing you and the other is facing the desk. Now, give the strip of paper a half-twist and glue or tape the ends together. Corner *C* should match to corner *B*, and corner *D* should match to corner *A*. Your Möbius Strip should look like the one at the right.

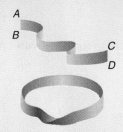
A
B
C
D

A Möbius Strip has only one surface. Prove it to yourself by drawing a straight line down the center of the strip. If your pencil returns to the same point, you have only one surface. Now cut along the pencil line. What happens?

Rotations

Objective

After studying this lesson, you should be able to:

- name and draw rotation images of figures with respect to intersecting lines.

So far you have studied two types of geometric transformations: reflections and translations. A third transformation is illustrated by the triangles at the right. The blue triangle is reflected onto the green triangle with respect to line *n*. Then the green triangle is reflected onto the red triangle with respect to line *l*.

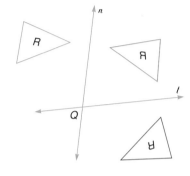

INVESTIGATION

Take a piece of notebook paper and place it over the diagram. Trace the blue triangle onto the paper. Placing your pencil point at point *Q*, rotate your piece of notebook paper about point *Q* until the tracing of the blue triangle coincides with the red triangle.

Since rotations are composites of two reflections, all properties preserved by reflections are preserved by rotations. A rotation is an isometry.

The composite of two reflections with respect to two intersecting lines is a transformation called a **rotation.** In the illustration, the red figure is the rotation image of the blue figure with respect to lines *n* and *l*. *Q*, the intersection of the two lines, is called the **center of rotation.**

Example 1

Suppose lines *n* and *l* intersect and △*ABC* is on one side of *n* and *l*. Draw the rotation image of △*ABC* with respect to *n* and then *l*.

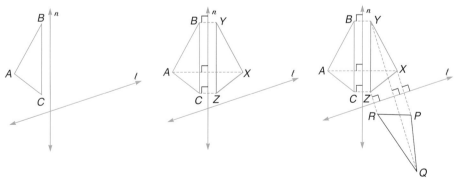

△*ABC* is on one side of *l* and *n*.

First reflect △*ABC* with respect to *n*. The image is △*XYZ*.

Then reflect △*XYZ* with respect to *l*. The image is △*PQR*.

Thus △*PQR* is the rotation image of △*ABC* with respect to *n* and *l*.

The figure at the right shows how two reflections can be used to find the rotation image of $\overline{AB}$ with respect to lines ℓ and m. The image is $\overline{PQ}$.

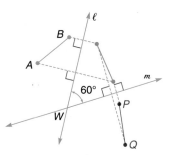

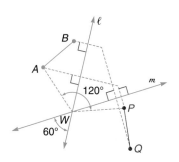

The same rotation image can be determined by using angles. Notice that lines ℓ and m form a 60° angle. It can be shown that $m\angle AWP = 2(60)$ or 120. Also, $\overline{AW} \cong \overline{WP}$. Likewise, it can be shown that if segments are drawn from B to W and from Q to W, then $m\angle BWQ = 120$ and $\overline{BW} \cong \overline{WQ}$.

The angles, $\angle AWP$ and $\angle BWQ$, are called **angles of rotation.** In both cases, the measure of the angles is 2(60) or 120.

Postulate 13-1	In a given rotation, if *A* is the preimage, *P* is the image, and *W* is the center of rotation, then the measure of the angle of rotation, $\angle AWP$, equals twice the measure of the angle formed by intersecting lines of reflection.

Example 2

The intersection of lines ℓ and m at P forms a 40° angle. Use the angles of rotation to find the rotation image of $\overline{XY}$.

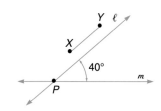

The angle of rotation has a measure of 2(40) or 80.

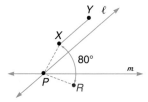

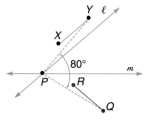

Construct $\angle XPR$ so that its measure is 80 and $\overline{XP} \cong \overline{PR}$. Then construct $\angle YPQ$ so that its measure is 80 and $\overline{YP} \cong \overline{PQ}$. The rotation image of $\overline{XY}$ is $\overline{RQ}$.

Example 3

Rex designs paddle fans for Functional Fans Inc. He designs one fan with 5 paddles. What is the measure of the angle of rotation if paddle *A* moves to the position of paddle *B*?

The fan has 5 paddles evenly spaced around the center. Since there are 360° in a complete rotation, the measure of the angle of rotation is 360 ÷ 5 or 72.

CHECKING FOR UNDERSTANDING

Communicating Mathematics

Read and study the lesson to answer these questions.

1. When is the composite of two line reflections a translation?
2. When is the composite of two line reflections a rotation?
3. Describe the relationship between the measure of the angle between the intersecting lines of reflection and the measure of the angle of rotation.
4. Describe two techniques you can use to locate a rotation image with respect to two intersecting lines.

Guided Practice

Use the figure to answer each problem.

5. Find the reflection image of quadrilateral *ABCD* with respect to line *m*.
6. Find the reflection image of quadrilateral *CDEF* with respect to line *ℓ*.
7. Find the rotation image of quadrilateral *ABCD* with respect to lines *m* and *ℓ*.
8. Find the reflection image of quadrilateral *JKHG* with respect to line *ℓ*.
9. Find the rotation image of quadrilateral *JKHG* with respect to lines *ℓ* and *m*.

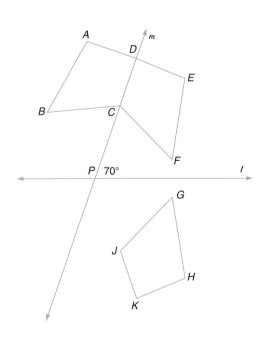

10. Find the measure of ∠*APH*.
11. Find the measure of ∠*BPG*.
12. Find the measure of ∠*CPJ*.
13. Find the measure of ∠*DPK*.
14. Find the reflection image of *C* with respect to line *m*.
15. Find the rotation image of $\overline{BD}$ with respect to lines *m* and *ℓ*.

Copy each figure. Then use the angles of rotation to find the rotation image of each geometric figure with respect to lines _n_ and _l_.

16.

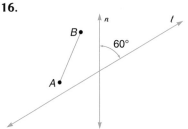

17.

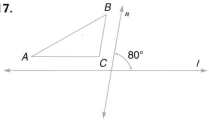

EXERCISES

Practice

For each of the following, determine whether the indicated composition of reflections is a rotation. Explain your answer.

18.

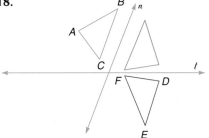

19.

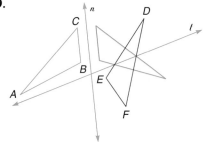

Two lines intersect to form an angle with the following measure. Find the measure of each angle of rotation.

20. 30°   **21.** 45°   **22.** 60°   **23.** 37°

Determine whether each of the following is preserved by a rotation. Write _yes_ or _no._

24. collinearity   **25.** betweenness of points

26. angle measure   **27.** distance measure

28. Draw a segment and two intersecting lines. Find the rotation image of the segment with respect to the two intersecting lines.

29. Draw a triangle and two intersecting lines. Find the rotation image of the triangle with respect to the two intersecting lines.

Copy each figure. Then use the angles of rotation to find the rotation image of each geometric figure with respect to lines ℓ and t.

30.
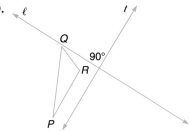

31.
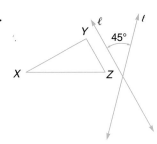

32.
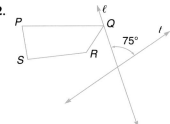

33.
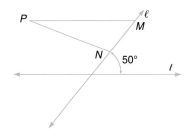

In the figure at the right, $\ell \parallel m$. Name the type of transformation for each mapping.

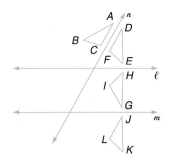

34. $\triangle GHI \rightarrow \triangle DEF$

35. $\triangle ABC \rightarrow \triangle GHI$

36. $\triangle DEF \rightarrow \triangle JKL$

37. Does a rotation preserve or reverse orientation? Explain.

Hexagon *ABCDEF* is a regular hexagon with diagonals as drawn. Point *C* is the rotation image of Point *A*.

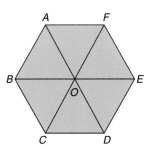

38. List three sets of possible intersecting lines of reflection for this rotation.

39. What is the measure of the angle of each set of intersecting lines of reflection? Explain your answer.

Critical Thinking   **40.** In $\triangle ABC$, $m\angle BAC = 40$. Triangle $AB'C$ is the reflection image of $\triangle ABC$ and $\triangle AB'C'$ is the reflection image of $\triangle AB'C$. How many such reflections would be necessary to map $\triangle ABC$ onto itself?

Applications

41. Computers Computer design specialist Wendell Schwartz forms various solids for large machine production by rotating a 2-dimensional geometric figure about a line.

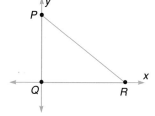

a. Describe the solid that would be formed by a rotation of △*PQR* about the *y*-axis.

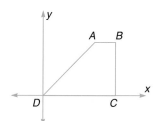

b. Describe the solid that Mr. Schwartz would form by a rotation of trapezoid *ABCD* about the *y*-axis.

42. Recreation What is the measure of the angle of rotation if seat number 1 of a 10-seat Ferris wheel moves to the position of seat number 3?

43. Recreation What is the measure of the angle of rotation if seat number 1 of a 10-seat Ferris wheel moves to the position of seat number 4?

Mixed Review

44. Copy the figure at the right. Then find the translation image of the geometric figure with respect to parallel lines ℓ and *m*. **(Lesson 13-6)**

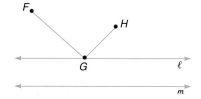

Find the value of *x*. (Lessons 9-2, 9-4, 9-6)

45.

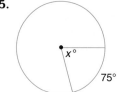

46.

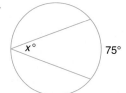

47.

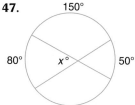

48. Can two angles be both complementary and supplementary? Explain. **(Lesson 1-8)**

Wrap-Up

49. Explain the difference between successive reflections with respect to two intersecting lines versus two parallel lines.

13-8 Dilations

Objectives

After studying this lesson, you should be able to:

- use scale factors to determine if a dilation is an enlargement, a reduction, or a congruence transformation,
- find the center and scale factor for a given dilation, and
- find the dilation image for a given center and scale factor.

We have already studied how translations, reflections, and rotations of geometric shapes produce figures congruent to each other. Such transformations are called isometries.

Application

A geometric figure can also be altered in size. For example the magnifying glass at the right enlarges the figure. Enlarging or reducing a figure will not change its shape. This type of transformation is called a **dilation** or a **similarity transformation.**

FYI...

In 1250, Roger Bacon of Oxford, England, invented the magnifying glass.

In the figure, $\triangle XYZ$ is the dilation image of $\triangle PQR$. The measure of the distance from C to a point on $\triangle XYZ$ is twice the measure of the distance from C to a point on $\triangle PQR$. For example, the following equations hold.

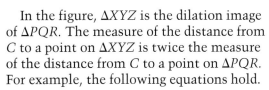

$$CX = 2(CP)$$
$$CY = 2(CQ)$$
$$CZ = 2(CR)$$

In this transformation, $\triangle PQR$ with **center** C and a **scale factor** of 2 is enlarged to $\triangle XYZ$.

This figure shows a dilation where the preimage $\overline{AB}$ is reduced to $\overline{ED}$ by a scale factor of $\frac{1}{3}$. Thus, $CE = \frac{1}{3}(CA)$. Therefore, $\frac{CE}{CA} = \frac{1}{3}$ and $\frac{CD}{CB} = \frac{1}{3}$. By proving $\triangle CAB \sim \triangle CED$, it can be shown that $ED = \frac{1}{3}(AB)$.

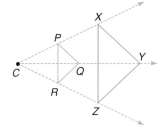
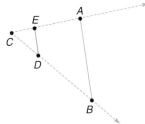

Theorem 13-1

If a dilation with center C and a scale factor k maps A onto E and B onto D, then $ED = k(AB)$. *You will be asked to prove this theorem in Exercise 57.*

Notice that when the scale factor is 2, the figure is enlarged. When the scale factor is $\frac{1}{3}$, the figure is reduced. In general, if k is the scale factor for a dilation with center C, then the following is true.

If $k > 0$, P', the image of point P, lies on $\overrightarrow{CP}$, and $CP' = k \cdot CP$.

If $k < 0$, P', the image of point P, lies on the ray opposite $\overrightarrow{CP}$, and $CP' = |k| \cdot CP$. *The center of a dilation is always its own image.*

If k = 1, the dilation is the identity transformation. That is, each point is mapped to itself.

If $|k| > 1$, the dilation is an enlargement. *The dilation is not an isometry.*

If $0 < |k| < 1$, the dilation is a reduction. *The dilation is not an isometry.*

If $|k| = 1$, the dilation is a congruence transformation. *The dilation is an isometry.*

Example 1

Given center C and a scale factor of $\frac{3}{4}$, find the dilation image of $\triangle PQR$.

Since the absolute value of the scale factor is less than 1, the dilation is a reduction.

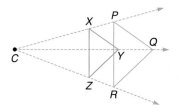

Draw $\overrightarrow{CP}$, $\overrightarrow{CQ}$, and $\overrightarrow{CR}$. Find X, Y, and Z so that $CX = \frac{3}{4}(CP)$, $CZ = \frac{3}{4}(CR)$, and $CY = \frac{3}{4}(CQ)$. $\triangle XYZ$ is the dilation image of $\triangle PQR$.

The following examples illustrate some of the basic properties of dilations.

Example 2

Given center C, $\angle EFG$, and its dilation image $\angle QRS$, examine EF and QR. Then examine $m\angle EFG$ and $m\angle QRS$. Determine whether the measures are greater, lesser, or equal.

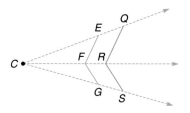

$\overline{EF}$ is enlarged to $\overline{QR}$. So, $EF < QR$.

$m\angle EFG = m\angle QRS$

The dilation preserves angle measure, but not the measure of a segment unless the scale factor is 1.

Example 3

For each figure, a dilation with center C produced the figure in red. What is the scale factor for each transformation?

Count the grids to compare relative sizes.

a.

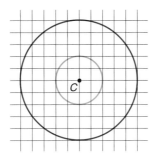

enlargement
The scale factor is $\frac{5}{2}$ or $-\frac{5}{2}$.

b.

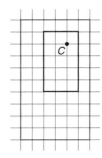

reduction
The scale factor is $\frac{3}{6}$ or $\frac{1}{2}$.

Since the image in Example 3a would be the same if $k = \frac{5}{2}$ or $-\frac{5}{2}$, k could be either one.

Example 4

Photography

A 4-inch by 5-inch photograph is enlarged to make an 8-inch by 10-inch photograph. Find the scale factor of this enlargement.

The 4-inch side of the original photograph corresponds to the 8-inch side of the enlargement. The scale factor is $\frac{8}{4}$ or 2. How do you know that the scale factor is 2 and not $\frac{1}{2}$?

CHECKING FOR UNDERSTANDING

Communicating Mathematics

Read and study the lesson to answer these questions.

1. How is a dilation different from the other transformations?
2. How can you determine if a dilation is a reduction or an enlargement?
3. Is it possible to have a scale factor of 0? Explain your answer.

Guided Practice

In the figure, $\triangle XYZ$ is the dilation image of $\triangle ABC$ with a scale factor of 8. Complete.

4. If $QB = 6$, then $QY = \underline{\ ?\ }$.
5. $\triangle ABC$ is $\underline{\ ?\ }$ to $\triangle XYZ$.
6. $BC \underline{\ ?\ } YZ$.
7. If $YZ = 32$, then $BC = \underline{\ ?\ }$.
8. If $m\angle BCA = 62$, then $m\angle YZX = \underline{\ ?\ }$.

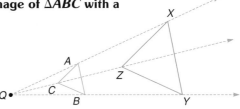

A dilation with center C and a scale factor k maps A onto D and B onto E. Find $|k|$ for each dilation. Then determine whether each dilation is an enlargement, a reduction, or a congruence transformation.

9. $CD = 10, CA = 5$ **10.** $CD = 6, CA = 4$ **11.** $AB = 16, DE = 4$

On a coordinate plane, graph the segment whose endpoints are given. Using (0, 0) as the center of dilation and for a scale factor of 2, draw the dilation image. Then repeat this using a scale factor of $-\frac{1}{2}$.

12. $(0, 2), (4, 0)$ **13.** $(3, -3), (-2, -2)$ **14.** $(-2, -1), (-2, -2)$

EXERCISES

Practice

For each of the following scale factors, determine whether the dilation is an enlargement, reduction, or a congruence transformation.

15. $4\frac{2}{5}$ **16.** $\frac{3}{8}$ **17.** $\frac{1}{6}$ **18.** $-\frac{3}{2}$

19. -0.61 **20.** -7 **21.** 1 **22.** 2.5

Find the measure of the image of $\overline{AB}$ with respect to a dilation with the given scale factor.

23. $AB = 5, k = -6$ **24.** $AB = \frac{2}{3}, k = \frac{1}{2}$ **25.** $AB = 16, k = 1\frac{1}{2}$

26. $AB = 3.1, k = -5$ **27.** $AB = 12, k = \frac{1}{4}$ **28.** $AB = 3\frac{1}{3}, k = -9$

For each scale factor, find the image of A with respect to a dilation with center P.

29. 1 **30.** 2 **31.** $\frac{2}{3}$ **32.** $1\frac{1}{3}$

33. $\frac{1}{2}$ **34.** $1\frac{1}{6}$ **35.** $1\frac{5}{6}$ **36.** $1\frac{1}{2}$

A dilation with center C and a scale factor k maps A onto D and B onto E. Find $|k|$ for each dilation. Then determine whether each dilation is an enlargement, a reduction, or a congruence transformation.

37. $CE = 18, CB = 9$ **38.** $CA = 2, CD = 10$ **39.** $AB = 3, DE = 1$

40. $AB = 3, DE = 4$ **41.** $CB = 28, CE = 7$ **42.** $DE = 12, AB = 4$

For each figure, a dilation with center C produced the figure in red. What is the scale factor for each transformation?

43.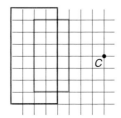

44.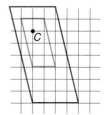

Draw and label a figure like the one shown at the right. Then draw the dilation image of △ABC for the given scale factor and center.

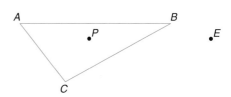

45. 3, center E 46. $\frac{1}{3}$, center E

47. -2, center P 48. $\frac{1}{2}$, center P

Graph each of the following ordered pairs. Then connect the points in order. Using (0, 0) as the center of dilation and a scale factor of 2, draw the dilation image. Then repeat this using a scale factor of $\frac{1}{2}$.

49. (3, 4), (6, 10), (-3, 5)   50. (6, 5), (4, 5), (3, 7)

51. (-1, 4), (0, 1), (2, 3)   52. (1, -2), (4, -3), (6, -1)

53. (1, 2), (3, 3), (3, 5), (1, 4)   54. (4, 2), (-4, 6), (-6, -8), (6, -10)

55. A dilation on a rectangle has a scale factor of 4.
   a. What is the effect of the dilation on the perimeter of the rectangle?
   b. What is the effect of the dilation on the area of the rectangle?

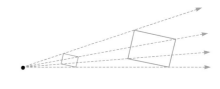

56. A dilation on a cube has a scale factor of 4.

   a. What is the effect of the dilation on the surface area of the cube?
   b. What is the effect of the dilation on the volume of the cube?

57. Write a paragraph proof for Theorem 13-1.

Critical Thinking

58. Draw △ABC with coordinates $A(3, 4)$, $B(4, 3)$, and $C(2, 1)$ and △RST with coordinates $R(7.5, 2.5)$, $S(10, 0)$, and $T(5, -5)$. If △RST is the dilation image of △ABC, find the coordinates of the center and the scale factor.

Applications

59. Art The picture at the right is Mauritz C. Escher's *Fish and Scales*. Explain how this piece of art relates to dilations.

©1959 M.C. Escher/Cordon Art-Baarn, Holland

60. Photography An 8-inch by 10-inch photograph is being reduced by the scale factor of $\frac{3}{4}$. What are the dimensions of the new photograph?

61. Publishing Tara is doing a layout for a new book. She has a drawing that is 12 centimeters by 10 centimeters, but the maximum space available for the drawing is 8 centimeters by 6 centimeters. If she wants the drawing to be as large as possible in the book, what scale factor should she use to reduce the original drawing?

Mixed Review

62. Copy the figure at the right. Then use the angles of rotation to find the rotation image of each geometric figure with respect to n and l. **(Lesson 13-7)**

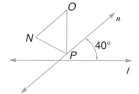

Find the value of x. (Lessons 8-1, 8-2, and 8-5)

63.
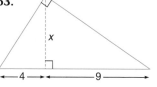

64.
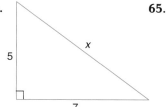

65.

CONSTRUCTION

66. Construct a rectangle with sides of 2 inches and 4 inches. **(Lesson 6-4)**

67. Describe an obtuse triangle. Can a triangle with a 90° angle be an obtuse triangle? **(Lesson 4-1)**

68. Two supplementary angles have the measures $20 + 5x°$ and $x + 10°$. Find the value of x. **(Lesson 1-8)**

Wrap-Up

69. Explain the difference between a dilation with a scale factor greater than 1 and a dilation with a scale factor between 0 and 1.

13-9 Problem-Solving Strategy: Make a Table

Objective

After studying this lesson, you should be able to:

- solve problems by making tables.

Application

Priscilla Adams manages The Clothes Closet. She needs to make decisions about what to order, what merchandise to display in front of the store, and who to schedule for work. Ms. Adams uses tables to help her solve some of these problems.

Example 1

APPLICATION

Management

Ms. Adams has five sales associates to cover the daytime hours. Rita and Tony are full-time employees and must work 5 days each week and have 2 consecutive days off. Beth, Cam, and Joshua each work 3 days a week, but Cam never works on Mondays. Ms. Adams knows that she needs 3 associates on Fridays, 4 associates on Saturdays and Sundays, and 2 associates the other days of the week. Make a work schedule for the 5 associates.

Ms. Adams starts by making a table with the days of the week across the top and the names of the associates down the right. Her job is to assign days for the associates to work and still meet all the criteria listed above. One possible solution is shown below.

Name	Sun.	Mon.	Tues.	Wed.	Thurs.	Fri.	Sat.
Rita			X	X	X	X	X
Tony	X	X			X	X	X
Beth	X			X			X
Cam	X		X			X	
Joshua	X	X					X

Ms. Adams checks her final schedule to make sure that all the criteria are met. Has Ms. Adams met all the criteria?

Sometimes tables are used to relay information to others.

Example 2

Make a table to summarize the properties preserved by the four transformations studied in this chapter.

Transformation	Collinearity	Betweenness	Angle Measure	Distance Measure	Orientation
Reflections	X	X	X	X	
Translations	X	X	X	X	X
Rotations	X	X	X	X	X
Dilations	X	X	X		X

CHECKING FOR UNDERSTANDING

Communicating Mathematics

Read and study the lesson to answer these questions.

1. Is making a table a good strategy for solving Example 1? Explain your answer.

2. In this lesson, the purpose for the table in Example 1 is different than the purpose for the table in Example 2. Explain the difference.

Guided Practice

Make a table to solve each problem.

3. Suppose that Rita in Example 1 needs a certain weekend off so she can go to her nephew's wedding. Make a possible work schedule for that week.

4. Tyron is studying for his geometry final exam. Help him to summarize the values for the sine, cosine, and tangent of a 30° angle, a 45° angle, and a 60° angle.

5. To determine a grade point average, 4 points are given for an A, 3 points for a B, 2 for a C, 1 for a D, and 0 for an F. The points for each course are multiplied by the credit value of the course to determine the total points. If each of Greg's five courses is worth one credit and he has a total of 13 points, what combinations of grades could he have?

EXERCISES

Practice

Solve. Use any strategy.

6. Nine dots are arranged as shown at the right. How many isosceles triangles can be drawn by using any three dots as the vertices of the triangle?

7. Find three positive integers x, y, and z such that $\frac{1}{x} + \frac{1}{y} + \frac{1}{z} = 1$.

8. To determine total bases in softball, 4 bases are given for a home run, 3 for a triple, 2 for a double, and 1 for a single. If Cathy has 15 total bases on 8 hits, what combinations of hits could she have?

9. At a recent student council meeting, 8 members were present. If each member shook hands with everyone else exactly once, how many handshakes occurred?

10. Bell, Thornton, Arnold, and Terri formed a band. The band has a lead guitar player, a rhythm guitar player, a keyboard player, and a drummer. Bell does not play a guitar. Arnold and the keyboard player are neighbors. Terri, the drummer, and the keyboard player like rap music. Thornton wants to learn to play the keyboard. Arnold, the drummer, and the lead guitar player are seniors. What instrument does each person play in the band?

Strategies

Look for a pattern.
Solve a simpler problem.
Act it out.
Guess and check.
Draw a diagram.
Make a chart.
Work backward.

11. Summarize information about regular polygons with 3 to 10 sides. Include information about the name of the figure, the number of sides, the number of diagonals, the measure of each interior angle, and the measure of each exterior angle.

12. The length of a rectangle is 8 feet more than the width. If the length were doubled and the width decreased by 12 feet, the area would be unchanged. Find the original dimensions.

13. A palindrome is a word, phrase, or number that reads the same backward as it does forward. For example, the word *pop* is a palindrome. How many 4-digit numbers are palindromes?

14. If a round cake is cut using five straight cuts, what is the maximum number of pieces that can be formed?

15. Find the sum of the first 50 even numbers.

16. In the figure, find the value of *x*.

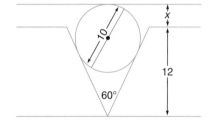

COOPERATIVE LEARNING PROJECT

Work in groups. Each person in the group must understand the solution and be able to explain it to any person in the class.

Each fall the winners of the American and National League Pennants meet in the World Series. They play until a team wins four games. The Series may last 4, 5, 6, or 7 games, and the team that wins the Series *always wins the final game.* Suppose the Boston Red Sox and the Chicago Cubs are playing. Use C to represent a Cubs' win and R to represent a Red Sox win, and arrange the letters to find combinations of wins in the Series.

- How many different combinations of wins are there if the Cubs win in 6 games?
- Make a table to show how many different ways the Cubs could win the Series in 4, 5, 6, and 7 games.
- How many different ways are there of either team winning the World Series?

VOCABULARY

Upon completing this chapter, you should be
familiar with the following terms:

angle of rotation	**660**	**646**	point of symmetry
center of dilation	**665**	**640**	preimage
center of rotation	**659**	**644**	reflection
composite of reflections	**653**	**659**	rotation
congruence transformation	**641**	**665**	scale factor
dilation	**665**	**641**	similarity transformation
image	**640**	**629**	system of equations
isometry	**641**	**640**	transformation
line of symmetry	**646**	**653**	translation
locus	**622**		

SKILLS AND CONCEPTS

OBJECTIVES AND EXAMPLES	REVIEW EXERCISES

Upon completing this chapter, you should
be able to:

Use these exercises to review and prepare
for the chapter test.

■ locate, draw, and describe a locus in a
plane or in space. **(Lesson 13-1)**

**Describe the locus of points for each
exercise.**

The locus of all points in a plane that are
2 feet from a given point is a circle with a
radius of 2 feet. The locus of all points in
space that are 2 feet from a given point is
a sphere with a radius of 2 feet.

1. all points in space that are equidistant
from two given points

2. all points in a plane that are less than 6
centimeters from a given point

■ find the locus of points that solve a
system of equations by graphing, by
substitution, or by elimination.
(Lesson 13-2)

**Find the locus of points that satisfy each
pair of equations.**

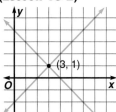

$x + y = 4$
$x - y = 2$

The locus of points
that satisfy both
equations is (3, 1).

3. $y = x - 2$
   $2x + y = 13$

4. $y = 2x - 1$
   $x + y = 7$

5. $3x - 4y = -1$
   $-2x + y = -1$

6. $3x + y = 5$
   $2x + 3y = 8$

OBJECTIVES AND EXAMPLES	REVIEW EXERCISES

■ solve locus problems that satisfy more than one condition. **(Lesson 13-3)**

The locus of points in a plane 2 inches from a line and 3 inches from a point on the line is the four points where the parallel lines of one locus intersect the circle of the other.

Describe all possible ways the given figures can intersect.

7. a circle and a line

8. a sphere and a line

9. two concentric circles and a plane

10. two parallel planes and a circle

■ name the image and preimage of a mapping. **(Lesson 13-4)**

$\triangle RST \rightarrow \triangle R'S'T'$

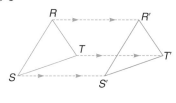

preimage　　　*image*

Suppose $\triangle ABE \rightarrow \triangle CBD$. Name the preimage for each of the following.

11. D

12. $\angle CBD$

13. B

14. $\triangle CBD$

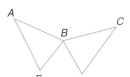

■ draw reflection images, lines of symmetry, and point of symmetry. **(Lesson 13-5)**

M is the reflection of X with respect to line ℓ.

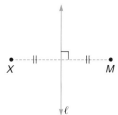

Copy each figure. Then draw the reflection image with respect to line ℓ.

15.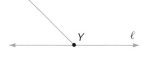

16.

■ name and draw translation images of figures with respect to parallel lines. **(Lesson 13-6)**

$\triangle XYZ$ is the translation image of $\triangle ABC$ with respect to parallel lines ℓ and m.

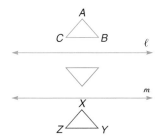

Copy each figure. Then find the translation image of each geometric figure with respect to the parallel lines n and ℓ.

17.

18.

- name and draw rotation images of figures with respect to intersecting lines. **(Lesson 13-7)**

$\triangle RST$ is the rotation image of $\triangle DEF$ with respect to lines r and ℓ.

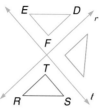

Use the figure to complete each exercise.

19. Find the measure of the angle of rotation.

20. Copy the figure at the right. Then draw the rotation image with respect to lines ℓ and m.

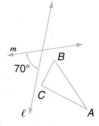

- find the dilation image for a given center and scale factor. **(Lesson 13-8)**

$\triangle XYZ$ is the dilation image of $\triangle PQR$ with center A and scale factor $\frac{1}{2}$. Since the scale factor is less than 1, the dilation is a reduction.

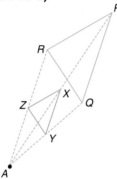

Use the figure to complete each exercise.

21. Copy the figure at the right. Then draw the dilation image with a scale factor of 3 and center at P.

22. Is the image for Exercise 21 an enlargement, a reduction, or a congruence transformation? Explain your answer.

APPLICATIONS AND CONNECTIONS

23. **Recreation**   Lucy is playing billiards. How can she hit the 5 ball with the cue ball (all white) if the cue ball must first hit a side of the billiard table? Draw a diagram. **(Lesson 13-5)**

24. **Pet Care**   Earl has two dogs. Part of each day, he ties the dogs in the back yard. Spot's 5-foot leash is tied to a long horizontal pole 4 feet above the ground so that it can slide along the pole. Bruno is tied to a stake with a 15-foot leash that can pivot around the stake. Explain how Earl should place the stake so that Spot and Bruno cannot tangle their leashes or get into a fight. **(Lesson 13-3)**

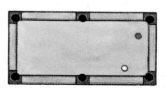

25. Make a table to solve the problem. Stephanie is buying a piece of poster board for 45¢. She gives the clerk a one-dollar bill. If the clerk gives her no pennies or half-dollars, what are the possible ways the clerk can give Stephanie her change? **(Lesson 13-9)**

Describe the locus of points for each exercise.

1. all the points in a plane that are 5 inches from a given line ℓ
2. all the points in a plane that are 4 centimeters from a given point A
3. all the points in space that are 3 meters from a given point C
4. all the points in space that are 10 feet from a given line n

Describe all the possible ways the figures can intersect.

5. two concentric spheres and a line   6. two parallel planes and a sphere

Determine the locus of points that satisfy each pair of equations.

7. $y = 4x$
   $x + y = 5$

8. $3x - 2y = 10$
   $x + y = 0$

9. $x - 4y = 7$
   $2x + y = -4$

Determine if each of the following is an isometry. Write *yes* or *no*.

10. reflection   11. translation   12. rotation   13. dilation

Describe each of the following as a reflection, a rotation, a translation, or a dilation.

14.   15. 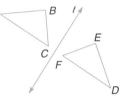   16.

17. Copy the figure at the right. Then draw all the possible lines of symmetry.

Copy the figure at the right. Then draw the dilation image with center *C* and the given scale factor. State whether the image is an enlargement, a reduction, or a congruence transformation.

18. $k = 3$ 19. $k = \frac{1}{3}$ 20. $k = 1$

Bonus

$A(2, -1)$, $B(4, 0)$, and $C(3, 4)$ are the coordinates of the vertices of a triangle. If this triangle is reflected over the *x*-axis and then rotated 90° in a clockwise direction with the center of rotation at the origin, what are the coordinates of the vertices of the resulting image?

Algebra Review

OBJECTIVES AND EXAMPLES

REVIEW EXERCISES

■ Solve rational equations.

$$\frac{4}{x} - \frac{1}{x-5} = \frac{1}{2x}$$
$$2x(x-5)\left(\frac{4}{x} - \frac{1}{x-5}\right) = 2x(x-5)\left(\frac{1}{2x}\right)$$
$$8(x-5) - 2x = x - 5$$
$$6x - 40 = x - 5$$
$$5x = 35$$
$$x = 7$$

Solve each equation. Check the solution.

1. $\frac{4x}{3} + \frac{7}{2} = \frac{7x}{12}$

2. $\frac{1}{h+1} + \frac{2}{3} = \frac{2h+5}{h-1}$

3. $\frac{3a+2}{a^2+7a+6} = \frac{1}{a+6} + \frac{4}{a+1}$

4. $\frac{3m-2}{2m^2-5m-3} - \frac{2}{2m+1} = \frac{4}{m-3}$

■ Solve systems of equations by the elimination method.

Use elimination to solve the system of equations $3r - 4s = 7$ and $2r + s = 1$.

$$\begin{array}{ll} 3r - 4s = 7 & \quad 3r - 4s = 7 \\ 2r + s = 1 \quad \times 4 & \quad (+)\, 8r + 4s = 4 \\ & \quad 11r = 11 \\ 2r + s = 1 & \quad r = 1 \\ 2(1) + s = 1 & \\ s = -1 & \end{array}$$

The solution is $(1, -1)$.

Use elimination to solve each system of equations.

5. $6x + 5y = 35$
$6x - 9y = 21$

6. $3x + 6 = 7y$
$x + 2y = 11$

7. $6a + 7b = 5$
$2a - 3b = 7$

8. $5m + 2n = -8$
$4m + 3n = 2$

■ Write a quadratic equation of the form $ax^2 + bx + c = 0$ given its roots.

Find a quadratic equation whose roots are 2 and $\frac{5}{3}$.

Sum of roots   **Product of roots**

$-\frac{b}{a} = 2 + \frac{5}{3}$ or $\frac{11}{3}$ $\frac{c}{a} = 2\left(\frac{5}{3}\right)$ or $\frac{10}{3}$

Try $a = 3$, $b = -11$, and $c = 10$. Then, the equation is $3x^2 - 11x + 10 = 0$.

Check to ensure that the roots of the equation are 2 and $\frac{5}{3}$.

Find a quadratic equation of the form $ax^2 + bx + c = 0$ having the given roots.

9. $1, -8$

10. $\frac{3}{2}, -4$

11. $-\frac{2}{3}, -\frac{3}{2}$

12. $3 + \sqrt{5}, 3 - \sqrt{5}$

■ Solve radical equations.

$$\sqrt{5 - 4x} = 13$$
$$5 - 4x = 13^2$$
$$-4x = 164$$
$$x = -41$$

Solve and check each equation. Check the solution.

13. $\sqrt{3x} = 6$

14. $\sqrt{7x - 1} = 5$

15. $\sqrt{\frac{4a}{3}} - 2 = 0$

16. $\sqrt{x + 4} = x - 8$

OBJECTIVES AND EXAMPLES

- Solve systems of inequalities by graphing.

  Solve the system of inequalities $x > -3$ and $y < x + 2$ by graphing.

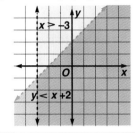

- Calculate the mean, median, and mode of a set of data.

  Find the mean, median, and mode of 18, 18, 19, 21, 24.

  mean: $\dfrac{18 + 18 + 19 + 21 + 24}{5} = \dfrac{100}{5}$ or 20

  median: The middle number is 19.

  mode: The number 18 occurs most often.

- Find the odds of a simple event.

  A soccer team consists of 8 seniors, 7 juniors, 3 sophomores, and 2 freshmen. The odds *(successes: failures)* that a player chosen at random is a senior is 8:12 or 2:3.

REVIEW EXERCISES

Solve each system of inequalities by graphing.

17. $y \le -3$
    $y > -x + 2$

18. $y > -x - 1$
    $y \le 2x + 1$

19. $2r + s < 9$
    $r + 11s < -6$

20. $|x + 2| \ge y$

Find the mean, median, and mode for each set of data.

21. 4, 5, 6, 8, 12

22. 9, 9, 9, 9, 8

23. 0, 2, 2, 2, 3, 3, 4, 5, 7, 8, 8

24. 9, 2, 17, 1, 9, 5, 12, 17

25. What are the odds that a student picked at random from the 30 students in Mr. Terry's English class will *not* have to recite one of the eight selections of poetry in class?

Applications and Connections

26. **Work** Kiko can clean the garage in 9 hours. Marcus can do the same job in 6 hours. If they work together, how many hours will it take them to clean the garage? Use the formula *rate of work* × *time* = *work done.*

27. **Law Enforcement** Lina told the police her speed was 55 mph when she applied the brakes and skidded. The skid marks were 240 feet long. Should Lina's car have skidded that far at 55 mph? Use the formula $speed = \sqrt{15 \times distance}$.

28. **Geometry** A rectangle is formed by increasing the length of one side of a square by 2 centimeters and by increasing an adjacent side by 3 centimeters. The area of the rectangle is twice the area of the square. What were the dimensions of the original square?

29. **Entertainment** The senior prom dance committee at Perry High School decided to award door prizes at the dance. They want the odds of winning a prize to be 1:5. If they plan for 180 tickets to be sold for the dance, how many door prizes do they need?

Appendix: Fractals

Objectives

After studying this appendix, you should be able to:

- recognize the structure present in the outcomes of random processes such as the chaos game,
- describe the fractal characteristic known as self-similarity, and
- recognize the presence of fractal geometry in the natural world.

Currently, one of the most exciting areas of modern mathematical exploration includes fractals, chaos, and dynamical systems. This area of study was virtually unknown until 1980 when Benoit B. Mandelbrot of the Physics Department at IBM's T. J. Watson Research Center in Yorktown Heights, New York, made a profound mathematical breakthrough. Mandelbrot discovered an incredibly complex geometric structure that is known as the Mandelbrot set. This recent event has brought geometry back to the forefront of mathematics.

Mandelbrot brought the concepts of chaos and geometry into a marriage that until 1980 was believed to be void of any connection. The Mandelbrot set is described by many in the field of science as one of the most complex and possibly the most beautiful objects ever seen in mathematics. Without the power of visualization afforded by our modern computers, many of the current discoveries being made in science and mathematics would be virtually impossible.

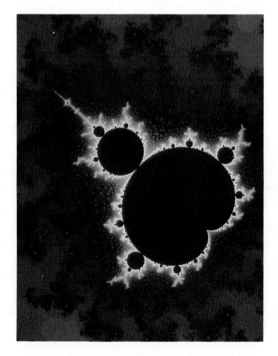

Mandelbrot Set

For centuries, people believed that the form of most of our natural world was strictly random, that the things we encountered of a structural nature such as trees or ferns, existed purely at random and were void of any sort of order.

Consider the following random process that surprisingly has incredible structure. Imagine that you have an equilateral triangle and a die. The vertices of the triangle are labeled *T, L,* and *R,* for top, left, and right. You start the process by randomly selecting a point on the triangle. Next, you roll the die and move halfway from your starting point to either of the three vertices according to the following rule.

> For 1 or 2, move halfway to *T.*
> For 3 or 4, move halfway to *L.*
> For 5 or 6, move halfway to *R.*

Mark the point that results. This is the midpoint of the segment whose endpoints are a vertex and your starting point as endpoints. Then repeat the process by rolling the die again and marking the midpoint of the segment from the next vertex to the last midpoint you marked. By repeating this process indefinitely, what pattern do you think emerges?

The results of simulating this random process on a personal computer are shown below.

Infinite Number of Rolls

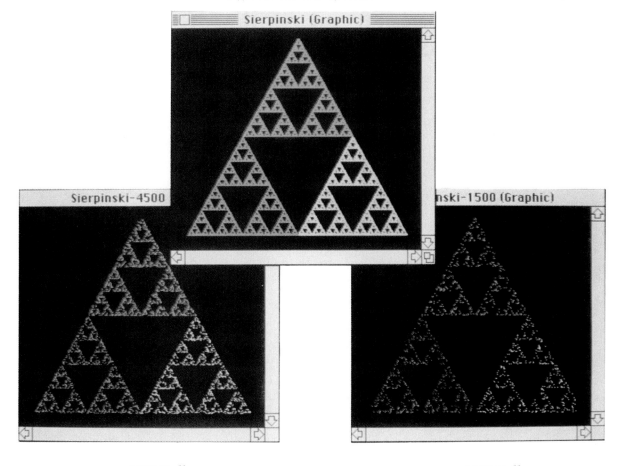

4500 Rolls   *1500 Rolls*

The Sierpinski Triangle, shown below, displays a common characteristic of all deterministic fractals. This characteristic is known as **self-similarity.** If any parts of a fractal image are replicas of the entire image, the image is self-similar. An image is *strictly self-similar* if the image can be broken down into parts that are exact copies of the entire image. That is, an arbitrary piece contains an exact copy of the entire image.

The images of the Sierpinski Triangle shown below reveal that it is strictly self-similar.

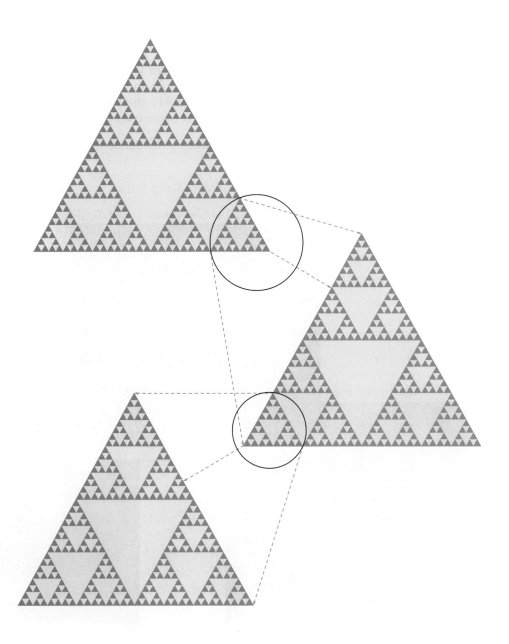

Sierpinski Triangle

A warning to the reader of this Appendix is eloquently stated by Michael Barnsley in the introduction to his book *Fractals Everywhere*.

> Fractal geometry will make you see everything differently. There is danger in reading further. You risk the loss of your childhood vision of clouds, forests, galaxies, leaves, feathers, flowers, rocks, mountains, torrents of water, carpets, bricks, and much else besides. Never again will your interpretation of these things be quite the same.

The self-similarity in fractal geometry that Barnsley refers to can be modeled mathematically on a desktop computer and at the same time seen in the structure of growing things. The figures below provide a model of a fractal that is found in nature.

Bracken Fern, Sierra Nevada, California

Notice the similarity of a computer-generated fractal bush and its natural counterpart.

Computer Generated Fractal Bush

Smoketree, Anza Borrego Desert, California

The same fractal self-similarity feature we saw in the fractal bush and its natural counterpart can be seen in this cast of the veins and arteries of a child's kidney.

Child's Kidney Venous and Arterial System

The fractal structure seen in the kidney can also be seen in this majestic oak.

Oak Tree, Arastradero Preserve, Palo Alto, California

A self-similar fractal branching or tree-like structure can be seen in these photos that were taken from the Gemini IV spaceprobe that was launched on June 3, 1965.

Wadi Hadramaut, Gemini IV Image

Dawn over the Himalayas, Gemini IV Image

Heinz-Otto Peitgen, Hartmut Jurgens, and Dietmar Saupe, in their new book *Fractals for the Classroom*, describe self-similarity in the following way.

Before we open our gallery of classical fractals and discuss in some detail several of these early masterpieces, let us introduce the concept of self-similarity. It will be an underlying theme in all fractals, more pronounced in some of them and in variations in others. In a way the word self-similarity needs no explanation, and at this point we merely give an example of a natural structure with that property, a cauliflower. It is not a classical mathematical fractal, but here the meaning of self-similarity is readily revealed without any math.

Each cauliflower head contains branches or buds off a main stem. Each main stem has other stems that branch off it. The heads or flowers on these stems all look similar to each other. So they are self-similar.

Cauliflower

The Mandelbrot set is one of the most complex and beautiful mathematical pictures. Let's take a tour of its incredible boundary and its beautiful embedded self-similarity. Study the series of sequential zoom-ins of the Mandelbrot set. The rectangular window in each figure shows the region for the next zoom-in.

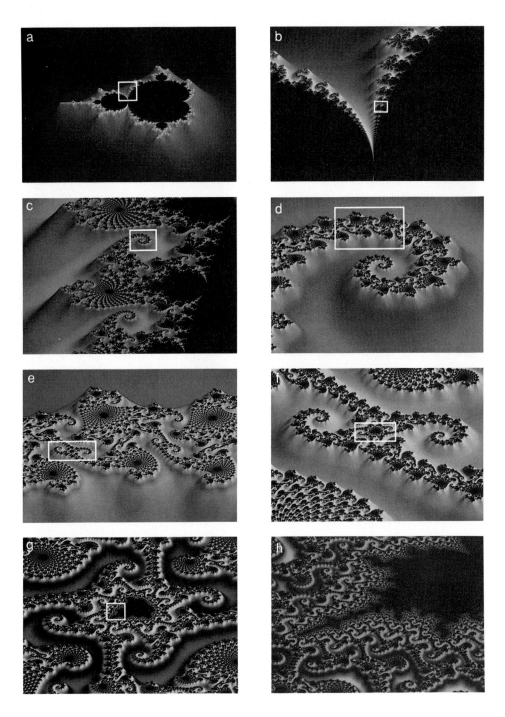

Blowups at the Boundary of the Mandelbrot Set

Study this close-up of the three-dimensional rendering of the Mandelbrot set. It is a high-resolution still picture that simulates a flight over the boundary of the Mandelbrot set.

Escalante 3D

The coloring of the rendering of the Mandelbrot set was inspired by winter scenes at Yellowstone Lake. The clouds in the background are created by a random fractal algorithm.

Yellowstone Lake

The beauty of nature may be random, but it is not without order. The field of fractal geometry is still young and there are many questions still unanswered. But, our understanding of the world around us is growing every day.

APPENDIX:
USING TABLES

A table of squares and approximate square roots and a table of values of trigonometric functions are provided for use in case a scientific calculator is not available. This guide will show you how to use the tables to find squares, square roots, and values of trigonometric functions.

Example 1

Find the square and square root of 51.

Read across the row labeled 51.

SQUARES AND APPROXIMATE SQUARE ROOTS

n	n^2	$\sqrt{n}$	n	n^2	$\sqrt{n}$
1	1	1.000	51	2601	7.141
2	4	1.414	52	2704	7.211
3	9	1.732	53	2809	7.280
4	16	2.000	54	2916	7.348
5	25	2.236	55	3025	7.416

The n^2 column shows that the square of 51 is 2601.
The $\sqrt{n}$ column shows that the square root of 51 to the nearest thousandth is 7.141.

Example 2

Find the sine, cosine, and tangent of 37° to the nearest ten thousandth.

Read across the row labeled 37°.

TRIGONOMETRIC RATIOS

Angle	sin	cos	tan	Angle	sin	cos	tan
36°	0.5878	0.8090	0.7265	81°	0.9877	0.1564	6.3138
37°	0.6018	0.7986	0.7536	82°	0.9903	0.1392	7.1154
38°	0.6157	0.7880	0.7813	83°	0.9925	0.1219	8.1443
39°	0.6293	0.7771	0.8098	84°	0.9945	0.1045	9.5144
40°	0.6428	0.7660	0.8391	85°	0.9962	0.0872	11.4301

The sin column shows that the sine of 37° is 0.6018.
The cos column shows that the cosine of 37° is 0.7986.
The tan column shows that the tangent of 37° is 0.7536.

SQUARES AND APPROXIMATE SQUARE ROOTS

n	n^2	$\sqrt{n}$	n	n^2	$\sqrt{n}$
1	1	1.000	51	2601	7.141
2	4	1.414	52	2704	7.211
3	9	1.732	53	2809	7.280
4	16	2.000	54	2916	7.348
5	25	2.236	55	3025	7.416
6	36	2.449	56	3136	7.483
7	49	2.646	57	3249	7.550
8	64	2.828	58	3364	7.616
9	81	3.000	59	3481	7.681
10	100	3.162	60	3600	7.746
11	121	3.317	61	3721	7.810
12	144	3.464	62	3844	7.874
13	169	3.606	63	3969	7.937
14	196	3.742	64	4096	8.000
15	225	3.873	65	4225	8.062
16	256	4.000	66	4356	8.124
17	289	4.123	67	4489	8.185
18	324	4.243	68	4624	8.246
19	361	4.359	69	4761	8.307
20	400	4.472	70	4900	8.367
21	441	4.583	71	5041	8.426
22	484	4.690	72	5184	8.485
23	529	4.796	73	5329	8.544
24	576	4.899	74	5476	8.602
25	625	5.000	75	5625	8.660
26	676	5.099	76	5776	8.718
27	729	5.196	77	5929	8.775
28	784	5.292	78	6084	8.832
29	841	5.385	79	6241	8.888
30	900	5.477	80	6400	8.944
31	961	5.568	81	6561	9.000
32	1024	5.657	82	6724	9.055
33	1089	5.745	83	6889	9.110
34	1156	5.831	84	7056	9.165
35	1225	5.916	85	7225	9.220
36	1296	6.000	86	7396	9.274
37	1369	6.083	87	7569	9.327
38	1444	6.164	88	7744	9.381
39	1521	6.245	89	7921	9.434
40	1600	6.325	90	8100	9.487
41	1681	6.403	91	8281	9.539
42	1764	6.481	92	8464	9.592
43	1849	6.557	93	8649	9.644
44	1936	6.633	94	8836	9.695
45	2025	6.708	95	9025	9.747
46	2116	6.782	96	9216	9.798
47	2209	6.856	97	9409	9.849
48	2304	6.928	98	9604	9.899
49	2401	7.000	99	9801	9.950
50	2500	7.071	100	10000	10.000

TRIGONOMETRIC RATIOS

Angle	sin	cos	tan	Angle	sin	cos	tan
0°	0.0000	1.0000	0.0000	45°	0.7071	0.7071	1.0000
1°	0.0175	0.9998	0.0175	46°	0.7193	0.6947	1.0355
2°	0.0349	0.9994	0.0349	47°	0.7314	0.6820	1.0724
3°	0.0523	0.9986	0.0524	48°	0.7431	0.6691	1.1106
4°	0.0698	0.9976	0.0699	49°	0.7547	0.6561	1.1504
5°	0.0872	0.9962	0.0875	50°	0.7660	0.6428	1.1918
6°	0.1045	0.9945	0.1051	51°	0.7771	0.6293	1.2349
7°	0.1219	0.9925	0.1228	52°	0.7880	0.6157	1.2799
8°	0.1392	0.9903	0.1405	53°	0.7986	0.6018	1.3270
9°	0.1564	0.9877	0.1584	54°	0.8090	0.5878	1.3764
10°	0.1736	0.9848	0.1763	55°	0.8192	0.5736	1.4281
11°	0.1908	0.9816	0.1944	56°	0.8290	0.5592	1.4826
12°	0.2079	0.9781	0.2126	57°	0.8387	0.5446	1.5399
13°	0.2250	0.9744	0.2309	58°	0.8480	0.5299	1.6003
14°	0.2419	0.9703	0.2493	59°	0.8572	0.5150	1.6643
15°	0.2588	0.9659	0.2679	60°	0.8660	0.5000	1.7321
16°	0.2756	0.9613	0.2867	61°	0.8746	0.4848	1.8040
17°	0.2924	0.9563	0.3057	62°	0.8829	0.4695	1.8807
18°	0.3090	0.9511	0.3249	63°	0.8910	0.4540	1.9626
19°	0.3256	0.9455	0.3443	64°	0.8988	0.4384	2.0503
20°	0.3420	0.9397	0.3640	65°	0.9063	0.4226	2.1445
21°	0.3584	0.9336	0.3839	66°	0.9135	0.4067	2.2460
22°	0.3746	0.9272	0.4040	67°	0.9205	0.3907	2.3559
23°	0.3907	0.9205	0.4245	68°	0.9272	0.3746	2.4751
24°	0.4067	0.9135	0.4452	69°	0.9336	0.3584	2.6051
25°	0.4226	0.9063	0.4663	70°	0.9397	0.3420	2.7475
26°	0.4384	0.8988	0.4877	71°	0.9455	0.3256	2.9042
27°	0.4540	0.8910	0.5095	72°	0.9511	0.3090	3.0777
28°	0.4695	0.8829	0.5317	73°	0.9563	0.2924	3.2709
29°	0.4848	0.8746	0.5543	74°	0.9613	0.2756	3.4874
30°	0.5000	0.8660	0.5774	75°	0.9659	0.2588	3.7321
31°	0.5150	0.8572	0.6009	76°	0.9703	0.2419	4.0108
32°	0.5299	0.8480	0.6249	77°	0.9744	0.2250	4.3315
33°	0.5446	0.8387	0.6494	78°	0.9781	0.2079	4.7046
34°	0.5592	0.8290	0.6745	79°	0.9816	0.1908	5.1446
35°	0.5736	0.8192	0.7002	80°	0.9848	0.1736	5.6713
36°	0.5878	0.8090	0.7265	81°	0.9877	0.1564	6.3138
37°	0.6018	0.7986	0.7536	82°	0.9903	0.1392	7.1154
38°	0.6157	0.7880	0.7813	83°	0.9925	0.1219	8.1443
39°	0.6293	0.7771	0.8098	84°	0.9945	0.1045	9.5144
40°	0.6428	0.7660	0.8391	85°	0.9962	0.0872	11.4301
41°	0.6561	0.7547	0.8693	86°	0.9976	0.0698	14.3007
42°	0.6691	0.7431	0.9004	87°	0.9986	0.0523	19.0811
43°	0.6820	0.7314	0.9325	88°	0.9994	0.0349	28.6363
44°	0.6947	0.7193	0.9657	89°	0.9998	0.0175	57.2900
45°	0.7071	0.7071	1.0000	90°	1.0000	0.0000	∞

POSTULATES, THEOREMS, AND COROLLARIES

If an abbreviation of a postulate, theorem, or corollary is used, it is listed after the number of the page on which the postulate, theorem, or corollary is given.

Chapter 1 The Language of Geometry

Postulate 1-1
Ruler Postulate
The points on any line can be paired with the real numbers so that, given any two points P and Q on the line, P corresponds to zero, and Q corresponds to a positive number. (24)

Postulate 1-2
Segment Addition Postulate
If Q is between P and R, then $PQ + QR = PR$.
If $PQ + QR = PR$, then Q is between P and R. (25)

Postulate 1-3
Protractor Postulate
Given $\overrightarrow{AB}$ and a number r between 0 and 180, there is exactly one ray with endpoint A, extending on each side of $\overrightarrow{AB}$, such that the measure of the angle formed is r. (38)

Postulate 1-4
Angle Addition Postulate
If R is in the interior of $\angle PQS$, then $m\angle PQR + m\angle RQS = m\angle PQS$. If $m\angle PQR + m\angle RQS = m\angle PQS$, then R is in the interior of $\angle PQS$. (39)

Chapter 2 Reasoning and Introduction to Proof

Postulate 2-1
Through any two points there is exactly one line. (77)
(Through any 2 pts. there is 1 line.)

Postulate 2-2
Through any three points not on the same line there is exactly one plane. (77) *(Through any 3 noncollinear pts. there is 1 plane.)*

Postulate 2-3
A line contains at least two points. (78) *(A line contains at least 2 pts.)*

Postulate 2-4
A plane contains at least three points not on the same line. (78) *(A plane contains at least 3 noncollinear pts.)*

Postulate 2-5
If two points lie in a plane, then the entire line containing those two points lies in that plane. (78) *(If 2 pts. are in a plane, then the line that contains them is in the plane.)*

Postulate 2-6
If two planes intersect, then their intersection is a line. (78) *(The intersection of 2 planes is a line.)*

Theorem 2-1
Congruence of segments is reflexive, symmetric, and transitive. (98) *(Congruence of segments is (reflexive/symmetric/ transitive).)*

Theorem 2-2 **Supplement Theorem**	If two angles form a linear pair, then they are supplementary angles. (105) *(If 2 ∡ form a linear pair, they are supp.)*
Theorem 2-3	Congruence of angles is reflexive, symmetric, and transitive. (106) *(Congruence of angles is (reflexive/symmetric/transitive).)*
Theorem 2-4	Angles supplementary to the same angle or to congruent angles are congruent. (106) *(∡ supp. (to the same ∠/ to ≅ ∡) are ≅.)*
Theorem 2-5	Angles complementary to the same angle or to congruent angles are congruent. (106) *(∡ comp. (to the same ∠/ to congruent ∡) are ≅.)*
Theorem 2-6	All right angles are congruent. (107) *(All rt. ∡ are ≅.)*
Theorem 2-7	Vertical angles are congruent. (107) *(Vertical ∡ are ≅.)*
Theorem 2-8	Perpendicular lines intersect to form four right angles. (107) *(⊥ lines form four rt. ∡.)*

Chapter 3 Parallels

Postulate 3-1 **Corresponding Angles Postulate**	If two parallel lines are cut by a transversal, then each pair of corresponding angles are congruent. (128) *(If 2 ∥ lines are cut by a transversal, corr. ∡ are ≅.)*
Theorem 3-1 **Alternate Interior Angle Theorem**	If two parallel lines are cut by a transversal, then each pair of alternate interior angles are congruent. (129) *(If 2 ∥ lines are cut by a transversal, alt. int. ∡ are ≅.)*
Theorem 3-2 **Consecutive Interior Angle Theorem**	If two parallel lines are cut by a transversal, then each pair of consecutive interior angles are supplementary. (129) *(If 2 ∥ lines are cut by a transversal, consec. int. ∡ are supp.)*
Theorem 3-3 **Alternate Exterior Angle Theorem**	If two parallel lines are cut by a transversal, then each pair of alternate exterior angles are congruent. (129) *(If 2 ∥ lines are cut by a transversal, alt. ext. ∡ are ≅.)*
Theorem 3-4 **Perpendicular Transversal Theorem**	In a plane, if a line is perpendicular to one of two parallel lines, then it is perpendicular to the other. (130) *(In a plane, if a line is ⊥ to one of 2 ∥ lines, then it is ⊥ to the other.)*
Postulate 3-2	If two lines are cut by a transversal so that corresponding angles are congruent, then the lines are parallel. (135) *(If 2 lines are cut by a transversal and corr. ∡ are ≅, then the lines are ∥.)*
Postulate 3-3 **Parallel Postulate**	If there is a line and a point not on the line, then there exists exactly one line through the point that is parallel to the given line. (136)
Theorem 3-5	If two lines in a plane are cut by a transversal so that a pair of alternate interior angles are congruent, then the two lines are parallel. (136) *(If 2 lines are cut by a transversal and alt. int. ∡ are ≅, then the lines are ∥.)*

Theorem 3-6	If two lines in a plane are cut by a transversal so that a pair of consecutive interior angles is supplementary, then the lines are parallel. (136) *(If 2 lines are cut by a transversal and consec. int. ∡ are supp., then the lines are ‖.)*
Theorem 3-7	If two lines in a plane are cut by a transversal so that a pair of alternate exterior angles is congruent, then the lines are parallel. (136) *(If 2 lines are cut by a transversal and alt. ext. ∡ are ≅, then the lines are ‖.)*
Theorem 3-8	In a plane, if two lines are perpendicular to the same line, then they are parallel. (136) *(In a plane, if 2 lines are ⊥ to the same line, they are ‖.)*
Postulate 3-4	Two lines have the same slope if and only if they are parallel and nonvertical. (143)
Postulate 3-5	Two nonvertical lines are perpendicular if and only if the product of their slopes is -1. (143)

Chapter 4 Congruent Triangles

Theorem 4-1 **Angle Sum Theorem**	The sum of the measures of the angles of a triangle is 180. (170) *(The sum of the ∡ in a Δ is 180.)*
Theorem 4-2 **Third Angle Theorem**	If two angles of one triangle are congruent to two angles of a second triangle, then the third angles of the triangles are congruent. (172) *(If 2 ∡ in a Δ are ≅ to 2 ∡ in another Δ, the third ∡ of the Δs are ≅.)*
Theorem 4-3 **Exterior Angle Theorem**	The measure of an exterior angle of a triangle is equal to the sum of the measures of the two remote interior angles. (172) *(The measure of an ext. ∠ of a Δ = the sum of the measures of the remote int. ∡.)*
Corollary 4-1	The acute angles of a right triangle are complementary. (173) *(The acute ∡ of a rt. Δ are comp.)*
Corollary 4-2	There can be at most one right or obtuse angle in a triangle. (173) *(There can be at most 1 rt. or obtuse ∠ in a Δ.)*
Theorem 4-4	Congruence of triangles is reflexive, symmetric, and transitive. (178) *(Congruence of triangles is (reflexive/symmetric/ transitive).)*
Postulate 4-1 **SSS Postulate**	If the sides of one triangle are congruent to the sides of a second triangle, then the triangles are congruent. (184)
Postulate 4-2 **SAS Postulate**	If two sides and the included angle of one triangle are congruent to two sides and an included angle of another triangle, then the triangles are congruent. (186)
Postulate 4-3 **ASA Postulate**	If two angles and the included side of one triangle are congruent to two angles and the included side of another triangle, the triangles are congruent. (186)

Theorem 4-5 **AAS**	If two angles and a non-included side of one triangle are congruent to the corresponding two angles and side of a second triangle, the two triangles are congruent. (192)
Theorem 4-6 **Isosceles Triangle** **Theorem**	If two sides of a triangle are congruent, then the angles opposite those sides are congruent. (202) *(If 2 sides of a △ are ≅ the ∡ opp. the sides are ≅.)*
Corollary 4-3	A triangle is equilateral if and only if it is equiangular. (203) *(An (equilateral/equiangular) △ is (equiangular/equilateral).)*
Corollary 4-4	Each angle of an equilateral triangle measures 60°. (203) *(Each ∠ of an equilateral △ measures 60°.)*
Theorem 4-7	If two angles of a triangle are congruent, then the sides opposite those angles are congruent. (204) *(If 2 ∡ of a △ are ≅, the sides opp. the ∡ are ≅.)*

Chapter 5 Applying Congruent Triangles

Theorem 5-1	A point on the perpendicular bisector of a segment is equidistant from the endpoints of the segment. (219) *(A pt. on the ⊥ bisector of a segment is equidistant from the endpts. of the segment.)*
Theorem 5-2	A point equidistant from the endpoints of a segment lies on the perpendicular bisector of the segment. (219) *(A pt. equidistant from the endpts. of a segment lies on the ⊥ bisector of the segment.)*
Theorem 5-3	A point on the bisector of an angle is equidistant from the sides of the angle. (219) *(A pt. on the bisector of an ∠ is equidistant from the sides of the ∠.)*
Theorem 5-4	A point in the interior of or on an angle and equidistant from the sides of an angle lies on the bisector of the angle. (219) *(A pt. in the int. of or on an ∠ and equidistant from the sides of an ∠ lies on the bisector of the ∠.)*
Theorem 5-5 **LL**	If the legs of one right triangle are congruent to the corresponding legs of another right triangle, then the triangles are congruent. (223)
Theorem 5-6 **HA**	If the hypotenuse and an acute angle of one right triangle are congruent to the hypotenuse and corresponding acute angle of another right triangle, then the two triangles are congruent. (224)
Theorem 5-7 **LA**	If one leg and an acute angle of one right triangle are congruent to the corresponding leg and acute angle of another right triangle, then the triangles are congruent. (225)
Postulate 5-1 **HL**	If the hypotenuse and a leg of one right triangle are congruent to the hypotenuse and corresponding leg of another right triangle, then the triangles are congruent. (226)

Theorem 5-8 *Exterior Angle* *Inequality Theorem*	If an angle is an exterior angle of a triangle, then its measure is greater than the measure of either of its corresponding remote interior angles. (234) *(If an ∠ is an ext. ∠ of a △, then its measure is greater than the measure of either of its corr. remote int. ∡.)*
Theorem 5-9	If one side of a triangle is longer than another side, then the angle opposite the longer side is greater than the angle opposite the shorter side. (240) *(If one side of a △ is longer than another side, then the ∠ opp. the longer side is greater than the ∠ opp. the shorter side.)*
Theorem 5-10	If one angle of a triangle is greater than another angle, then the side opposite the greater angle is longer than the side opposite the lesser angle. (240) *(If one ∠ of a △ is greater than another ∠, then the side opp. the greater ∠ is longer than the side opp. the lesser ∠.)*
Theorem 5-11	The perpendicular segment from a point to a line is the shortest segment from the point to the line. (242)
Corollary 5-1	The perpendicular segment from a point to a plane is the shortest segment from the point to the plane. (242)
Theorem 5-12 *Triangle Inequality* *Theorem*	The sum of the lengths of any two sides of a triangle is greater than the length of the third side. (246)
Theorem 5-13 *SAS Inequality* *(Hinge Theorem)*	If two sides of one triangle are congruent to two sides of another triangle, and the included angle in one triangle is greater than the included angle in the other, then the third side of the first triangle is longer than the third side in the second triangle. (252)
Theorem 5-14 *SSS Inequality*	If two sides of one triangle are congruent to two sides of another triangle and the third side in one triangle is longer than the third side in the other, then the angle between the pair of congruent sides in the first triangle is greater than the corresponding included angle in the second triangle. (252)

Chapter 6 Quadrilaterals

Theorem 6-1	Opposite sides of a parallelogram are congruent. (267) *(Opp. sides of a ▱ are ≅.)*
Theorem 6-2	Opposite angles of a parallelogram are congruent. (267) *(Opp. ∡ of a ▱ are ≅.)*
Theorem 6-3	Consecutive angles in a parallelogram are supplementary. (267) *(Consec. ∡ in a ▱ are supp.)*
Theorem 6-4	The diagonals of a parallelogram bisect each other. (267) *(Diagonals of a ▱ bisect each other.)*

Theorem 6-5	If both pairs of opposite sides of a quadrilateral are congruent, then the quadrilateral is a parallelogram. (275) *(If opp. sides of a quad. are ≅, it is a ▱.)*
Theorem 6-6	If one pair of opposite sides of a quadrilateral are both parallel and congruent, then the quadrilateral is a parallelogram. (276) *(If a pair of opp. sides of a quad. are ≅ and ∥, it is a ▱.)*
Theorem 6-7	If the diagonals of a quadrilateral bisect each other, then the quadrilateral is a parallelogram. (276) *(If the diagonals of a quad. bisect, it is a ▱.)*
Theorem 6-8	If both pairs of opposite angles in a quadrilateral are congruent, then the quadrilateral is a parallelogram. (276) *(If both pairs of opp. ∡ of a quad. are ≅, it is a ▱.)*
Theorem 6-9	If a parallelogram is a rectangle, then its diagonals are congruent. (282) *(If a ▱ is a rect. then its diagonals are ≅.)*
Theorem 6-10	The diagonals of a rhombus are perpendicular. (288) *(Diagonals of a rhom. are ⊥.)*
Theorem 6-11	Each diagonal of a rhombus bisects a pair of opposite angles. (288) *(Each diagonal of a rhom. bisects opp. ∡.)*
Theorem 6-12	Both pairs of base angles of an isosceles trapezoid are congruent. (294) *(Base ∡ of an iso. trap. are ≅.)*
Theorem 6-13	The diagonals of an isosceles trapezoid are congruent. (295) *(The diagonals of an isos. trap. are ≅.)*
Theorem 6-14	The median of a trapezoid is parallel to the bases and its measure is one half the sum of the measures of the bases. (295) *(Median of a trap. is ∥ to the bases. Length of median of a trap. = $\frac{1}{2}$ (sum of the lengths of bases))*

Chapter 7 Similarity

Postulate 7-1 *AA Similarity*	If two angles of one triangle are congruent to two angles of another triangle, then the triangles are similar. (329)
Theorem 7-1 *SSS Similarity*	If the measures of the corresponding sides of two triangles are proportional, then the triangles are similar. (329)
Theorem 7-2 *SAS Similarity*	If the measures of two sides of a triangle are proportional to the measures of two corresponding sides of another triangle, and the included angles are congruent, then the triangles are similar. (330)
Theorem 7-3	Similarity of triangles is reflexive, symmetric, and transitive. (331) *(Similarity of triangles is (reflexive/symmetric/ transitive).)*
Theorem 7-4 *Triangle* *Proportionality*	If a line is parallel to one side of a triangle and intersects the other two sides in two distinct points, then it separates these sides into segments of proportional lengths. (336)

Theorem 7-5	If a line intersects two sides of a triangle and separates the sides into corresponding segments of proportional lengths, then the line is parallel to the third side. (337)
Theorem 7-6	A segment whose endpoints are the midpoints of two sides of a triangle is parallel to the third side of the triangle and its length is one-half the length of the third side. (338)
Corollary 7-1	If three or more parallel lines intersect two transversals, then they cut off the transversals proportionally. (338)
Corollary 7-2	If three or more parallel lines cut off congruent segments on one transversal, then they cut off congruent segments on every transversal. (338)
Theorem 7-7 *Proportional* *Perimeters*	If two triangles are similar, then the perimeters are proportional to the measures of corresponding sides. (342)
Theorem 7-8	If two triangles are similar, then the measures of the corresponding altitudes are proportional to the measures of the corresponding sides. (343)
Theorem 7-9	If two triangles are similar, then the measures of the corresponding angle bisectors of the triangles are proportional to the measures of the corresponding sides. (344)
Theorem 7-10	If two triangles are similar, then the measures of corresponding medians are proportional to the measures of the corresponding sides. (344)
Theorem 7-11 *Angle Bisector* *Theorem*	An angle bisector in a triangle separates the opposite side into segments that have the same ratio as the other two sides. (345)

Chapter 8 Right Angles and Trigonometry

Theorem 8-1	If the altitude is drawn from the vertex of the right angle of a right triangle to its hypotenuse, then the two triangles formed are similar to the given triangle and to each other. (360)
Theorem 8-2	The measure of the altitude drawn from the vertex of the right angle of a right triangle to its hypotenuse is the geometric mean between the measures of the two segments of the hypotenuse. (361)
Theorem 8-3	If the altitude is drawn to the hypotenuse of a right triangle, then the measure of a leg of the triangle is the geometric mean between the measures of the hypotenuse and the segment of the hypotenuse adjacent to that leg. (362)
Theorem 8-4 *Pythagorean Theorem*	In a right triangle, the sum of the squares of the measures of the legs equals the square of the measure of the hypotenuse. (365)

Theorem 8-5 *Converse of the* *Pythagorean Theorem*	If the sum of the squares of the measures of two sides of a triangle equals the square of the measure of the longest side, then the triangle is a right triangle. (366)
Theorem 8-6	In a 45°-45°-90° triangle, the hypotenuse is $\sqrt{2}$ times as long as a leg. (371)
Theorem 8-7	In a 30°-60°-90° triangle, the hypotenuse is twice as long as the shorter leg and the longer leg is $\sqrt{3}$ times as long as the shorter leg. (372)

Chapter 9 Circles

Postulate 9-1 *Arc Addition Postulate*	The measure of an arc formed by two adjacent arcs is the sum of the measures of the two arcs. That is, if Q is a point on $\overset{\frown}{PR}$, then $m\overset{\frown}{PQ} + m\overset{\frown}{QR} = m\overset{\frown}{PR}$. (417)
Theorem 9-1	In a circle, if a diameter is perpendicular to a chord, then it bisects the chord and its arc. (422) *(In a ⊙, if a diameter is ⊥ to a chord, then it bisects the chord and its arc.)*
Theorem 9-2	In a circle or in congruent circles, two minor arcs are congruent if and only if their corresponding chords are congruent. (423) *(In a ⊙ or in ≅ ⊙s, 2 minor arcs are ≅ if and only if their corr. chords are ≅.)*
Theorem 9-3	In a circle or in congruent circles, two chords are congruent if and only if they are equidistant from the center. (424) *(In a ⊙ or in ≅ ⊙s, 2 chords are congruent if and only if they are equidistant from the center.)*
Theorem 9-4	If an angle is inscribed in a circle, then the measure of the angle equals one-half the measure of the intercepted arc. (428) *(If an ∠ is inscribed in a ⊙, then the measure of the ∠ = $\frac{1}{2}$ the measure of the intercepted arc.)*
Theorem 9-5	If two inscribed angles of a circle or congruent circles intercept congruent arcs or the same arc, then the angles are congruent. (429) *(If 2 inscribed ∠ of a ⊙ or ≅ ⊙s intercept ≅ arcs or the same arc, then the ∠ are ≅.)*
Theorem 9-6	If an angle is inscribed in a semicircle, then the angle is a right angle. (429) *(If an ∠ is inscribed in a semicircle, then the ∠ is a rt. ∠.)*
Theorem 9-7	If a quadrilateral is inscribed in a circle, then its opposite angles are supplementary. (430) *(If a quad. is inscribed in a ⊙, then its opp. ∠ are supp.)*
Theorem 9-8	If a line is tangent to a circle, then it is perpendicular to the radius drawn to the point of tangency. (434)
Theorem 9-9	In a plane, if a line is perpendicular to a radius of a circle at the endpoint on the circle, then the line is a tangent of the circle. (435)

Theorem 9-10	If two segments from the same exterior point are tangent to a circle, then they are congruent. (436)
Theorem 9-11	If a secant and a tangent intersect at the point of tangency, then the measure of each angle formed is one-half the measure of its intercepted arc. (440)
Theorem 9-12	If two secants intersect in the interior of a circle, then the measure of an angle formed is one-half the sum of the measures of the arcs intercepted by the angle and its vertical angle. (441)
Theorem 9-13	If two secants, a secant and a tangent, or two tangents intersect in the exterior of a circle, then the measure of the angle formed is one-half the positive difference of the measures of the intercepted arcs. (441)
Theorem 9-14	If two chords intersect in a circle, then the products of the measures of the segments of the chords are equal. (447)
Theorem 9-15	If two secant segments are drawn to a circle from an exterior point, then the product of the measures of one secant segment and its external secant segment is equal to the product of the measures of the other secant segment and its external secant segment. (448)
Theorem 9-16	If a tangent segment and a secant segment are drawn to a circle from an exterior point, then the square of the measure of the tangent segment is equal to the product of the measures of the secant segment and its external secant segment. (449)

Chapter 10 Polygons and Area

Theorem 10-1 *Interior Angle* *Sum Theorem*	If a convex polygon has n sides and S is the sum of the measures of its angles, then $S = 180(n - 2)$. (474)
Theorem 10-2 *Exterior Angle* *Sum Theorem*	If a polygon is convex, then the sum of the measures of the exterior angles, one at each vertex, is 360. (475)
Postulate 10-1	The area of a region is the sum of the areas of all of its nonoverlapping parts. (483)
Postulate 10-2	Congruent figures have equal areas. (489)
Postulate 10-3 *Length Probability* *Postulate*	If a point on $\overline{AB}$ is chosen at random and C is between A and B, then the probability that the point is on $\overline{AC}$ is $\dfrac{\text{length of } \overline{AC}}{\text{length of } \overline{AB}}$. (507)
Postulate 10-4 *Area Probability* *Postulate*	If a point in region A is chosen at random, then the probability that the point is in region B, which is in the interior of region A, is $\dfrac{\text{area of region } B}{\text{area of region } A}$. (508)

Chapter 12 More on Coordinate Geometry

Theorem 12-1
Slope-Intercept Form

If the equation of a line is written in the form $y = mx + b$, m is the slope of the line and b is the y-intercept.   (575)

Theorem 12-2

Given two points $A(x_1, y_1, z_1)$ and $B(x_2, y_2, z_2)$ in space, the distance between A and B is given by the following equation.
(608) $AB = \sqrt{(x_2 - x_1)^2 + (y_2 - y_1)^2 + (z_2 - z_1)^2}$

Chapter 13 Loci and Transformations

Postulate 13-1

In a given rotation, if A is the preimage, P is the image, and W is the center of rotation, then the measure of the angle of rotation, $\angle AWP$, equals twice the measure of the angle formed by intersecting lines of reflection.   (660)

Theorem 13-1

If a dilation with center C and a scale factor k maps A onto E and B onto D, then $ED = k(AB)$.   (665)

GLOSSARY

absolute value   The absolute value of a number is the number of units that it is from zero on a number line.   (24)

acute angle   An acute angle is one whose degree measure is less than 90.   (44)

acute triangle   An acute triangle is a triangle with all acute angles.   (164)

adjacent angles   Two angles in the same plane are adjacent if they have a common side and a common vertex, but no interior points in common.   (50)

adjacent arcs   Adjacent arcs are arcs of a circle that have exactly one point in common.   (417)

alternate exterior angles   In the figure, transversal l intersects lines ℓ and m. $\angle 5$ and $\angle 3$, and $\angle 6$ and $\angle 4$ are alternate exterior angles.   (124)

alternate interior angles   In the figure, transversal l intersects lines ℓ and m. $\angle 1$ and $\angle 7$, and $\angle 2$ and $\angle 8$ are alternate interior angles.   (124)

altitude of a cone   The altitude of a cone is the segment from the vertex perpendicular to the base.   (544)

altitude of a parallelogram   An altitude of a parallelogram is any perpendicular segment between parallel sides.   (484)

altitude of a prism   An altitude of a prism is a segment perpendicular to the base planes with an endpoint in each plane. The length of an altitude is called the height of the prism.   (536)

altitude of a pyramid   The altitude of a pyramid is the segment from the vertex perpendicular to the base.   (542)

altitude of a triangle   A segment is an altitude of a triangle if the following conditions hold.

1. Its endpoints are a vertex of a triangle and a point on the line containing the opposite side.
2. It is perpendicular to the line containing the opposite side.   (216)

angle   A figure is an angle if and only if it consists of two noncollinear rays with a common endpoint. The rays are the sides of the angle. The endpoint is the vertex of the angle. An angle separates a plane into three parts, the interior of the angle, the exterior of the angle, and the angle itself.   (36)

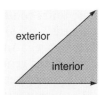

angle bisector   A ray, QS, is the bisector of $\angle PQR$ if S is in the interior of the angle and $\angle PQS \cong \angle RQS$.   (45)

angle of depression   An angle of depression is formed when a person must look down from a tower or cliff to see an object below. The angle of depression is the angle the person's line of sight forms with a horizontal line.   (385)

angle of elevation   When a person must look up to see an object, an angle of elevation is formed. The angle of elevation is the angle the person's line of sight forms with a horizontal line.   (384)

angle of rotation   The angle of rotation, $\angle ABC$, is determined by A, the preimage; B, the center of rotation; and C, the rotation image.   (660)

apothem   A segment that is drawn from the center of a regular polygon perpendicular to a side of the polygon is called an apothem. (495)

arc   An arc is an unbroken part of a circle. (416)

arc measure The degree measure of a minor arc is the degree measure of its central angle. The degree measure of a major arc is 360 minus the degree measure of its central angle. The degree measure of a semicircle is 180. (417)

arc of a chord A minor arc that has the same endpoints as a chord is called an arc of the chord. (422)

area The area of a figure is the number of square units contained in the interior of the figure. (483)

auxiliary figure An auxiliary figure is included on a geometric figure in order to prove a given theorem. (170)

axis **1.** In a coordinate plane, the x-axis is the horizontal number line and the y-axis is the vertical number line. (8)
2. The axis of a cylinder is the segment whose endpoints are the centers of the bases. (538)
3. The axis of a cone is the segment whose endpoints are the vertex and the center of the base. (544)

base **1.** In an isosceles triangle, the side opposite the vertex angle is called the base. (165)
2. The parallel sides of a trapezoid are called bases. (294)
3. Any side of a parallelogram can be called a base. (484)
4. In a prism, the bases are the two faces formed by congruent polygons that lie in parallel planes. (536)
5. In a cylinder, the bases are the two congruent circles that lie in parallel planes. (538)
6. In a pyramid, the base is the face that does not intersect the other faces at the vertex. The base is a polygon. (542)
7. In a cone, the base is the flat circular side. (544)

base angle **1.** In an isosceles triangle, either angle formed by the base and one of the legs is called a base angle. (165)

2. In the trapezoid at the right, ∠A and ∠D, and ∠B and ∠C are pairs of base angles. (294)

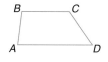

between In the figure, point B is between A and C. (23)

Cavalieri's Principle If two solids have the same cross-sectional area at every level and the same height, then they have the same volume. (555)

center of a regular polygon A point is a center of a regular polygon if it is the common center of its inscribed and circumscribed circles. (495)

central angle **1.** A central angle of a circle is an angle formed by two rays coplanar with the circle. The vertex of the angle is the center of the circle. (416)
2. An angle formed by two radii drawn to consecutive vertices of a regular polygon is called a central angle of the polygon. (497)

chord **1.** A chord of a circle is a segment whose endpoints are points on the circle. (410)
2. A chord of a sphere is a segment whose endpoints are on the sphere. (560)

circle A figure is a circle if it is the set of all points in a plane that are a given distance from a given point in the plane, called the center. A circle separates a plane into three parts, the interior, the exterior, and the circle itself. (410)

circumference The circumference of a circle is the limit of the perimeter of the inscribed regular polygons as the number of sides increases. (502)

circumscribed polygon A polygon is circumscribed about a circle if each side of the polygon is tangent to the circle. (436)

collinear points Points are collinear if they lie on the same line. (9)

common tangent   A line that is tangent to two circles that are in the same plane is called a common tangent of the two circles. A common tangent that does not intersect the segment whose endpoints are the centers of the circles is a common external tangent. A common tangent that intersects the segment whose endpoints are the centers of the circles is a common internal tangent.   (413)

compass   A compass is an instrument used to draw circles and arcs of circles.   (25)

complementary angles   Two angles are complementary if the sum of their degree measure is 90.   (52)

complete network   In graph theory, a complete network has at least one path between each pair of nodes.   (514)

composite of reflections   Two successive reflections are called a composite of reflections.   (653)

concave polygon   A polygon is concave if a line containing a side of the polygon contains a point in the interior of the polygon.   (466)

concentric circles   Concentric circles are circles that lie in the same plane and have the same center.   (418)

conclusion   In a conditional statement, the part immediately following *then* is called the conclusion.   (76)

conditional statement   A conditional statement is a statement that can be written in if-then form. The part following *if* is called the hypothesis. The part following *then* is called the conclusion. Conditional statements may be true or false.   (76)

cone   A cone may be thought of as a pyramid with a circular base. The figure at the right is a cone. It has a circular base and a vertex at *V*.   (544)

congruence transformation   When a geometric figure and its transformation image are congruent, the mapping is called a congruence transformation or isometry.   (641)

congruent angles   Two angles are congruent if they have the same measurement.   (44)

congruent segments   Two segments are congruent if they have exactly the same length.   (32)

congruent triangles   Two triangles are congruent if their corresponding parts are congruent.   (177)

conjecture   A conjecture is an educated guess.   (70)

consecutive interior angles   In the figure, transversal *l* intersects lines ℓ and *m*. ∠8 and ∠1, and ∠7 and ∠2 are consecutive interior angles.   (124)

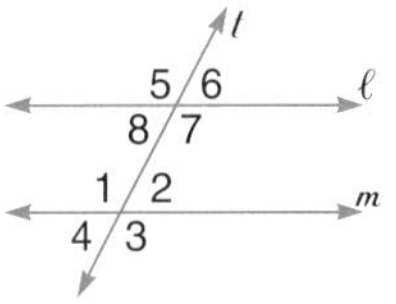

converse   The converse of a conditional statement is formed by interchanging the hypothesis and conclusion.   (77)

convex polygon   A polygon is convex if any line containing a side of the polygon does not contain a point in the interior of the polygon.   (466)

coordinate   In an ordered pair, the first component is called the *x*-coordinate and the second component is called the *y*-coordinate.   (8)

coordinate plane   A coordinate plane is a number plane formed by two perpendicular number lines that intersect at their zero points.   (8)

coordinate proof   A coordinate proof is a format used for proof. The figure is placed on a coordinate plane as an algebraic argument is used. Justification of statements is provided as needed.   (595)

coplanar points   Points are coplanar if they lie in the same plane.   (14)

corollary   A statement that can be easily proven using a theorem is called a corollary of that theorem.   (173)

corresponding angles   In the figure, transversal *l* intersects lines ℓ and *m*. ∠5 and ∠1, ∠8 and ∠4, ∠6 and ∠2, and ∠7 and ∠3 are corresponding angles.   (124)

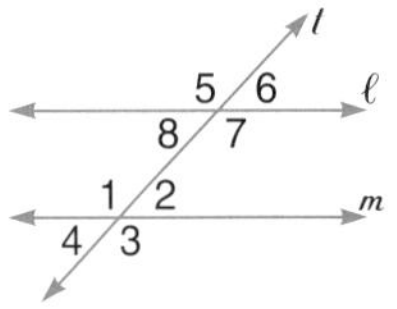

cosine   A ratio is the cosine of an acute angle of a right triangle if it is the ratio of the measure of the leg adjacent to the acute angle to the measure of the hypotenuse.   (376)

counterexample   A counterexample is an example used to show that a statement does not agree with or confirm a given idea.   (71)

cross products   Every proportion has two cross products. In the proportion $\frac{a}{b} = \frac{c}{d}$, where $b \neq 0$ and $d \neq 0$, the cross products are ad and bc. The cross products of a proportion are equal. (309)

cylinder   A cylinder is a figure whose bases are formed by congruent circles in parallel planes. The segment whose endpoints are the centers of the circles is called the axis of the cylinder. The altitude is a segment perpendicular to the base planes with an endpoint in each plane.   (538)

deductive reasoning   Deductive reasoning is a system of reasoning used to reach logical conclusions.   (82)

degree   A degree is one of the units of measure used in measuring angles.   (37)

degree of a node   In a network, the number of edges meeting at each node is called the degree of the node.   (513)

diagonal   A segment joining two nonconsecutive vertices of a polygon is called a diagonal of the polygon.   (266)

diameter   **1.** A diameter of a circle is a chord that contains the center of the circle.   (410)
2. A diameter of a sphere is a segment that contains the center, and whose endpoints are points on the sphere. (560)

dilation   A dilation is a transformation in which size is altered based on a center, C, and a scale factor, k. If $k > 1$, the dilation is an enlargement. If $0 < k < 1$, the dilation is a reduction. If $k = 1$, the dilation is a congruence transformation. (665)

distance   The absolute value of the difference of the coordinates of two points on a number line represents the measure of the distance between the two points.   (24)

distance between a point and a line   The distance between a point and a line is the length of the segment perpendicular to the line from the point. The measure of the distance between a line and a point on the line is zero.   (148)

distance between two parallel lines   The distance between two parallel lines is the distance between one of the lines and any point on the other line.   (149)

dot product   If $a = (x_1, y_1)$ and $b = (x_2, y_2)$, then $a \cdot b = x_1x_2 + y_1y_2$ is the product of a and b.   (606)

edge   **1.** In a polyhedra, pairs of faces intersect at line segments called edges. (468)
2. In graph theory, the path connecting two nodes is called an edge.   (512)

equiangular triangle   An equiangular triangle is a triangle with all angles congruent.   (164)

equilateral triangle   An equilateral triangle is a triangle with all sides congruent. (165)

exterior   **1.** Any point that is not on the angle or in the interior of the angle is in the exterior of the angle.   (37)
2. A point is in the exterior of a circle if the measure of the segment joining the point to the center of the circle is greater than the measure of the radius.   (411)

exterior angle   An angle is an exterior angle of a polygon if it forms a linear pair with one of the angles of the polygon. In the figure, $\angle 2$ is an exterior angle. (172)

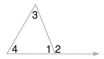

exterior angles In the figure, transversal t intersects lines ℓ and m. $\angle 5$ and $\angle 6$ are exterior angles. (124)

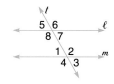

external secant segment An external secant segment is the part of the secant segment that is exterior to the circle. *See secant segment.* (448)

extremes In the proportion $\frac{a}{b} = \frac{c}{d}$, a and d are called the extremes. (309)

face In a polyhedra, flat surfaces formed by polygons and their interiors are called faces. (468)

fractals A fractal is a shape that is irregular or broken. (7)

geometric mean For any positive numbers a and b, x is the geometric mean between a and b if $\frac{a}{x} = \frac{x}{b}$ and x is positive. (360)

geometric probability Geometric probability involves using the principles of length and area to find the probability of an event. (507)

graph theory A representation of a real-world situation involving points that are connected. (512)

great circle If a plane intersects a sphere in more than one point and contains the center of the sphere, the intersection of the plane and the sphere is called a great circle. (561)

height The length of an altitude is called the height. (484)

hemisphere Each great circle separates a plane into two congruent parts called hemispheres. (561)

hypotenuse In a right triangle, the side opposite the right angle is called the hypotenuse. (165)

hypothesis In a conditional statement, the part immediately following *if* is called the hypothesis. (76)

if-then statement An if-then statement is a sentence that states if something is true then something else is true. (76)

image The result of a transformation is called the image. If A is mapped onto A', then A' is called the image of A. The preimage of A' is A. (640)

incenter The center of the circle inscribed in a given triangle is called the incenter. The incenter of a triangle is the point of intersection of the angle bisectors. (436)

included angle In a triangle, the angle formed by two given sides is called the included angle. (185)

incomplete network In a graph theory, an incomplete network has at least one pair of nodes that are not connected by an edge. (514)

indirect proof The steps for writing an indirect proof are listed below.
1. Assume that the hypothesis is true and the conclusion is false.
2. Show that the assumption leads to a contradiction of the hypothesis or some other fact, such as a postulate, theorem, or corollary.
3. Point out that the assumption must be false, and therefore, the conclusion must be true. (233)

indirect reasoning In indirect reasoning, you assume the opposite of what you want to prove. Then, show that this assumption leads to a contradiction. (233)

inductive reasoning Looking at several specific situations to arrive at a conjecture is called inductive reasoning. (70)

inscribed angle   An angle is an inscribed angle if its vertex lies on a circle and its sides contain chords of the circle.   (428)

inscribed polygon   A polygon is inscribed in a circle if each of its vertices lie on the circle.   (430)

intercepted arc   An angle intercepts an arc if and only if each of the following conditions holds.
1. The endpoints of the arc lie on the angle.
2. All points of the arc, except the endpoints, are in the interior of the circle.
3. Each side of the angle contains an endpoint of the arc.   (428)

intercepts method   The intercepts method is a technique for graphing a linear equation that locates the x- and y-intercepts.   (574)

interior   **1.** A point is in the interior of an angle if it does not lie on the angle itself and it lies on a segment whose endpoints are on each side of the angle.   (37)
2. A point is in the interior of a circle if the measure of the segment joining the point to the center of the circle is less than the measure of the radius.   (411)

interior angles   In the figure, transversal t intersects lines ℓ and m. $\angle 1$, $\angle 2$, $\angle 7$, and $\angle 8$ are interior angles.   (124)

intersection   The intersection of two figures is the set of points that are in both figures.   (14)

isometry   When a geometric figure and its transformation image are congruent, the mapping is called an isometry or congruence transformation.   (641)

isosceles trapezoid   An isosceles trapezoid is a trapezoid in which the legs are congruent.   (294)

isosceles triangle   An isosceles triangle is a triangle with at least two sides congruent. The congruent sides are called legs. The angles opposite the legs are base angles. The angle formed by two legs is the vertex angle. The third side is the base.   (165)

lateral area   The lateral area of a prism is the area of all the lateral faces.   (537)

lateral edge   **1.** In a prism, lateral edges are the intersection of two adjacent lateral faces. Lateral edges are parallel segments. (536)
2. In a pyramid, lateral edges are the edges of the lateral faces and have the vertex as an endpoint.   (542)

lateral faces   **1.** In a prism, the lateral faces are the faces that are not bases. The lateral faces are formed by parallelograms.   (536)
2. In a pyramid lateral faces are faces that intersect at the vertex.   (542)

law of cosines   Let $\triangle ABC$ be any triangle with a, b, and c representing the measures of sides opposite angles with measures A, B, and C respectively. Then, the following equations hold true.
$$a^2 = b^2 + c^2 - 2bc \cos A$$
$$b^2 = a^2 + c^2 - 2bc \cos B$$
$$c^2 = a^2 + b^2 - 2bc \cos C \quad (394)$$

law of detachment   If $p \rightarrow q$ is a true conditional and p is true, then q is true.   (82)

law of sines   Let $\triangle ABC$ be any triangle with a, b, and c representing the measures of sides opposite angles with measures A, B, and C respectively. Then, $\dfrac{\sin A}{a} = \dfrac{\sin B}{b} = \dfrac{\sin C}{c}$.   (389)

law of syllogism   If $p \rightarrow q$ and $q \rightarrow r$ are true conditionals then $p \rightarrow r$ is also true.   (83)

leg   **1.** In a right triangle, the sides opposite the acute angles are legs.   (165)
2. In an isosceles triangle, the congruent sides are called legs.   (165)
3. In a trapezoid, the nonparallel sides are called legs.   (294)

line   Line is one of the basic undefined terms of geometry. Lines extend indefinitely and have no thickness or width. Lines are represented by double arrows and named by lower case script letters. A line also can be named using double arrows over capital letters representing two points on the line.   (13)

linear equation   An equation is linear if it can be written in the form $Ax + By = C$, where A, B, and C are any real numbers, and A and B are not both 0.   (574)

linear pair   Two angles form a linear pair if they are adjacent and their noncommon sides are opposite rays.   (50)

line of reflection   Line ℓ is a line of reflection if it is the perpendicular bisector of the segment drawn from point X to its reflection image point A.   (644)

line of symmetry   A line of symmetry is a line that can be drawn through a plane figure so that the figure on one side is the reflection image of the figure on the opposite side.   (646)

line symmetry   A figure has line symmetry if it can be folded along the lines of symmetry and the two halves match exactly.   (202)

locus   In geometry, a figure is a locus if it is the set of all points and only those points that satisfy a given condition. (622)

midpoint   A point M is the midpoint of a segment, $\overline{PQ}$, if M is between P and Q, and $PM = MQ$.   (30)

minor arc   If $\angle APB$ is a central angle of circle P, then points A and B and all points of the circle interior to the angle form a minor arc called $\overarc{AB}$.   (416)

net   A two-dimensional figure that when folded forms the surfaces of a three-dimensional object.   (530)

network   A network is a figure consisting of nodes and edges.   (512)

n-gon   A polygon with n sides is called an n-gon.   (467)

node   The points in graph theory are called nodes.   (512)

noncollinear points   Points are noncollinear if they do not lie on the same line.   (9)

major arc   If $\angle APB$ is a central angle of circle P, and C is any point on the circle and in the exterior of the angle, then points A and B and all points of the circle exterior to $\angle APB$ form a major arc called ACB. Three letters are needed to name a major arc.   (416)

means   In the proportion $\frac{a}{b} = \frac{c}{d}$, b and c are called the means.   (309)

measure   **1.** The measure of $\overline{AB}$, written AB, is the distance between A and B. (23)
2. A protractor can be used to find the measure of an angle.   (37)

median   **1.** A median of a triangle is a segment that joins a vertex of the triangle and the midpoint of the side opposite the vertex.   (216)
2. The median of a trapezoid is the segment that joins the midpoints of the legs.   (295)

oblique cone   A cone that is not a right cone is called an oblique cone.   (544)

oblique cylinder   A cylinder that is not a right cylinder is called an oblique cylinder.   (538)

oblique prism   A prism that is not a right prism is called an oblique prism.   (536)

obtuse angle   An obtuse angle is one whose degree measure is greater than 90. (44)

obtuse triangle   An obtuse triangle is a triangle with one obtuse angle.   (164)

opposite angles of a parallelogram   In a parallelogram, any pair of nonconsecutive angles are opposite angles. Opposite angles of a parallelogram are congruent.   (267)

opposite rays $\overrightarrow{PQ}$ and $\overrightarrow{PR}$ are opposite rays if P is between Q and R.
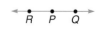
(36)

ordered pair   An ordered pair is a pair of numbers in which the order is specified. Ordered pairs are used to locate points in a plane.   (8)

ordered triple   An ordered triple is three numbers in which the order is specific. Ordered triples are used to locate points in space.   (607)

origin   The point of intersection of the *x*-axis and *y*-axis in a coordinate plane is called the origin and named *O*.   (8)

paragraph proof   In a paragraph proof, the statements and reasons are written informally in a paragraph.   (129)

parallel lines   Two lines are parallel if they lie in the same plane and do not intersect.   (122)

parallelogram   A quadrilateral is a parallelogram if both pairs of opposite sides are parallel. Any side of a parallelogram may be called a base. For each base there is a corresponding segment called the altitude that is perpendicular to the base and has its endpoints on the lines containing the base and the opposite side.   (117)

parallelogram law   The parallelogram law is a method for adding two vectors. The two vectors with the same initial point form part of a parallelogram. The resultant or sum of the two vectors is the diagonal of the parallelogram.   (602)

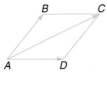

parallelpiped   A parallelpiped is a three-dimensional figure whose six faces are parallelograms.   (117)

parallel planes   Two planes are parallel if they do not intersect.   (122)

perpendicular bisector   A segment bisector is a perpendicular bisector if the bisector is perpendicular to the segment.   (218)

perpendicular lines   Two lines are perpendicular if they intersect to form a right angle.   (56)

plane   Plane is one of the basic undefined terms of geometry. Planes extend indefinitely in all directions and have no thickness. A plane is represented by a four-sided figure and is named by a capital script letter or by three points of the plane not on the same line.   (13)

Platonic solid   A Platonic solid is any one of the five regular polyhedrons: tetrahedron, hexahedron, octahedron, dodecahedron, or icosohedron.   (468)

point   Point is one of the basic undefined terms of geometry. Points have no dimension, are represented by dots, and are named by capital letters.   (13)

point of reflection   Point *S* is the reflection of point *R* with respect to point *Q*, the point of reflection, if *Q* is the midpoint of the segment drawn from *R* to *S*.   (644)

point of symmetry   A point of symmetry is the point of reflection for all points on a figure.   (646)

point of tangency   The point of the intersection of a circle and a tangent to the circle is the point of tangency.   (434)

point-slope form   The equation of a line passing through a point whose coordinates are (x_1, y_1) and that has a slope *m* is $y - y_1 = m(x - x_1)$.   (581)

polygon   A figure is a polygon if it meets each of the following conditions.
1. It is formed by three or more coplanar segments called sides.
2. Sides that have a common endpoint are noncollinear.
3. Each side intersects exactly two other sides, but only at their endpoints called vertices.   (266)

polyhedron   A solid with flat surfaces that form polygons is called a polyhedron. The flat surfaces formed by the polygons and their interiors are called faces. Pairs of faces intersect at edges. Three or more edges intersect at a vertex.   (468)

postulate   A postulate in geometry is a statement that describes a fundamental property of the basic terms. Postulates are accepted as being true.   (77)

preimage In a transformation if A is mapped onto A', then A is the preimage of A'. (640)

prism A solid with the following characteristics is a prism.
1. Two faces, called bases, are formed by congruent polygons that lie in parallel planes.
2. The faces that are not bases, called lateral faces, are formed by parallelograms.
3. The intersections of two adjacent lateral faces are called lateral edges and are parallel segments. (536)

probability Probability is a ratio or fraction that compares the chances of an event happening (or not happening) to all possible outcomes in a given situation. (507)

proof A proof is a method of constructing a valid argument. (89)

proportion A proportion is an equation of the form $\frac{a}{b} = \frac{c}{d}$ that states that two ratios are equivalent. (308)

protractor A protractor is a tool used to find the degree measure of a given angle. (37)

pyramid A solid with the following characteristics is a pyramid.
1. All the faces, except one face, intersect at a point called the vertex.
2. The face that does not intersect at the vertex is called the base and forms a polygon.
3. The faces meeting at the vertex are called lateral faces and form triangles. (542)

Pythagorean Theorem In a right triangle, the sum of the squares of the measures of the legs equals the square of the measure of the hypotenuse. (365)

Pythagorean triple This is a set of numbers that satisfy the equation $a^2 + b^2 = c^2$. (369)

quadrant One of the four regions into which two perpendicular number lines separate the plane is a quadrant. (8)

quadrilateral A four-sided polygon is called a quadrilateral. (266)

radius
1. A radius of a circle is a segment whose endpoints are the center of the circle and a point on the circle. (410)
2. A segment is a radius of a regular polygon if it is a radius of a circle circumscribed about the polygon. (495)
3. A radius of a sphere is a segment whose endpoints are the center and a point on the sphere. (560)

rate A rate is the ratio of two measurements that may have different types of units. (315)

ratio A ratio is a comparison of two numbers using division. (308)

ray PQ is a ray if it is the set of points PQ and all points S for which Q is between P and S. (36)

rectangle A rectangle is a quadrilateral with four right angles. (282)

reflection A reflection is a transformation that flips a figure over a line called the line of reflection. *Also see lines of reflection and points of reflection.* (644)

regular polygon A polygon is regular if it is a convex polygon with all sides congruent and all angles congruent. (467)

regular polyhedron A regular polyhedron is a polyhedron in which all faces are congruent regular polygons. (468)

regular pyramid A pyramid is a regular pyramid if its base is regular, and the segment is the center of the base, and the vertex is perpendicular to the base. This segment is called the altitude. (542)

regular tessellation A tessellation is a regular tessellation if it is formed by regular polygons. (478)

remote interior angles The angles in a triangle that are not adjacent to a given exterior angle are called remote interior angles. (172)

resultant   The sum of two or more vectors is called the resultant of the vectors. (602)

rhombus   A quadrilateral is a rhombus if all four sides are congruent.   (288)

right angle   A right angle is an angle whose degree measure is 90.   (44)

right circular cone   A solid figure that has a circular base and an axis from the vertex that is perpendicular to the base is a right circular cone. The axis is also the altitude of the cone.   (544)

right cylinder   A cylinder whose axis is also an altitude of the figure is a right cylinder.   (538)

right prism   If the lateral edges of a prism are also altitudes, then the prism is a right prism.   (536)

right triangle   A right triangle is a triangle with one right angle. The side opposite the right angle is called the hypotenuse. The other two sides are called legs. (164)

rotation   The composite of two reflections with respect to two intersecting lines is a transformation called a rotation. The intersection of the two lines is called the center of rotation.   (621)

ruler postulate   The point on any line can be paired with real numbers so that, given any two points P and Q on the line, P corresponds to zero, and Q corresponds to a positive number.   (24)

scalar multiplication   Multiplying a vector by a constant is called scalar multiplication.   (602)

scale factor   The ratio of the lengths of two corresponding sides of two similar polygons is called the scale factor.   (321)

scalene triangle   A scalene triangle is a triangle with no two sides congruent. (165)

scatter plot   In a scatter plot, two sets of data are plotted as ordered pairs in a coordinate plane.   (586)

secant   A secant is a line that intersects a circle in exactly two points.   (412)

secant segment   A secant segment is a segment that contains a chord of a circle. The part or parts of a secant segment that are exterior to the circle are called external secant segments.   (448)

sector   A sector of a circle is a region bounded by a central angle and the intercepted arc.   (509)

segment   A segment is a part of a line that consists of two points, called endpoints, and all the points between them.   (23)

segment bisector   A segment bisector is a segment, line, or plane that intersects a segment at its midpoint.   (31)

self-similarity   If any parts of a fractal image are replicas of the entire image, the image is self-similar.   (7)

semicircle   A line containing the diameter of a circle separates the circle into two semicircles.   (416)

side   1. The sides of an angle are the two rays that form the angle.   (36)
2. Any segment that forms part of a polygon is a side of the polygon. (164)

similar figures   Figures that have the same shape but that may differ in size are called similar figures.   (7)

similarity transformation   When a geometric figure and its transformation image are similar, the mapping is called a similarity transformation.   (641)

similar polygons   Two polygons are similar if there is a correspondence such that their corresponding angles are congruent and the measures of their corresponding sides are proportional.   (321)

sine   A ratio is the sine of an acute angle of a right triangle if it is the ratio of the measure of the leg opposite the acute angle to the measure of the hypotenuse. (376)

skew lines   Two lines are skew if they do not intersect and are not in the same plane.   (123)

slant height   1. In a regular pyramid, the slant height is the height of a lateral face. (542)
2. In a right circular cone, the slant height is the measure of any segment joining the vertex to the edge of the circular base.   (544)

slope   The slope of a line containing two points with coordinates (x_1, y_1) and (x_2, y_2) is given by the following formula.
$m = \frac{y_2 - y_1}{x_2 - x_1}$ where $x_2 \neq x_1$.   (142)

slope-intercept form   The equation of the line having a slope m and y-intercept b is $y = mx + b$.   (575)

solid   A solid is a three-dimensional figure consisting of all of its surface points and all of its interior points.   (529)

solving the triangle   Finding the measures of all the angles and sides of a triangle is called solving the triangle.   (389)

space   Space is the set of all points.   (14)

sphere   In space, a figure is a sphere if it is the set of all points that are a given distance from a given point, called the center.   (560)

spreadsheets   Spreadsheets are computer programs designed especially for creating charts involving many calculations.   (501)

square   A square is a quadrilateral with four right angles and four congruent sides.   (289)

standard form   The standard form of a linear equation is $Ax + By = C$.   (574)

straight angle   A straight angle is one whose degree measure is 180.   (37)

straightedge   Any instrument used as a guide to draw a line is a straightedge.   (25)

supplementary angles   Two angles are supplementary if the sum of their degree measures is 180.   (51)

surface area   The sum of the areas of the faces of a solid figure is the surface area.   (529)

system of equations   A set of equations with the same variables is a system of equations.   (629)

tangent   **1.** A ratio is the tangent of an acute angle of a right triangle if it is the ratio of the measure of the leg opposite the acute angle to the measure of the leg adjacent to the acute angle.   (376)

2. A tangent is a line in a plane that intersects a circle in the plane in exactly one point. The point of intersection is the point of tangency.   (412)
3. A tangent to a sphere is a line that intersects the sphere in exactly one point.   (560)

tangent segment   A tangent segment is a segment that intersects a circle in exactly one point and lies on a tangent.   (436)

tessellation   Tiled patterns formed by repeating shapes to fill a plane without gaps or overlaps are tessellations.   (478)

theorem   A theorem is a statement that must be proven before it is accepted as true.   (98)

traceability   A network is traceable if it can be traced in one continuous path without retracting any edge. This can be done if the graph has modes with even degrees or the graph has exactly two nodes with odd degrees.   (512)

transformation   In a plane, a mapping is a transformation if each point has exactly one image point and each image point has exactly one preimage point.   (621)

translation   A composite of two reflections over two parallel lines is a translation. A translation is also the result of sliding a given figure a given distance.   (653)

transversal   A line that intersects two or more lines in a plane at different points is a transversal.   (124)

trapezoid   A quadrilateral is a trapezoid if it has exactly one pair of parallel sides. The parallel sides of a trapezoid are called bases. The nonparallel sides are called legs. The angles formed by bases and the legs are called base angles. The line segment joining the midpoints of the legs of a trapezoid is called the median. The altitude is a segment perpendicular to both bases with its endpoints on the bases.   (294)

triangle   A triangle is a figure formed by three noncollinear segments called sides. Each endpoint of a side is an endpoint of exactly one other side. The endpoints are the vertices of the triangle. A triangle separates a plane into three parts, the triangle, its interior, and its exterior.   (164)

trigonometric ratio   A ratio of the measures of two sides of a regular triangle is called a trigonometric ratio. Trigonometry means triangle measurement.   (376)

two-column proof   A two-column proof is a formal proof in which statements are listed in one column and the reasons for each statement are listed in a second column.   (89)

undefined term   An undefined term is a word that has a meaning that is readily understood. The basic undefined terms of geometry are point, line, and plane.   (14)

vector   A vector is a directed segment. It is a quantity which possesses both magnitude (length) and direction.   (601)

vertex   **1.** The vertex of an angle is the common endpoint of the two rays that form the angle.   (37)
2. In a polygon, each endpoint of a side is called a vertex.   (164)
3. In a polyhedron, three or more edges intersect at a point called a vertex. (468)

4. In a pyramid, the vertex is the point where all but one of the faces intersect. (542)

5. In the figure, the vertex of the cone is point *V*. (544)

vertex angle   In an isosceles triangle, the angle formed by the congruent sides (legs) is called the vertex angle.   (165)

vertical angles   Two angles are vertical if they are two nonadjacent angles formed by two intersecting lines.   (50)

volume   The measure of the amount of space a figure encloses is the volume of the figure.   (548)

x-axis   The x-axis is the horizontal number line in a coordinate plane.   (8)

x-coordinate   The x-coordinate is the first component in an ordered pair.   (8)

y-axis   The y-axis is the vertical number line in a coordinate plane.   (8)

y-coordinate   The y-coordinate is the second component in an ordered pair. (8)

SELECTED ANSWERS

CHAPTER 1 THE LANGUAGE OF GEOMETRY

Pages 10–12 Lesson 1-1
5. III **7.** *y*-axis **9.** R; (-3, 4) **11.** Find the lengths of each side using the distance formula and then add the sides together. **13.** (3, -1) **15.** (-2, 4)
17, 19, 21.

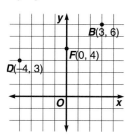

23. (0, -11) **25.** (15, 0)
27. (15, -11) **29.** no
31. yes
33a, b. **c.** (-3, -4)

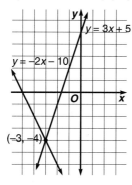

35a. Magic Kingdom **b.** (E, 3)
c. (D, 5), (D, 6)
d. Epcot Center Dr.
e. (C, 1), (D, 1), (D, 2), (E, 3), (E, 4), (F, 4), (F, 5), (F, 6)

Pages 15–18 Lesson 1-2
7. **9.**

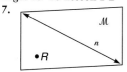 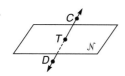

11. true **13.** true **15.** *E* **17.** no; yes **19.** line *TU*
21. *R*, *U* **23.** no; infinitely many **25.** plane **27.** line
29. plane
31. **35.**

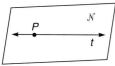 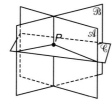

39. **41.** no

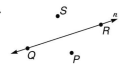

43. planes *PDC* and *PDA* **45.** parts of the figure that are hidden from view **47.** This figure is not possible unless *P*, *Q*, and *R* are collinear. If points *P*, *Q*, and *R* determine a unique plane, it is not possible for *P*, *Q*, and *R* to be in a plane that does not contain *S* and a different plane that does contain *S*.

49.

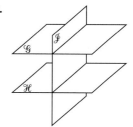

57–59.

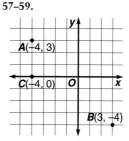

55. Since three points determine a plane, he should use three legs.

60. (0, 3) **61.** (-1, -2) **62.** 5
63. -2 **65.** 31 **66.** 55
67. 27.84 **68.** 80.22

Pages 21–22 Lesson 1-3
5. 10 combinations **7.** 9 **9.** 22 pieces **11.** Sample answer: $3 \times (3 + 3 \div 3) - 3 \div 3 = 11$ **13.** about $265

Pages 26–29 Lesson 1-4
5. 4 **7.** 6 **9.** 10 **11.** 10 **13.** 10 **15.** 6.32 **17.** 7; 15
19. 1 **21.** 10 **23.** 4 **25.** 17 **27.** 23 **29.** 11.5 **31.** 5
33. 13 **35.** 5 **37.** 1.41 **39.** 9.49 **41.** 1; 5 **43.** 7; 21
45. **47.**

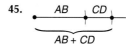 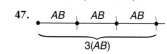

49. 29.31 units **51.**

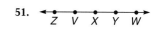

55a. the distance formula **60.**
b. 10 **c.** ≈ 3.07 **d.** ≈ 16.55

56. (-1, 3) **57.** *E* **58.** *A*

59. *D*, *A*, *E*, *O*

61. 18 area codes

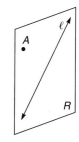

Pages 32–35 Lesson 1-5
5. 3 **7.** 1 **9.** $4\frac{1}{2}$ **11.** (-1, -1) **13.** $(2, -4\frac{1}{2})$ **15.** $(5\frac{1}{2}, -2)$
17. *x* = 4; 34 **19.** true **21.** true **23.** false **25.** true
27. false **29.** *B*(5, 5) **31.** *A*(2, 8) **33.** 8; 54 **35.** 5; 52
37. 8; 18 **39.** no **43.** *P*(4, -1), *Q*(6, -5) **45.** *x* = 2; yes
49. (-1, 5) or (-1, -1) **50a.** *P*, *Q*, *R*, *S* **b.** yes, since any three points are coplanar **51.** 12 ways **52.** 25 **53.** *F* is between *D* and *E*.

Page 35 Mid-Chapter Review
1. (1, 3) **2.** (-1, 1) **3.** (4, -2) **4.** true **5.** -2 **6.** *U, V, W,* or *X* **7.** $\overrightarrow{RS}, \overrightarrow{TS}, \overrightarrow{VS}$ **8.** $\overline{XY}$ **9.** true **10.** 3 unicycles, 4 bicycles, and 23 tricycles **11.** *T* is between *R* and *S*.
12. *x* = 5; 13 **13.** 10; (3, 4) **14.** *x* = 8

Pages 39–42 Lesson 1-6
5.  **7.**

9. $\overrightarrow{QR}, \overrightarrow{QT}$ **11.** false **13.** *Y, R, S,* or *T* **15.** *N, S,* or *T*
17. 10 **19.** 105 **21.** 80 **23.** 95 **25.** *J* **27.** *V* **29.** no

31.  **33.**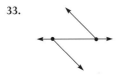

35. 29; 122 **37.** 12; 118 **39.** 12 angles **41.** 24 angles
43. 1; 2 if *PQ* ≥ 2, 1 if *PQ* < 2 **45a.** As the loft of the club increases, the ball will travel higher in the air and for a shorter distance, as long as the ball is struck with the same amount of force.
b.

9-iron
3-iron
47. (2, -1)

48.
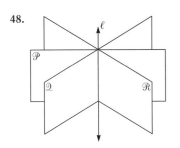
49. greater than 0
50. 5
51. (-1, 2)
52. *x* = 7

Pages 46–48 Lesson 1-7
11. acute **13.** straight **15.** right **17.** 0 < 3*x* + 12 < 90 or -4 < *x* < 26 **19.** acute **21.** obtuse **23.** straight
25. acute **27.** ∠*MHA* and ∠*HMT*, ∠*MAH* and ∠*TAH*, ∠*ATM* and ∠*HTM* **29.** 29 **31.** 56 **33.** 94; 63 **35.** -7; 19
37. 18 **39.** yes **41.** *x* = 6, *y* = 8 **45.** *m*∠*IMP* = 40, *m*∠*IMN* = 50 **46.** point **47.** 3 **48.** 12 **49.** False, since *R* is not in the interior of ∠*TPM*. **50.** 58

Pages 52–55 Lesson 1-8
5. 52, 142 **7.** no complement, 70 **9.** ∠*NML* and ∠*PMK*
11. ∠*JMP* and ∠*JML* **13.** Sample answers: ∠*NMP* and ∠*PMJ*, ∠*PMJ* and ∠*JMK*, ∠*JMK* and ∠*KML* **15.** 7; 53; 37
17. 33 **19.** 20 **21.** adjacent, complementary
23. adjacent, supplementary, linear pair **25.** supplementary
27. 13; 89 **29.** 112, 68 **31.** 45 **33.** 28, 62

35. 72, 108, 18 **37.** 12, 36, 144 **41.** 170° **42.** 4 **43.** 1
44. 21 **45.** yes; *m*∠1 = *m*∠2 **46.** 6

Pages 59–61 Lesson 1-9
5. $\overline{CQ}$ **7.** yes; ∠*BQD* and ∠*CQE* are right angles, so *m*∠*BQD* = 90 and *m*∠*CQE* = 90. Therefore *m*∠*BQD* = *m*∠*CQE*. *m*∠*BQD* = *m*∠*BQC* + *m*∠*CQD* and *m*∠*CQE* = *m*∠*CQD* + *m*∠*DQE* by the angle addition postulate. By substitution *m*∠*BQC* + *m*∠*CQD* = *m*∠*CQD* + *m*∠*DQE*. *m*∠*BQC* = *m*∠*DQE* **9.** yes **11.** no **13.** no **15.** yes
17. no **19.** no **21.** yes **23.** yes **25.** 25; no **27.** no
29. yes **31.** no **33.** yes **35.** a square **37.** *x* = 60, *y* = 30 **39.** *x* = 15, *y* = 10 **41.** 45 **42.** 4 **43.** $\overline{CF}, \overline{AD}, \overline{AF}, \overline{FD}, F$ **44.** ∠*ACE*, ∠*FCE*, ∠*CEB* **45.** *x* = 6 **46.** 23

Pages 62–64 Chapter 1 Summary and Review
1, 3.
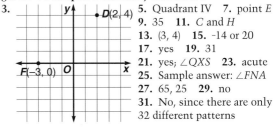
5. Quadrant IV **7.** point *E*
9. 35 **11.** *C* and *H*
13. (3, 4) **15.** -14 or 20
17. yes **19.** 31
21. yes; ∠*QXS* **23.** acute
25. Sample answer: ∠*FNA*
27. 65, 25 **29.** no
31. No, since there are only 32 different patterns

Pages 66–67 Algebra Review
1. 8 **3.** -1.3 **5.** $\frac{3}{8}$ **7.** 11 **9.** -9 **11.** 135 **13.** 9*r* − 5*s*
15. $\frac{5ab}{3} - \frac{a^2}{4}$ **17.** $\frac{18}{7}$ **19.** 4*a* + 7 **21.** 3
23. 12 − *a*² = *b* **25.** *xy*² = *t* **27.** -24 **29.** 22 **31.** -9
33. 4.5 **35.** 35 **37.** $\frac{9}{10}$ **39.** 10.2375 miles **41.** 1.25 liters

CHAPTER 2 REASONING AND INTRODUCTION TO PROOF

Pages 71–74 Lesson 2-1
5. False; *A, B,* and *C* could be as shown. They are not collinear.

7. ∠*A* ≅ ∠*B*

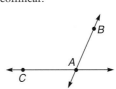

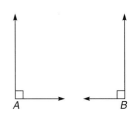

9. Sample answer:

$\overline{PQ} \cong \overline{RQ}$

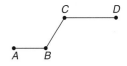

11. True; any three noncollinear points can be the vertices of a triangle.

13. false; counterexample:

17. Points Q, P, and R are collinear.

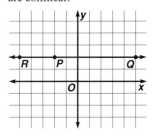

19. $1111^2 = 1,234,321$
21. $\overline{DE}$ is parallel to $\overline{BC}$.

23. false; counterexample:

27. $ABCDE$ is a pentagon.

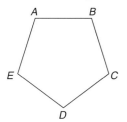

33b. Yes, the fungus is killing the plants.
c. Introduce the fungus to some healthy plants and see if they droop.

34. yes **36.** 6 **37.** plane

Pages 78–81 Lesson 2-2

7. Hypothesis: two lines are perpendicular; Conclusion: they intersect. **9.** Hypothesis: $x = 4$; Conclusion: $x^2 = 16$ **11.** If an angle is acute, then it is 37°. False; an angle that measures 58° is acute. **13.** If he or she may serve as President, then he or she is a native born United States citizen who is at least thirty-five years old. true **15.** If two planes intersect, then the intersection is a line. **17.** If an aluminum can is recycled, then it is remelted and back in the store within six weeks. **19.** false **21.** false **23.** Hypothesis: a candy bar is a Milky Way™; Conclusion: it contains caramel. **25.** true **27.** false **29.** true **31.** If a vehicle is a car, then it has four wheels. **33.** If an angle is acute, then it has a measure less than 90°. **35.** If two lines are parallel, then they do not intersect. **37.** If the distance of a race is about 6.2 miles, then it is 10 kilometers. true **39.** 2-1 **41.** 2-5 **43.** 2-6 **45.** one **47.** six **49.** fifteen **51.** one **53.** ten

55.

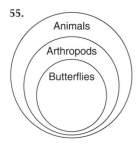

57a. 92 **b.** 1023
59. false; A, B, and P are not necessarily collinear.
60. 72, 108 **61.** 15
62. (1, 4)
63. 51

64.

65. no

Pages 84–87 Lesson 2-3

5. yes; detachment **7.** no **9.** no **11.** $\overline{CD} \cong \overline{CD}$; detachment **13.** Planes $\mathcal{M}$ and $\mathcal{N}$ intersect in a line; detachment **15.** Bobby Rahal is a professional race car driver; detachment **17.** yes; syllogism **19.** yes; detachment **21.** no **23.** A, B, and C are collinear; detachment **25.** p and q have a point in common; detachment **27.** If an ordered pair for a point has 0 as its x-coordinate, then it is not contained in any of the four quadrants; syllogism **29.** no conclusion **31.** no conclusion **33.** Basalt was formed by volcanos; syllogism **35a.** Sample answer: (2) Angles X and Y are adjacent and supplementary. (3) Angles X and Y are right angles. **b.** Sample answer: (2) Right angles measure 90°. (3) Angles that are adjacent and supplementary measure 90°. **37a.** Sample answer: (2) Dr. Garcia is a physician. (3) Dr. Garcia has graduated from medical school. **b.** Sample answer: (2) All medical school graduates have studied chemistry. (3) All physicians have studied chemistry. **39.** If a mineral sample is quartz, then it can scratch glass; syllogism **42.** If a geometry test score is 89, then it is above average. **43.** acute **44.** $(\frac{1}{2}, 4)$ **45.** $\sqrt{29}$

46.

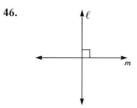

Pages 90–94 Lesson 2-4

5. Subtraction **7.** Division **9.** Substitution **11a.** Given **b.** Reflexive property of equality **c.** Addition property of equality **d.** Angle addition postulate **e.** Substitution property of equality **13.** Substitution **15.** Substitution **17.** Reflexive **19.** Subtraction **21.** Addition **23a.** Given **b.** Symmetric property of equality **c.** Multiplication property of equality **d.** Distribution property **e.** Subtraction property of equality **f.** Multiplication property of equality **25a.** Given **b.** Angle addition postulate **c.** Substitution property of equality **d.** Substitution property of equality **e.** Subtraction property of equality

27. Given: $m\angle M = m\angle P$
$m\angle N = m\angle P$
Prove: $m\angle M = m\angle N$

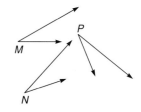

Statements	Reasons
a. $m\angle M = m\angle P$ $m\angle N = m\angle P$	**a.** Given
b. $m\angle M = m\angle N$	**b.** Transitive property of equality

29. Given: $m\angle ABC = 90$
$m\angle EDC = 90$
$m\angle 1 = m\angle 3$
Prove: $m\angle 2 = m\angle 4$

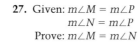

Statements	Reasons
a. $m\angle ABC = 90$ $m\angle EDC = 90$ $m\angle 1 = m\angle 3$	a. Given
b. $m\angle ABC = m\angle EDC$	b. Substitution property of equality
c. $m\angle ABC = m\angle 1 + m\angle 2$ $m\angle EDC = m\angle 3 + m\angle 4$	c. Angle addition postulate
d. $m\angle 1 + m\angle 2 = m\angle 3 + m\angle 4$	d. Substitution property of equality
e. $m\angle 1 + m\angle 2 = m\angle 1 + m\angle 4$	e. Substitution property of equality
f. $m\angle 2 = m\angle 4$	f. Subtraction property of equality

31. Given: $A = p + prt$

Prove: $p = \frac{A}{1 + rt}$

Statements	Reasons
a. $A = p + prt$	a. Given
b. $A = p(1 + rt)$	b. Distributive property
c. $\frac{A}{1 + rt} = p$	c. Division property of equality
d. $p = \frac{A}{1 + rt}$	d. Symmetric property of equality

34. $7 = 7$; detachment   **35.** $\frac{5}{9} = 0.\overline{5}$ **36.** $(3, \frac{11}{2})$

37.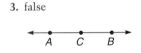

Page 94 Mid-Chapter Review
1. true, since a midpoint divides a segment into two congruent segments   **2.** True, since $x^3 = x \cdot x \cdot x$ and the reals are closed under multiplication.

3. false

4. True, since all right angles have a measure of 90 and congruent angles are by definition angles with the same measure.
5. If two lines are perpendicular, then they form four right angles.   **6.** If a substance is ordinary table sugar, then its chemical formula is $C_{12}H_{22}O_{11}$.   **7.** If it is a non-leap year, then the month of February has 28 days.   **8.** no conclusion
9. The diagonals of $ABCD$ are congruent; detachment
10. If two angles are right angles, then they have the same measure; syllogism
11. Given: $x = 7$

Prove: $4x^2 = 196$

Statements	Reasons
a. $x = 7$	a. Given
b. $x \cdot x = 7 \cdot x$	b. Multiplication property of equality
c. $x^2 = 7 \cdot 7$	c. Substitution property of equality
d. $4x^2 = 196$	d. Multiplication property of equality

12. Given: $AC = AB$, $AC = 4x + 1$, $AB = 6x - 13$

Prove: $x = 7$

Statements	Reasons
a. $AC = AB$ $AC = 4x + 1$ $AB = 6x - 13$	a. Given
b. $4x + 1 = AB$	b. Substitution property of equality
c. $4x + 1 = 6x - 13$	c. Substitution property of equality
d. $1 = 2x - 13$	d. Subtraction property of equality
e. $14 = 2x$	e. Addition property of equality
f. $7 = x$	f. Division property of equality
g. $x = 7$	g. Symmetric property of equality

Pages 96–97 Lesson 2-5
3. 9th - Anthony, 10th - Erin, 11th - Brad, 12th - Lisa   **5.** 19
7. Umeko - Drama Club, delivery person; Jim - Spanish Club, tutor; Gwen - marching band, lifeguard   **9.** The jeans box contained T-shirts and jeans, the T-shirt box contained jeans, and the jeans and T-shirts box contained T-shirts.

Pages 100–104 Lesson 2-6
5. Theorem 2-1 (Congruence of segments is reflexive.)
7. Theorem 2-1 (Congruence of segments is symmetric.)
9. Substitution property of equality   **11.** Addition property of equality

13. Given: $\angle A$ is a right angle.
Prove: $m\angle A = 90$

15. Given: $\angle 1$ and $\angle 2$ are vertical angles.
Prove: $\angle 1 \cong \angle 2$

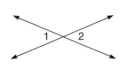

17. Given: ΔQED is a triangle.
Prove: $m\angle Q + m\angle E + m\angle D = 180$

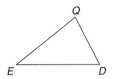

19a. $\overline{LE} \cong \overline{MR}$; $\overline{EG} \cong \overline{RA}$ **b.** Definition of congruent segments   **c.** Segment addition postulate   **e.** $LG = MA$ **f.** Definition of congruent segments   **21a.** Given   **b.** Definition of congruent segments   **c.** $AM = MB$; $CN = ND$ **d.** Segment addition postulate   **e.** Substitution property of equality   **f.** Substitution property of equality   **h.** Division property of equality   **i.** Definition of congruent segments

23. Given: $MP = NP$
$PO = PL$
Prove: $MO = NL$

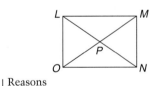

Statements	Reasons
1. $MP = NP$ $PO = PL$	1. Given
2. $MP + PO = NP + PL$	2. Addition property of equality
3. $MO = MP + PO$ $NL = NP + PL$	3. Segment addition postulate
4. $MO = NL$	4. Substitution property of equality

25. Given: $\overline{SA} \cong \overline{ND}$
Prove: $\overline{SN} \cong \overline{AD}$

Statements	Reasons
1. $\overline{SA} \cong \overline{ND}$	1. Given
2. $SA = ND$	2. Definition of congruent segments
3. $SA + AN = AN + ND$	3. Addition property of equality
4. $SN = SA + AN$ $AD = AN + ND$	4. Segment addition postulate
5. $SN = AD$	5. Substitution property of equality
6. $\overline{SN} \cong \overline{AD}$	6. Definition of congruent segments

27. Given: $\overline{QT} \cong \overline{RT}$
$\overline{TS} \cong \overline{TP}$
Prove: $\overline{QS} \cong \overline{RP}$

Statements	Reasons
1. $\overline{QT} \cong \overline{RT}$ $\overline{TS} \cong \overline{TP}$	**1.** Given
2. $QT = RT$ $TS = TP$	**2.** Definition of congruent segments
3. $QT + TS = RT + TP$	**3.** Addition property of equality
4. $QS = QT + TS$ $RP = RT + TP$	**4.** Segment addition postulate
5. $QS = RP$	**5.** Substitution property of equality
6. $\overline{QS} \cong \overline{RP}$	**6.** Definition of congruent segments

29. Given: $AB = CD$
M is the midpoint of $\overline{AB}$.
N is the midpoint of $\overline{CD}$.
Prove: $AM = MB = CN = ND$

Statements	Reasons
1. $AB = CD$ M is the midpoint of $\overline{AB}$. N is the midpoint of $\overline{CD}$.	**1.** Given
2. $AB = AM + MB$ $CD = CN + ND$	**2.** Segment addition postulate
3. $AM = MB$ $CN = ND$	**3.** Definition of midpoint
4. $AB = AM + AM$ $CD = CN + CN$	**4.** Substitution property of equality
5. $2AM = 2CN$	**5.** Substitution property of equality
6. $AM = CN$	**6.** Division property of equality
7. $AM = ND$	**7.** Transitive property of equality
8. $AM = MB = CN = ND$	**8.** Transitive property of equality

31. Statements	Reasons
1. The defendant drove through a red traffic light at the corner of Washington and Elm.	**1.** The defendant was seen.
2. The signal was not down.	**2.** The traffic computer shows no indication that the signal was down.
3. The defendant is subject to a $50 fine.	**3.** The law states that if a driver proceeds through a red traffic light that is in proper working order, that driver is subject to a $50 fine.

33. Jan. - Pablo, Feb. - Timothy, Aug. - Emma, Sept. - Amy
34. Division or multiplication property of equality   **35.** If a student maintains a C average, then he or she is eligible to play a varsity sport.   **36.** a satellite   **37.** $m\angle AND = 72$, $m\angle NOR = 18$ **38a.** 5 units   **b.** 2 units   **c.** 7 units   **d.** 8 units

Pages 107–109 Lesson 2-7
3. sometimes   **5.** sometimes   **7.** sometimes   **9.** always
11. $m\angle 1 = 112$, $m\angle 2 = 112$ **13.** $m\angle 1 = 140$, $m\angle 2 = 40$
15. (1) Given   (4) $m\angle ABC = m\angle 1 + m\angle 2$ (5) Substitution property of equality   (6) Definition of supplementary
17. $\angle MLN$ or $\angle PLQ$ **19.** 110   **21.** 30
23. Given: $\angle ABC \cong \angle EFG$
$\angle ABD \cong \angle EFH$
Prove: $\angle DBC \cong \angle HFG$

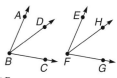

Statements	Reasons
1. $\angle ABC \cong \angle EFG$ $\angle ABD \cong \angle EFH$	**1.** Given
2. $m\angle ABC = m\angle EFG$ $m\angle ABD = m\angle EFH$	**2.** Definition of congruent angles
3. $m\angle ABC = m\angle ABD + m\angle DBC$ $m\angle EFG = m\angle EFH + m\angle HFG$	**3.** Angle addition postulate
4. $m\angle ABD + m\angle DBC = m\angle EFH + m\angle HFG$	**4.** Substitution property of equality
5. $m\angle DBC = m\angle HFG$	**5.** Subtraction property of equality
6. $\angle DBC \cong \angle HFG$	**6.** Definition of congruent angles

25. Given: $\angle A$ is an angle.
Prove: $\angle A \cong \angle A$

Statements	Reasons
1. $\angle A$ is an angle.	**1.** Given
2. $m\angle A = m\angle A$	**2.** Reflexive property of equality
3. $\angle A \cong \angle A$	**3.** Definition of congruent angles

27. Given: $\angle X$ and $\angle Y$ are right angles.
Prove: $\angle X \cong \angle Y$

Statements	Reasons
1. $\angle X$ and $\angle Y$ are right angles.	**1.** Given
2. $m\angle X = 90$ $m\angle Y = 90$	**2.** Definition of right angle
3. $m\angle X = m\angle Y$	**3.** Substitution property of equality
4. $\angle X \cong \angle Y$	**4.** Definition of congruent angles

29. Given: $\ell \perp m$
Prove: $\angle 1$, $\angle 2$, and $\angle 3$, and $\angle 4$ are right angles.

Statements	Reasons
1. $\ell \perp m$	**1.** Given
2. $\angle 1$ is a right angle.	**2.** Definition of perpendicular lines
3. $m\angle 1 = 90$	**3.** Definition of right angle
4. $\angle 1 \cong \angle 3$	**4.** Vertical $\angle$s are $\cong$.
5. $m\angle 3 = 90$	**5.** Definition of congruent angles
6. $\angle 3$ is a right angle.	**6.** Definition of right angle

7. ∠1 and ∠4 form a linear pair. ∠1 and ∠2 form a linear pair.	7. Definition of linear pair
8. ∠1 and ∠4 are supplementary. ∠1 and ∠2 are supplementary.	8. If 2 ∠s form a linear pair, they are supp.
9. $m∠1 + m∠4 = 180$ $m∠1 + m∠2 = 180$	9. Definition of supplementary
10. $90 + m∠4 = 180$ $90 + m∠2 = 180$	10. Substitution property of equality
11. $m∠4 = 90$ $m∠2 = 90$	11. Subtraction property of equality
12. ∠4 is a right angle. ∠2 is a right angle.	12. Definition of right angle

31. Given: ∠S ≅ ∠T
∠S and ∠T are supplementary.
Prove: ∠S and ∠T are right angles.

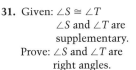
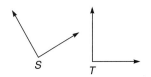

Statements	Reasons
1. ∠S ≅ ∠T ∠S and ∠T are supplementary.	1. Given
2. $m∠S = m∠T$	2. Definition of congruent angles
3. $m∠S + m∠T = 180$	3. Definition of supplementary
4. $2m∠S = 180$	4. Substitution property of equality
5. $m∠S = 90$	5. Division property of equality
6. $m∠T = 90$	6. Substitution property of equality
7. ∠S and ∠T are right angles.	7. Definition of right angles

35. Reflexive property of equality
36. $F = \frac{9}{5}C + 32$ Given
$F - 32 = \frac{9}{5}C$ Subtraction property of equality
$\frac{5}{9}(F - 32) = C$ Multiplication property of equality
37. no conclusion   38. no complement, 21

Pages 110–113 Chapter 2 Summary and Review
1. true   3. If something is a cloud, then it has a silver lining.   5. If a rock is obsidian, then it is a glassy rock produced by a volcano.   7. ∠A and ∠B have measures with a sum of 90; detachment   9. The Sun is in constant motion; syllogism
11. Given: $MN = PN$
$NL = NO$
Prove: $ML = PO$

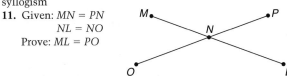

Statements	Reasons
1. $MN = PN$ $NL = NO$	1. Given
2. $MN + NL = PN + NO$	2. Addition property of equality
3. $ML = MN + NL$ $PO = PN + NO$	3. Segment addition postulate
4. $ML = PO$	4. Substitution property of equality

13. Given: $\overline{AM} ≅ \overline{CN}$
$\overline{MB} ≅ \overline{ND}$
Prove: $\overline{AB} ≅ \overline{CD}$

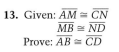
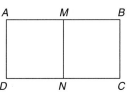

Statements	Reasons
1. $\overline{AM} ≅ \overline{CN}$ $\overline{MB} ≅ \overline{ND}$	1. Given
2. $AM = CN$ $MB = ND$	2. Definition of congruent segments
3. $AM + MB = CN + ND$	3. Addition property of equality
4. $AB = AM + MB$ $CD = CN + ND$	4. Segment addition postulate
5. $AB = CD$	5. Substitution property of equality
6. $\overline{AB} ≅ \overline{CD}$	6. Definition of congruent segments

15. Given: ∠1 and ∠2 form a linear pair.
∠1 ≅ ∠2
Prove: ∠1 and ∠2 are right angles.

Statements	Reasons
1. ∠1 and ∠2 form a linear pair. ∠1 ≅ ∠2	1. Given
2. $m∠1 = m∠2$	2. Definition of congruent angles
3. ∠1 and ∠2 are supplementary.	3. If 2 ∠s form a linear pair, they are supp.
4. $m∠1 + m∠2 = 180$	4. Definition of supplementary
5. $m∠1 + m∠1 = 180$	5. Substitution property of equality
6. $2m∠1 = 180$	6. Substitution property of equality
7. $m∠1 = 90$	7. Division property of equality
8. $m∠2 = 90$	8. Substitution property of equality
9. ∠1 and ∠2 are right angles.	9. Definition of right angle

17. A sponge remains permanently attached to a surface for all of its adult life; syllogism
19. $t = 35d + 20$ Given
$t - 20 = 35d$ Subtraction property of equality
$\frac{t - 20}{35} = d$ Division property of equality

CHAPTER 3 PARALLELS

Pages 119–121 Lesson 3-1
5. 21
7. Given: ∠AED and ∠BEC are supplementary.
Prove: ∠AED is a right angle.
∠BEC is a right angle.

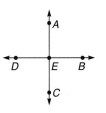

9. They are the same.   **11.** 1,111,088,889   **13.** 6   **17.** 11

Pages 124–127 Lesson 3-2
5. parallel   **7.** parallel, intersecting, skew   **9.** intersecting or parallel   **11.** false; could be skew   **13.** true   **15.** false   **17.** false   **19.** true   **21.** true   **23.** $\overrightarrow{RT}$; alternate interior angles   **25.** $\overrightarrow{RV}$; consecutive interior angles   **27.** m; corresponding angles   **29.** a; corresponding angles   **31.** ℓ; consecutive interior angles   **33.** parallel and intersecting   **35.** intersecting   **37.** intersecting

39.

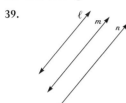

43.

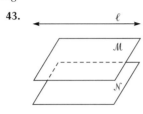

45. $\overline{CV}, \overline{BW}, \overline{ZX}, \overline{AY}, \overline{RU}$ **47.** plane BWX **49.** If plane $\mathcal{A}$ is parallel to plane $\mathcal{B}$ and plane $\mathcal{B}$ is parallel to plane $\mathcal{C}$, then plane $\mathcal{A}$ is parallel to plane $\mathcal{C}$. The basement floor is parallel to the ground-level floor and the ground-level floor is parallel to the upstairs floor, so the basement floor is parallel to the upstairs floor.   **51a.** 2,000 feet   **b.** easy to keep track of which airplanes are eastbound and which are westbound; less worry about collisions
52. Given: $\overrightarrow{BA}$ and $\overrightarrow{BD}$ are opposite rays.
  Prove: $\angle ABC$ and $\angle CBD$
      are supplementary.

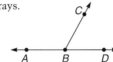

53. Given: $\angle 1 \cong \angle 2$
  Prove: $\angle 1 \cong \angle 3$

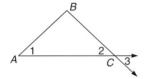

Statements	Reasons
1. $\angle 1 \cong \angle 2$	1. Given
2. $\angle 2 \cong \angle 3$	2. Vertical $\angle$s are $\cong$.
3. $\angle 1 \cong \angle 3$	3. Congruence of angles is transitive.

54. Symmetric property of equality   **55.** If two lines lie in the same plane and do not intersect, then they are parallel.
56. $(2, -2)$ **57.** 13

Pages 130–134 Lesson 3-3
7. $\angle 5$ and $\angle 2$ are supplementary; $\angle 1$ and $\angle 4$ are supplementary; $\angle 3 \cong \angle 2$; $\angle 3$ and $\angle 5$ are supplementary.
9. 82   **11.** 82   **13.** 40   **15.** 140   **17.** $x = 16$, $y = 11$
19. 90   **21.** 125   **23.** 35   **25.** $x = 52$, $y = 13$ **27.** $x = 7$, $y = 16$ **29.** 77   **31.** 62   **33.** 103   **35.** $x = 16$, $y = 41$, $z = 42$ **37.** $x = 127$, $y = 5$, $z = 31$ **39.** 35   **41.** 125   **43.** 35
45. a. Given   **b.** $\perp$ lines form 4 rt. $\angle$s.   **c.** $m\angle 1 = 90$
d. If 2 $\parallel$ lines are cut by a trasversal, corr. $\angle$s are $\cong$.   **e.** $m\angle 2$ $= 90$ **f.** Definition of right angle   **g.** $m \perp p$

47. Given: $\ell \parallel m$
  Prove: $\angle 3$ and $\angle 5$ are
      supplementary.
      $\angle 4$ and $\angle 6$ are
      supplementary.

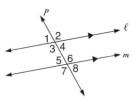

Proof: Since $\ell \parallel m$, $\angle 1 \cong \angle 5$ and $\angle 2 \cong \angle 6$ because they are corresponding angles. $\angle 1$ and $\angle 3$ form a linear pair, as do $\angle 2$ and $\angle 4$, by definition. Hence $\angle 1$ and $\angle 3$ are supplementary, as are $\angle 2$ and $\angle 4$, because if 2 $\angle$s form a linear pair, they are supplementary. By definition of supplementary $m\angle 1 + m\angle 3$ $= 180$ and $m\angle 2 + m\angle 4 = 180$. Then by substitution, $m\angle 5 + m\angle 3 = 180$ and $m\angle 6 + m\angle 4 = 180$. Hence $\angle 3$ and $\angle 5$ and $\angle 4$ and $\angle 6$ are supplementary by definition.

49. Given: $\overline{MQ} \parallel \overline{NP}$
      $\angle 1 \cong \angle 5$
  Prove: $\angle 4 \cong \angle 3$

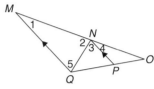

Statements	Reasons
1. $\overline{MQ} \parallel \overline{NP}$ $\angle 1 \cong \angle 5$	1. Given
2. $\angle 4 \cong \angle 1$	2. If 2 $\parallel$ lines are cut by a transversal, corr. $\angle$s are $\cong$.
3. $\angle 4 \cong \angle 5$	3. Congruence of angles is transitive.
4. $\angle 5 \cong \angle 3$	4. If 2 $\parallel$ lines are cut by a transversal, alt. int. $\angle$s are $\cong$.
5. $\angle 4 \cong \angle 3$	5. Congruence of angles is transitive.

53. parallel

54. Given: $\overline{AB} \cong \overline{FE}$
      $\overline{BC} \cong \overline{ED}$
  Prove: $\overline{AC} \cong \overline{FD}$

Statements	Reasons
1. $\overline{AB} \cong \overline{FE}$ $\overline{BC} \cong \overline{ED}$	1. Given
2. $AB = FE$ $BC = ED$	2. Definition of congruent segments
3. $AC = AB + BC$ $FD = FE + ED$	3. Segment addition postulate
4. $AB + BC = FE + ED$	4. Addition property of equality
5. $AC = FD$	5. Substitution property of equality
6. $\overline{AC} \cong \overline{FD}$	6. Definition of congruent segments

55. Lines p and q never meet; detachment   **56.** 7   **57.** -3 or 9   **58.** III

Page 134 Mid-Chapter Review
1. Given: $a \parallel b$
      $\angle 1 \cong \angle 2$
  Prove: $a \perp \ell$
      $b \perp \ell$

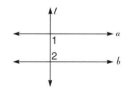

2. $\overline{EF}$ **3.** $\overline{AC}$, $\overline{EF}$ **4.** plane *DEF* **5.** none **6.** 65 **7.** 65
8. 115 **9.** 65 **10.** 115 **11.** 115 **12.** 65

Pages 137–141 Lesson 3-4
7. 13 **9.** 10 **11.** 20 **13.** $\overleftrightarrow{AB} \parallel \overleftrightarrow{DC}$, Theorem 3-5
15. false; They must be in a plane. **17.** false; The alternate
interior angles would be congruent. **19.** $\overleftrightarrow{GK} \parallel \overleftrightarrow{HL}$,
corresponding angles congruent (Postulate 3-2) **21.** $\overleftrightarrow{FG} \parallel \overleftrightarrow{JK}$,
alternate interior angles congruent (Theorem 3-5) **23.** none
25. $c \parallel d$, corresponding angles congruent (Postulate 3-2)
27. none **29.** none **31.** $x = 7$, $y = 138$ **33.** $\angle 2$ and $\angle 3$;
$\angle 6$ and $\angle 7$; $\angle 2$ and $\angle 8$; $\angle 6$ and $\angle 4$; $\angle 3$ and $\angle 5$; $\angle 7$ and
$\angle 1$; $\angle 5$ and $\angle 8$; $\angle 1$ and $\angle 4$ **35.** $\overleftrightarrow{AE} \parallel \overleftrightarrow{DF}$ since in a plane, if
2 lines are $\perp$ to the same line they are $\parallel$. $\overleftrightarrow{EB} \parallel \overleftrightarrow{FH}$ since if 2
lines in a plane are cut by a transversal and alt. int. $\angle$s are $\cong$,
then the lines are $\parallel$. **37.** **a.** Given **b.** $\angle 1$ and $\angle 2$ form a
linear pair. **c.** If 2 $\angle$s form a linear pair, they are supp.
d. $\angle 1 \cong \angle 3$ **e.** If 2 lines in a plane are cut by a transversal
and corr. $\angle$s are $\cong$, the lines are $\parallel$.
39. Given: $\angle RQP \cong \angle PSR$
　　　　$\angle SRQ$ and $\angle PSR$ are
　　　　supplementary.
　　Prove: $\overline{QP} \parallel \overline{RS}$

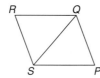

Statements	Reasons
1. $\angle RQP \cong \angle PSR$ $\angle SRQ$ and $\angle PSR$ are supplementary.	1. Given
2. $m\angle RQP = m\angle PSR$	2. Definition of congruent angles
3. $m\angle SRQ + m\angle PSR = 180$	3. Definition of supplementary
4. $m\angle SRQ + m\angle RQP = 180$	4. Substitution property of equality
5. $\angle SRQ$ and $\angle RQP$ are supplementary.	5. Definition of supplementary
6. $\overline{QP} \parallel \overline{RS}$	6. If 2 lines are cut by a transversal and consec. int. $\angle$s are supp., then the lines are $\parallel$.

41. Given: $\angle ABD \cong \angle BEF$
　　　　$\overrightarrow{BC}$ bisects $\angle ABD$.
　　　　$\overrightarrow{EH}$ bisects $\angle BEF$.
　　Prove: $\overrightarrow{BC} \parallel \overrightarrow{EH}$

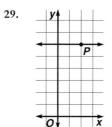

Statements	Reasons
1. $\angle ABD \cong \angle BEF$ $\overrightarrow{BC}$ bisects $\angle ABD$. $\overrightarrow{EH}$ bisects $\angle BEF$.	1. Given
2. $m\angle 1 + m\angle 2 = m\angle ABD$ $m\angle 3 + m\angle 4 = m\angle BEF$	2. Angle addition postulate
3. $m\angle 1 + m\angle 2 = m\angle 3 + m\angle 4$	3. Substitution property of equality
4. $m\angle 1 = m\angle 2$ $m\angle 3 = m\angle 4$	4. Definition of angle bisector
5. $m\angle 1 + m\angle 1 = m\angle 3 + m\angle 3$	5. Substitution property of equality

6. $2m\angle 1 = 2m\angle 3$

7. $m\angle 1 = m\angle 3$

8. $\overrightarrow{BC} \parallel \overrightarrow{EH}$

6. Substitution property of equality	
7. Division property of equality	
8. If 2 lines are cut by a transversal and corr. $\angle$s are $\cong$, then the lines are $\parallel$.	

43. Postulate 3-2 **44.** true **45.** intersecting or skew
46.

Statements	Reasons
1. $A = 2\pi r^2 + 2\pi rh$	1. Given
2. $A - 2\pi r^2 = 2\pi rh$	2. Subtraction property of equality
3. $\frac{A - 2\pi r^2}{2\pi r} = h$	3. Division property of equality

47. If something is a cloud, then it is composed of millions
of water droplets. **48.** yes **49.** 6

Pages 144–147 Lesson 3-5
7. $-\frac{1}{2}$, falling **9.** 1, rising **11.** undefined, vertical
13. perpendicular **15.** $\frac{3}{4}$, $-\frac{4}{3}$ **17.** $-\frac{3}{2}$, $\frac{2}{3}$ **19.** $\frac{11}{3}$, $-\frac{3}{11}$ **21.** 2
23. $\frac{3}{2}$ **25.** undefined **27.** $-\frac{2}{3}$

29. 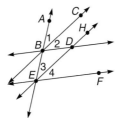　　**33.**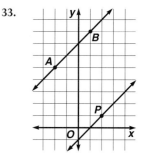

35. slope of $\overleftrightarrow{AB} = -\frac{1}{3}$; slope of $\overleftrightarrow{BC} = -\frac{1}{3}$ Either $\overleftrightarrow{AB} \parallel \overleftrightarrow{BC}$ or
$\overleftrightarrow{AB}$ and $\overleftrightarrow{BC}$ are the same line. Since B is a common point,
$\overleftrightarrow{AB}$ is not parallel to $\overleftrightarrow{BC}$. Thus $\overleftrightarrow{AB}$ and $\overleftrightarrow{BC}$ are the same line
and A, B, and C are collinear. **37.** slope of $\overleftrightarrow{AB} = 3$; slope
of $\overleftrightarrow{CD} = 3$. No, $\overleftrightarrow{AB}$ and $\overleftrightarrow{CD}$ are the same line since A, B, C,
and D are collinear. **43.** (7, 5), (5, -1), or (-3, 3) **45.** slope
between (5, 5) and (9, 1) = -1; slope between (5, 5) and (0, 10)
= -1. So the lines between (5, 5) and (9,1) and (5, 5) and (0, 10)
are either parallel or the same line. But the lines share the
point (5, 5). So, they must be the same line. Therefore, the
line through (5, 5) and (9, 1) crosses the y-axis at (0, 10).
47a. 7500 **b.** The rate of change is the slope of the line that
relates population and time. **c.** 300,000

49a.　　　　　　　　　　　　　　　**b.** 11,080 feet

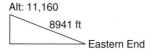

Alt: 11,160

8941 ft

Eastern End

51. Given: $\angle 1 \cong \angle 2$
$\overline{PQ} \perp \overline{QR}$
Prove: $\overline{ST} \perp \overline{PQ}$

Statements	Reasons
1. $\angle 1 \cong \angle 2$ $\overline{PQ} \perp \overline{QR}$	**1.** Given
2. $\overline{ST} \parallel \overline{QR}$	**2.** If 2 lines in a plane are cut by a transversal and corr. $\angle$s are $\cong$, the lines are $\parallel$.
3. $\overline{ST} \perp \overline{PQ}$	**3.** In a plane, if a line is $\perp$ to one of 2 $\parallel$ lines, it is $\perp$ to the other.

52. Sample answers: $\angle 1$ and $\angle 2$ are supplementary; $\angle 1 \cong \angle 3$; $\angle 2$ and $\angle 3$ are supplementary; $\angle 4$ and $\angle 5$ are supplementary; $\angle 4 \cong \angle 6$; $\angle 5$ and $\angle 6$ are supplementary.

53.

Statements	Reasons
1. $5x - 7 = x + 1$	**1.** Given
2. $4x - 7 = 1$	**2.** Subtraction property of equality
3. $4x = 8$	**3.** Addition property of equality
4. $x = 2$	**4.** Division property of equality

54. $(\frac{1}{2})^2 = \frac{1}{4}$, and $\frac{1}{2} \not< \frac{1}{4}$ **55.** $(2, 9)$

56.

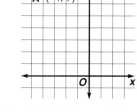

Pages 151–154 Lesson 3-6

5.

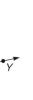

7.

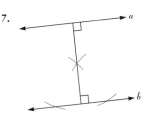

9. true   **11.** true   **13.** yes; equidistant at all points

15.

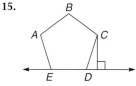

17.

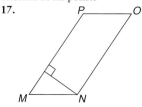

19. $2\sqrt{10}$ **21.** 0   **23.** 5

25.

27.

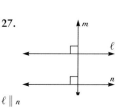

$\ell \parallel n$

$\angle CBA$ and $\angle DBA$ are right angles. AB is the distance from A to $\overline{CD}$.

29. $\overline{PS}$ **31.** $\overline{QR}$ **33.** Not necessarily. The lines may be skew.

35. yes   **37a.** E. 69th St.; it is the perpendicular from the point to the line.   **b.** Sample answers: heavy traffic, one-way streets   **39a.** 2.1 kilometers   **b.** 2.7 kilometers   **c.** 4.1 kilometers   **d.** Riverboat; bird; bird is not restricted to traveling on the roads or in the rivers.   **40.** $-\frac{2}{5}$

41.

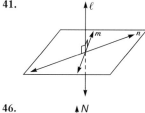

42. true   **43.** 43
44. yes; detachment
45. Hypothesis—two lines are parallel; conclusion—they are everywhere equidistant.

46.

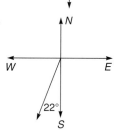

Pages 156–158 Chapter 3 Summary and Review
1. l, n **3.** l, n, ℓ **5.** l, m **7.** $\angle FAC, \angle ACD; \angle GAD, \angle ADC$
9. $\angle EBC$ **11.** $\overline{DE}$ is parallel to $\overline{CF}$; The alternate interior angles are congruent.   **13.** 6   **15.** $-\frac{1}{3}$ **17.** $-\frac{9}{7}, \frac{7}{9}$
19. $\frac{1}{6}, -6$ **21.** $\overline{PS}$ **23.** $\overline{RQ}$ or $\overline{MP}$ **25.** 36   **27.** 65 mph

Pages 160–161 Algebra Review
1. 10   **3.** -18   **5.** 69   **7.** 18   **9.** 22   **11.** 16   **13.** 48
15. 87.5%
17.

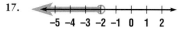

19. $\{n | n < 13\}$
21. $\{a | a \geq -5.5\}$

23. $\{x | x \leq 7\}$ **25.** y^8 **27.** $20a^5x^5$ **29.** $\frac{-y^2}{w}$ **31.** 4   **33.** 3
35. $2mr$ **37.** xyz^3 **39.** 3.7 hours   **41.** 7 weeks

CHAPTER 4 CONGRUENT TRIANGLES

Pages 166–169 Lesson 4-1
13. $\Delta BCD, \Delta ABD$ **15.** ΔBCD **17.** $\overline{BD}, \overline{AD}$ **19.** 33 units
21. right isosceles   **23.** scalene   **25.** obtuse isosceles
27. $\angle R, \angle O, \angle M$ **29.** $\angle O, \angle M$ **31.** $\overline{RO}$ and $\overline{RM}$ **33.** true
35. true   **37.** false   **39.** equiangular, equilateral
41. equiangular, equilateral   **43.** ΔOAT and ΔRYE, and ΔWHT and ΔCAR; ΔOAT and ΔRYE **45.** obtuse, isosceles
47. Given: $m\angle NMO = 20$
Prove: ΔLMN is an obtuse triangle.

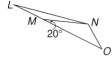

Statements	Reasons
1. $m\angle NMO = 20$	**1.** Given
2. $\angle LMN$ and $\angle NMO$ form a linear pair.	**2.** Definition of linear pair
3. $\angle LMN$ and $\angle NMO$ are supplementary.	**3.** If 2 $\angle$s form a linear pair, they are supp.
4. $m\angle LMN + m\angle NMO = 180$	**4.** Definition of supplementary

5. $m\angle LMN + 20 = 180$ | **5.** Substitution property of equality
6. $m\angle LMN = 160$ | **6.** Subtraction property of equality
7. $\angle LMN$ is obtuse. | **7.** Definition of obtuse angle
8. ΔLMN is obtuse. | **8.** Definition of obtuse triangle

49. Write an inequality for the value of the perimeter.
$$23 < 2x + 2 + 10 + x + 4 < 32$$
$$23 < 3x + 16 < 32$$
$$7 < 3x < 16$$
$$\tfrac{7}{3} < x < \tfrac{16}{3}$$
Since ΔDEF is isosceles, one of the following is true.

$EF = ED$	or	$EF = FD$	or	$FD = ED$
$2x + 2 = 10$		$2x + 2 = x + 4$		$x + 4 = 10$
$2x = 8$		$x = 2$		$x = 6$
$x = 4$				

The only case that satisfies the inequality is $x = 4$, so $EF = ED$. The vertex angle is the angle between the congruent sides, so $\angle DEF$ is the vertex angle.   **51.** Isosceles triangle; the segments from the vertex A to B, C, D, and E will be congruent.

55a. 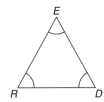   **b.** 36   **56.** Yes; the definition of parallel is that they are everywhere equidistant.

57. 0.1 or $\frac{1}{10}$ **58.** No; if they were parallel, $m\angle 1 = m\angle 5$.

59. 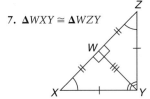   **60.** Substitution property of equality   **61.** 17

Pages 174–176 Lesson 4-2
5. 42   **7.** 73   **9.** 65   **11.** 51
13.

exterior angle	remote interior angles
$\angle 1$	$\angle 3$ and $\angle 4$
$\angle 5$	$\angle 3$ and $\angle 4$
$\angle 6$	$\angle 2$ and $\angle 4$
$\angle 7$	$\angle 2$ and $\angle 4$
$\angle 8$	$\angle 2$ and $\angle 3$
$\angle 9$	$\angle 2$ and $\angle 3$

15. yes   **17.** Sample answer: $\angle S$ and $\angle SUV$ are complementary and $\angle R$ and $\angle RUT$ are complementary.
19. 90   **21.** 70   **23.** 17   **25.** 55   **27.** 55   **29.** 55   **31.** 90
33. 30   **35.** 30   **37.** 15   **39.** $m\angle BTD = 140$, $m\angle CTD = 40$, $m\angle ATB = 40$, $m\angle B = 20$, $m\angle A = 120$, $m\angle D = 120$

41. Given: ΔRED is equiangular.
Prove: $M\angle R = m\angle E = m\angle D = 60$

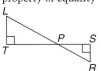

Statements	Reasons
1. ΔRED is equiangular.	1. Given
2. $\angle R \cong \angle E \cong \angle D$	2. Definition of equiangular
3. $m\angle R = m\angle E = m\angle D$	3. Definition of congruent angles
4. $m\angle R + m\angle E + m\angle D = 180$	4. The sum of the $\angle$s in a Δ is 180.
5. $m\angle R + m\angle R + m\angle R = 180$	5. Substitution property of equality
6. $3m\angle R = 180$	6. Substitution property of equality
7. $m\angle R = 60$	7. Division property of equality
8. $m\angle R = m\angle E = m\angle D = 60$	8. Substitution property of equality

43. Given: ΔABC 3
Prove: $m\angle DCB = m\angle CAB + m\angle CBA$

Statements	Reasons
1. ΔABC	1. Given
2. $\angle DCB$ and $\angle BCA$ form a linear pair.	2. Definition of linear pair
3. $\angle DCB$ and $\angle BCA$ are supplementary.	3. If 2 $\angle$s form a linear pair, they are supp.
4. $m\angle DCB + m\angle BCA = 180$	4. Definition of supplementary
5. $m\angle BCA + m\angle CAB + m\angle CBA = 180$	5. The sum of the $\angle$s in a Δ is 180.
6. $m\angle DCB + m\angle BCA = m\angle BCA + m\angle CAB + m\angle CBA$	6. Substitution property of equality
7. $m\angle DCB = m\angle CAB + m\angle CBA$	7. Subtraction property of equality

45. Given: $\overline{LT} \perp \overline{TS}$
$\overline{ST} \perp \overline{SR}$
Prove: $\angle TLR \cong \angle LRS$

Statements	Reasons
1. $\overline{LT} \perp \overline{TS}$ $\overline{ST} \perp \overline{SR}$	1. Given
2. $\angle LTP$ is a right angle. $\angle RSP$ is a right angle.	2. $\perp$ lines form 4 rt. $\angle$s
3. $\angle LTP \cong \angle RSP$	3. All rt. $\angle$s are $\cong$.
4. $\angle LPT \cong \angle RPS$	4. Vertical $\angle$s are $\cong$.
5. $\angle TLR \cong \angle LRS$	5. If 2 $\angle$s in a Δ are $\cong$ to 2 $\angle$s in another Δ, the third $\angle$s are $\cong$ also.

47. $\angle 2 \cong \angle 3$ **49.** 360   **53.** yes   **54.** $-\frac{1}{9}$

55. Congruence of angles is reflexive.
56. 45   **57.** 2 or 8

Pages 179–182 Lesson 4-3
5. ΔPSK **7.** $\Delta WXY \cong \Delta WZY$

9. ΔEFD **11.** $\angle K$ **13.** $\overline{BR}$ **15.** $\Delta MIT \cong \Delta NIT$

17. The congruent parts are not corresponding. **19.** $\triangle EAD$
21. $\triangle ABR \cong \triangle ABS$, $\triangle ARO \cong \triangle ASO \cong \triangle BRO \cong \triangle BSO$,
$\triangle ASR \cong \triangle BSR$ **23.** 11 **25.** $\overline{BD} \cong \overline{AE}$ is true because the
segments are corresponding parts of congruent triangles.
27. $\overline{BC} \cong \overline{AC}$ is not necessarily true. They are not
corresponding parts of the triangles. **29.** $\angle CBA \cong \angle CED$ is
true. The angles are corresponding parts of congruent
triangles.
35. Given: $\triangle XYZ$
     Prove: $\triangle XYZ \cong \triangle XYZ$

Statements	Reasons
1. $\angle X \cong \angle X$ $\angle Y \cong \angle Y$ $\angle Z \cong \angle Z$	**1.** Congruence of angles is reflexive.
2. $\overline{XY} \cong \overline{XY}$ $\overline{YZ} \cong \overline{YZ}$ $\overline{XZ} \cong \overline{XZ}$	**2.** Congruence of segments is reflexive.
3. $\triangle XYZ \cong \triangle XYZ$	**3.** Definition of congruent triangles

37. Given: $\triangle MNO \cong \triangle ONM$
     Prove: $\triangle MNO$ is isosceles.

Statements	Reasons
1. $\triangle MNO \cong \triangle ONM$	**1.** Given
2. $\overline{MN} \cong \overline{ON}$	**2.** CPCTC
3. $\triangle MNO$ is isosceles.	**3.** Definition of isosceles triangle

42. 133 **43.** 105, 33, 42 **44.** true **45.** -2 **46.** 12

Pages 187–191 Lesson 4-4
5. SSS **7a.** Given **b.** Definition of a midpoint **c.** Given
d. If 2 ∥ lines are cut by a traversal, alt. int. ∠s are ≅.
e. Vertical ∠s are ≅. **f.** ASA **9.** SAS **11.** true
13. could be true **15.** could be true **17.** could be true
19. SAS **21.** SSS **23.** $\triangle GHF$; SSS
25. Given: $\angle A \cong \angle D$
       $AO \cong OD$
     Prove: $\triangle AOB \cong \triangle DOC$

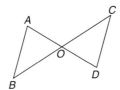

Statements	Reasons
1. $\angle A \cong \angle D$ $\overline{AO} \cong \overline{OD}$	**1.** Given
2. $\angle AOB \cong \angle DOC$	**2.** Vertical ∠s are ≅.
3. $\triangle AOB \cong \triangle DOC$	**3.** ASA

27. Given: $\overline{MO} \cong \overline{PO}$
       $\overline{NO}$ bisects $\overline{MP}$.
     Prove: $\triangle MNO \cong \triangle PNO$

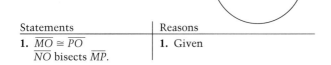

Statements	Reasons
1. $\overline{MO} \cong \overline{PO}$ $\overline{NO}$ bisects $\overline{MP}$.	**1.** Given

	2. $\overline{MN} \cong \overline{PN}$	**2.** Definition of bisector
	3. $\overline{NO} \cong \overline{NO}$	**3.** Congruence of segments is reflexive.
	4. $\triangle MNO \cong \triangle PNO$	**4.** SSS

29.

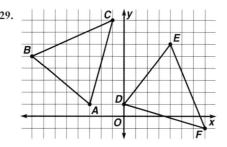

$AB = \sqrt{(-3-(-8))^2 + (1-5)^2} = \sqrt{41}$
$DE = \sqrt{(0-4)^2 + (1-6)^2} = \sqrt{41}$
$BC = \sqrt{(-8-(-1))^2 + (5-8)^2} = \sqrt{58}$
$EF = \sqrt{(4-7)^2 + (6-(-1))^2} = \sqrt{58}$
$AC = \sqrt{(-3-(-1))^2 + (1-8)^2} = \sqrt{53}$
$DF = \sqrt{(0-7)^2 + (1-(-1))^2} = \sqrt{53}$

Since corresponding sides are congruent, the triangles are
congruent by SSS.
31. Given: $\angle 3 \cong \angle 4$
       $\overline{DC} \cong \overline{BA}$
     Prove: $\angle 1 \cong \angle 6$

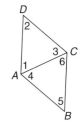

Statements	Reasons
1. $\angle 3 \cong \angle 4$ $\overline{DC} \cong \overline{BA}$	**1.** Given
2. $\overline{AC} \cong \overline{AC}$	**2.** Congruence of segments is reflexive.
3. $\triangle ADC \cong \triangle CBA$	**3.** SAS
4. $\angle 1 \cong \angle 6$	**4.** CPCTC

33a. $\triangle DCF$ **b.** $\triangle EFC$ **c.** $\triangle EDF$ **d.** $\triangle FCG$ **e.** $\triangle ECG$
f. $\triangle CGA$ **34.** $\angle B \cong \angle T$; $\angle I \cong \angle O$; $\angle G \cong \angle P$; $\overline{BI} \cong \overline{TO}$;
$\overline{IG} \cong \overline{OP}$; $\overline{BG} \cong \overline{TP}$ **35.** 12 units **36.** Hypothesis: you
want a great pizza; Conclusion: go to Katie's **37.** a straight
angle

Page 191 Mid-Chapter Review
1. False, the hypotenuse must be
longer than the legs.
2. true **3.** true **4.** true **5.** yes **6.** no **7.** yes **8.** yes

9. $\angle O$ **10.** $\overline{TP}$ **11.** $\angle T$

12. Given: $\overline{QP} \cong \overline{ST}$
       $\angle P$ and $\angle T$ are
       right angles.
       R is the midpoint
       of $\overline{PT}$.
     Prove: $\overline{QR} \cong \overline{SR}$

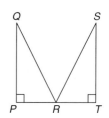

Statements	Reasons
1. $\overline{QP} \cong \overline{ST}$ $\angle P$ and $\angle T$ are right angles. R is the midpoint of PT.	1. Given
2. $\angle P \cong \angle T$	2. All rt. $\angle$s are $\cong$.
3. $\overline{PR} \cong \overline{TR}$	3. Definition of midpoint
4. $\triangle QPR \cong \triangle STR$	4. SAS
5. $\overline{QR} \cong \overline{SR}$	5. CPCTC

Pages 194–197 Lesson 4-5

7. $\overline{CE} \cong \overline{AB}$ **9.** $\angle ACB \cong \angle CAE$
11. Given: $\overline{AB} \cong \overline{CD}$
$\overline{AB} \parallel \overline{CD}$
Prove: $\triangle AOB \cong \triangle DOC$

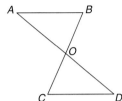

Proof: We are given that $\overline{AB} \cong \overline{CD}$ and $\overline{AB} \parallel \overline{CD}$. $\angle A \cong \angle D$ and $\angle B \cong \angle C$ since if parallel lines are cut by a transversal, alternate interior angles are congruent. Therefore, $\triangle AOB \cong \triangle DOC$ by ASA.
13. $\triangle ADO \cong \triangle CBO$ by AAS or ASA; $\triangle ADB \cong \triangle CBD$ by SAS; $\triangle ABD \cong \triangle CAD$ by SAS **15.** $\triangle AOD \cong \triangle COB$ by SAS; $\triangle AOB \cong \triangle COD$ by SAS **17.** $\angle E \cong \angle S$ **19.** $\overline{DE} \cong \overline{RS}$ or $\overline{EF} \cong \overline{ST}$ **21.** $\triangle ABD \cong \triangle CDB$ by AAS; $\triangle AOD \cong \triangle CDB$ by AAS and CPCTC **23.** One order of steps: 3, 1, 5, 9, 8, 2, 7, 6, 4. (1) $\perp$ lines form rt. $\angle$s. (2) Given (3) Given (4) CPCTC (5) Given (6) ASA (7) Given (8) All rt. $\angle$s are $\cong$. (9) $\perp$ lines form four rt. $\angle$s. **25.** not valid because the conditions of the theorems or postulates that prove the congruence of triangles are not met **27.** $\overline{ST} \cong \overline{QN}$, $\angle S \cong \angle Q$, and $\overline{PS} \cong \overline{PQ}$ is given. $\triangle TSP \cong \triangle NQP$ by SAS. $\overline{TP} \cong \overline{NP}$ by CPCTC. So $\triangle TPN$ is isosceles by the definition of isosceles triangle.

29. Given: $\overline{AB} \cong \overline{AC}$
D is the midpoint
of $\overline{BC}$.
Prove: $\triangle ABD \cong \triangle ACD$

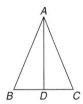

Statements	Reasons
1. $\overline{AB} \cong \overline{AC}$ D is the midpoint of $\overline{BC}$.	1. Given
2. $\overline{BD} \cong \overline{CD}$	2. Definition of midpoint
3. $\overline{AD} \cong \overline{AD}$	3. Congruence of segments is reflexive.
4. $\triangle ABD \cong \triangle ACD$	4. SSS

33.

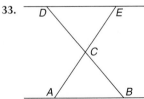

$\overline{DE}$ represents the ironing board and AB represents the floor. Since the legs bisect each other, $\overline{DC} \cong \overline{BC}$ and $\overline{CE} \cong \overline{CA}$. $\angle DCE \cong \angle BCA$ because vertical angles are congruent.

Therefore, $\triangle DCE \cong \triangle BCA$ by SAS. Since corresponding parts of congruent triangles are congruent, $\angle D \cong \angle B$. Since if two lines are cut by a transversal so that alt. int. $\angle$s are congruent, the lines are parallel, $\overline{DE} \parallel \overline{AB}$.
34. Given: $\overline{PR} \cong \overline{TR}$
$\angle 1 \cong \angle 2$
$\angle P$ and $\angle T$ are
right angles.
Prove: $\overline{QR} \cong \overline{SR}$

Statements	Reasons
1. $\overline{PR} \cong \overline{TR}$ $\angle 1 \cong \angle 2$ $\angle P$ and $\angle T$ are right angles.	1. Given
2. $\angle P \cong \angle T$	2. All rt. $\angle$s are $\cong$.
3. $\triangle QPR \cong \triangle STR$	3. ASA
4. $\overline{QR} \cong \overline{SR}$	4. CPCTC

35. $\angle ACK$ **36.** Alternate interior angles are congruent; corresponding angles are congruent; alternate exterior angles are congruent; consecutive interior angles are supplementary; two lines in a plane are perpendicular to a third line.
37. $2\sqrt{29}$; $(2, 6)$

Pages 199–201 Lesson 4-6

5. $2\frac{7}{12}$ **7.** $\triangle BCG \cong \triangle FCD$; $\overline{BG} \cong \overline{FD}$; $\triangle ABG \cong \triangle EFD$; $\angle A \cong \angle E$ **9.** $4624 = 68^2$

11. Subgoals: $\triangle ABD \cong \triangle ACB$, $\angle DAB \cong \angle CAB$
Proof:
Given: $\overline{AB} \perp$ plane BCD
$\overline{DB} \cong \overline{CB}$
Prove: $\angle DAB \cong \angle CAB$

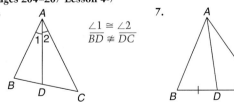

Statements	Reasons
1. $\overline{AB} \perp$ plane BCD	1. Given
2. $\overline{AB} \perp \overline{DB}$ $\overline{AB} \perp \overline{CB}$	2. Definition of perpendicular plane
3. $\angle ABD$ and $\angle ABC$ are right angles.	3. $\perp$ lines form four rt. $\angle$s.
4. $\angle ABD \cong \angle ABC$	4. All rt. $\angle$s are $\cong$.
5. $\overline{DB} \cong \overline{CB}$	5. Given
6. $\overline{AB} \cong \overline{AB}$	6. Congruence of segments is reflexive.
7. $\triangle ABD \cong \triangle ABC$	7. SAS
8. $\angle DAB \cong \angle CAB$	8. CPCTC

13. 3; 1 dog, 1 cat, and 1 hamster **15.** 12

Pages 204–207 Lesson 4-7

5.

$\angle 1 \cong \angle 2$
$\overline{BD} \not\cong \overline{DC}$

7.

9. isosceles **11.** $\angle EAB$ and $\angle 4$ **13.** $\angle 5$ and $\angle 6$ or $\angle EFA$ and $\angle BGA$ **15.** 67 **17.** 21

19. Given: $\overline{AB} \cong \overline{BC}$
Prove: $\angle 3 \cong \angle 5$

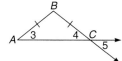

Statements	Reasons
1. $\overline{AB} \cong \overline{BC}$	1. Given
2. $\angle 3 \cong \angle 4$	2. If 2 sides of a **Δ** are ≅ the ∡ opp. the sides are ≅.
3. $\angle 4 \cong \angle 5$	3. Vertical ∡ are ≅.
4. $\angle 3 \cong \angle 5$	4. Congruence of angles is transitive.

21. 6 **23.** 3 **25.** 18 **27.** ΔTBR is isosceles—Definition of isosceles triangle; $\angle RQB \cong \angle BCR$—$\angle RQB$ and $\angle 1$ form a linear pair and so do $\angle BCR$ and $\angle 2$, so these pairs are supplementary. Since $\angle 1 \cong \angle 2 \angle RQB \cong \angle BCR$. $\Delta TQR \cong \Delta TCB$—AAS ($\angle 1 \cong \angle 2$, $\angle T \cong \angle T$, and $\overline{TB} \cong \overline{TR}$) $\angle TBR \cong \angle TRB$—If two sides of a **Δ** are ≅, then the ∡ opp. the sides are ≅.

29. Given: $\overline{PS} \cong \overline{QR}$
$\angle 3 \cong \angle 4$
Prove: $\angle 1 \cong \angle 2$

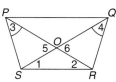

Statements	Reasons
1. $\overline{PS} \cong \overline{QR}$ $\angle 3 \cong \angle 4$	1. Given
2. $\angle 5 \cong \angle 6$	2. Vertical ∡ are ≅.
3. $\Delta POS \cong \Delta QOR$	3. AAS
4. $\overline{OS} \cong \overline{OR}$	4. CPCTC
5. $\angle 1 \cong \angle 2$	5. If 2 sides of a **Δ** are ≅, the ∡ opp. the sides are ≅.

31. Given: ΔABC is isosceles.
$\overline{DE} \parallel \overline{AB}$
$\overline{AC} \cong \overline{BC}$
Prove: ΔDEC is isosceles.

Statements	Reasons
1. $\overline{AC} \cong \overline{BC}$ $\overline{DE} \parallel \overline{AB}$	1. Given
2. $\angle A \cong \angle B$	2. If 2 sides of a **Δ** are ≅, the ∡ opp. the sides are ≅.
3. $\angle A \cong \angle EDC$ $\angle B \cong \angle DEC$	3. If 2 ∥ lines are cut by a transversal, corr. ∡ are ≅.
4. $\angle EDC \cong \angle DEC$	4. Congruence of angles is transitive.
5. $\overline{CD} \cong \overline{CE}$	5. If 2 ∡ of a **Δ** are ≅, the sides opp. the ∡ are ≅.
6. ΔDEC is isosceles.	6. Definition of isosceles triangle

33. Given: $\overline{AB} \cong \overline{AC}$
$\overline{BX}$ bisects $\angle ABC$.
$\overline{CX}$ bisects $\angle ACB$.
Prove: $\overline{BX} \cong \overline{CX}$

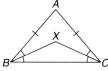

Statements	Reasons
1. $\overline{AB} \cong \overline{AC}$ $\overline{BX}$ bisects $\angle ABC$. $\overline{CX}$ bisects $\angle ACB$.	1. Given
2. $\angle ABX \cong \angle XBC$ $\angle ACX \cong \angle XCB$	2. Definition of angle bisector
3. $\angle ABC \cong \angle ACB$	3. If 2 sides of a **Δ** are ≅, the ∡ opp. the sides are ≅.
4. $m\angle ABX = m\angle XBC$ $m\angle ACX = m\angle XCB$ $m\angle ABC = m\angle ACB$	4. Definition of congruent angles
5. $m\angle ABC = m\angle ABX$ $+\angle XBC$ $m\angle ACB = m\angle ACX$ $+ m\angle XCB$	5. Angle addition postulate
6. $m\angle ABC = 2m\angle XBC$ $m\angle ACB = 2m\angle XCB$	6. Substitution property of equality
7. $2m\angle XBC = 2m\angle XCB$	7. Substitution property of equality
8. $m\angle XBC = m\angle XCB$	8. Division property of equality
9. $\angle XBC = mXCB$	9. Definition of congruent angles
10. $\overline{BX} \cong \overline{CX}$	10. If 2 ∡ of a **Δ** are ≅, the sides opp. the ∡ are ≅.

35. $m\angle 1 = 20$, $m\angle 2 = 140$, $m\angle 3 = 20$, $m\angle 4 = 40$, $m\angle 5 = 100$, $m\angle 6 = 40$ $m\angle 7 = 140$

37. Given: ΔMNO is equilateral.
Prove: $m\angle M$, $m\angle N$, and $m\angle O = 60$

Statements	Reasons
1. ΔMNO is equilateral.	1. Given
2. ΔMNO is equiangular.	2. An equilateral **Δ** is equiangular.
3. $m\angle M = m\angle N = m\angle O$	3. Definition of equiangular
4. $m\angle M + m\angle N$ $+m\angle O = 180$	4. The sum of the ∡ in a **Δ** is 180.
5. $3m\angle M = 180$	5. Substitution property of equality
6. $m\angle M = 60$	6. Division property of equality
7. $m\angle M$, $m\angle N$, $m\angle O = 60$	7. Substitution property of equality

39. Given: ΔPQR
$\angle P \cong \angle R$
Prove: $\overline{PQ} \cong \overline{RQ}$

Statements	Reasons
1. Let $\overline{QT}$ bisect $\angle PQR$.	1. Through any 2 pts. there is 1 line.
2. $\angle P \cong \angle R$	2. Given
3. $\angle 1 \cong \angle 2$	3. Definition of an angle bisector
4. $\overline{QT} \cong \overline{QT}$	4. Congruence of segments is reflexive.
5. $\Delta PQT \cong \Delta RQT$	5. AAS
6. $\overline{PQ} \cong \overline{RQ}$	6. CPCTC

41. 135 **43a.** 77 **b.** 30 **c.** 74 **d.** 51 **e.** 47.5 **f.** 39.5

44. $\triangle BLS \cong \triangle BES$; $\angle LBS \cong \angle EBS$; $\triangle LBU \cong \triangle EBU$; $\angle LUS \cong \angle EUS$ **45.** $\angle NTI \cong \angle NCA$, $\angle TIN \cong \angle CAN$, and $\overline{TN} \cong \overline{CN}$; $\angle NTI \cong \angle NCA$; $\angle TIN \cong \angle CAN$, and $\overline{IN} \cong \overline{AN}$; $\angle NTI \cong \angle NCA$, $\angle INT \cong \angle ANC$, and $\overline{TI} \cong \overline{CA}$; $\angle NTI \cong \angle NCA$, $\angle INT \cong \angle ANC$, and $\overline{IN} \cong \overline{AN}$; $\angle TIN \cong \angle CAN$, $\angle INT \cong \angle ANC$, and $\overline{TI} \cong \overline{CA}$; $\angle TIN \cong \angle CAN$, $\angle INT \cong \angle ANC$, and $\overline{TN} \cong \overline{CN}$
46. 31, 52, 97
47. $\angle 1 \cong \angle 4 \cong \angle 5 \cong \angle 8$; $\angle 2 \cong \angle 3 \cong \angle 6 \cong \angle 7$; Angles that are supplementary because the lines are parallel: $\angle 2$ and $\angle 5$, $\angle 4$ and $\angle 6$, $\angle 1$ and $\angle 7$, $\angle 3$ and $\angle 8$; Angles that are supplementary because they form linear pairs: $\angle 1$ and $\angle 2$, $\angle 2$ and $\angle 4$, $\angle 4$ and $\angle 3$, $\angle 3$ and $\angle 1$, $\angle 5$ and $\angle 7$, $\angle 7$ and $\angle 8$, $\angle 8$ and $\angle 6$, $\angle 6$ and $\angle 5$.
48. Don will receive an A on the Geometry test; detachment

Pages 208–210 Summary and Review
1. $\triangle ABE$, $\triangle DBE$ **3.** $\overline{BD}$ **5.** $\triangle ABF$, $\triangle BFD$, $\triangle DFE$, $\triangle AFE$, $\triangle BCD$ **7.** 53 **9.** 53 **11.** 120 **13.** 60 **15.** 55 **17.** 25 **19.** $\angle I$ **21.** $\overline{HG}$ **23.** $\overline{IG}$
25. Given: E is the midpoint of $\overline{AC}$.
$\angle 1 \cong \angle 2$
Prove: $\angle 3 \cong \angle 4$

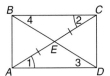

Statements	Reasons
1. E is the midpoint of $\overline{AC}$. $\angle 1 \cong \angle 2$	1. Given
2. $\overline{AE} \cong \overline{CE}$	2. Definition of midpoint
3. $\angle BEC \cong \angle DEA$	3. Vertical $\angle$s are $\cong$.
4. $\triangle BEC \cong \triangle DEA$	4. ASA
5. $\angle 3 \cong \angle 4$	5. CPCTC

27. We are given that $\overline{KL} \cong \overline{ML}$, $\angle J \cong \angle N$, and $\angle 1 \cong \angle 2$. Therefore, $\triangle JKL \cong \triangle NML$ by AAS. So, $\overline{JK} \cong \overline{NM}$ by CPCTC.
29. 12
31. Given: $\angle A \cong \angle D$
$\overline{AB} \cong \overline{DC}$
E is the midpoint of $\overline{AD}$.
Prove: $\angle 3 \cong \angle 4$

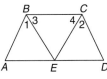

Statements	Reasons
1. $\angle A \cong \angle D$ $\overline{AB} \cong \overline{DC}$ E is the midpoint of AD.	1. Given
2. $\overline{AE} \cong \overline{DE}$	2. Definition of midpoint
3. $\triangle ABE \cong \triangle DCE$	3. SAS
4. $\overline{BE} \cong \overline{CE}$	4. CPCTC
5. $\angle 3 \cong \angle 4$	5. If 2 sides of a $\triangle$ are $\cong$, the $\angle$s opp. the sides are $\cong$.

33. 729

CHAPTER 5 APPLYING CONGRUENT TRIANGLES

Pages 219–222 Lesson 5-1
5. $\overline{AD}$ **7.**

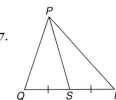

11.

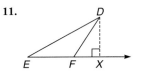

13. Given: $\overline{AB} \cong \overline{CB}$
$\overline{BD}$ is a median of $\triangle ABC$.
Prove: $\overline{BD}$ is an altitude of $\triangle ABC$.

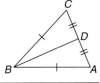

Statements	Reasons
1. $\overline{AB} \cong \overline{CB}$	1. Given
2. $\overline{BD}$ is a median of $\triangle ABC$.	2. Given
3. D is the midpoint of $\overline{CA}$.	3. Definition of median
4. $\overline{CD} \cong \overline{AD}$	4. Definition of midpoint
5. B is on the perpendicular bisector of $\overline{AC}$. D is on the perpendicular bisector of $\overline{AC}$.	5. A pt. equidistant from the endpts. of a segment lies on the $\perp$ bisector of the segment.
6. $\overline{BD}$ is the perpendicular bisector of $\overline{AC}$.	6. Through any 2 pts. there is 1 line.
7. $\overline{BD} \perp \overline{AC}$	7. Definition of perpendicular bisector
8. $\overline{BD}$ is an altitude of $\triangle ABC$.	8. Definition of altitude

15. **17.**

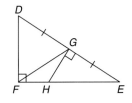

19. always **21.** sometimes **23a.** $(7, 4)$ **b.** 3 **c.** yes **25.** 2

27. Given: $\overline{BD}$ is a median of $\triangle ABC$.
$\overline{BD}$ is an altitude of $\triangle ABC$.
Prove: $\triangle ABC$ is isosceles.

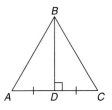

Statements	Reasons
1. $\overline{BD}$ is a median of $\triangle ABC$.	1. Given
2. D is the midpoint of $\overline{AC}$.	2. Definition of median
3. $\overline{AD} \cong \overline{CD}$	3. Definition of midpoint
4. $\overline{BD}$ is an altitude of $\triangle ABC$.	4. Given
5. $\overline{BD} \perp \overline{AC}$	5. Definition of altitude
6. $\angle BDA$ and $\angle BDC$ are right angles.	6. $\perp$ lines form 4 rt. $\angle$s.
7. $\angle BDA \cong \angle BDC$	7. All rt. $\angle$s are $\cong$.
8. $\overline{BD} \cong \overline{BD}$	8. Congruence of segments is reflexive.
9. $\triangle BDA \cong \triangle BDC$	9. SAS
10. $\overline{BA} \cong \overline{BC}$	10. CPCTC
11. $\triangle ABC$ is isosceles.	11. Definition of isosceles triangle

29. Given: C is equidistant from A and B.
E is the midpoint of $\overline{AB}$.
Prove: C lies on the perpendicular bisector of $\overline{AB}$.

Statements	Reasons
1. C is equidistant from A and B. E is the midpoint of $\overline{AB}$.	1. Given
2. $\overline{CA} \cong \overline{CB}$	2. Definition of equidistant
3. Draw $\overleftrightarrow{CE}$.	3. Through any 2 pts. there is 1 line.
4. $\overline{CE} \cong \overline{CE}$	4. Congruence of segments is reflexive.
5. $\overline{AE} \cong \overline{BE}$	5. Definition of midpoint
6. $\triangle CAE \cong \triangle CBE$	6. SSS
7. $\angle CEA \cong \angle CEB$	7. CPCTC
8. $\angle CEA$ and $\angle CEB$ form a linear pair.	8. Definition of linear pair
9. $\angle CEA$ and $\angle CEB$ are supplementary.	9. If 2 $\angle$ s form a linear pair, they are supp.
10. $m\angle CEA + m\angle CEB = 180$	10. Definition of supplementary
11. $m\angle CEA + m\angle CEA = 180$	11. Substitution property of equality
12. $2m\angle CEA = 180$	12. Substitution property of equality
13. $m\angle CEA = 90$	13. Division property of equality
14. $m\angle CEB = 90$	14. Substitution property of equality
15. $\angle CEA$ and $\angle CEB$ are right angles.	15. Definition of right angle
16. $\overleftrightarrow{CE} \perp \overline{AB}$	16. Definition of perpendicular
17. $\overleftrightarrow{CE}$ is the perpendicular bisector of $\overline{AB}$.	17. Definition of perpendicular bisector
18. C is on the perpendicular bisector of $\overline{AB}$.	18. A line contains at least 2 pts.

31. Given: $\overline{AB} \cong \overline{AC}$
$\overline{AD}$ is an altitude of $\triangle ABC$.
Prove: $\overline{AD}$ is a median of $\triangle ABC$.

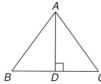

Statements	Reasons
1. $\overline{AD}$ is an altitude of $\triangle ABC$.	1. Given
2. $\overline{AD} \perp \overline{BC}$	2. Definition of altitude.
3. $\angle ADB$ and $\angle ADC$ are right angles.	3. $\perp$ lines form 4 rt. $\angle$s.
4. $\angle ADB \cong \angle ADC$	4. All rt. $\angle$s are $\cong$.
5. $\overline{AB} \cong \overline{AC}$	5. Given
6. $\angle ABD \cong \angle ACD$	6. If 2 sides of a $\triangle$ are $\cong$, the $\angle$s opp. the sides are $\cong$.
7. $\triangle ADB \cong \triangle ADC$	7. AAS
8. $\overline{BD} \cong \overline{CD}$	8. CPCTC
9. $\overline{AD}$ is a median of $\triangle ABC$.	9. Definition of median

33. Given: $\triangle ABC \cong \triangle DEF$
$\overline{BG}$ is an angle bisector of $\triangle ABC$.
$\overline{EH}$ is an angle bisector of $\triangle DEF$.
Prove: $\overline{BG} \cong \overline{EH}$

Statements	Reasons
1. $\triangle ABC \cong \triangle DEF$	1. Given
2. $\angle A \cong \angle D$ $\overline{AB} \cong \overline{DE}$ $\angle ABC \cong \angle DEF$	2. CPCTC
3. $\overline{BG}$ is an angle bisector of $\triangle ABC$. EH is an angle bisector of $\triangle DEF$.	3. Given
4. $\overline{BG}$ bisects $\angle ABC$. $\overline{EH}$ bisects $\angle DEF$.	4. Definition of angle bisector
5. $m\angle ABG = \frac{1}{2}m\angle ABC$ $m\angle DEH = \frac{1}{2}m\angle DEF$	5. Definition of bisect
6. $m\angle ABC = m\angle DEF$	6. Definition of congruent angles
7. $\frac{1}{2}m\angle ABC = \frac{1}{2}m\angle DEF$	7. Multiplication property of equality
8. $m\angle ABG = m\angle DEH$	8. Substitution property of equality
9. $\angle ABG \cong \angle DEH$	9. Definition of congruent angles
10. $\triangle ABG \cong \triangle DEH$	10. ASA
11. $\overline{BG} \cong \overline{EH}$	11. CPCTC

38. 7   **39.** 107   **40.** 8   **41.** yes; both have slope $-\frac{2}{3}$

42. $\overline{AX}$ is an altitude of $\triangle ABC$; detachment.

Pages 226–229 Lesson 5-2

5. none   **7.** yes, HA or AAS   **9.** none   **11.** $x = 6$

13. Given: $\angle Q$ and $\angle S$ are right angles.
$\angle 1 \cong \angle 2$
Prove: $\triangle PQR \cong \triangle RSP$

Statements	Reasons
1. $\angle Q$ and $\angle S$ are right angles. $\angle 1 \cong \angle 2$	1. Given
2. $\triangle QRP$ and $\triangle SPR$ are right triangles.	2. Definition of right triangle
3. $\overline{PR} \cong \overline{RP}$	3. Congruence of segments is reflexive.
4. $\triangle PQR \cong \triangle RSP$	4. HA

15. No; there is no AA or AAA congruence theorem.

17. Yes; HA   **19.** $x = 5$, $y = 3$ **21.** $x = 7$, $y = 20$

23. Given: $\overline{QP} \cong \overline{SR}$
$\angle Q$ and $\angle S$ are right angles.
Prove: $\angle 1 \cong \angle 2$

Statements	Reasons
1. $\overline{QP} \cong \overline{SR}$ $\angle Q$ and $\angle S$ are right angles.	1. Given
2. $\triangle RQP$ and $\triangle PSR$ are right triangles.	2. Definition of right triangle

3. $\overline{PR} \cong \overline{RP}$ **3.** Congruence of segments is reflexive.
4. $\triangle RQP \cong \triangle PSR$ **4.** HL
5. $\angle 1 \cong \angle 2$ **5.** CPCTC

25. Given: $\triangle ABY$ and $\triangle CBY$ are right triangles.
$\overline{AB} \cong \overline{CB}$
$\overline{YX} \perp \overline{AC}$
Prove: $\overline{AX} \cong \overline{CX}$

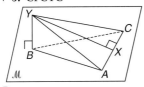

Statements	Reasons
1. $\triangle ABY$ and $\triangle CBY$ are right triangles. $\overline{AB} \cong \overline{CB}$	1. Given
2. $\overline{YB} \cong \overline{YB}$	2. Congruence of segments is reflexive.
3. $\triangle YBA \cong \triangle YBC$	3. LL
4. $\overline{YA} \cong \overline{YC}$	4. CPCTC
5. $\overline{YX} \perp \overline{AC}$	5. Given
6. $\angle YXA$ and $\angle YXC$ are right angles.	6. $\perp$ lines form four rt. $\angle$s.
7. $\triangle YXA$ and $\triangle YXC$ are right triangles.	7. Definition of right triangle
8. $\overline{YX} \cong \overline{YX}$	8. Congruence of segments is reflexive.
9. $\triangle YXA \cong \triangle YXC$	9. HL
10. $\overline{AX} \cong \overline{CX}$	10. CPCTC

27. Given: $\triangle ABC$ and $\triangle DEF$ are right triangles.
$\overline{AB} \cong \overline{DE}$
$\overline{BC} \cong \overline{EF}$
Prove: $\triangle ABC \cong \triangle DEF$

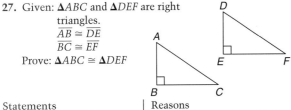

Statements	Reasons
1. $\triangle ABC$ and $\triangle DEF$ are right triangles.	1. Given
2. $\angle ABC$ and $\angle DEF$ are right angles.	2. Definition of right triangle
3. $\angle ABC \cong \angle DEF$	3. All rt. $\angle$s are $\cong$.
4. $\overline{AB} \cong \overline{DE}$ $\overline{BC} \cong \overline{EF}$	4. Given
5. $\triangle ABC \cong \triangle DEF$	5. SAS

29. Given: $\triangle ABC \cong \triangle DEF$
$\overline{AG}$ is the altitude to $\overline{BC}$.
$\overline{DH}$ is the altitude to $\overline{EF}$.
Prove: $\overline{AG} \cong \overline{DH}$

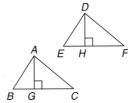

Statements	Reasons
1. $\triangle ABC \cong \triangle DEF$	1. Given
2. $\overline{AC} \cong \overline{DF}$ $\angle C \cong \angle F$	2. CPCTC
3. $\overline{AG}$ is the altitude to $\overline{BC}$. $\overline{DH}$ is the altitude to $\overline{EF}$.	3. Given
4. $\overline{AG} \perp \overline{BC}$ $\overline{DH} \perp \overline{EF}$	4. Definition of altitude
5. $\angle AGC$ is a right angle. $\angle DHF$ is a right angle.	5. $\perp$ lines form four rt. $\angle$s.
6. $\triangle AGC$ is a right triangle. $\triangle DHF$ is a right triangle.	6. Definition of right triangle
7. $\triangle AGC \cong \triangle DHE$	7. HA
8. $\overline{AG} \cong \overline{DH}$	8. CPCTC

33. The brace represents the hypotenuse of a right triangle as shown in the diagram, (assuming the deck will be attached at a right angle to the wall). If each brace is attached at the same distance from the wall, this distance represents a leg of the triangle. HL says that the four triangles will be congruent and hence all will be attached at the same distance on the wall since that distance represents the other leg of the triangle.

34. $\angle 6 \cong \angle 1$ **35.** $\angle 5$ and $\angle 4$; $\angle 1$ and $\angle 6$ **36.** $d = 11$
37. If a segment is a median of a triangle, then it bisects one side of the triangle.

Pages 231–232 Lesson 5-3
5. 61   **7.** 5   **9.** \$136   **11.** $O = 0$, $N = 5$, $G = 2$ or $O = 2$, $N = 3$, $G = 9$ **13.** 225 square units

Pages 235–239 Lesson 5-4
5. $\triangle ABC$ is right or obtuse.   **7.** If two parallel lines are cut by a transversal, then alternate exterior angles are not congruent.
9. >   **11.** <   **13.** division   **15.** transitive
17. Given: $\overline{PQ} \cong \overline{PR}$
      $\angle 1 \not\cong \angle 2$
   Prove: $\overline{PZ}$ is not a median of $\triangle PQR$.

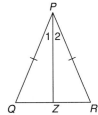

Proof: Assume that $\overline{PZ}$ is a median of $\triangle PQR$. We are given that $\overline{PQ} \cong \overline{PR}$ and that $\angle 1 \not\cong \angle 2$. Since $\overline{PZ}$ is a median of $\triangle PQR$, $\overline{QZ} \cong \overline{RZ}$ since Z is the midpoint of $\overline{QR}$ by the definition of median. If two sides of a triangle are congruent, then the angles opposite the sides are congruent, so $\angle Q \cong \angle R$. Therefore, $\triangle PZQ \cong \triangle PZR$ by SAS. Then $\angle 1 \cong \angle 2$ by CPCTC. This is a contradiction of a given fact. Therefore, our assumption that $\overline{PZ}$ is a median of $\triangle PQR$ must be false, which means $\overline{PZ}$ is not a median of $\triangle PQR$.
19. >   **21.** <

23. Given: $\triangle KNL$
      $\overline{NM} \cong \overline{OM}$
   Prove: $m\angle 1 > m\angle 2$

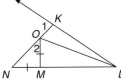

Statements	Reasons
1. $\overline{NM} \cong \overline{OM}$	1. Given
2. $\angle 2 \cong \angle N$	2. If 2 sides of a $\triangle$ are $\cong$, the $\angle$s opp. the sides are $\cong$.
3. $m\angle 1 > m\angle N$	3. If an $\angle$ is an ext. $\angle$ of a $\triangle$, then its measure is greater than the measure of either of its corr. remote int. $\angle$s.
4. $m\angle 2 = m\angle N$	4. Definition of congruent angles
5. $m\angle 1 > m\angle 2$	5. Substitution property of equality

25. Given: △ABC
m∠ABC = m∠BCA
Prove: x < y

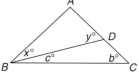

Statements	Reasons
1. m∠ABC = m∠BCA	**1.** Given
2. m∠ABC = x + c	**2.** Angle addition postulate
3. m∠ABC > x	**3.** Definition of inequality
4. m∠BCA > x	**4.** Substitution property of equality
5. m∠ADB > m∠BCA	**5.** If an ∠ is an ext. ∠ of a △, then its measure is greater than the measure of either of its curr. remote int. ∡s.
6. m∠ADB > x	**6.** Transitive property of inequality
7. y > x	**7.** Substitution property of equality

27. Given: ∠2 ≇ ∠1
Prove: ℓ is not parallel to m.

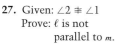

Proof: We are given that ∠2 ≇ ∠1. We assume that ℓ is parallel to m. Since ∠1 and ∠2 are corresponding angles, the corresponding angles postulate says that ∠1 ≅ ∠2. But this is a contradiction of our given fact. Therefore, our assumption that ℓ is parallel to m must be false, and hence ℓ is not parallel to m.

29. Given: Intersecting lines ℓ and m.
Prove: ℓ and m intersect in no more than one point.

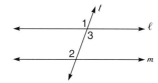

Proof: Assume lines ℓ and m intersect in more than one point. This means a point P is on both lines ℓ and m, and a point Q is on both lines ℓ and m. Therefore points P and Q determine both lines ℓ and m. But this contradicts our postulate that states through any two points there is exactly one line. So, our assumption must be false, which means lines ℓ and m intersect in no more than one point.

31. Given: $\overline{CD}$, $\overline{BE}$, and $\overline{AF}$ are altitudes of △ABC. $\overline{CD} \cong \overline{BE} \cong \overline{AF}$
Prove: △ABC is scalene.

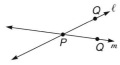

Proof: Assume △ABC is not scalene. Then at least two sides must be congruent by the definition. Say $\overline{AB} \cong \overline{AC}$. Now consider △ABE and △ACD. Since $\overline{CD}$ and $\overline{BE}$ are altitudes and altitudes are perpendicular to the opposite side by definition, and perpendicular lines form right angles, we have that △ABE and △ACD are right triangles. By the reflexive property of congruent angles, ∠BAC ≅ ∠BAC, and hence △ABE ≅ △ACD by HA. Then by CPCTC, $\overline{BE} \cong \overline{CD}$. But this contradicts our given statement and hence our assumption must be false. $\overline{AB} \not\cong \overline{BC}$ and $\overline{AC} \not\cong \overline{BC}$ can be proved in a similar manner. Therefore, △ABC is scalene.

33. Given: X is in the interior of △PQR.
Prove: m∠X > m∠Q

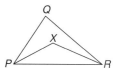

Statements	Reasons
1. m∠X + m∠XPR + m∠XRP = 180 m∠Q + m∠QPR + m∠QRP = 180	**1.** The sum of the ∡s in a △ is 180.
2. m∠X + m∠XPR + m∠XRP = m∠Q + m∠QPR + m∠QRP	**2.** Substitution property of equality
3. m∠QPR = m∠QPX + m∠XPR m∠QRP = m∠QRX + m∠XRP	**3.** Angle addition postulate
4. m∠X + m∠XPR + m∠XRP = m∠Q + (m∠QPX + m∠XPR) + (m∠QRX + m∠XRP)	**4.** Substitution property of equality
5. m∠X = m∠Q + (m∠QPX + m∠QRX)	**5.** Subtraction property of equality
6. m∠X > m∠Q	**6.** Definition of inequality

35. Mr. Sopher; Photos and Tebbe each have alibis and Bloom couldn't shoot an arrow with a sling on. **36.** S; RS + ST = RT **37.** Distributive, Division **38.** x = 4; The figure is a square because the two triangles are congruent. **39.** 3000 bacteria

Page 239 Mid-Chapter Review

1. 3 **2.** 9 **3.** yes by LL or SAS **4.** $\overline{XY} \cong \overline{XZ}$ **5.** <; Exterior angle inequality theorem **6.** a > 2 **7.** We are given that we have two lines that are noncoplanar and that do not intersect. Assume the lines are not skew. Then we have two possibilities for the lines.
Case I: The lines intersect. This contradicts the given statement that the lines do not intersect.
Case II: The lines are parallel. By the definition of parallel, this contradicts the given statement that the lines are noncoplanar.
In each case, we are led to a contradiction. Hence our assumption must be false and therefore the lines are skew.
8. Eric

Pages 242–245 Lesson 5-5

5. ∠H, ∠G, ∠I **7.** $\overline{MN}$, $\overline{LN}$, $\overline{LM}$ **9.** $\overline{PQ}$, $\overline{RQ}$, $\overline{RP}$ **11.** $\overline{HT}$ **13.** $\overline{PA}$ **15.** ∠CBA; ∠A **17.** $\overline{QT}$ **19.** ∠EAD, ∠ADE, ∠DEA **21.** ∠ADB, ∠ABD, ∠BAD **23.** $\overline{QR}$, $\overline{PQ}$, $\overline{PR}$ **25.** Given: QR > QP $\overline{PR} \cong \overline{PQ}$
Prove: m∠P > m∠Q

Statements	Reasons
1. QR > QP	**1.** Given
2. m∠P > m∠R	**2.** If one side of a △ is longer than another side, then the ∠ opp. the longer side is greater than the ∠ opp. the shorter side.
3. $\overline{PR} \cong \overline{PQ}$	**3.** Given
4. ∠Q ≅ ∠R	**4.** If 2 sides of a △ are ≅, the ∡s opp. the sides are ≅.

5. $m\angle Q = m\angle R$ **5.** Definition of congruent angles
6. $m\angle P > m\angle Q$ **6.** Substitution property of equality
27. Given: $TE > AE$
$\qquad\qquad m\angle P > m\angle PAE$
$\qquad$ Prove: $TE > PE$

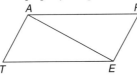

Statements	Reasons
1. $m\angle P > m\angle PAE$	**1.** Given
2. $AE > PE$	**2.** If one angle of a triangle is greater than another angle, then the side opp. the greater $\angle$ is longer than the side opp. the lesser $\angle$.
3. $TE > AE$	**3.** Given
4. $TE > PE$	**4.** Transitive property of inequality

29. Given: $\overline{PQ} \perp$ plane $\mathcal{M}$
$\qquad$ Prove: $\overline{PQ}$ is the shortest
$\qquad\qquad$ segment from P
$\qquad\qquad$ to plane $\mathcal{M}$.

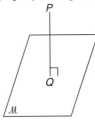

Proof: By definition, $\overline{PQ}$ is perpendicular to plane $\mathcal{M}$ if it is perpendicular to every line in $\mathcal{M}$ that intersects it. But by Theorem 5-11, that perpendicular segment is the shortest segment from the point to each of these lines. Therefore, $\overline{PQ}$ is the shortest segment from P to $\mathcal{M}$.

31. C; distance from C to line is greatest.

32. $-3 < x < \frac{21}{4}$ **33.** 5

34.

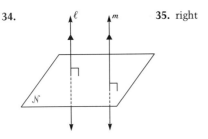

35. right

Pages 248–250 Lesson 5-6
5. no   **7.** yes   **9.** 9 and 17   **11.** no; the points are collinear.   **13.** yes   **15.** no   **17.** no   **19.** yes   **21.** no
23. no   **25.** yes   **27.** no   **29.** no   **31.** yes

33. $\{x \mid x > \frac{8}{3}\}$ **35.** 1

37. Given: $\overline{AD}$, $\overline{BE}$, $\overline{CF}$
$\qquad\qquad$ are altitudes
$\qquad\qquad$ of $\triangle ABC$.
$\qquad$ Prove: $AB + BC + AC >$
$\qquad\qquad AD + BE + CF$

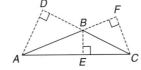

Statements	Reasons
1. $\overline{AD}$ is the altitude to $\overleftrightarrow{DC}$. $\overline{BE}$ is the altitude to $\overleftrightarrow{AC}$. $\overline{CF}$ is the altitude to $\overleftrightarrow{AB}$.	**1.** Given
2. $\overline{AD} \perp \overleftrightarrow{DC}$ $\overline{BE} \perp \overleftrightarrow{AC}$ $\overline{CF} \perp \overleftrightarrow{AB}$	**2.** Definition of altitude

3. $AB > AD$
$\quad\, BC > BE$
$\quad\, AC > CF$
3. The perpendicular segment from a point to a line is the shortest segment from the point to the line.

4. $AB = AD + x$
$\quad\, BC = BE + y$
$\quad\, AC = CF + z$
4. Definition of inequality

5. $AB + BC + AC = AD + BE + CF + x + y + z$
5. Addition property of equality

6. $AB + BC + AC > AD + BE + CF$
6. Definition of inequality

41a. 1 in., 4 in., 4 in.; 2 in., 3 in., 4 in.; 3 in., 3 in., 3 in.
b. The shortest side is 1 in. and the triangle is 1 in., 4 in., 4 in., an isosceles triangle   **43.** $\overline{XZ}$ **44.** acute, isosceles
45. -10

46. Given: $\overline{PQ}$ bisects $\overline{AB}$.
$\qquad$ Prove: $\overline{AM} \cong \overline{MB}$

Statements	Reasons
1. $\overline{PQ}$ bisects $\overline{AB}$.	**1.** Given
2. M is the midpoint of $\overline{AB}$.	**2.** Definition of bisect
3. $AM = MB$	**3.** Definition of midpoint
4. $\overline{AM} \cong \overline{MB}$	**4.** Definition of congruent segments

47. 73, 17

Pages 254–257 Lesson 5-7
5. $m\angle ALK < m\angle NLO$ **7.** $m\angle KLO = m\angle ALN$ **9.** $m\angle B$

11. Given: $\overline{PQ} \cong \overline{SQ}$
$\qquad$ Prove: $PR > SR$

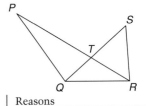

Statements	Reasons
1. $\overline{PQ} \cong \overline{SQ}$	**1.** Given
2. $\overline{QR} \cong \overline{QR}$	**2.** Congruence of segments is reflexive.
3. $m\angle PQR = m\angle PQS + m\angle SQR$	**3.** Angle addition postulate
4. $m\angle PQR > m\angle SQR$	**4.** Definition of inequality
5. $PR > SR$	**5.** SAS Inequality

13. $AB > AC$ **15.** $m\angle 1 < m\angle 2$ **17.** $m\angle DFE > m\angle DFG$
19. $x > 4$

21. Given: $\overline{PQ} \cong \overline{RS}$
$\qquad\qquad QR < PS$
$\qquad$ Prove: $m\angle 3 < m\angle 1$

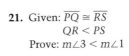

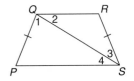

Statements	Reasons
1. $\overline{PQ} \cong \overline{RS}$	**1.** Given
2. $\overline{QS} \cong \overline{QS}$	**2.** Congruence of segments is reflexive.
3. $QR < PS$	**3.** Given
4. $m\angle 3 < m\angle 1$	**4.** SSS Inequality

23. Given: $\triangle TER$
$\overline{TR} \cong \overline{EU}$
Prove: $TE > RU$

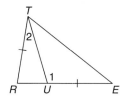

Statements	Reasons
1. $\overline{TR} \cong \overline{EU}$	1. Given
2. $\overline{TU} \cong \overline{TU}$	2. Congruence of segments is reflexive.
3. $m\angle 1 > m\angle 2$	3. If an $\angle$ is an ext. $\angle$ of a $\triangle$, then its measure is greater than the measure of either of its corr. remote int. $\angle s$.
4. $TE > RU$	4. SAS Inequality

25. Given: $\overline{ED} \cong \overline{DF}$
$m\angle 1 > m\angle 2$
D is the midpoint of CB.
$\overline{AE} \cong \overline{AF}$
Prove: $AC > AB$

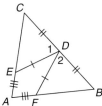

Statements	Reasons
1. $\overline{ED} \cong \overline{DF}$ D is the midpoint of $\overline{CB}$.	1. Given
2. $CD = BD$	2. Definition of midpoint
3. $\overline{CD} \cong \overline{BD}$	3. Definition of congruent segments
4. $m\angle 1 > m\angle 2$	4. Given
5. $CE > FB$	5. SAS Inequality
6. $AC = EA + CE$ $AB = AF + FB$	6. Segment addition postulate
7. $CE = AC - EA$ $FB = AB - AF$	7. Subtraction property of equality
8. $AC - EA > AB - AF$	8. Substitution property of equality
9. $\overline{AE} \cong \overline{AF}$	9. Given
10. $AE = AF$	10. Definition of congruent segments
11. $AC > AB$	11. Addition property of inequality

27. Given: $\overline{AC} \cong \overline{DF}$
$\overline{BC} \cong \overline{EF}$
$m\angle F > m\angle C$
Prove: $DE > AB$

We are given that $\overline{AC} \cong \overline{DF}$ and $\overline{BC} \cong \overline{EF}$. We also know that $m\angle F > m\angle C$. Now draw auxiliary ray $\overrightarrow{FZ}$ such that $m\angle DFZ = m\angle C$ and that $\overline{ZF} \cong \overline{BC}$. This leads to two cases.
Case I: If Z lies on $\overline{DE}$, then $\triangle FZD \cong \triangle CBA$ by SAS. Hence $ZD = BA$ by CPCTC and the definition of congruent segments. By the Segment Addition Postulate, $DE = EZ + ZD$ and hence $DE > ZD$ by the definition of inequality, then $DE > AB$ by substitution property of equality.

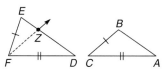

Case II: If Z does not lie on $\overline{DE}$, then let the intersection of $\overrightarrow{FZ}$ and $\overline{ED}$ be point T.

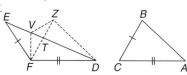

Now draw another auxiliary segment $\overline{FV}$ such that V is on $\overline{DE}$ and $\angle EFV \cong \angle VFZ$. Since $\overline{FZ} \cong \overline{BC}$ and $\overline{BC} \cong \overline{EF}$, we have $\overline{FZ} \cong \overline{EF}$ by the transitive theorem. Also $\overline{VF}$ is congruent to itself by the reflexive theorem. Hence, $\triangle EFV \cong \triangle ZFV$ by SAS. Then by CPCTC, $\overline{EV} \cong \overline{ZV}$. In $\triangle VZD$ the Triangle Inequality Theorem gives $VD + VZ > ZD$ and so by substitution $VD + EV > ZD$. By the Segment Addition Postulate $ED > ZD$. We also have $\triangle FZD \cong \triangle CBA$ by SAS which gives $\overline{ZD} \cong \overline{AB}$ by CPCTC. Making the substitution, we get $ED > BA$ or $DE > AB$.
31. yes; the measures satisfy the triangle inequality.

32. Given: $\overline{AC} \cong \overline{BD}$
$\overline{AD} \cong \overline{BC}$
Prove: $\triangle AXC \cong \triangle BXD$

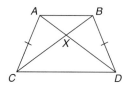

Statements	Reasons
1. $\overline{AC} \cong \overline{BD}$ $\overline{AD} \cong \overline{BC}$	1. Given
2. $\overline{AB} \cong \overline{BA}$	2. Congruence of segments is reflexive.
3. $\triangle ACB \cong \triangle BDA$	3. SSS
4. $\angle ACB \cong \angle BDA$	4. CPCTC
5. $\angle AXC \cong \angle BXD$	5. Vertical $\angle s$ are $\cong$.
6. $\triangle AXC \cong \triangle BXD$	6. AAS

33. 9 **34.** $m\angle A = 36$, $m\angle B = 54$

Pages 258–260 Summary and Review
1. $m\angle BDC = m\angle BCD = 57$, $m\angle DBC = 66$ **3.** no; $m\angle AED \neq 90$ **5.** $x = 11$, $y = 26$ **7.** $x = 2$, $y = 1$

9. Given: $\triangle QXP \cong \triangle QXR$
Prove: $\overline{QX}$ is an altitude of $\triangle PQR$.

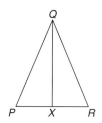

Proof: We assume $\overline{QX}$ is not an altitude of $\triangle PQR$. This means that $\overline{QX} \not\perp \overline{PR}$ by the definition of altitude. So, by the definition of perpendicular this means $\angle QXR$ is not a right angle. Therefore, $\angle QXR$ is either acute or obtuse. Since $\angle QXP$ and $\angle QXR$ form a linear pair, these two angles are supplementary. Hence, if $\angle QXP$ is acute, $\angle QXR$ is obtuse (or vice versa), by the definition of supplementary. In either case, $\angle QXP \not\cong \angle QXR$. But we are given that $\triangle QXP \cong \triangle QXR$ and by CPCTC this would mean $\angle QXP \cong \angle QXR$. This is a contradiction and hence our assumption must be false. Therefore $\overline{QX}$ is an altitude of $\triangle PQR$.
11. $>$ **13.** $<$ **15.** $\overline{SP}$ **17.** $\angle D$, $\angle CBD$, $\angle BCD$ **19.** 6 and 16 **21.** yes **23.** 3 **25.** $m\angle PSN < m\angle NQP$

27. Given: $AD = BC$
Prove: $DB < AC$

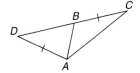

Statements	Reasons
1. $AD = BC$	1. Given
2. $\overline{AD} \cong \overline{BC}$	2. Definition of congruent segments
3. $\overline{BA} \cong \overline{BA}$	3. Congruence of segments is reflexive.
4. $m\angle CBA > m\angle DAB$	4. If an $\angle$ is an ext. $\angle$ of a $\triangle$ then its measure is greater than the measure of either of its corr. remote int. $\angle$s.
5. $AC > DB$	5. SAS Inequality

29. $(2, -1)$

Pages 262–263 Algebra Review
1. 7   **3.** 3   **5.** 25%   **7.** $48   **9.** 15   **11.** $\{x|x \le -4\}$
13. $\{t|t > 1.2\}$ **15.** $\{k|k \ge \frac{1}{5}\}$ **17.** $64a^6b^3$ **19.** $-\frac{432d^{10}}{c^5}$
21. 2.4×10^5 **23.** 3.14×10^{-4} **25.** $-3x^3 + x^2 - 5x + 5$
27. $16m^2n^2 - 2mn + 11$ **29.** $(y + 3)(y + 4)$
31. $(a - b)(a - 9b)$ **33.** $(2x - 3)(3x + 2)$ **35.** 6.4 cm
37. 7 at $2.99, 9 at $2.79

CHAPTER 6 QUADRILATERALS

Pages 268–271 Lesson 6-1
5. $\overline{DC}$; definition of parallelogram
7. $\triangle CBA$; SAS or SSS   **9.** $\overline{EB}$; Th. 6-4
11. $A(1, 1)$, $B(3, 6)$, $C(8, 8)$, and $D(6, 3)$
slope of $\overline{AB} = \frac{1-6}{1-3}$ or $\frac{5}{2}$

slope of $\overline{BC} = \frac{6-8}{3-8}$ or $\frac{2}{5}$

slope of $\overline{CD} = \frac{8-3}{8-6}$ or $\frac{5}{2}$

slope of $\overline{DA} = \frac{3-1}{6-1}$ or $\frac{2}{5}$

Since the opposite sides have the same slope, they are parallel and $ABCD$ is a parallelogram.
13. Given: parallelogram $MNOP$
Prove: $\triangle MNO \cong \triangle OPM$

Statements	Reasons
1. $MNOP$ is a parallelogram.	1. Given
2. $\overline{MN} \cong \overline{OP}$ $\overline{NO} \cong \overline{PM}$	2. Opp. sides of a $\square$ are $\cong$.
3. $\angle MNO \cong \angle OPM$	3. Opp. $\angle$s of a $\square$ are $\cong$.
4. $\triangle MNO \cong \triangle OPM$	4. SAS

15. definition of parallelogram   **17.** Th. 6-2   **19.** 64
21. true; Th. 6-4; vertical angles are congruent and SAS
23. false   **25.** true; Th. 6-4   **27.** $x = 30$, $y = 45$, $z = 75$
29. $m\angle R = 32$, $m\angle S = 148$ **31.** $(9, 4)$, $(5, -2)$, $(-3, 4)$
33. yes; $\overline{JU} \parallel \overline{YL}$ and $\overline{JY} \parallel \overline{UL}$ since if two lines are cut by a tranversal so that corresponding angles are congruent, then the lines are parallel, so $JULY$ is a parallelogram by definition.   **35.** The opposite angles should be congruent.

37. The opposite sides should be parallel, and if they were parallel, then $\angle ADC$ would be congruent to $\angle BCE$.   **39.** 36
41. Given: $ABCD$ is a parallelogram.
Prove: $\angle BAD \cong \angle DCB$
$\angle ABC \cong \angle CDA$

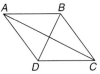

Statements	Reasons
1. $ABCD$ is a parallelogram.	1. Given
2. $\overline{AD} \cong \overline{BC}$ $\overline{AB} \cong \overline{CD}$	2. Opp. sides of a $\square$ are $\cong$.
3. $\overline{BD} \cong \overline{BD}$ $\overline{AC} \cong \overline{AC}$	3. Congruence of segments is reflexive.
4. $\triangle BAD \cong \triangle DCB$ $\triangle ABC \cong \triangle CDA$	4. SSS
5. $\angle BAD \cong \angle DCB$ $\angle ABC \cong \angle CDA$	5. CPCTC

43. Given: $\square EAST$
Prove: $\overline{ES}$ bisects $\overline{AT}$.
$\overline{AT}$ bisects $\overline{ES}$.

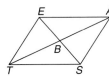

Statements	Reasons
1. $\square EAST$	1. Given
2. $\overline{EA} \cong \overline{ST}$	2. Opp. sides of a $\square$ are $\cong$.
3. $\overline{EA} \parallel \overline{ST}$	3. Definition of $\square$
4. $\angle AEB \cong \angle TSB$ $\angle EAB \cong \angle STB$	4. If 2 $\parallel$ lines are cut by a transversal, alt. int. $\angle$s are $\cong$.
5. $\triangle EBA \cong \triangle SBT$	5. ASA
6. $\overline{EB} \cong \overline{SB}$ $\overline{AB} \cong \overline{TB}$	6. CPCTC
7. $\overline{ES}$ bisects $\overline{AT}$. $\overline{AT}$ bisects $\overline{ES}$.	7. Definition of segment bisector

45. Given: $PQST$ is a parallelogram.
$\overline{RP}$ bisects $\angle QPT$.
$\overline{VS}$ bisects $\angle QST$.
Prove: $\overline{RP} \cong \overline{VS}$

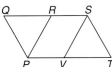

Statements	Reasons
1. $PQST$ is a parallelogram.	1. Given
2. $\angle Q \cong \angle T$ $\angle QST \cong \angle QPT$	2. Opp. $\angle$s of a $\square$ are $\cong$.
3. $m\angle QST = m\angle QPT$	3. Definition of congruent angles
4. $m\angle QST = m\angle QSV + m\angle VST$ $m\angle QPT = m\angle QPR + m\angle RPT$	4. Angle addition postulate
5. $\overline{RP}$ bisects $\angle QPT$. $\overline{VS}$ bisects $\angle QST$.	5. Given
6. $m\angle QSV = m\angle VST$ $m\angle QPR = m\angle RPT$	6. Definition of angle bisector
7. $m\angle QST = 2m\angle VST$ $m\angle QPT = 2m\angle QPR$	7. Substitution property of equality
8. $2m\angle VST = 2m\angle QPR$	8. Substitution property of equality
9. $m\angle VST = m\angle QPR$	9. Division property of equality
10. $\angle VST \cong \angle QPR$	10. Definition of congruent angles
11. $\overline{QP} \cong \overline{ST}$	11. Opp. sides of a $\square$ are $\cong$.
12. $\triangle QPR \cong \triangle TSV$	12. ASA
13. $\overline{RP} \cong \overline{VS}$	13. CPCTC

49. = ; > **50.** HA, LL, LA, and HL **51.** $\overline{TA}$ **52.** right; yes; no **53.** $-\frac{8}{7}$ **54.** parallel **55.** If a quadrilateral has opposite sides parallel, then it is a parallelogram.

Pages 273–274 Lesson 6-2
5. 29 **7.** 7 tables **9.** 56 pairs **11.** 47 **13.** 123, 454, 321

Pages 277-280 Lesson 6-3
7. No; the top and bottom segments are parallel, but the other pair may not be.
9. Yes; both pairs of opposite angles are congruent.
11. $x = 8$ or -2, $y = 5$ or -5
13. midpoint A of $\overline{JK} = (\frac{3+8}{2}, \frac{-2+(-2)}{2})$ or $(\frac{11}{2}, -2)$

midpoint B of $\overline{KL} = (\frac{8+7}{2}, \frac{-2+(-4)}{2})$ or $(\frac{15}{2}, -3)$

midpoint C of $\overline{LM} = (\frac{7+3}{2}, \frac{-4+(-6)}{2})$ or $(5, -5)$

midpoint D of $\overline{MJ} = (\frac{3+3}{2}, \frac{-6+(-2)}{2})$ or $(3, -4)$

slope of $\overline{AB} = \frac{-2-(-3)}{\frac{11}{2}-\frac{15}{2}}$ or $-\frac{1}{2}$

slope of $\overline{BC} = \frac{-3-(-5)}{\frac{15}{2}-5}$ or $\frac{4}{5}$

slope of $\overline{CD} = \frac{-5-(-4)}{5-3}$ or $-\frac{1}{2}$

slope of $\overline{DA} = \frac{-4-(-2)}{3-\frac{11}{2}}$ or $\frac{4}{5}$

Since the opposite sides are parallel, $ABCD$ is a parallelogram.

15. Given: $\square PQRS$
$\overline{XS} \cong \overline{QY}$
Prove: $PYRX$ is a parallelogram.

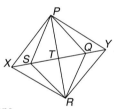

Statements	Reasons
1. $\square PQRS$ $\overline{XS} \cong \overline{QY}$	**1.** Given
2. $\overline{PT} \cong \overline{TR}$ $\overline{ST} \cong \overline{TQ}$	**2.** The diagonals of a $\square$ bisect each other.
3. $ST = TQ$ $XS = QY$	**3.** Definition of congruent segments
4. $XS + ST = TQ + QY$	**4.** Addition property of equality
5. $XT = XS + ST$ $TY = TQ + QY$	**5.** Segment addition postulate
6. $XT = TY$	**6.** Substitution property of equality
7. $\overline{XT} \cong \overline{TY}$	**7.** Definition of congruent segments
8. $\overline{XY}$ bisects $\overline{PR}$. $\overline{PR}$ bisects $\overline{XY}$.	**8.** Definition of bisector
9. $PYRX$ is a parallelogram.	**9.** If the diagonals of a quad. bisect, it is a $\square$.

17. Given: $\overline{SR} \cong \overline{TA}$
$\overline{SR} \parallel \overline{TA}$
Prove: $STAR$ is a parallelogram.

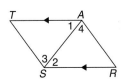

Statements	Reasons
1. $\overline{SR} \cong \overline{TA}$ $\overline{SR} \parallel \overline{TA}$	**1.** Given
2. $\angle 1 \cong \angle 2$	**2.** If 2 $\parallel$ lines are cut by a transversal, alt. int. $\angle$s are $\cong$.
3. $\overline{SA} \cong \overline{AS}$	**3.** Congruence of segments is reflexive.
4. $\triangle RSA \cong \triangle TAS$	**4.** SAS
5. $\angle 3 \cong \angle 4$	**5.** CPCTC
6. $\overline{ST} \parallel \overline{RA}$	**6.** If 2 lines are cut by a transversal and alt. int. $\angle$s are $\cong$, then the lines are $\parallel$.
7. $STAR$ is a parallelogram.	**7.** Definition of parallelogram

19. yes; Theorem 6-8 **21.** no **23.** 8 **25.** 3 **27.** 49
29. $x = 1$, $y = 4$

31. no; sample drawing:

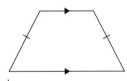

35. slope of $\overline{AB} = \frac{6-8}{-2-2} = \frac{-2}{-4} = \frac{1}{2}$

slope of $\overline{BC} = \frac{8-8}{2-3} = \frac{0}{-1} = 0$

slope of $\overline{CD} = \frac{8-3}{3-(-1)} = \frac{5}{4}$

slope of $\overline{DA} = \frac{3-6}{-1-(-2)} = \frac{-3}{1} = -3$

This is not a parallelogram because the opposite sides are not parallel.

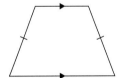

37.

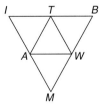

39. Given: $\triangle TWA$ is equilateral.
$TBWA$ is a parallelogram.
$TWAI$ is a parallelogram.
Prove: $\triangle IBM$ is equilateral.

Statements	Reasons
1. $\triangle TWA$ is equilateral. $TBWA$ is a parallelogram. $TWAI$ is a parallelogram.	**1.** Given
2. $m\angle ATW = m\angle TWA$ $= m\angle WAT = 60$	**2.** Each $\angle$ of an equilateral triangle measures $60°$.
3. $\angle TAW \cong \angle IBW$ $\angle TWA \cong \angle BIA$	**3.** Opp. $\angle$s of a $\square$ are $\cong$.
4. $m\angle TAW = m\angle IBW$ $m\angle TWA = m\angle BIA$	**4.** Definition of congruent angles
5. $m\angle IBW = 60$ $m\angle BIA = 60$	**5.** Substitution property of equality
6. $m\angle IBW + m\angle BIA$ $+ m\angle IMB = 180$	**6.** The sum of the $\angle$s in a $\triangle$ is 180.

7. $60 + 60 + m\angle IMB = 180$	7. Substitution property of equality
8. $m\angle IMB = 60$	8. Subtraction property of equality
9. $\triangle IBM$ is equiangular.	9. Definition of equiangular
10. $\triangle IBM$ is equilateral.	10. An equiangular $\triangle$ is equilateral.

41. Given: $\overline{BD}$ bisects $\overline{AC}$.
$\overline{AC}$ bisects $\overline{BD}$.
Prove: $ABCD$ is a parallelogram.

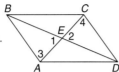

By definitions of segment bisector and midpoint, $\overline{AE} \cong \overline{CE}$ and $\overline{BE} \cong \overline{DE}$. Since they are vertical angles, $\angle 1 \cong \angle 2$. By SAS, $\triangle BEA \cong \triangle DEC$. Since they are corresponding parts in congruent triangles, $\angle 3 \cong \angle 4$ and $\overline{AB} \cong \overline{CD}$. Since $\angle 3$ and $\angle 4$ are congruent alternate interior angles, $\overline{AB} \parallel \overline{CD}$. So, since a pair of opposite sides are congruent and parallel, $ABCD$ is a parallelogram.   **43.** The legs are made so that they bisect each other, so the quadrilateral formed by the ends of the legs is a parallelogram. So the table top is parallel to the floor.
45. 37   **46.** 30   **47.** no; Triangle inequality $30 + 35 \not> 66$
48. obtuse; one obtuse angle   **49.** $0 < y < 5$

Pages 284–287 Lesson 6-4
5. 2   **7.** 2 or 5   **9.** no; sample drawing:

11. Given: $\square WXYZ$
$\angle 1$ and $\angle 2$ are complementary.
Prove: $WXYZ$ is a rectangle.

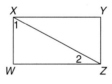

Statements	Reasons
1. $\square WXYZ$ $\angle 1$ and $\angle 2$ are complementary.	1. Given
2. $m\angle 1 + m\angle 2 = 90$	2. Definition of complementary
3. $m\angle 1 + m\angle 2 + m\angle W = 180$	3. The sum of the $\angle$s in a $\triangle$ is 180.
4. $90 + m\angle W = 180$	4. Substitution property of equality
5. $m\angle W = 90$	5. Subtraction property of equality
6. $\angle W \cong \angle Y$	6. Opp. $\angle$s of a $\square$ are $\cong$.
7. $m\angle W = m\angle Y$	7. Definition of congruent angles
8. $m\angle Y = 90$	8. Substitution property of equality
9. $\angle W$ and $\angle WXY$ are supplementary. $\angle W$ and $\angle YZW$ are supplementary.	9. Consec. int. $\angle$s in a $\square$ are supp.
10. $m\angle W + m\angle WXY = 180$ $m\angle W + m\angle YZW = 180$	10. Definition of supplementary
11. $90 + m\angle WXY = 180$ $90 + m\angle YZW = 180$	11. Substitution property of equality
12. $m\angle WXY = 90$ $m\angle YZW = 90$	12. Subtraction property of equality
13. $\angle W$, $\angle Y$, $\angle WXY$, and $\angle YZW$ are right angles.	13. Definition of right angle
14. $WXYZ$ is a rectangle.	14. Definition of rectangle

13. 12   **15.** 6   **17.** 70

19.

21.

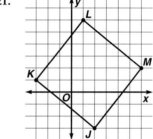

a. Two possible ways are to use slopes to see if opposite sides are parallel and adjacent sides are perpendicular, or to use slopes to see if opposite sides are parallel and the distance formula to see if the diagonals are congruent.
b. midpoint of $\overline{JL} = (\frac{2+1}{2}, \frac{-3+6}{2})$ or $(\frac{3}{2}, \frac{3}{2})$
midpoint of $\overline{KM} = (\frac{-3+6}{2}, \frac{1+2}{2})$ or $(\frac{3}{2}, \frac{3}{2})$

Since they have the same midpoint, $\overline{JL}$ and $\overline{KM}$ bisect each other. Therefore, $JKLM$ is a parallelogram.
$JL = \sqrt{(2-1)^2 + (-3-6)^2}$ or $\sqrt{82}$
$KM = \sqrt{(-3-6)^2 + (1-2)^2}$ or $\sqrt{82}$
Since the diagonals are congruent, $JKLM$ is a rectangle.
23. 3, 10   **25.** 32, 58, 58   **27.** 32, 58, 58   **29.** no; not all right angles   **31.** yes; opposite sides parallel and all right angles   **33.** yes   **35.** no   **37.** yes   **41.** Sample answer: Since all of the angles and the opposite sides are congruent, the bricks are interchangeable and can be installed in rows easily.   **43.** no   **44.** 58   **45.** 7   **46.** If a quadrilateral is a rectangle, then it is a parallelogram.   **47.** 16

Page 287 Mid-Chapter Review
1. $\overline{FE}$; definition of parallelogram
2. $\angle GFE$; Theorem 6-2   **3.** $\overline{HD}$; Theorem 6-4   **4.** $\triangle HDG$; Theorem 6-4 and SAS   **5.** $\overline{FG}$; Theorem 6-1   **6.** $\overline{HE}$; definition of parallelogram   **7.** 21   **8.** 125   **9.** 32
10. $-\frac{3}{2}$ **11.** definition of parallelogram   **12.** Theorem 6-6
13. Theorem 6-8   **14.** Theorem 6-7

15. Given: rectangle $JKLM$
$\overline{KF} \cong \overline{MH}$
$\overline{JE} \cong \overline{LG}$
Prove: $EFGH$ is a parallelogram.

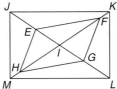

Statements	Reasons
1. rectangle $JKLM$ $\overline{KF} \cong \overline{MH}$ $\overline{JE} \cong \overline{LG}$	1. Given
2. $\overline{KI} \cong \overline{MI}$ $\overline{JI} \cong \overline{LI}$	2. Diagonals of a $\square$ bisect each other.
3. $KI = MI$ $JI = LI$ $KF = MH$ $JE = LG$	3. Definition of congruent segments
4. $KI = KF + FI$ $MI = MH + HI$ $JI = JE + EI$ $LI = LG + GI$	4. Segment addition postulate
5. $KF + FI = MH + HI$ $JE + EI = LG + GI$	5. Substitution property of equality
6. $FI = HI$ $EI = GI$	6. Subtraction property of equality
7. $\overline{FI} \cong \overline{HI}$ $\overline{EJ} \cong \overline{GI}$	7. Definition of congruent segments
8. $EFGH$ is a parallelogram.	8. If the diagonals of a quad bisect, it is a $\square$.

Pages 290–293 Lesson 6-5
5. parallelogram **7.** parallelogram

9.

Property	Parallelogram	Rectangle	Rhombus	Square
The diagonals bisect each other.	yes	yes	yes	yes
The diagonals are congruent.	no	yes	no	yes
Each diagonal bisects a pair of opposite angles.	no	no	yes	yes
The diagonals are perpendicular.	no	no	yes	yes

11. 51.8 **13.** 6 or -4 **15.** parallelogram, rectangle, rhombus, square **17.** parallelogram, rectangle, rhombus, square **19.** False; the diagonals of a rhombus are not congruent unless it is also a square. **21.** True; the diagonals of a rhombus are perpendicular. **23.** False; the consecutive angles of a rhombus are not congruent unless it is also a square. **25.** parallelogram, rectangle **27.** parallelogram, rectangle, rhombus, square **29.** 28, 62, 90 **31.** 14 **33.** 8 **35.** The diagonals are perpendicular.

Given: $KITE$
$\overline{KI} \cong \overline{KE}$
$\overline{IT} \cong \overline{ET}$
Prove: $\overline{KT} \perp \overline{IE}$

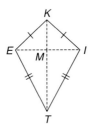

Statements	Reasons
1. $\overline{KI} \cong \overline{KE}$ $\overline{IT} \cong \overline{ET}$	1. Given
2. $\overline{KT} \cong \overline{KT}$	2. Congruence of segments is reflexive.
3. $\Delta KIT \cong \Delta KET$	3. SSS
4. $\angle IKM \cong \angle EKM$	4. CPCTC
5. $\overline{KM} \cong \overline{KM}$	5. Congruence of segments is reflexive.
6. $\Delta IKM \cong \Delta EKM$	6. SAS
7. $\angle IMK \cong \angle EMK$	7. CPCTC
8. $\angle IMK$ and $\angle EMK$ form a linear pair.	8. Definition of linear pair
9. $\angle IMK$ and $\angle EMK$ are supplementary.	9. If 2 $\angle$s form a linear pair, they are supp.
10. $m\angle IMK + m\angle EMK = 180$	10. Definition of supplementary
11. $m\angle IMK = m\angle EMK$	11. Definition of congruent angles
12. $2m\angle IMK = 180$	12. Substitution property of equality
13. $m\angle IMK = 90$	13. Division property of equality
14. $\angle IMK$ is a right angle.	14. Definition of right angle
15. $\overline{KT} \perp \overline{IE}$	15. Definition of perpendicular

37. Given: $ABCD$ is a rhombus.
$\overline{AF} \cong \overline{BG}$
$\overline{BG} \cong \overline{CH}$
$\overline{CH} \cong \overline{DE}$
$\overline{DE} \cong \overline{AF}$
Prove: $EFGH$ is a parallelogram.

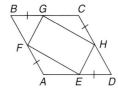

Statements	Reasons
1. $ABCD$ is a rhombus. $\overline{AF} \cong \overline{BG}$; $\overline{BG} \cong \overline{CH}$; $\overline{CH} \cong \overline{DE}$; $\overline{DE} \cong \overline{AF}$	1. Given
2. $\overline{AB} \cong \overline{BC}$; $\overline{BC} \cong \overline{CD}$; $\overline{CD} \cong \overline{AD}$; $\overline{AD} \cong \overline{AB}$	2. Definition of rhombus
3. $AF = BG$, $BG = CH$, $CH = DE$, $DE = AF$, $AB = BC$, $BC = CD$, $CD = AD$, $AD = AB$	3. Definition of congruent segments
4. $AB = AF + FB$, $BC = BG + GC$, $CD = CH + HD$, $AD = DE + EA$	4. Segment addition postulate
5. $AF + FB = BG + GC$ $BG + GC = CH + HD$ $CH + HD = DE + EA$ $DE + EA = AF + FB$	5. Substitution property of equality
6. $FB = GC$, $GC = HD$, $HD = EA$, $EA = FB$	6. Subtraction property of equality
7. $\overline{FB} \cong \overline{GC}$, $\overline{GC} \cong \overline{HD}$ $\overline{HD} \cong \overline{EA}$, $\overline{EA} \cong \overline{FB}$	7. Definition of congruent segments
8. $\angle A \cong \angle C$, $\angle B \cong \angle D$	8. Opp. angles of a $\square$ are $\cong$.
9. $\overline{FB} \cong \overline{HD}$, $\overline{BG} \cong \overline{ED}$, $\overline{GC} \cong \overline{AE}$, $\overline{CH} \cong \overline{FA}$	9. Transitive property of equality
10. $\Delta FBG \cong \Delta HDE$, $\Delta GCH \cong \Delta EAF$	10. SAS

11. $\overline{FG} \cong \overline{HE}$, $\overline{FE} \cong \overline{GH}$
12. Quad *EFGH* is a parallelogram.
11. CPCTC
12. If opp. sides of a quad are $\cong$, it is a $\square$.

39a. $\overline{AC}$ becomes shorter and $\overline{BD}$ becomes longer. **b.** The base of the jack and the plate that supports the car are parallel to the rod between points A and C and perpendicular to the diagonal $\overline{BD}$. The diagonals of a rhombus are perpendicular. Changing the lengths of the diagonals doesn't affect the level of the car. **c.** Since the diagonals of other parallelograms are not perpendicular, the load would shift as the jack was raised. **40.** 32, 58, 58 **41.** $AB = 31$, $BC = 35$, $CD = 31$, and $AD = 35$ **42.** no solution
43. yes, LA or AAS **44.** Transitive property of equality
45. plane

Pages 296–299 Lesson 6-6

5. 21 **7.** 57 **9.** $\frac{9\sqrt{2}}{2} \approx 6.364$ units **11.** $x = 11\frac{1}{2}$

13. 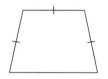 **17.**

19. cannot be drawn; it would be a parallelogram **21.** 32.1
23. 57 **25.** 1, 8; 2, 7; 3, 6; 4, 5 **27.** isosceles trapezoid

29. Given: Trapezoid *RSPT* is isosceles.
Prove: $\triangle RSQ$ is isosceles.

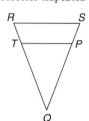

Statements	Reasons
1. Trapezoid *RSPT* is isosceles.	1. Given
2. $\angle R \cong \angle S$	2. Base $\angle$ of an isos. trap. are $\cong$.
3. $\overline{RQ} \cong \overline{SQ}$	3. If 2 $\angle$ of a $\triangle$ are $\cong$, the sides opp. the $\angle$ are $\cong$.
4. $\triangle RSQ$ is isosceles.	4. Definition of isosceles triangle

31a. 3; *JKZX, IJXY, IKZY* **b.** Yes; the bases and legs must be congruent if they are isosceles.

33a.

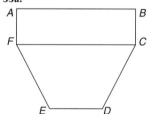 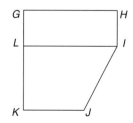

ABCF - rectangle, *FCDE* - trapezoid, *GHIL* - rectangle, *LIKJ* - trapezoid **b.** Yes; *LIJK* from the end view is congruent to the trapezoid on the other end of the box. **35a.** 6.18
b. 4.74 **36.** false; any rhombus that is not a square **37.** 30
38. yes; detachment **39.** obtuse

Pages 300–302 Summary and Review
1. $\overline{ED}$; Th. 6-4 **3.** $\angle ABC$; Th. 6-2 **5.** $\triangle DAB$; SAS or SSS
7. $\overline{EA}$; Th. 6-4
9. Given: $\square PRSV$
$\triangle PQR \cong \triangle STV$
Prove: Quadrilateral *PQST* is a parallelogram.

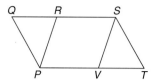

Statements	Reasons
1. $\square PRSV$ $\triangle PQR \cong \triangle STV$	1. Given
2. $\overline{RS} \cong \overline{PV}$	2. Opp. sides of a $\square$ are $\cong$.
3. $\overline{QS} \parallel \overline{PT}$	3. Definition of parallelogram
4. $\overline{QR} \cong \overline{VT}$	4. CPCTC
5. $QR + RS = QS$ $PV + VT = PT$	5. Segment addition postulate
6. $QR + RS = PV + VT$	6. Addition property of equality
7. $QS = PT$	7. Substitution property of equality
8. $\overline{QS} \cong \overline{PT}$	8. Definition of congruent segments
9. Quadrilateral *PQST* is a parallelogram.	9. If a pair of opp. sides of a quad. are $\cong$ and $\parallel$, it is a $\square$.

11. 6 **13.** 2 **15.** 3 or –3 **17.** 46 **19.** 7 **21.** rectangle, parallelogram **23.** 35 **25.** 47 **27.** 30 **29.** Yes; the corners will each have a 90° angle.

CHAPTER 7 SIMILARITY

Pages 310–313 Lesson 7-1
7. 0.542 **9.** 0.976 **11.** yes **13.** yes **15.** 4.5 **17.** 0.4
19. 0.50 **21.** 0.25 **23.** 11 **25.** 0.405 **27.** 12 **29.** $\frac{1}{1}$
31. $\frac{2}{3}$ **33.** 37.5% **35.** 325%
37. $\frac{a-b}{b} = \frac{c-d}{d}$
$(a-b)\,d = (c-d)\,b$
$ad - bd = cb - db$
$ad - bd = cb - bd$
$ad - bd + bd = cb - bd + bd$
$ad = cb$
$\frac{a}{b} = \frac{c}{d}$

39. $x = 1.25$ $y = 40$ **41.** Sample answer: $\frac{x}{2} = \frac{11}{y}$
45. 2984 **47.** about 1.3 **48.** yes; if it is a square **49.** no; fails the triangle inequality
50.

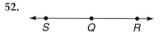

51. If a quadrilateral is a trapezoid, then it has exactly two opposite sides parallel.

52.

Pages 316–320 Lesson 7-2
5. a **7.** about 5.06 **9.** no **11.** 487 soft drinks **13.** yes
15. no **17.** 5 **19a.** about 643 people **b.** about 47,174 people **21.** 10; 8 **23.** about 18.5 ft **25.** $A(1, -1)$ and $T(1, 5)$ or $A(-5, -1)$ and $T(-5, 5)$ **27.** 50 points

31. $166.67 for CD component, $500 for a receiver, $333.33 for speakers   **33.** 20.8 in.   **35.** 25%   **37a.** 1, 1, 2, 3, 5, 8, 13, 21, 34, 55, 89, 144, 233, 377, 610, 987, 1597, 2584, 4181, 6765   **b.** A term is the sum of the two previous terms. **c.** 1, 2, 1.5, 1.666, 1.6, 1.625, 1.61538, 1.619047, 1.617647, 1.6181818, 1.6179775, 1.618055556, 1.6180258, 1.618037135, 1.618032787, 1.618034448, 1.618033813, 1.618032056, 1.618033963; They are closer and closer approximations of the golden ratio.   **38.** 15   **39.** 38, 38

40. 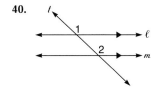   **41.**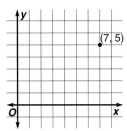

Pages 323–326 Lesson 7-3

5. $\angle A \cong \angle D$, $\angle B \cong \angle E$, $\angle C \cong \angle F$, $\frac{AB}{DE} = \frac{BC}{EF} = \frac{AC}{DF}$ **7.** Yes; corresponding angles are congruent and $\frac{2.0}{3.0} = \frac{1.6}{2.4} = \frac{1.8}{2.7}$.

9. No; their two pairs of acute angles may not be congruent.
11. No; they may not have sides with proportional measures.
13. $\frac{2}{3}$ **15.** 15   **17a.** 43   **b.** 64.5   **c.** $\frac{43}{64.5} = \frac{2}{3}$ **19.** 12, 12

21. 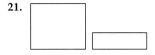   **25.**

27. $\frac{6.2}{17.6}$ or $\frac{31}{88}$ or about 0.35

31. 6 times   **33.** $M(0, -4)$, $N(3, -2)$ or $M(0,4)$, $N(3,2)$
39. 90 cm   **41.** 3.75 in.   **43.** 14 seniors   **44.** trapezoid
45. It is less than 13 and greater than 3.   **46.** intersecting, parallel, skew   **47.** If it is cool, then I will wear a sweater.
48. the angle itself, the interior, the exterior

Pages 331–335 Lesson 7-4

5. $\overline{RA}$, $\overline{OF}$; $\overline{AT}$, $\overline{FT}$; $\overline{RT}$, $\overline{OT}$ **7.** no   **9.** no   **11.** yes; AA Similarity   **13.** yes; AA Similarity; $x = 3\frac{1}{3}$, $y = 3$
15. yes; AA Similarity   **17.** yes; SSS Similarity   **19.** 11
21. $\triangle ABC \sim \triangle ADB$; AA Similarity; $\triangle ABC \sim \triangle BDC$; AA Similarity; $\triangle ADB \sim \triangle BDC$ (Th. 7-3)   **23.** 12, 22.5, 7.5; $\triangle AEB \approx \triangle ADC$ by AA Similarity
25. Given: $\angle D$ is a right angle.
$\overline{BE} \perp \overline{AC}$
Prove: $\triangle ADC \sim \triangle ABE$

Statements	Reasons
1. $\angle D$ is a right angle. $\overline{BE} \perp \overline{AC}$	**1.** Given
2. $\angle EBA$ is a right angle.	**2.** $\perp$ lines form 4 rt. $\angle$
3. $\angle D \cong \angle EBA$	**3.** All rt. $\angle$ are $\cong$.
4. $\angle A \cong \angle A$	**4.** Congruence of angles is reflexive.
5. $\triangle ADC \sim \triangle ABE$	**5.** AA Similarity

27. Given: $\angle B \cong \angle E$
$\frac{AB}{DE} = \frac{BC}{EF}$
Prove: $\triangle ABC \sim \triangle DEF$

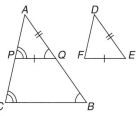

Statements	Reasons
1. Draw $\overline{QP} \parallel \overline{BC}$ so that $\overline{QP} \cong \overline{EF}$.	**1.** Parallel postulate
2. $\angle APQ \cong \angle C$ $\angle AQP \cong \angle B$	**2.** If 2 $\parallel$ lines are cut by a transversal, corr. $\angle$ are $\cong$.
3. $\angle B \cong \angle E$	**3.** Given
4. $\angle AQP \cong \angle E$	**4.** Congruence of angles is transitive.
5. $\triangle ABC \sim \triangle AQP$	**5.** AA Similarity
6. $\frac{AB}{AQ} = \frac{BC}{QP}$	**6.** Definition of similar polygons
7. $\frac{AB}{DE} = \frac{BC}{EF}$	**7.** Given
8. $AB \cdot QP = AQ \cdot BC$; $AB \cdot EF = DE \cdot BC$	**8.** Equality of cross products
9. $QP = EF$	**9.** Definition of congruent segments
10. $AB \cdot EF = AQ \cdot BC$	**10.** Substitution property of equality
11. $AQ \cdot BC = DE \cdot BC$	**11.** Substitution property of equality
12. $AQ = DE$	**12.** Division property of equality
13. $\overline{AQ} \cong \overline{DE}$	**13.** Definition of congruent segments
14. $\triangle AQP \cong \triangle DEF$	**14.** SAS
15. $\angle APQ \cong \angle F$	**15.** CPCTC
16. $\angle C \cong \angle F$	**16.** Congruence of angles is transitive.
17. $\triangle ABC \sim \triangle DEF$	**17.** AA Similarity

29. Reflexive property
Given: $\triangle ABC$
Prove: $\triangle ABC \sim \triangle ABC$

Statements	Reasons
1. $\triangle ABC$	**1.** Given
2. $\angle A \cong \angle A$ $\angle B \cong \angle B$	**2.** Congruence of angles is reflexive.
3. $\triangle ABC \sim \triangle ABC$	**3.** AA Similarity

Symmetric property
Given: $\triangle ABC \sim \triangle DEF$
Prove: $\triangle DEF \sim \triangle ABC$

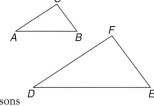

Statements	Reasons
1. $\triangle ABC \sim \triangle DEF$	**1.** Given
2. $\angle A \cong \angle D$ $\angle B \cong \angle E$	**2.** Definition of similar polygons
3. $\angle D \cong \angle A$ $\angle E \cong \angle B$	**3.** Congruence of angles is symmetric.
4. $\triangle DEF \sim \triangle ABC$	**4.** AA Similarity

Transitive property
Given: $\triangle ABC \sim \triangle DEF$
$\qquad \triangle DEF \sim \triangle GHI$
Prove: $\triangle ABC \sim \triangle GHI$

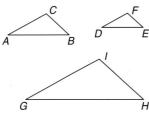

Statements	Reasons
1. $\triangle ABC \sim \triangle DEF$ $\triangle DEF \sim \triangle GHI$	1. Given
2. $\angle A \cong \angle D$ $\angle B \cong \angle E$ $\angle D \cong \angle G$ $\angle E \cong \angle H$	2. Definition of similar polygons
3. $\angle A \cong \angle G$ $\angle B \cong \angle H$	3. Congruence of angles is transitive.
4. $\triangle ABC \sim \triangle GHI$	4. AA Similarity

33. 12 m **35.** false **36.** true **37.** true **38.** true
39. false **40.** true

Page 335 Mid-Chapter Review

1. $\frac{12}{16}$ or $\frac{3}{4}$ **2.** about \$2.22 **3.** about 93.1 gal **4.** 2.5

5. Two polygons are similar if and only if their corresponding angles are congruent and the measures of their corresponding sides are proportional. **6.** yes; AA Similarity **7.** David; the right angles are congruent, but the other two pairs of angles may not be.

Pages 339–341 Lesson 7-5

7. true **9.** true **11.** no **13.** no **15.** 2

17.

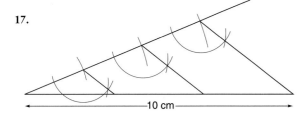

19. AR **21.** AE **23.** DE **25.** 2, 12 **27.** 2.5 **29.** 3
31. 70; 1:2

33.

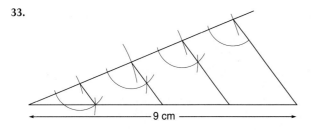

35. Given: D is the
midpoint of $\overline{AB}$.
E is the
midpoint of $\overline{AC}$.
Prove: $\overline{DE} \parallel \overline{BC}$
$\qquad DE = \frac{1}{2}BC$

Statements	Reasons
1. D is the midpoint of $\overline{AB}$. E is the midpoint of $\overline{AC}$.	1. Given
2. $\overline{AD} \cong \overline{DB}$ $\overline{AE} \cong \overline{EC}$	2. Definition of midpoint
3. $AD = DB$ $AE = EC$	3. Definition of congruent segments
4. $AB = AD + DB$ $AC = AE + EC$	4. Segment addition postulate
5. $AB = AD + AD$ $AC = AE + AE$	5. Substitution property of equality
6. $AB = 2AD$ $AC = 2AE$	6. Substitution property of equality
7. $\frac{AB}{AD} = 2$ $\frac{AC}{AE} = 2$	7. Division property of equality
8. $\frac{AB}{AD} = \frac{AC}{AE}$	8. Transitive property of equality
9. $\angle A \cong \angle A$	9. Congruence of angles is reflexive.
10. $\triangle ADE \sim \triangle ABC$	10. SAS Similarity
11. $\angle ADE \cong \angle ABC$	11. Definition of similar polygons
12. $\overline{DE} \parallel \overline{BC}$	12. If 2 lines are cut by a transversal so that corr. $\angle$s are $\cong$, the lines are $\parallel$.
13. $\frac{BC}{DE} = \frac{AB}{AD}$	13. Definition of similar polygons
14. $\frac{BC}{DE} = 2$	14. Substitution property of equality
15. $2DE = BC$	15. Multiplication property of equality
16. $DE = \frac{1}{2}BC$	16. Division property of equality

37. (4, 6) or (6, 9) **41.** $w = 74.1$ ft; $x = 80.2$ ft; $y = 86.4$ ft;
$z = 92.6$ ft; $v = 98.7$ ft **43.** 25 cm **44.** $\angle R$, $\angle Q$, $\angle P$
45. obtuse **46.** $-\frac{7}{6}$ **47.** Addition property of equality

Pages 345–348 Lesson 7-6
5. false **7.** true **9.** 6 **11.** true **13.** false **15.** true
17. $8\frac{1}{3}$ **19.** 6.75 **21.** 6 **23.** 3 or $\frac{1}{3}$ **25.** 8; 18.75 **27.** $\frac{1}{1}$

29. Given: $\triangle ABC \sim \triangle RST$
$\overline{AD}$ is a median
of $\triangle ABC$.
$\overline{RU}$ is a median
of $\triangle RST$.
Prove: $\frac{AD}{RU} = \frac{AB}{RS}$

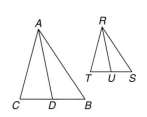

We are given that $\triangle ABC \sim \triangle RST$, $\overline{AD}$ is a median of $\triangle ABC$, and $\overline{RU}$ is a median of $\triangle RST$. So, by the definition of median, $CD = DB$ and $TU = US$. According to the definition of similar polygons, $\frac{AB}{RS} = \frac{CB}{TS}$. $CB = CD + DB$ and $TS = TU + US$ by the segment addition postulate.

Substituting, $\frac{AB}{RS} = \frac{CD + DB}{TU + US}$

$\frac{AB}{RS} = \frac{DB + DB}{US + US}$

$\frac{AB}{RS} = \frac{2DB}{2US}$

$\frac{AB}{RS} = \frac{DB}{US}$

$\angle B \cong \angle S$ by the definition of similar polygons, and $\triangle ABD \sim \triangle RSU$ using SAS Similarity. Therefore, $\frac{AD}{RU} = \frac{AB}{RS}$ by the definition of similar polygons.

31. It is a trapezoid.
Given: $\overline{SV}$ bisects $\angle RST$
$\overline{RA} \cong \overline{RV}$
$\overline{BT} \cong \overline{VT}$
Prove: $ABTR$ is a trapezoid.

Statements	Reasons
1. $\overline{SV}$ bisects $\angle RST$. $\overline{RA} \cong \overline{RV}$ $\overline{BT} \cong \overline{VT}$	**1.** Given
2. $\frac{RV}{VT} = \frac{SR}{ST}$	**2.** An angle bisector in a triangle separates the opposite side into segments that have the same ratio as the other sides.
3. $\frac{ST}{VT} = \frac{SR}{RV}$	**3.** Property of proportions
4. $ST = SB + BT$ $SR = SA + AR$	**4.** Segment addition postulate
5. $\frac{SB + BT}{VT} = \frac{SA + AR}{RV}$	**5.** Substitution property of equality
6. $RA = RV$ $BT = VT$	**6.** Definition of congruent segments
7. $\frac{SB + BT}{BT} = \frac{SA + AR}{AR}$	**7.** Substitution property of equality
8. $\frac{SB}{BT} = \frac{SA}{AR}$	**8.** Property of proportions
9. $\overline{AB} \parallel \overline{RT}$	**9.** If a line intersects two sides of a triangle into corresponding segments of proportional lengths then the line is parallel to the third side.
10. $ABTR$ is a trapezoid.	**10.** Definition of trapezoid

33. 0.78 cm **35.** 6 ft **36.** 8 **37.** 5 **38.** $\angle 2 \cong \angle 1$, $\angle 2 \cong \angle N$, $\overline{PQ} \parallel \overline{LN}$, $\angle 1 \cong \angle L$, $\angle 3 \cong \angle 4$, $\angle L \cong \angle N$, $\angle 2 \cong \angle L$, $\triangle MPQ \sim \triangle MLN$, $\frac{MP}{ML} = \frac{PQ}{LN} = \frac{QM}{NM}$, $\triangle MPQ$ and $\triangle MLN$ are isosceles **39.** $\overline{PQ} \perp \overline{QS}$, $\triangle RPQ$ and $\triangle RSQ$ are isosceles, $\overline{RP} \cong \overline{QP}$, $\overline{RS} \cong \overline{QS}$ **40.** $\triangle AEF$ and $\triangle BDC$ are right triangles, $\overline{CD} \perp \overline{AD}$, $\overline{AF} \perp \overline{AD}$, $\overline{CD} \parallel \overline{AF}$ **41.** $\overline{AB} \parallel \overline{DC}$, $\angle 1 \cong \angle 3$, $\triangle ADC \cong \triangle CBA$, $\angle 1$ and $\angle 2$ are complementary, $\angle 3$ and $\angle 4$ are complementary

Pages 350–351 Lesson 7-7
5. 9 socks **7.** 1869 **9.** 24 guests

11. The series of fractions simplifies to $\frac{1}{2} + 1 + \frac{3}{2} + 2 \cdots + \frac{99}{2}$. So, find half of the sum of the series $1 + 2 + 3 + \cdots + 99$. $1 + 2 + 3 + \cdots + 99 = 49(100) + 50$ or 4950, since there are 49 pairs of addends with a sum of 100, and 50 has no match.

The sum of the series of fractions is $\frac{4950}{2}$ or 2475.

Pages 352–354 Summary and Review
1. 7.5 **3.** 1.25 **5.** false **7.** $9\frac{1}{3}$ **9.** true **11.** 8

13. Given: $\frac{RP}{QS} = \frac{RS}{QP}$
$\overline{QR} \parallel \overline{PS}$
isosceles trapezoid $PQRS$
Prove: $\triangle PQR \sim \triangle SRQ$

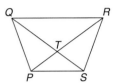

Statements	Reasons
1. isosceles trapezoid $PQRS$ $\overline{QR} \parallel \overline{PS}$	**1.** Given
2. $\overline{QP} \cong \overline{RS}$	**2.** Definition of isosceles trapezoid
3. $\overline{RQ} \cong \overline{RQ}$	**3.** Congruence of segments is reflexive.
4. $QP = RS$ $RQ = RQ$	**4.** Definition of congruent segments
5. $\frac{RQ}{RQ} = 1$, $\frac{QP}{RS} = 1$	**5.** Division property of equality
6. $\frac{RQ}{RQ} = \frac{QP}{RS}$	**6.** Substitution property of equality
7. $\frac{RP}{QS} = \frac{RS}{QP}$	**7.** Given
8. $\angle PQR \cong \angle SRQ$	**8.** Base $\angle$s of an iso. trap. are $\cong$.
9. $\triangle PQR \sim \triangle SRQ$	**9.** SAS Similarity

15. 4 **17.** $9\frac{1}{3}$ **19.** 29 **21.** $AT = TC = 4$ and $DT = TB = 7$ **23.** 2

Pages 356–357 Algebra Review
1. $7x^3y + 28x^2y^2 - 56xy^3$ **3.** $-2x^3 + 56x^2 - 9x$ **5.** $3x^2 + 13x - 10$ **7.** $6x^2 + 25xy - 9y^2$ **9.** $(x + 9)^2$
11. $\frac{1}{4}\left(n + \frac{3}{2}\right)\left(n - \frac{3}{2}\right)$ **13.** $(4p - 9r^2)(4p + 9r^2)$ **15.** $\frac{x}{4y^2z}$;
$x = 0, y = 0, z = 0$ **17.** $\frac{a-5}{a-2}$; $a = -5$ or 2 **19.** Domain: $\{-2, -1, 0\}$; Range: $\{-1, 0, 2\}$; Inverse: $\{(-1, -2), (0, -1), (2, 0)\}$
21. Domain: $\{4\}$; Range: $\{-2, -1, 1, 7\}$; Inverse: $\{(1,4), (-2, 4), (7, 4), (-1, 4)\}$ **23.** $\{y \mid y \le -\frac{9}{2}\}$ **25.** $\{d \mid d \ge 20\}$ **27.** $\frac{245}{3}$ or 81.6 **29.** $cd - y$ **31.** $\frac{14b - 9}{a}$ **33.** $62.40 **35.** 14 in. by 11 in.

CHAPTER 8 RIGHT TRIANGLES AND TRIGONOMETRY

Pages 362–364 Lesson 8-1
5. $\sqrt{45} \approx 6.7$ **7.** $\sqrt{154} \approx 12.4$ **9.** $\sqrt{40} \approx 6.3$ **11.** $\sqrt{44} \approx 6.6$, $\sqrt{28} \approx 5.3$ **13.** $\sqrt{15} \approx 3.9$ **15.** $\frac{3}{2}$ **17.** 1 **19.** $\sqrt{45} \approx 6.7$ **21.** $\sqrt{30} \approx 5.5$ **23.** $\sqrt{32} \approx 5.7$ **25.** 4, 5 **27.** 8, $\sqrt{1280} \approx 35.8$ **29.** 5, $\sqrt{20} \approx 4.5$ **31.** $PQ = 9$, $PR = 13$, $PV = 3\sqrt{13}$, $VR = 2\sqrt{13}$

33. Given: $\triangle ADC$
$\angle ADC$ is a right angle.
$\overline{DB}$ is an altitude of $\triangle ADC$.
Prove: $\frac{AB}{DB} = \frac{DB}{CB}$

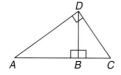

We are given that $\angle ADC$ is a right angle and $\overline{DB}$ is an altitude of $\triangle ADC$. $\triangle ADC$ is a right triangle by the definition of a right triangle. Therefore, $\triangle ADB \sim \triangle DCB$ since if the altitude is drawn from the vertex of the right angle to the hypotenuse of a right triangle, then the two triangles formed are similar to the given triangle and to each other. So $\frac{AB}{DB} = \frac{DB}{CB}$ by the definition of similar polygons. **37.** 127 pages **38.** 30, 60, 90 **39.** Distributive property **40.** No; complementary angles have measures with a sum of 90 and the measure of an obtuse angle is greater than 90.
41. $AB + BC = AC$

Pages 367–370 Lesson 8-2

5. yes **7.** yes **9.** yes **11.** $\sqrt{27} \approx 5.2$ **13.** 3 **15.** 4.5 miles **17.** yes **19.** yes **21.** no **23.** 1 **25.** 13.6 **27.** 9.8 **29.** 68 cm **31.** $32 + 4\sqrt{241}$ units or about 94.1 units **33.** 19.2 ft **35.** 13 feet

37. Given: $\triangle ABC$ with sides of measure a, b, and c where $a^2 + b^2 = c^2$.
Prove: $\triangle ABC$ is a right triangle.

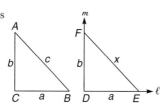

Draw $\overline{DE}$ on line ℓ with measure equal to a. At D, draw line $m \perp \overline{DE}$. Locate point F on m so that $DF = b$. Draw $\overline{FE}$ and call its measure x. Because $\triangle FED$ is a right triangle, $a^2 + b^2 = x^2$. But $a^2 + b^2 = c^2$, so $x^2 = c^2$ or $x = c$. Thus, $\triangle ABC \cong \triangle FED$ by SSS. This means $\angle C \cong \angle D$. Therefore, $\angle C$ must be a right angle, making $\triangle ABC$ a right triangle. **39.** 13 ft **41.** the area of a trapezoid **43a.** yes **b.** yes **c.** The conjecture is true. **d.** No; 1, 2, and 30 have a product of 60, but $1^2 + 2^2 \neq 30^2$. **44.** $3\sqrt{15}$ **45.** 15 m **46.** $\overrightarrow{XT} \| \overrightarrow{WY}$; $\overrightarrow{TZ} \| \overrightarrow{SY}$ **47.** 4 **48.** A duck-billed platypus is a mammal that lays eggs; syllogism

Pages 373–375 Lesson 8-3

5. $16, 8\sqrt{3}$ **7.** $31.2\sqrt{2} \approx 44.1$ m **9.** $2\sqrt{3} \approx 3.5$ ft **11.** $\frac{\sqrt{3}}{3} \approx 0.6$ yd **13.** $3\sqrt{3} \approx 5.2$ **15.** $\frac{7\sqrt{2}}{2} \approx 4.9$ **17.** $7.5\sqrt{3} \approx 13$
19. $\sqrt{37} \approx 6.083$ **21.** 27 **23.** $12 + 12\sqrt{3}$ or about 32.8 units **25.** $\sqrt{147} \approx 12.1$ cm **27.** $\sqrt{3}$ **29.** $\sqrt{5}$ **31.** $\sqrt{7}$ **33.** $m\angle BFD = 60$ **37.** about 9.5 feet **38.** yes; SAS Similarity **39.** 78 in.

40. Given: $m\angle BAC = 90$,
$m\angle ABC = 30$,
$m\angle EDC = 60$
Prove: ℓ is parallel to m.

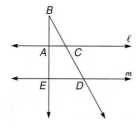

Proof: Since $m\angle BAC = 90$, $\triangle BAC$ is a right triangle. Thus, $m\angle BAC + m\angle ABC + m\angle ACB = 180$, and since $m\angle ABC = 30$, it follows that $90 + 30 + m\angle ACB = 180$. By subtraction, $m\angle ACB = 60$. Since $\angle ACB$ and $\angle EDC$ are congruent corresponding angles, ℓ is parallel to m.

Pages 378–382 Lesson 8-4

5. $\frac{15}{17} \approx 0.882$ **7.** $\frac{15}{8} \approx 1.875$ **9.** $\frac{15}{17} \approx 0.882$ **11.** tan Q **13.** sin P **15.** 0.174 **17.** 0.781 **19.** $m\angle S = 35$ **21.** about 15.4 miles **23.** $\frac{20}{29} \approx 0.690$ **25.** $\frac{21}{29} \approx 0.724$ **27.** $\frac{3}{4} \approx 0.750$ **29.** cos T **31.** tan T **33.** $\frac{1}{2}$ **35.** $\frac{1}{2}$ **37.** $\frac{\sqrt{3}}{3}$ **39.** 9.3 **41.** 42, 26.8 **43.** 30, 3.1 **45.** 12.9, 17.6 **47.** about 122 feet **49.** about 43.6 in. **53.** about 2733 feet **55.** 16 units; $8\sqrt{3}$ or about 13.9 units **56.** $x = \sqrt{104} \approx 10.2$, $y = \sqrt{65} \approx 8.1$ **57.** no; $\frac{AB}{AC} \neq \frac{AE}{AD}$ **58.** 16 in. **59.** False; the diagonals of a rhombus bisect opposite angles.

60. Given: $\overline{GA} \cong \overline{AI}$
$GL < IL$
Prove: $m\angle 1 < m\angle 2$

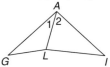

Proof: $\triangle GAL$ and $\triangle IAL$ satisfy the SSS inequality. That is, $\overline{GA} \cong \overline{AI}$, $\overline{AL} \cong \overline{AL}$, and $GL < IL$. It follows that $m\angle 1 < m\angle 2$. **61.** 100 **62.** 10

Page 382 Mid-Chapter Review

1. $\sqrt{112} \approx 10.6$ **2.** $\sqrt{135} \approx 11.6$ **3.** $\sqrt{77} \approx 8.8$ **4.** $\sqrt{128} \approx 11.3$ **5.** yes **6.** no **7.** yes **8.** no **9.** yes **10.** yes **11.** 72 units **12.** $8\sqrt{2} \approx 11.3$ cm **13.** 51°

Pages 386–388 Lesson 8-5

5. E: $\angle ZXY$; D: $\angle WYX$ **7.** E: $\angle KHJ$; D: $\angle IJH$ **9.** sin 47° $= \frac{10}{PQ}$; 13.7 **11.** tan 72° $= \frac{13}{QR}$; 4.2 **13.** cos 24° $= \frac{43.7}{PQ}$; 47.8 **15.** 19 feet **17.** 27 **19.** 61 **21.** 2 **23.** 19 **25.** 132.63 meters **27.** 8° **29.** 31.07 meters **31.** 938.22 feet **35.** 11.7 cm **37.** cos $A = \frac{1}{3}$ or 0.333; sin $A = \frac{2\sqrt{2}}{3}$ or 0.943; tan $A = 2\sqrt{2}$ or 2.828 **38.** 254 miles **39.** 15.2

40. Given: $\overline{CI} \cong \overline{MI}$
$\overline{IT}$ is a median of $\triangle CIM$.
Prove: $\angle CIT \cong \angle MIT$

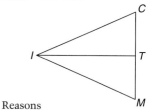

Statements	Reasons
1. $\overline{CI} \cong \overline{MI}$ $\overline{IT}$ is a median of $\triangle CIM$.	1. Given
2. $\overline{IT} \cong \overline{IT}$	2. Congruence of segments is reflexive.
3. $\overline{TC} \cong \overline{TM}$	3. Definition of median
4. $\triangle CTI \cong \triangle MTI$	4. SSS
5. $\angle CIT \cong \angle MIT$	5. CPCTC

Pages 391–393 Lesson 8-6

5.

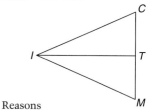

$\frac{\sin 50°}{14} = \frac{\sin B}{10}$

7.

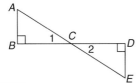

$\frac{\sin 42°}{16} = \frac{\sin C}{12}$

25. 536 feet **27.** 21.8 miles **31.** 109.6 feet **32.** 50.50 meters **33.** about 9.7

35. Given: $\overline{AB} \perp \overline{BD}$
$\overline{DE} \perp \overline{DB}$
$\overline{DB}$ bisects $\overline{AE}$.
Prove: $\angle A \cong \angle E$

Statements	Reasons
1. $\overline{AB} \perp \overline{BD}$ $\overline{DE} \perp \overline{DB}$	**1.** Given
2. $\angle B$ is a right angle. $\angle D$ is a right angle.	**2.** $\perp$ lines form four rt. $\angle$s.
3. $\angle B \cong \angle D$	**3.** All rt. $\angle$s are $\cong$.
4. $\angle 1 \cong \angle 2$	**4.** Vertical $\angle$s are $\cong$.
5. $\overline{DB}$ bisects $\overline{AE}$.	**5.** Given
6. $\overline{AC} \cong \overline{EC}$	**6.** Definition of bisector
7. $\Delta ABC \cong \Delta EDC$	**7.** AAS
8. $\angle A \cong \angle E$	**8.** CPCTC

Pages 396–398 Lesson 8-7
5. law of cosines; $a \approx 6.1$, $m\angle B \approx 54$, $m\angle C \approx 71$ **7.** law of cosines; $m\angle A \approx 54$, $m\angle B \approx 59$, $m\angle C \approx 67$ **9.** law of cosines; $m\angle A \approx 44$, $m\angle B \approx 56$, $m\angle C \approx 80$ **11.** law of cosines; $c \approx 22.7$, $m\angle A \approx 68$, $m\angle B \approx 34$ **13.** law of cosines; $c \approx 6.5$, $m\angle A \approx 76$, $m\angle B \approx 69$ **15.** law of sines; $a \approx 23.1$, $m\angle B \approx 98$, $b \approx 27.6$ **17.** $m\angle A \approx 23$, $m\angle B \approx 67$, $m\angle C \approx 90$ **19.** $m\angle C \approx 81$, $a \approx 9.1$, $b \approx 12.1$ **21.** $a = 2.5$, $m\angle B \approx 76$, $m\angle C \approx 75$ **23.** $m\angle A \approx 103$, $m\angle B \approx 49$, $m\angle C \approx 28$ **25.** $m\angle A \approx 15$, $m\angle B \approx 131$, $m\angle C \approx 34$ **27.** 36 **29.** 26 cm **31.** 61.9 ft **35.** about 67 nautical miles **37.** about 74 yards **38.** 24.56 cm **39.** 12 **40.** 13 **41.** Yes; the lengths of the segments satisfy the triangle inequality.
42. Given: $\overline{AB} \cong \overline{BC}$
Prove: $\angle 3 \cong \angle 4$

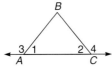

We are given that $\overline{AB} \cong \overline{BC}$. So $\angle 1 \cong \angle 2$ since if two sides of a triangle are congruent, then the angles opposite those sides are congruent. $\angle 1$ and $\angle 3$ and $\angle 2$ and $\angle 4$ form linear pairs. The angles in a linear pair are supplementary, so $\angle 1$ and $\angle 3$ are supplementary and $\angle 2$ and $\angle 4$ are supplementary. If two angles are supplementary to the same or congruent angles, then they are congruent. So $\angle 3$ and $\angle 4$ are congruent.

Pages 400–401 Lesson 8-8
3. look for a pattern, act it out, make a chart; 220 cans
5. 2730; multiply the previous term by 4 and add 2 **7.** Start both timers. When the 3-minute timer runs out, start boiling the spaghetti. When, after 4 minutes, the 7-minute timer runs out, start it over and cook for 7 more minutes.

9. $144,000 **11.** 2660 cm^3 **13.** 70; It is the only number divisible by 5 ($20\% = \frac{1}{5}$) and 7 between 50 and 100.

Pages 402–404 Summary and Review
1. 18 **3.** $\sqrt{3120} \approx 55.9$ **5.** 2 **7.** $\sqrt{11,979} \approx 109.4$ **9.** $\sqrt{mn}$ **11.** $LM = \sqrt{24}$ or 4.9 **13.** $LM = \sqrt{30}$ or 5.5 **15.** $KM \approx 25.8$ **17.** 9.8 **19.** 17.0 **21.** no **23.** yes **25.** $1.55\sqrt{3} \approx 2.7$ **27.** $\frac{14.2}{\sqrt{2}} \approx 10.0$ **29.** $\frac{15}{17} \approx 0.882$ **31.** $\frac{15}{17} \approx 0.882$ **33.** $m\angle A = 51$, $c \approx 89.7$, $a \approx 70.2$ **35.** $m\angle A \approx 13$, $m\angle C \approx 145$, $c \approx 10.4$ **37.** $b \approx 30.6$, $m\angle A \approx 89$, $m\angle C \approx 47$ **39.** $m\angle A \approx 41$, $m\angle B \approx 79$, $m\angle C \approx 60$ **41.** about 4° **43.** about 1675 km

CHAPTER 9 CIRCLES

Pages 413–415 Lesson 9-1
5. P **7.** $\overline{DB}$ **9.** $\overrightarrow{HB}$ **11.** G, P **13.** A, B, C, D, E **15.** $(-2, -7), 9$ **17.** true **19.** false **21.** false **23.** 7.6 **25.** x **27.** $x^2 + y^2 = 49$ **29.** $x^2 + y^2 = 14$

31.

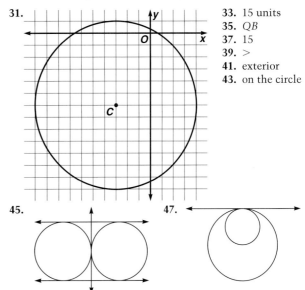

33. 15 units **35.** QB **37.** 15 **39.** > **41.** exterior **43.** on the circle

45.

47.

49. $(x + 2)^2 + (y - 11)^2 = 32$ **51.** $5\frac{1}{4}$ in. **53.** 16 **54.** true **55.** false **56.** true **57.** false **58.** false **59.** false

Pages 418–421 Lesson 9-2
7. major; 320 **9.** semicircle; 180 **11.** major; 270 **13.** minor; 40 **15.** minor; 130 **17.** minor; 90 **19.** 28 **21.** 246 **23.** 180 **25.** 152 **27.** 180 **29.** 114 **31.** 40 **33.** 85 **35.** 135 **37.** 85 **39.** 140 **41.** 320 **43.** false **45.** true **47.** false **49.** true **51.** 144 **53.** 32 **55.** 112 **57.** 216 **59.** 30 **61.** 13 **63.** 60 **65.** 46 **67.** $(x - 1)^2 + (y - 2)^2 = 9$ **68.** 4.5 **69.** -3, subtract 2 from the previous term **70.** 8 **71.** Obtuse triangles have one obtuse angle and all the angles in an acute triangle are acute. **72.** If circles are concentric, then they have the same center. **73.** $\sqrt{65} \approx 8.1$

Pages 425–427 Lesson 9-3
5. Theorem 9-2 **7.** Theorem 9-3 **9.** Theorem 9-1 **11.** Theorem 9-1 **13.** 75 **15.** $\overline{QV}$ **17.** V **19.** $\overarc{YT}$

21. $\overline{WA}$ **23.** no **25.** 16 **27.** Yes, because in a circle, if a diameter is perpendicular to a chord, then it bisects the chord and its arc. **29.** 13 in. **31.** 6 cm **33.** longer chord **35.** 31 **37.** Given: $\odot O$
$\overline{OS} \perp \overline{RT}$
$\overline{OV} \perp \overline{UW}$
$\overline{OS} \cong \overline{OV}$
Prove: $\overline{RT} \cong \overline{UW}$

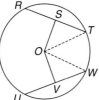

Proof: Draw radii $\overline{OT}$ and $\overline{OW}$. Since $\overline{OT} \cong \overline{OW}$ and $\overline{OS} \cong \overline{OV}$, $\Delta STO \cong \Delta VWO$ by HL. Then $\overline{ST} \cong \overline{VW}$ and $ST = VW$. Since a diameter perpendicular to a chord bisects a chord, $\overline{OS}$ bisects $\overline{RT}$ and $\overline{OV}$ bisects $\overline{UW}$. So $RT = 2ST$ and $UW = 2VW$. Therefore, $RT = UW$ and $\overline{RT} \cong \overline{UW}$. **39.** $20\sqrt{3}$ or about 34.6 units **43.** about 7.1 in. **45.** 57 **46.** $\sqrt{189} \approx 13.7$ **48.** Division or multiplication property of equality **49.** 5

Pages 431–433 Lesson 9-4
5. no **7.** no **9.** 60 **11.** 30 **13.** 40 **15.** 100 **19.** 47 **21.** 52 **23.** 99 **25.** 38 **27.** 99 **29.** 94 **31.** 68 **33.** 34 **35.** 47 **37.** 95 **39.** 105 **41.** 190 **43.** 150 **45.** yes; yes **47.** Given: $\overline{MH} \parallel \overline{AT}$
Prove: $\widehat{AM} \cong \widehat{HT}$

Proof: Draw $\overline{MT}$. Since $\overline{MH} \parallel \overline{AT}$, $\angle HMT \cong \angle MTA$ and $m\angle HMT = m\angle MTA$. But $m\angle HMT = \frac{1}{2}m\widehat{HT}$ and $m\angle MTA = \frac{1}{2}m\widehat{MA}$. Therefore, $\frac{1}{2}m\widehat{HT} = \frac{1}{2}m\widehat{MA}$ and $m\widehat{HT} = m\widehat{MA}$. The arcs are in the same circle and $\widehat{AM} \cong \widehat{HT}$.
49. Given: T lies inside $\angle PRQ$.
Prove: $m\angle PRQ = \frac{1}{2}m\widehat{PQ}$
Proof: $m\angle PRQ = m\angle PRK + m\angle KRQ$
$= \frac{1}{2}(m\widehat{PK}) + \frac{1}{2}(m\widehat{KQ})$
$= \frac{1}{2}(m\widehat{PK} + m\widehat{KQ})$
$= \frac{1}{2}m\widehat{PQ}$

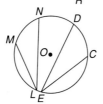

51. Give: $\odot O$
$\widehat{MN} \cong \widehat{DC}$
Prove: $\angle MLN \cong \angle DEC$

Proof: Since $\widehat{MN} \cong \widehat{DC}$, $m\widehat{MN} = m\widehat{DC}$ and $\frac{1}{2}m\widehat{MN} = \frac{1}{2}m\widehat{DC}$, $m\angle MLN = \frac{1}{2}m\widehat{MN}$ and $m\angle DEC = \frac{1}{2}m\widehat{DC}$. Therefore, $m\angle MLN = m\angle DEC$ and $\angle MLN \cong \angle DEC$.
53. Given: quadrilateral inscribed in $\odot O$.
Prove: $\angle DCB$ and $\angle DAB$ are supplementary.

Proof: In $\odot O$ $m\widehat{DCB} + m\widehat{DAB} = 360$. Since $m\angle DAB = $

$\frac{1}{2}m\widehat{DCB}$ and $m\angle DCB = \frac{1}{2}m\widehat{DAB}$, $m\angle DAB + m\angle DCB = \frac{1}{2}m\widehat{DCB} + \frac{1}{2}m\widehat{DAB}$ or $m\angle DAB + m\angle DCB = \frac{1}{2}(m\widehat{DCB} + m\widehat{DAB}) = \frac{1}{2}(360)$ or 180. Since $m\angle DCB + m\angle DAB = 180$, the angles are supplementary by definition.
55a. 60 **b.** 120 **56.** 65 cm **57.** no, $6^2 + 9^2 \neq 11^2$ **58.** yes, all angles 90° and the sides are proportional **59.** 7 inches **60.** hypothesis: an angle is inscribed in a semicircle; conclusion: the angle is a right angle
61.

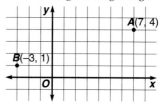

$AB = \sqrt{109} \approx 10.4$

Page 433 Mid-Chapter Review
1.

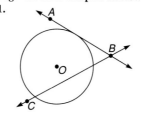

2. 12 **3.** 180 **4.** 64 **5.** 148 **6.** 116 **7.** 116 **8.** 3 cm **9.** 20.8 units **10.** 12 units **11.** 120

Pages 436-439 Lesson 9-5
5. neither **7.** 12 **9.** 14 **11.** 8 **13.** 45 **15.** 45 **17.** 90 **19.** 45 **21.** 5 **23.** 13 **25.** $5\sqrt{2} \approx 7.1$ **27.** $\sqrt{329} \approx 18.1$ **29.** $\Delta OKM \sim \Delta CLM$; Sample answer: A line parallel to a side of a triangle forms a triangle similar to the original triangle by AA Similarity. **31.** 15
33. Sample answer:
$$TB = TS$$
$$TA = TR$$
$$TA - TB = TR - TS$$
$$AB = RS$$
35.

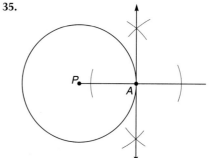

37. Given: $\overleftrightarrow{CA}$ is tangent to the circle at A.
Prove: $\overline{XA} \perp \overleftrightarrow{CA}$

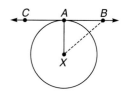

Proof: Pick any point on $\overleftrightarrow{CA}$ other than A and call it B. Now, draw $\overline{XB}$. From the definition of tangent, we know that $\overleftrightarrow{CA}$ intersects $\odot X$ at exactly one point, A, and that B lies in the exterior of $\odot X$. As a result, $XA < XB$. Thus, since $\overline{XA}$ is the shortest segment from X to $\overleftrightarrow{CA}$, it follows that $\overline{XA} \perp \overleftrightarrow{CA}$.

39. Given: $\ell \perp \overline{AB}$
$\overline{AB}$ is a radius of $\odot A$.
Prove: ℓ is tangent to $\odot A$.

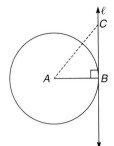

Proof: Assume that ℓ is not tangent to the circle. Since ℓ touches $\odot A$ at B, it must touch the circle in another point. Call this point C. Then $AB = AC$. But if $\overline{AB}$ is perpendicular to ℓ, AB must be the shortest distance between A and ℓ There is a contradiction. Therefore, ℓ is tangent to $\odot A$.
44. 84   **45.** 68 in.   **46.** yes; AA Similarity   **47.** yes; isosceles   **48.** -6

Pages 442–446 Lesson 9-6
5. 59   **7.** 105   **9.** 40   **11.** $35 = \frac{1}{2}[(360 - x) - x]$; 145
13. 114   **15.** 66   **17.** 138   **19.** 174   **21.** 49   **23.** 198
25. 236   **27.** 38   **29.** 44   **31.** 44   **33.** 200   **35.** 46
37. 134   **39.** 144   **41.** 160   **43.** 26   **45.** 144   **47.** 15
49. $x = 46$, $y = 64$, $z = 40$

51. Given: Secants $\overrightarrow{AC}$ and $\overrightarrow{BD}$ intersect at X in the interior of $\odot P$.
Prove: $m\angle AXB = \frac{1}{2}(m\,\widehat{AB} + m\,\widehat{CD})$.

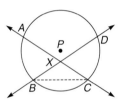

Proof: We are given that secants $\overrightarrow{AC}$ and $\overrightarrow{BD}$ intersect at X inside circle P. Draw $\overline{BC}$. Because an angle inscribed has the measure of $\frac{1}{2}$ the measure of its intercepted arc, $m\angle XBC = \frac{1}{2}m\widehat{CD}$ and $m\angle XCB = \frac{1}{2}m\widehat{AB}$. By the Exterior Angle Theorem, $m\angle AXB = m\angle XCB + m\angle XBC$. By substitution, $m\angle AXB = \frac{1}{2}m\widehat{AB} + \frac{1}{2}m\widehat{CD}$. Then by use of the distributive property, $m\angle AXB = \frac{1}{2}(m\widehat{AB} + m\widehat{CD})$.

53. Give: $\overrightarrow{DG}$ is a tangent to the circle.
$\overrightarrow{DF}$ is a secant to the circle.
Prove: $m\angle FDG = \frac{1}{2}(m\,\widehat{FG} - m\,\widehat{EG})$

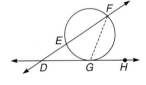

Proof: We are given that $\overrightarrow{DG}$ is a tangent to a circle, and $\overrightarrow{DF}$ is a secant to that circle. Draw $\overline{FG}$: $m\angle DFG = \frac{1}{2}m\widehat{EG}$; $m\angle FGH = \frac{1}{2}m\widehat{FG}$ because the measure of an inscribed angle equals $\frac{1}{2}$ the measure of the intercepted arc. By the Exterior Angle Theorem, $m\angle FGH = mDFG + m\angle FDG$. Then by substitution $\frac{1}{2}m\widehat{FG} = \frac{1}{2}m\widehat{EG} + m\angle FDG$, and by the subtraction property of equality $\frac{1}{2}m\widehat{FG} - \frac{1}{2}m\widehat{EG} = m\angle FDG$. Finally, by the distributive property of equality, $\frac{1}{2}(m\widehat{FG} - m\widehat{GE} = m\angle FDG)$.

55. Case I: The secant contains the center of the circle.
Given: Secant $\overleftrightarrow{AB}$ contains the center of the circle P. $\overleftrightarrow{CB}$ is tangent to $\odot P$ at B.
Prove: $m\angle CBA = \frac{1}{2}m\,\widehat{ADB}$

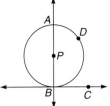

Proof: By Theorem 9-8, $\overline{BA} \perp \overline{BC}$. Thus $\angle CBA$ is a right angle and $m\angle CBA = 90$. $\widehat{ADB}$ is a semi-circle, thus $m\,\widehat{ADB} = 180$.
Substituting, $m\angle CBA = 90 = \frac{1}{2}(180) = m\,\widehat{ADB}$

Case II: The secant does not contain the center of the circle.
Given: Secant $\overleftrightarrow{AB}$; $\overleftrightarrow{CE}$ is tangent to $\odot P$ at B. D is a point on the major arc BFA.
Prove: $m\angle CBA = \frac{1}{2}m\,\widehat{AB}$
$m\angle EBA = \frac{1}{2}m\,\widehat{BDA}$

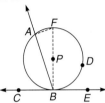

Proof: Draw diameter $\overline{BF}$ and chord $\overline{AF}$.
$m\angle CBA + m\angle ABF = m\angle CBF$ by the Angle Addition Postulate and, subtracting, $m\angle CBA = m\angle CBF - m\angle ABF$. $m\angle ABF = \frac{1}{2}m\widehat{AF}$ by Theorem 9-4. $m\widehat{FAB} = 180$ as $\widehat{FAB}$ is a semi-circle. $\overline{BF} \perp \overline{CB}$ by Theorem 9-8, making $\angle CBF$ a right angle and $m\angle CBF = 90$. Substituting $m\angle CBA = 90 - \frac{1}{2}m\widehat{AF} = \frac{1}{2}(180) - \frac{1}{2}m\,\widehat{AF} = \frac{1}{2}(180 - m\,\widehat{AF}) = \frac{1}{2}(m\,\widehat{FAB} - m\,\widehat{AF})$. Using the Arc Addition Postulate, $m\,\widehat{BA} + m\,\widehat{AF} = m\,\widehat{FAB}$. Subtracting, $m\,\widehat{BA} = m\,\widehat{FAB} - m\,\widehat{AF}$. Substituting, $m\angle CBA = \frac{1}{2}m\,\widehat{AB}$. Using the Angle Addition Postulate, $m\angle EBA = m\angle EBF + m\angle FBA$. $m\angle FBA = \frac{1}{2}m\widehat{FA}$ by Theorem 9-4. $m\,\widehat{BDF} = 180$ as $\widehat{BDF}$ is a semi-circle. $\overline{BF} \perp \overline{CB}$ by Theorem 9-8 making $\angle EBF$ a right angle, thus $m\angle EBF = 90$. Substituting, $m\angle EBA = 90 + m\,\widehat{FA} = \frac{1}{2}(180) + \frac{1}{2}m\,\widehat{FA} = \frac{1}{2}(180 + m\,\widehat{FA}) = \frac{1}{2}(m\,\widehat{BDF} + m\,\widehat{FA})$. Using the Arc Addition Postulate, $m\,\widehat{BDA} = m\,\widehat{BDF} + m\,\widehat{FA}$. Therefore, $m\angle EBA = \frac{1}{2}m\,\widehat{BDA}$.

57a. yes   **b.** no   **58.** It is tangent to the circle.   **59.** 17.5
60. yes   **61.** 116   **62.** 50   **63.** Congruence of angles is reflexive.

64.

66. 8, -1   **67.** $\frac{1}{3}$, -5
68. $\frac{13 + 3\sqrt{33}}{2}$, $\frac{13 - 3\sqrt{33}}{2}$
69. $\frac{4 - \sqrt{29}}{2}$, $\frac{4 + \sqrt{29}}{2}$

Pages 450–453 Lesson 9-7
7. $3x = 7 \cdot 2$; $4\frac{2}{3}$ **9.** $4(x + 4) = 3 \cdot 8$; 2
11. $3x = 7 \cdot 3$; 7   **13.** $x^2 = 8 \cdot 16$; $8\sqrt{2} \approx 11.31$
15. $\frac{17}{8}$ **17.** 0.46   **19.** 1.2
21. $\frac{-5 + \sqrt{73}}{2} \approx 1.77$ **23.** $\frac{\sqrt{5}}{4} \approx 0.56$
25. 15.3   **27.** 26.3   **29.** 12.25   **31.** $\sqrt{51.25} \approx 7.16$

33. Given: $\overline{RP}$ and $\overline{RT}$ are secant segments.
Prove: $RQ \cdot RP = RS \cdot RT$

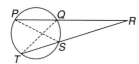

Statements	Reasons
1. Draw $\overline{PS}$ and $\overline{TQ}$.	1. Through any 2 pts there is 1 line.
2. $\angle SRP \cong \angle QRT$	2. Congruence of angles is reflexive.
3. $\angle RPS \cong \angle RTQ$	3. If 2 inscribed $\angle$ of a $\odot$ intercept the same arc, then the $\angle$ are $\cong$.
4. $\triangle PSR \approx \triangle TQR$	4. AA Similarity
5. $\frac{RQ}{RS} = \frac{RT}{RP}$	5. Definition of similar polygons
6. $RQ \cdot RP = RS \cdot RT$	6. Cross products

37. about 1106 mi   **39.** 925 ft   **41.** about 2.6   **43.** about 1
45. 25   **46.** 25   **47.** about 0.85 ft. or 10.2 in.

48. yes;

49. 2.5

Pages 456–457 Lesson 9-8
7.

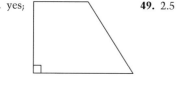

9a.

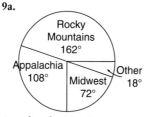

b. Sample answer: quality and price of coal   **11.** 40;
Dividing by 2, 3, 4, 5, . . .   **13.** about 18.0%

Pages 458–460 Summary and Review
1. no
3.

5. 180   **7.** 117   **9.** 123
11. 34.2 cm   **13.** 144
15. 36   **17.** 12   **19.** 42
21. 18.5   **23.** 138   **25.** 6
27. $15\frac{1}{3}$ **29.** 36
31. 22.1 m

Pages 462–463 Algebra Review
1. $\{x|\ x < -1 \text{ or } x \geq 3\}$
3. $\{r|\ 7 < r \leq 11\}$ **5.** $\{0, -17\}$
7. $\{\frac{3}{2}, -8\}$ **9.** $\{0, -\frac{2}{5}\}$ **11.** $(3, 5)$ **13.** $6\sqrt{3}$ **15.** $\frac{\sqrt{11}}{11}$
17. $4x^2\sqrt{6}$ **19.** $\frac{3axy}{10}$ **21.** $b + 7$ **23.** $\frac{7ab(x + 9)}{3(x - 5)}$ **25.** 3
27. $\frac{3}{4}$ **29.** 36 ounces   **31.** $16, $9   **33.** about 14.7 seconds

CHAPTER 10 POLYGONS AND AREA

Pages 469–472 Lesson 10-1
7. not a polygon   **9.** concave polygon   **11.** concave

polygon   **13.** not a polygon   **15.** faces: quadrilaterals $ABFE$, $FBCG$, $HGCD$, $EHDA$, $ABCD$, and $EFGH$; edges: $\overline{AE}$, $\overline{EF}$, $\overline{FB}$, $\overline{AB}$, $\overline{FG}$, $\overline{GC}$, $\overline{CB}$, $\overline{HG}$, $\overline{CD}$, $\overline{DH}$, $\overline{EH}$, and $\overline{AD}$; vertices: A, B, C, D, E, F, G, H **17.** triangle   **19.** 20-gon   **21.** hexagon
23. not a polygon   **25.** convex pentagon   **27.** not a polygon
29. M, N, O, P, Q **31.** $\overline{MN}$, $\overline{NO}$, $\overline{OP}$, $\overline{PQ}$, $\overline{QM}$
33. pentagon   **35.** not regular; not all of the sides are congruent   **37.** regular; It is convex, all the sides are congruent, and all of the angles are congruent.
39. a square

41. not possible   **43a.** triangles
b. squares   **c.** triangles
d. pentagons   **e.** triangles
45a. 4, 4, 6   **b.** 6, 8, 12   **c.** 8, 6, 12
d. 12, 20, 30   **e.** 20, 12, 30

49.

50. 6.9   **51.** not enough information   **52.** undefined
53. If a polygon is regular, then it is convex and has all sides congruent.

Pages 476–479 Lesson 10-2
5. 720   **7.** 4140   **9.** $180(x - 2)$ **11.** $128\frac{4}{7}$
13. $\frac{180x - 360}{x}$ **15.** 5   **17.** $\frac{360}{n}$ **19.** 10   **21.** 20
23. 3240   **25.** 2340   **27.** $360t - 360$ **29.** 6   **31.** 24
33. 360   **35.** 135, 45   **37.** 162, 18   **39.** $\frac{180(x - 2)}{x}$, $\frac{360}{x}$
41. 21   **43.** 26; 36   **45.** 36   **47a.** 43   **b.** 7380   **49a.** 1010
b. 110   **c.** They don't always work; no
51a.

Regular Polygon	triangle	square	pentagon
Does it tessellate the plane?	yes	yes	no
Measure, m, of one interior angle	60	90	108
Is m a factor of 360?	*yes*	yes	no

Regular Polygon	hexagon	heptagon	octagon
Does it tessellate the plane?	yes	no	no
Measure, m, of one interior angle	120	$128\frac{4}{7}$	135
Is m a factor of 360?	yes	no	no

b. If the measure of an interior angle of a regular polygon is a factor of 360, the polygon will tessellate the plane.
52. concave polygon; heptagon   **53.** 13 in.   **54.** 45 feet
55. yes   **56.** 46   **57.** 61

Pages 481–482 Lesson 10-3
3. 120

5. Sample answer:

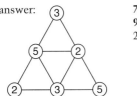

7. 35
9. 0.1; 4, 5; 12, 13; 24, 25; 40, 41

Pages 486–488 Lesson 10-4
5. 56 in^2 **7.** 100 mm^2 **9.** 56 ft^2 **11.** 4 m   **13.** 31.75 mm
15.

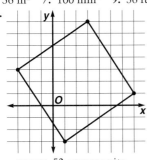

square; 52 square units

17.

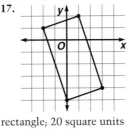

rectangle; 20 square units

19. $x + 4$ cm   **21.** 11 in.   **23.** The triangles can be assembled to form a square that is congruent to *MNOP*. So the area of the new square = the area of *MNOP*. Since the area of *ABCD* is the sum of these two areas, it is twice the area of *MNOP*.   **25.** 1577.8 pounds per square inch
27. $4 \cdot (5 - 2) + 7 = 19$ **28.** 18
29. Given: $\overline{AB} \parallel \overline{DC}$
Prove: $\angle A$ and $\angle D$ are
supplementary.

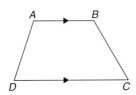

Statements	Reasons
1. $\overline{AB} \parallel \overline{DC}$	1. Given
2. $\angle A$ and $\angle D$ are supplementary.	2. If 2 ∥ lines are cut by a transversal, consec. int. ⦤ are supp.

Page 488 Mid-Chapter Review
1. concave; pentagon   **2.** not a polygon   **3.** convex; quadrilateral   **4.** not a polygon   **5.** 120, 60   **6.** 157.5, 22.5
7. 168, 12   **8.** 727   **9.** Yes, the area is 650 sq yd.

Pages 492–494 Lesson 10-5
5. 52.5 ft^2 **7.** 24 ft^2 **9.** $9\sqrt{3}$ or about 15.6 m^2 **11.** 30 in^2
13. 114 cm^2 **15.** 148.5 m^2 **17.** 120 square units
19. 3.5 in.   **21.** 4 cm   **23.** The base of the triangle is twice as long as the base of the parallelogram.

25. The diagonals of a rhombus are perpendicular, so $\overline{AE} \perp \overline{BD}$ and $\overline{CE} \perp \overline{DB}$. Therefore $\overline{AE}$ is an altitude of $\triangle ABD$ and $\overline{CE}$ is an altitude of $\triangle BCD$. Since the diagonals of a rhombus bisect each other, $\overline{AE} \cong \overline{EC}$.

So $AE = \frac{1}{2}AC$ and $EC = \frac{1}{2}AC$.
area of $ABCD$ = area of $\triangle ABD$ + area of $\triangle BCD$
area of $ABCD = \frac{1}{2}(BD)(AE) + \frac{1}{2}(BD)(EC)$
area of $ABCD = \frac{1}{2}(BD)(\frac{1}{2}AC) + \frac{1}{2}(BD)(\frac{1}{2}AC)$
area of $ABCD = \frac{1}{4}(BD)(AC) + \frac{1}{4}(BD)(AC)$
area of $ABCD = \frac{1}{2}(BD)(AC)$
The area is one-half the product of the diagonals.

29a. 35 sq units   **b.** 27.125 sq units   **30.** $288\sqrt{2}$ or about 407.3 square inches   **31.** $\sqrt{98}$ or about 9.9   **32.** $\frac{2}{27}$
33. No; $4 + 9 < 21$; it fails the triangle inequality.

Pages 498–500 Lesson 10-6
5. 30 cm   **7.** $30° - 60° - 90°$; Central angle COA would measure 120; $\overline{OT}$ bisects $\angle COA$, so $m\angle TOA = 60$.
9. OT, $\frac{5}{\sqrt{3}}$ or about 2.89 cm; OA, $\frac{10}{\sqrt{3}}$ or about 5.77 cm
11. 18 cm; 120; $\sqrt{3}$ cm; $9\sqrt{3}$ or about 15.6 cm^2
13. 10 m; 72; 1.4 m; 7m^2 **15.** 256 in^2 **17.** 289.3 mi^2
19. 186 ft^2 **21.** 88 units^2 **23.** $12 + 4\sqrt{3} \approx 18.9$ units^2
25. $108\sqrt{3} \approx 187.1$ units^2 **27.** 60 cm; $5\sqrt{3}$; $150\sqrt{3}$ cm^2
29. $4\sqrt{3}$ in.   **31.** $P = 24.72$ ft, $A = 47.02$ ft^2 **33a.** square
b. 2.91 in^2 **c.** 1.91 in^2 **35a.** 54 ft   **b.** 140.3 ft^2
37. 204 ft^2 **38.** no; $18 + 32 \leq 67$ so these lengths fail the triangle inequality.   **39.** If two figures are congruent, then they have equal areas; two figures are congruent; they have equal areas.

Pages 504–506 Lesson 10-7
5. $r = 2.4$ km; $C = 4.8\pi$ km; $A = 5.76\pi$ km^2 **7.** $r = 9$ ft;
$d = 18$ ft; $C = 18\pi$ ft   **9.** 62.8 m   **11.** 44.0 yd   **13.** 6.9 mm
15. 1017.9 in^2 **17.** 295.6 km^2 **19.** 34.9 yd^2 **21.** 11; 121π
23. $\sqrt{11}$; 11π **25.** 25; 625π **27.** 6π m; 9π m^2 **29.** $18\pi - 36$ units^2; 2055 units^2 **31.** $12.25\pi - 2.25\sqrt{3}$ units^2; 34.59 units^2 **33.** 50π units^2; 157.08 units^2 **35.** 150π units^2; 471.24 units^2 **37.** $56.25\pi - 108$ units^2; 68.71 units^2
41. about 7 square miles   **43.** about 7.7 ounces   **45.** 259.81 in^2 **46.** 10 cm   **47.** 10.3 feet; 31, 46   **48.** acute   **49.** 6

Pages 509–511 Lesson 10-8
5. $\frac{8}{10} = \frac{4}{5}$ **7.** $\frac{2}{10} = \frac{1}{5}$ **9.** $\frac{7}{10}$ **11.** 0.57   **13.** 0.625
15b. 0.08   **c.** No; as long as the dart is randomly thrown and the cards do not overlap, the area of the board and the cards is always the same.   **17.** $\frac{2}{3}$ **21.** 0.0625   **22.** 5.0 in., 2.0 in^2 **23.** $\sqrt{1144} \approx 33.8$

Pages 514–517 Lesson 10-9
5. $A - 1$, $M - 1$, $C - 0$ **7.** $A - 1$, $B - 2$, $C - 2$, $D - 2$, $E - 2$, $F - 1$ **9.** no   **11.** Not complete; add edges between I and T and between C and M.   **13.** complete   **15.** $A - 3$, $B - 2$, $C - 3$, $D - 2$, $E - 5$, $F - 2$, $G - 3$ **17.** edges between: P and E, E and N, P and N **19.** edges between: R and U, R and N, R and K, E and U, E and N, and K and U **21.** traceable
23. traceable

25. Sample answer:

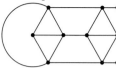

27a. yes **b.** no **c.** No; if a network has an Euler circuit, each node has an even degree, so any path will return to its starting node.

29a.

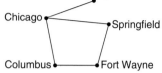

b. yes; through Chicago or Fort Wayne

31a. complete **b.** 21 **32.** 0.21
33. Given: Lines ℓ and m intersect at P.
Prove: Plane $\mathcal{R}$ contains both ℓ and m.
Assume: Plane $\mathcal{R}$ does not contain both ℓ and m.

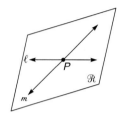

34. reflexive
35.

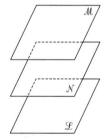

Pages 518–520 Summary and Review
1. $L, M, N, O, P; \overline{LM}, \overline{LP}, \overline{LN}, \overline{LO}, \overline{OP}, \overline{PM}, \overline{MN},$ and $\overline{NO}$
3. No; all of the faces are not congruent. **5.** 13 **7.** 20.3 ft^2
9. 9 m **11.** 27.09 cm^2 **13.** 411.6 in^2 **15.** 10,641.7 mm^2
17. 44.0 mm; 153.9 mm^2 **19.** 5.7 ft; 2.5 ft^2 **21.** $A, B, C,$
$D, E; \overline{AB}, \overline{BC}, \overline{BE}, \overline{CD}, \overline{BD}$ **23.** Edges need to be drawn between A and C, A and D, A and E, C and E, and D and E
25. Yes; the total area is 272 ft^2.

CHAPTER 11 SURFACE AREA AND VOLUME

Pages 527–528 Lesson 11-1
5. 31 games **7.** $12 - 34 + 56 - 7 + 89 = 116$ **9.** 25 bacteria **11.** 15

13.

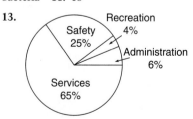

15. 7¢, 12¢, 16¢, 27¢, 31¢, 36¢, 40¢

Pages 531–534 Lesson 11-2
5.

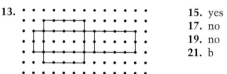

7.

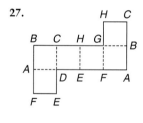

9. 2 pentagons, 5 rectangles **11.** c

13.

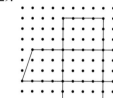

15. yes
17. no
19. no
21. b

23. 2 pentagons, 5 rectangles
25. 2 hexagons, 6 rectangles

27.

29.

31. yes **33.** yes **35.** yes **39.** Wright brothers': 1.41, Four-passenger: 9.83, Supersonic: 97.40 **40.** 1656 in^2
41. 12π or 37.7 in. **42.** 7ft^2 **43a.** < **b.** >

Pages 539–541 Lesson 11-3
7. 569.5 cm^2 **9.** about 412.7 cm^2 **11.** about 1016.9 cm^2
13. right prism **15.** 42 units **17.** 6π cm **19.** 9π cm^2
21. $L = 432$ in^2; $T = 432 + 108\sqrt{3} \approx 619.1$ in^2
23. $L = 150 + 50\sqrt{3} \approx 236.6$ ft^2; $T = 150 + 150\sqrt{3} \approx 409.8$ ft^2 **25.** 384 units^2 **27.** 96π or about 301.6 m^2
29. about 559.8 in^2 **33.** about 2010.6 ft^2

35.

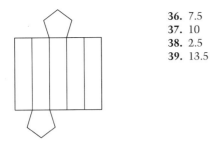

36. 7.5
37. 10
38. 2.5
39. 13.5

Pages 545–547 Lesson 11-4
5. pyramid   **7.** prism   **9.** 27 cm^2 **11.** 143.5 cm^2
13. 736 cm^2 **15.** prism   **17.** neither   **19.** pyramid
21. about 144.5 units^2 **23.** about 1154.0 units^2
25. 144 cm^2 **27.** about 282.7 cm^2 **29.** 420 ft^2
31. 864 in^2 **33.** about 1382.3 m^2 **35.** about 255,161.7 ft^2
37. about 46 ft   **38.** false   **39.** 4π or about 12.6 yd^2
40. The measure of the inscribed angle is half the measure of
the central angle.   **41.** They are perpendicular.

Page 547 Mid-Chapter Review
1. T **2.** $\overline{TO}$ **3.** 6
4.

5. 108π in^2
6. 176π cm^2
7. 720 units^2

Pages 550–553 Lesson 11-5
5. 48 m^3 **7.** about 77.0 in^3 **9.** 32π or about 100.5 m^3
11. about 1950.9 ft^3 **13.** 272 cm^3 **15.** 18.5 cm
17. 67.2 ft^3 **19.** 9π or about 28.3 ft^3 **21.** about 754.0 cm^3
23. 600 m^3 **25.** about 2598.1 cm^3 **27a.** about 1011.5 mm^3
b. about 9 grams   **29.** 360 cm^3 **31.** no   **33.** 6 units
37. almost 3 cords   **39a.** 84 feet   **b.** 139 ft^3 or about
5.15 yd^3 **c.** $368.55   **40.** 52.3 mm^2 **41.** 3 cm   **42.** 9
43. 12   **44.** 42

Pages 556–559 Lesson 11-6
5. about 314.2 in^3 **7.** about 536.2 in^3 **9.** about 1322.3
units^3 **11.** 35 ft^3 **13.** 190.3 m^3 **15.** about 1407.4 units^3
17. about 134.0 cm^3 **19.** about 5178.8 m^3 **21.** 192 units^3
23a. 108 units^3 **b.** 1 to 6   **25.** 58.9 in^3 **27.** about 48.9
units^3 **29.** about 5730.3 units^3 **33.** 90,187,500 ft^3
35. The formula for the volume of a cone rounded to the
nearest unit.   **38.** 120 ft^3 **39.** 0.82   **40.** They are
complementary.   **41.** $\sqrt{73} \approx 8.5$

Pages 562–565 Lesson 11-7
5. true   **7.** true   **9.** 5   **11.** $T \approx 200.96$ cm^2; $V \approx 267.9$ cm^3
13. about 448.9 in^3 **15.** circle   **17.** circle   **19.** circle
21. neither   **23.** neither   **25.** true   **27.** true   **29.** true
31. true   **33.** 12   **35.** $T \approx 50,265,482$ ft^2;
$V \approx 33,510,321,640$ ft^3 **37.** 2 in.   **39.** 2:1   **41.** about 4701
cm^3 **43.** Sample answer: Labrador Sea, Glasgow, Scotland;
Essen, Germany   **45.** about 620,561.5 yd^3

47. about 10,053.1 ft^2 **49.** about 418.9 ft^3 **50.** 60.35 ft^2
51. no; $2.7^2 + 3.0^2 \neq 5.3^2$ **52.** no   **53.** 10

Pages 566–568 Summary and Review
1. c   **3.** b   **5.** $L = 264$ in^2; $T = 312$ in^2 **7.** $L \approx 197.9$ ft^2;
$T \approx 325.2$ ft^2 **9.** $L = 48$ in^2; $T = 84$ in^2 **11.** $L \approx 52.3$ mm^2;
$T \approx 84.4$ mm^2 **13.** about 6283.2 cm^3
15. about 175.4 cm^3 **17.** about 1570.8 mm^3 **19.** about
14,657,415 mi^2 **21.** 64 blocks   **23.** 2145 lb

Pages 570–571 Algebra Review
1. $\{y \mid -4 \le y \le 6\}$ **3.** $\{p \mid p < -2 \text{ or } p > \frac{5}{2}\}$ **5.** $\frac{1}{x+2}$
7. $\frac{x^2 - 7x - 15}{x^2 + x - 6}$ **9.** (4, 2)   **11.** (0,2)   **13.** $\frac{11}{20}$ or 0.55
15.

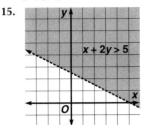

17.

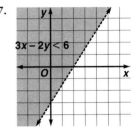

19. $5\sqrt{13} + 5\sqrt{15}$ **21.** $-6\sqrt{2} - 12\sqrt{7}$ **23.** 10, 2
25. $\frac{2 \pm \sqrt{5}}{3}$; 1.41, -0.08   **27.** $2000 at 6%, $8000 at 8%
29. $y = 1.65x + 3$; $4.65

CHAPTER 12 MORE COORDINATE GEOMETRY

Pages 576–579 Lesson 12-1
7. No; y is squared.   **9.** yes   **11.** yes   **13.** 1; -4   **15.** 4; no
y-intercept   **17.** 2; -1   **19.** neither; 1, $\frac{1}{2}$ **21.** parallel; 0, 0
23. parallel; -1, -1
25.

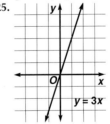

29.

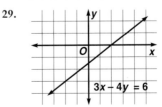

31. parallel   **33.** $m = 1$, $b = 0$ **35.** $m = -\frac{3}{4}$, $b = 2$
37. m is undefined, no y-intercept   **39.** $y = -\frac{1}{3}x + 2$
41. $y = 4x - 3$ **43.** $y = 2x + 3$ **45.** $y = \frac{1}{3}x + 4$
47. $x = 2$ **49.** All are of the form $y = mx + 3$, but all have
a different value for m.

53a. 200 **b.** 1400 **c.**

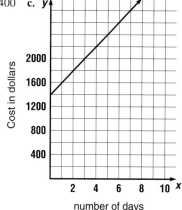

54. about 523.6 in^3 **55.** about 113.1 ft^2 **56.** $-\frac{2}{5}$
58. $m\angle 1 = 49$, $m\angle 2 = 131$ **59.** If it snows Saturday, then we will have to rent skis; law of syllogism

Pages 582–585 Lesson 12-2
5. $y = 4x + 2$ **7.** $y = 5x + 8$ **9.** $y = -5x - 31$ **11.** $y = 3x - 2$ **13.** 2, -3; $y = 2x - 3$ **15.** 1, -1; $y = x - 1$
17. $y = \frac{1}{6}x - 5$ **19.** $y = -4x - 14$ **21.** $y = 7$
23. $y = -2x + 1$ **25.** $y = -\frac{1}{2}x + \frac{11}{2}$ **27.** $y = -4x - 11$
29. $y = 2$ **31.** $y = -x + 5$ **33.** $y = -5x + 4$
35. $y = \frac{3}{4}x + 8$ **39.** $y = 12x + 900$

41. $\frac{3}{2}$; $-\frac{3}{4}$

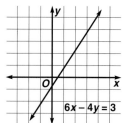

42. $m\angle C \approx 23.9$, $m\angle A \approx 40.4$, $AC \approx 11.1$

43. Given: $\angle 4 \cong \angle 6$
Prove: $\angle 3 \cong \angle 5$

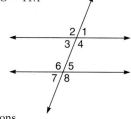

Statements	Reasons
1. $\angle 4 \cong \angle 6$	1. Given
2. $\angle 3$ and $\angle 4$ form a linear pair. $\angle 5$ and $\angle 6$ form a linear pair.	2. Definition of linear pair
3. $\angle 3$ and $\angle 4$ are supplementary. $\angle 5$ and $\angle 6$ are supplementary.	3. If 2 $\angle$s form a linear pair, they are supp.
4. $m\angle 3 + m\angle 4 = 180$ $m\angle 5 + m\angle 6 = 180$	4. Definition of supplementary
5. $m\angle 3 + m\angle 4 = m\angle 5 + m\angle 6$	5. Substitution property of equality
6. $m\angle 3 = m\angle 5$	6. Subtraction property of equality
7. $\angle 3 \cong \angle 5$	7. Definition of congruent angles

44. $\sqrt{74} \cong 8.6$ **45.** (-1.5, 1)

Pages 588–591 Lesson 12-3
5. $y = -x - 1$, $y = 3x + 7$, $y = 1$ **7.** no **9.** yes
11b. Sample answer: $y = 0.15x + 3$ **c.** Answer based on equation in 11b: $1950.30 **13.** $y = -2$, $y = -2x + 6$, $y = 2x + 6$ **15.** $y = \frac{1}{9}x + \frac{44}{9}$ **17.** $y = x - 2$, $y = -\frac{1}{2}x + 7$, $x = -6$ **19.** $y = -x + 4$, $y = 2x + 4$, $y = 4$
21. $(3 + 1)^2 + (5 - 2)^2 \stackrel{?}{=} 25$
$(4)^2 + (3)^2 \stackrel{?}{=} 25$
$16 + 9 \stackrel{?}{=} 25$
$25 = 25$ ✔

25b. Sample answer: $y = -\frac{5}{6}x + 142$ **c.** Answer based on equation in 25b: about 126 **26.** $428 **27.** 248 cm^2
28. $RS > QR$; SAS Inequality

Pages 593–594 Lesson 12-4
5. 3 hours **7.** 8 adult tickets and 12 student tickets
9. $10\pi \approx 31.4$ feet **11.** 89 mph and 99 mph **13.** 46

Pages 597–600 Lesson 12-5
5. $E(d - f, r)$
7. Midpoint A of $\overline{TS}$ is $\left(\frac{2d + 2a}{2}, \frac{2e + 2c}{2}\right)$ or $(d + a, e + c)$.
Midpoint B of $\overline{SR}$ is $\left(\frac{2a + 2b}{2}, \frac{2c + 0}{2}\right)$ or $(a + b, c)$.
Midpoint C of $\overline{VR}$ is $\left(\frac{0 + 2b}{2}, \frac{0 + 0}{2}\right)$ or $(b, 0)$.
Midpoint D of $\overline{TV}$ is $\left(\frac{0 + 2d}{2}, \frac{0 + 2e}{2}\right)$ or (d, e).
Slope of $\overline{AB}$ is $\frac{e + c - c}{d + a - (a + b)}$ or $\frac{e}{d - b}$.
Slope of $\overline{DC}$ is $\frac{e - 0}{d - b}$ or $\frac{e}{d - b}$.
Slope of $\overline{DA}$ is $\frac{e + c - e}{d + a - d}$ or $\frac{c}{a}$.
Slope of $\overline{CB}$ is $\frac{c - 0}{a + b - b}$ or $\frac{c}{a}$.

Since opposite sides are parallel, $ABCD$ is a parallelogram.

9.

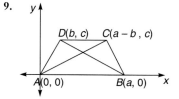

$DB = \sqrt{(a - b)^2 + (0 - c)^2} = \sqrt{(a - b)^2 + c^2}$
$AC = \sqrt{((a - b) - 0)^2 + (c - 0)^2} = \sqrt{(a - b)^2 + c^2}$
$DB = AC$ and $\overline{DB} \cong \overline{AC}$

11. $A(0, 0)$, $Y(b, 0)$ **13.** $R(-b, 2b)$
15. $AB = \sqrt{(a - 0)^2 + (b - 0)^2} = \sqrt{a^2 + b^2}$
$BC = \sqrt{(2a - a)^2 + (0 - b)^2} = \sqrt{a^2 + b^2}$
$AB = BC$ and $\triangle ABC$ is isosceles.

17. $DE = \sqrt{(a-0)^2 + (a\sqrt{3} - 0)^2} = \sqrt{a^2 + 3a^2} = \sqrt{4a^2} = 2a$

$EF = \sqrt{(2a-a)^2 + (0-a\sqrt{3})^2} = \sqrt{a^2 + 3a^2} = \sqrt{4a^2} = 2a$

$DF = \sqrt{(2a-0)^2 + (0-0)^2} = \sqrt{4a^2} = 2a$

$DE = EF = DF$ and $\overline{DE} \cong \overline{EF} \cong \overline{DF}$

Δ DEF is equilateral.

19.

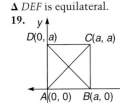

Slope of $\overline{AC}$ is $\frac{a-0}{a-0}$ or $\frac{a}{a}$ or 1.

Slope of $\overline{BD}$ is $\frac{a-0}{0-a}$ or $\frac{a}{-a}$ or -1.

$\overline{AC} \perp \overline{BD}$

21.

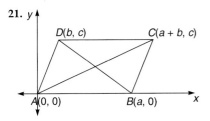

Midpoint of $\overline{AC}$ is $\left(\frac{(a+b)+0}{2}, \frac{c+0}{2} \right)$ or $\left(\frac{a+b}{2}, \frac{c}{2} \right)$.

Midpoint of $\overline{DB}$ is $\left(\frac{a+b}{2}, \frac{0+c}{2} \right)$ or $\left(\frac{a+b}{2}, \frac{c}{2} \right)$.

$\overline{AC}$ and $\overline{DB}$ bisect each other.

23.

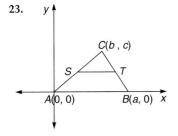

Midpoint S is $\left(\frac{b+0}{2}, \frac{c+0}{2} \right)$ or $\left(\frac{b}{2}, \frac{c}{2} \right)$.

Midpoint T is $\left(\frac{a+b}{2}, \frac{0+c}{2} \right)$ or $\left(\frac{a+b}{2}, \frac{c}{2} \right)$.

Slope of $\overline{ST}$ is $\dfrac{\frac{c}{2} - \frac{c}{2}}{\frac{a+b}{2} - \frac{b}{2}}$ or $\frac{0}{\frac{a}{2}}$ or 0.

Slope of $\overline{AB}$ is $\frac{0-0}{a-0}$ or $\frac{0}{a}$ or 0.

$\overline{ST} \parallel \overline{AB}$

27.

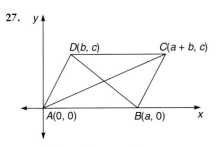

Slope of $\overline{BD}$ is $\frac{c-0}{b-a}$ or $\frac{c}{b-a}$.

Slope of $\overline{AC}$ is $\frac{c-0}{(a+b)-0}$ or $\frac{c}{a+b}$.

But $\overline{BD} \perp \overline{AC}$ and $\frac{c}{b-a} = -\frac{a+b}{c}$. $\frac{c}{b-a} = \frac{a+b}{-c}$

$-c^2 = b^2 - a^2$

$a^2 = b^2 + c^2$ and $\sqrt{a^2} = \sqrt{b^2 + c^2}$

$AD = \sqrt{(b-0)^2 + (c-0)^2} = \sqrt{b^2 + c^2}$

$DC = \sqrt{((a+b)-b)^2 + (c-c)^2} = \sqrt{a^2}$

$BC = \sqrt{((a+b)-a)^2 + (c-0)^2} = \sqrt{b^2 + c^2}$

$AB = \sqrt{(a-0)^2 + (0-0)^2} = \sqrt{a^2}$

$AD = DC = BC = AB$ and $\overline{AD} \cong \overline{DC} \cong \overline{BC} \cong \overline{AB}$

$ABCD$ is a rhombus.

29. $x = 20$ **31.** $\sqrt{58} \cong 7.6$ km **32.** false **33.** false **34.** true **35.** true **36.** true **37.** false **38.** false **39.** false

Page 600 Mid-Chapter Review

1. 3; -4 **2.** no x-intercept; 15 **3.** $y = -\frac{2}{3}x + \frac{10}{3}$

4. $y = 3x$ **5.** $y = -\frac{1}{2}x + 3$ **6.** yes **7.** no **8.** 10:33

9. $A(-b, 0)$, $C(b, 2b)$, $D(-b, 2b)$ **10.** $E(0, 0)$, $G(a, b+c)$

11.

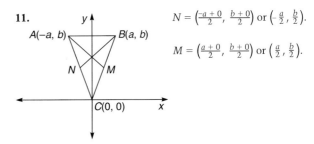

$N = \left(\frac{-a+0}{2}, \frac{b+0}{2} \right)$ or $\left(-\frac{a}{2}, \frac{b}{2} \right)$.

$M = \left(\frac{a+0}{2}, \frac{b+0}{2} \right)$ or $\left(\frac{a}{2}, \frac{b}{2} \right)$.

$BN = \sqrt{\left(-\frac{a}{2} - a \right)^2 + \left(\frac{b}{2} - b \right)^2}$ $AM = \sqrt{\left(-a - \frac{a}{2} \right)^2 + \left(b - \frac{b}{2} \right)^2}$

$= \sqrt{\left(\frac{-3a}{2} \right)^2 + \left(\frac{-b}{2} \right)^2}$ $= \sqrt{\left(\frac{-3a}{2} \right)^2 + \left(\frac{b}{2} \right)^2}$

$= \sqrt{\frac{9a^2}{4} + \frac{b^2}{4}}$ or $\frac{\sqrt{9a^2 + b^2}}{2}$ $= \sqrt{\frac{9a^2}{4} + \frac{b^2}{4}}$ or $\frac{\sqrt{9a^2 + b^2}}{2}$

Therefore, $\overline{BN} \cong \overline{AM}$.

Pages 604–606 Lesson 12-6

9. $\sqrt{73} \cong 8.5$, 69° **11.** $\overrightarrow{AB}$ and $\overrightarrow{DC}$ **13.** $\overrightarrow{AB}$ and $\overrightarrow{BC}$

15.

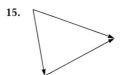

17.

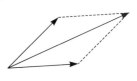

19. (1, 10) **21.** (-1, 9) **23.** $6\sqrt{2} \approx 8.5$ units **25.** $4\sqrt{2} \approx 5.7$ units, 45° **27.** $\overrightarrow{BC}$ and $\overrightarrow{AD}$ **29.** $\overrightarrow{RT}$ **31.** $\overrightarrow{PR}$ **33.** (9, 6) **35.** (20, 17) **37.** 11 **39.** $\overrightarrow{AB}$ and $2\overrightarrow{AB}$, $\overrightarrow{AB} + \overrightarrow{BL}$ and $\overrightarrow{SR}$ **41.** $\vec{v}$ and $\vec{w}$, $\vec{v}$ and $\vec{t}$ **43.** The resulting speed is 250 km/h and the direction is about 16° south of west. **45.** about 86 newtons, about 54° west of north

46.

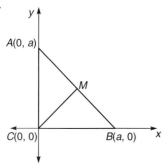

Midpoint M is $\left(\frac{0+a}{2}, \frac{a+0}{2}\right)$ or $\left(\frac{a}{2}, \frac{a}{2}\right)$.

Slope of $\overline{AB}$ is $\frac{0-a}{a-0}$ or $\frac{-a}{a}$ or -1.

Slope of $\overline{CM}$ is $\dfrac{\frac{a}{2}-0}{\frac{a}{2}-0}$ or $\dfrac{\frac{a}{2}}{\frac{a}{2}}$ or 1.

Since $-1 \cdot 1 = -1$, $\overline{CM} \perp \overline{AB}$.
47. about 157.1 m^2 **48.** about 50.3 in. **49.** 18, 22

Pages 610–612 Lesson 12-7
7. 5 **9.** (1.5, -2, 2) **11.** (0, 3, 4), 9 **13.** $(x-4)^2 + (y-1)^2 + (z+2)^2 = 36$ **15.** $11 + \sqrt{21} \approx 15.6$ units
17.

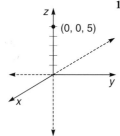

19.

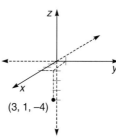

21. $\sqrt{94} \approx 9.7$ units **23.** $\sqrt{134} \approx 11.6$ units **25.** (0, 0, 0)
27. (11, 1, 2.5) **29.** (-2, -3, 2), 10 **31.** (-8, 0, -4), $\sqrt{18} \approx 4.2$
33. $(x-6)^2 + (y+1)^2 + (z-3)^2 = 144$ **35.** $(x+2)^2 + (y-4)^2 + (z-1)^2 = 13$

37a. (1, 2, 6)
b. $\sqrt{26} \approx 5.1$ units
c. $(x-1)^2 + (y-2)^2 + (z-6)^2 = 26$
d.

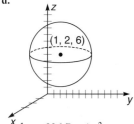

e. about 326.7 units^2
f. about 555.3 units^3

39. 2 or $\frac{4}{13}$ **41.** 3 to 5 **43.** 27 to 125 **45.** $\sqrt{147} \approx 12.1$ miles **47.** (-2, -8) **48.** about 78.5 cm^2 **49.** 110 cm^2
50. 10.5 in^2 **51.** Two lines intersect and more that one plane contains them. **52.** 100

Pages 614–616 Summary and Review
1.

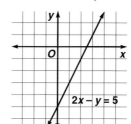

3. perpendicular
5. $y = x + 8$
7. $y = 2$
9. $y = \frac{1}{2}x - 2$, $y = \frac{2}{3}x - 3$, $y = x - 3$
11. $y = -2x - 3$, $y = -\frac{3}{2}x + 2$, $y = -x + 7$

13.

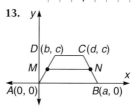

Midpoint M is $\left(\frac{b+0}{2}, \frac{c+0}{2}\right)$ or $\left(\frac{b}{2}, \frac{c}{2}\right)$.
Midpoint N is $\left(\frac{a+d}{2}, \frac{c+0}{2}\right)$ or $\left(\frac{2a+d}{2}, \frac{c}{2}\right)$.
Slope of $\overline{DC}$ is $\frac{c-c}{d-b}$ or $\frac{0}{d-b}$ or 0.

Slope of $\overline{MN}$ is $\dfrac{\frac{c}{2}-\frac{c}{2}}{\frac{a+d}{2}-\frac{b}{2}}$ or $\dfrac{0}{\frac{a+d-b}{2}}$ or 0.

Slope of $\overline{AB}$ is $\frac{0-0}{a-0}$ or $\frac{0}{a}$ or 0.

$\overline{DC} \parallel \overline{MN} \parallel \overline{AB}$
15. $\sqrt{50} \approx 7.1$ units; about 8.1° **17.** (7, 1) **19.** $4\sqrt{2} \approx 5.7$ units; (5, -3, 3) **21.** $x^2 + y^2 + z^2 = 25$ **25.** (20, 20)

CHAPTER 13 LOCUS AND TRANSFORMATIONS

Pages 624–627 Lesson 13-1
7.

The perpendicular bisector of the line segment

11.

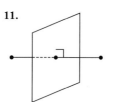

a plane that is the perpendicular bisector of the given line segment

15. a line segment perpendicular to the floor at the intersection of the diagonals of the floor of the classroom with endpoints on the floor and ceiling **17.** the 50-yard line of the football field **19.** a cylindrical surface with line ℓ as the axis and a radius of 4 inches

21. 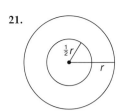 a circle, concentric to the given circle with a radius that is half the radius of the given circle

25. 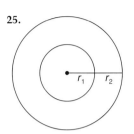 the circle with radius equal to the sum of the radii of the two gears and center at the center of the large gear

29. 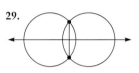 the line that is the perpendicular bisector of the line segment joining the two points

33.  two planes parallel to the given plane, one on each side of the given plane and the given distance from the plane

37. a line perpendicular to the plane of the circle at the center of the circle

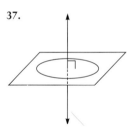

39. a line perpendicular to $\mathscr{R}$ at the point of intersection of the diagonals of the square **41.** two planes perpendicular to $\mathscr{R}$ and passing through two perpendicular lines, one parallel to $\overline{AB}$ and $\overline{DC}$ and the other parallel to $\overline{AD}$ and $\overline{BC}$ and intersecting at the center of the square **43.** all points on a circle with center at the vertex and radius half the length of the hypotenuse except the points on the given line and the points on the line perpendicular to the given line at the vertex **45.** a curve which repeats itself every time the wheel makes one complete rotation

49. a great circle crossing the great circle passing through Chicago and London such that their tangents at the point of intersection are perpendicular to each other **51.** the points on the perpendicular bisector of the line segment between the two students **52.** slope **53.** about 1256.6 cm^2 **54.** 90 **55.** diameters **56.** right triangle **57.** skew

Pages 631–633 Lesson 13-2
5. (6, 5) **7.** (-3, -9) **9.** (0, 4) **11.** (-2, 2) **13.** (6, 2) **15.** (0, -1) **17.** a, c, d **19.** a, b, d **21.** (1, 3) **23.** $(5\frac{1}{2}, -\frac{1}{2})$ **25.** (-1, -1) **27.** (14, 4) **29.** (-9, -7) **31.** (-1, -1) **33.** (7, 4) **35a.** (2, 2), (12, 2), (6, 6) **b.** 20 units^2 **37.** (-1, -1) and (4, 4) **39.** $\frac{9\sqrt{10}}{10} \approx 2.8$ **41.** about 5.58 years or 5 years 7 months **43.** yes; their areas of coverage overlap **45.** two circles, concentric to the given circle, one with radius of 2 inches and the other with a radius of 8 inches **46.** $x = \sqrt{76}$ or 8.7, $y = \sqrt{92}$ or 9.6, $z = \sqrt{437}$ or 20.9 **48.** If three points lie in the same plane, then they are coplanar. **49.** 36

Pages 636–639 Lesson 13-3
5. a sphere with radius 7 and center (2, -6, 5)
7. a cylinder with radius 3 and an axis passing through (5, 7, z)
9. They may have no points of intersection; the plane may be tangent to the sphere or it may intersect the sphere in a circle; numbers of possible points of intersection: 0, 1, infinite **11.** They may have no points of intersection; one or both lines may be tangent to the outer circle; one or both lines may be tangent to the inner circle and intersect the outer circle in 2 places; one or both lines may be secants of just the outer circle; one or both lines may be secants of both circles; numbers of possible points of intersection: 0, 1, 2, 3, 4, 5, 6, 7, or 8

15. 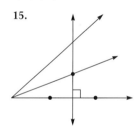 the point inside the angle where the ray that bisects the angle intersects the line that is the perpendicular bisector of the segment that has the given points as endpoints

17. a cylinder with radius 6 whose axis is perpendicular to the xy-plane through the point (2, -4, 0) **19.** a sphere with radius 4 and center (3, 4, 5) **21.** $(x + 1)^2 + (y + 6)^2 = 16$ **23.** They may have no points of intersection; the line may be tangent to either circle; the line may be secant of the outer circle; the line may be secant of both circles; numbers of possible points of intersection: 0, 1, 2, 3, or 4 **25.** They may have no points of intersection; one or both lines may be tangent to the sphere; one or both lines may be secants of the sphere; numbers of possible points of intersection: 0, 1, 2, 3, or 4 **27.** They may have no points of intersection; one or both planes may be tangent to the sphere; one or both planes may intersect the sphere in a circle; numbers of possible points of intersection 0, 1, 2, or infinite

33. 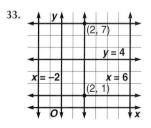 **37.**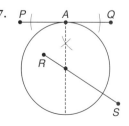

points (2, 1) and (2, 7)

39. ∅ **41.** 5 points **43.** 0 points
47.

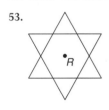

1 point   2 points

49. sometimes **50.** never **51.** always **52.** never
53. always **54.** sometimes **55.** sometimes **56.** always

Pages 641–643 Lesson 13-4
5. $\overline{DC}$ **7.** ∠S **9.** $\overline{RS}$ **11.** B **13.** ∠DEB **15.** $\overline{ED}$
17. B **19.** $\overline{BA}$ **21.** ∠CBA **23.** ∠T **25.** ∠B **27.** ∆RQS
29. ∆NMW **31.** ∆XZY **33.** CDAB **35.** $\overline{WY}$
37. ∆XYW → ∆ZYW **41.** Angel D is rotated or reflected
to form angel C. **43.** For each frame of the cartoon, the
picture of the ball slides just slightly in one direction. **44.** a
circle **45.** 315 cm² **46.** 5 **47.** $\sqrt{558} \approx 23.6$ **48.** If a
mapping is a transformation, then it is one-to-one.

Pages 647–651 Lesson 13-5
7. C   **9.**

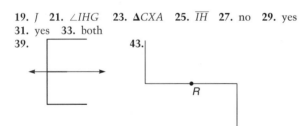

13.   **17.**

19. J **21.** ∠IHG **23.** ∆CXA **25.** $\overline{IH}$ **27.** no **29.** yes
31. yes **33.** both
39.   **43.**

47. none **49.** none **51.** none

53.   **57.** 168 cm **58.** ∠S **59.** $\overline{CD}$
          60. HA **61.** SSS **62.** AAS
          63. not enough information
          64. HL **65.** not enough
          information

Page 651 Mid-Chapter Review
1.a. Read the problem carefully. **b.** Draw the given figure.
c. Locate the points that satisfy the given conditions.
d. Draw a smooth curve or line. **e.** Describe the locus.
2. a line which is the perpendicular bisector of the segment
joining the two given points **3.** a plane which is the

perpendicular bisector of the segment joining the two given
points **4.** a **5.** b **6.** a doughnut shape 6 cm wide **7.** If
the two points are more than 8 units apart, there are no
points in the locus of points. If the 2 points are 8 units apart,
the locus of points is one point at the point of tangency of the
two spheres. If the two points are less than 8 units apart, the
locus of points is a circle formed by the intersection of the
two spheres.
8. A   **9.** $\overline{BC}$
          10. ∠B

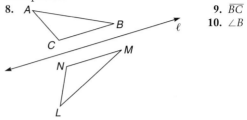

Pages 654–658 Lesson 13-6
5. no **7.** none **9.** none **13.** F **15.** C **17.** P
19. Q **21.** C **23.** P **25.** yes **27.** yes **29.** ∆HGI
31. pentagon AHGFE **33.** ∆JKL **35.** pentagon IJKLM

37.   **39.**

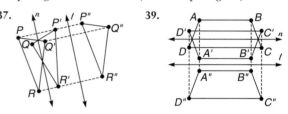

41. ∆STU **43.** ∆LMN **45.** Plan: A translation is
composed of two successive reflections over parallel lines.
The first reflection with respect to ℓ preserves collinearity.
The second reflection with respect to m preserves
collinearity. Therefore, by transitivity, collinearity is
preserved from preimage to image. **47.** Plan: A translation
is composed of two successive reflections over parallel lines.
The first reflection with respect to ℓ preserves angle and
distance measure. The second reflection with respect to m
preserves angle and distance measure. Therefore, by
transitivity, angle and distance measure is preserved from
preimage to image.
49.

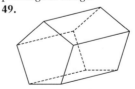

53.   **54.** 72 ft²
          55. about 166.3 cm²
          56. 6 in²
          57. ∠P
          58. 56.5

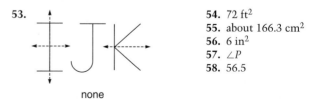

Pages 661–664 Lesson 13-7
5. quadrilateral EFCD **7.** quadrilateral HGJK
9. quadrilateral CDAB **11.** 140 **13.** 140 **15.** $\overline{GK}$

17.

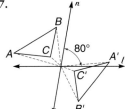

19. yes **21.** 90° **23.** 74°
25. yes **27.** yes

31.

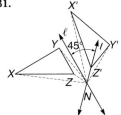

33.

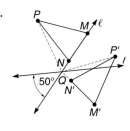

35. rotation **37.** Sample answer: It preserves orientation, because if vertices of the preimage are in clockwise order, the vertices of the image will be in clockwise order.
39. $m\angle AOC$ is 120 so the measure of the angle of intersecting lines will be 60. **41a.** right circular cone **b.** right cylinder with hollowed out right cone. **43.** 108
45. 75 **46.** 37.5 **47.** 65 **48.** No; the sum of their measures would have to be both 90 and 180.

Pages 667–670 Lesson 13-8

5. similar **7.** 4 **9.** 2; enlargement **11.** $\frac{1}{4}$; reduction

13.

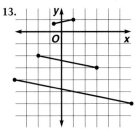

15. enlargement
17. reduction
19. reduction
21. congruence
23. 30
25. 24 **27.** 3
29. A **31.** T
33. S **35.** F

37. 2; enlargement **39.** $\frac{1}{3}$; reduction **41.** $\frac{1}{4}$; reduction

45.

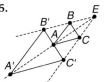

49.

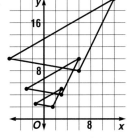

53.

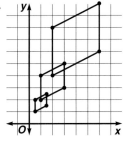

55a. The perimeter of the dilation image will be 4 times the perimeter of the preimage. **b.** The area of the dilation image will be 16 times the area of the preimage.

57. Given: Dilation with center C and scale factor k.
Prove: $ED = k(AB)$

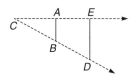

$CE = k(CA)$ and $CD = k(CB)$ by the definition of a dilation.
$\frac{CE}{CA} = k$ and $\frac{CD}{CB} = k$ so, $\frac{CE}{CA} = \frac{CD}{CB}$ by substitution.
$\angle ACB \cong \angle ECD$ since congruence of angles is reflexive. Therefore, by SAS Similarity, $\Delta ACB \approx \Delta ECD$. The corresponding sides of similar triangles are proportional, so $\frac{ED}{AB} = \frac{CE}{CA}$. We know that $\frac{CE}{CA} = k$, so $\frac{ED}{AB} = k$ by substitution. Therefore, $ED = k(AB)$ by the multiplication property of equality. **59.** Sample answer: The fish are dilations of each other. **61.** $\frac{3}{5}$ **63.** 6 **64.** $\sqrt{74} \approx 8.6$ **65.** 21.4
67. a triangle with one obtuse angle; no **68.** 25

Pages 672–673 Lesson 13-9
5. AAADF, AABCF, AABDD, AACCD, ABBBF, ABBCD, ABCCC, BBBBD, BBBCC **7.** 2, 3, 6 **9.** 28
13. 90 **15.** 2550

Pages 674–676 Summary and Review
1. a circle in the plane with the given point as its center and radius 11 inches **3.** (5, 3) **5.** (1,1) **7.** They may have no points of intersection; the line may be tangent to the circle; the line may intersect the circle at two points; numbers of possible points of intersection: 0, 1, or 2 **9.** They may have no points of intersection; the plane may be tangent to the outside circle; the plane may intersect the outside circle at two points and not touch the inside circle; the plane may be tangent to the inside circle and intersect the outside circle at two points; the plane may intersect both circles at two points; the circles may lie in the plane; numbers of possible points of intersection: 0, 1, 2, 3, 4, or infinite **11.** E **13.** B

15.

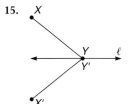

17.

19. 140
23. Imagine the reflection of the 5 ball with respect to the line formed by one side of the billiard table and aim for the reflection.

Pages 678–679 Algebra Review
1. $-\frac{14}{3}$ **3.** $-\frac{23}{2}$ **5.** (5, 1) **7.** (2, -1) **9.** $x^2 + 7x - 8 = 0$
11. $6x^2 + 13x + 6 = 0$ **13.** 12 **15.** 3
21. 7; 6; no mode **23.** 4; 3; 2 **25.** 22:8 or 11:4
27. No; it would skid 240 ft at 60 mph. **29.** 30

INDEX

parallel, 122–154, 156–158, 266, 276, 294, 307, 336–341, 574–579, 599, 623, 634–635, 653–658, 675

perpendicular, 56–61, 64, 107, 119–120, 130–131, 134, 136, 142–145, 148–153, 157–158, 241–242, 288–289, 422–423, 434–436, 495, 576–579, 598–599

of reflection, 644–653, 675

secants, 412–414, 440–445, 458–460

skew, 123–126, 156

slopes of, 142–147, 157, 217, 383, 574–588, 614–615, 628

of symmetry, 292, 646–649

tangents, 412–415, 434–445, 458–459, 560

transversals, 124–141

vertical, 143–145, 383, 575

Line symmetry, 202, 292, 646–649

Loci, 622–639, 674–675
and equations, 628–633, 636, 674
intersections of, 629, 634–639, 674–675

LOGO, 49, 535, 652

Lovelace, Ada Byron, 230

Magnitude, 601–606, 615

Major arcs, 416–419, 442

Mandelbrot, Benoit B., 6–7, 680

Mandelbrot set, 680, 689–691

Maoris, 127

Mappings, 640–671, 675–676

Means, 309–310
arithmetic, 364
geometric, 360–365, 402

Measures,
of angles, 37–44, 50–54, 89–91, 416–419, 428–433, 440–445, 458–459, 473–478, 518, 645, 654, 671
of arcs, 417–420, 428–433, 440–445, 458–459
of segments, 23–28, 89–91, 645, 654, 671

Medians,
of trapezoids, 295–297, 299, 492
of triangles, 216–217, 258, 344–347, 354, 597, 598, 600

Nets, 530–534, 538, 543, 566

Networks, 512–517, 520
complete, 514–515
incomplete, 514
traceable, 512–516, 520

Newton, Isaac, 591

N-gon, 467, 475

Nodes, 512–516, 520

Nonagons, 467

Noncollinear points, 9–10, 13

Noncollinear rays, 36

Number lines
distance between points, 24
midpoints on, 30, 63
x-axis, 8, 601, 607, 613
y-axis, 8, 383, 607, 613
z-axis, 607, 613

Oblique cones, 544–545, 556

Oblique cylinders, 538, 552, 556

Oblique prisms, 536, 552, 555–556

Oblique pyramids, 556

Obtuse angles, 44, 164, 173

Obtuse triangles, 164, 208, 216, 234

Octagons, 467, 473–474

Octahedra, 468, 471

Open graphs, 512

Opposite rays, 36–37, 50, 120

Mid-chapter reviews, 35, 94, 134, 191, 239, 287, 335, 382, 433, 488, 547, 600, 651

Midpoints, 30–34, 63, 216–217, 295, 338, 582, 596–599, 608, 616, 644, 647

Minor arcs, 416–419, 423

Mobius strip, 658

Morgan, Garrett, 69

Multiplication property, 88–89, 235

Ordered pairs, 8–11, 595–607

Ordered triples, 607–613, 616

Origins, 8, 601

Palindromes, 673

Parabolas, 573

Paraboloids, 573

Paragraph proofs, 129, 193, 240–241, 267, 283, 422, 424, 429

Parallel lines, 122, 154, 156–158, 266, 276, 294, 307, 336–341, 574–579, 599, 623, 634–635, 653–658, 675
constructing, 135
distance between, 148–151, 158
slopes of, 143–147, 157, 276

Parallelogram law, 602

Parallelograms, 13, 75, 117, 266–271, 275–293, 300–301, 596–599
angles of, 267–270, 276–280, 301
area of, 484–493, 519
diagonals of, 267–270, 276–277, 300, 599
rectangles, 18–20, 282–287, 301, 483–488, 492–493
rhombi, 288–293, 301, 599
sides of, 266–270, 275–280
squares, 289–291, 483–484, 492, 598

Parallelpipeds, 117

Parallel planes, 122–123, 156

Parallel postulate, 136

Parallel vectors, 602

Pei, I.M., 464–465

Peitgen, Heinz-Otto, 688

Pentagons, 467, 473, 475, 497–499

Percents, 312

Perimeters, 496–502, 537, 543
of rectangles, 19–20, 526, 592–593
of similar triangles, 342–343, 354

Perpendicular bisectors, 218–222, 423, 435, 582, 615, 634, 644–646

Perpendicular lines, 56–61, 64, 107, 119–120, 130–131, 134, 136, 142–145, 148–153, 157–158, 241–242, 288–289, 422–423, 434–436, 495, 576–579, 598–599

constructing, 57–58, 149
   slopes of, 143–145, 157, 217
Perpendicular transversal theorem,
   130
Perpendicular vectors, 606
Peruvians, 410
Planes, 13–18, 77–80
   coordinate, 8–12, 14
   horizontal, 17
   intersecting, 78, 123
   lines perpendicular to, 58
   parallel, 122–123, 156
   vertical, 17
Plato, 468
Platonic solids, 468
Points, 13–18, 77–80
   betweenness, 23, 36, 59, 71–72,
      645, 654, 671
   collinear, 9–11, 14–16, 59, 70,
      587, 644, 654, 671
   on coordinate planes, 8–11, 14,
      62
   coplanar, 14–17, 59
   distance between, 23–26, 63,
      166, 185
   incenters, 436
   locus of, 622–639, 674–675
   midpoints, 30–34, 63, 216–217,
      295, 338, 582, 596–599, 608,
      616, 644, 647
   nodes, 512–516, 520
   noncollinear, 9–10, 13
   origin, 8, 601
   power of, 454
   of reflections, 644, 646
   of symmetry, 646
   of tangency, 434
Point-slope form, 581–582,
   586–588, 614
Point symmetry, 646–651
Polygons, 164, 266, 466–479, 512,
   518–519
   angles of, 473–479, 518, 590
   area of, 483–502, 519–520
   circumscribed, 436–438
   concave, 466–467
   convex, 466–479, 518
   decagons, 467
   diagonals of, 350
   dodecagons, 467
   heptagons, 467, 473–474
   hexagons, 467, 473, 475,
      495–496, 519
   inscribed, 430–432
   n-gons, 467, 475
   nonogons, 467
   octagons, 467, 473–474
   pentagons, 467, 473, 475,
      497–499

quadrilaterals, 264–271, 275–303,
   430–432, 467, 473, 483–494,
   519
   regular, 467–469, 474–478,
      495–502, 519
   sides of, 164, 266, 466–467, 473
   similar, 321–348, 353–354,
      360–362, 376
   triangles, 163–197, 199–210,
      217–229, 233–261, 327–348,
      353–354, 360–398, 403–404,
      467, 473, 489–493, 496–500,
      519
   vertices of, 164–165, 266
Polyhedra, 468–472, 529–537,
   542–559, 566–568
   dodecahedra, 468, 471
   hexahedra, 468, 471
   icosahedra, 468, 471
   octahedra, 468, 471
   prisms, 17, 121, 123, 126, 359,
      529–537, 548–556, 566–567,
      607, 609, 654
   pyramids, 17, 169, 525, 529–531,
      542–547, 554–559, 562,
      567–568
   regular, 468–469
   rhombohedra, 539
   tetrahedra, 468, 471
Polynesians, 127
Postulates, 24, 77–78, 98
   angle addition, 38–39, 57
   angle-angle similarity, 329, 353
   angle-side-angle, 186–187, 192,
      225
   arc addition, 417–418
   area probability, 508, 520
   corresponding angles, 128
   hypotenuse-leg, 226–227
   length probability, 507, 520
   parallel, 136
   protractor, 38
   ruler, 24
   segment addition, 25, 38
   side-angle-side, 186–188, 194,
      209, 223
   side-side-side, 184–185
Power of points, 454
Preimages, 640–643, 665, 675
Primitive triples, 370
Prisms, 17, 121, 123, 126, 359,
   529–537, 548–556, 607, 609,
   654
   lateral area of, 537, 567
   oblique, 536, 552, 555–556
   right, 536, 548–556, 567
   surface area of, 537, 567
   volume of, 548–556, 567
Probability, 507–511, 520

Problem-solving strategies
   decision making, 399–401
   draw a diagram, 118–121
   guess and check, 480–482
   identify subgoals, 198–201
   list the possibilities, 19–22
   look for a pattern, 272–274
   make a model, 526–528
   make a table, 671–673
   process of elimination, 95–97
   solve a simpler problem,
      349–351
   using graphs, 455–457
   work backward, 230–232, 364,
      399
   write an equation, 592–594
Projections, 613
Proofs, 89
   coordinate, 595–600, 615
   indirect, 233–239, 253, 259
   paragraph, 129, 193, 240–241,
      267, 283, 422, 424, 429
   two column, 89–94, 98–103,
      106–109, 111–112, 129, 150,
      186–188, 192–194, 199, 204,
      242–244, 253, 275, 288, 447
Properties
   addition, 88–92, 235
   comparison, 234–235
   distributive, 88–90, 537
   division, 88–92, 235
   multiplication, 88–89, 235
   reflexive, 88–91, 98–101, 106,
      111, 178, 331
   substitution, 88–91
   subtraction, 88, 235
   symmetric, 88–92, 98–99, 106,
      111, 178, 331
   transitive, 88–90, 98–101, 106,
      111, 129, 178, 235, 331
Proportional perimeters theorem,
   342
Proportions, 308–346, 352–354,
   360–362, 365
Protractor postulate, 38
Ptolemy, 29
Pyramids, 17, 169, 525, 529–531,
   542–547
   lateral area of, 543–547, 567
   oblique, 556
   regular, 542–547, 567
   right, 554–559
   surface area of, 543–546, 567
   volume of, 554–559, 568
Pythagoras, 365
Pythagorean Theorem, 365–372,
   403, 423–424, 434–435, 491,
   607
Pythagorean triples, 369–370

Photo Credits

Cover, Romilly Lockyer/The Image Bank; vii, Pictures Unlimited; x(t), NASA, (ct), Tom Mareschal/The Image Bank, (cb), Frank A. Cezus/FPG, (b), Art Montes De Oca/FPG; xii, xiv, Pictures Unlimited; 6, Adrian Baker/FPG, (inset), Courtesy Benoit Mandelbrot; 7, A. M. Rosario/The Image Bank; 8, Todd Gray/LGI Photo Agency; 9,12(t), Pictures Unlimited, 12(b), David Frazier; 13, Jerry Schad/Photo Researchers; 16, Bob Peterson/FPG; 18, Steve Lissau; 19, William H. Allen, Jr.; 20,21, Elaine Shay; 23(t), E. A. McGee/FPG, (b),24, PEANUTS reprinted by permission of UFS, Inc; 28, Michel Tcherevkoff/The Image Bank; 30,34, Pictures Unlimited; 36, Courtesy United Airlines; 41, Pictures Unlimited; 42, Larry West/FPG; 43(l), Phillip Hayson/Photo Researchers, (r), William Rivelli/The Image Bank; 45,50, Pictures Unlimited; 55, T. Zimmerman/FPG; 56, Pictures Unlimited; 67, Elaine Shay; 68, Marc Romanelli/The Image Bank, (inset), The Western Reserve Historical Society Library; 69, James Westwater; 70, Drawing by Ed Fisher, ©1966 Saturday Review, Inc; 73, Crown Studio; 74(t), Robert A. Isaacs/Photo Researchers, (b), Historical Pictures Service, Chicago; 76(t), Susan Biddle/The White House, (b), Elaine Shay; 78, United Negro College Fund; 79, The White House; 80(t), Courtesy Chrysler Corporation, (b), Alan Carey; 82(l), Crown Studio, (r), Larry Hamill; 83(t), Elaine Shay, (b), Crown Studio; 84, Pictures Unlimited; 85(t), Elaine Shay, (b), Richard Dole/DUOMO; 86, MAK-1, 87, William Weber; 88, By permission of Johnny Hart and NAS, Inc; 93, Courtesy of Kennywood; 94,95,96, Crown Studio; 103, Doug Martin; 104, Crown Studios; 105, L. D. Franga; 109, Ruth Dixon; 113, Randy Scheiber; 116,117, Courtesy Kenzo Tange Associates; 118, Studiohio; 119, C. J. Zimmerman/FPG; 120, Pictures Unlimited; 120, Tom Campbell/FPG; 121, Bob Winsett/Tom Stack & Assoc; 122, Charles Feil/FPG; 123, Jim Pickerell/FPG; 125,126, Pictures Unlimited; 128, By permission of Johnny Hart and NAS, Inc; 129,133, Pictures Unlimited; 135, D & P Valenti/H. Armstrong Roberts; 137, Doug Martin; 140, Pictures Unlimited; 141, ©1960 M. C. Escher/Cordon Art-Baarn, Holland; 142, Historical Pictures Service, Chicago; 146, First Image; 147, Tom Stack/Tom Stack & Assoc; 151,153,154, Pictures Unlimited; 162, Scala/Art Resource; 163, David Lyle Millard, STILL LIFE PAINTING TECHNIQUES, Watson-Guptill Publications; 164, Benn Mitchell/The Image Bank; 165, Studiohio; 166, 168, Pictures Unlimited; 170, Doug Martin; 171,172, Pictures Unlimited; 177, Courtesy Ford Motor Company; 178, Pictures Unlimited; 182(t), First Image, (b), Museum of Modern Art/Rosenthal Art Slides; 184, Pictures Unlimited/Glencoe; 186, Allen Zak; 191, Pictures Unlimited; 192, 193, Doug Martin; 200,201, First Image; 202(t), Doug Martin, (b), Larry Hamill; 206, Frank Cezus; 213, Pictures Unlimited; 214, Courtesy U.S. Embassy, Tokyo, Japan, (inset), Courtesy Norma Sklarek; 215(t), Robert Frerck/Odyssey/Chicago, (b), file photo; 216, Janet Adams; 223,225, Leo Mason/The Image Bank; 229(t), Steve Dunwell/The Image Bank, (b), Historical Pictures Service, Chicago; 230, Doug Martin; 231, First Image; 232,233,235, Pictures Unlimited; 239, file photo; 245, Aaron Haupt/Glencoe; 246,247,249, Pictures Unlimited; 250, R. Krubner/H. Armstrong Roberts; 252,254, Pictures Unlimited; 257, Zefa/H. Armstrong Roberts; 264, David Heald © Solomon R. Guggenheim Foundation, New York, (inset), Philadelphia Museum of Art/Rosenthal Art Slides; 265, David Heald © Solomon R. Guggenheim Foundation, New York; 266,271, 272,273, Pictures Unlimited; 274(t), James Westwater, (b), Pictures Unlimited; 275,277, Pictures Unlimited; 278, Lee E. Yunker; 280(t), Anne Van DerVaeren/The Image Bank, (b), Pictures Unlimited; 282(t), Ken Frick, (top inset), Andrea Pistolesi/The Image Bank, (b), Studiohio, (bottom inset), Steve Proehl/The Image Bank; 283, Pictures Unlimited; 286, Mark Gibson; 287, Tim Courlas; 288, Denise morris Curt/The Connecticut Limner; 291, Pictures Unlimited; 292, Doug Martin; 293, Jean Kugler/FPG; 294, Hedrich-Blessing, Ltd; 298, H. Armstrong Roberts; 302, Pictures Unlimited; 306, Art Resource/Giraudon; (inset), Art Resource/Prado, Madrid; 307, The Bettmann Archive; 308,309, Pictures Unlimited; 310(l), Richard Kane/Sportschrome East/West, (r),311,312, Pictures Unlimited; 313(t), Mark Godfrey, (b), NOAO; 314, Don C. Nieman; 315, Frank A. Cezus/FPG; 316,317, Pictures Unlimited; 318, David Hiser/The Image Bank; 319, Doug Martin; 320, Pictures Unlimited; 321(l), Doug Martin, (r), Eric Grave/Science Source/Photo Researchers; 325, Pictures Unlimited; 334, M. Thonig/H. Armstrong Roberts; 335, Pictures Unlimited; 336, Tim Courlas; 337, Folger Shakespeare Library; 350, Pictures Unlimited; 351, Mak-1; 358, 359, Robert Frerck/Odyssey/Chicago; 360, Craig Kramer; 364, Susan Snyder; 366, Pictures Unlimited; 367, Mark Gibson; 369, Doug Martin; 370, Animals Animals/Hans & Judy Beste; 371, Lindsay Gerard/Glencoe; 375,376, file photo; 377, Courtesy Carnival Cruise Line; 378, William Weber; 381, Tim Slattery/Harbor Reflections; 384, Ed & Chris Kumler; 387, Aaron Haupt/Glencoe; 388, Susan Snyder; 391, Joseph DiChello; 392, Gary Gay/The Image Bank; 393,394, Pictures Unlimited; 396, Metro Dade Tourism/Walter Marks; 397, Jeff Smith/The Image Bank; 398, Ken Frick; 400, Pictures Unlimited; **408,** Nyina Warwinu/Picture Group, **(inset),** Jerry Sinkovec/Artwerks; **409,** Nyina Warwinu/Picture Group; **410,** David Frazier; **415(l),** Doug Martin; **(r),417,418,420,** Pictures Unlimited; **421,** Perez Siquier/Spanish Tourist Office, Chicago; **427,430,** Pictures Unlimited; **433,** Mazda of America, Inc; **436,** Doug Martin; **439,** Tim Courlas; **440,** file photo; **442,** Doug Martin; **445,** David Frazier; **447(l),** Gerard Photography, **(r),** Pictures Unlimited; **452,** NASA; **455,** Pictures Unlimited; **456,** Aaron Haupt/Glencoe; **460,** David L. Perry; **464,** Paul Warchol, **(inset),** Serge Hambourg; **465,** Paul Warchol; **466,** Matt Meadows; **467,** George Matchneer; **468,** Bob Mullenix; **472,** Kathleen O'Donnell/Stockphotos/The Image Bank; **473,** Mak-1; **475,** Department of Defense; **478,** Pictures Unlimited; **479,** John Shaw/Tom Stack & Assoc; **482,** Pictures Unlimited; **483,484,** John Colwell from Grant Heilman; **487,** Matt Meadows/OSU, Department of Engineering Mechanics; **488,** First Image; **489,494,** Bob Mullenix; **495,496,** Pictures Unlimited; **501,** Crown Studio; **502,** Matt Meadows; **503,** Crown Studios; **504,** John Banagan/The Image Bank; **506,** Bob Mullenix; **507,** Blair Seitz/Photo Researchers; **508,** Crown Studios; **510,** Comstock, Inc/Bonnie Camin; **511,** Crown Studios; **512,** W. Cody/Westlight; **517,** Crown Studios; **520(t),** David Germon, **(b),** Paul Brown; **524,** Eugene G. Schulz, **525,** Farrell Grehan/FPG; **526,** Doug Martin; **527,** Chris Michaels/FPG; **528,** Visual Horizons/FPG; **529,** Harald Sund/The Image Bank; **534,** Crown Studios; **534,** Matt Meadows; **536,** Bob Mullenix; **538,** Bill Ross/Westlight; **541,542(t),** Lee Yunker, **(b),** Salt Institute; **546,** Bob Mullenix; **547,** Mark Stephensen/Westlight; **548(l),** Pictures Unlimited, **(r),** Robert Winslow/Tom Stack & Assoc; **549,** Pictures Unlimited; **552,** Bob Mullenix; **553,** Alex Bartel/FPG; **553,** Pictures Unlimited; **554,** ©1989 Tully/Ballenger; **554,** Neal & Mary Jane Mishler/FPG; **555,** Crown Studios; **559,** Tom Tracy/FPG; **560,** Sportschrome East/West; **561,** Doug Martin; **563,** NASA; **564,** Studiohio; **572,573,** Robert Frerck/Odyssey/Chicago; **574(t),** Aaron Haupt/Glencoe, **(b),** Thomas Kitchin/Tom Stack & Assoc; **575,** Pictures Unlimited; **578,** Ken Frick; **579,** Courtesy Navy Research Laboratory, Washington DC; **580,581,** Mak-1; **584,** Pictures Unlimited; **585,** Mak-1; **586,** Norbert Wu/TSW; **587,** Pictures Unlimited; **588,** George Obremski/The Image Bank; **589,** Pictures Unlimited; **590,** Rich Buzzelli/Tom Stack & Assoc; **591,** The Bettmann Archive; **592,** Comstock, Inc/ Billy Brown; **593,** Mak-1; **594,595,** Pictures Unlimited; **599,** Mak-1; **601,** Pictures Unlimited; **602,** Bill Ross/Westlight; **606,** Don & Pat Valenti/Tom Stack & Assoc; **607,** F. Stuart Westmorland/Tom Stack & Assoc; **608,** William Rivelli/The Image Bank; **616,** Jim Zuckerman/Westlight; **620,** Star Trek II: The Wrath of Kahn, Paramount, 1982; **621,** Stock Imagery; **622(t),** NASA, **(b),** Pete Turner/The Image Bank; **624,** Melchior Digiacomo/The Image Bank; **625(t),** Steve Proehl/The Image Bank, **(b),** Murray Alcosser/The Image Bank; **626,** Greg Vaughn/Tom Stack & Assoc; **627,** Courtesy Christine M. Darden/Advanced Vehicles Division/NASA; **628,** Harald Sund/The Image Bank; **629,** Andrea Pistolesi/ The Image Bank; **633,** Travelpix/FPG; **634,** Eddie Hironaica/The Image Bank; **639,** Pictures Unlimited; **640,** ©1938 M. C. Escher/Cordon Art-Baarn, Holland; **643,** ©1941 M. C. Escher/Cordon Art-Baarn, Holland; **644,** Cradoc Bagshaw/Westlight; **646(t),** Pictures Unlimited, **(bl),** Geoffrey Gove/The Image Bank, **(br),** Runk/Schoenberger from Grant Heilman; **650,** Pictures Unlimited; **653,** Stock Imagery; **654,** Pictures Unlimited; **657,** Robin Smith/FPG; **658,** Doug Martin; **665,** Elaine Shay; **667,** Pictures Unlimited; **670,** ©1959 M. C. Escher/Cordon Art-Baarn, Holland; **672(t),** Paul J. Sutton/DUOMO, **(b),** Jim Cummins/ALLSTOCK; **679,** Pictures Unlimited; **680,** Courtesy Benoit Mandelbrot, IBM, Watson Research Center; **681,682,** Images by H. JÜrgens, H.-O. Peitgen, D. Saupe, from: THE BEAUTY OF FRACTALS by H.-O. Peitgen, P. Richter. Springer-Verlag, Heidelberg, 1986, and FRACTALS FOR THE CLASSROOM, H.-O.Peitgen, H. JÜrgens, D. Saupe. Springer-Verlag, New York, 1991; **683,684,** Michael D. McGuire; **685,** Image by Manfred Kage, Institut fur wissenschaftliche Fotografie; from FRACTALS FOR THE CLASSROOM, H.-O. Peitgen, H. JÜrgens, D. Saupe. Springer-Verlag, New York, 1991; **686,** Michael D. McGuire; **687,** Dr. Vehrenberg KG, from FRACTALS FOR THE CLASSROOM, H.-O. Peitgen, H. JÜrgens, D. Saupe. Springer-Verlag, New York 1991; **688,** FRACTALS FOR THE CLASSROOM by H.-O. Peitgen, H. JÜrgens, D. Saupe. Springer-Verlag, New York, 1991; **689,** Images by H. JÜrgens, H.-O. Peitgen from FRACTELE: GEZAHMTES CHAOS, Carl Freidrich Von Siemens Stiftung, 1988; **690,691,** Images by H. JÜrgens, H.-O. Peitgen, D. Saupe, from: THE BEAUTY OF FRACTALS by H.-O. Peitgen, P. Richter. Springer-Verlag, Heidelberg, 1986, and FRACTALS FOR THE CLASSROOM, H.-O.Peitgen, H. JÜrgens, D. Saupe. Springer-Verlag, New York, 1991.